PHOTOCHEMICAL SPLITTING OF WATER

PHOTOCHEMICAL SPLITTING OF WATER

Fundamentals to Applications

Edited by

SONEY C. GEORGE

Center for Nanoscience and Technology, Amal Jyothi College of Engineering, Kanjirappally, Kottayam, Kerala, India

LUÍS P.M. SANTOS

Graduate Program of Materials Science and Engineering, Federal University of Ceará, Campus of PICI, Fortaleza, Ceará, Brazil

SAJITH KURIAN

Department of Chemistry, Mar Ivanios College (Autonomous), Thiruvananthapuram, Kerala, India

ELSEVIER

Elsevier
Radarweg 29, PO Box 211, 1000 AE Amsterdam, Netherlands
125 London Wall, London EC2Y 5AS, United Kingdom
50 Hampshire Street, 5th Floor, Cambridge, MA 02139, United States

ISBN: 978-0-443-29064-0

For Information on all Elsevier publications
visit our website at https://www.elsevier.com/books-and-journals

Publisher: Megan Ball
Acquisitions Editor: Edward Payne
Editorial Project Manager: Rupinder Heron
Production Project Manager: Kamesh R.
Cover Designer: Vicky Pearson Esser

Typeset by MPS Limited, Chennai, India

Contents

List of contributors ix
Preface xiii

1. Potential of water splitting in the global energy future

Sajith Kurian, Luís P.M. Santos and Soney C. George

1.1 Introduction—global energy challenge 1
1.2 Hydrogen economy 3
1.3 Water: the potential source of hydrogen 6
1.4 Role of artificial intelligence and additive manufacturing in water electrolysis 14
1.5 Future prospects 16
1.6 Conclusion 17
References 17

2. Concepts, kinetics, and mechanism of water splitting

Rohit Gupta and Rahul Gupta

2.1 Introduction 23
2.2 Kinetics of splitting water techniques 27
2.3 Comparison of the mechanisms of splitting water by electrolysis method and photochemical water splitting 38
2.4 Case study 38
2.5 Learning and knowledge outcomes 45
2.6 Conclusion 46
References 47

3. Technologies for water splitting

Rakhi Khandelwal

3.1 Introduction 51
3.2 Water-splitting technologies 52
3.3 Challenges of water-splitting technologies 63
3.4 Comparative study of various water-splitting technologies 65
3.5 Future aspects of water-splitting technologies 65
3.6 Case study: Seawater to Green Hydrogen: A Prototype by IIT Madras Scientists 67
3.7 Conclusion 71
References 71
Further reading 73

4. Instrumentation and experimental setup for photochemical splitting of water

R. Anjana, Nithya S. George, Ayona K. Jose, Jibi Kunjumon, P.A. Aleena and D. Sajan

4.1 Introduction 75
4.2 Cell design 78
4.3 Electrode considerations 83
4.4 Cell setup in a photochemical water-splitting reaction 86
4.5 j–V (photocurrent) analysis 91
4.6 Determination of flat band potential (E_{fb}) 92
4.7 Two-electrode j–V measurement 92
4.8 Conclusion 93
References 93

5. Photocatalytic water splitting

Shalini Viswanathan, A.C. Swathi and K. Aparna

5.1 Introduction 97
5.2 Material design for photocatalytic water splitting 100
5.3 Reaction setup of water-splitting devices 106
5.4 Recent progress in photocatalytic water splitting 108
5.5 Case study 110
5.6 Constraints in photocatalytic water splitting 112
5.7 Conclusions 112
References 114

6. Photoelectrochemical water splitting

Seema Rani, Soumyadip Sharangi, Subhabrata Das, Mansi Pahuja and Kaushik Ghosh

6.1 Introduction 119
6.2 Photoelectrochemical water splitting: working principles 120
6.3 Materials for photoelectrochemical water splitting 126
6.4 Role of electrolyte 130
6.5 Advancements in photoelectrochemical water splitting 133
6.6 Case Study 136
6.6 Conclusion 139
6.7 Future outlook 139
References 140

7. Selection of materials and cell design components for photochemical splitting of water

Sapna Balayan, Mamta Shandilya, Itika Kainthla and Gun Anit Kaur

7.1 Introduction 143
7.2 Background of water-splitting 145
7.3 Material selection for photochemical splitting of water 151
7.4 Cell Design 155
7.5 Future perspective and recommendations 157
7.6 Conclusion 158
References 159

8. Noble metal-based catalysts for water splitting

Lekshmi Krishna, Paulose Thomas, Angel Susan Cherian and Anju K. Nair

8.1 Introduction 163
8.2 Fundamentals of water splitting 165
8.3 Catalysis in water splitting 166
8.4 Noble metals: properties and characteristics 168
8.5 Noble metals in hydrogen evolution reaction 171
8.6 Noble metal-based photocatalysts 179
8.7 Noble metal-free alternatives 181
8.8 Comparative analysis of noble metals versus noble metal-free catalysts 182
8.9 Challenges and limitations 184
8.10 Recent trends and future directions 186
8.11 Future prospects for noble metals in water-splitting technologies 187
8.12 Conclusion 188
References 189

9. Noble metal-free catalysts – oxides, chalcogenides, nitrides, oxynitrides for water splitting

Swapnali Walake, Pragati Thakur and Yogesh Jadhav

9.1 Introduction 193
9.2 Electrochemical and photo-electrochemical water splitting 194
9.3 Photocatalytic water splitting 196
9.4 Transition metal oxides, chalcogenides, nitrides, oxynitrides catalysts 197
9.5 Mechanism of water splitting 197
9.6 Strategies to improve water splitting 206
9.7 Conclusion and future prospects 212
References 213

10. Metal ion incorporated metal oxides for water splitting

Anju John, Jain Maria Thomas and Jesty Thomas

10.1 Introduction 217
10.2 Mechanism of metal oxide–assisted photochemical water splitting 219
10.3 TiO_2-based photocatalysts 221
10.4 ZnO-based photocatalysts 226
10.5 WO_3-based photocatalysts 229
10.6 Case studies 232
10.7 Conclusion and future perspective 237
References 237

11. Plasmonic material and quantum dot sensitized photocatalysts for water splitting

Nikita Chaudhary, Jyoti Jyoti and Kaushik Ghosh

11.1 Introduction 243
11.2 Plasmonic materials for PEC 244
11.3 Surface plasmon resonance 244
11.4 Design 250
11.5 PEC system based on quantum dots 250
11.6 Photoanode based on quantum dots 251

11.7 Surface customization of quantum dots 256
11.8 Photocathode based on quantum dot 257
11.9 Quantum dots assembly 258
11.10 Tandem quantum dot-based PEC architecture 259
11.11 Case study for TiO_2-based photoanode 260
11.12 Conclusion 260
References 262

12. Carbon-based materials for water-splitting

Mansi Pahuja, Mohd Afshan, Daya Rani, E.M. Harini and Kaushik Ghosh

12.1 Introduction 269
12.2 Carbon nanomaterials 270
12.3 Case study 286
References 290

13. Metal organic frameworks water splitting

Olugbenga Akande and Jude. A. Okolie

13.1 Introduction 295
13.2 Overview of different metal organic frameworks used in water splitting 296
13.3 Advances in metal-organic framework photocatalysts for efficient water-splitting 298
13.4 Types of water-splitting methods for extraction of heavy metals 307
13.5 Extraction of heavy metals using hydrogen evolution reaction 309
13.6 Analyzing chemical bonding via raman spectroscopy 310
13.7 Case study 310
13.8 Conclusion 313
References 313

14. 2D materials for water splitting

Misbah Mirza, Muhammad Safdar, Nazia Yasmin and Sadia Yasmin

14.1 Introduction 319
14.2 2D materials 321
14.3 2D materials superiority 323
14.4 Synthesis techniques of 2D materials 325
14.5 Characterization techniques for 2D materials 326
14.6 Mechanism of water splitting 329
14.7 2D materials for water splitting 333
14.8 SACs supported on diverse 2D materials 341
14.9 Conclusions 348
References 349

15. Conjugated polymer materials for water-splitting

Benjamin Tawiah, Emmanuel A. Ofori, Yang Ming and Bin Fei

15.1 Introduction 361
15.2 The basics of water-splitting 363
15.3 CP material synthesis and characterization 368
15.4 Photocatalytic water-splitting with CPs 379
15.5 Challenges and future perspectives 383
15.6 Case study: practical application of CP for water-splitting 385
15.7 Conclusion 387
References 387

16. Role of cocatalyst in water splitting

Sumesh George and Jain Maria Thomas

16.1 Introduction 395
16.2 Approaches to enhance the photocatalytic activity 397
16.3 Significance of cocatalysts in water splitting 398
16.4 Factors influencing the performance of cocatalyst 399
16.5 Types of cocatalysts 399
16.6 Conclusion and future perspective 417
References 418

17. Nanostructured materials for water-splitting

Suman Thakur and Sharda Bharti

17.1 Introduction 427
17.2 Conventional catalysts 429
17.3 Synthesis and properties of nanostructured materials 431
17.4 Metal-based nanostructured materials for water-splitting 437
17.5 2D materials and heterostructures 440

17.6 Nanostructured materials for photoelectrochemical water-splitting 442
17.7 Commercial prospects and technological challenges 446
17.8 Case study on nanostructured materials for water-splitting 446
17.9 Conclusion and future perspectives 450
References 451

18. Photovoltaic electrochemical integration for water-splitting

Princemon Philip, Ajesh K. Zachariah and Anju K. Nair

18.1 Introduction 457
18.2 Water-splitting 459
18.3 Photovoltaic systems for water-splitting 462
18.4 Electrochemical systems for water-splitting 465
18.5 Efficiency of water-splitting 469
18.6 Photovoltaic electrocatalysis 470
18.7 Case study 471
18.8 Conclusion 482
References 482

19. Challenges for photoelectrochemical water-splitting technology for the near-future

Gaweł Sołowski and Fethi Ahmet Özdemir

19.1 Introduction 487
19.2 Methodology 488
19.3 Discussion 489
19.4 Learning, knowledge outcomes, and conclusions 499
References 499

Index 507

List of contributors

Mohd Afshan Quantum Materials and Devices Unit, Institute of Nano Science and Technology, Mohali, Punjab, India

Olugbenga Akande Department of Computer Science and Electrical Engineering, Handong Global University, Pohang, Gyeongsangbukdo, Republic of Korea

P.A. Aleena Department of Physics, Centre for Sustainable Energy and Environment Technologies (CE^2 T), Bishop Moore College, Alappuzha, Kerala, India

R. Anjana Department of Physics, M.S.M College, Alappuzha, Kerala, India

K. Aparna Department of Chemical Engineering, Materials Science and Environmental Sustainability Group, National Institute of Technology, Calicut, Kerala, India

Sapna Balayan Amity Institute of Nanotechnology (AINT), Amity University Uttar Pradesh (AUUP), Noida, Uttar Pradesh, India

Sharda Bharti Department of Biotechnology, National Institute of Technology (NIT) Raipur, Raipur, Chhattisgarh, India

Nikita Chaudhary Quantum Materials and Devices Unit, Institute of Nano Science and Technology, Mohali, Punjab, India

Angel Susan Cherian Department of Physics, Mar Thoma College, Tiruvalla, Kerala, India

Subhabrata Das Quantum Materials and Devices Unit, Institute of Nano Science and Technology, Mohali, Punjab, India

Luís P.M. Santos Graduate Program of Materials Science and Engineering, Federal University of Ceará, Campus of PICI, Fortaleza, Ceará, Brazil

Bin Fei Department of Industrial Art (Fashion and Textiles), Kwame Nkrumah University of Science and Technology, Kumasi, Ghana

Nithya S. George Department of Physics, Centre for Sustainable Energy and Environment Technologies (CE^2 T), Bishop Moore College, Alappuzha, Kerala, India

Soney C. George Center for Nanoscience and Technology, Amal Jyothi College of Engineering, Kanjirappally, Kottayam, Kerala, India

Sumesh George Department of Physics, St. George's College, Aruvithura, Kerala, India

Kaushik Ghosh Quantum Materials and Devices Unit, Institute of Nano Science and Technology, Mohali, Punjab, India

Rahul Gupta Department of Physics, Faculty of Physics, G.D. Goenka Public Higher Secondary School, Jammu, Jammu and Kashmir, India

Rohit Gupta Department of Applied Sciences, Faculty of Physics, Yogananda College of Engineering and Technology, Jammu, Jammu and Kashmir, India

E.M. Harini Quantum Materials and Devices Unit, Institute of Nano Science and Technology, Mohali, Punjab, India

Yogesh Jadhav Symbiosis Centre for Nanoscience and Nanotechnology (SCNN), Symbiosis International University, Pune, Maharashtra, India; BacPlexTechnologies Private Limited, C/O AIC IISER Pune SEED Foundation IISER Pune Campus, Pashan, Pune, Maharashtra, India

Anju John Research Department of Chemistry, Kuriakose Elias College, Kottayam, Kerala, India

Ayona K. Jose Department of Physics, Centre for Sustainable Energy and Environment Technologies (CE^2 T), Bishop Moore College, Alappuzha, Kerala, India

Jyoti Jyoti Quantum Materials and Devices Unit, Institute of Nano Science and Technology, Mohali, Punjab, India

Itika Kainthla School of Physics and Materials Science, Shoolini University of Biotechnology and Management Sciences, Solan, Himachal Pradesh, India

Gun Anit Kaur School of Physics and Materials Science, Shoolini University of Biotechnology and Management Sciences, Solan, Himachal Pradesh, India

Rakhi Khandelwal Government Mahila Engineering College, Ajmer, Rajasthan, India

Lekshmi Krishna Department of Physics, Mar Thoma College, Tiruvalla, Kerala, India

Jibi Kunjumon Department of Physics, Centre for Sustainable Energy and Environment Technologies (CE^2 T), Bishop Moore College, Alappuzha, Kerala, India

Sajith Kurian Department of Chemistry, Mar Ivanios College (Autonomous), Thiruvananthapuram, Kerala, India

Yang Ming Department of Industrial Art (Fashion and Textiles), Kwame Nkrumah University of Science and Technology, Kumasi, Ghana

Misbah Mirza Department of Physics, The Women University Multan, Multan, Pakistan

Anju K. Nair Department of Physics, Mar Thoma College, Tiruvalla, Kerala, India

Emmanuel A. Ofori Department of Materials Science and Engineering, Hubei University of Automotive Technology, Hubei, P.R. China

Jude. A. Okolie Engineering Pathways, Gallogly College of Engineering, University of Oklahoma, Norman, OK, United States

Fethi Ahmet Özdemir Department of Molecular Biology and Genetics, Faculty of Science and Art, Bingöl University, Bingöl, Turkey

Mansi Pahuja Quantum Materials and Devices Unit, Institute of Nano Science and Technology, Mohali, Punjab, India

Princemon Philip St. Alosyius College, Edathua, Alappuzha, Kerala, India

Daya Rani Quantum Materials and Devices Unit, Institute of Nano Science and Technology, Mohali, Punjab, India

Seema Rani Quantum Materials and Devices Unit, Institute of Nano Science and Technology, Mohali, Punjab, India

Muhammad Safdar Department of Basic Sciences and Humanities, Khawaja Fareed University of Engineering and Information Technology, Rahim Yar Khan, Pakistan

D. Sajan Department of Physics, Centre for Sustainable Energy and Environment Technologies (CE^2 T), Bishop Moore College, Alappuzha, Kerala, India

Mamta Shandilya School of Physics and Materials Science, Shoolini University of Biotechnology and Management Sciences, Solan, Himachal Pradesh, India

Soumyadip Sharangi Quantum Materials and Devices Unit, Institute of Nano Science and Technology, Mohali, Punjab, India

Gaweł Sołowski Department of Molecular Biology and Genetics, Faculty of Science and Art, Bingöl University, Bingöl, Turkey

A.C. Swathi Department of Physics, National Institute of Technology, Calicut, Kerala, India

Benjamin Tawiah School of Fashion and Textiles (SFT), The Hong Kong Polytechnic University, Hung Hom, Kowloon, Hong Kong SAR, P.R. China; Department of Industrial Art (Fashion and Textiles), Kwame Nkrumah University of Science and Technology, Kumasi, Ghana

Pragati Thakur Department of Chemistry, Savitribai Phule Pune University, Pune, Maharashtra, India

Suman Thakur Department of Biotechnology, National Institute of Technology (NIT) Raipur, Raipur, Chhattisgarh, India

Jain Maria Thomas Department of Chemistry, St. George's College, Aruvithura, Kottayam, Kerala, India

Jesty Thomas Research Department of Chemistry, Kuriakose Elias College, Kottayam, Kerala, India

Paulose Thomas Department of Physics, Mar Thoma College, Tiruvalla, Kerala, India; Department of Physics, Mar Thoma College for Women, Perumbavoor, Kerala, India

Shalini Viswanathan Department of Chemical Engineering, Materials Science and Environmental Sustainability Group, National Institute of Technology, Calicut, Kerala, India

Swapnali Walake Symbiosis Centre for Nanoscience and Nanotechnology (SCNN), Symbiosis International University, Pune, Maharashtra, India

Nazia Yasmin Department of Physics, The Women University Multan, Multan, Pakistan

Sadia Yasmin National Transmission and Despatch Company, Multan, Pakistan

Ajesh K. Zachariah Department of Physics, Mar Thoma College, Tiruvalla, Kerala, India

Preface

Splitting of water is an area of active research for the last few decades. This research field is remarkably improved day by day as the researchers are focused on developing photocatalysts based on nanomaterials to enhance the efficiency of water splitting. Nanostructured materials and nanocomposites based on noble and non-noble metals, semiconductors, transition metal oxides, MOF, 2D materials, conjugated polymers, quantum dots, and carbonaceous materials have shown promise in improving light absorption and enhancing electron transfer processes. Some new designs such as tandem photoelectrochemical cells where two or more photocatalysts are stacked together to capture a broader spectrum of light, increasing overall efficiency, are also in progress The photochemical water splitting technique contains great potential for clean hydrogen production although several challenges are there. Low efficiency is one of the most important challenges and researchers are concentrating on developing extraordinary catalysts that raise efficiency several times higher than the currently available ones.

Many efficient photocatalysts are made from rare or expensive materials, which raise the cost of large-scale hydrogen production. Long-term stability of the photocatalysts under operating conditions is another major challenge. Many materials degrade over time when exposed to light and water. Harnessing the full spectrum of sunlight and optimizing the absorption of visible light (instead of just UV light) is a key area of research. While significant challenges remain in terms of efficiency, cost, and stability, ongoing research into new materials, better catalyst design, and improved systems could make this process a viable technology for sustainable hydrogen production in the future.

The advancement of research in the field of splitting water is discussed in this book. Chapter 1 provides the potential of water splitting in the global energy future. Chapter 2 focuses on the concepts, kinetics, and mechanism of water splitting. Chapter 3 discusses the technologies adopted for water splitting and Chapter 4 guides us to instrumentation and experimental setup for photochemical splitting of water. Various aspects of photocatalytic water splitting are discussed in Chapter 5. Chapter 6 discusses photoelectrochemical water splitting whereas the selection of materials and cell design is described in Chapter 7. Chapter 8 narrates on noble metal-based catalysts and Chapter 9 on noble metal-free catalysts, including oxides, chalcogenides, nitrides, and oxynitrides for water splitting. Chapter 10 illustrates the metal ion incorporated metal oxides for water splitting. Plasmonic material and quantum dots sensitized photocatalysts for water splitting are discussed in Chapter 11 and Chapter 12 brings us to a tour of carbon-based materials for water splitting. MOF-assisted water splitting is narrated in Chapter 13 and in Chapter 14, 2D materials for water splitting are concentrated. Chapter 15 leads us into a deep discussion on conjugated polymer materials for water splitting. Chapter 16 discusses the role of cocatalyst in water splitting, and Chapter 17 on nanostructured materials for water splitting. Chapter 18 throws a light on photovoltaic electrochemical integration for water splitting and Chapter 19 concludes with challenges for photoelectrochemical water splitting technology for the near future.

The immense contributions of all authors to the book are admired. The reviewers who reviewed the chapters within the agreed time and gave valuable suggestions are recognized with appreciation. We are deeply grateful to the editorial team members of Elsevier Publishers for their guidance and sustained support in this venture. We are profoundly indebted to the support, guidance, and motivation of our management, colleagues, and family members. I hope that this book gives a delightful experience to the readers who focus on water splitting-related research.

Editors
Dr. Soney C. George
Dr. Luís P. M. Santos
Dr. Sajith Kurian

CHAPTER

1

Potential of water splitting in the global energy future

Sajith Kurian[1], *Luís P.M. Santos*[2], *and Soney C. George*[3]

[1]Department of Chemistry, Mar Ivanios College (Autonomous), Thiruvananthapuram, Kerala, India [2]Graduate Program of Materials Science and Engineering, Federal University of Ceará, Campus of PICI, Fortaleza, Ceará, Brazil [3]Center for Nanoscience and Technology, Amal Jyothi College of Engineering, Kanjirappally, Kottayam, Kerala, India

1.1 Introduction—global energy challenge

The 21st century faces an unprecedented global energy challenge. As the world's population continues to grow and economies expand, the demand for energy is increasing at an alarming rate. Conventional fossil fuels, such as coal, oil, and natural gas, accounted for around 82% of the world's energy consumed in 2023 (Outlook, 2023). While they have historically been crucial for industrial development and modern living, their detrimental environmental impact is undeniable (Hoegh-Guldberg et al., 2018). Greenhouse gas emissions, air pollution, and the geopolitical tensions arising from fossil fuel dependence necessitate a paradigm shift in how we produce and consume energy.

The World Energy Outlook presents a challenging backdrop due to conflict and uncertainty, with instability in the Middle East potentially disrupting energy markets and prices (IEA, 2023). Clean energy projects face challenges, such as cost inflation, supply chain bottlenecks, and higher borrowing costs. However, clean energy is the most dynamic aspect of global energy investment. The three main scenarios for the future of clean energy include the stated policies scenario, announced pledges scenario (APS), and net zero emissions (NZE) by 2050 scenario. The stated policies scenario predicts an average annual growth rate of 0.7% in total energy demand by 2030, while the APS flattens demand owing to improved efficiency and the advantages of electricity-powered technologies. The analysis also examines uncertainties, such as China's economic growth and the potential for rapid solar photovoltaic (PV) deployment. The report emphasizes the importance of affordable, reliable, and resilient supply, especially in price-sensitive developing economies. Emerging markets and developing economies account for almost 80% of global electricity demand in the stated policies scenario and over two-thirds in other scenarios.

Photochemical Splitting of Water
DOI: https://doi.org/10.1016/B978-0-443-29064-0.00005-X

Renewable energy sources such as solar, wind, and hydroelectric power have made significant strides, but their intermittent nature and the limitations of current storage technologies pose significant hurdles (Brouwer et al., 2014; Notton et al., 2018). In this context, a sustainable, efficient, and clean energy carrier is crucial for meeting the global energy demand without compromising environmental integrity (Capurso et al., 2022). This is where hydrogen, the most abundant element in the universe, emerges as a promising solution. As a cutting-edge energy source, hydrogen fuel has the power to revolutionize how we use and switch to sustainable energy sources, supporting sustainable development. This adaptable fuel is a great option to meet our energy needs in an eco-friendly manner because it has a number of important qualities. Because of its adaptability, hydrogen may be integrated with a variety of energy systems. It allows for an easy transition for current infrastructure because it may be used as a stand-alone fuel or combined with natural gas. It can also be used in conjunction with renewable energy sources to store excess electricity produced during periods of peak production, which helps to mitigate the intermittent problems that come with renewable energy (Dincer and Aydin, 2023). According to projections, the world's need for hydrogen is expected to surpass 500 million metric tons by 2070. Notably, as automakers rapidly move away from traditional combustion engines and petroleum-based motor fuels, the transportation sector is expected to become the main consumer of hydrogen. As shown in Fig. 1.1, projections indicate that by 2070, this industry alone will need about 158.2 million metric tons of hydrogen (Adeli et al., 2023; Statista, 2021).

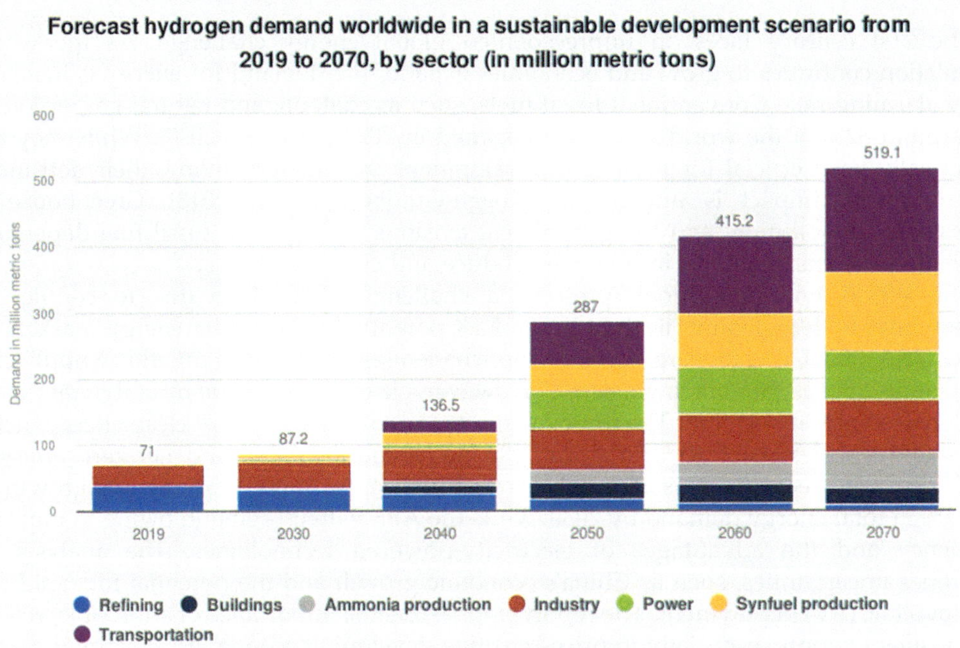

FIGURE 1.1 Projected global demand for hydrogen in different sectors (2019–70) in million metric tons (Statista, 2021). *Source: From IEA, Statista.*

Hydrogen production, utilization, and storage are making much-needed progress. The currently available data show the projects that are already pipelined estimates more than 400 GW from hydrogen by electrolysis. If all of the planned projects are completed, this might be able to fulfil the APS targets; but, supply chain constraints and cost inflation could make progress difficult (Feldmann et al., 2023).

1.2 Hydrogen economy

The concept of a hydrogen economy envisions a world where molecular hydrogen (H_2) serves as a primary energy carrier, replacing fossil fuels for various applications—from powering vehicles to heating homes and fueling industrial processes. Hydrogen's appeal lies in its versatility and clean-burning properties and its high gravimetric energy density, superior to conventional fossil sources, such as methane, gasoline, and diesel (Kim et al., 2019). Furthermore, when used in fuel cells, H_2 combines with oxygen gas from the air to produce electricity, with water vapor as the only byproduct (Yu et al., 2019). During the 26th United Nation (UN) Climate Change Conference of the Parties (COP26) held in Glasgow in 2021, hundreds of nations reaffirmed the Paris Agreement's temperature goal. This goal aims to keep the increase in the global average temperature well below 2°C above preindustrial levels, with efforts to limit the rise to 1.5°C. Furthermore, to achieve this target, it was emphasized that global greenhouse gas emissions must be significantly reduced. This includes reducing global carbon dioxide (CO_2) emissions by 45% by 2030 relative to 2010 levels and achieving NZE around mid-century, along with substantial reductions in other greenhouse gases (UNFCCC, 2022).

The hydrogen economy proposes an integrated system whegasesre H_2 production, storage, distribution, and utilization form a cohesive and sustainable energy infrastructure. This system would significantly reduce carbon emissions, enhance energy security, and support economic growth. Countries around the world are investing heavily in hydrogen technologies, recognizing their potential to transform the energy landscape, for instance, Japan (Japan, 2023), Australia (COAG Energy Council, 2019), United Kingdom (Lachlan Carey, 2021), and South Korea (Nakano, 2021) have already started implementing regulatory policies to decarbonize their economies and incorporate hydrogen as a key alternative energy source. Hydrogen has been considered not only as a sustainable energy carrier but also as an important clear feedstock for the production of important industrial chemicals, such as alcohols, acids, and ammonia (Li et al., 2020; Seh et al., 2017). Fig. 1.2 shows possible free-carbon pathways for utilizing hydrogen as an energy source and feedstock for a decarbonized economy.

1.2.1 Properties of hydrogen

Hydrogen is the smallest atom and the lightest gas. The energy released when hydrogen is burned completely is measured as the higher heating value, HHV (gross calorific value), and the lower heating value, LHV (net calorific value). The HHV and LHV are approximately 141.8 MJ/kg (39.4 kWh/kg) and 120 MJ/kg (33.3 kWh/kg), respectively (Osman et al., 2022; Raveendran et al., 2023). This makes it one of the most energy-dense

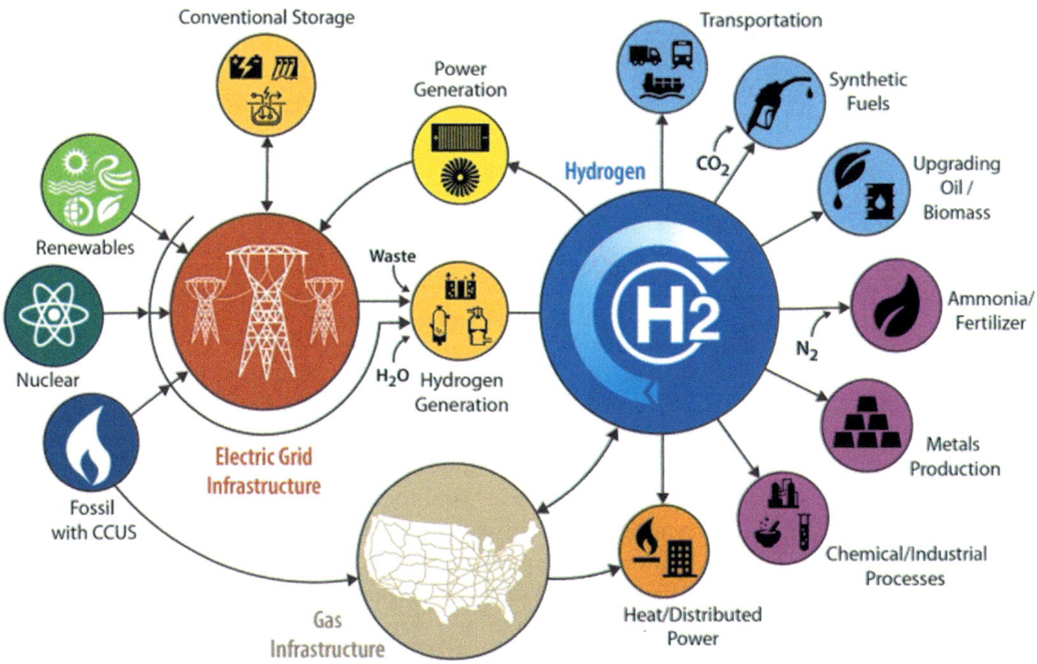

FIGURE 1.2 A schematic way to the utilization of hydrogen as sustainable energy source for decarbonized economy (McQueen et al., 2020). *Source: From McQueen, S., Stanford, J., Satyapal, S., Miller, E., Stetson, N., Papageorgopoulos, D., et al., 2020. US-DOE Hydrogen Program Plan.*

fuels by weight. When hydrogen is burned, it primarily produces water vapor as a byproduct. This makes it an environmentally friendly alternative to fossil fuels, which produce CO_2 and other pollutants. Hydrogen is the most abundant element in the universe, though on Earth it is usually found in compounds like water and hydrocarbons. It can be extracted from these compounds through various processes.

1.2.2 Methods adapted for hydrogen generation—colors of hydrogen

Fig. 1.3 shows different technologies used in the manufacturing of hydrogen gas (Adeli et al., 2023; Ji and Wang, 2021). It consists of both sustainable and nonsustainable methods. Steam methane reforming (SMR) is the most common pathway for H_2 production

in refineries and ammonia production facilities. Around 96% of hydrogen worldwide produced comes from SMR (Adeli et al., 2023; Lee et al., 2009). This technique uses heat and steam to convert natural gas into H_2 and CO_2. On the other hand, autothermal reforming (ATR), which is similar to SMR uses methane and oxygen with either carbon dioxide or steam to produce hydrogen and carbon monoxide (syngas) (Osman et al., 2022). The release of greenhouse gas into the atmosphere makes it a nonviable method in the current net-zero scenario. Carbon capture technology can be employed for low-emission hydrogen production but it escalates the cost of production. The ability to capture CO_2 relative to SMR makes the ATR a more attractive process.

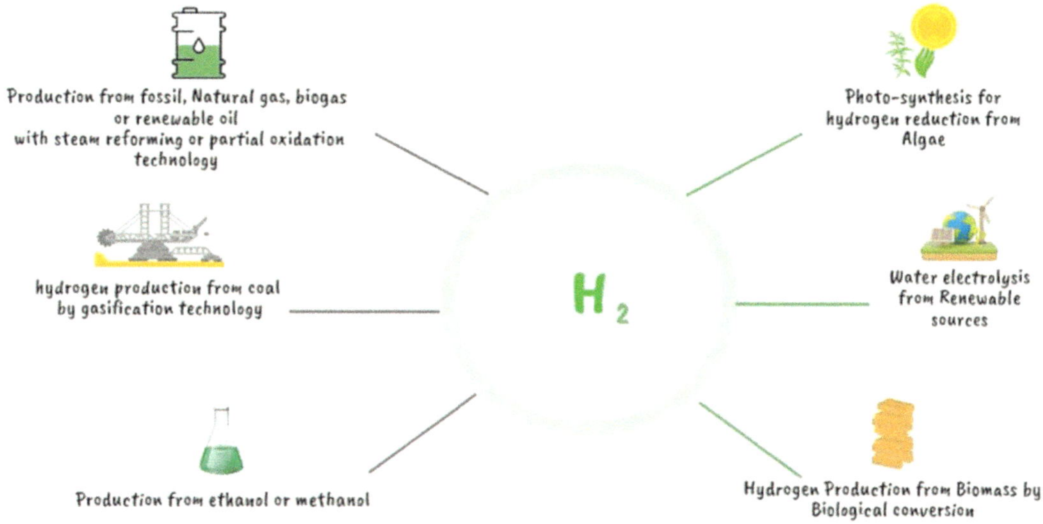

FIGURE 1.3 Sustainable and nonsustainable hydrogen production techniques (Adeli et al., 2023; Ji and Wang, 2021).

Coal gasification is a process that produces syngas, which can be converted into pure hydrogen (Feldmann et al., 2023; Osman et al., 2022). This process is common in countries like China and India. Gasification plants can use carbon capture technology to reduce CO_2 emissions. Biomass gasification is similar but requires more steps due to the increased number of intermediate products (Feldmann et al., 2023; Osman et al., 2022). In this approach, the net emission can be negative when carbon capture is added. Methane is thermally broken down into hydrogen and solid carbon by a process called methane pyrolysis (Feldmann et al., 2023; Osman et al., 2022). Due to its lower production costs and emissions, this technology has the potential to close the gap between hydrogen derived from fossil fuels and renewable sources. However, technology is still developing.

Water electrolysis by using clean electricity produces clean energy in the form of clean hydrogen (Seh et al., 2017; Zhang et al., 2024a). Greenhouse gas emissions are negligible or minimized and that too results from materials manufacturing and construction (e.g., solar panels, wind turbines, electrolysis plants) (Kim et al., 2019). It is essential to use dedicated renewable energy rather than the use of today's fuels for the production of hydrogen to achieve the APS and NZE by 2050.

The generation, delivery, and storage of H_2 must be low-carbon processes for hydrogen or hydrogen-based fuels to be considered low-carbon fuel options in freight transportation or any other area. The manufacturing route and any extra abatement techniques, such as carbon capture and storage and hydrogen leak detection and prevention, will determine the environmental impact of hydrogen production. Fig. 1.4 shows the different methods adopted for hydrogen generation and the color coding of hydrogen depends on the source and method which helps to understand the environmental impacts of each method. It is the green hydrogen that we look for by the electrolysis of water by using renewable energy sources for a sustainable energy future.

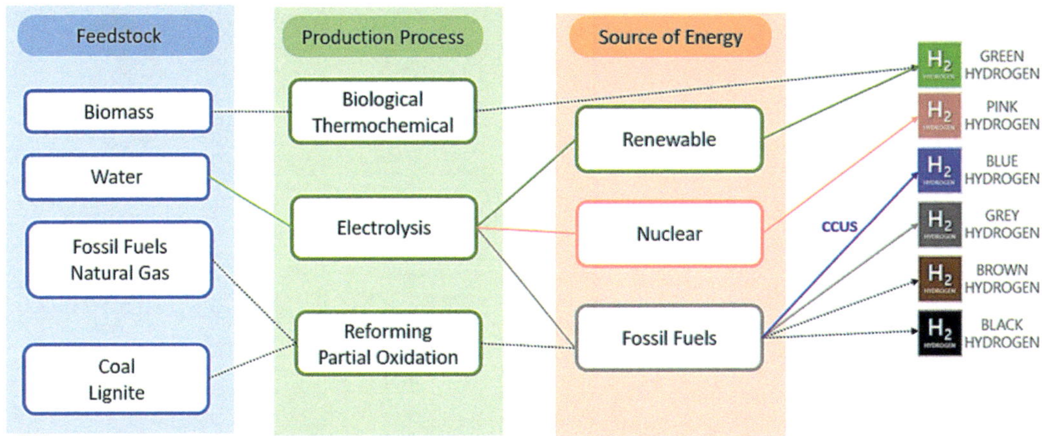

FIGURE 1.4 Colors of hydrogen based on the different methods of hydrogen generation (Adeli et al., 2023).

Global warming potential of hydrogen

Although hydrogen leakage to the atmosphere is not included in most life cycle assessments as a greenhouse gas, it indirectly contributes to warming by increasing stratospheric water vapor, forming tropospheric ozone, and extending the lifetime of atmospheric methane. This results in a 20-year global warming potential (GWP) that is about 33 times greater than CO_2. This value for the global warming potential of hydrogen is around seven times greater than earlier predictions. If hydrogen is to provide either low- or zero-carbon solutions, more investigation and technological advancement in the areas of hydrogen leakage detection, prevention, and mitigation will be required (Derwent et al., 2020; Fan et al., 2022; Ocko and Hamburg, 2022; Warwick et al., 2022).

1.3 Water: the potential source of hydrogen

One of the most compelling sources of hydrogen is water. Water is abundant and can be found in almost every part of the globe, making it a reliable and sustainable feedstock for hydrogen production. The chemical composition of H_2O consists of two hydrogen atoms bonded to one oxygen atom, providing a direct source of hydrogen.

The electrolysis of H_2O is a well-known procedure for extracting hydrogen. In this electrochemical process, the H_2O molecule splits into its constituents following the hydrogen evolution reaction (HER) and oxygen evolution reaction (OER). This process can be powered by renewable energy sources, ensuring that the hydrogen produced is a truly green and sustainable method. By tapping into water as a hydrogen source, we can leverage an inexhaustible resource to meet our energy needs without depleting natural reserves or harming the environment.

> Ammonia (NH_3) can be used in a fuel cell as a carbon-free fuel to produce energy. Due to its higher boiling point, −33°C as against −253°C for hydrogen, NH_3 can be stored much more easily than hydrogen. Nevertheless, NH_3 is a toxic compound and its combustion produces environmentally unfriendly gases like N_2O and NO_x and raises great concern over its use in fuel cells. Much research is going on in this field to use NH_3 as a hydrogen carrier and storage and, industry and research is exploring the possibilities of ammonia-mediated hydrogen economy (Aziz et al., 2020; Chang et al., 2021; Christensen et al., 2006; Klerke et al., 2008; Little et al., 2015; Negro et al., 2023; Sun et al., 2022).

1.3.1 Splitting of water into H_2 and O_2

The current requirement of hydrogen for various industrial needs is met by different processes outlined above and all of them may lead to the emission of CO_2 in one way or another. The use of sunlight (photons) in the photocatalytic process is a very interesting pathway for splitting water to create H_2 in photoelectrochemical cells. Despite being more environmentally friendly and producing pure hydrogen, poor solar to hydrogen (STH) conversion efficiencies (< 10%) (Kim et al., 2019) mean that they cannot replace bulk and instantaneous generation since they produce insufficient amounts per unit of time. Reactive metal and metal hydrides can be hydrolyzed to quickly create huge amounts of H_2. However, they cannot be produced using a greener way because their precursors are often dangerous metals that are created through dirty chemical industries that pollute the environment. Water electrolysis is the only environmentally friendly way to produce H_2 on a big scale and instantly, hence research is being done to improve the performance of water electrolyzers by creating inexpensive, highly efficient electrocatalysts for the breakdown of water (Raveendran et al., 2023; Yan et al., 2016; Yu et al., 2019; Zhang et al., 2024a).

Water splitting, or electrolysis, involves passing an electric current through water to break it down into hydrogen and oxygen gases. At ambient temperature (298 K) and atmospheric pressure (1 atm), the fundamental reaction request overcome +237.2 kJ/mol of free energy, as can be represented using the following reaction:

$$\mathbf{H_2O} \quad (l) \rightarrow \mathbf{H_2}(g) + \frac{\mathbf{1}}{\mathbf{2}}\mathbf{O_2}(g) \tag{1.1}$$

$\Delta G^0 = +237.2$ kJ/mol (at 298 K and 1 atm.), $\Delta E^0 = 1.23$V (vs RHE)

The process of electrolysis converts one mole of water into one mole of hydrogen and half a mole of oxygen following the thermodynamic potentials and the first law of thermodynamics. The positive value for the change in free energy indicates that the electrolysis is a nonspontaneous process and the energy for the dissociation and expansion of gases must be supplied by the process and is accounted for in the change in enthalpy (285.83 kJ), entropy (163.4 J), and internal energy (282.1 kJ) at 298 K and 1 atm (Raveendran et al., 2023).

The splitting of water is affected by two main processes: water oxidation and water reduction. At the cathode occurs the HER by the reduction of protons as in Eq. (1.2):

$$\mathbf{2H^+(aq.) + 2e^- \rightarrow H_2(g)} \tag{1.2}$$

On the other hand, the OER occurs at the anode by the water oxidation reaction and proceeds as in Eq. (1.3) (Morales-Guio et al., 2014):

$$\mathbf{2H_2O \quad (l) \rightarrow O_2(g) + 4H^+(aq.) + 4e^-} \tag{1.3}$$

In water splitting by the electrolysis of water, the OER is the bottleneck reaction for the process and it is responsible for the high over-potential request to drive electrochemical reaction. Therefore the development of new electrocatalyst materials with high performance, stability, and cost-effectiveness is the one main issue for large-scale hydrogen generation.

Electrolytes increase the conductivity and thus water splitting may occur in the acidic, basic, and neutral media (Zou and Zhang, 2015).

The reactions in acidic media follow Eqs. (1.4) and (1.5) (Shi and Zhang, 2016):

$$\text{Cathode:} \quad 4H^+(aq.) + 4e^- \rightarrow \quad 2H_2(g), \quad E^0 = 0.0 \quad V \tag{1.4}$$

$$\text{Anode:} \quad 2H_2O \quad (l) \rightarrow O_2(g) + 4H^+(aq.) + 4e^-, \quad E^0 = 1.23 \quad V \tag{1.5}$$

In a basic medium, the oxidation and reduction reactions are as follows:

$$\text{Cathode:} \quad 4H_2O \quad (aq.) + 4e^- \rightarrow \quad 2H_2(g) + 4 \quad OH^-, \quad E^0 = -0.828 \quad V \tag{1.6}$$

$$\text{Anode:} \quad 4OH^- \quad (l) \rightarrow O_2(g) + 2H_2O(aq.) + 4e^-, \quad E^0 = 0.401 \quad V \tag{1.7}$$

In neutral medium:

$$\text{Cathode:} \quad 4H_2O \quad (aq.) + 4e^- \rightarrow \quad 2H_2(g) + 4 \quad OH^-, \quad E^0 = 0.413 \quad V \tag{1.8}$$

$$\text{Anode:} \quad 2H_2O(aq.) \rightarrow O_2(g) + 4H^+(aq.) + 4e^-, \quad E^0 = 0.817 \quad V \tag{1.9}$$

Electrolysis in an acidic medium has a better efficiency of HER as plenty of H^+ ions are available for charge and discharge. As there is no instantaneous supply of H^+ ions at the cathode, the efficiency of HER is low in the basic medium. On the other hand, OER efficiency is higher in a basic medium and lower in an acidic medium (Anantharaj et al., 2016b; Wang et al., 2012).

HER follows a multistep mechanism irrespective of the pH of the medium. Adsorption of hydrogen on the surface of the catalyst is the first step and is known as the Volmer reaction. The second step is the desorption of H_2 from the cathode and occurs either through an electrochemical or the chemical route, as depicted in Fig. 1.5 (Gupta et al., 2020). The electrochemical pathway is known as the Heyrovsky step and the chemical one is known as the Tafel step (de Chialvo and Chialvo, 1994a). Both reactions occur simultaneously over the electrode surface, however, depending on the electrocatalyst nature, one reaction is favored over another. It is well accepted that the Tafel reaction is more facile in comparison to

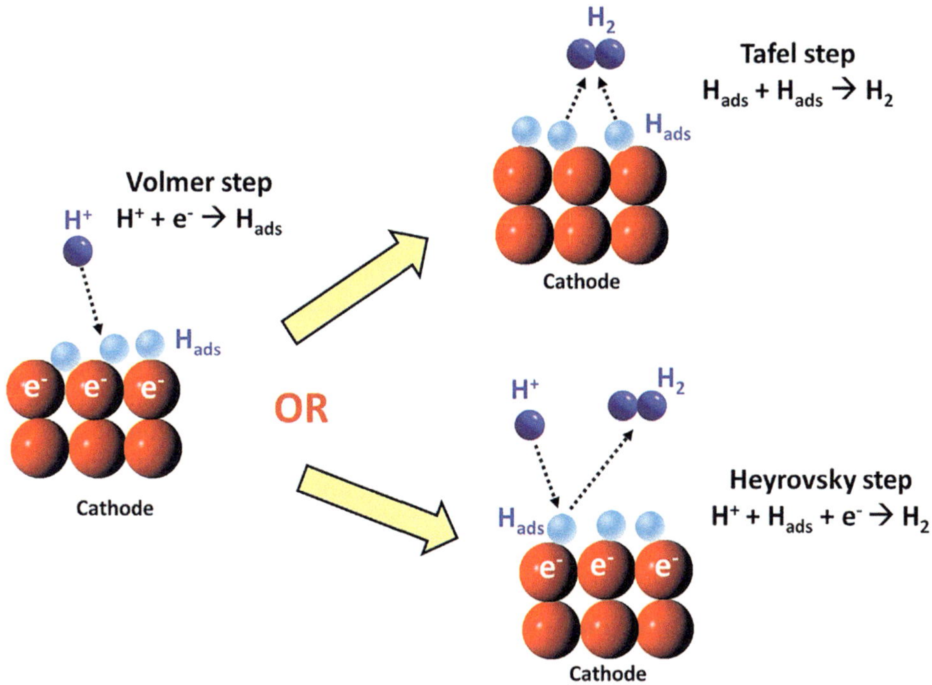

FIGURE 1.5 Schematic illustration of hydrogen evolution reaction pathways. *Source: Reproduced from Gupta, S., Patel, M.K., Miotello, A., Patel, N., 2020. Metal boride-based catalysts for electrochemical water-splitting: a review. Advanced Functional Materials 30, 1906481.*

Heyrovsky one (Yu et al., 2019). The reactions in different steps in acidic media are as follows (de Chialvo and Chialvo, 1994a, 1994b; Zeng and Li, 2015):

$$H^+ + e^- \rightarrow \quad H^* \quad \text{(Volmer)} \tag{1.10}$$

$$H^+ + e^- + H^* \rightarrow H_2 \text{ (Heyrovsky) or } 2H^+ \rightarrow H_2 \text{ (Tafel)} \tag{1.11}$$

And in alkaline media,

$$H_2O + e^- \rightarrow H^* + OH^- \quad \text{(Volmer)} \tag{1.12}$$

$$H_2O + e^- \rightarrow H_2 + OH^- \text{ (Heyrovsky) or } 2H^+ \rightarrow H_2 \text{ (Tafel)} \tag{1.13}$$

Based on these reaction pathways and the values of Tafel slopes, the thermodynamic property, ΔG_{H^*} for the adsorption of hydrogen, can be taken as an indicative measure of the HER performance of the electrocatalyst (Peter, 2002). A catalyst with ΔG_{H^*} value of approximately zero is most efficient. Too negative and too positive values leads to Heyrovsky step and Volmer step, respectively, and arises from difficulty in desorbing in the former case and due to the low interaction of hydrogen with the surface of the electrocatalyst in the latter case (Anantharaj et al., 2016a; Bard and Faulkner, 2001; Gupta et al., 2020; Zhang et al., 2020; Zhu et al., 2015).

The OER is a multistep four-electron transfer process occurring at the anode. Along with the well-known electrochemical oxide and oxide pathway Krasil' shchkov path and Wade and Hackerman's path in acidic medium and Krasil' shchkov path, Yeager's path, and Bockris path in alkaline medium are well studied in literature (Bockris and Otagawa, 1984; Fabbri et al., 2014; Man et al., 2011; Matsumoto and Sato, 1986; Raveendran et al., 2023). In all the mechanisms (Fig. 1.6), both in acidic and basic mediums, active sites on the catalyst undergo cyclic oxidation and reduction. So metals with stable and variable oxidation states will be a good catalyst for OER.

While looking at the chemical reactions proposed for all the mechanisms, it is understood that the formation of intermediates, O*, OH*, and OOH*, the * corresponding to the activity sites on the electrocatalyst surface, is a prerequisite and the free energy for the adsorption value has an impact on the OER activity (Seh et al., 2017; Zhang et al., 2020). Too strong and weak bonds between the intermediate and the catalyst surface cause sluggish reactions and thus OER mechanism varies for different oxide materials and with the compositions of the oxide and even with the differences in synthesis methods that lead to different morphological features.

1.3.2 Current scenario: industrial and laboratory (research)

1.3.2.1 Hydrogen produced at industry

Industrially, hydrogen is already in use, primarily for refining petroleum, producing ammonia for fertilizers, and in various chemical processes. However, most of this hydrogen is produced from natural gas through a process called SMR, which is carbon-intensive. The shift toward green hydrogen—produced via water electrolysis powered by renewable energy—is gaining momentum. Several countries have launched ambitious hydrogen strategies. For instance, the European Union has committed to producing 10 million tonnes of renewable hydrogen by 2030. Similarly, countries like Japan, South Korea, and Australia

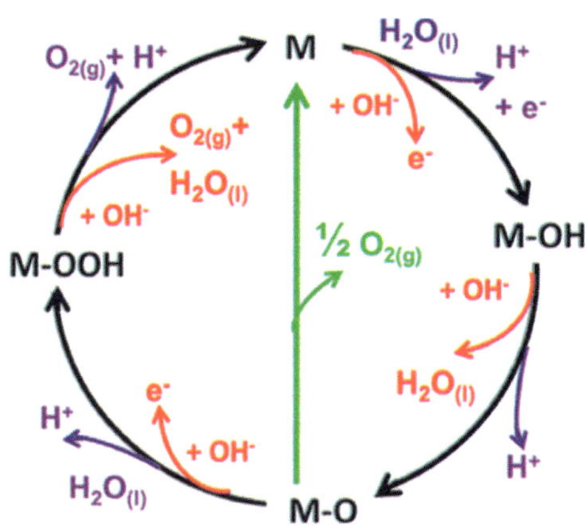

FIGURE 1.6 Mechanisms of oxygen evolution reaction in acidic and basic mediums (Suen et al., 2017). *Source: Adapted from Suen, N.-T., Hung, S.-F., Quan, Q., Zhang, N., Xu, Y.-J., Chen, H.M., 2017. Electrocatalysis for the oxygen evolution reaction: recent development and future perspectives. Chemical Society Reviews 46, 337–365.*

are investing in hydrogen infrastructure, aiming to become global leaders in hydrogen technology and export.

1.3.2.2 Laboratory research

Significant efforts directed toward improving the efficiency and cost-effectiveness of water electrolysis are underway in research laboratories worldwide. Scientists are exploring new catalysts, materials, and electrolysis techniques to enhance performance and reduce energy consumption. Research is also focused on integrating hydrogen production with renewable energy systems and developing advanced storage solutions.

Innovations such as nanoengineered catalysts, advanced membrane materials, and hybrid electrolyzer systems are showing promising results, bringing us closer to realizing the full potential of the hydrogen economy.

Platinum (Pt) metal which has almost zero ΔG_{H*} has low overpotential (almost zero) and tafel slop is about 30 mV/dec (Qadeer et al., 2024). However, owing to its uneconomical nature researchers are focused on developing new cost-effective materials for electrocatalytic water splitting. Copper (Cu)-based materials are used to replace noble metals. Cu-W alloy electrocatalyst is found to be good for HER activity (Jian et al., 2024), and the same is the case for copper oxide (CuO, Cu_2O) (Sajid et al., 2024), copper sulfide (Mottakin et al., 2024), copper phosphide (Jiang et al., 2024). Cu-based alloys and nanocomposites are also widely employed in HER (Chen et al., 2024). In addition to Cu-based materials, other transition metals, including cobalt (Co), iron (Fe), and nickel (Ni) are used as electrocatalysts for water splitting in alkaline medium (Lakhan et al., 2024). Currently, researchers are focused on improving the efficiency and stability of transition metal-based material which has the potential to be implemented as catalyst materials for large-scale hydrogen production (Fan et al., 2024; Sun et al., 2021; Wu et al., 2024). The improvement strategies are summarized in Fig. 1.7.

Renewable energy can be used as an external potential for the electrolysis of water. Abundantly available solar energy can be integrated into the process of electrolyzing water into H_2 and O_2. In photocatalytic water splitting, light energy is used to split water into hydrogen and oxygen (He et al., 2024). In this approach, semiconductor materials are used as photocatalysts. TiO_2 is the widely used semiconductor photocatalyst because of its low cost, high photocatalytic activity, nontoxicity, high stability, and high surface area, despite that it has some limitations including the large band gap energy and high charge recombination rate (Ruan et al., 2024). To improve the photocatalytic activity, researchers are exploring the different approaches for the synthesis of TiO_2 nanoparticles and their composites with various other low band gap semiconductor photocatalysts. Doping with transition metals, such as silver (Ag) (Rani et al., 2024), Ni (Thabet et al., 2024), Fe (Mishra et al., 2024), Cu (Fazil et al., 2024), Ce (Zhang et al., 2021b), etc., is also conducted to improve the water-splitting efficiency of TiO_2. Furthermore, semiconductor photocatalysts based on tungsten (W), iron, Cu, Mo, etc., are also found to have good photocatalytic activity and profound research is going on in this field. Inorganic oxide perovskites, inorganic halide perovskites, organic-inorganic halide perovskites are found to be efficient in photocatalytic water splitting and intense research is being carried out in these fields (Gurudayal et al., 2015; Hoang et al., 2016; Jayaraman et al., 2021; Kim et al., 2015). Rather than using a single

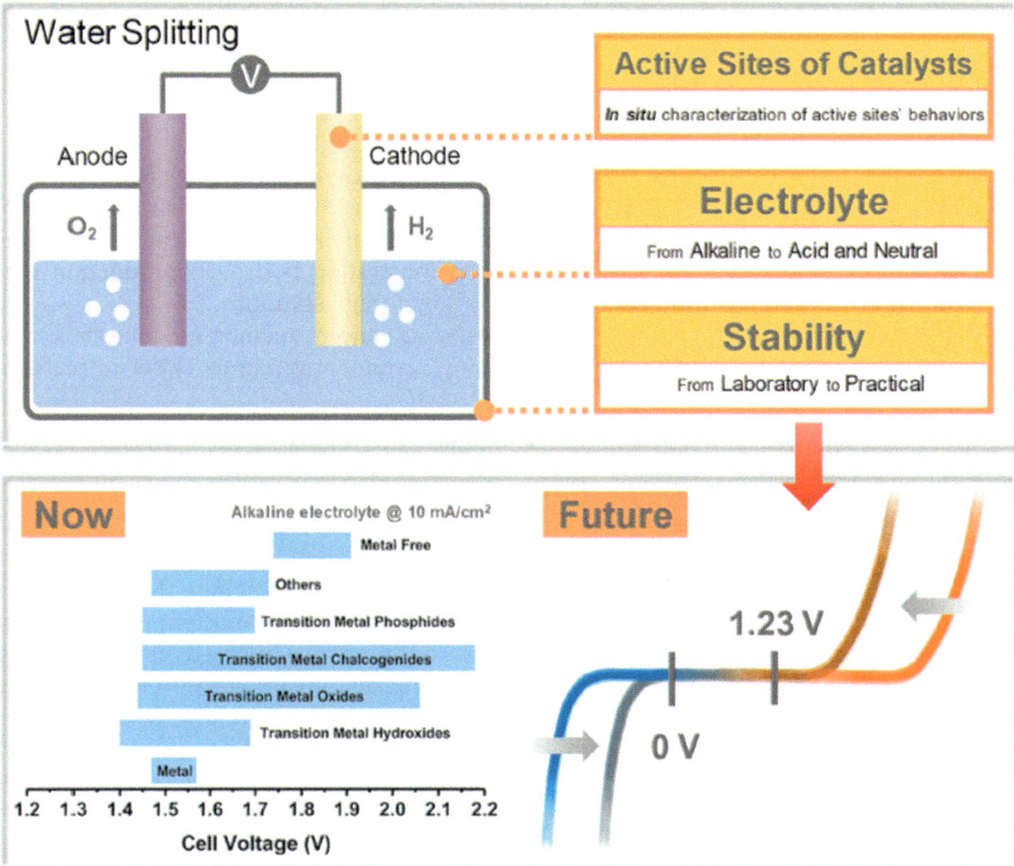

FIGURE 1.7 Future perspective for enhancing the electrocatalytic water splitting (Sun et al., 2021).

catalyst in the photocatalytic water-splitting reaction, the use of composite material is found to have gained momentum in this research area due to comparatively high efficiency and stability through proper band gap management (Huang et al., 2022; Song et al., 2022b; Vidhya et al., 2023).

The use of photovoltaic-electrochemical cells (PV-EC) is another area where intense research is going on. PV-EC, available now, use commercial silicon-based solar cells and thus requires a high capital investment (Report on Hydrogen Production in India, 2016; Samsudin, 2023; Sharma et al., 2023; Song et al., 2022a). Integration of solar cells especially the perovskite-based solar cells (PSC) which have achieved over 26.7% efficiency comparable to the commercial silicon solar cells may lead to an efficient PSC-EC system (NREL, Best Research-Cell Efficiency Chart, 2024). Research in using tandem cells is also undergoing widely in worldwide research labs (Gurudayal et al., 2015; Kim et al., 2015).

In the thermochemical method, heat energy is used to split water molecules at high temperatures up to 2000°C (Kumar and Sahu, 2024). Aluminum is an efficient, inexpensive, and easily produced material for thermochemical water-splitting cycles (Samidin et al., 2024). Thermochemical water-splitting reactions can be effectively conducted at lower temperatures using Sn (Pan et al., 2024), Zn (Bhosale, 2024), Fe (Kumar and Sahu, 2024), Ce (Liu et al., 2024), and Mg (Boretti, 2022). Transition metal-based perovskites are another promising material for thermochemical water-splitting cycles (Zhang et al., 2023). A schematic representation of the perovskite thermochemical water-splitting catalyst is shown in Fig. 1.8. The foremost disadvantage of this method is the essentiality of high temperature and lower hydrogen yield.

In a photobiological water-splitting reaction, an algae-based reactor is facilitated for the generation of hydrogen and oxygen (Ramadhani and Said, 2024). However, in this case, also the yield is the main challenge (Goria et al., 2024). Innovative approaches to enhancing hydrogen production through bio-photo-electro-chemical cells, microbial electrolysis, and cellular bionic systems are underway. Thylakoid membranes extracted from spinach, when incorporated into bio-photo-electrochemical cells, facilitate water splitting with a solar-to-hydrogen efficiency of 0.3%. Furthermore, bioelectrochemical systems integrating bacteria like *Ralstonia eutropha* demonstrate scalable solar-to-fuels conversions, yielding hydrogen and oxygen produced during water splitting and biomass and alcohols from CO_2. The engineering of living algae interfaced with conductive polymers also enables sustained photosynthetic-independent hydrogen production. Microbial electrolysis cell combined with algal biomass production offers a resource-efficient method for a circular economy and sustainable environment. These advancements highlight the integration of biotic and abiotic systems for sustainable hydrogen production, crucial for the development of future carbon-neutral energy technologies (Gautam et al., 2024; Pinhassi et al., 2016; Torella et al., 2015; Xu et al., 2023).

Now the most important technique for water splitting is the photoelectrochemical method (Marwat et al., 2021). This method uses light energy and a low external electrical bias to improve the efficiency of water-splitting reactions. This method is considered one of the clean methods

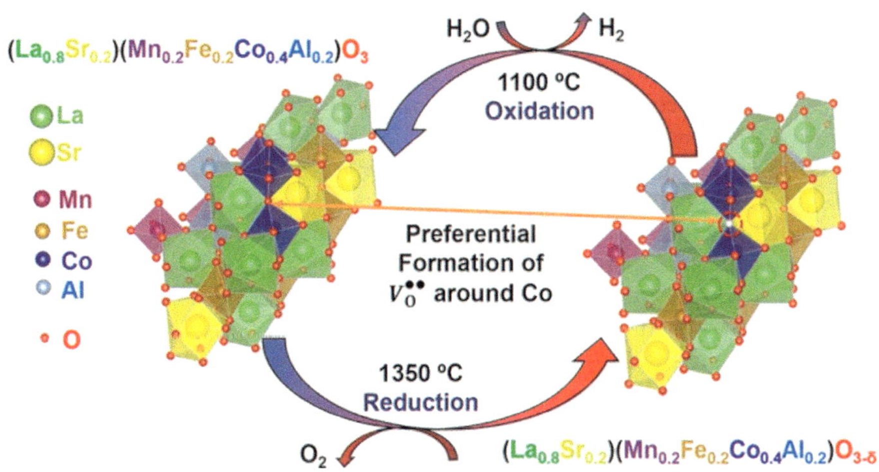

FIGURE 1.8 Schematic representation of the perovskite thermochemical water-splitting catalyst (Zhang et al., 2023).

that can be adopted for water-splitting reactions for the generation of hydrogen. This method is one of the simplest and most efficient approaches for the conversion of solar energy to hydrogen fuel. This method is superior to the thermochemical method because high temperature is not required in this process (Imran and Hussain, 2024). In this technique, a low input bias is used to overcome the overpotential of the water-splitting reaction. The solar-to-hydrogen conversion efficiency is higher in the case of photoelectrochemical water-splitting reaction, but highly efficient photocathodes and photoanodes are essential for HER and OER. For efficient PEC performance, uniform surface and bulk engineering of a nanocatalyst is essential with coupling of p- and n-type material (Marwat et al., 2021).

By a similar approach, Yoo et al. (2019) were able to generate a photocurrent of 30 mA/cm^2 at 1 V versus Ag/AgCl electrode using a p-n junction type TiO_2–NR/CdS-NiOx catalyst using a solar simulator. A favorable band edge alignment resulted in an efficient p-n junction formation, which underscores the enhanced activity. The corresponding band edge alignment diagrams are shown in Fig. 1.9. The efficiency of the current PEC catalysts can be modified with surface and defect engineering, efficient heterojunction formation, morphological and bulk compositional modifications, etc. (Marwat et al., 2021). However, no stable PEC catalyst for industrial application has yet been developed. Modification of current catalysts with suitable electrocatalysts including NiCoOOH, NiOOH, CoO_x, etc., significantly enhances PEC water-splitting efficiency (Benavente Llorente et al., 2023; Luo et al., 2022).

1.4 Role of artificial intelligence and additive manufacturing in water electrolysis

Despite the years of dedicated research in water electrolysis by catalysts for green hydrogen production, the efficiency, cyclability, and material stability have not been improved much except in the case of expensive metal catalysts like platinum and palladium. However, the massive amount of data and different models originating from research worldwide can be used efficiently by the use of artificial intelligence (AI) and additive manufacturing (AM). AI can be used in optimizing the production process to maximize efficiency and reduce costs (Lim et al., 2023). By analyzing vast amounts of data and providing insights and recommendations, AI has the potential to improve the quantity and quality of information available to human decision-makers, ultimately resulting in better and more informed choices (Bilgiç and Öztürk, 2023; Kheirrouz et al., 2022; Nasir et al., 2020). Tools for AI are becoming more widely acknowledged as a promising method for bioprocess modelling and optimization. These instruments have the power to greatly enhance industrial equipment's functionality and design, as well as anticipate and stop equipment breakdowns. Fig. 1.10 represents a schematic representation of the usage of AI in water splitting (Kheirrouz et al., 2022).

AM, also referred to as 3D printing, has drawn a lot of interest recently as a potent technique for precisely and accurately constructing intricate shapes. This is very useful when creating catalysts and reactors. Researchers have created intricately shaped microreactors using 3D printing and these microreactors exhibit increased conversion rates as well as selectivity in comparison to conventional reactors (Hegde et al., 2018). Additionally, additive manufacturing can be utilized to create unique catalysts for producing green

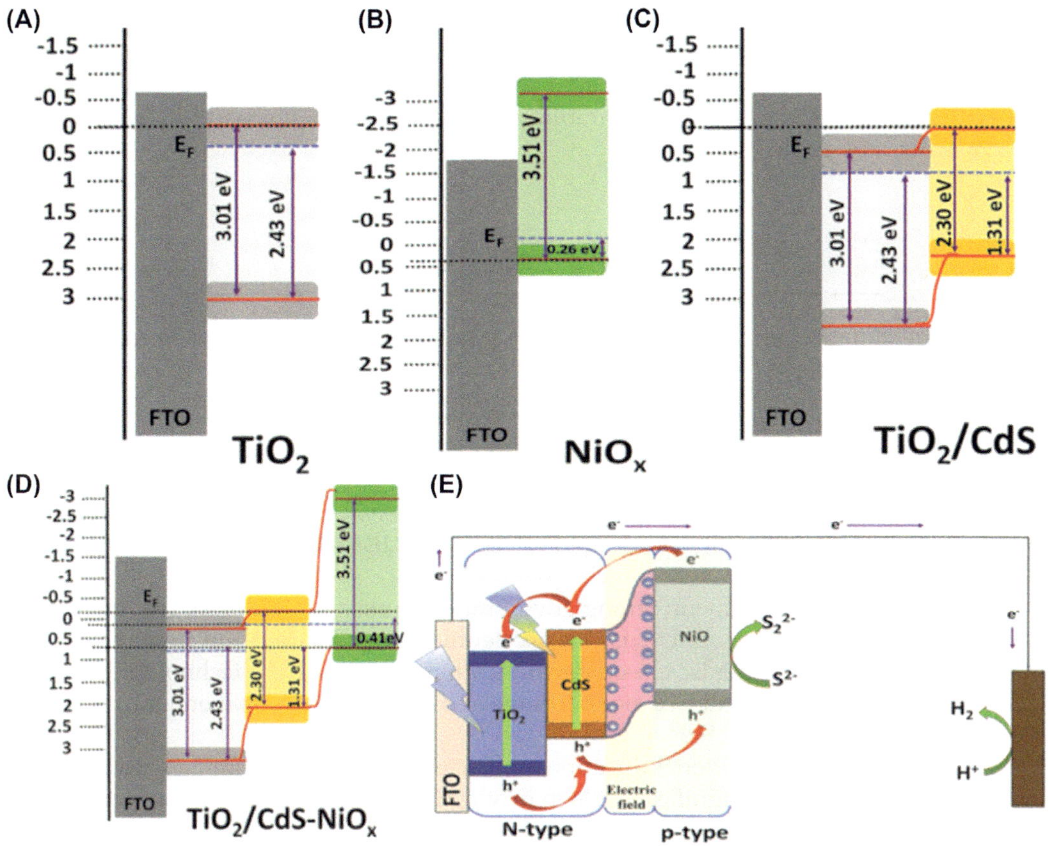

FIGURE 1.9 Band alignment in p-n junction type TiO_2–NR/CdS-NiOx catalyst (Yoo et al., 2019).

FIGURE 1.10 Schematic representation of AI in water splitting (Adeli et al., 2023; Kheirrouz et al., 2022).

hydrogen. For instance, scientists have created nickel foam catalysts that can be 3D printed and used to electrolyze water to produce hydrogen (Al-Othman et al., 2022; Chiriță et al., 2019; Mandal et al., 2018). When compared to conventional catalysts, these catalysts have demonstrated increased activity and durability. It might result in the synthesis of hydrogen more cheaply and efficiently.

1.5 Future prospects

The future of hydrogen as a cornerstone of the global energy system looks promising but hinges on overcoming several key challenges:

1. Cost reduction: The high cost of electrolyzers and the electricity required for water splitting must be reduced. Continued research and economies of scale are expected to drive down costs.
2. Infrastructure development: Building a comprehensive hydrogen infrastructure, including production facilities, storage systems, and distribution networks, is essential for widespread adoption.
3. Policy support and investment: Strong governmental policies, incentives, and investment in hydrogen technologies are crucial for accelerating development and deployment.
4. Technological advancements: On-going innovation in electrolysis technologies, renewable energy integration, and hydrogen storage and transportation will enhance efficiency and feasibility.
5. Public and industrial adoption: Education and awareness programs, along with industrial partnerships, will play a significant role in driving the adoption of hydrogen technologies across various sectors.

The book "Photochemical Splitting of Water: Fundamentals to Applications" provides a comprehensive understanding of the photochemical water-splitting process, from its basic concepts to experimental setups and results analysis. It reviews different materials used as catalysts, their recent advancements, and future prospects. The entire book is divided into three sections. A basic understanding of the theories behind photochemical water splitting, instrumentation, experimental setup, and the criteria for the selection of materials are discussed in the first section. The second section is dedicated to the materials used as active materials in photocatalytic and photoelectrocatalytic water-splitting processes. It is discussed categorically chapterwise under 10 titles. Recent advances in each of these titles are discussed. The third and last section consists of the future prospective of this research area by including a discussion on integrating the photovoltaic cell with an electrocatalytic hydrogen evolution and the challenges for photoelectrochemical water-splitting technology, scale-up, and limitations. Furthermore, the third section discusses emerging trends, such as using AI and machine learning for catalyst design and optimization of the water-splitting process.

1.6 Conclusion

In conclusion, hydrogen holds immense potential to revolutionize the global energy landscape. By harnessing water as a sustainable source of hydrogen and advancing the technologies for its efficient production and utilization, we can move toward a cleaner, more resilient, and sustainable energy future. Advanced technologies like AI and AM will help optimize the production process, thereby increasing efficiency while reducing the cost of green hydrogen production. The journey toward a hydrogen economy is challenging, but the rewards—a sustainable planet and a secure energy future—make it a worthy pursuit.

References

Adeli, K., Nachtane, M., Faik, A., Saifaoui, D., Boulezhar, A., 2023. How green hydrogen and ammonia are revolutionizing the future of energy production: a comprehensive review of the latest developments and future prospects. Applied Sciences 13, 8711.

Al-Othman, A., Tawalbeh, M., Martis, R., Dhou, S., Orhan, M., Qasim, M., et al., 2022. Artificial intelligence and numerical models in hybrid renewable energy systems with fuel cells: advances and prospects. Energy Conversion and Management 253, 115154.

Anantharaj, S., Ede, S.R., Sakthikumar, K., Karthick, K., Mishra, S., Kundu, S., 2016a. Recent trends and perspectives in electrochemical water splitting with an emphasis on sulfide, selenide, and phosphide catalysts of Fe, Co, and Ni: a review. ACS Catalysis 6, 8069–8097.

Anantharaj, S., Karthik, P.E., Subramanian, B., Kundu, S., 2016b. Pt nanoparticle anchored molecular self-assemblies of DNA: an extremely stable and efficient HER electrocatalyst with ultralow Pt content. ACS Catalysis 6, 4660–4672.

Aziz, M., Wijayanta, A.T., Nandiyanto, A.B.D., 2020. Ammonia as effective hydrogen storage: a review on production, storage and utilization. Energies 13, 3062.

Bard, A.J., Faulkner, L.R., 2001. Fundamentals and applications: electrochemical methods. Electrochemical Methods 2 (482), 580–632.

Benavente Llorente, V., Jenewein, K.J., Bierling, M., Körner, A., Hutzler, A., Kormányos, A., et al., 2023. Photocorrosion of hematite photoanodes in neutral and alkaline electrolytes. The Journal of Physical Chemistry C 127, 19687–19697.

Bhosale, R.R., 2024. H_2 producing hybrid solar thermochemical $ZnSO_4/ZnO$ water splitting cycle: thermodynamic efficiency analysis. International Journal of Hydrogen Energy 49, 1584–1592.

Bilgiç, G., Öztürk, B., 2023. Modeling of artificial neural networks for hydrogen production via water electrolysis. El-Cezeri 10, 137–146.

Bockris, J.O., Otagawa, T., 1984. The electrocatalysis of oxygen evolution on perovskites. Journal of the Electrochemical Society 131 (2), 290.

Boretti, A., 2022. Which thermochemical water-splitting cycle is more suitable for high-temperature concentrated solar energy? International Journal of Hydrogen Energy 47, 20462–20474.

Brouwer, A.S., Van Den Broek, M., Seebregts, A., Faaij, A., 2014. Impacts of large-scale intermittent renewable energy sources on electricity systems, and how these can be modeled. Renewable and Sustainable Energy Reviews 33, 443–466.

Capurso, T., Stefanizzi, M., Torresi, M., Camporeale, S.M., 2022. Perspective of the role of hydrogen in the 21st century energy transition. Energy Conversion and Management 251, 114898.

Chang, F., Gao, W., Guo, J., Chen, P., 2021. Emerging materials and methods toward ammonia-based energy storage and conversion. Advanced Materials 33, 2005721.

Chen, Y., Zhao, C., Wang, S., Zhang, L., Liu, G., Dai, W., et al., 2024. Bridging sustainability and catalytic efficiency: Cu_7Co alloy-decorated biomass-derived carbon as a highly efficient electrocatalyst for hydrogen generation. New Journal of Chemistry 48, 12166–12173.

Chiriță, A.-P., Bere, P.-P., Radoi, R.I., Dumitrescu, L., 2019. Aspects regarding the use of 3D printing technology and composite materials for testing and manufacturing vertical axis wind turbines. Materiale Plastice 56 (4), 910–917.

Christensen, C.H., Johannessen, T., Sørensen, R.Z., Nørskov, J.K., 2006. Towards an ammonia-mediated hydrogen economy? Catalysis Today 111, 140–144.

COAG Energy Council, 2019. Commonwealth of Australia Australia's National Hydrogen Strategy. COAG Energy Council.

de Chialvo, M.R.G., Chialvo, A.C., 1994a. The resolution of the volmer-heyrovsky-tafel mechanism with a normal distribution of the standard gibbs energy of adsorption. Journal of the Brazilian Chemical Society 5, 137–143.

de Chialvo, M.R.G., Chialvo, A.C., 1994b. Hydrogen evolution reaction: analysis of the Volmer-Heyrovsky-Tafel mechanism with a generalized adsorption model. Journal of Electroanalytical Chemistry 372, 209–223.

Derwent, R.G., Stevenson, D.S., Utembe, S.R., Jenkin, M.E., Khan, A.H., Shallcross, D.E., 2020. Global modelling studies of hydrogen and its isotopomers using STOCHEM-CRI: likely radiative forcing consequences of a future hydrogen economy. International Journal of Hydrogen Energy 45, 9211–9221.

Dincer, I., Aydin, M.I., 2023. New paradigms in sustainable energy systems with hydrogen. Energy Conversion and Management 283, 116950.

Fabbri, E., Habereder, A., Waltar, K., Kötz, R., Schmidt, T.J., 2014. Developments and perspectives of oxide-based catalysts for the oxygen evolution reaction. Catalysis Science and Technology 4, 3800–3821.

Fan, C., Zang, Z., Zhang, X., 2024. Non-metal doping regulation in transition metal and their compounds for electrocatalytic water splitting. International Journal of Hydrogen Energy 56, 1273–1283.

Fan, Z., Sheerazi, H., Bhardwaj, A., Corbeau, A.-S., Longobardi, K., Castañeda, A., et al., 2022. Hydrogen leakage: a potential risk for the hydrogen economy. Center on Global Energy Policy (CGEP). Columbia University, pp. 1–33.

Fazil, M., Alshehri, S.M., Mao, Y., Ahmad, T., 2024. Enhanced photo/electrocatalytic hydrogen evolution by hydrothermally derived Cu-doped TiO_2 solid solution nanostructures. Langmuir 40, 4063–4076.

Feldmann, J., Byrum, Z., Cyrs, T., 2023. Clean Hydrogen: Outlook for Freight Transport in the United States. World Resources Institute, Washington, DC.

Gautam, R., Steinberger Wilckens, R., Ghosh, U.K., 2024. Hydrogen production in microbial electrolysis cell and valorization of reactor effluent for algal biomass. ACS Sustainable Resource Management 1 (8), 1738–1749.

Goria, K., Singh, H.M., Singh, A., Kothari, R., Tyagi, V.V., 2024. Insights into biohydrogen production from algal biomass: challenges, recent advancements and future directions. International Journal of Hydrogen Energy 52, 127–151.

Gupta, S., Patel, M.K., Miotello, A., Patel, N., 2020. Metal boride-based catalysts for electrochemical water-splitting: a review. Advanced Functional Materials 30, 1906481.

Gurudayal, Sabba, D., Kumar, M.H., Wong, L.H., Barber, J., Grätzel, M., et al., 2015. Perovskite–hematite tandem cells for efficient overall solar driven water splitting. Nano Letters 15, 3833–3839.

He, L., Zeng, X., Chen, H., Zhao, L., Huang, Z., Wang, D., et al., 2024. A hybrid photocatalytic system splits atmospheric water to produce hydrogen. Advanced Functional Materials 34, 2313058.

Hegde, C., Yan, Q., Li, H., 2018. 3D printing electro-catalysts for hydrogen production. Progress in Additive Manufacturing (Pro-AM 2018, 625–630.

Hoang, M.T., Pham, N.D., Han, J.H., Gardner, J.M., Oh, I., 2016. Integrated photoelectrolysis of water implemented on organic metal halide perovskite photoelectrode. ACS Applied Materials and Interfaces 8, 11904–11909.

Hoegh-Guldberg, O., Jacob, D., Bindi, M., Brown, S., Camilloni, I., Diedhiou, A., Djalante, R., Ebi, K., Engelbrecht, F., Guiot, J. and Hijioka, Y., 2018. Impacts of 1.5 C global warming on natural and human systems. In Global warming of 1.5 C.: An IPCC Special Report (pp. 175-311). IPCC Secretariat.

Huang, Q., Guo, Y., Chen, J., Lou, Y., Zhao, Y., 2022. NiCoP modified lead-free double perovskite $Cs_2AgBiBr_6$ for efficient photocatalytic hydrogen generation. New Journal of Chemistry 46, 7395–7402.

IEA, 2023. World Energy Outlook 2023. IEA, Paris.

Imran, S., Hussain, M., 2024. Emerging trends in water splitting innovations for solar hydrogen production: analysis, comparison, and economical insights. International Journal of Hydrogen Energy 77, 975–996.

Japan, 2023. Basic Hydrogen Strategy. The Ministerial Council on Renewable Energy, Hydrogen and Related Issues.

Jayaraman, J.P., Hamdan, M., Velpula, M., Kaisare, N.S., Chandiran, A.K., 2021. $BiVO_4/Cs_2PtI_6$ vacancy-ordered halide perovskite heterojunction for panchromatic light harvesting and enhanced charge separation in photoelectrochemical water oxidation. ACS Applied Materials and Interfaces 13, 16267–16278.

Ji, M., Wang, J., 2021. Review and comparison of various hydrogen production methods based on costs and life cycle impact assessment indicators. International Journal of Hydrogen Energy 46, 38612–38635.

Jian, X., Zhang, W., Yang, Y., Li, Z., Pan, H., Gao, Q., et al., 2024. Amorphous Cu–W Alloys as stable and efficient electrocatalysts for hydrogen evolution. ACS Catalysis 14, 2816–2827.

Jiang, H., Shi, J., Liu, X., Tang, J., 2024. Highly dispersed copper phosphide nanoparticles accelerate the electrolytic water oxidation process. Green Chemistry 26, 3388–3396.

Kheirrouz, M., Melino, F., Ancona, M.A., 2022. Fault detection and diagnosis methods for green hydrogen production: a review. International Journal of Hydrogen Energy 47, 27747–27774.

Kim, J.H., Jo, Y., Kim, J.H., Jang, J.W., Kang, H.J., Lee, Y.H., et al., 2015. Wireless solar water splitting device with robust cobalt-catalyzed, dual-doped $BiVO_4$ photoanode and perovskite solar cell in tandem: a dual absorber artificial leaf. ACS Nano 9, 11820–11829.

Kim, J.H., Hansora, D., Sharma, P., Jang, J.-W., Lee, J.S., 2019. Toward practical solar hydrogen production–an artificial photosynthetic leaf-to-farm challenge. Chemical Society Reviews 48, 1908–1971.

Klerke, A., Christensen, C.H., Nørskov, J.K., Vegge, T., 2008. Ammonia for hydrogen storage: challenges and opportunities. Journal of Materials Chemistry 18, 2304–2310.

Kumar, S., Sahu, R.K., 2024. Fe–Al core-shell structure as an efficient catalyst for dual hydrogen production and storage by thermochemical water splitting: a reactive molecular dynamic simulation. International Journal of Hydrogen Energy 52, 1498–1506.

Lachlan Carey, 2021. The United Kingdom's Hydrogen Strategy [WWW Document]. Center for Strategic and International Studies https://www.csis.org/analysis/united-kingdoms-hydrogen-strategy.

Lakhan, M.N., Hanan, A., Hussain, A., Ali Soomro, I., Wang, Y., Ahmed, M., et al., 2024. Transition metal-based electrocatalysts for alkaline overall water splitting: advancements, challenges, and perspectives. Chemical Communications 60, 5104–5135.

Lee, M., Noah, M., Park, J., Seong, M.J., Kwon, Y.K., Hong, S., 2009. Textured network devices: overcoming fundamental limitations of nanotube/nanowire network-based devices. Small 5, 1642–1648.

Li, X., Zhao, L., Yu, J., Liu, X., Zhang, X., Liu, H., et al., 2020. Water splitting: from electrode to green energy system. Nano-Micro Letters 12 (1), 1–29.

Lim, C.Y.J., Made, R.I., Khoo, Z.H.J., Ng, C.K., Bai, Y., Wang, J., et al., 2023. Machine learning-assisted optimization of multi-metal hydroxide electrocatalysts for overall water splitting. Materials Horizons 10, 5022–5031.

Little, D.J., Smith III, M.R., Hamann, T.W., 2015. Electrolysis of liquid ammonia for hydrogen generation. Energy and Environmental Science 8, 2775–2781.

Liu, Y., Gu, T., Bu, C., Liu, D., Piao, G., 2024. Investigation on the activity of Ni doped $Ce_{0.8}Zr_{0.2}O_2$ for solar thermochemical water splitting combined with partial oxidation of methane. International Journal of Hydrogen Energy 62, 1077–1088.

Luo, Y., Wu, Y., Huang, C., Menon, C., Feng, S., Chu, P.K., 2022. Plasma modified and tailored defective electrocatalysts for water electrolysis and hydrogen fuel cells. EcoMat 4, e12197.

Man, I.C., Su, H., Calle-Vallejo, F., Hansen, H.A., Martínez, J.I., Inoglu, N.G., et al., 2011. Universality in oxygen evolution electrocatalysis on oxide surfaces. ChemCatChem 3, 1159–1165.

Mandal, M., Valls, A., Gangnus, N., Secanell, M., 2018. Analysis of inkjet printed catalyst coated membranes for polymer electrolyte electrolyzers. Journal of the Electrochemical Society 165 (7), F543.

Marwat, M.A., Humayun, M., Afridi, M.W., Zhang, H., Abdul Karim, M.R., Ashtar, M., et al., 2021. Advanced catalysts for photoelectrochemical water splitting. ACS Applied Energy Materials 4 (11), 12007–12031.

Matsumoto, Y., Sato, E., 1986. Electrocatalytic properties of transition metal oxides for oxygen evolution reaction. Materials Chemistry and Physics 14, 397–426.

McQueen, S., Stanford, J., Satyapal, S., Miller, E., Stetson, N., Papageorgopoulos, D., 2020. US-DOE Hydrogen Program Plan.

Mishra, D., Rathore, D., Bansal, P., Garg, M., Pande, S., Chatterjee, S., 2024. Novel Fe-doped TiO_2 metal-organic framework for electrocatalytic hydrogen evolution reaction. Chemical Engineering Research and Design 207, 268–280.

Morales-Guio, C., Stern, L., X, H., 2014. Nanostructured hydrotreating catalysts for electrochemical hydrogen evolution. Chemical Society Reviews 43, 6555–6569.

Mottakin, M., Su'ait, M.S., Selvanathan, V., Ibrahim, M.A., Abdullah, H., Akhtaruzzaman, M., 2024. Integrating multiphasic CuS_x/FeS_x Nanostructured Electrocatalyst For Enhanced Oxygen And Hydrogen Evolution Reactions In Saline Water Splitting. Journal of Alloys and Compounds 1002, 175351.

Nakano, J., 2021. South Korea's Hydrogen Industrial Strategy [WWW Document]. Center for Strategic and International Studies https://www.csis.org/analysis/south-koreas-hydrogen-industrial-strategy.

Nasir, T., Asmaela, M., Zeeshana, Q., Solyalib, D., 2020. Applications of machine learning to friction stir welding process optimization. Jurnal Kejuruteraan 32 (1), 171–186.

Negro, V., Noussan, M., Chiaramonti, D., 2023. The potential role of ammonia for hydrogen storage and transport: a critical review of challenges and opportunities. Energies 16, 6192.

Notton, G., Nivet, M.L., Voyant, C., Paoli, C., Darras, C., Motte, F., et al., 2018. Intermittent and stochastic character of renewable energy sources: consequences, cost of intermittence and benefit of forecasting. Renewable and Sustainable Energy Reviews 87, 96–105.

NREL, Best Research-Cell Efficiency Chart [WWW Document], 2024. NREL, National Renewable Energy Laboratory https://www.nrel.gov/pv/cell-efficiency.html

Ocko, I.B., Hamburg, S.P., 2022. Climate consequences of hydrogen emissions. Atmospheric Chemistry and Physics 22, 9349–9368.

Osman, A.I., Mehta, N., Elgarahy, A.M., Hefny, M., Al-Hinai, A., Al-Muhtaseb, A.H., et al., 2022. Hydrogen production, storage, utilisation and environmental impacts: a review. Environmental Chemistry Letters 1–36.

Outlook, B.P.E., 2023. edition, BP Energy Economics

Pan, H., Lu, Y., Hu, B., 2024. Materials used for solar thermal/thermochemical processes for CO_2/H_2O dissociation/conversion. Conversion of Water and CO_2 to Fuels Using Solar Energy: Science, Technology and Materials. pp. 203–231.

Peter, L., 2002. Comprehensive modern electrochemistry. Platinum Metals Review 46 (1), 15–17.

Pinhassi, R.I., Kallmann, D., Saper, G., Dotan, H., Linkov, A., Kay, A., et al., 2016. Hybrid bio-photo-electro-chemical cells for solar water splitting. Nature Communications 7, 12552.

Qadeer, M.A., Zhang, X., Farid, M.A., Tanveer, M., Yan, Y., Du, S., et al., 2024. A review on fundamentals for designing hydrogen evolution electrocatalyst. Journal of Power Sources 613, 234856.

Ramadhani, R., Said, A., 2024. A review of insights into algae as a sustainable bio-based photocatalyst for environmental remediation. International Journal of Hydrological and Environmental for Sustainability 3, 18–47.

Rani, S., Singh, S., Pal, B., 2024. Core-shell structure of reduced graphene oxide@ Ag–TiO_2 for photocatalytic H_2O splitting and CH_3OH dehydrogenation under UV light irradiation. International Journal of Hydrogen Energy 49, 910–924.

Raveendran, A., Chandran, M., Dhanusuraman, R., 2023. A comprehensive review on the electrochemical parameters and recent material development of electrochemical water splitting electrocatalysts. RSC Advances 13, 3843–3876.

Report on Hydrogen Production in India, 2016.

Ruan, X., Li, S., Huang, C., Zheng, W., Cui, X., Ravi, S.K., 2024. Catalyzing artificial photosynthesis with TiO_2 heterostructures and hybrids: emerging trends in a classical yet contemporary photocatalyst. Advanced Materials 36, 2305285.

Sajid, M., Qayyum, W., Farhan, A., Qamar, M.A., Nawaz, H., 2024. Progress in the development of copper oxide-based materials for electrochemical water splitting. International Journal of Hydrogen Energy 62, 209–227.

Samidin, S., Isahak, W.N.R.W., Ahmad, K.N., Mijan, N.A., Yusop, M.R., Samsuri, A., et al., 2024. Improving green hydrogen production: microwave-assisted synthesis of diverse Ni-based nanocatalysts on Al_2O_3 for thermochemical water splitting. Fuel 373, 132370.

Samsudin, M.F.R., 2023. Photovoltaic-assisted photo (electro) catalytic hydrogen production: a review. Energies 16, 5743.

Seh, Z.W., Kibsgaard, J., Dickens, C.F., Chorkendorff, I.B., Nørskov, J.K., Jaramillo, T.F., 2017. Combining theory and experiment in electrocatalysis: insights into materials design. Science 355 (6321), eaad4998 (1979).

Sharma, A., Longden, T., Catchpole, K., Beck, F.J., 2023. Comparative techno-economic analysis of different PV-assisted direct solar hydrogen generation systems. Energy and Environmental Science 16, 4486–4501.

Shi, Y., Zhang, B., 2016. Recent advances in transition metal phosphide nanomaterials: synthesis and applications in hydrogen evolution reaction. Chemical Society Reviews 45, 1529–1541.

Song, H., Luo, S., Huang, H., Deng, B., Ye, J., 2022a. Solar-driven hydrogen production: recent advances, challenges, and future perspectives. ACS Energy Letters 1043–1065.

Song, K., Gou, J., Wu, L., Zeng, C., 2022b. Two-dimensional black phosphorus-modified $Cs_2AgBiBr_6$ with efficient charge separation for enhanced visible-light photocatalytic H_2 evolution. Journal of Materials Chemistry C 10, 15386–15393.

Statista, 2021. Forecast Global Hydrogen Sector Demand in Sustainable Development Scenario 2019–2070 [WWW Document] https://www.statista.com/statistics/760001/global-hydrogen-demand-by-sector-sustainable-scenario/.

Suen, N.-T., Hung, S.-F., Quan, Q., Zhang, N., Xu, Y.-J., Chen, H.M., 2017. Electrocatalysis for the oxygen evolution reaction: recent development and future perspectives. Chemical Society Reviews 46, 337–365.

Sun, L., Luo, Q., Dai, Z., Ma, F., 2021. Material libraries for electrocatalytic overall water splitting. Coordination Chemistry Reviews 444, 214049.

Sun, S., Jiang, Q., Zhao, D., Cao, T., Sha, H., Zhang, C., et al., 2022. Ammonia as hydrogen carrier: advances in ammonia decomposition catalysts for promising hydrogen production. Renewable and Sustainable Energy Reviews 169, 112918.

Thabet, S.M., Abdelhamid, H.N., Ibrahim, S.A., El-Bery, H.M., 2024. Boosting photocatalytic water splitting of TiO_2 using metal (Ru, Co, or Ni) co-catalysts for hydrogen generation. Scientific Reports 14, 10115.

Torella, J.P., Gagliardi, C.J., Chen, J.S., Bediako, D.K., Colón, B., Way, J.C., et al., 2015. Efficient solar-to-fuels production from a hybrid microbial–water-splitting catalyst system. Proceedings of the National Academy of Sciences 112, 2337–2342.

UNFCCC, 2022. Report of the conference of the parties serving as the meeting of the parties to the Paris Agreement on its third session, held in Glasgow from 31 October to 13 November 2021. Decisions adopted by the conference of the parties serving as the meeting of the parties to the Paris Agreement 03276, pp. 1–46.

Vidhya, C., Meera, B., Nair, R.B., Kurian, S., 2023. Enhanced photocatalytic hydrogen evolution through suppressed electron-hole recombination in $Cs_2AgBiBr_6$-NC/g-C_3N_4 nanocomposites. Catalysis Science and Technology 14 (3), 746–757.

Wang, C., Dekrafft, K.E., Lin, W., 2012. Pt nanoparticles@photoactive metal-organic frameworks: efficient hydrogen evolution via synergistic photoexcitation and electron injection. Journal of the American Chemical Society 134, 7211–7214.

Warwick, N., Griffiths, P., Keeble, J., Archibald, A., Pyle, J., Shine, K., 2022. Policy paper. Atmospheric Implications of Increased Hydrogen Use.

Wu, K., Lyu, C., Cheng, J., Ding, W., Wu, J., Wang, Q., et al., 2024. Defect engineering in transition-metal (Fe, Co, and Ni)-based electrocatalysts for water splitting. Carbon Energy 6, e485.

Xu, Z., Qi, J., Wang, S., Liu, X., Li, M., Mann, S., et al., 2023. Algal cell bionics as a step towards photosynthesis-independent hydrogen production. Nature Communications 14, 1872.

Yan, Y., Xia, B.Y., Zhao, B., Wang, X., 2016. A review on noble-metal-free bifunctional heterogeneous catalysts for overall electrochemical water splitting. Journal of Materials Chemistry A 4, 17587–17603.

Yoo, I., Kalanur, S.S., Seo, H., 2019. A nanoscale p–n junction photoelectrode consisting of an NiOx layer on a TiO_2/CdS nanorod core-shell structure for highly efficient solar water splitting. Applied Catalysis B 250, 200–212.

Yu, P., Wang, F., Shifa, T.A., Zhan, X., Lou, X., Xia, F., et al., 2019. Earth abundant materials beyond transition metal dichalcogenides: a focus on electrocatalyzing hydrogen evolution reaction. Nano Energy 58, 244–276.

Zeng, M., Li, Y., 2015. Recent advances in heterogeneous electrocatalysts for the hydrogen evolution reaction. Journal of Materials Chemistry A 3 (29), 14942–14962.

Zhang, C., Nan, H., Tian, H., Zheng, W., 2020. Recent advances in pentlandites for electrochemical water splitting: a short review. Journal of Alloys and Compounds 838, 155685.

Zhang, D., De Santiago, H.A., Xu, B., Liu, C., Trindell, J.A., Li, W., et al., 2023. Compositionally complex perovskite oxides for solar thermochemical water splitting. Chemistry of Materials 35, 1901–1915.

Zhang, C., Low, J., Xiong, Y., 2024a. Integration of green hydrogen production and storage via electrocatalysis. Precision Chemistry 2, 229–238.

Zhang, P., Chen, L., Kuo, D.-H., Wu, B., Su, Z., Lu, D., et al., 2024b. Photocatalytic fixation of nitrogen to ammonia with a Ce/S co-doped TiO_2 catalyst: synergistic tuning of heterovalent metal states and oxygen vacancy defects. Journal of Materials Chemistry A 12, 7163–7177.

Zhu, Y.-P., Xu, X., Su, H., Liu, Y.-P., Chen, T., Yuan, Z.-Y., 2015. Ultrafine metal phosphide nanocrystals in situ decorated on highly porous heteroatom-doped carbons for active electrocatalytic hydrogen evolution. ACS Applied Materials and Interfaces 7, 28369–28376.

Zou, X., Zhang, Y., 2015. Noble metal-free hydrogen evolution catalysts for water splitting. Chemical Society Reviews 44, 5148–5180.

CHAPTER

2

Concepts, kinetics, and mechanism of water splitting

Rohit Gupta[1], *and Rahul Gupta*[2]

[1]Department of Applied Sciences, Faculty of Physics, Yogananda College of Engineering and Technology, Jammu, Jammu and Kashmir, India [2]Department of Physics, Faculty of Physics, G.D. Goenka Public Higher Secondary School, Jammu, Jammu and Kashmir, India

2.1 Introduction

Splitting water is a synthetic process where water (H_2O) is broken up into oxygen gas (O_2) and hydrogen gas (H_2) (El-Shafie, 2023). In this process, an external source (electricity, solar energy, etc.) is required to run the chemical reaction (Santos et al., 2012). The process of splitting water produces hydrogen gas, which is a good, ethical, and energetic source of fuel (Rashid et al., 2015). This hydrogen can be utilized in energy cells (i.e., fuel cells) to turn out electricity. Moreover, it can be utilized as an immediate fuel for vehicles and industrial purposes (Kumar and Lim, 2022). Splitting water is an important step toward viable energy production and bringing down the reliance on traditional fuels (Miller et al., 2020). The concept of water splitting has its sources in the basic comprehension of electrochemistry as well as photochemistry (Lasia, 2019). It originates from early experiments in the 18th and 19th centuries investigating the breaking up of water into H_2 and O_2 gases by electrolysis method (Brauns and Turek, 2020). The revolutionary efforts of scientists like Alessandro Volta and Michael Faraday put down the basis for grasping the principles of electrolysis (Buelvas et al., 2014). In this process, water molecules disintegrate into protons (i.e., H^+ ions) and hydroxide ions (i.e., OH^- ions) by passing electric current (Wang et al., 2021). Moreover, the exposure of the photochemical technique of splitting water in the early 20th century by Johannes Wilhelm Mouton paved the way for exploring ways to split water using sunlight (photocatalysis) (Amira and Ibrahem, 2022). This technique became very important as interest in sources of green energy grew throughout the 20th and 21st centuries (Cheng and Jiang, 2015). In present days, water-splitting techniques are actively investigated and enlarged to harness imperishable sources of energy, especially solar energy, for turning out hydrogen as an ethical fuel (Zeng and Zhang, 2010). This progress is important for viable energy solutions, aiming to bring down greenhouse gas emissions and

Photochemical Splitting of Water
DOI: https://doi.org/10.1016/B978-0-443-29064-0.00013-9

alleviate the environs impact of energy output (Dragica LJ et al., 2003). In a nutshell, the process of splitting water is an important technique for moving forward toward a viable energy future, bringing down environs impact, intensifying energy security, and bringing up technological innovation in imperishable energy technologies (Naimi and Antar, 2018). The generation of hydrogen as an ethical or eco-friendly fuel contributes to imperishable or green energy solutions and the sustainability of the environment (Kumar and Himabindu, 2019.). Higher efficiency brings down energy costs and also increases determination compared to traditional fossil fuels (Gómez and Castro, 2024). The future scope of the techniques of splitting water is balanced for remarkable growth and impression, furnishing a cleaner, more feasible energy prospect and stimulating new measures in related fields (Brauns and Turek, 2020). Hydrogen produced from the process of splitting water can take part in an important contribution to imperishable energy storage, remove carbon from transportation and industry, and bring down emissions of greenhouse gases (Nie et al. 2008; Sinha et al., 2022; Engge and Maulana, 2020; U.S. Department of Energy, Office of Energy, 2009; Wang et al., 2019).

2.1.1 Objectives

The objectives of the chapter: 'concepts, kinematics, and mechanisms of water splitting' are as follows:

1. The study of concepts of water-splitting techniques helps scientists and engineers learn the fundamental principles involved in the process of splitting water. This includes learning the electrolysis reaction, electrode kinetics, and the contribution of catalysts in increasing reaction rates.
2. The study of kinematics of the water-splitting technique helps to learn how to improve the design and functioning of electrolysis cells to maximize its efficiency. The maximization of efficiency results in the lowering of input energy and higher rates of production of hydrogen per unit of input energy.
3. The study of mechanisms of water-splitting technique helps to learn the thorough engineering designs and layout of electrolysis cells. Scientists, engineers, and researchers aim to evolve new mechanisms that can enhance its potency. These new mechanisms may include how to improve the separation of gases, reduce side reactions, and enlarge the longevity of catalysts and electrodes.
4. The knowledge of concepts, kinematics, and mechanisms of water-splitting technique helps to advance the water-splitting technology from laboratory to industrial purposes. This includes designing wide-ranging electrolysis systems that are economically viable and can integrate with imperishable energy sources for sustainable production of hydrogen.
5. In general, the objective of the "chapter" is to label energy challenges at the global level. Hydrogen produced from the process of splitting water can take part in an important contribution to imperishable energy storage, remove carbon from transportation and industry, and bring down emissions of greenhouse gases. It is aimed at promoting the scientific knowledge, technological abilities, and empirical use of the process of splitting water in the transformation toward a viable energy future.

2.1.2 Concept of water splitting

Splitting water is a synthetic technique where H_2O is broken up into O_2 and H_2 gases as shown in Fig. 2.1. In this process, an external source (electricity, solar energy, etc.) is required to run the chemical reaction. The process of splitting water produces hydrogen gas, which is a good and energetic source of fuel (Gilmore and Gouma, 2022; Terlouw and Bauer, 2022; Sun and Xu, 2023).

2.1.2.1 Important processes and reactions

Some of the important processes and reactions in the process of splitting water are as follows:

1. Electrolysis process
   The process of splitting water by electrolysis, or electrolytic water splitting, takes place in an electrolytic cell where two electrodes (i.e., one is anode and the other is cathode) are employed in water. These electrodes are connected to an external direct current (DC) source.
2. Electrode reactions
   The oxidation of H_2O molecules takes place at the anode. Due to this, O_2, electrons, and hydrogen ions (H^+) are released.
   This reaction is written as

$$2H_2O\text{ (liquid)} \rightarrow O_2\text{(gas)} + 4\,H^+\text{(aqueous)} + 4\,e^-$$

The reduction of H_2O molecules takes place at the cathode. Due to this, H_2 and OH^- ions are produced.

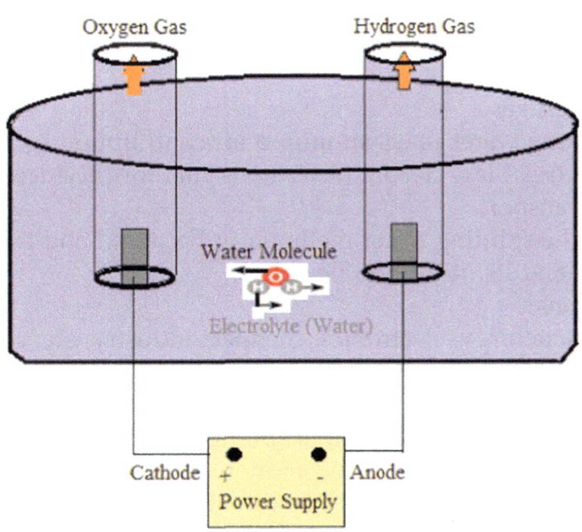

FIGURE 2.1 Electrolysis of water.

This reaction is written as

$$4H_2O\ (\text{liquid}) + 4\,e^- \rightarrow 2H_2(\text{gas}) + 4\,OH^-(\text{aqueous})$$

The general reaction for the water splitting process or water electrolysis process is written as

$$4H_2O\ (\text{liquid}) + \rightarrow 2H_2(\text{gas}) + O_2(\text{gas})$$

3. Energy requirement
   The splitting water method needs energy. The least possible amount of energy needed per mole of water is known as the thermodynamic minimum. It is approximately equal to 237.2 kJ per mole. This energy corresponds to a decomposition potential of about 1.2 V.

2.1.2.2 *Applications of water splitting*

The applications of water splitting are as follows:

1. Production of H_2
   H_2 obtained from the splitting water method can be cast off as a natural fuel in fuel cells, ignition engines, and various industrial processes.
2. Storing energy
   H_2 obtained from the process of splitting water can be used for storing energy obtained from imperishable sources (i.e., solar power plant, wind power plant, etc.) when they are not producing electricity.
3. Chemical industry
   H_2 obtained from the process of splitting water may be an important raw material for various chemical processes (i.e., for the production of ammonia [NH_3] for fertilizers).
4. Economy of hydrogen gas
   The process of splitting water plays an important contribution to the economy of H_2. H_2 produced by this process is a clean and virtuous fuel for producing energy.
5. Easy to store and transport
   H_2 produced from the splitting water method can be stored and transported. This makes energy distribution and its utilization flexible.
6. Industrial applications
   In industries (i.e., metallurgy, electronics, chemical industry, etc.), the H_2 produced from the splitting water method can be casted-off for various industrial processes.
7. Research and development
   Advances in materials (i.e., platinum, iridium, nickel alloys, etc.) required for the development of anode and cathode and modernization in catalysts (i.e., metal oxides, materials made from carbon, etc.) can modify the competence and durability of the splitting water technique and minimize its cost.

2.1.2.3 *Key challenges of splitting water technique*

The key challenges of splitting water technique for the yield of hydrogen gas are as follows:

1. Efficiency of water-splitting process
The electrolysis process often requires significant energy input. This makes the process of splitting water less economical when compared to customary fossil fuels.
2. Materials required for water-splitting process
The electrode materials required should be stable, highly effective, and cost-effective. This is again an important challenge.
3. Boosting of water-splitting process
Boosting the electrolysis process to industrial levels while sustaining efficiency and cost-effectiveness is a key challenge.
4. Mechanism and efficiency
The rate and efficiency of the splitting water technique are determined by the potential difference (i.e., voltage) put in across the electrodes. A higher voltage enhances the rate, but it consumes more energy.
5. Efficiency considerations
The efficiency of the splitting water method is worked out by the energy needed for producing one unit of H_2.

2.1.2.4 *Types of water splitting*

The two types of splitting of water are as follows:

1. Water splitting by electrolysis or electrolytic water splitting
In this technique, H_2O is broken up into O_2 and H_2 by casting off electricity. This action can take place in acidic, neutral, or alkaline conditions. These conditions affect the competence and requirements of the process of splitting water.
2. Photochemical or photocatalytic water splitting
In this process, light energy obtained from sunlight in the form of photons is used to run the process of splitting water. In the process, semiconductors act as photocatalysts. These semiconductors absorb photons from sunlight and initiate the transfer of electrons. The transferred electrons help H_2O molecules to break up into O_2 and H_2.

2.2 Kinetics of splitting water techniques

In the kinetics of splitting water technique, reaction rates and the factors that affect them are studied. In this chapter, the kinetics of water splitting by electrolysis, or electrolytic water splitting process, and photochemical water splitting, or photocatalytic technique of splitting water, are explained.

2.2.1 Kinetics of splitting water by electrolysis or electrolytic splitting water technique

In the anodic reaction, the oxidation of H_2O molecules comes into being. Due to this reaction, O_2, electrons, and H^+, that is, protons are released.

This reaction is written as

$$2H_2O\ (\text{liquid}) \rightarrow O_2(\text{gas}) + 4\,H^+(\text{aqueous}) + 4\,e^-$$

In this reaction, the oxidation of H_2O molecules is taking place. This oxidation process is usually slow in the absence of a catalyst. To accelerate this reaction, platinum or noble elements are usually used as catalysts. The rate of this reaction relies on the potential difference (i.e., voltage) put in across the electrodes employed in the electrolytic cell. The higher the voltage put in across the electrodes, the faster the rate of evolution of O_2. The gap between the thermodynamic potential and the actual potential difference (i.e., voltage) put in across the electrodes is known as overpotential. This overpotential affects the kinetics of water-splitting process significantly. So, it should be minimized to enhance the competence of the splitting water by electrolysis method.

In the cathodic reaction, the reduction of H_2O molecules comes into being. Due to this, H_2 and OH^- ions are released in the electrolytic solution of water.

This reaction is written as

$$4H_2O\ (\text{liquid}) + 4\,e^- \rightarrow 2H_2(\text{gas}) + 4\,OH^-(\text{aqueous})$$

In this reaction, the reduction of water molecules is taking place. This reduction process is also usually slow in the absence of a catalyst. To accelerate this reaction, nickel, cobalt, or iron are usually used as catalysts to increase the evolution of H_2. The rate of this reaction relies on the voltage put in across the electrodes of the electrolytic cell. The lower the voltage put in across the electrodes employed in the electrolytic cell, the faster the rate of evolution of H_2. The gap between the thermodynamic potential and the actual voltage put in across the electrodes is known as overpotential. The excessive overpotential affects the kinetics of water splitting significantly. So, it should be minimized to improve competence and to decrease the consumption of electrical energy by process of splitting water by electrolysis.

The general reaction for process of splitting water by the electrolysis, or electrolytic water splitting, is written as

$$2H_2O\ (\text{liquid}) \rightarrow 2H_2(\text{gas}) + O_2(\text{gas})$$

2.2.1.1 Factors affecting the kinetics of water splitting by the electrolysis process

The factors affecting the kinetics of the process of splitting water by electrolysis, or electrolytic water splitting, are as follows (El-Shafie, 2023; Santos et al., 2012; Rashid et al., 2015; Kumar and Lim, 2022):

1. Nature of electrode material
   The nature of electrode material significantly affects the kinetics of evolutions of O_2 and H_2 at the electrodes. Moreover, catalysts make a prime contribution to lowering the overpotential and increasing the reactions' rates at the electrodes.
2. Composition of electrolyte
   The nature and concentration of electrolytes (acids, bases, or salts) affect the conductivity of the solution of electrolyte. This, in turn, affects the transfer of electric charge (i.e.,

positive and negative ions) from one electrode to the other electrode, and also, affects the reactions' kinetics at the electrodes.

3. Effect of temperature
   The reactions' rates can be enhanced by enhancing the temperature. This is because the mobility of ions is increased at elevated temperatures, and the reactions' kinetics at the electrodes is also increased at elevated temperatures. However, the rates of reactions can be controlled by making use of stable electrodes and also by the composition of the electrolyte.
4. Effect of surface area of electrode
   The active sites can be increased for reactions to run by increasing the area of electrodes. This, in turn, significantly improves the overall rates of reactions of splitting water by electrolysis. Nanostructured electrodes with an elevated ratio of surface to volume can increase the activities of catalysts and decrease the drawbacks of the transport of mass.
5. Effect of pressure
   Pressure also affects the rates of reactions at the electrodes employed in the splitting water method. It has a prime effect on the evolution of gases at the electrodes.
6. Effect of overpotential
   The competence of catalysts and electrode materials can decrease overpotential to improve the competence of electrical energy consumed.
7. Effect of durability of catalysts
   A stable catalyst (i.e., a catalyst that can last for a long time) under rough electrochemical conditions can ensure its performance for a long time.
8. Integration of splitting water method and renewable energy
   The process of splitting water by electrolysis, or electrolytic water splitting, can be optimized to integrate with the sources of imperishable energy (i.e., solar power and wind power). This engages in an important contribution to the viable production of H_2.

2.2.1.2 *Activity of catalysts*

Catalysts play an important contribution in minimizing overpotentials. They provide active sites for electrochemical reactions in electrolytic cells during the process of splitting water. The electrochemical reaction's kinetics during the splitting water process by electrolysis or electrolytic water splitting are enhanced by enhancing the superficial area of catalysts, using stable catalysts, and also by controlling the electronic properties of material of catalysts. By adding promoters, one can further enhance the activity of catalysts and optimize their performance in water-splitting process.

2.2.1.3 *Kinetic aspects of water splitting by electrolysis*

A brief of kinetic aspects of the process of splitting water by electrolysis or electrolytic water splitting are as follows:

1. Reaction mechanism
   The overall water-splitting reaction involves the adsorption of H_2O molecules, transfer of electrons, and desorption of products (i.e., O_2 and H_2) of reactions at the electrodes.

2. Volmer reaction
   This reaction is written as $2H_2O+2e^- \rightarrow 2H_{ad}+2HO^-$, where H_{ad} is the adsorbed hydrogen atom on an active site. The first step in the anodic reaction as well as in the cathodic reaction includes the adsorption of H_2O molecules onto the facet of the electrodes during the splitting water procedure.
3. Heyrovsky reaction
   This reaction is written as $H_{ad}+H_2O+e^- \rightarrow H_2+HO^-$. This reaction includes the generation of intermediates (i.e., HO^- ions) during the evolution of hydrogen gas.
4. Tafel reaction
   This reaction is written as $H_{ad}+H_{ad} \rightarrow H_2$. This reaction describes the overall reaction's kinetics of splitting water by electrolysis technique. In the overall reaction, the rate of reactions at the electrodes and the overpotential are related exponentially.
5. Transition or conversion states
   The transition (i.e., conversion) states and the intermediates formed are considered important in explaining the mechanism of reactions and for designing highly effective catalysts.
6. Overpotential reduction
   It is the additional potential required beyond the thermodynamic potential to run the desired electrochemical reactions at the electrodes. It should be minimized to improve the competence of process of splitting water by electrolysis.

2.2.2 Mechanism of water splitting by electrolysis process

The mechanism of the process of splitting water by electrolysis or electrolytic water splitting involves some complicated steps that occur at the surfaces of the electrodes of electrolytic cells. In this section, a precise explanation of the mechanism of the process of splitting water by electrolysis or electrolytic water splitting is provided. In the explanation, anodic reaction (i.e., release of O_2) and cathodic reaction (i.e., release of H_2) is specially focused.

2.2.2.1 *Anodic reaction (i.e., release of O_2)*

This reaction takes place at the anode. In this reaction, the oxidation of H_2O molecules takes place. Due to this, O_2, electrons, and H^+ are released.

This reaction is written as

$$2H_2O\ (\text{liquid}) \rightarrow O_2(\text{gas}) + 4\,H^+(\text{aqueous}) + 4\,e^-$$

The precise mechanism of anodic reaction is as follows:

Adsorption of H_2O molecules:
H_2O molecules adsorb onto the facet of the anode, which is a noble metal (i.e., platinum, etc.) or a catalyst of oxide (i.e., iridium oxide, etc.)

Oxidation of H_2O molecules:
The adsorbed H_2O molecules onto the facet of the anode lose electrons and engage in the decomposition of H_2O molecules.
Formation of hydroxyl radicals:
The oxidation or decomposition of H_2O molecules proceeds through several intermediate steps and yields hydroxyl radicals and intermediates containing O_2.
O_2 gas formation:

At the instant of oxidation or decay of H_2O molecules, O_2 gas is set free as a product, and hydrogen ions (H^+ ions or protons) are released into the electrolyte of the electrolytic cell (i.e., in water-electrolyte).

2.2.2.2 *Factors affecting anodic reaction*

The key factors affecting the anodic reaction are as follows (Miller et al., 2020; Lasia, 2019; Brauns and Turek, 2020):

1. Electrode potential
   The rate of evolution of oxygen relies on the voltage put in across electrodes. The anodic reaction can be accelerated by putting in a higher positive potential put in across the anode. In addition, putting in a higher positive voltage put in across the anode provides the required electrical energy to prevail over activation barriers, thereby reducing overpotential.
2. Catalysts
   The kinetics of the evolution of oxygen is increased by using catalysts (i.e., iridium oxide [IrO_2] or ruthenium oxide [RuO_2]). These catalysts decrease the overpotential needed for highly effective production of O_2.

2.2.2.3 *Cathodic reaction (i.e., evolution of H_2)*

The process of cathodic reaction takes place at the cathode. In this reaction, the disintegration of H_2O molecules takes place. As a result, H_2 gas and hydroxide (OH^-) ions are produced.

This reaction is written as

$$4H_2O\,(\text{liquid}) + 4\,e^- \rightarrow 2\,H_2(\text{gas}) + 4\,OH^-(\text{aqueous})$$

The precise mechanism of cathodic reaction is as follows:

Adsorption of H^+ ions (i.e., protons):
H^+ ions from the solution in the electrolytic cell adsorb onto the facet of the cathode.
Reduction of H^+ ions (i.e., protons):
Adsorbed H^+ ions (i.e., protons) gain electrons (i.e., electrical supply) and form hydrogen atoms.
Formation of H_2 gas:

The H atoms fuse and form H_2 gas. This H_2 gas is then set free into the electrolyte.

2.2.2.4 *Factors affecting cathodic reaction*

The key factors affecting the cathodic reaction are as follows:

1. Electrode potential
The rate of evolution of hydrogen relies on the voltage put in across the electrodes. The cathodic reaction can be accelerated by putting in a higher negative voltage put in across the cathode. Moreover, putting in a higher negative voltage put in across the cathode provides the required electrical energy to prevail over activation barriers, thereby reducing overpotential.
2. Catalysts
The kinetics of the evolution of hydrogen is increased by using catalysts (i.e., platinum [Pt], nickel [Ni], or cobalt [Co]). These catalysts lower the activation energy, provide active sites for the assimilation of H^+ ions and electrons, and increase the rate of cathodic reaction needed for the highly effective production of H_2 gas.

2.2.3 Overall reaction of water splitting by electrolysis process

The general reaction for the process of splitting water by electrolysis, or electrolytic water splitting, is written as

$$2H_2O\ (\text{liquid}) \rightarrow 2\ H_2(\text{gas}) + O_2(\text{gas})$$

In this process, an external electrical source (i.e., a power supply) is required to run the water splitting by electrolysis process. This source of electrical energy provides electrons (i.e., negatively charged particles) required for the release of H_2 at the cathode and the release of O_2 at the anode. The electrolyte can be an acidic solution, a neutral solution, or an alkaline solution. It mainly provides ions to conduct electricity between the electrodes of the electrolytic cell. It also helps in the transport of ions. Moreover, it stabilizes the intermediates of reactions at the anode as well as at the cathode.

2.2.4 Techniques for optimizing water splitting by electrolysis process

The techniques for optimizing water splitting by electrolysis process for viable production of H_2 gas are as follows (Kailai et al. 2021; Molaei, 2024; Chen et al., 2021):

1. Reduction of overpotential
Reduction of overpotentials (i.e., the additional electrical potential needed to run the reactions) is important for increasing the energy competence of water splitting by the electrolysis process.
2. Development of catalysts
Highly effective and durable catalysts are needed to improve the kinetics of splitting water by the electrolysis technique and also, to minimize the cost of electrolytic cells.
3. Study of operating conditions
Modern techniques are used to study the dynamic behavior of anodic and cathodic reactions under operating conditions. These conditions help in designing catalysts and optimizing the process of splitting water by the electrolysis.

4. Effect of mass transport
   The mass transport of reactants (i.e., H_2O molecules and H^+ ions) and products (i.e., O_2 gas and H_2 gas) onto the facets of electrodes is important for optimizing the kinetics of water splitting by the electrolysis process and for minimizing overpotential.
5. Catalysts composed of single atoms
   Catalysts composed of single atoms possess high competence and help significantly in overcoming the limitations of traditional catalysts. In recent years, catalysts composed of single atoms have developed. They are highly active and can be used as active sites in the process of splitting water. They can remove the limitations of traditional catalysts (i.e., metal nanoparticles).
6. Integration of the splitting water by electrolysis method or electrolytic water splitting and renewable energy
   The integration of the splitting water by electrolysis method with sources of green energy (i.e., solar power and wind power) is very important for the viable yield of H_2 gas and the storage of electrical energy.
7. Adoption of the process of splitting water by electrolysis or electrolytic water splitting technique
   The durability, cost-effectiveness, and integration of the process of splitting water by electrolysis or electrolytic water splitting with imperishable energy sources are essential for the widespread adoption of techniques of splitting water.

2.2.5 Introduction of photochemical splitting of water

Photochemical water splitting refers to the process by which light energy (i.e., photons) is utilized to split up H_2O molecules into H_2 and O_2 as shown in Fig. 2.2 (Molaei, 2024; Tentu and Basu, 2017; Chen and Shen, 2010).

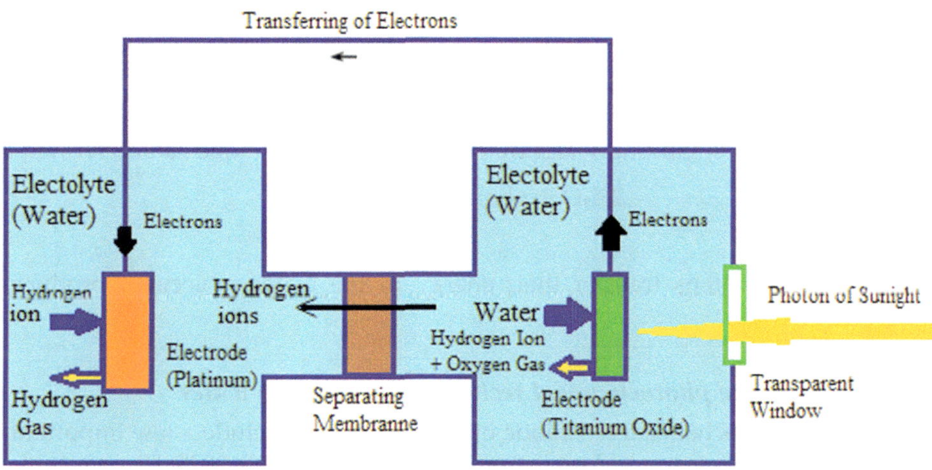

FIGURE 2.2 Photochemical or photocatalytic technique of splitting water.

A detailed explanation of the photochemical or photocatalytic technique of splitting water is as follows:

1. Absorption of photons
   The photochemical or photocatalytic technique of splitting water starts with the assimilation of photons by the H_2O molecules. Each photon carries a fixed amount of energy, corresponding to its frequency or wavelength. In the process of splitting water, the wavelength or frequency of photons falls in the visible or UV range.
2. Excitation of electrons
   When a photon of energy of the sunlight is absorbed by a water molecule, it excites the electrons of the molecule to a higher energy state. This excitation causes the molecule to become unstable or undergo a structural change.
3. Breaking of O-H bonds
   The absorbed photons break the O-H bonds of water molecules.
4. Radicals or ion formation
   Due to the disintegration of O-H bonds in H_2O molecules, radicals or ions (highly reactive species), are formed.

$$H_2O \text{ (water molecule)} \rightarrow H^{\cdot}\text{(hydrogen radical)} + OH^{\cdot}\text{(hydroxyl radical)}$$

5. Redox reactions
   The radicals or ions formed participate in reduction and oxidation reactions.

$$2H_2O \text{ (water molecules)} \rightarrow O_2\text{(oxygen gas)} + 4\,H^{+}\text{(hydrogen ions)} + 4e^{-}\text{(electrons)}$$

   This reaction is known as oxidation half-reaction.
   In this reaction, the H_2O molecule is disintegrated into O_2, protons (i.e., positively charged H^+ ions), and negatively charged electrons (e^-).

$$2H^{+}\text{(hydrogen ions)} + 2e^{-}\text{(electrons)} \rightarrow H_2\text{(hydrogen gas)}$$

   This reaction is known as reduction half-reaction.
6. Overall reaction
   The overall process of splitting water by photochemical technique can be written as

$$2H_2O + \text{photons} \rightarrow O_2 + 2H_2$$

   In this reaction, photons transfer their energy to the water molecules, producing O_2 and H_2.

2.2.5.1 *Kinetics of the photochemical technique of splitting water*

The kinetics of photochemical technique of splitting water include a few important steps that determine the competence and rate of reactions (Mohsin et al., 2023; Vijayan et al., 2023; Kazuhiro, 2017; Di Paola et al. 2013).

1. Absorption of photons
   The process starts with the absorption of photons by a photosensitive material (i.e., a semiconductor or a catalyst). The frequency or wavelength of photons lies in the visible or ultraviolet range. The process of absorption of photons generates excited states within the photosensitive material. Due to this, electrons make a move from the lower valence band (VB) to the upper conduction band (CB).
2. Separation of charge
   In a photosensitive material, the moved electrons (i.e., negatively charged carrier) in CB and the resulting holes (i.e., vacancies or positively charged carrier) in VB are separated by a band gap (E_G). This separation is important for preventing the reassimilation of the pairs of the electrons and holes. As a result, a loss of energy takes place.
3. Evolution of hydrogen
   The moved electrons in the CB are gained by protons (i.e., H^+ ions) to form H_2 gas. This reaction occurs at the facet of a catalyst or electrode. $2H^+ + 2e^- \rightarrow H_2$.
4. Evolution of oxygen
   At the matching instant or moment, holes (i.e., produced vacancies or positively charged carriers) in the lower VB can split up H_2O molecules to produce O_2.

$$2H_2O \rightarrow O_2 + 4\,H^+ + 4e^-$$

These reactions proceed concurrently and are interconnected through the movement of H^+ and e^-.

5. Superficial reactions at catalysts
   The kinetics at the facet of the catalyst or semiconductor are very significant. The rate of evolution of gases is affected by the facet area and structure of the catalyst, as well as by active sites. Catalysts or semiconductors are often designed to optimize the rates of evolutions of H_2 and O_2 and, hence, to increase the kinetics of the photochemical or photocatalytic technique of splitting water.
6. Dynamics of electron transfer
   The overall kinetics of the photochemical or photocatalytic technique of splitting water depends upon the rate at which electrons move from the higher energy state to the facet of the catalyst, and in turn, take part in the reduction and oxidation reactions. This process of transfer of electrons can be controlled by the electric characteristics (forbidden energy gap and positions of the band edge) of the material of the catalyst or semiconductor.
7. Overpotential
   The photochemical or photocatalytic technique of splitting water needs to overcome overpotentials. So, it should be minimized to improve the competence of the photochemical or photocatalytic technique of splitting water.
8. Quantum competence
   Quantum Competence is defined as the efficiency with which photons absorbed by photosensitive material (i.e., a semiconductor or a catalyst) are converted into necessary products. It can be enhanced by enhancing the intensity of light, increasing the wavelength or frequency of light, and the ability of photosensitive material (i.e., a semiconductor or a catalyst) to absorb the photons of incident light effectively.

9. **Reaction mechanisms**
   The kinetics of photochemical or photocatalytic water splitting can be optimized by knowing reaction mechanisms, which include the knowledge of intermediate species involved evolution of O_2 and H_2.

2.2.5.2 *Mechanism of photochemical technique of splitting water*

The mechanism of the photochemical or photocatalytic technique of splitting water includes some steps that occur at the interface of photosensitive material (i.e., a semiconductor or a catalyst) sensitive to photons of incident light (Zhou, 2020; Ismael, 2021). A detailed explanation of the mechanism of the photochemical or photocatalytic technique of splitting water is as follows:

1. **Absorption of photons of incident light**
   The process starts with the absorption of photons by a photosensitive material (i.e., a semiconductor [titanium oxide] or a molecular catalyst). The frequency or wavelength of photons of sunlight lies in the visible or ultraviolet range. The process of absorption of photons creates excited states within the photosensitive material. Due to this, electrons make a move from the lower VB to the higher CB of the semiconductor, or upper energy states in the molecular catalyst.
2. **Production of pairs of electrons and holes**
   When photons of light are captivated by the photosensitive material, pairs of electrons and holes are created within the body of photosensitive material. Electrons (i.e., negatively charged carriers) make a move to the CB, whereas holes (i.e., vacancies or positively charged carriers) are produced in the VB.
3. **Charge separation**
   The electrons (i.e., negatively charged carriers) that excite in the CB are displaced from their actual positions and move toward the facet of the photosensitive material. These delocalized electrons run reduction reactions. At the matching instant or moment, the holes (i.e., positively charged carriers) in the VB move toward the facet of photosensitive material or into electrolyte and run oxidation reactions.
4. **Oxidation reaction (evolution of oxygen)**
   Holes (i.e., positively charged carriers) in the lower VB oxidize H_2O molecules to turn out O_2 gas.

$$2H_2O \rightarrow O_2 + 4\,H^+ + 4e^-$$

   The protons and electrons produced by this reaction run the reduction reaction.
5. **Reduction reaction (evolution of hydrogen)**
   Excited electrons in the CB combine with protons (H^+) to form H_2.

$$2H^+ + 2e^- \rightarrow H_2$$

   This reaction occurs at the facet of photosensitive material.
6. **Photosensitive material**
   The competence as well as kinetics of the photochemical or photocatalytic technique of splitting water relies on the superficial characteristics of the photosensitive material (i.e., a

semiconductor or catalyst). Photosensitive material provides active sites for the evolution of O_2 and H_2, and hence, increases their rates of evolution.

7. Applied potential
The photochemical water-splitting process requires the applied potential to run the reactions. This applied potential promotes the transferring of charge carriers toward the electrodes. These charge carriers start the reduction and oxidation reactions.
8. Dynamics of reaction
The competence and rate of photochemical water splitting can be determined by the transferring of charge carriers, reactions occurring at the facet of photosensitive material, and the transferring of charge carriers to the boundaries of photosensitive material.

2.2.5.3 *Factors affecting photochemical water splitting*

The photochemical or photocatalytic technique of splitting water employs sunlight energy to split up H_2O into H_2 and O_2. It is affected by the following factors (Lu and Yates, 1995; Singh, 2018; Maulana et al., 2023; Nakata and Fujishima, 2012; Aspasia and Christos, 2017; Kemppainen et al., 2015; Liu et al., 2024):

1. Properties of the catalysts
Catalysts take part in a prime role in the photochemical method of splitting water. They bring down the activation energy needed for the splitting water technique. The configuration of the catalyst seriously affects its competence and stability when it is exposed to light. The relevant band gap of the catalyst can adeptly take up light and assist in the uncoupling of charge carriers. A large facet of the catalyst imparts more active sites for catalytic activities to take place.
2. Properties of light source
The characteristics of the source of light are also important. The photochemical reactions depend on the energy (i.e., frequency) of the photons of the sunlight. A photon of higher frequency is usually more effective for initiating the reaction. The frequency of photons of the incident light measures the amount of energy taken up by the catalyst and its capability to excite electrons. Moreover, light of higher intensities can boost the rates of reactions to an extent where the saturation of the catalyst occurs.
3. Environmental and experimental conditions of reaction
The photochemical water-splitting process is affected by the various environmental and experimental conditions that impact the competence and stability of the process. The reaction's kinetics and the stability of the catalyst are affected by the acidity or alkalinity of the electrolytic solution. Optimal temperatures can boost the rates of reaction by providing essential thermal energy without adjusting the stability of the catalyst. The kind of electrolyte used and its concentration can have an effect on the conductivity as well as the redox potential of the system.
4. Dynamics of charge carriers
The charge carriers' separation and their transport by an efficient method are important for the effective photochemical process of splitting water. The electrons and holes generated by photons of light should competently separate to avert coalesce and maximize the number of charge carriers convenient for the reactions. Moreover, a considerate conductivity and

appropriate band alignment between catalyst and semiconductor enable quick transport of charge carriers to the catalytic sites.

5. Loading and dispersal of catalysts
   The catalytic activity is affected by the procedure when catalysts are diffused on the facet of the semiconductor. The suitable catalytic sites are ensured by the highest loading density without obstructing the absorption of light or the routes through which the charge is transported. A consistent process of splitting water across the semiconductor is ensured by the constant dispersal of catalysts.
6. Photostability and durability of the catalysts and semiconductors
   The performance of a photochemical splitting water system must be maintained under light exposure for a long time. Catalysts and semiconductors should resist deterioration caused by exposure to light and conditions of the reaction.
7. Material design
   Progress in material science takes part in an important contribution to increase the competence of the photochemical process of splitting water. Nanostructured materials can add to absorption of light, and elevate the dynamics of charge carriers. The techniques of facet engineering can modify catalytic activity and stability toward particular routes of reaction.
8. External bias or photovoltaic assistance
   The competence of the process of splitting water can be enhanced by putting in the external bias or by combining photovoltaic components. The external bias aids in lowering the energy barrier. The integration of photovoltaic cells with the photoelectrochemical system can directly convert solar energy into electricity. This energy is then employed for the splitting water technique and therefore, enhances its overall competence.

2.3 Comparison of the mechanisms of splitting water by electrolysis method and photochemical water splitting

A comparison of the mechanisms of splitting water by electrolysis method and photochemical water splitting is given in Table 2.1.

2.4 Case study

2.4.1 Objective

The objective is to advance the scientific understanding of water splitting and to improve its practical applications for energy production and environmental sustainability.

2.4.2 Scope

The scope of the study aims to provide a comprehensive understanding of water splitting, from fundamental principles to practical applications and future advancements.

TABLE 2.1 Comparison of the mechanisms of splitting water by electrolysis method and photochemical water splitting.

S. no.	Splitting water by electrolysis method	Photochemical water splitting
1.	Process: It uses electrical energy to run the reaction within an electrochemical cell.	Process: It uses energy of sunlight to run splitting water procedure with the help of a photocatalyst.
2.	Electrochemical cell employed in the technique consists of electrodes (i.e., anode and cathode) dipped in a solution.	Photocatalyst employed in the technique is a substance that absorbs light and produces electron–hole pairs.
3.	Anode reaction: In this reaction, H_2O is oxidized to turn out O_2, H^+ ions, and electrons. $2H_2O$ (liquid) $\rightarrow O_2$ (gas) + $4H^+$(aqueous)+$4e^-$	Oxidation reaction: Photogenerated holes oxidize water to turn out O_2, H^+ ions, and electrons. $2H_2O \rightarrow O_2 + 4H^+ + 4e^-$
4.	Cathode reaction: In this reaction, H^+ ions combine with electrons to turn out H_2. $4H^+$ (aqueous)+$4e^- \rightarrow 2H_2$(gas)	Reduction reaction: Photogenerated electrons reduce H^+ ions to turn out H_2. $4H^+ + 4e^- \rightarrow 2H_2$
5.	The efficiency of alkaline electrolysis method varies from 60% to 70% efficiency. The efficiency of proton exchange membrane electrolysis ranges from 70% to 80% efficiency. Here, efficiency is influenced by input electrical power and the design of the system (Shit et al. 2021; Grigoriev et al., 2020).	Its efficiency ranges from 5% to 18%. Here, efficiency is influenced by photocatalyst properties, semiconductor materials and design of the system, and light absorption (Burton et al., 2021).

2.4.3 Audience

Undergraduate, graduate and scholars, scientists and industries.

2.4.4 Rationale

The rationale for exploring the "Concepts, Kinetics, and Mechanism of Water Splitting" lies in its potential to address critical energy and environmental challenges, improve the efficiency and cost-effectiveness of renewable energy technologies, and foster innovation in energy systems.

2.4.5 Expected results and deliverables

1. Detailed understanding: Clear explanations of fundamental concepts, reaction mechanisms, and different methods of water splitting.
2. Kinetic data: Explained effects of variables on the kinetics of water splitting.
3. Catalyst performance: Explained the efficiency, stability, and properties of catalysts used in water splitting.
4. Technological insights: Information on scalability, cost-effectiveness, and environmental impact of water-splitting technologies.

5. Publications: Research papers and presentations to share findings with the scientific community.
6. Recommendations: Suggestions for future research and practical applications to enhance water-splitting technologies.

2.4.6 Safety considerations

1. Chemical safety
   a. Hydrogen and oxygen gases: Water splitting typically produces hydrogen and oxygen gases. Both gases are highly flammable and can form explosive mixtures in air. Ensure proper ventilation, avoid open flames, and use explosion-proof equipment.
   b. Corrosive agents: Some water-splitting processes involve corrosive chemicals like acids or bases. Handle these substances with appropriate protective equipment (gloves, goggles, lab coats) and follow proper storage and disposal guidelines.
   c. Pressure concerns: In electrolysis or other high-pressure methods, the buildup of gas pressure can be hazardous. Use pressure-rated equipment and safety valves to prevent explosions.
2. Electrical safety
   a. Electrolysis: When performing electrolysis, you are dealing with electrical currents and potentially high voltages. Ensure that electrical equipment is properly insulated and grounded. Use equipment designed for the voltages and currents you are working with.
   b. Short circuits and overheating: Avoid short circuits and overheating by ensuring that connections are secure and equipment is in good working order. Regularly inspect and maintain electrical setups.
3. Thermal safety
   a. Heat generation: Water-splitting processes, especially those involving high temperatures, can generate significant heat. Use heat-resistant materials and keep flammable materials away from heat sources. Implement cooling systems or thermal shields if necessary.
   b. Thermal stress: Rapid changes in temperature can cause stress on materials and equipment. Gradually increase or decrease temperatures to prevent damage or accidents.
4. Mechanical safety
   a. Equipment integrity: Ensure that all equipment, such as electrolysis cells or reactors, is well-maintained and free from cracks or defects. Use equipment that can withstand the pressures and temperatures involved in the process.
   b. Handling and installation: When installing or handling equipment, follow manufacturer instructions and use appropriate tools. Avoid manual handling of heavy or awkward items to prevent injury.
5. Personal safety
   a. Protective gear: Always wear appropriate personal protective equipment (PPE) including lab coats, gloves, safety goggles, and face shields.
   b. Training: Ensure that all personnel involved are trained in the specific procedures and safety protocols related to water-splitting experiments.

6. Emergency preparedness
   a. Spill and accident response: Have protocols in place for dealing with chemical spills, gas leaks, or electrical failures. Ensure that emergency equipment, such as fire extinguishers and eye wash stations, is readily accessible and in working order.
   b. First aid: Be familiar with first aid procedures for exposure to chemicals or electrical accidents. Ensure that first aid kits are available and well-stocked.
7. Waste management
   a. Chemical disposal: Properly dispose of any chemicals, including residues and by-products, according to local regulations. Use appropriate waste containers and label them clearly.
   b. Gas handling: Manage and dispose of any residual gases safely. Use gas scrubbers or other methods to handle and neutralize gases before releasing them into the environment.

By adhering to these safety considerations, one can mitigate risks and ensure a safe environment when studying or working with water-splitting technologies.

2.4.7 Actions taken/workflow/tools used/simulations and analyses

The workflow for studying water splitting involves a combination of experimental techniques, computational modeling, and data analysis. By utilizing the right tools and following a structured approach, researchers can gain insights into the kinetics, mechanisms, and efficiencies of water-splitting processes.

2.4.8 Challenges and Solutions

1. Material and Catalyst Challenges

Challenges:
- Catalyst Efficiency: Finding catalysts that are highly efficient, stable, and cost-effective for water splitting is a major hurdle. Many catalysts degrade or lose efficiency over time.
- Material Durability: Catalysts and materials used in water splitting must withstand harsh conditions (e.g., high temperatures, corrosive environments).

Solutions:
- Development of Novel Catalysts: Research into new materials, such as advanced metal alloys, transition metal or non-precious metal-based catalysts, can improve efficiency and reduce costs.
- Protective Coatings: Applying protective coatings or designing composite materials can enhance the durability of catalysts.
- Optimization and Surface Engineering: Tailoring the surface properties of catalysts to increase their activity and stability.

2. Energy Efficiency

Challenges:
- Energy Input vs. Output: Achieving high energy efficiency is challenging. The energy required for water splitting often exceeds the energy obtained from the hydrogen produced.

Solutions:

- Optimizing Process Conditions: Fine-tuning operational parameters (e.g., temperature, pressure, pH) to enhance efficiency.
- Integration with Renewable Energy: Coupling water splitting processes with renewable energy sources (e.g., solar or wind) to improve overall energy efficiency and sustainability.

3. Kinetics and Reaction Mechanism

Challenges:

- Complex Kinetics: The kinetics of water splitting can be complex, involving multiple steps and intermediates, making it difficult to fully understand and optimize.
- Mechanistic Uncertainties: Uncertainty about the detailed reaction mechanisms and the role of intermediates can hinder the design of more efficient systems.

Solutions:

- Advanced Analytical Techniques: Using techniques like spectroscopy (e.g., FTIR, UV- Vis) and electrochemical analysis (e.g., impedance spectroscopy) to better understand reaction mechanisms.
- Computational Modeling: Employing computational tools and simulations to model reaction pathways and predict kinetic behaviors.

4. Scalability and Practical Implementation

Challenges:

- Scale-Up Issues: Translating laboratory results to industrial-scale processes can be challenging due to differences in scale and operational conditions.
- Cost of Implementation: High costs associated with materials, catalysts, and energy requirements can be prohibitive for large-scale deployment.

Solutions:

- Pilot Plants: Developing and testing pilot-scale systems to address scale-up challenges and optimize processes for larger scales.
- Cost Reduction Strategies: Research into cheaper materials and methods, as well as process optimization, to make water splitting more economically viable.

5. Safety and Environmental Impact

Challenges:

- Handling of Gases: Hydrogen and oxygen gases are highly flammable and can form explosive mixtures.
- Environmental Concerns: Ensuring that the production and use of catalysts and other materials do not have negative environmental impacts.

Solutions:

- Safety Protocols: Implementing rigorous safety protocols for handling gases, including proper ventilation, explosion-proof equipment, and regular safety checks.
- Sustainable Materials: Developing and using environmentally friendly materials and methods to minimize the environmental footprint of water splitting processes.

6. Interdisciplinary Collaboration

Challenges:

- Complexity of Research: Water splitting involves aspects of chemistry, physics, materials science, and engineering, requiring expertise from multiple disciplines.

Solutions:

- Cross-Disciplinary Teams: Forming teams with diverse expertise to address various aspects of water splitting research.
- Collaborative Research: Engaging in collaborative projects and partnerships with academic, industrial, and research institutions to leverage collective knowledge and resources.

By addressing these challenges with targeted solutions, researchers can advance the development of more efficient, scalable, and practical water splitting technologies.

2.4.9 Learning and Knowledge Outcomes

1 Understanding Fundamental Concepts
Learning Outcomes:

- Water Splitting Principles: Gain a thorough understanding of the basic principles behind water splitting, including the electrochemical and photochemical processes involved.
- Energy Requirements: Learn about the energy inputs required for splitting water into hydrogen and oxygen and the role of different energy sources (electrical, solar, thermal).

Knowledge Outcomes:

- Electrolysis: Comprehend the process of electrolysis, including the roles of the anode and cathode, and the overall chemical reaction.
- Photo-catalysis: Understand how light-driven processes can facilitate water splitting, including the functioning of photo-catalysts and the nature of the photo-excited states.

2 Kinetics of Water Splitting
Learning Outcomes:

- Reaction Rates: Learn the impact of different variables such as temperature, pressure, and catalyst type.
- Kinetic Models: Understand and apply kinetic models to describe the behavior of water splitting reactions, including factors that influence reaction rates and mechanisms.

Knowledge Outcomes:

- Catalytic Mechanisms: Understand how different catalysts affect the kinetics of the water splitting process and the mechanisms through which they operate.

3 Mechanisms of Water Splitting
Learning Outcomes:

- Reaction Pathways: Study the detailed mechanisms of water splitting, including the formation and role of intermediates and transition states.
- Catalyst Functionality: Learn how catalysts facilitate water splitting at the molecular level, including electron transfer, bond breaking, and formation processes.

Knowledge Outcomes:

- Mechanistic Insights: Develop a deeper understanding of the step-by-step processes involved in water splitting, including the identification of key reaction intermediates.
- Catalyst Design: Gain knowledge about how to design and optimize catalysts based on their mechanistic behavior and interaction with reactants.

4 Material Science and Engineering

Learning Outcomes:

- Material Properties: Learn about the properties of materials used in water splitting, such as stability, conductivity, and catalytic activity.
- Optimization Techniques: Study techniques for optimizing materials and catalysts for improved performance and durability in water splitting applications.

Knowledge Outcomes:

- Material Selection: Understand the criteria for selecting materials based on their performance in water splitting processes.
- Engineering Design: Gain insights into the design considerations for developing and scaling up water splitting technologies.

5 Practical Applications and Innovations

Learning Outcomes:

- Application Scenarios: Study the practical applications of water splitting technology, including its potential in renewable energy production and hydrogen storage.
- Innovative Solutions: Explore innovative solutions and emerging technologies related to water splitting, including new materials, processes, and integration with other technologies.

Knowledge Outcomes:

- Technology Integration: Understand how water splitting technology can be integrated with renewable energy systems and other industrial processes.
- Commercialization: Gain insights into the challenges and opportunities for commercializing water splitting technologies.

6 Sustainability and Environmental Impact

Learning Outcomes:

- Sustainability Issues: Learn about the environmental and sustainability aspects of water splitting technology, including resource use and waste management.
- Lifecycle Analysis: Understand the lifecycle of water splitting technologies, from material extraction and manufacturing to end-of-life disposal.

Knowledge Outcomes:

- Environmental Considerations: Develop an understanding of the environmental impact of water splitting technologies and strategies for minimizing their footprint.
- Sustainable Practices: Gain knowledge about sustainable practices and policies related to the development and deployment of water splitting technologies.

By achieving these learning and knowledge outcomes, researchers and practitioners can significantly advance the field of water splitting, contributing to the development of efficient and sustainable technologies for hydrogen production and renewable energy.

Authors: Rohit Gupta and Rahul Gupta

2.5 Learning and knowledge outcomes

1. Understanding fundamental concepts learning outcomes:

a. Water-splitting principles: Gain a thorough understanding of the basic principles behind water splitting, including the electrochemical and photochemical processes involved.
b. Energy requirements: Learn about the energy inputs required for splitting water into hydrogen and oxygen and the role of different energy sources (electrical, solar, thermal).

Knowledge outcomes:

a. Electrolysis: Comprehend the process of electrolysis, including the roles of the anode and cathode, and the overall chemical reaction.
b. Photocatalysis: Understand how light-driven processes can facilitate water splitting, including the functioning of photocatalysts and the nature of the photo-excited states.

2. Kinetics of water-splitting learning outcomes:

a. Reaction rates: Learn the impact of different variables such as temperature, pressure, and catalyst type.
b. Kinetic models: Understand and apply kinetic models to describe the behavior of water-splitting reactions, including factors that influence reaction rates and mechanisms.

Knowledge outcomes:

a. Catalytic mechanisms: Understand how different catalysts affect the kinetics of the water-splitting process and the mechanisms through which they operate.

3. Mechanisms of water-splitting learning outcomes:

a. Reaction pathways: Study the detailed mechanisms of water splitting, including the formation and role of intermediates and transition states.
b. Catalyst functionality: Learn how catalysts facilitate water splitting at the molecular level, including electron transfer, bond breaking, and formation processes.

Knowledge outcomes:

a. Mechanistic insights: Develop a deeper understanding of the step-by-step processes involved in water splitting, including the identification of key reaction intermediates.
b. Catalyst design: Gain knowledge about how to design and optimize catalysts based on their mechanistic behavior and interaction with reactants.

4. Material science and engineering learning outcomes:

a. Material properties: Learn about the properties of materials used in water splitting, such as stability, conductivity, and catalytic activity.
b. Optimization techniques: Study techniques for optimizing materials and catalysts for improved performance and durability in water-splitting applications.

Knowledge outcomes:

a. Material selection: Understand the criteria for selecting materials based on their performance in water-splitting processes.
b. Engineering design: Gain insights into the design considerations for developing and scaling up water-splitting technologies.

5. Practical applications and innovations learning outcomes:

a. Application scenarios: Study the practical applications of water-splitting technology, including its potential in renewable energy production and hydrogen storage.
b. Innovative solutions: Explore innovative solutions and emerging technologies related to water splitting, including new materials, processes, and integration with other technologies.

Knowledge outcomes:

a. Technology integration: Understand how water-splitting technology can be integrated with renewable energy systems and other industrial processes.
b. Commercialization: Gain insights into the challenges and opportunities for commercializing water-splitting technologies.

6. Sustainability and environmental impact learning outcomes:

a. Sustainability issues: Learn about the environmental and sustainability aspects of water-splitting technology, including resource use and waste management.
b. Lifecycle analysis: Understand the lifecycle of water-splitting technologies, from material extraction and manufacturing to end-of-life disposal.

Knowledge outcomes:

a. Environmental considerations: Develop an understanding of the environmental impact of water-splitting technologies and strategies for minimizing their footprint.
b. Sustainable practices: Gain knowledge about sustainable practices and policies related to the development and deployment of water-splitting technologies.

By achieving these learning and knowledge outcomes, researchers and practitioners can significantly advance the field of water splitting, contributing to the development of efficient and sustainable technologies for hydrogen production and renewable energy.

2.6 Conclusion

It is concluded from the chapter that the kinetics of the process of splitting water by electrolysis, or electrolytic water splitting, call for complicated and disordered electrochemical processes affected by the employed electrodes' nature, the nature of catalysts, the conditions of the composition of the electrolytic cell, the factors affecting the rate of reactions, and also by external factors, that is, temperature and pressure. Moreover, the kinetics of the process of splitting water by electrolysis, or electrolytic water splitting, can be improved by using advanced materials and the proper design of catalysts. This improvement is significant for increasing the competence as well as scalability of the splitting water technique in the coming days. The water splitting or electrolysis mechanism should be acknowledged and optimized for the viable production of H_2 gas.

In the photochemical or photocatalytic technique of splitting water, sunlight energy is absorbed by water molecules. This causes the bond-breaking of water molecules and the generation of extremely reactive ions or radicals. This process makes a prime contribution to turn out the oxygen and hydrogen. The photochemical or photocatalytic water-splitting technology is being explored for synthetic photosynthesis and the production of solar fuel (Gupta et al., 2023; Marian, 2022). In this technology, sunlight energy is used to generate hydrogen gas, which is an ethical source of imperishable energy. Solar fuels are those fuels that are produced directly or indirectly from solar energy. This energy can be used to run the chemical reactions that produce storable fuels, that is, hydrogen and hydrocarbons (Kazemi and Manteghi, 2024; Serpone et al., 1992). The photochemical or photocatalytic technique of splitting water includes the sucking up of photons of sunlight by a photosensitive material, the turning out and partition of pairs of electrons and holes, and the successive evolution of O_2 and H_2 gases at the facet of the photosensitive material. This mechanism aims to exploit solar energy for the green or eco-friendly generation of hydrogen as an ethical fuel. The generation of hydrogen as an ethical or eco-friendly fuel contributes to imperishable or imperishable or green energy solutions and the sustainability of the environment (Smolinka, 2009). It is observed that the efficiency of the photochemical technique of splitting water is 5%–18%, depending on the material of the semiconductor or catalyst, and also, the design of the system employed in the process (Ouabi and Lajouad, 2024). Also, the efficiency of electrolysis employing alkaline solution is 60%–70% and that employing proton exchange membrane is 70%–80%.

It is inferred from the chapter that higher efficiency brings down energy costs and also increases determination compared to traditional fossil fuels. Current research and technological advancements are important to enhance efficiency, bring down costs, and increase the viability of splitting water techniques for the ethical turning out of hydrogen. The future scope of the splitting water techniques is balanced for remarkable growth and impression, furnishing a cleaner, more feasible energy prospect and stimulating new measures in related fields (Hota et al., 2023).

References

Amira, M., Ibrahem, H., 2022. Optimization of proton exchange membrane electrolyze cell design using machine learning. Energies 15 (18), 6657. Available at https://doi.org/10.3390/en15186657.

Aspasia, N., Christos, T., 2017. Alternative photocatalysts to TiO_2 for the photocatalytic reduction of CO_2. Applied Surface Science 391 (B), 149–174. Available at https://doi.org/10.1016/j.apsusc.2016.06.172.

Brauns, J., Turek, T., 2020. Alkaline water electrolysis powered by renewable energy: a review. Processes 8 (2), 248. Available at https://doi.org/10.3390/pr8020248.

Buelvas, W.L., Ávila, K.C.P., Jiménez, A.R., 2014. Temperature as a factor determining on water electrolysis. International Journal of Engineering Trends and Technology 7 (1), 5–9. Available at https://doi.org/10.14445/22315381/IJETT-V7P217.

Burton, N.A., Padilla, R.V., Rose, A., Habibullah, H., 2021. Increasing the efficiency of hydrogen production from solar powered water electrolysis. Renewable and Sustainable Energy Reviews 135, 110255–110275. Available at https://doi.org/10.1016/j.rser.2020.110255.

Chen, X., Shen, S., 2010. Semiconductor-based photocatalytic hydrogen generation. Chemical Reviews 110 (11), 6503–6570. Available at https://doi.org/10.1021/cr1001645.

Chen, Z., Wei, W., Ni, B.-J., 2021. Cost-effective catalysts for renewable hydrogen production via electrochemical water splitting: recent advances. Current Opinion in Green and Sustainable Chemistry 27, 100398. Available at https://doi.org/10.1016/j.cogsc.2020.100398.

Cheng, Y., Jiang, S.P., 2015. Advances in electrocatalysts for oxygen evolution reaction of water electrolysis-from metal oxides to carbon nanotubes. Progress in Natural Science 25 (6), 545–553. Available at https://doi.org/10.1016/j.pnsc.2015.11.008.

Di Paola, A., Bellardita, M., Palmisano, Brookite, L., 2013. The least known TiO_2 photocatalyst. Catalysts 3, 36–73. Available at https://doi.org/10.3390/catal3010036.

Dragica LJ, S., Milica, P.M., Sofija, P., 2003. Hydrogen generation from water electrolysis possibilities of energy saving. Journal of Power Sources 118 (1–2), 315–319. Available at https://doi.org/10.1016/S0378-7753(03)00077-6.

El-Shafie, M., 2023. Hydrogen production by water electrolysis technologies: a review. Results in Engineering 20, 101426. Available at https://doi.org/10.1016/j.rineng.2023.101426.

Engge, Y., Maulana, F., 2020. Dissociation of water into hydrogen and oxygen through a combination of electrolysis and photocatalyst. In: IOP Conference Series: Earth and Environmental Science, Volume 743, The 11th International Conference on Global Resource Conservation 28-29 July 2020, East Java, Indonesia.

Gilmore, T., Gouma, P.I., 2022. Photochemical water splitting via transition metal oxides. Catalysts 12, 1303. Available at https://doi.org/10.3390/catal12111303.

Grigoriev, S.A., Fateev, V.N., Bessarabov, D.G., Millet, P., 2020. Current status, research trends, and challenges in water electrolysis science and technology. International Journal of Hydrogen Energy 45 (49), 26036–26058. Available at https://doi.org/10.1016/j.ijhydene.2020.03.109.

Gupta, R., Gupta, R., Pandita, N., 2023. Solar energy: an ideal solution for the energy crisis in the union territories of Jammu and Kashmir and Ladakh. Engineering and Scientific International Journal 10 (1), 10–13.

Gómez, J., Castro, R., 2024. Green Hydrogen energy systems: a review on their contribution to a renewable energy system. Energies 17, 3110. Available at https://doi.org/10.3390/en17133110.

Hota, P., Das, A., Dilip, K., Maiti, 2023. A short review on generation of green fuel hydrogen through water splitting. International Journal of Hydrogen Energy 48 (2), 523–541. Available at https://doi.org/10.1016/j.ijhydene.2022.09.264.

Ismael, M., 2021. Latest progress on the key operating parameters affecting the photocatalytic activity of TiO_2-based photocatalysts for hydrogen fuel production: a comprehensive review. Fuel 303, 121207. Available at https://doi.org/10.1016/j.fuel.2021.121207.

Kailai, Z., Haijun, H., Litong, S., Baohua, J., 2021. Strategies for optimizing the photocatalytic water-splitting performance of metal–organic framework-based materials. Small Science 1 (12), 1–34. Available at https://doi.org/10.1002/smsc.202100060 2100060.

Kazemi, A., Manteghi, F., 2024. Metal electrocatalysts for hydrogen production in water splitting. ACS Omega 9 (7), 7310–7335. Available at http://doi.org/10.1021/acsomega.3c07911.

Kazuhiro, T., 2017. Photocatalytic water splitting: quantitative approaches toward photocatalyst by design. ACS Catalysis 7 (11), 8006–8022. Available at https://doi.org/10.1021/acscatal.7b02662.

Kemppainen, E., Halme, J., Lund, P., 2015. Physical modeling of photoelectrochemical hydrogen production devices. The Journal of Physical Chemistry C 119 (38), 21747–21766. Available at http://doi.org/10.1021/acs.jpcc.5b04764.

Kumar, S.S., Himabindu, V., 2019. Hydrogen production by PEM water electrolysis – a review. Materials Science for Energy Technologies 2 (3), 442–454. Available at https://doi.org/10.1016/j.mset.2019.03.002.

Kumar, S.S., Lim, H., 2022. An overview of water electrolysis technologies for green hydrogen production. Energy Reports 8, 13793–13813. Available at https://doi.org/10.1016/j.egyr.2022.10.127.

Lasia, A., 2019. Mechanism and kinetics of the hydrogen evolution reaction. International Journal of Hydrogen Energy 44 (36), 19484–19518. Available at https://doi.org/10.1016/j.ijhydene.2019.05.183.

Liu, L., Ma, H., Khan, M., Hsiao, B.S., 2024. Recent advances and challenges in anion exchange membranes development/application for water electrolysis: a review. Membranes 14, 85. Available at https://doi.org/10.3390/membranes14040085.

Lu, G., Yates, J.T., 1995. Photocatalysis on TiO_2 surfaces: principles, mechanisms, and selected results. Chemical Reviews 95 (3), 735–758. Available at https://doi.org/10.1021/cr00035a013.

Marian, C., 2022. Water electrolysis: from textbook knowledge to the latest scientific strategies and industrial developments. Chemical Society Reviews 51, 4583–4762. Available at https://doi.org/10.1039/D0CS01079K.
Maulana, F., Engge, Y., Nurhuda, M., Istiroyah, I., 2023. Hydrogen production of the water splitting process by electrolysis-photocatalysis reactor. AIP Conference Proceedings 2913, 040010. Available at https://doi.org/10.1063/5.0171889.
Miller, H.A., Bouzek, K., Hnat, J., Loos, S., 2020. Green hydrogen from anion exchange membrane water electrolysis: a review of recent developments in critical materials and operating conditions. Sustainable Energy and Fuels 4 (5), 2114–2133. Available at https://doi.org/10.1039/C9SE01240K.
Mohsin, M., Ishaq, T., Bhatti, I.A., Maryam, Jilani, A., 2023. Semiconductor nanomaterial photocatalysts for water-splitting hydrogen production: the holy grail of converting solar energy to fuel. Nanomaterials 13, 546. Available at https://doi.org/10.3390/nano13030546.
Molaei, M.J., 2024. Recent advances in hydrogen production through photocatalytic water splitting: a review. Fuel 365, 131159. Available at https://doi.org/10.1016/j.fuel.2024.131159.
Naimi, Y., Antar, A., 2018. Hydrogen generation by water electrolysis. Advances in Hydrogen Generation Technologies. InTech. Available at http://doi.org/10.5772/intechopen.76814.
Nakata, K., Fujishima, A., 2012. TiO_2 photocatalysis: Design and applications. Journal of Photochemical and Photobiology C 13 (3), 169–189. Available at https://doi.org/10.1016/j.jphotochemrev.2012.06.001.
Nie, J.H., Chen, Y., Boehm, R.F., Katukota, S., 2008. A Photoelectrochemical model of proton exchange water electrolysis for hydrogen production. Journal of Heat Transfer 130 (4), 042409. Available at https://doi.org/10.1115/1.2789722 042409.
Ouabi, H., Lajouad, R., 2024. Hydrogen production by water electrolysis driven by a photovoltaic source: a review, e-Prime - advances in electrical engineering. Electronics and Energy 8, 100608. Available at https://doi.org/10.1016/j.prime.2024.100608.
Rashid, M.D., Al Mesfer, M.K., Naseem, H., Danish, M., 2015. Hydrogen production by water electrolysis: a review of alkaline water electrolysis, PEM water electrolysis and high temperature water electrolysis. International Journal of Engineering and Advanced Technology 4 (3), 80–93 ISSN: 2249 – 8958. C3749024315/2013©BEIESP.
Santos, D.M.F., Sequeira, C.A.C., Figueiredo, J.L., 2012. Hydrogen production by alkaline water electrolysis. Química Nova 36 (8), 1176–1193. Available at https://doi.org/10.1590/S0100-40422013000800017.
Serpone, N., Lawless, D., Terzian, R., 1992. Solar fuels: status and perspectives. Solar Energy 49 (4), 221–234. Available at https://doi.org/10.1016/0038-092X(92)90001-Q.
Shit, S., Bolar, S., Murmu, N.C., Kuila, T., 2021. An account of the strategies to enhance the water splitting efficiency of noble-metal-free electrocatalysts. Journal of Energy Chemistry 59, 160–190. Available at https://doi.org/10.1016/j.jechem.2020.10.022.
Singh, K.S.S., 2018. Reduction in fuel consumption and GHG emission by using Hydrogen gas in gasoline engine. Journal of Emerging Technologies and Innovative Research 5 (7), 435–444. Available at https://www.jetir.org.
Sinha, A.S.K., Ojha, U., Alam, Z., 2022. Production of hydrogen via water splitting using photocatalytic and photo-electrocatalytic route. Clean Energy Technologies - Hydrogen and Gasification Processes. Intech Open. Available at 10.5772/intechopen.106915.
Smolinka, T., 2009. Fuels-Hydrogen Production-Water Electrolysis, Encyclopedia of Electrochemical Power Sources. Elsevier, pp. 394–413. Available at https://doi.org/10.1016/B978-044452745-5.00315-4.
Sun, H., Xu, X., 2023. Electrochemical water splitting: bridging the gaps between fundamental research and industrial applications. Energy and Environmental Materials 6 (5), 1–21. Available at https://doi.org/10.1002/eem2.12441.
Tentu, R.D., Basu, S., 2017. Photocatalytic water splitting for hydrogen production. Current Opinion in Electrochemistry 5 (1), 56–62. Available at https://doi.org/10.1016/j.coelec.2017.10.019.
Terlouw, T., Bauer, C., 2022. Large-scale hydrogen production *via* water electrolysis: a techno-economic and environmental assessment. Energy and Environmental Science 15, 3583–3602. Available at https://doi.org/10.1039/D2EE01023B.
U.S. Department of Energy, Office of Energy, 2009. Efficiency and Renewable Energy, "Current (2009) State-of-the-Art Hydrogen Production Cost Estimate Using Water Electrolysis - Independent Review - Published for the U.S. Department of Energy Hydrogen Program," National Renewable Energy Laboratory, Golden, CO, 80401-3393.
Vijayan, D.S., Koda, E., Sivasuriyan, A., 2023. Advancements in solar panel technology in civil engineering for revolutionizing renewable energy solutions—a review. Energies 16, 6579. Available at https://doi.org/10.3390/en16186579.

Wang, S., Lu, A., Zhong, C.J., 2021. Hydrogen production from water electrolysis: role of catalysts. Nano Convergence 8 (4), 1–23. Available at https://doi.org/10.1186/s40580-021-00254-x.

Wang, Z., Li, C., Domen, K., 2019. Recent developments in heterogeneous photocatalysts for solar-driven overall water splitting. Chemical Society Reviews 48 (7), 2109–2125. Available at https://doi.org/10.1039/C8CS00542G15.

Zeng, K., Zhang, D., 2010. Recent progress in alkaline water electrolysis for hydrogen production and applications. Progress in Energy and Combust Science 36, 307–326. Available at https://doi.org/10.1016/j.pecs.2009.11.002.

Zhou, H., 2020. Increasing the efficiency of photocatalytic reactions via surface microenvironment engineering. Journal of the American Chemical Society 142 (6), 2738–2743. Available at https://doi.org/10.1021/jacs.9b12247.

CHAPTER

3

Technologies for water splitting

Rakhi Khandelwal

Government Mahila Engineering College, Ajmer, Rajasthan, India

3.1 Introduction

Water splitting is a chemical process that entails dividing water (H_2O) into its component constituents, oxygen (O_2) and hydrogen (H_2).

$$2H_2O \rightarrow H_2 + O_2$$

Since the bond between hydrogen and oxygen is strong, therefore, a huge amount of energy is required for H-O bond breaking. This energy requirement may be accomplished by the use of an external power source or by sunlight. The hydrogen and oxygen gases produced may be collected and employed in several applications, including producing and storing renewable energy.

The key significance of water splitting lies in the production of hydrogen gas, which can be used as a source of sustainable and environmentally friendly fuel (Dincer and Acar, 2015). Hydrogen gas is considered a versatile energy carrier because it can be combusted. Alternatively, it may be used in fuel cells to create energy while producing just water vapor as a resultant product. Energy and environmental challenges might be greatly improved by the use of water-splitting technology by offering an efficient means of storing and using hydrogen fuel with the utilization of sources of renewable energy like wind energy and solar energy.

The objective of this chapter is to provide a comprehensive overview of the technologies used for water splitting, including an in-depth examination of the underlying principles, the various methods employed, and their limitations and advancements. In this chapter, we will discuss the practical applications and implications of water-splitting technologies in various fields, including energy production, environmental sustainability, and industrial processes. The chapter also includes the key challenges facing the development and commercialization of water-splitting technologies. To illustrate the potential and impact of water-splitting technologies being implemented, one case study has also been discussed at the end of this chapter.

Photochemical Splitting of Water
DOI: https://doi.org/10.1016/B978-0-443-29064-0.00014-0

3.2 Water-splitting technologies

The various technologies and methods that enable water splitting are as follows:

1. Water electrolysis
2. Photochemical water splitting
3. Thermochemical water splitting
4. High temperature steam electrolysis
5. Biological water splitting

The different types of water-splitting technologies are shown in Fig. 3.1.

3.2.1 Water electrolysis

The process in which electricity is used as a source of energy for splitting of water molecules into hydrogen and oxygen is known as water electrolysis. The underlying idea of water electrolysis is that when electricity is passed through water, which is as an electrolyte, then electrochemical reactions take place in it that causes its splitting. The two primary types of processes involved in water splitting are the oxygen evolution reaction (OER) and the hydrogen evolution reaction (HER) as shown in Fig. 3.2. These reactions take place during the process of breaking down molecules of water into their constituent elements, hydrogen and oxygen.

Hydrogen evolution reaction: In HER, reduction of water molecules (H_2O) resulted in the formation of hydrogen (H_2) and oxygen gas (O_2).

The half-reaction occurring at the cathode is

$$2H_2O_{(l)} + 2e^- \rightarrow H_{2(g)} + 2OH^-_{(aq)}$$

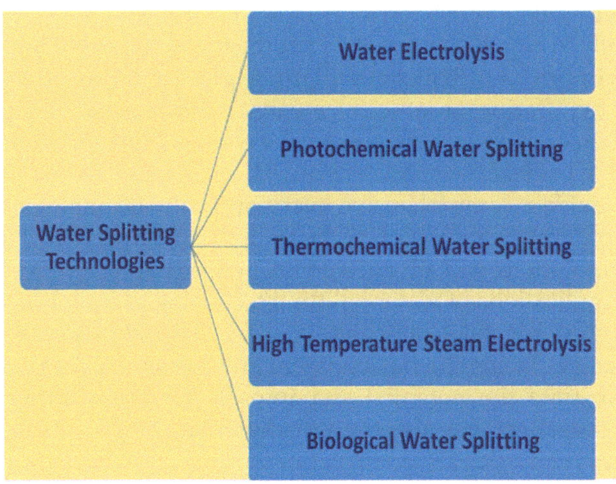

FIGURE 3.1 Different types of water-splitting technologies.

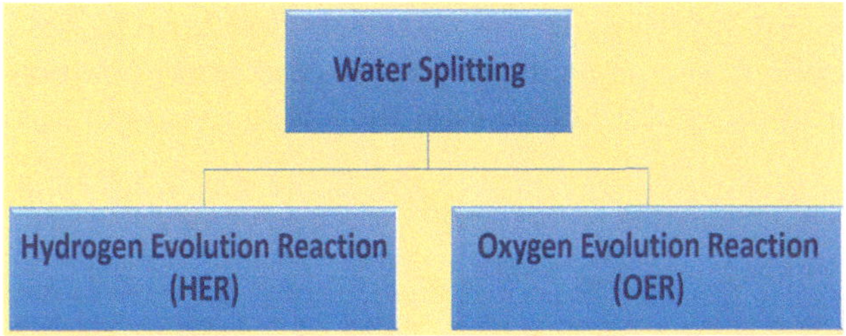

FIGURE 3.2 Two main types of reactions involved in water splitting.

At the anode, the half-reaction occurs:

$$2H_2O_{(l)} \rightarrow O_2(g) + 4H^+_{(aq)} + 4e^-$$

The overall reaction is

$$2H_2O_{(l)} \rightarrow H_{2(g)} + O_2(g)$$

Enormous effort has been made to design active catalysts for HER efficiently, utilizing transition metals and related derivatives (Chhowalla et al., 2016; Kibsgaard et al., 2012; Lukowski et al., 2013; Xie et al., 2013; Kong et al., 2014).

Oxygen evolution reaction: In OER, water molecules are oxidized to produce oxygen gas (O_2) and protons (H^+).

The half-reaction at the anode is

$$2H_2O_{(l)} \rightarrow O_2(g) + 4H^+_{(aq)} + 4e^-$$

The half-reaction takes place at cathode:

$$4H^+_{(aq)} + 4e^- \rightarrow 2H_{2(g)}$$

The overall reaction is

$$2H_2O_{(l)} \rightarrow O_2(g) + 2H_{2(g)}$$

HER and OER both are fundamental processes in water splitting. The hydrogen and oxygen gases released in these processes can be collected and used for various applications, including clean energy production and storage. Conventional water electrolysis produces both H_2 and O_2 concurrently due to the close coupling of HER and OER (You and Sun, 2018).

The process takes place in an electrolysis cell, which is known as electrolyzer. The main components of an electrolysis cell are as follows:

3.2.1.1 *Electrolyte*

An electrolyte is a material that, when dissolved in water, conducts electricity. In water electrolysis, an alkali metal hydroxide solution in diluted form, such as dilute potassium hydroxide (KOH) or dilute sodium hydroxide (NaOH), is commonly used as the electrolyte. The electrolyte allows the flow of ions, which is essential for the electrochemical reactions to occur.

3.2.1.2 *Anode*

The electrolysis cell's positive electrode is known as the anode. Electric current induces oxidation at the anode. In water electrolysis, water molecules at the anode release oxygen gas (O_2) through the process of OER.

Reaction at anode:

$$2H_2O_{(l)} \rightarrow O_2(g) + 4H^+_{(aq)} + 4e^-$$

3.2.1.3 *Cathode*

The electrolysis cell's negative electrode is known as the cathode. The reduction reaction takes place at the cathode. In HER at the cathode, reduction of water molecules takes place resulting in the formation of hydrogen gas (H_2).

Reaction at cathode:

$$H_2O + 2e^- \rightarrow 2H_2 + 2OH^-$$

3.2.1.4 *Transporter of the electrolyte*

The transporter of the electrolyte has a significant impact in facilitating the movement of ions within the electrolyzer and thereby enhancing the efficiency of the electrolyzer.

3.2.1.5 *Electrolysis process*

When the anode and cathode are connected to an external direct current (DC) electrical source, an external circuit allows electrons to go from the anode to the cathode. Simultaneously, ions (H^+, OH^-) flow through the electrolyte to maintain electrical neutrality. An illustration of the electrolysis cell for water splitting is depicted in Fig. 3.3.

Hydrogen and oxygen collection: Separate collections are made of the gases produced at the anode (oxygen) and the cathode (hydrogen). These gases may then be stored and used for a variety of purposes, including renewable energy generation in the form of fuel cells.

The following balanced chemical equation may be used to represent the whole water electrolysis process:

Overall reaction:

$$2H_2O \rightarrow 2H_2 + O_2$$

Because it needs an outside electrical energy source to fulfill the energy requirements of the reactions, the water electrolysis is an energy-intensive process.

For the electrolysis of pure water, it is often necessary to use excess energy in the form of over potential (to overcome different activation barriers). This extra energy is very crucial because, without it, the process would go very slowly or even not at all. Furthermore, pure water has approximately one millionth the electrical conductivity of seawater. In such cases, the effectiveness of electrolysis can be enhanced by combining a suitable electrolyte, such as an acid, base, or salt, with an electrocatalyst.

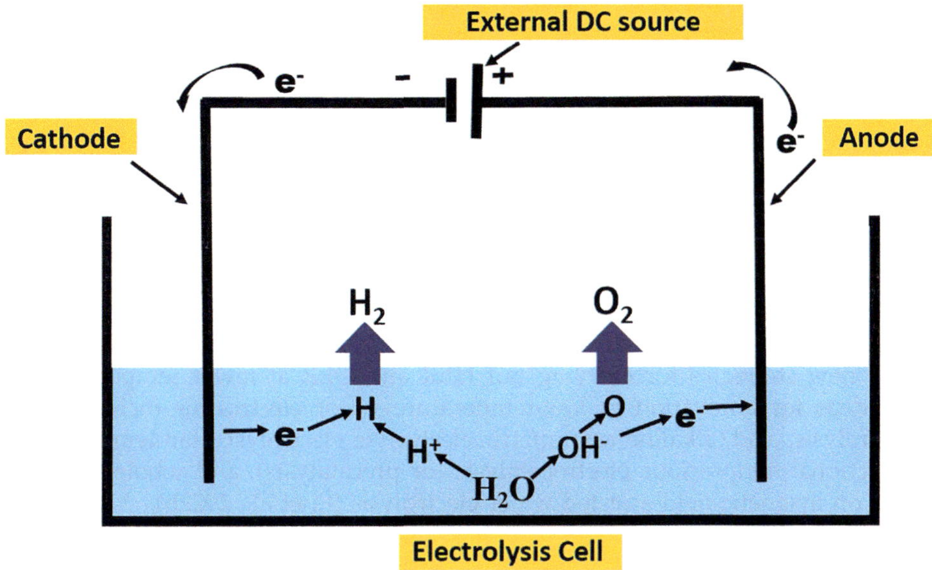

FIGURE 3.3 Illustration of an electrolysis cell for water splitting.

3.2.1.6 *Types of electrolyzers*

Different types of electrolyzers use various electrolyte materials and transporters, and the choice of electrolyte is typically influenced by the specific requirements of the electrolyzer design, the target application, and the desired performance characteristics. Based on the type of transporter of the electrolyte, there are three main types of electrolyzers (Chisholm et al., 2022):

1. Polymer electrolyte membrane electrolyzer (PEM)
2. Alkaline electrolyzers
3. Solid oxide electrolyzer

1. PEM electrolyzer: The electrodes of this electrolyzer are separated by a perfluorosulfonic acid polymer called Nafion that permits the flow of hydrogen ions created by the oxidation of water at the anode through it and into the cathode compartment, where they are released to evolve hydrogen gas. PEM electrolysis is efficient and has a rapid response time, making it appropriate for on-demand hydrogen production and integration with renewable energy sources.
2. Alkaline electrolyzers: In the alkaline electrolyzer, dilute aqueous NaOH or KOH is used in the electrolysis. This allows hydroxide ions to travel to the anode, where they produce oxygen. It is a mature technology and has been used for decades. While it is less efficient than PEM electrolysis, it is more cost-effective.
3. Solid oxide electrolyzer: Ceramic oxide is used in this electrolyzer to keep the electrodes apart. At the cathode, reduction of water takes place resulting in the formation of hydrogen and oxide ions. Oxide ions go through the ceramic oxide to the anode and get oxidized to become oxygen gas. To reduce the external voltage needed for electrolysis, the process is carried out at high temperature of 700°C–800°C.

Water electrolysis efficiency is affected by electrolyte choice, electrolysis cell design, and the quality of electrodes.

3.2.1.7 *Recent advancements in water electrolysis*

Low-temperature electrolysis (LTE): This is the method in which electrolysis of water is performed at relatively mild temperatures, typically around room temperature or slightly higher. Therefore LTE systems are characterized by their simplicity, ease of operation, and potential for cost-effectiveness. But at the same time, since the whole process is to be carried out at lower temperatures, very specific and effective efficient catalysts and electrolyzers are required for LTE. In the past few years, researchers have continuously worked on developing new strategies for carrying out HER and OER at lower temperatures. These emerging ideas for innovation in low-temperature water electrolysis include decoupled water electrolysis, acid/alkaline asymmetric electrolyte electrolysis, tandem water electrolysis, bifunctional nonprecious electrocatalysts (or precatalysts), and whole water electrolysis based on magnetic field-assisted water electrolysis (Li et al., 2022b).

3.2.2 Photochemical water splitting

In this type of water splitting, a photocatalyst is used to facilitate a water-splitting reaction. Actually, in this process, photocatalysts harness the solar energy and this energy is used for splitting the water into hydrogen (H_2) and oxygen (O_2) (Hassan et al., 2023). It is based on the principle of using a semiconductor material (the photocatalyst) to absorb light energy and create electron–hole pairs.

The absorption of light with greater energy than the semiconductor's bandgap causes electrons to move from the valence band to the conduction band, resulting in the formation of electron–hole pairs. Subsequently, electron–hole pairs take part in the oxidation and reduction processes necessary for the splitting of water.

The key components of photocatalytic water splitting are as follows:

3.2.2.1 *Photocatalyst*

A photocatalyst is typically a semiconductor material that can absorb photons (light particles) and generate electron–hole pairs. Common photocatalysts include titanium dioxide (TiO_2) (Galvão et al., 2022), strontium titanate ($SrTiO_3$) (Shuai et al., 2023), bismuth vanadate ($BiVO_4$) (Lei et al., 2023), and tungsten trioxide (WO_3) (Gong et al., 2022). Photocatalysts for this type of water splitting are chosen based on their bandgap energy, light absorption capacity, and their stability (Gupta et al., 2022).

3.2.2.2 *Bandgap energy*

The photocatalyst's bandgap energy determines its ability to absorb light at specific wavelengths. It should match the energy of the incident photons for efficient absorption and electron excitation.

3.2.2.3 *Light sources*

Photocatalytic water splitting can be driven by various light sources, including sunlight, ultraviolet (UV) light, or artificial sources such as LEDs or lamps. The choice of light source depends on the photocatalyst's absorption characteristics.

3.2.2.4 *Photocatalytic water-splitting process*

The process of photochemical water splitting is depicted in Fig. 3.4. It involves the following steps:

Absorption of photons: Photocatalysts absorb photons from a light source, often sunlight or artificial light.
Generation of electron–hole pairs: Due to the absorption of photons, electrons in the photocatalyst get excited leading to the formation of electron–hole pairs.
Redox reactions: These excited electrons and holes participate in reduction and oxidation reactions with water.
Reduction: Electrons are used for the reduction of water which results in the formation of hydrogen ions (H^+).
Oxidation: Holes are used to oxidize the water to form oxygen gas (O_2).
Formation of hydrogen and oxygen: The generated hydrogen ions combine to form molecular hydrogen (H_2), while oxygen (O_2) gas is released (Jiang et al., 2021).

3.2.2.5 *Recent advancements in photochemical water splitting*

To make photochemical water splitting more scalable and cost-effective, researchers have been keenly involved in the development of various strategies like the development of low bandgap semiconductors (Chavas et al., 2020), heterojunctions (Meena et al., 2022), doping

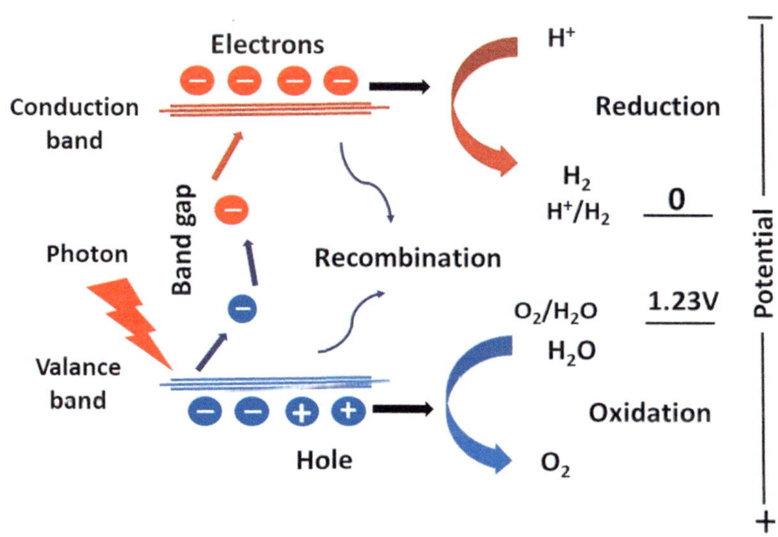

FIGURE 3.4 Photochemical water splitting.

(Tay et al., 2018), incorporating transition metal nanoparticles for surface plasmon resonance (Kumar et al., 2022), heterogeneous cocatalyst (Fang et al., 2020), and change in surface morphology (Shuchi et al., 2023). A lot of research has been done on various photocatalysts such as MoS_2 (Ibrahim et al., 2024), dip catalyst (Othmane et al., 2024), and ZnV_2O_6 nanosheet (El Ouardi et al., 2024).

3.2.3 Thermochemical water splitting

High temperatures (500°C–2000°C) are used in thermochemical water splitting to generate hydrogen (H_2) and oxygen (O_2). The basic principle of thermal water splitting involves supplying sufficient heat for breaking the H-O bonds in water molecules, causing them to dissociate into oxygen and hydrogen (Li et al., 2022a). Such high temperatures might be achieved with concentrated solar power or the waste heat from nuclear power processes, which has the potential to start a series of chemical reactions structured into cycles. Different cycles have been proposed and researched, each with its unique set of reactions and materials. Among all those cycles, the sulfur-iodine and cerium oxide cycles are well-studied examples. These cycles involve multiple steps, such as water decomposition, chemical reactions with intermediate compounds, and the regeneration of reactants, ultimately resulting in the formation of hydrogen gas.

3.2.3.1 *Sulfur-iodine (S-I) cycle*

The sulfur-iodine cycle is one of the utmost extensively studied thermochemical water-splitting cycles.

It involves three main chemical reactions: sulfuric acid (H_2SO_4) decomposition, hydrogen iodide (HI) production from sulfur dioxide (SO_2) and water, and the final step of HI electrolysis to yield hydrogen gas (Colette Maatouk et al., 2009). This cycle is known for its high efficiency but requires a complex set of reactions and corrosive chemicals. Various reactions and steps involved in S-I cycle are shown in Fig. 3.5.

3.2.3.2 *Cerium oxide cycle*

The cerium oxide cycle utilizes cerium oxide (CeO_2) as a redox material. The cycle includes two key steps: first is the reduction of cerium oxide by high-temperature heat to produce cerium oxide with a lower oxidation state and second is the oxidation of cerium oxide back to its original state using water vapor to release oxygen and regenerate cerium oxide (Cho et al., 2021). Reactions involved in this cycle are shown in Fig. 3.6. This cycle is promising due to the availability of cerium and its high oxygen storage capacity.

However, there are some challenges also associated with thermal water splitting, such as the high energy requirements, the need for advanced materials to withstand high temperatures and corrosive environments, and the overall efficacy of the process. Current research and development efforts aim to address these challenges and make thermal water splitting a more practical and efficient method for hydrogen production in the future.

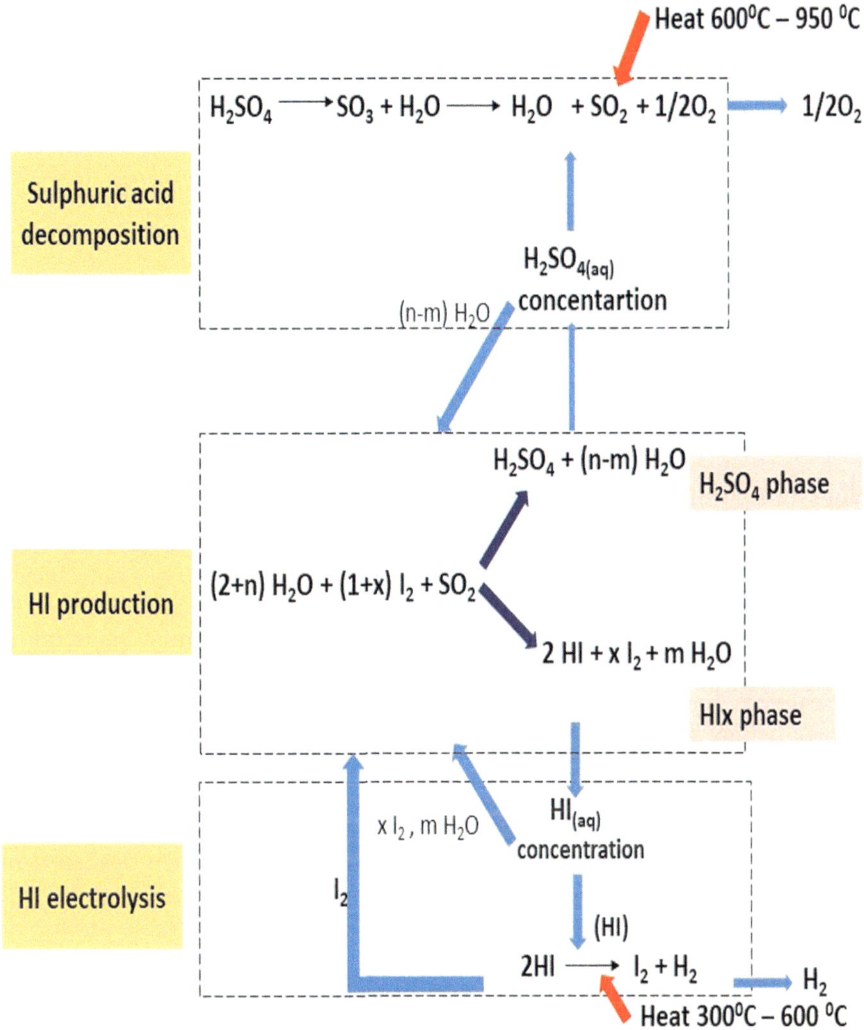

FIGURE 3.5 Thermochemical water splitting: sulfur-iodine cycle (S-I cycle).

3.2.3.3 *Recent advancements in thermochemical water splitting*

Since these two-step oxidation–reduction cycles require a large amount of heat, therefore, to make out the balance between thermodynamic and kinetic characteristics of the thermochemical water-splitting process and to optimize the solar thermochemical hydrogen production (STCH) performance, various new materials have been investigated. Such as new-fangled class of compositionally complex perovskite oxides (Zhang et al., 2023b), $(La_{0.8}Sr_{0.2})(Mn_{(1-x)/3}Fe_{(1-x)/3}CoxAl_{(1-x)/3})O_3$, $Ca(Ti_{0.5}Mn_{0.5})O_3$ (Qian et al., 2021) and $(_{La1-x}Srx)(Mn_{1-y}Al_y)O_3$ (McDaniel et al., 2013).

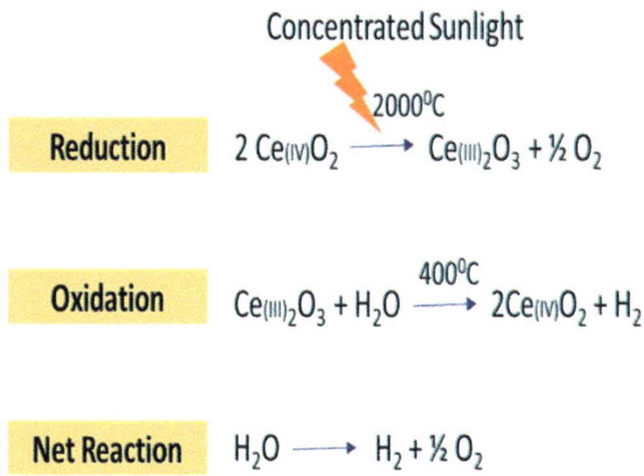

FIGURE 3.6 Thermochemical water splitting: cerium oxide cycle.

3.2.4 High-temperature steam electrolysis

This is another potential method for producing hydrogen using thermal energy from nuclear reactors. This method capitalizes on the increased efficiency of electrolysis at higher temperatures (Pandiyan et al., 2019). At elevated temperatures, ranging between 800°C and 1000°C, the electrical conductivity of water increases significantly, allowing for more effective electrolysis. Additionally, thermal energy can be harnessed more efficiently to drive the reaction, thereby potentially reducing the overall energy input required for hydrogen production. This strategy makes use of the reverse reaction of solid oxide fuel cell (SOFC) technology, which is an electrochemical conversion device that generates electricity through redox reactions and uses a solid oxide electrolyte to transport oxygen ions, while in the high-temperature steam electrolysis (HTSE) method, using nuclear thermal energy instead of electricity, water is first converted to steam. Steam is then broken down at the cathode to produce hydrogen molecules and oxygen ions, which move through the solid oxide electrolyte material to produce oxygen molecules at the anode surface. The basic features and components of HTSE are as follows:

3.2.4.1 *High operating temperature*

HTSE is conducted at temperatures significantly higher than traditional water electrolysis. Typical operating temperatures range from 700°C to 1000°C, or even higher in some cases. These elevated temperatures speed up the water-splitting reaction and lower the need for electricity in the process.

3.2.4.2 *Solid oxide electrolysis cells*

Solid oxide electrolysis cells (SOECs) are the primary electrochemical cells used in HTSE. These cells consist of a solid oxide electrolyte, often made of zirconia-based materials, sandwiched between the cathode and the anode. The electrolyte is a ceramic material that conducts oxygen ions at high temperatures.

3.2.4.3 Electrochemical reactions

The key electrochemical reactions that occur in a HTSE cell are as follows:

Anode reaction: At the anode, oxygen ions (O^{2-}) move through the solid oxide electrolyte to the anode side. These oxygen ions get the electrons from the external circuit at the anode and are oxidized into oxygen gas (O_2).
Cathode reaction: At the cathode, steam (H_2O) is supplied, and the high temperature causes the steam to dissociate into oxygen ions (O^{2-}) and hydrogen gas (H_2). These generated oxygen ions move through the solid oxide electrolyte toward the anode (Guo et al., 2020). HTSE process is shown in Fig. 3.7.

3.2.4.4 External electrical power

Similar to traditional electrolysis, to provide the required potential difference (voltage) for the electrochemical reactions to occur, an external electrical power source is needed. However, because of the elevated operating temperature, the energy efficiency of HTSE is higher than that of room-temperature electrolysis.

3.2.4.5 Hydrogen and oxygen separation

The gases generated at the anode and cathode, hydrogen and oxygen, are collected separately. Various methods can be used for gas separation and purification.

3.2.4.6 Advantages of high-temperature steam electrolysis

- Higher energy efficiency as compared to traditional water electrolysis at lower temperatures.
- Rapid reaction kinetics, leading to faster hydrogen production rates.
- Enhanced energy integration through compatibility with high-temperature heat sources like concentrated solar power and nuclear energy.
- Possibility of coelectrolysis, in which the synthesis of synthetic fuels like syngas—a blend of hydrogen and carbon monoxide—can be coupled with steam electrolysis.

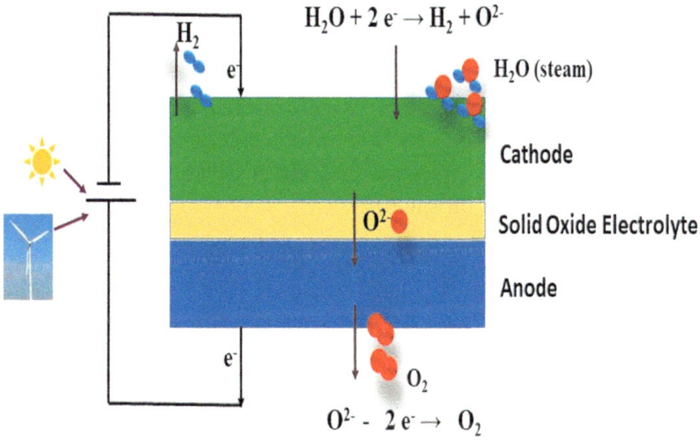

FIGURE 3.7 High-temperature steam electrolysis.

3.2.4.7 *Challenges and limitations*

- Materials and components must withstand high-temperature operation, which can lead to materials degradation over time.
- System complexity and cost, especially in the development of robust and durable SOECs.
- The need for steam as a feedstock, which may require additional energy for steam generation.
- HTSE is an area of active research and development, with the potential to contribute to clean and efficient hydrogen production, particularly when coupled with high-temperature heat sources or industrial processes that can provide the required heat.

3.2.4.8 *Recent advancements in high-temperature steam electrolysis*

To improve the efficiency of the HTSE process, researchers have focused mainly in two domains. First is the development of a more efficient cathodic material and the second one is the exploration of new oxides for SOEC. In the past few years, several novel nanoengineered cathode materials have been synthesized and studied for their potential to work as the cathode in the HTSE process (Saeed et al., 2023; Biswas et al., 2023; Luo et al., 2023). The development of novel SOEC materials must take into account important performance metrics such as thermal, chemical, and mechanical stability, as well as an application-oriented assessment (cost-effectiveness, ease of manufacture). In recent years, researchers worked on the SOFCs with electrolytic membranes based on ZrO_2-Sc_2O_3 systems that showed the highest power characteristics compared to devices using membranes based on ZrO_2-Y2O_3. Reducing the operating temperature of the fuel cell by using electrolytic membranes made of zirconium dioxide stabilized by scandium oxide enhances the stability and dependability of electrochemical devices (Agarkov et al., 2023).

3.2.5 Biological water splitting

Biological water splitting involves using living organisms, such as algae, bacteria, or engineered microorganisms, to break water molecules into oxygen and hydrogen. This process often mimics natural photosynthesis, where these organisms utilize sunlight to drive the conversion of water into its elemental components. These methods aim to harness the natural ability of organisms to use sunlight and perform water splitting while optimizing and enhancing their capabilities for hydrogen production. Developments in synthetic biology, genetic engineering, and cultivation methods are accelerating advances in biological water splitting for sustainable hydrogen generation. Many strategies or techniques have been researched to do water splitting biologically.

3.2.5.1 *Algae and cyanobacteria*

These photosynthetic organisms naturally perform photosynthesis and can split water. They use sunlight to power the process, converting water and carbon dioxide into oxygen and sugars (Li et al., 2020). Researchers work with these organisms or modify them to increase their hydrogen production while minimizing the production of sugars.

3.2.5.2 *Genetic engineering*

Scientists manipulate the genetic makeup of microorganisms to enhance their capability for hydrogen production. By modifying enzymes or introducing specific genes (Srirangan et al., 2011), they aim to optimize the process of water splitting within these organisms to produce more hydrogen.

3.2.5.3 *Biological systems with synthetic components*

Synthetic biology techniques involve incorporating artificial or engineered components into biological systems to improve efficiency. This can include introducing synthetic catalysts or proteins to enhance the water-splitting process (Chong et al., 2016).

3.2.5.4 *Bioreactors and cultivation systems*

To scale up hydrogen production, bioreactors or cultivation systems are used to cultivate large quantities of these organisms under controlled conditions (Ben et al., 2007). This allows for controlled optimization of environmental factors like light, temperature, and nutrients to maximize hydrogen production.

3.2.5.5 *Enzymatic catalysis*

Splitting of water molecules can be done with the help of enzymes. Researchers study and modify these enzymes to improve their efficiency in separating water into hydrogen and oxygen (Böhmer et al., 2017).

Biological water splitting is an active area of research, facing challenges like stability, efficiency, and scalability. Researchers are working on improving the efficiency of catalysts, enhancing light absorption, increasing stability, and reducing costs to make these systems more viable for large-scale energy production.

By combining principles from biology, chemistry, and materials science, scientists aim to develop efficient and sustainable methods for artificial biological water splitting to produce renewable energy, particularly hydrogen, as a clean fuel source.

3.3 Challenges of water-splitting technologies

Water-splitting technology, particularly through electrolysis, faces several challenges:

3.3.1 Energy efficiency

Electrolysis needs a large amount of electrical energy to split the molecules of water into hydrogen and oxygen. Improving the efficiency of this method to minimize energy input is a major challenge.

3.3.2 Costs

The materials used in electrolysis, especially those for catalysts (like platinum), are expensive. Reducing costs associated with catalysts and other components is crucial for making the technology more economically viable.

3.3.3 Corrosion and durability

Electrolysis involves tough conditions (such as high temperatures and corrosive environments) that can degrade the materials used in the process. Developing durable materials that can withstand these conditions and maintain efficiency over time is a challenge.

3.3.4 Scaling up

While water-splitting technologies have been demonstrated at a laboratory scale, scaling up these processes to industrial levels without compromising efficiency and cost-effectiveness remains a challenge.

3.3.5 Storage and transportation

Transporting and storing hydrogen, which is the main byproduct of water splitting, is a challenging task. For the extensive use of this green hydrogen, safe, cost-effective, and efficient techniques for its storage and transportation are essential.

3.3.6 Integration of renewable energy sources

Fluctuations in the energy availability from renewable energy sources such as wind and solar power creates complexities in the integration of water-splitting technologies with renewable energy sources. Overcoming this intermittency challenge is a significant problem in ensuring a regular and consistent supply of green hydrogen.

3.3.7 Environmental impact

Since the catalyst used in water-splitting techniques contains rare or ecologically harmful elements, this may cause serious concerns about their environmental impact in the long term and their sustainability.

Researchers and scientists are actively working to address these challenges by exploring new materials, improving the efficacy of electrolysis, developing better catalysts, optimizing processes, and finding innovative solutions to make water-splitting technologies more viable for widespread adoption as a clean and sustainable energy source.

3.4 Comparative study of various water-splitting technologies

Hydrogen generation by water-splitting technology varies greatly in terms of cost, maturity, and efficiency. A well-known technique with a high efficiency (60%–80%), water electrolysis is costly since it requires expensive catalysts and a lot of energy. Although photochemical water splitting uses sunlight, it has stability problems and poor efficiency (10%–20%), which makes it less practical for industrial use. Thermochemical water splitting, which uses high temperatures from solar or nuclear power, is an emerging technology with moderate efficiency (40%–60%) but significant material and operating expenses. HTSE has a high initial cost but has good integration with current heat sources and excellent efficiency (70%–85%). Although biological water splitting, which uses an organism's natural processes, is inexpensive and ecologically benign, it confronts substantial issues with efficiency and scalability. Its efficiency ranges from 1% to 5%. Every technique has distinct benefits and drawbacks that affect which sizes and applications it is appropriate for when producing hydrogen.

A comparative study of all the above water-splitting technologies is summarized in Table 3.1.

3.5 Future aspects of water-splitting technologies

The future of water-splitting technology holds promise for advancements that could significantly impact various sectors, especially in the realm of clean energy and sustainability. Following are some anticipated future aspects:

3.5.1 Improved efficiency and cost reduction

To improve the efficiency of existing water-splitting technologies, researchers now focus on developing more efficient and cost-effective catalysts for both HER and OER. This includes the exploration of abundant materials and innovative nanostructures that can enhance catalytic activity.

3.5.2 Renewable energy integration

The energy input problem can be solved by combining water splitting with renewable energy sources like solar and wind power. This integration can utilize surplus energy during peak production times to perform electrolysis, enhancing overall efficiency and reducing reliance on fossil fuels. This can lead to more sustainable and large-scale hydrogen production.

3.5.3 Scaling up for commercialization

Efforts are being made to scale up water-splitting technologies for industrial use.

TABLE 3.1 Comparative study of various water-splitting technologies.

Technology	Water electrolysis	Photochemical water splitting	Thermochemical water splitting	High temperature steam electrolysis	Biological water splitting
Efficiency in %	60–80	10–20	40–60	70–85	1–5
Cost	High	Moderate to high	Moderate	High	Moderate to low
Technology maturity	Well established	Emerging	Developing	Advanced	Experimental
Key advancement	Mature technology, high purity hydrogen	Uses sunlight directly, low emissions	High efficiency at high temperatures	High efficiency, can use nuclear heat	Low cost, renewable
Key challenges	Energy- intensive, expensive catalysts	Low efficiency, stability issues	High temperature requirement, material stability	Expensive, requires high-temperature materials	Very low efficiency, scalability issues

3.5.4 Technological innovations

Continuous advancements in engineering and materials science will likely lead to the development of more efficient electrolysis systems. These innovations could involve improvements in cell designs, membrane materials, and electrode configurations, resulting in higher performance and lower energy requirements.

3.5.5 Hybrid systems and storage solutions

Combining water-splitting technology with energy storage solutions could resolve problem of the uneven nature of renewable energy sources. Hydrogen produced by electrolysis can work as an energy storage tool, enabling a more reliable and stable renewable energy grid.

3.5.6 Environmental sustainability

Focus on environmentally friendly materials and processes is increasing. Research aims to develop water-splitting technologies that rely on abundant, nontoxic materials, minimizing environmental impact throughout the production cycle.

3.5.7 Market growth and commercial viability

As technology advances and production costs decrease, water splitting could become more economically viable, attracting investment and driving market growth. This could lead to broader adoption across industries, including transportation, energy production, and manufacturing.

3.5.8 Policy support and collaboration

Supportive policies, government incentives, and collaborative efforts among academia, industry, and governments can enhance the research, innovation, and set out of water-splitting knowledge and technology.

As these advancements continue, water splitting could become a cornerstone in the evolution toward a sustainable energy future by providing hydrogen as a clean and adaptable energy source.

3.6 Case study: Seawater to Green Hydrogen: A Prototype by IIT Madras Scientists

3.6.1 Objective

This case study aims to survey and evaluate many facets of the technology created by scientists at the Indian Institute of Technology Madras (IIT Madras) to split saltwater and make hydrogen.

3.6.2 Scope

Along with India, efforts are being made worldwide to research and develop technology to exploit the plentiful saltwater for the generation of green hydrogen. India is ambitious in creating green hydrogen, but the lack of the essential input, freshwater, is a hurdle. Parts for an alkaline water electrolyzer that can directly synthesize green hydrogen from saltwater have been developed by IIT Madras researchers. This is recommended as an alternative to freshwater, which is currently the primary ingredient utilized to make green hydrogen. This prototype addresses the combined issues of water scarcity and cost optimization.

3.6.3 Audience

1. **Energy industry professionals:** Individuals working in the energy sector, including executives, engineers, and researchers, who are interested in exploring innovative approaches to sustainable energy production and decarbonization.
2. **Government officials and policymakers:** Government representatives and policymakers involved in energy policy development, renewable energy promotion, and climate change mitigation strategies, seeking insights into successful green hydrogen initiatives and their policy implications.
3. **Investors and financial institutions:** Investors, venture capitalists, and financial institutions interested in financing sustainable energy projects, evaluating the economic viability and potential returns of green hydrogen production initiatives.
4. **Academia and research institutions**: Scholars, educators, and students in the fields of energy, environmental science, and sustainability, seeking case studies and real-world examples to enhance their understanding of green hydrogen technologies and their applications.
5. **General public:** Individuals interested in learning about innovative solutions to address climate change, reduce carbon emissions, and promote the enactment of renewable energy, contributing to broader awareness and engagement in sustainable energy transition efforts.

3.6.4 Rationale

This case study aims to highlight the innovative approach taken by scientists at IIT Madras in developing a prototype that utilizes seawater for green hydrogen production. This pioneering technology represents a noteworthy progress in the field of renewable energy and has the potential to revolutionize hydrogen production methods. This study seeks to showcase the scientific, environmental, and socio-economic significance of the technology developed, while also providing valuable insights for stakeholders interested in advancing renewable energy solutions.

3.6.5 Expected results and deliverables

Currently, commercial water electrolyzers employ high-purity freshwater to produce green hydrogen. However, as freshwater is a finite resource, over 80% of the world's population faces a high-risk water shortage. In contrast, saltwater is almost an endless supply because it makes up over 96% of the world's water storage and covers over 71% of the planet's surface. Therefore if seawater can be successfully utilized in a water electrolyzer for green hydrogen production, then it will prove as a game changer in this domain. The prototype electrolyzer created at IIT Madras in India demonstrated outstanding selectivity for OER over hypochlorite generation at normal temperature (26°C), low total seawater splitting voltage of 1.71 V, and high stability for 50 hours in salt water.

3.6.6 Workflow

The goal of the new technology and study from IIT Madras is to directly produce clean hydrogen from the plentiful ocean, potentially lowering the need for freshwater altogether. The alternate electrolyzer is presently in the prototype stage and may require some time to reach the market. This is shown in Fig. 3.8. The anodic chlorine corrosion and delayed cathodic HER limit the amount of hydrogen that may be produced using seawater electrolysis. This approach seems simple for developing cobalt iron (CoFe)-based Prussian blue analog with a nitrogen-doped carbon nanotube composite (CoFePBA/NCNT) and nickel-encapsulated carbon tubules (Ni/NCT) as OER and HER catalysts for overall

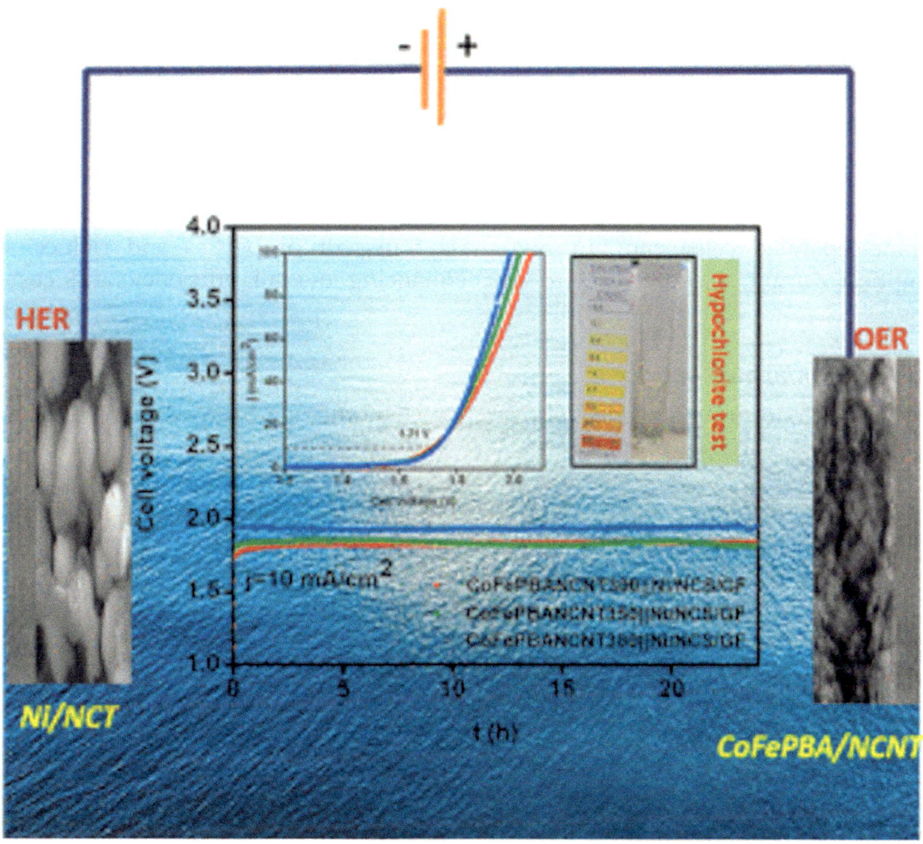

FIGURE 3.8 Prototype of seawater electrolyzer developed at IIT Madras (Ghosh et al., 2023).

seawater electrolysis employing graphite felt or carbon paper as the catalyst substrate, thus avoiding chlorine corrosion of the metal substrate. A simple brush coating method was employed for catalyst coating instead of a generally known hydrothermal method, which is achievable for large-scale electrolyzer fabrication (Ghosh et al., 2023).

3.6.7 Challenges and solutions

3.6.7.1 Corrosion and scaling

Seawater contains chlorides and other corrosive ions. These can lead to corrosion of electrodes and other components. Additionally, the formation of scales on electrode surfaces can reduce performance.

3.6.7.1.1 Potential solution

Utilizing corrosion-resistant electrode materials, like titanium-based alloys or coated electrodes, can mitigate corrosion and fouling issues associated with seawater electrolysis, ensuring long-term stability and performance of the electrolysis system.

3.6.7.2 High energy consumption

Seawater electrolysis requires higher energy input compared to freshwater electrolysis due to the higher concentration of ions and increased electrical conductivity, leading to increased operational costs.

3.6.7.2.1 Potential solution

Implementing advanced electrolysis system designs, such as flow-through reactors or membrane-separated cells, can improve mass transport properties and reduce energy consumption during seawater electrolysis, enhancing overall efficiency and cost-effectiveness.

3.6.7.3 Environmental impact

Disposal of concentrated brine (leftover seawater after electrolysis) can harm marine ecosystems if not managed properly.

3.6.7.3.1 Potential solution

Proper management of the concentrated brine is crucial. Options include reinjection into deep wells or utilizing it for other industrial processes.

3.6.7.4 Oxygen evolution reaction kinetics

OER at the anode during seawater electrolysis typically exhibits slow kinetics, limiting the overall efficiency and hydrogen production rate of the electrolysis process.

3.6.7.4.1 Potential solution

The development of robust and effective electrocatalysts, specifically for OER, can enhance the overall performance of seawater electrolysis systems and speed up reaction kinetics, leading to higher rates of hydrogen generation and reduced energy consumption.

3.6.7.5 *Learning and knowledge outcomes*

Researchers have developed components that enable efficient electrolysis of seawater, splitting it into hydrogen and oxygen. By leveraging seawater, the process becomes more sustainable and less resource-intensive.

3.6.7.6 *Future prospects*

While the prototype is still in its early stages, it holds promise for India's green hydrogen goals. Collaborative efforts between academia, industry, and policymakers are essential for further development and commercialization.

3.7 Conclusion

Water-splitting technologies play a highly promising role in the pursuit of sustainable energy solutions while acknowledging the need for continued innovation and collaboration across scientific, technological, and policy domains to fully harness its potential.

This chapter has covered a wide range of water-splitting technologies, each with its own benefits and drawbacks. Conventional techniques, like electrolysis, have been honed over many years, offering a fundamental comprehension of the process of splitting water into hydrogen and oxygen. More sophisticated methods of using solar and thermal energy for water splitting have surfaced, such as thermochemical reactions and photoelectrochemical cells.

Due to its well-established infrastructure and relative simplicity, electrolysis is still the most extensively used technology; nevertheless, its usage is restricted by its dependency on energy and the requirement for effective catalysts. However, there are also promising opportunities to directly incorporate renewable energy sources into the water-splitting process using photoelectrochemical and thermochemical processes, which might lower prices and increase efficiency.

To make water splitting a practical and sustainable way to produce hydrogen, research is still needed to improve catalyst performance, scale up production, and lower the total energy consumption of these technologies. Future advancements will probably concentrate on making these processes more efficient, integrating them with renewable energy sources, and taking economic and environmental factors into account.

In the end, there is a great deal of potential for the development and application of water-splitting technologies to contribute to a sustainable energy future. These technologies have the potential to significantly contribute to solving the world's energy problems and lowering dependency on fossil fuels by offering a clean and sustainable supply of hydrogen.

References

Agarkov, D., et al., 2023. Structure and physical properties of ceramic materials based on ZrO_2-Sc_2O_3 for SOFC electrolytic membranes obtained from powders of melted solid solutions with a similar composition. Membranes 13, 717.

Ben, H., Lehr, F., Rupprecht, J., Mussgnug, J.H., Posten, C., Kruse, O., 2007. Photosynthetic biomass and H_2 production by green algae: from bioengineering to bioreactor scale-up. Physiologia Plantarum 131 (1), 10–21.

Biswas, S., Kaur, G., Paul, G., Giddey, S., 2023. A critical review on cathode materials for steam electrolysis in solid oxide electrolysis. International Journal of Hydrogen Energy 48 (34), 12541–12570.

Böhmer, et al., 2017. Enzymatic oxyfunctionalization driven by photosynthetic water-splitting in the *Cyanobacterium synechocystis* sp. PCC 6803. Catalysts 7, 240.

Chavas, A., et al., 2020. Bandgap engineering of two dimensional semiconductor materials. Npj 2D Materials and Applications 4 (29).

Chhowalla, M., Shin, H.S., Eda, G., Li, L.J., Loh, K.P., Zhang, H., 2016. Photoanode current of large-area MoS_2 ultrathin nanosheets with vertically mesh-shaped structure on indium tin oxide. Nature Chemistry 5, 263–275.

Chisholm, G., Zhao, T., Cronin, L., 2022. Hydrogen from water electrolysis. In: third ed. In: Letcher, T.M., (Ed.), Storing Energy 24. Elsevier, pp. 559–591.

Cho, H.-S., Kodama, T., Gokon, N., Bellan, S., Kim, J.-K., 2021. Development of synthesis and fabrication process for Mn-CeO_2 foam via two-step water-splitting cycle hydrogen production. Energies 14, 6919.

Colette Maatouk, S., Brijou Mokrani, N., Tabarant, M., Fleche, J.-L., Carles, P., 2009. Study of the miscibility gap in $H_2SO_4/HI/I_2/H_2O$ mixtures produced by the Bunsen reaction – Part I: preliminary results at 308K. International Journal of Hydrogen Energy 34 (17), 7155–7161.

Dincer, I., Acar, C., 2015. A review on clean energy solutions for better sustainability. International Journal of Energy Research 39, 585–606.

El Ouardi, M., et al., 2024. Synthesis of ZnV_2O_6 nanosheet photocatalysts for efficient photodegradation of Rhodamine B: Experimental and RSM modelling. Journal of Environmental Chemical Engineering 12 (5), 113505.

Fang, G., Liu, Z., Han, C., 2020. Enhancing the PEC water splitting performance of $BiVO_4$ co-modifying with NiFeOOH and Co-Pi double layer cocatalysts. Applied Surface Science 515, 146095.

Galvão, R.A., Silva, G.M.M., Ferreira Coêlho, N.C., Santa-Cruz, L.A., Machado, G., 2022. Hydrogen production by the layer-by-layer assembled films of PAni-TiO_2-AuNPs. Materials Today Chemistry 26, 101072.

Ghosh, A., T K, S.F., Ganguly, D., Sundara, R., 2023. CoFePBA nanosheets on carbon nanotubes coupled with nickel-encapsulated carbon tubules for efficient and highly stable overall seawater electrolysis. ACS Applied Energy Materials 6 (11), 6080–6090.

Gong, L., Chu, Q., Liua, X., Tan, Y., 2022. Plasmonic platinum nanoparticles–tungsten oxide nanoarchitectures as visible light photocatalysts for highly efficient overall water splitting. Journal of Materials Chemistry A 10, 21161–21176.

Guo, M., Ru, X., Lin, Z., Xiao, G., Wang, J., 2020. Optimization design of rib width and performance analysis of solid oxide electrolysis cell. Energies 13, 5468.

Gupta, A., Likozar, B., Jana, R., Chanu, W.C., Singh, M.K., 2022. A review of hydrogen production processes by photocatalytic water splitting – from atomistic catalysis design to optimal reactor engineering. International Journal of Hydrogen Energy 47 (78), 33282–33307.

Hassan, I.U., Naikoo, G.A., Salim, H., Awan, T., Tabook, M.A., Pedram, M.Z., et al., 2023. Advances in photochemical splitting of seawater over semiconductor nano-catalysts for hydrogen production: a critical review. Journal of Industrial and Engineering Chemistry 121, 1–14.

Ibrahim, A., Hafeez, H.Y., Mohammed, J., Abdu, S., Suleiman, A.B., Ndikilar, C.E., 2024. A recent progress and advancement on MoS_2-based photocatalysts for efficient solar fuel (hydrogen) generation via photocatalytic water splitting. International Journal of Hydrogen Energy 71, 1006–1025.

Jiang, S., Liua, Y., Xu, J., 2021. Rare earth oxynitrides: promising visible-light-driven photocatalysts for water splitting. Materials Advances 2, 1190–1203.

Kibsgaard, J., Chen, Z.B., Reinecke, B.N., Jaramillo, T.F., 2012. Engineering the surface structure of MoS_2 to preferentially expose active edge sites for electrocatalysis. Nature Materials 11, 963–969.

Kong, D., Wang, H.T., Lu, Z.Y., Cui, Y., 2014. $CoSe_2$ nanoparticles grown on carbon fiber paper: an efficient and stable electrocatalyst for hydrogen evolution reaction. Journal of the American Chemical Society 136, 4897–4900.

Kumar, M., Ghosh, C.C., Meena, B., Ma, T., Subrahmanyam, C., 2022. Plasmonic Au nanoparticle sandwiched $CuBi_2O_4/Sb_2S_3$ photocathode with multi-mediated electron transfer for efficient solar water splitting. Sustainable Energy Fuels 6, 3961–3974.

Li, Y., Yao, R., Chen, Y., Xu, B., Chen, C., Zhang, C., 2020. Mimicking the catalytic center for the water-splitting reaction in photosystem II. Catalysts 10, 185.

Lukowski, M.A., Daniel, A.S., Meng, F., Forticaux, A., Li, L.S., Jin, S., 2013. Enhanced hydrogen evolution catalysis from chemically exfoliated metallic MoS_2 nanosheets. Journal of the American Chemical Society 135, 10274–10277.

Luo, D., Qian, J., Zhang, H., Lin, C., Xiao, Q., Li, X., 2023. Nanoarchitecturing hydrogen electrode with intimately coupled ion-conducting and catalytic phases for enhanced intermediate-temperature steam electrolysis. Journal of Alloys and Compounds 969, 172362.

McDaniel, A.H., Miller, E.C., Arifin, D., Ambrosini, A., Coker, E.N., O'Hayre, R., Chueh, W.C., Tong, J., 2013. Sr- and Mn-doped $LaAlO_3-\delta$ for solar thermochemical H_2 and CO production. Energy and Environmental Science 6, 2424–2428.

Meena, B.R., Kumar, M., Gupta, S., Sinha, L., Subramanyam, P., Subrahmanyam, C., 2022. Rational design of $TiO_2/BiSbS_3$ heterojunction for efficient solar water splitting. Sustainable Energy Technologies and Assessments 49, 101775.

Othmane, et al., 2024. Polyester fabric supported graphene oxide/Cu-Ag bimetallic nanoparticles as a "dip catalyst" for the reduction of p-nitrophenol and organic dyes. Colloids and Surfaces A 697, 134314.

Pandiyan, A., Uthayakumar, A., Subrayan, R., Cha, S.W., 2019. Review of solid oxide electrolysis cells: a clean energy strategy for hydrogen generation. Nanomaterials and Energy 8 (1).

Qian, X., et al., 2021. Outstanding properties and performance of $CaTi_{0.5}Mn_{0.5}O_3-\delta$ for solar-driven thermochemical hydrogen production. Matter 4, 688–708.

You, B., Sun, Y., 2018. Innovative strategies for electrocatalytic water splitting. Accounts of Chemical Research 51 (7), 1571–1580.

Xie, J.F., Zhang, H., Li, S., Wang, R.X., Sun, X., Zhou, M., et al., 2013. Defect-rich MoS_2 ultrathin nanosheets with additional active edge sites for enhanced electrocatalytic hydrogen evolution. Advanced Materials 25, 5807–5813.

Further reading

Li, W., Tian, H., Ma, L., Wang, Y., Liu, X., Gao, X., 2022b. Low-temperature water electrolysis: fundamentals, progress, and new strategies. Materials Advances 3, 5598–5644.

Li, X., Sun, X., Song, Q., Yang, Z., Wang, H., Duan, Y., 2022a. A critical review on integrated system design of solar thermochemical water-splitting cycle for hydrogen production. International Journal of Hydrogen Energy 47 (79), 33619–33642.

Srirangan, K., Pyne, M.E., Perry Chou, C., 2011. Biochemical and genetic engineering strategies to enhance hydrogen production in photosynthetic algae and cyanobacteria. Bioresource Technology 102 (18), 8589–8604.

Tay, Y.F., et al., 2018. Solution-processed Cd-substituted CZTS photocathode for efficient solar hydrogen evolution from neutral water. Joule 2 (3), 537–548.

Zhang, D., et al., 2023b. Compositionally complex perovskite oxides for solar thermochemical water splitting. Chemistry of Materials 35 (5), 1901–1915.

CHAPTER

4

Instrumentation and experimental setup for photochemical splitting of water

R. Anjana[1], Nithya S. George[2], Ayona K. Jose[2], Jibi Kunjumon[2], P.A. Aleena[2], and D. Sajan[2]

[1]Department of Physics, M.S.M College, Alappuzha, Kerala, India [2]Department of Physics, Centre for Sustainable Energy and Environment Technologies (CE[2] T), Bishop Moore College, Alappuzha, Kerala, India

4.1 Introduction

4.1.1 Renewable energy and hydrogen production

Establishing energy security and sustaining environmental safety standards are critical priorities and the major challenge in the 21st century for all countries. Power consumption is expected to hit 100% increase by 2050 globally hence, this abrupt exhaustion of fossil fuel reserves driven by the excessive use of coal and other petroleum products coupled with widespread environmental damages has emphasized the urgency of this issue and has propelled this issue to the forefront of academic and public discussions all over the world. Consequently, considerable amount of research endeavors are aimed on establishing technologies having environmentally friendly energy carrier for the postfossil fuel era which can efficiently and economically tackle these concerns. In this scenario, H_2, a carbon-free clean energy vector, produces an additional 2.5 times energy per unit mass and releases H_2O along with the large amount of energy ($\Delta H^0 = -284\,kJ/mol$) while burning has attracted big attention. Currently, the commercially available H_2 production primarily relies on conventional and nonrenewable fossil fuels, therefore, developing a clean and renewable solution for hydrogen production is a crucial objective (Chen et al., 2012). The advantages of hydrogen include (Ma et al., 2015):

1. Better resource utilization
2. Good energy conversion effectiveness

Photochemical Splitting of Water
DOI: https://doi.org/10.1016/B978-0-443-29064-0.00015-2

3. Pollution/carbon free while burning
4. Different storage options available (e.g., as liquid or gas)
5. Conversion into different varieties of fuels
6. Ease in transportation over long distances
7. Better energy security and environment.

A process called photocatalytic water splitting, that uses energy from the sun to split H_2 and O_2 from water, is considered as a highly promising method for the generation of green and renewable hydrogen energy, leveraging sunlight as the most abundant sustainable energy source (Ma et al., 2015). Hydrogen can be stored in a fuel tank and can be made available for fuel cell-powered cars in which only water is eliminated as the waste product (Chen et al., 2010; Xiang et al., 2016; Xiang et al., 2016; Fujishima and Honda, 1972). In this point of view, exploring diverse photocatalyst, which supports in water splitting, and comprehending the underlying mechanism behind water splitting requires detailed investigation. This technique was first used to evolve hydrogen via the splitting of water photoelectrochemically using n-type TiO_2 electrodes (Jiang et al., 2017). Thus much research has gone into creating the required semiconductor photocatalysts. In the ultraviolet (UV) light range, numerous photocatalysts have allegedly demonstrated high photocatalytic activity over the past 40 years for separating water into H_2 and O_2 (2:1 by molar ratio). At present, most popular water-splitting photocatalysts are metal oxides, sulfides, and nitrides.

This chapter intends to provide an understanding of the various tools and methodologies employed in the experiment and optimization of photochemical water-splitting processes. This section delves into the essential instruments required for efficient experimentation, including different types of cell configurations. It describes the setup configurations that maximize photon absorption, enhance reaction kinetics, and ensure accurate analysis of various parameters. Emphasis is placed on the design, as well as the integration of various components to create a coherent experimental framework. The chapter also addresses the importance of electrode considerations, which is crucial for replicating and analyzing results. By providing detailed guidance on cell setup, current photocurrent (j–V) analysis, and flat band potential analysis, this chapter aims to equip researchers with the knowledge needed to advance the field of photochemical water splitting.

4.1.2 Principle of photocatalytic water splitting

The different steps in photocatalytic water-splitting process consists of the harvesting of light, its separation and then the movement of photogenerated electrons and holes. Photocatalytic water-splitting reaction on the surface of a photocatalytically active material involves oxygen evolution and hydrogen evolution. Around 237 kJ/mol of Gibbs free energy is required for the splitting of water into H_2 and O_2.

$$H_2O \rightarrow H_2 + \frac{1}{2}O_2 \tag{4.1}$$

Half-cell reactions are as follows:

$$2H^+ + 2e^- \rightarrow H_2 \tag{4.2}$$

$$H_2O + 2h^+ \rightarrow \frac{1}{2}O_2 + 2H^+ \tag{4.3}$$

Bandgaps acts as the crucial factor for considering and assessing the activity of a material in the photocatalytic water splitting. In precise, to ensure the optimal performance and efficiency in a water-splitting phenomenon, the redox potentials of water must lie within the photocatalyst's bandgaps. In order for the desired redox reactions to happen in a water-splitting process, one crucial requirement is that the maximum of the valence band (VB) needs to be very much lower than the oxidation potential of H_2O/O_2 and the conduction band (CB) minimum must lie higher than the reduction potential of H^+/H_2 simultaneously. Since the least energy requirement to separate the water into hydrogen and oxygen is 1.23 eV, the bandgap of a photocatalytically active material should be larger than the same, which determines the effectiveness of the photocatalytic reactions. In the view of above-mentioned facts, the photocatalyst transferred the photogenerated carriers to the water molecules which have been absorbed on the photocatalyst surface to facilitate the process of hydrogen and oxygen evolution. In practical applications, by considering the facts like energy loss in electron transfer process and the need to sustain a kinetic overpotentials to overcome the challenges in hydergen evolution reaction (HER) and oxygen evolution reaction (OER), it is imperative to consider photocatalysts with bandgaps usually larger than 1.8 eV. Noting that the efficiency of a water-splitting photocatalyst is significantly influenced by several parameters including solar adsorption, capturing the radiations, electron–hole recombination stimulated by irradiation, and the surface activities of photocatalyst in relation to HER and OER (Wang et al., 2014). Fig. 4.1 illustrates the photocatalytic water-splitting mechanism using a semiconductor-based photocatalyst.

Keeping in mind that the solar spectrum encompasses a range of wavelengths, including UV light, which constitutes a small fraction, less than 400 nm, accounting for approximately 6.8% of the solar power. Meanwhile, visible light, spanning from 400 to 700 nm, contributes about 38.9% of the solar power. However, the majority of sunlight, approximately 54.3%, falls within the near-infrared (IR) range, ranging from 760 to 3000 nm. Hence, a wide bandgap photocatalyst cannot effectively harness the long range wavelength in the sunlight, consequently restricts the efficiency of solar-to-hydrogen conversion. Here comes the challenges in exploring the photocatalyst particularly in tailoring its narrow bandgaps with achieving appropriate band alignments involves minimizing the VB and CB, improving the separation between the photogenerated carriers and minimizing the electron–hole carriers recombination during migration. Different strategies are widely used for developing photocatalyst with superior properties including introduction of impurity states inside the wide bandgap, so that it could create intermediate energy levels and can utilize the long wavelengths without making a modification in its VB and CB. Tuning wider bandgaps, functionalizing the surface of photocatalyst by intrinsic electric fields or building heterojunctions and also reducing the dimension of the prepared

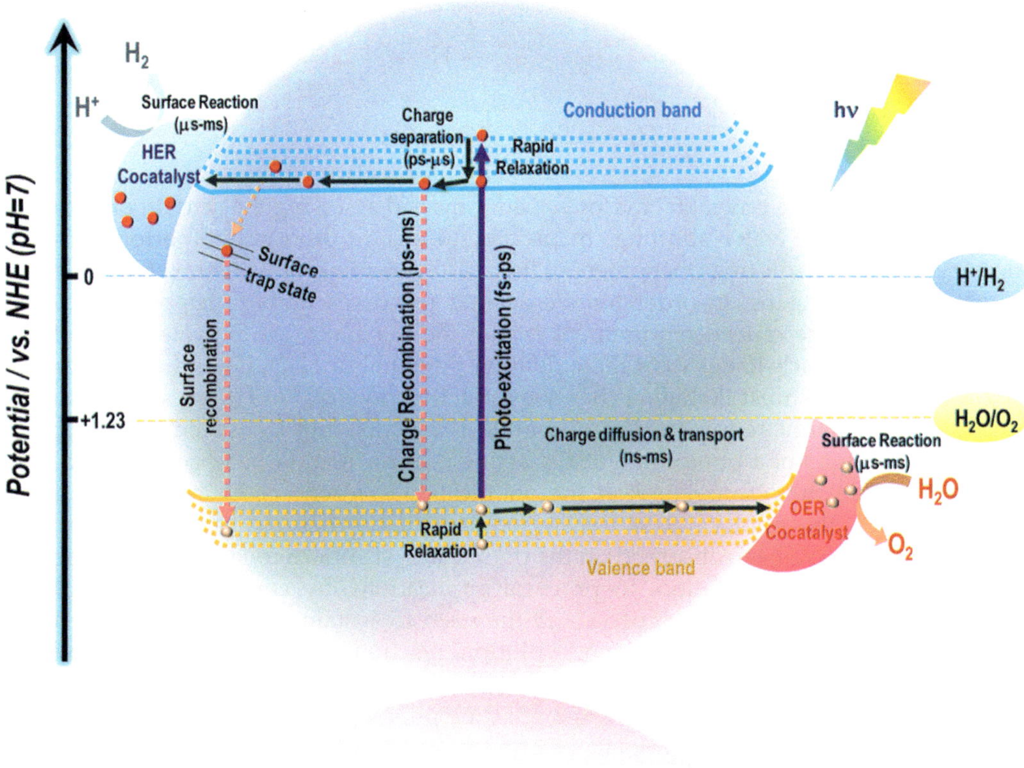

FIGURE 4.1 Illustration of photocatalytic water-splitting mechanism on a semiconductor-based photocatalyst (Tao et al., 2022).

photocatalyst which can shortenen migration pathways help in improving the performance of a photocatalytically active material (Wang et al., 2014).

4.2 Cell design

Photoelectrochemical (PEC) cell is required for a PEC reaction to happen. Minimum requirement of a PEC cell is a photo anode, a photo cathode, and electrolyte solution. The conceptual dissimilarity between a PEC cell and a normal electrochemical cell is the requirement of one or more photoactive electrodes, and absorption of light is required to initiate the chemical reactions associated with water splitting. Therefore in a PEC cell, there is the requirement of at least one window that is transparent so that light may penetrate within the electrochemical cell and one or both electrodes absorb the light. Contact between anode and cathode can be established in two methods. A conductive wire may be used as an electrical contact between anode and cathode or a monolithic cell with anode and cathode integrated on the same conductive substrate is also viable.

4.2.1 Monilithic photoelectrochemical cell

Fig. 4.2 represents a monolithic cell design. Electrode that absorbs light is known as the working electrode (either photocathode or photoanode) and the electrode which is not photoactivated and that initiates the other half reaction is referred as the counter electrode. However, if both the electrodes are photoactivated, they may be referred as (working) photoanode and (working) photocathode. In addition to the aforementioned components, a reference electrode is required for the accurate determination of working electrode potential. In order to ensure reactant/product mobility and to maintain uniform electrolyte concentration, gas circulation as well as magnetic stirrer are recommended. Generation of hydrogen and oxygen can be carried out in separate compartments if a membrane is used in between. PEC cell with membrane and without membrane is demonstrated in Figs. 4.3 and 4.4. Generally monolithic cell with a single substrate is widely preferred.

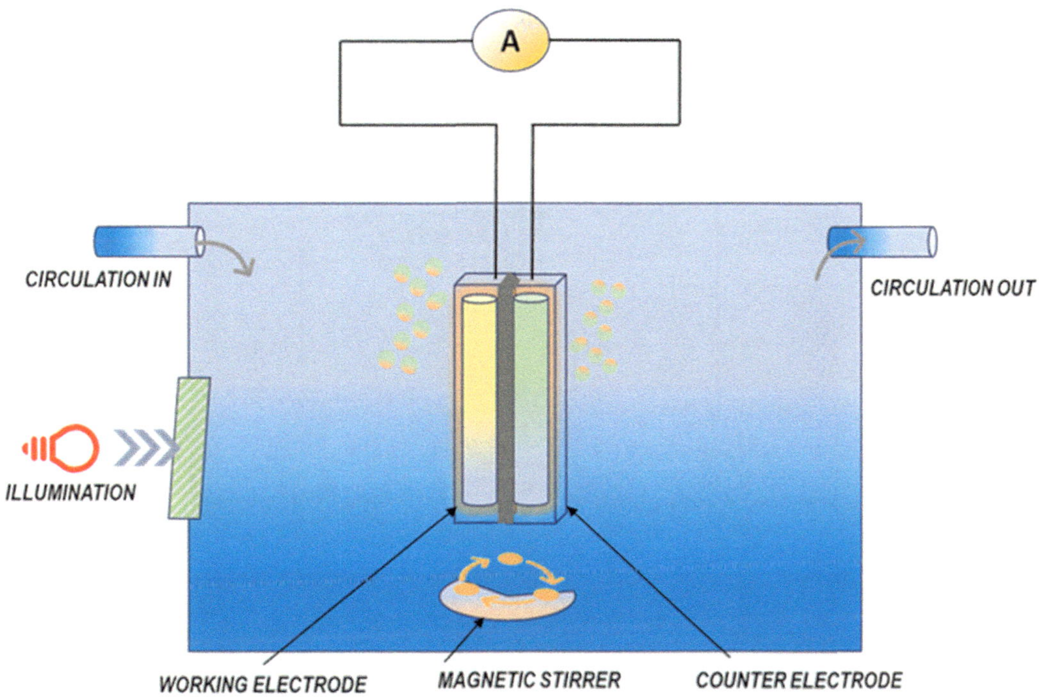

FIGURE 4.2 Monilithic photoelectrochemical cell.

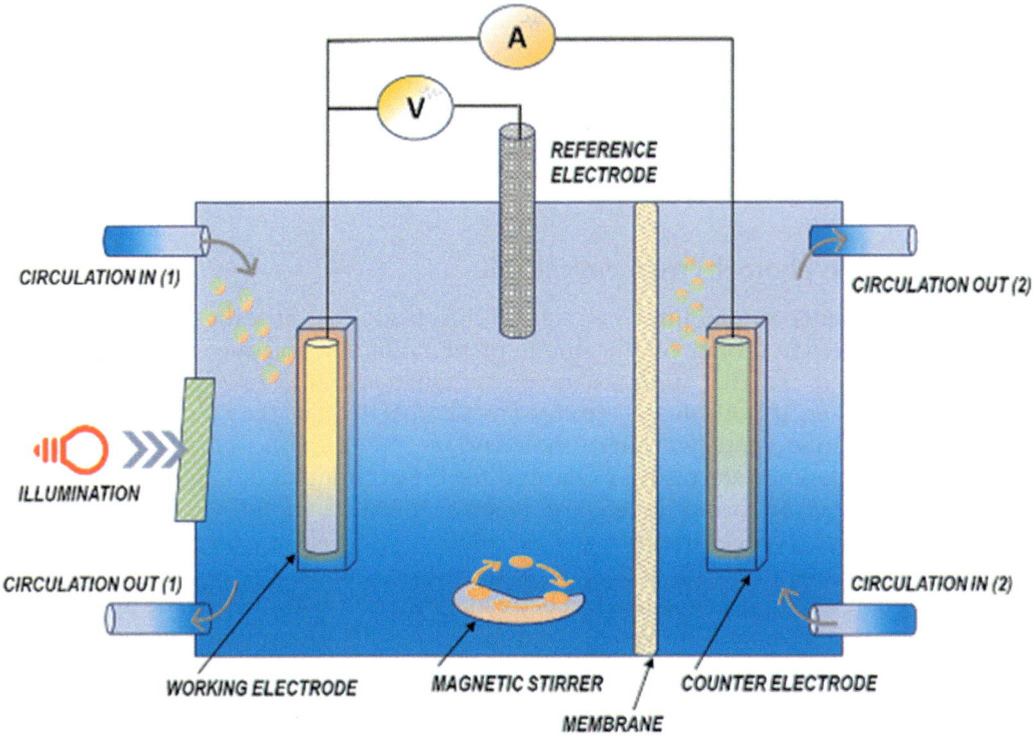

FIGURE 4.3 Photoelectrochemical (PEC) with separator membrane.

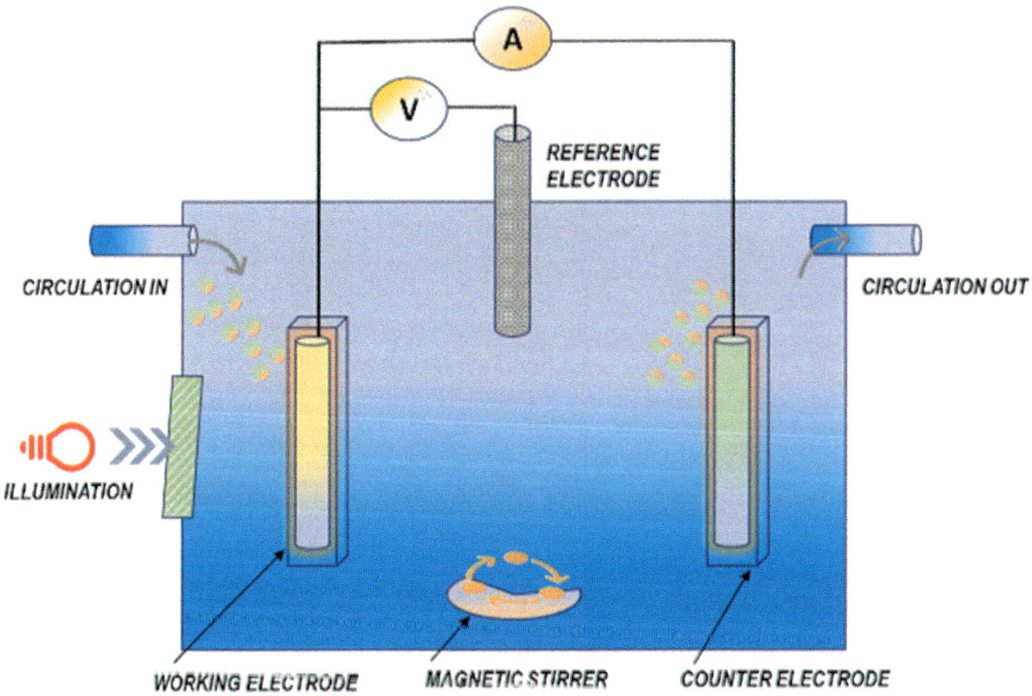

FIGURE 4.4 Photoelectrochemical (PEC) without separator membrane.

4.2.2 PEC/PEC tandem cells

For further enhancement in the overall efficiency of water splitting, a dual absorber, that is, four photon (D4) approach may be utilized. In this concern, a PEC tandem cell may be designed. Here, both anode and cathode can be photoactivated. Availability of two bandgaps make greater spectrum utilization of incident light which can increase the efficiency. A PEC/PEC tandem cell is demonstrated in Fig. 4.4. In this configuration, the top electrode (photoanode) absorbs photons of shorter wavelength while longer wavelength photons are absorbed by the bottom electrode (photocathode). That ensures the maximum utilization of energy in the incident solar radiation. Thermodynamically semiconductors with smaller bandgap are required when compared with cell with single photoelectrode (PE) since one electrode is required to supply only part of the potential requirement to drive the overall process. Smaller bandgap material can absorb more photons from the incident radiation. As shown in Fig. 4.5, photoanode conduction band is more negative than photocathode valence band. However, since the water-splitting potential is entirely generated at the interface of semiconductor–electrode, it should be carefully selected.

4.2.3 PEC/PV tandem cells

PEC/PV tandem cell is another setup for unassisted water splitting. Here, PV device is employed to voltage bias PEC. Fig. 4.6 shows a PEC/PV-integrated cell. If an n-type semiconductor is used as the photoanode, majority carriers, that is, the electrons travel through the external circuit toward the PV cell where recombination takes place. Hydrogen is generated

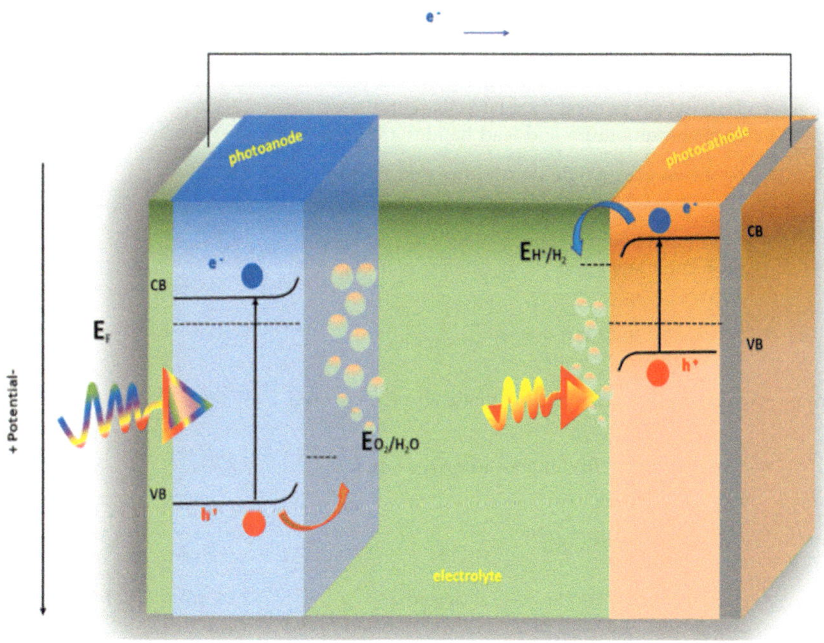

FIGURE 4.5 PEC/PEC tandem cell.

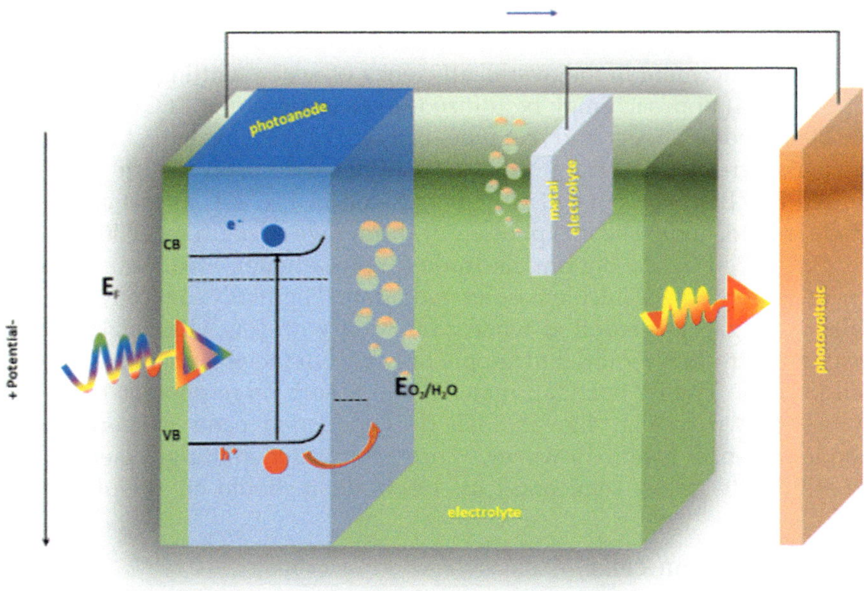

FIGURE 4.6 PEC/PV tandem cell.

TABLE 4.1 Efficiency comparison of various cell configurations.

Configuration	Peculiarities	STH efficiency/ current density	References
Tandem PV cell	• Two hybrid perovskite solar cells with the cation methyl ammonium and lead halides are used.	12.3%	Luo et al. (2014)
Monolithic tandem PV cell	• Based on lead halide perovskites and $BiVO_4$	2.5%	Chen et al. (2015)
Multijunction monolithic PEC cells	• Multifunctional crystalline titania as corrosion protection layer and Rh as catalyst is used	19.3%	Cheng et al. (2018)
PV cell	• Perovskite solar cell is used	14.4%	Turan et al. (2016)
PEC–PV tandem cell	• Perovskite–silicon photocathode absorber is used	17.6%	Karuturi et al. (2020)
Tandem cell	• Used modified $BiVO_4$ and α-Fe_2O_3 as dual photoanodes	7.7%	Kim et al. (2016)
PEC/solar tandem device	• Photon recycling is done	7%	Shi et al. (2016)
Dual photoelectrode tandem cell	• TiO_2/PANi electrode was fabricated and assembled in a dual photoelectrochemical system combined with $WO_3/BiVO_4$ in tandem and parallel illumination modes	2.29 mA/cm^2 at 1.23 V_{RHE}	Keshavarzi et al. (2023)

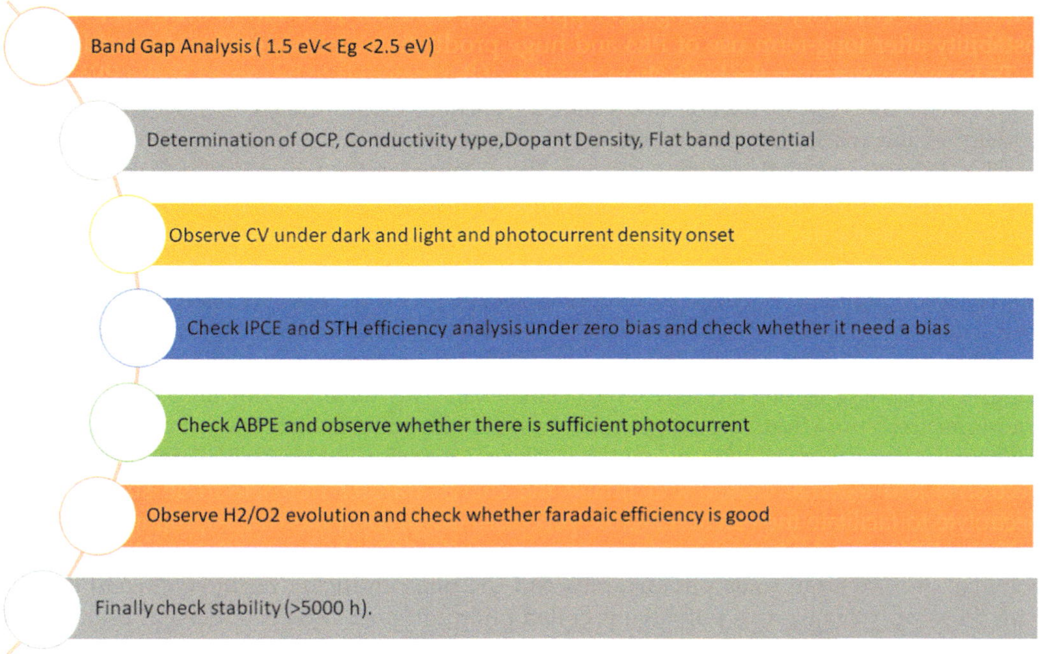

FIGURE 4.7 Analysis parameters and check list for a good photoelectrochemical (PEC) material.

at the counter electrode (metal electrode) as shown in Fig. 4.6. Here, the additional power excluding the power produced by minority carriers are supplied by the PV cell. This relaxes the requirement of bandgap matching which thereby ensures flexibility in choosing semiconductor materials. Table 4.1 summarizes the efficiency of various popular cell configurations.

4.2.4 Working principle for photoelectrochemical water splitting

In PEC water splitting, the electrochemical energy requirement for water splitting is satisfied by the semiconductor that absorbs photons having energy higher than its bandgap (Hamdani and Bhaskarwar, 2021). Minority charge carriers, that is, holes for n-type semiconductor and electrons for p-type semiconductor have a crucial role in the process. Minority charge carriers are generated upon illumination and they initiate both the oxidation reaction and reduction reaction at the electrode surface. Fig. 4.7 represents a quick reference for various analysis parameters for a PEC material.

4.3 Electrode considerations

In PEC devices, absorption of light and the electrolysis of water take place at the electrodes. Design of low-cost electrodes without compromising stability in aqueous solution and efficiency is challenging. Appropriate material selection in order to overcome instability after long-term use of PEs and huge production cost is significant consideration.

To be more specific on the basic characteristics of the materials to be served as the PEs, urgent requirements are optical function and catalytic function. Optical function is a measure of the capacity of the material to absorb the visible region of the solar energy spectrum as far as possible. Whereas, catalytic function refers to the potential of the material to perform HER and OER. Appreciable bandgap for these semiconductor materials should be 1.6–2.5 eV which has to be obviously above the thermodynamic potential requirement of water-splitting reaction which is 1.23 V versus reversible hydrogen electrode (RHE) (Hamdani and Bhaskarwar, 2021). Appreciable long-term photostability, abundance, nontoxicity, and selectivity for HER and OER has to be taken into considerations. Apart from these, high charge carrier mobility, lower recombination rate, and charge transfer between semiconductor–electrolyte interface play an important role in designing efficient electrode materials. Any material with all these properties is not yet realized but efforts are being made to make the suitable materials with some of these qualities. In general, absorption of photons, charge separation, transport toward electrodes via electrolyte and external circuit, and finally the charge transfer from electrode surface to the electrolyte to facilitate the reaction are the processes involved in the water-splitting reaction.

If the PE is producing a positive current density in response to the applied positive potential, it is referred to as photoanodes and the one producing negative current density with the supply of negative potential is called photocathode. O_2 is produced at photoanode and H_2 at photocathode. Band bending at the electrode–electrolyte interface is an indication of whether it is p-type or n-type. n-Type semiconductors, which is having a band bending in upward direction, act as the photoanode and the p-type materials having a band bending in the downward direction serves as the photocathode as illustrated in Fig. 4.8. Electrons are transferred from conduction band to the electrolyte in p-type materials.

Metal oxides have been widely explored as the PE materials owing to their abundance, sustainability, and ease of fabrication. Metal oxides are best known for their photostability in the aqueous solutions but its wide application is hindered by less light

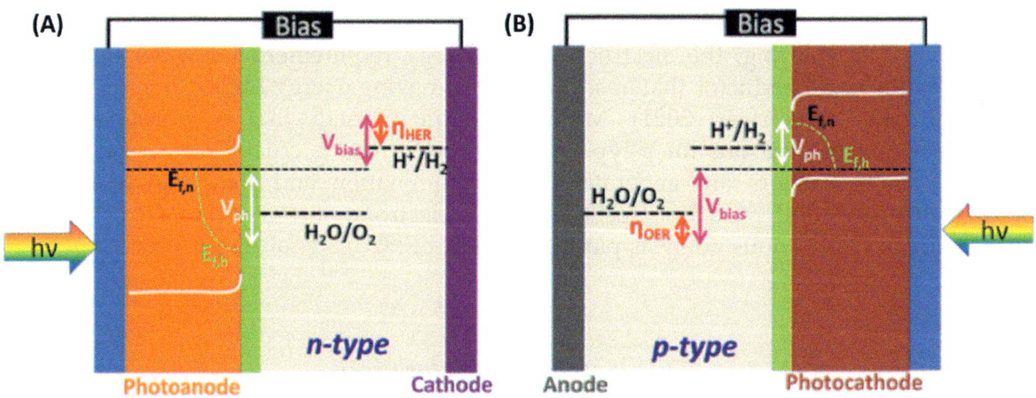

FIGURE 4.8 Electrode/electrolyte interface during band bending: (A) Upward band bending at the n-type semiconductor interface, (B) downward band bending in p-type semiconductor interface (Thakur et al., 2020).

absorption ability and poor electrical conductivity whereas nonoxide semiconductors is famous for higher efficiency but are expensive. Transition metal oxides, metal chalcogenides, nitrides, and perovskite oxide materials havealso been investigated as the electrode materials. For the fabrication of efficient electrode materials, synthesis methods, tuning the morphology, plasmonic enhancement, the creation of p–n junctions, etc. are the popular strategies to be adapted. Electrode surface area has a crucial role since it has direct relationship with the electrochemical reaction rates. This fact confirms the advantages of polycrystallinity of the electrode materials and also single crystals with higher surface area. Doping of the materials for bandgap tuning to ensure maximum visible light absorption, formation of hetrojunctions to limit the charge carrier recombination, modification of reaction kinetics by using surface cocatalysts, and defining the conductivity of the material by precisely controlling the growth conditions are all techniques to be adopted for the successful designing of PEs (Moss et al., 2021). Charge separation efficiency of a material depends on the charge generation and transfer. Strategies to prevent recombination are of importance. Separated charges can be collected by a substrate before recombination (Fig. 4.9). Slow charge transfer is another issue to be addressed and it can be arising from the lower concentrations of redox species. Hence higher concentrations of ion to be oxidized and reduced is preferable and the reaction product of the charge transfer has to be removed from the electrode as soon as possible.

STH (Solar to hydrogen conversion) efficiency specifies the ability of the material for hydrogen production. This is a standard for comparing the efficiency of PEs. Photocurrent density, applied bias, efficiency of photon-to-current conversion and efficiency of incident photon-to-current conversion are the other parameters for analyzing the effectiveness of PEs.

PEs can be broadly classified into four according to its STH efficiency. They are poor PE having STH < 1%, marginal PE with STH > 1%, acceptable PE having STH 4%–10%, and highly efficient PE possessing STH > 10%. Low efficiency can be omitted by proper control

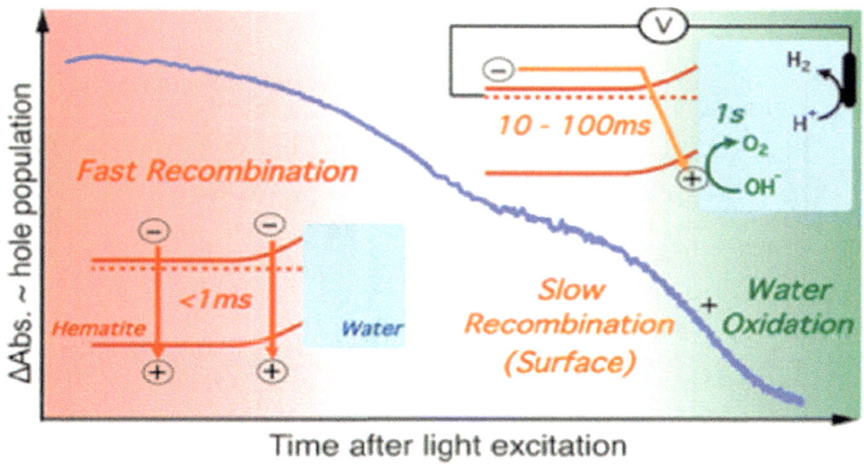

FIGURE 4.9 Various processes for charge recombination, like (direct charge) fast recombination and (back) surface recombination process on hematite photoanode (Wang and Wang, 2018).

over the synthesis conditions. Characteristics of the material viz. size, shape, crystallinity, and also the optical and electrical properties affect the efficiency and STH value. Highly efficient PEs are also reported but not many. Their significant performance arises from the narrow bandgaps and ideal charge carrier mobility. Certain materials like GaAs, InGaP, and their alloys have the potential as highly efficient PEs.

4.4 Cell setup in a photochemical water-splitting reaction

The efficiency of the photochemical water splitting depends mainly on the configuration of cell setup. The PEC splitting of water can be carried out on two different ways. One is particulate photocatalytic system (PPS) and the other is PEC cell. PPS is a system where the photocatalyst is directly dispersed in the electrolytic solution. PEC cell contains a n-type semiconductor-based photoanode, p-type semiconductor-based cathode, a counter electrode, and an electrolyte. The PEC cell can be configured in various ways other than the PEs. The PEs (photoanode and photocathode) are combined with counter electrodes which together operate the PEC cell using the external bias (Chen et al., 2018). The tandem PEC cell consists of cathode and anode semiconductor-based electrodes.

The basic PEC characterization principle is to evaluate the photocurrent generated between the working and counter electrode described as a function of the potential at the working electrode (e.g., TiO_2, Fe_2O_3, WO_3, Si, $BiVO_4$) with respect to the reference electrode (Rajaitha et al., 2022). The PEC cell is filled with electrolyte, an aqueous electrolytic solution where the electrolysis will occur (Pinhassi et al., 2016). Some examples include the NaOH, H_2SO_4, KOH, Na_2SO_4. Mainly the cell setup in a PEC water-splitting reaction includes two electrodes and three electrodes assembly that is presented in Fig. 4.10.

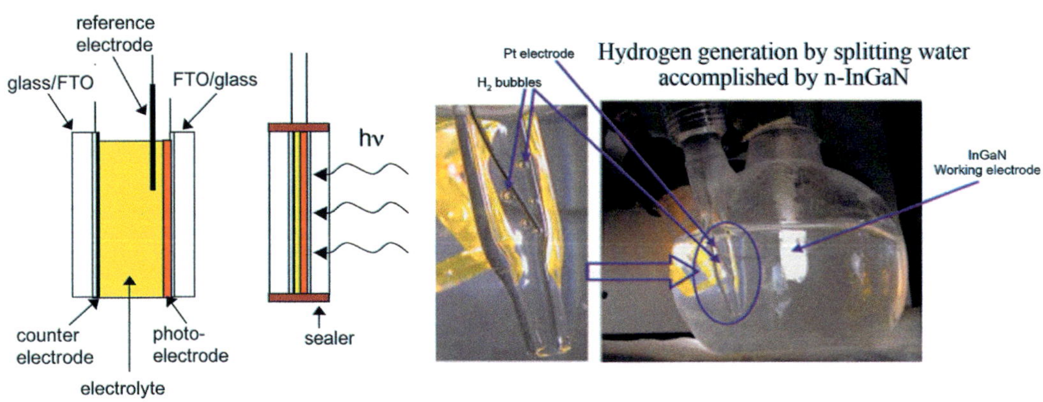

FIGURE 4.10 Scheme for a three- and two-electrode photoelectrochemical cell (left) and photograph of generation of hydrogen gas at the platinum-based counter electrode during water-splitting process with n-type InGaN as the working electrode under illumination (Li et al., 2008).

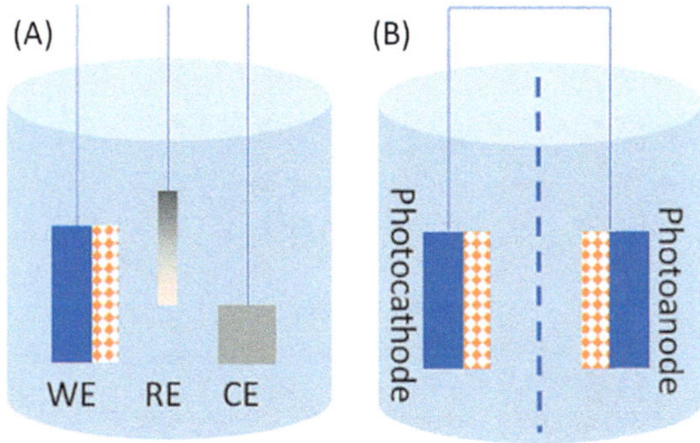

FIGURE 4.11 Typical PEC cell setup: (A) Three-electrode assembly with photoelectrode as working electrode, counter electrode, and reference electrode; (B) two-electrode unbiased system with photoanode and photocathode (Liu and Kuang, 2024).

4.4.1 Three-electrode configuration

Single semiconductor PE setup determines the charge transfer kinetics, stability, and surface reactions without any interference from the counter electrode. Other than cell efficiency, the three-electrode setup (Fig. 4.11) will provide the PE efficiencies. The three-electrode setup uses a reference electrode (Ag/AgCl) for setting a reference potential ($V_{Ag/AgCl}$=+0.197 vs NHE). The working electrode (anode) is set with a fixed potential using a potentiostat. Normally, the PE (cathode) is also used as a counter electrode (Pt wire, graphite, nickel, glassy carbon) for getting the counter electrode polarization (Oka et al., 2018). The PE potential is maintained by an external potentiostat and gets the current at that particular voltage. The counter electrode will be able to conduct the current as long as the potentiostat is able to provide the required voltage. Thus the three-electrode assembly can only measure the PE and will not consider the counter electrode polarization (Hodes, 2012).

4.4.2 Two-electrode configuration

Two-electrode is arranged using a photoelectrode and a counter electrode which evaluated the efficiency of the whole PEC system. A typical PEC consists of a photoactive semiconductor-based working electrode and a counter electrode with the supporting electrolyte. A schematic illustration showing PEC water-splitting cells along with its energy diagram is presented in Fig. 4.12. Normally platinum (Pt) is used as the counter electrode in these cells (Zhang et al., 2019). A n-type semiconductor—photoanode (VB positive than H_2O/O_2 potential) and p-type semiconductor function—photocathode (CB negative than H^+/H_2 potential). Also, in a two-electrode setup, the photoanode and cathode can be connected without a counter electrode. Here in the water-splitting reaction, oxygen and hydrogen will respectively be evolved at photoanode and photocathode, along the light irradiation without any external voltage. For instance, a standard PEC cell includes a working electrode (e.g., InGaN), platinum-based counter as the cathode, and NaCl-saturated calomel silver chloride as the other electrode. These electrodes will be immersed in an electrolytic solution. The potential difference between the working electrode and the counter measured as V_{CE} and recorded and the applied voltage and photocurrent I_p are measured as a function of V_{CE} (Li et al., 2008).

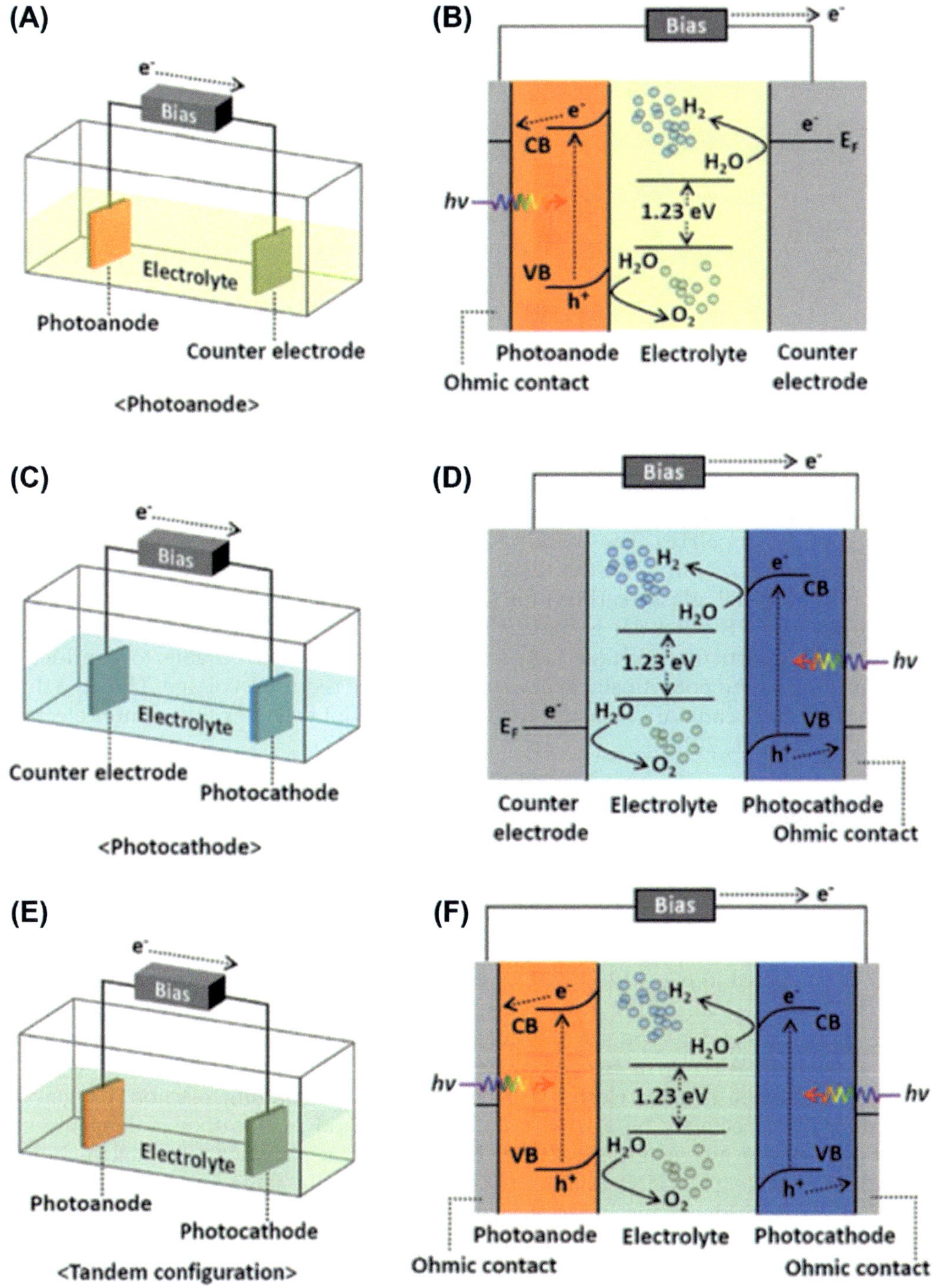

FIGURE 4.12 A schematic representation of PEC water splitting cells and related energy diagram. Schematic representation of (A) photoanode, (C) photocathode, and (E) Z-scheme (self-biased) PEC cell configurations with detailed description of their respective mechanisms (B), (D), and (F) in energy band (Ishaq et al., 2021).

In contrast with the typical three-electrode PEC setup with single PE, the PEC tandem cell will evaluate the material property of both the electrodes, provides a synergetic effect for getting enhanced performance. This PEC can also be coupled with a photovoltaic cell which allows the solar water-splitting reaction which is unbiased. Here the photovoltaic cell must have enough supply for photovoltage for balancing the PEC cell voltage deficiency (Chiu et al., 2019).

4.4.3 Role of counter electrode

The electrons accumulated in the counter electrode results in reduction of water (Yang et al., 2019; Zhang et al., 2021). Normally the electrode is made of a metal-based material (Xing et al., 2013). In a typical PEC system, the electrolyte acts as the ionic media for the charge transport while these counter electrodes complete the external circuit of a three-electrode cell which strongly affects the efficiency of the cell (Liu and Kuang, 2024). The counter has an important role in determining the stability of the PEC system. Sometimes, the counter electrodes will be dissolving in the electrolyte during the electrochemical reaction and will deposit in the PE making the PEC measurement more complicated (Fig. 4.13). Therefore choosing stable counter electrode is important for preventing these problems (Vanka et al., 2022). During HER, the current density of the platinum electrode may get influenced by the type of electrocatalyst, the area of working electrode (catalytic area) with respect to counter, range of potential, and the scan rate of working electrode. During the electrochemical reaction, the counter electrode balances the charge by accepting or donating the photogenerated electrons. Also, one chooses the counter with high surface area for efficient participation in the reaction and thereby reduces the overpotential and improving the efficiency. It is used to keep half cell reaction rate on the electrode surface faster than the other half cell reaction rate at the working electrode. So, the counter electrode must be having larger electrical conductivity and also larger surface area. The

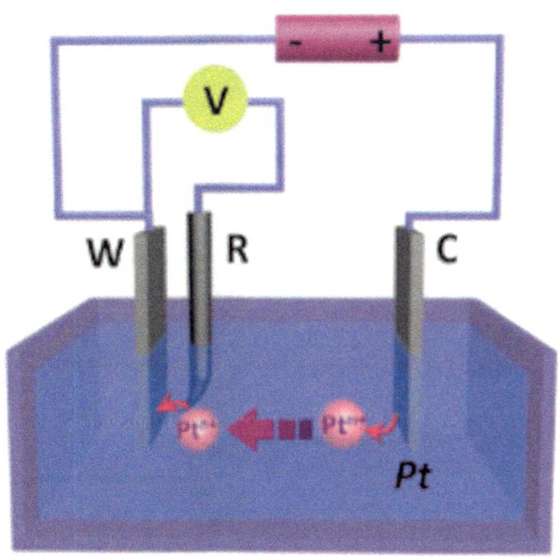

FIGURE 4.13 Schematic diagram showing Pt dissolution and redeposition during the hydrogen evolution reaction with Pt-based counter electrode.

surface area must be ideally 10 times larger than the working electrode surface area for achieving fast reaction kinetics on the counter electrode. Thus for this purpose, the platinum electrode area is kept as the same by the working electrode surface area. If the area of working electrode increases, the overpotential will be decreasing and it will cause Pt dissolution and then it will be deposited at the working electrode. Therefore platinum-based counter electrode is coupled with ion exchange membrane for preventing the Pt deposition during the reaction (Chen et al., 2017). Ti mesh is also having excellent stability as a counter electrode as per literature (Hasan and Allam, 2022).

4.4.4 Role of reference electrode

The significance of using reference electrode in a PEC system is to control the applied voltage. For the proper design and evaluation of electrochemical device, it is important to get correct understanding of the electrode potential. During the current flow in an electrochemical system, the chemical reaction can vary the composition and electronic states of the electrode and will result in variation of electrode potential. Thus a standard and constant reference is essential for evaluating the potential difference between the PEs because the absolute single electrode potential cannot be experimentally determined. So, for determining the electrode potential, the potential between the working and reference electrodes are evaluated. The reference electrode is placed near the photoanode to make sure

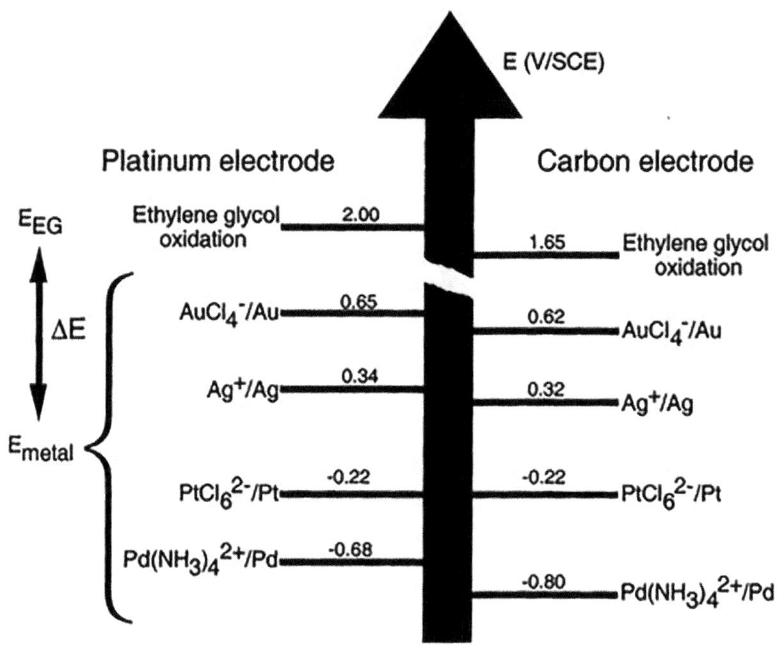

FIGURE 4.14 Schematic representation of an electrochemical series of the reduction potential of metal species in ethylene glycol at both a glassy carbon electrode and a platinum electrode at room temperature.

the potential measurements are accurate. In order to evaluate the solar to photocurrent efficiency, reference electrode is required. Normally, the electrochemical potential is measured with respect to standard hydrogen electrode (SHE). A diagram representing various redox systems on solid state scale versus the vacuum (eV) and on electrochemical scale versus SHE (V) is represented in Fig. 4.14.

Ideally, the reference electrodes are made of nonpolarizable type materials, which stay unchanged during the current flow. The electrode is selected with respect to the experimental conditions, that is, according to the current flow and type of electrolyte. Reference electrodes can be saturated calomel electrode (SCE), silver/silver chloride (Ag/AgCl), or mercury/mercury oxide (Hg/HgO). Among this, usually Ag/AgCl reference electrode is used in aqueous-based electrolytes and is also the standard RHE.

Generally, for any reference electrode used, the conversion with respect to RHE potential expressed is as follows (Harris-Lee et al., 2023):

$$E_{RHE} = E_{Ref} + (pH \times 0.059\ V) + E_{Ref\ vs\ SHE} \tag{4.4}$$

The relative potential of Ag/AgCl is converted to the RHE using the formula:

$$E_{RHE} = E_{Ag/AgCl} + (pH \times 0.059\ V) + 0.21\ V \tag{4.5}$$

where 0.21 V is the Ag/AgCl reference electrode potential with respect to SHE in 3 M KCl. The RHE potential depends on the electrolytic pH.

4.5 j–V (photocurrent) analysis

The j–V analysis is used to determine the fundamental characteristics of a PE like its photocurrent onset potential, and the flat band potential (E_{fb}) of the material under consideration. Before performing the j–V analysis, the semiconductor's conductivity type (n-type or p-type), which dictates the potential scan direction which is to be swept, must be investigated. This should be done in a potential between the E_{fb} and open circuit potential. Also, precautions should be taken while forward biasing the corresponding n-type or p-type electrodes to the potentials more cathodic or anodic than their E_{fb} to avoid cathodic corrosion reaction and degradation from anodic stripping, respectively.

The suitability of the PE materials for the water-redox reactions was determined from the magnitude of the photocurrent generated.

At the interface between electrode and electrolyte, the minority hole carriers in the n-type electrode facilitates the OER, while HER is initiated by the minority electron carriers in the p-type electrodes, under light irradiation (forward biasing) with an energy which is equal to or greater than semiconductor bandgap (region III). Photocurrent onset potential (E_{onset}) is the potential at which this begins to occur. The difference between photocurrent onset potential and reversible redox potential of interest ($E°$) gives the onset voltage (V_{onset}). Under higher reverse biasing condition, a dark current is generated and the semiconductor surface will mimic a metal. The potential at which this process begins is the $E_{dark\text{-}onset}$. This is evaluated by sweeping the potential from E_{onset} to the reverse bias conditions. Reverse bias breakdown of semiconductor or the reduction of Schottky barrier height (region IV) causes the dark current to

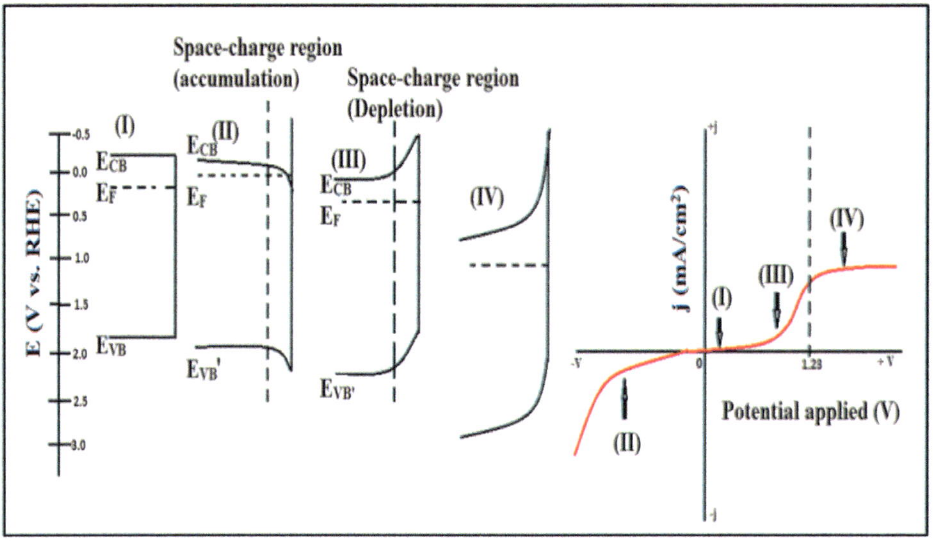

FIGURE 4.15 Effect of the band-edge biasing for photoanode. (I) Electrode potential is kept at the flat band potential, (II) potential of the electrode < flat band potential, (III) potential of electrode > flat band potential, (IV) high reverse-bias potential is given at the electrode, resulting in the saturation of photocurrent. In regions (III) and (IV), band bending in downward direction can be observed. Plot to the right side of the figure (Hamdani and Bhaskarwar, 2021) is the j–V curve.

appear. Under this condition, the bending of band is observed and the charge carrier transport occurs in the space charge region. The forward biasing causes the formation of accumulation layer within the space charge region (region II) while higher reverse bias results in the depletion layer. The j–V measurements are done between the $E_{\text{dark-onset}}$ and E_{onset} (Fig. 4.15).

4.6 Determination of flat band potential (E_{fb})

Photocurrent density (j) which is estimated as a function of potential (V) versus reference electrode is employed to determine the flat band potential (E_{fb}). Also, the dark and photocurrent scans are measured. Total current excluding the dark current gives the measure of photocurrent from the j–V curve. The flat band potential (E_{fb}) is the potential at which the semiconductor alters from accumulation to depletion which coincides with the onset of photocurrent at ideal conditions.

4.7 Two-electrode j–V measurement

One of the major drawback of the three-electrode j–V analysis is that, it is not able to check the efficiency power conversion. This happens because the three electrode configuration evaluates only the half-cell voltage in between the working and reference electrodes. For a two-electrode j–V measurement, an external bias is applied connecting the working electrode and counter electrode. In this situation, the conductivity relies on the direction of voltage scan as well as its endpoints, where its scan rate varies from 10 to 100 mV/s (Hamdani and Bhaskarwar, 2021; Chen et al., 2013).

4.8 Conclusion

In conclusion, this chapter has thoroughly examined the instrumentation and experimental setup required for the efficient photochemical splitting of water, offering invaluable insights into the tools and methodologies crucial for advancing research in this field. By detailing the various cell configurations, the chapter highlights how careful design and integration of components can significantly enhance photon absorption, optimize reaction kinetics, and ensure precise measurement and analysis. Considering literature on different cell configurations, the PEC–PV device, which coins the tandem cell with a PV cell, is the most viable tandem configuration. Emphasis on electrode considerations underscores its pivotal role in achieving efficiency. The STH conversion efficiency is a crucial parameter to index the competitiveness of PEC systems. An efficiency bench mark of 25% to reach market competitiveness of PEC water-splitting systems is set as the minimum requirement as per the United States Department of Energy. In short cell configuration, electrodes and device design should be well optimized for analyzing its suitability in practical water-splitting devices.

References

Chen, Q., Fan, G., Fu, H., Li, Z., Zou, Z., 2018. Tandem photoelectrochemical cells for solar water splitting. Advances in Physics: X 3, 1487267. Available at https://doi.org/10.1080/23746149.2018.1487267.

Chen, R., Yang, C., Cai, W., Wang, H.Y., Miao, J., Zhang, L., et al., 2017. Use of platinum as the counter electrode to study the activity of nonprecious metal catalysts for the hydrogen evolution reaction. ACS Energy Letters 2, 1070–1075. Available at https://doi.org/10.1021/acsenergylett.7b00219.

Chen, X., Li, C., Grätzel, M., Kostecki, R., Mao, S.S., 2012. Nanomaterials for renewable energy production and storage. Chemical Society Reviews 41, 7909. Available at https://doi.org/10.1039/c2cs35230c.

Chen, X., Shen, S., Guo, L., Mao, S.S., 2010. Semiconductor-based photocatalytic hydrogen generation. Chemical Reviews 110, 6503–6570. Available at https://doi.org/10.1021/cr1001645.

Chen, Y.-S., Manser, J.S., Kamat, P.V., 2015. All solution-processed lead halide perovskite-$BiVO_4$ tandem assembly for photolytic solar fuels production. Journal of the American Chemical Society 137, 974–981. Available at https://doi.org/10.1021/ja511739y.

Chen, Z., Dinh, H.N., Miller, E., 2013. Photoelectrochemical water splitting: standards, experimental methods, and protocols. SpringerBriefs in Energy. Springer New York, New York, NY. Available at https://doi.org/10.1007/978-1-4614-8298-7.

Cheng, W-.H., Richter, M.H., May, M.M., Ohlmann, J., Lackner, D., Dimroth, F., Hannappel, T., Atwater, H.A., 2018. Monolithic photoelectrochemical device for direct water splitting with 19% efficiency. ACS Energy Letters 3 (8), 1795–1800. Available at https://doi.org/10.1021/acsenergylett.8b00920.

Chiu, Y.-H., Lai, T.-H., Kuo, M.-Y., Hsieh, P.-Y., Hsu, Y.-J., 2019. Photoelectrochemical cells for solar hydrogen production: challenges and opportunities. APL Materials 7. Available at https://doi.org/10.1063/1.5109785.

Fujishima, A., Honda, K., 1972. Electrochemical photolysis of water at a semiconductor electrode. Nature 238, 37–38. Available at https://doi.org/10.1038/238037a0.

Hamdani, I.R., Bhaskarwar, A.N., 2021. Recent progress in material selection and device designs for photoelectrochemical water-splitting. Renewable and Sustainable Energy Reviews 138, 110503. Available at https://doi.org/10.1016/j.rser.2020.110503.

Harris-Lee, T.R., Marken, F., Bentley, C.L., Zhang, J., Johnson, A.L., 2023. A chemist's guide to photoelectrode development for water splitting – the importance of molecular precursor design. EES Catalysis 1, 832–873. Available at https://doi.org/10.1039/d3ey00176h.

Hasan, M.M., Allam, N.K., 2022. An alternative, low-dissolution counter electrode to prevent deceptive enhancement of HER overpotential. Scientific Reports 12, 1–6. Available at https://doi.org/10.1038/s41598-022-13385-w.

Hodes, G., 2012. Photoelectrochemical cell measurements: getting the basics right. The Journal of Physical Chemistry Letters 3, 1208–1213. Available at https://doi.org/10.1021/jz300220b.

Ishaq, T., Yousaf, M., Bhatti, I.A., Batool, A., Asghar, M.A., Mohsin, M., et al., 2021. A perspective on possible amendments in semiconductors for enhanced photocatalytic hydrogen generation by water splitting. International Journal of Hydrogen Energy 46, 39036–39057. Available at https://doi.org/10.1016/j.ijhydene.2021.09.165.

Jiang, C., Moniz, S.J.A., Wang, A., Zhang, T., Tang, J., 2017. Photoelectrochemical devices for solar water splitting – materials and challenges. Chemical Society Reviews 46, 4645–4660. Available at https://doi.org/10.1039/C6CS00306K.

Karuturi, S.K., Shen, H., Sharma, A., Beck, F.J., Varadhan, P., Duong, T., et al., 2020. Over 17% efficiency stand-alone solar water splitting enabled by perovskite-silicon tandem absorbers. Advanced Energy Materials 10, 2000772. Available at https://doi.org/10.1002/aenm.202000772.

Keshavarzi, R., Mousavian, M., Omrani, M., Mirkhani, V., Afzali, N., Mesa, C.A., et al., 2023. Photoelectrochemical water splitting with dual-photoelectrode tandem and parallel configurations: enhancing light harvesting and carrier collection efficiencies. Surfaces and Interfaces 38, 102813. Available at https://doi.org/10.1016/j.surfin.2023.102813.

Kim, J.H., Jang, J.-W., Jo, Y.H., Abdi, F.F., Lee, Y.H., Van De Krol, R., et al., 2016. Hetero-type dual photoanodes for unbiased solar water splitting with extended light harvesting. Nature Communications 7, 13380. Available at https://doi.org/10.1038/ncomms13380.

Li, J., Lin, J.Y., Jiang, H.X., 2008. Direct hydrogen gas generation by using InGaN epilayers as working electrodes. Applied Physics Letters 93, 2008–2011. Available at https://doi.org/10.1063/1.3006332.

Liu, D., Kuang, Y., 2024. Particle-based photoelectrodes for PEC water splitting: concepts and perspectives. Advanced Materials 19, 1–30. Available at https://doi.org/10.1002/adma.202311692.

Luo, J., Im, J.-H., Mayer, M.T., Schreier, M., Nazeeruddin, M.K., Park, N.-G., et al., 2014. Water photolysis at 12.3% efficiency via perovskite photovoltaics and Earth-abundant catalysts. Science 345, 1593–1596. Available at https://doi.org/10.1126/science.1258307.

Ma, G., Hisatomi, T., Domen, K., 2015. Semiconductors for photocatalytic and photoelectrochemical solar water splitting. In: Rozhkova, E.A., Ariga, K. (Eds.), From Molecules to Materials. Springer International Publishing, Cham, pp. 1–56. Available at https://doi.org/10.1007/978-3-319-13800-8_1.

Moss, B., Babacan, O., Kafizas, A., Hankin, A., 2021. A review of inorganic photoelectrode developments and reactor scale-up challenges for solar hydrogen production. Advanced Energy Materials 11, 2003286. Available at https://doi.org/10.1002/aenm.202003286.

Oka, K., Tsujimura, O., Suga, T., Nishide, H., Winther-Jensen, B., 2018. Light-assisted electrochemical water-splitting at very low bias voltage using metal-free polythiophene as photocathode at high pH in a full-cell setup. Energy and Environmental Science 11, 1335–1342. Available at https://doi.org/10.1039/c7ee03669h.

Pinhassi, R.I., Kallmann, D., Saper, G., Dotan, H., Linkov, A., Kay, A., et al., 2016. Hybrid bio-photo-electro-chemical cells for solar water splitting. Nature Communications 7, 1–10. Available at https://doi.org/10.1038/ncomms12552.

Rajaitha, P.M., Hajra, S., Mistewicz, K., Panda, S., Sahu, M., Dubal, D., et al., 2022. Multifunctional materials for photo-electrochemical water splitting. Journal of Materials Chemistry A 10, 15906–15931. Available at https://doi.org/10.1039/d2ta01869a.

Shi, X., Jeong, H., Oh, S.J., Ma, M., Zhang, K., Kwon, J., et al., 2016. Unassisted photoelectrochemical water splitting exceeding 7% solar-to-hydrogen conversion efficiency using photon recycling. Nature Communications 7, 11943. Available at https://doi.org/10.1038/ncomms11943.

Tao, X., Zhao, Y., Wang, S., Li, C., Li, R., 2022. Recent advances and perspectives for solar-driven water splitting using particulate photocatalysts. Chemical Society Reviews 51, 3561–3608. Available at https://doi.org/10.1039/D1CS01182K.

Thakur, A., Ghosh, D., Devi, P., Kim, K.-H., Kumar, P., 2020. Current progress and challenges in photoelectrode materials for the production of hydrogen. Chemical Engineering Journal 397, 125415. Available at https://doi.org/10.1016/j.cej.2020.125415.

Turan, B., Becker, J.-P., Urbain, F., Finger, F., Rau, U., Haas, S., 2016. Upscaling of integrated photoelectrochemical water-splitting devices to large areas. Nature Communications 7, 12681. Available at https://doi.org/10.1038/ncomms12681.

Vanka, S., Zeng, G., Deutsch, T.G., Toma, F.M., Mi, Z., 2022. Long-term stability metrics of photoelectrochemical water splitting. Frontiers in Energy Research 10, 1–10. Available at https://doi.org/10.3389/fenrg.2022.840140.

Wang, J., Guan, Z., Huang, J., Li, Q., Yang, J., 2014. Enhanced photocatalytic mechanism for the hybrid g-C_3N_4/MoS_2 nanocomposite. Journal of Materials Chemistry A 2, 7960–7966. Available at https://doi.org/10.1039/C4TA00275J.

Wang, Z., Wang, L., 2018. Progress in designing effective photoelectrodes for solar water splitting. Chinese Journal of Catalysis 39, 369–378. Available at https://doi.org/10.1016/S1872-2067(17)62998-X.

Xiang, C., Weber, A.Z., Ardo, S., Berger, A., Chen, Y., Coridan, R., et al., 2016. Modeling, simulation, and implementation of solar-driven water-splitting devices. Angewandte Chemie International Edition 55, 12974–12988. Available at https://doi.org/10.1002/anie.201510463.

Xing, Z., Zong, X., Pan, J., Wang, L., 2013. On the engineering part of solar hydrogen production from water splitting: photoreactor design. Chemical Engineering Science 104, 125–146. Available at https://doi.org/10.1016/j.ces.2013.08.039.

Yang, W., Prabhakar, R.R., Tan, J., Tilley, S.D., Moon, J., 2019. Strategies for enhancing the photocurrent, photovoltage, and stability of photoelectrodes for photoelectrochemical water splitting. Chemical Society Reviews 48, 4979–5015. Available at https://doi.org/10.1039/c8cs00997j.

Zhang, S., Ye, H., Hua, J., Tian, H., 2019. Recent advances in dye-sensitized photoelectrochemical cells for water splitting. EnergyChem 1, 100015. Available at https://doi.org/10.1016/j.enchem.2019.100015.

Zhang, Y., Guo, W., Zhang, Y., Wei, W.D., 2021. Plasmonic photoelectrochemistry: in view of hot carriers. Advanced Materials 33, 1–16. Available at https://doi.org/10.1002/adma.202006654.

CHAPTER

5

Photocatalytic water splitting

Shalini Viswanathan[1], *A.C. Swathi*[2], *and K. Aparna*[1]

[1]Department of Chemical Engineering, Materials Science and Environmental Sustainability Group, National Institute of Technology, Calicut, Kerala, India [2]Department of Physics, National Institute of Technology, Calicut, Kerala, India

5.1 Introduction

The escalating energy demands of the modern world have intensified efforts to discover sustainable energy sources. Among these, solar energy stands out as a promising solution due to its abundant availability. Photocatalytic water splitting, a technology capable of generating clean energy from light and water, has garnered significant attention within various light-to-energy conversion methodologies (Huang et al., 2023; Tachibana et al., 2012; Walter et al., 2010). This process, akin to natural photosynthesis, harnesses light energy to drive the decomposition of water molecules into hydrogen and oxygen (Jin et al., 2023; Ye et al., 2021). Beyond producing clean hydrogen energy, photocatalytic water splitting contributes to sustainable energy management by reducing greenhouse gas emissions (Finke et al., 2021). The ability of this technology to capture and utilize solar energy marks a substantial advance in renewable energy production (Hasija et al., 2021). If successfully scaled to industrial levels, solar photocatalytic water splitting could offer a safe, clean, and sustainable alternative to traditional energy production methods, which often deplete natural resources and produce harmful emissions (Ishaq et al., 2021; Jeffry et al., 2021; Zhang et al., 2022). Thus a comprehensive understanding of the fundamentals, as well as innovative materials and strategies to address current technical challenges, is essential for the efficient and cost-effective application of photocatalytic water-splitting technology.

The overall performance of photocatalytic water splitting is influenced by several factors, including the catalytic material's physicochemical and optoelectronic properties, the type and intensity of the light source, and the reaction conditions (Ali et al., 2021; Rahman et al., 2020; Shi et al., 2020). Photocatalytic technology relies on the development of suitable materials that produce photo-induced charge carriers when exposed to light with energy equal to or greater than their bandgap. These charge carriers are the chemical species that initiate the redox reactions on the catalyst's surface, driving the water-splitting process. The landmark discovery by Fujishima and Honda in 1972, demonstrating photocatalytic water splitting on TiO_2 electrodes, laid the foundation for semiconductor-based photocatalytic

Photochemical Splitting of Water
DOI: https://doi.org/10.1016/B978-0-443-29064-0.00003-6

water-splitting research (Fujishima and Honda, 1972). Since then, numerous materials have been explored to create an economically viable and energy-efficient photocatalytic water-splitting device. Titanium dioxide (TiO_2), due to its affordability, availability, and ease of fabrication, remains a highly preferred photocatalytic material (Arifin et al., 2021; Ismael, 2020; Ma et al., 2022). However, TiO_2 has a wide bandgap and relies mainly on the UV fraction of sunlight, limiting its application in industrial settings (Moridon et al., 2022). The limited use of TiO_2 within the visible light spectrum has led to ongoing research into performance-enhancing techniques, including doping, coupling with narrow-bandgap materials, and surface modification (Aldosari and Hussain, 2024; Peiris et al., 2021). The quest for innovative materials capable of facilitating water splitting via photocatalysis has identified a variety of photocatalysts with unique properties, applications, and challenges, including metal oxides, two-dimensional carbon-based materials, nitrides, and sulfides (Kundu et al., 2020; Sahani et al., 2022; Yaseen et al., 2022; Zhang et al., 2021). For optimal light absorption and utilization, a water-splitting photocatalyst should exhibit strong light absorptive properties, particularly in the visible spectrum. This requirement has driven the development of materials like tungsten trioxide (WO_3) and bismuth vanadate ($BiVO_4$) (Costa et al., 2022; Rather et al., 2021). Furthermore, repeated exposure to light may lead to photocatalyst degradation, resulting in photocorrosion and loss of catalytic activity, which could impede large-scale hydrogen production. Continuous research efforts aim to develop materials resistant to such photocorrosion.

This chapter presents a detailed analysis of the core principles of photocatalytic water splitting, explores practical techniques for process optimization, and explains the mechanisms driving hydrogen and oxygen generation through water splitting. It also examines various experimental setups used in photocatalytic water-splitting studies and highlights recent advancements in the field.

5.1.1 Basics of semiconductor-mediated photocatalytic water splitting

The overall splitting of H_2O into H_2 and half O_2 is a thermodynamically unfavorable process involving a large positive change in Gibbs free energy ($\Delta G > 0$, 238 kJ/mol), typically known as an "uphill reaction" (Maeda, 2011). This photochemical reaction is facilitated by light and the half-reactions can be expressed as follows:

$$2H^+ + 2e^- \rightarrow H_2 \quad (5.1)$$

$$2H_2O + 4h^+ \rightarrow O_2 + 4H^+ \quad (5.2)$$

The overall reaction is given by

$$2H_2O \rightarrow 2H_2 + O_2 \quad (5.3)$$

In photocatalytic water splitting over a semiconductor particle, the photocatalyst is suspended in water under specified reaction conditions. For low-cost, large-scale hydrogen production, a setup with dispersed semiconductor particles is particularly advantageous (Liu et al., 2019). Fig. 5.1 provides a basic schematic of this reaction setup, illustrating the key principles of overall water splitting via a semiconductor photocatalyst. Upon exposure to light, photocatalytic water splitting over the catalyst proceeds through multiple stages:

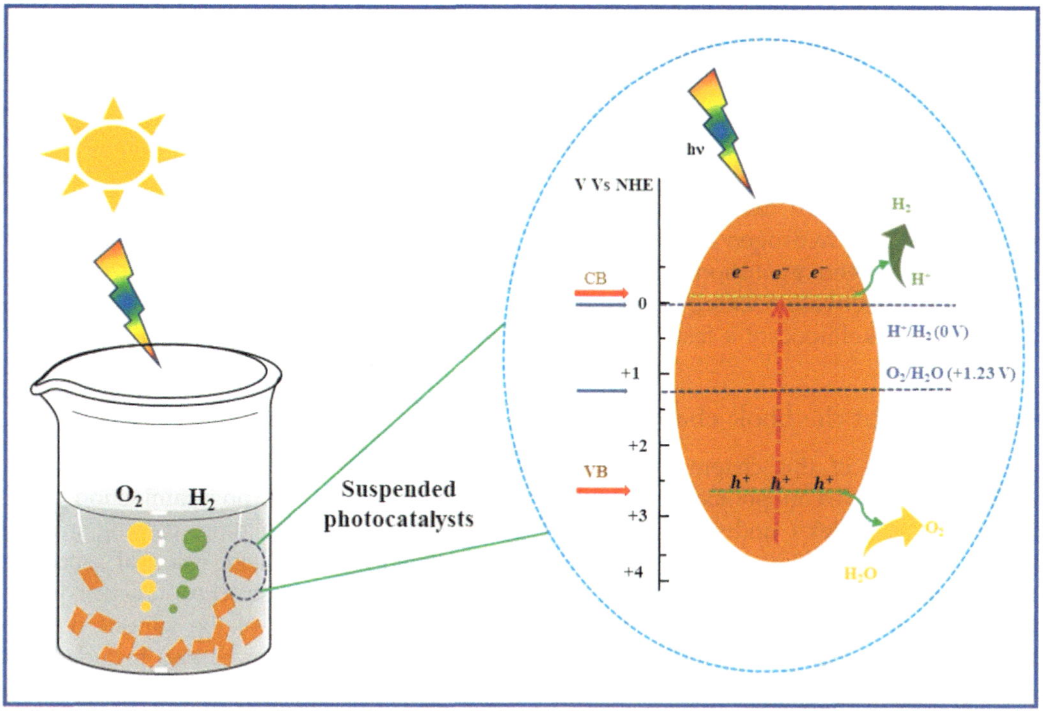

FIGURE 5.1 Schematic representation of the photocatalytic reaction setup with the principle of overall water splitting over semiconductor photocatalyst.

(1) the catalyst absorbs photon energy when it meets or exceeds the material's bandgap; (2) photoinduced charge carriers are then separated from the material's surface; and (3) the redox reaction, facilitated by these charge carriers, occurs across the catalytic surface, generating H_2 and O_2.

The photocatalyst's optoelectronic properties are critical for light absorption and the separation of photoinduced charge carriers. High crystallinity and enhanced light absorptivity improve the efficiency of photocatalytic water splitting, while crystal defects, which act as recombination centers for photoinduced carriers, reduce efficiency (Sheng Chua et al., 2016; Song et al., 2020). Since faster charge separation is positively correlated with shorter charge diffusion lengths from the bulk to the catalytic surface, reducing photocatalytic particle size has generally shown to increase overall efficiency (Cesar et al., 2009). Additionally, incorporating solid cocatalysts has been found to enhance the photocatalytic production of H_2 and O_2 over dispersed semiconductor catalysts, lowering the activation energy required for gas generation (Rosman et al., 2022). Noble metals, such as platinum or ruthenium, are commonly used for this purpose, though recent studies also highlight the use of transition metal oxides like RuO_2 and NiO_x (Shangguan, 2007). Selecting suitable cocatalysts is crucial to promote photocatalytic water splitting, as most existing photocatalysts struggle to activate H_2 on the catalytic surface. Beyond these factors, the relative alignment of the photocatalyst's energy levels with the thermodynamic requirements of water splitting significantly influences the overall process efficiency. To satisfy the thermodynamic criteria, a photocatalyst typically requires a minimum bandgap of 1.23 eV to perform water splitting under standard conditions (Weber and Dignam, 1986). Consequently, in the potential versus normal hydrogen electrode (NHE), which is a reference electrode used to establish a zero-point for electrode potentials in electrochemistry diagram, the valence and conduction bands must be positioned at +1.23 V and 0 V, respectively, as depicted in Fig. 5.1, which further illustrates these thermodynamic considerations. Additionally, for hydrogen and oxygen evolution via the hydrogen evolution reaction (HER) and oxygen evolution reaction (OER), the overpotential must be sufficient to overcome the activation energy barrier (Tsai et al., 2020). A high enough overpotential enhances the likelihood of a faster rate for the water-splitting reaction.

5.1.2 Scope of the book chapter

The objective of this chapter is to provide a comprehensive overview of the multiple dimensions involved in photocatalytic water splitting. It begins with an examination of the fundamental principles underpinning overall photocatalytic water splitting, delving into the intricacies of the process. Additionally, recent advancements in the field, essential aspects of material design, and the development of effective strategies for constructing efficient photocatalytic water-splitting technologies are discussed. A brief overview of the various types of water-splitting systems, along with their characteristics and applications, is also included. The primary focus of this chapter is on visible light-driven photocatalytic water splitting to enhance sustainability and efficiency. The chapter further explores the individual steps of photocatalytic HER and OER from water splitting, offering an in-depth understanding of the kinetics and mechanisms involved. Emphasis is placed on the importance of an ideal reaction setup to achieve energy-efficient and sustainable outcomes, with discussions covering a variety of device types used in recent photocatalytic water-splitting experiments. In addition to addressing the relevance of optimal reaction setups for achieving desired results, this chapter examines the diverse water-splitting devices utilized in recent studies. Moreover, the chapter thoroughly reviews current advancements and innovations in photocatalytic water splitting, highlighting ongoing research efforts and improvements within this rapidly advancing field. It also covers the challenges and limitations associated with photocatalytic water splitting, shedding light on the obstacles faced by researchers and practitioners in realizing the full potential of this technology. In conclusion, the chapter offers an extensive summary of photocatalytic water splitting, emphasizing key findings and suggesting potential directions for future research and innovation. It underscores the vast potential of photocatalytic water splitting as a sustainable, clean energy source and highlights the importance of continued innovation and collaboration to propel this field forward.

5.2 Material design for photocatalytic water splitting

The effectiveness of photocatalytic water splitting is significantly influenced by several factors, including sunlight absorption, the retention and recombination of excited electron–hole pairs, and the surface reactivity of photocatalysts toward HER and OER. These factors interrelate in practice, as the process strongly depends on the photocatalyst's surface chemical reactivity, requiring that irradiated electrons or holes possess sufficient energy to facilitate both HER and OER. A large bandgap value implies a limited capacity for the photocatalyst to capture long-wavelength photons from sunlight, reducing the solar-to-hydrogen (STH) conversion efficiency (Chen et al., 2012; Jiang et al., 2017). However, a wide bandgap in a photocatalyst does fulfill this energy requirement.

It is important to note that the solar spectrum includes only a small fraction of ultraviolet (UV) light at wavelengths below 400 nm, accounting for approximately 6.8% of solar energy. In comparison, around 54.3% of sunlight resides within the near-infrared (IR) region (760–3000 nm), while visible light (400–700 nm) comprises about 38.9% of solar energy. Consequently, a primary challenge is to develop photocatalytic materials with narrow bandgaps and optimal alignment of the valence band maximum (VBM) and conduction band minimum (CBM). Additionally, enhancing electron–hole separation and minimizing carrier recombination during migration are beneficial for improved efficiency. To address these challenges, band engineering in semiconductors has become a prominent approach for enhancing the water-splitting capabilities of photocatalysts. This involves introducing impurity states within the broad bandgaps of photocatalysts to capture long-wavelength light, while carefully maintaining the VBM and CBM positions through doping. Alternatively, increasing the photocatalyst's chemical reactivity allows for adjustments to VBM and CBM locations, effectively reducing the bandgap. Intrinsic electric fields within photocatalysts can be generated through surface chemical functionalization or heterojunction formation, improving electron–hole separation induced by irradiation. Furthermore, efforts to minimize the migration distance of photogenerated charge carriers are exploring the reduction of material dimensionality (Fu et al., 2018). In summary, current research is focused on manipulating electronic structural properties to meet photocatalytic water-splitting requirements. However, achieving highly efficient photocatalytic systems also necessitates a comprehensive understanding of three core processes: light absorption, electron–hole separation, and the surface redox reactions within photocatalysts.

5.2.1 Types of photocatalysts for water splitting

The most basic form of a water-splitting photocatalyst consists of a single semiconductor particle capable of promoting H_2 and O_2 evolution by absorbing photons. Typically, this reaction necessitates a secondary nanoparticle cocatalyst to facilitate the evolution of H_2. Although a variety of photocatalysts can generate H_2 or O_2 in the presence of electron acceptors or donors, energy losses constrain the range of visible-light photocatalysts that can drive overall water splitting to produce H_2 and O_2 without relying on sacrificial agents (Galińska and Walendziewski, 2005; Kudo et al., 2004; Ni et al., 2007).

In contrast, energy losses are more manageable in Z-scheme (or tandem) water-splitting systems, which are characterized by two distinct photocatalysts operating in sequence to generate H_2 and O_2 (Sasaki et al., 2009). This configuration, analogous to the photosynthetic processes observed in green plants, comprises two photocatalysts arranged in series to separately conduct water reduction and oxidation. The connection between these photocatalysts can be established either through direct electrical contact or via a reversible redox mediator, such as I_3^-/I^- or Fe^{3+}/Fe^{2+} (Fig. 5.2). Effective tandem photocatalysts must exhibit selective redox capabilities while simultaneously inhibiting H_2-O_2 recombination to facilitate

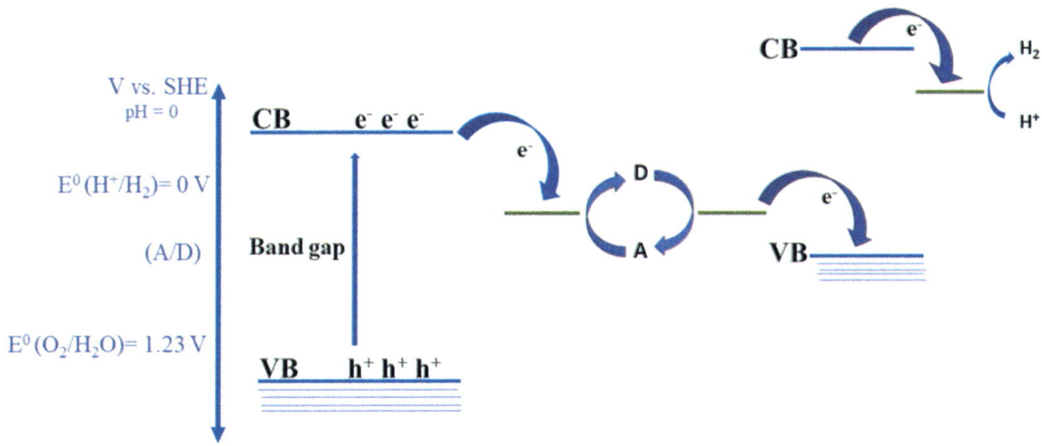

FIGURE 5.2 Schematic diagram of two-step photoexcitation system.

the water-splitting reaction. For instance, I_3^- ions are preferentially reduced over H_2O (or H^+), making it essential to suppress this reduction to enable efficient catalysis by the H_2 evolution photocatalyst (Yuan et al., 2016).

5.2.2 Importance of visible light-driven photocatalytic water-splitting systems

A particularly promising approach to renewable hydrogen generation, which is critical for sustainable energy systems, is photocatalytic water splitting powered by sunlight. Recent advancements in this field have notably emphasized the exploration of both one-step and two-step excitation overall water-splitting systems. Currently, the development of systems capable of visible light-driven overall water splitting is a primary objective, as achieving high STH conversion efficiency is essential for practical application (Lin et al., 2020).

Since the 1980s, numerous UV-driven water-splitting systems have been reported (Fu et al., 2020; Fujiwara et al., 2020; Wang et al., 2018). Among these, $SrTiO_3$ is recognized for its responsiveness to UV light (Mu et al., 2016; Yoshinaga et al., 2018), with Al-doped $SrTiO_3$ emerging recently as a promising photocatalyst due to its enhanced stability and efficiency (Takata et al., 2020). However, these UV-based systems face inherent limitations in meeting cost competitiveness with fossil fuels, primarily because UV radiation constitutes only about 5% of the total energy in the solar spectrum. In contrast, visible light, spanning wavelengths from 400 to 800 nm, accounts for more than 50% of solar energy. The potential to harness visible light, particularly wavelengths below 600 nm, represents a significant advancement, as it could theoretically yield a STH conversion efficiency of approximately 16%. Thus for efficient, renewable hydrogen production, developing photocatalysts responsive to longer wavelengths is critical (Fu et al., 2020; Hisatomi et al., 2014; Pinaud et al., 2013). Significant progress has been made recently in developing visible light-driven overall water-splitting systems. However, the efficiency of current systems remains below the levels required for large-scale, practical applications. Furthermore, considerable challenges persist, particularly in optimizing photocatalytic materials and overall system design. The primary objective is to engineer photocatalysts that respond effectively to long-wavelength light, a crucial step toward improving STH efficiency. Insights from previous studies on UV-responsive photocatalysts may inform the development of visible light-responsive alternatives, notably through optimized cocatalyst loading and doping strategies. Efficiency improvements can also be achieved by employing coating layers on both cocatalysts and photocatalysts to inhibit reverse reactions. Recent studies indicate that several photocatalysts previously deemed inactive for overall water splitting may be activated through refined synthesis methods, advanced modification strategies, and optimized reaction conditions. While current advanced visible light-responsive photocatalysts may be limited in one-step excitation configurations, they show promise in two-step excitation overall water splitting. In such cases, selecting appropriate electron mediators is essential to minimize reverse reactions and facilitate effective charge transfer between the hydrogen evolution photocatalyst (HEP) and the oxygen evolution photocatalyst (OEP).

5.2.3 Photocatalytic hydrogen evolution from water splitting

The same parameters that control the entire water-splitting process also influence photocatalytic water splitting and the production of green hydrogen from the process. Here, the electrons and holes that are photoexcited can go through a number of processes, including recombination, migration to the catalyst's surface, trapping by different agents like scavenging species or system defects, or transfer into water or other molecules (Zhang et al., 2012). A schematic illustration of the hydrogen energy system created by photocatalytic water splitting is shown in Fig. 5.3.

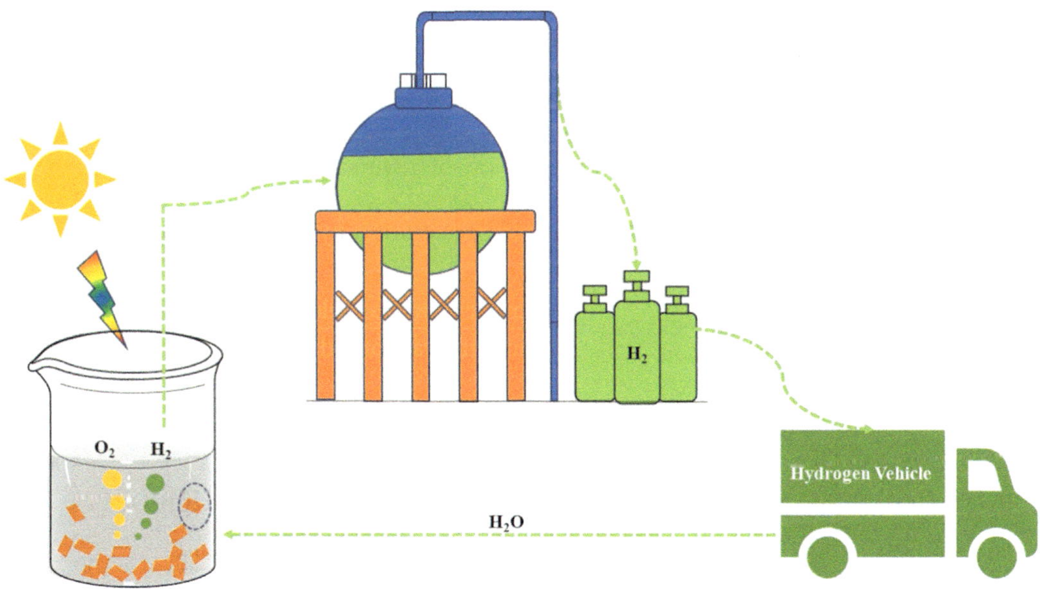

FIGURE 5.3 Schematic illustration of the hydrogen energy system generated from photocatalytic water splitting.

The effective production of hydrogen through photocatalysis is critically dependent on the abundant supply of free electrons capable of facilitating the reduction of water. Consequently, several key factors must be meticulously considered when selecting a photocatalyst and designing the corresponding experimental setup:

1. Bandgap and energy alignment: The photocatalyst must possess an appropriate bandgap that aligns favorably with the relative positions of the valence and conduction bands, as well as an adequate overpotential to enable the reduction of water at the catalytic surface under the specified experimental conditions.
2. Light absorption: Ideally, the selected photocatalyst should exhibit minimal light transmission and scattering, along with strong absorption capabilities in the visible light spectrum to maximize the utilization of solar energy.
3. Charge carrier generation: The photocatalytic material should be designed to generate a maximum number of electrons and holes while minimizing the recombination of these charge carriers within the material, which can significantly hinder efficiency.
4. Charge separation and migration: Efficient charge separation and the rapid migration of charge carriers to the catalyst surface are essential for optimal performance. Any delays in this process can lead to reduced efficiency in hydrogen production.
5. Electron trapping: The presence of species that could trap photoinduced electrons should be minimized to ensure a high concentration of free electrons available for water reduction.
6. Surface back reaction: It is crucial to mitigate surface back reactions (SBR), wherein the hydrogen (H_2) and oxygen (O_2) produced during the water-splitting process recombine to form water (H_2O), thereby decreasing the overall yield of hydrogen.

To comprehensively address the aforementioned parameters, careful synthesis and selection of the photocatalyst are essential. Modifying the nature of surface reaction sites and optimizing surface morphology can significantly influence critical characteristics, including the suppression of surface back reactions (SBR) (Zeng et al., 2022). Additionally, the introduction of sacrificial agents or the physical separation of active sites for H_2 and O_2 generation can effectively inhibit the progression of reverse reactions that lead to water formation.

The efficiency of the photocatalytic hydrogen generation by water splitting is generally expressed as:

$$\text{Overall quantum yield (\%)} = \frac{\text{Number of reacted electrons}}{\text{Number of absorbed photons}} \times 100$$

$$\text{Apparent quantum yield (\%)} = \frac{\text{Number of reacted electrons}}{\text{Number of incident photons}} \times 100$$

In terms of hydrogen evolution, the apparent quantum yield can be expressed as:

$$\text{Apparent quantum yield (\%)} = \frac{2 \times \text{Number of evolved H}_2\text{ molecules}}{\text{Number of incident photons}} \times 100$$

The apparent quantum yield is generally smaller than the overall quantum yield because the number of absorbed photons is lesser than the number of incident photons. In terms of solar energy conversion, photocatalytic efficiency can also be expressed as:

$$\text{Solar energy conversion (\%)} = \frac{\text{Output energy of } H_2 \text{ generated}}{\text{Energy of incident solar light}} \times 100$$

5.2.4 Photocatalytic oxygen evolution from water splitting

Recent advancements in photocatalytic water splitting have drawn significant attention to the crucial role of oxygen (O_2) evolution, primarily due to its intricate four-electron reaction mechanism. The O_2 evolution reaction is identified as the rate-limiting step in the overall photocatalytic water-splitting process (Lin et al., 2021). Consequently, achieving optimal efficiency in water oxidation photocatalysts necessitates the fulfillment of several critical criteria:

1. Valence band potential: The potentials of the valence band (VB) in the photocatalysts must exceed those of the O_2/H_2O couple, specifically requiring values greater than 1.23 eV.
2. Bandgap optimization: To maximize solar energy utilization, it is imperative that the bandgaps of the photocatalysts fall within the range of 1.23–3 eV.
3. Charge carrier dynamics: Efficient separation of photogenerated electron–hole pairs within the photocatalysts is essential to facilitate prompt redox reactions at the photocatalyst surface upon illumination.
4. Active site density: Maintaining a sufficient density of active sites for water oxidation on the photocatalyst surface is critical for minimizing the rapid recombination of photogenerated electrons and holes.

Currently, heterogeneous photocatalysts are the focus of extensive research due to their versatility and effectiveness in promoting photocatalytic reactions (Walter et al., 2010). Over the past decade, a variety of OEP have been developed. However, substantial challenges remain before these technologies can be effectively implemented in industrial applications. The photocatalysts face ongoing issues, including low efficiency, inadequate physicochemical stability, and a dependence on sacrificial agents and cocatalysts. The interplay of these factors significantly heightens the economic constraints associated with industrial deployment. To advance the field, it is essential to address these challenges through targeted measures:

1. Development of effective photocatalysts: Current research primarily focuses on enhancing the development of efficient water oxidation photocatalysts. Promising results have been observed with Sillén–Aurivillius perovskites, such as Bi_4NbO_8Cl, which exhibit high photochemical stability. The valence bands of these perovskites are predominantly filled with O 2p orbitals, rendering them resistant to corrosion from photogenerated holes. Moreover, the availability of diverse synthesis methods—including solid-state reactions, two-step polymerized complex methods, conventional single-step solid-state reactions, and flux techniques—enhances the adaptability and optimization of photocatalytic performance throughout the preparation process. The varied compositions within the Sillén–Aurivillius perovskite family also offer opportunities for tailoring light absorption and bandgap characteristics, further improving photocatalytic O_2 evolution activity. Thus

employing rational design strategies focused on crystal and band structures presents considerable promise for the development of high-performance perovskites specifically engineered for efficient O_2 evolution in future applications.

2. Enhancement of surface catalytic reactions: Current initiatives aimed at improving O_2 evolution performance predominantly concentrate on increasing light absorption and charge separation efficiency, while relatively less emphasis has been placed on surface catalytic reactions. The reactive sites of photocatalysts are critical in determining reactant adsorption and the activation energy required for reactions. Surface defects can significantly enhance the active sites of photocatalysts, indicating that exploring catalytic active sites for O_2 evolution is likely to become a prominent focus in forthcoming studies.
3. Stability and robustness: While significant attention has been dedicated to enhancing the separation efficiency of photogenerated charge carriers to improve O_2 evolution performance, less focus has been directed toward maintaining the physicochemical stability of photocatalysts during the photocatalytic process. The stability of most synthesized materials can be adversely affected by various factors, including synthesis methods and conditions. Furthermore, certain O_2 evolution photocatalysts may experience self-poisoning due to the accumulation of photogenerated holes, leading to inactivation following photoreactions. Therefore developing robust strategies to enhance the chemical and physicochemical stability of O_2 evolution photocatalysts is essential.

Additionally, it is crucial to consider the spatial segregation of reduction and oxidation reactions. This spatial distinction can effectively inhibit the reverse reactions that commonly occur on the photocatalyst surface. An illustrative approach is the hydrogen farm strategy proposed by Li's group, which demonstrates considerable potential for advancing the field, and it is anticipated that further progress will be made in this area.

5.3 Reaction setup of water-splitting devices

Effective photocatalysts for comprehensive water splitting should have the ability to produce hydrogen and oxygen in the presence of air at atmospheric pressure. Photocatalytic water-splitting experiments are generally performed in air-free environments, utilizing argon gas atmospheres or degassed conditions. This method enables accurate quantification of the generated H_2 and O_2 gases, thereby validating the reaction stoichiometry ($H_2/O_2=2$). For these experiments, two primary reactor types have been established: closed systems and flow systems, each designed to accommodate different types of reaction cells. Fig. 5.4 presents the experimental configurations utilized for photocatalytic water splitting.

5.3.1 Closed system

In closed systems, two primary types of reaction cells are utilized for photocatalytic processes. The first category encompasses batch-type cells, which include test tubes or cuvettes typically sealed with a rubber septum. The second category consists of custom-

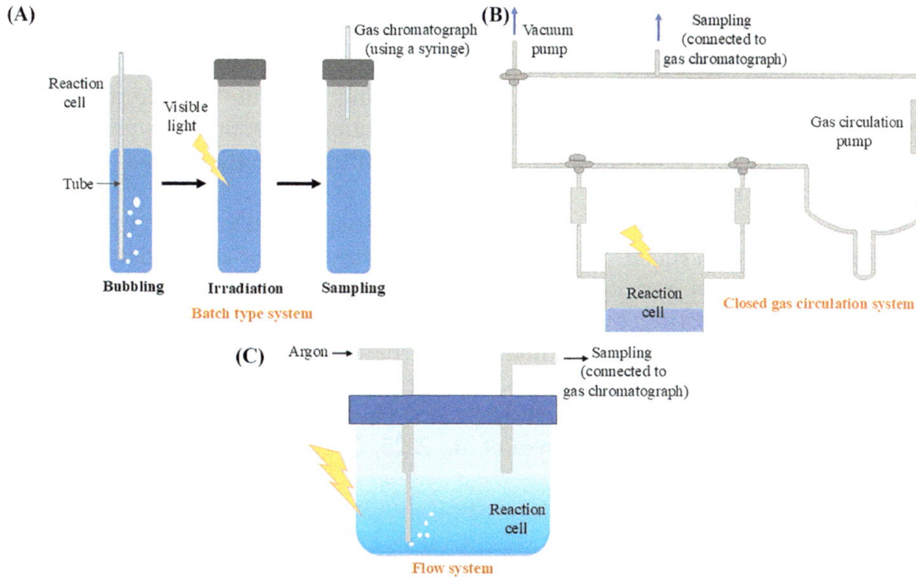

FIGURE 5.4 Experimental reaction setups for photocatalytic water-splitting processes (A) Batch type system (B) closed gas circulation system (C) flow system.

designed reaction cells integrated into a closed gas circulation system, which features a vacuum pump. These cells are required to be optically transparent to incoming light and are generally constructed from glass materials such as quartz or Pyrex.

In a batch-type configuration, the reaction cell may take the form of a test tube, an optical cuvette, or a transparent plastic thin-film bag, such as those made from polypropylene. Prior to irradiation, the reaction system is purged with argon gas to remove impurities present in both the gas and liquid phases (Pinaud et al., 2013). It is advisable to avoid using nitrogen (N_2) for purging, as it could lead to confusion between oxygen (O_2) produced through photocatalysis and potential air leakage. The reaction cells can be irradiated using either a comprehensive illuminator or an external light source, such as a merry-go-round system. Gas samples for analysis are extracted with a gas-tight syringe; however, accurately measuring O_2 levels poses challenges due to the risk of air contamination during the sampling process.

In a closed gas circulation system, the gases generated from the water-splitting reaction are continuously recirculated and analyzed at regular intervals using a gas chromatograph that is directly linked to the system. This sealed gas circulation setup allows for precise measurements of hydrogen (H_2) and oxygen (O_2) levels. The system accommodates a variety of reaction cell configurations, including top, inner, and side irradiation cells. Experiments involving water splitting in this configuration are conducted under degassed conditions, utilizing saturated water vapor alongside a minimal amount of carrier gas, such as argon. The low-pressure environment facilitates the identification of evolved gases throughout the reaction; however, it also introduces a pressure bias that favors the forward reaction (Osterloh, 2017).

5.3.2 Flow system

Flow systems, characterized by the continuous bubbling of a liquid phase with a carrier gas, provide a notable benefit for efficient water splitting by inhibiting the reverse reaction (the recombination of H_2 and O_2 to generate water). Nonetheless, flow systems can result in product dilution, which requires additional energy input later in the process to isolate pure H_2 and O_2 from the carrier gas.

In both flow and closed systems, maintaining the suspension of photocatalyst particles with a magnetic stirrer is crucial. This guarantees maximum light absorption and facilitates the effective diffusion of reactants and products, thereby improving reproducibility. Nonetheless, extended stirring over a period of months may adversely affect the photocatalyst, resulting in diminished performance. Alternatively, photocatalyst particles may be attached to a designated substrate, such as an inert glass surface, resulting in the formation of a photocatalyst panel. This method could provide benefits, particularly when expanding for more extensive applications (Nishioka et al., 2023; Ohno et al., 2012).

5.4 Recent progress in photocatalytic water splitting

Over the years, a steady increase in photocatalytic water-splitting literature has been observed and such statistics for the keyword "photocatalytic water splitting" from the Web of Science database are shown in Fig. 5.5.

Recent extensive investigations have focused on the design and development of effective photocatalysts for water splitting. Various materials, including metal and nonmetal oxides, (oxy)sulfides, nitride compounds, and chalcopyrites, have been extensively studied for their potential to achieve promising photocatalytic efficiencies (Han et al., 2018; Salomao et al., 2021; Zhang and Wang, 2019). Notably, there has been a marked increase in interest in solar

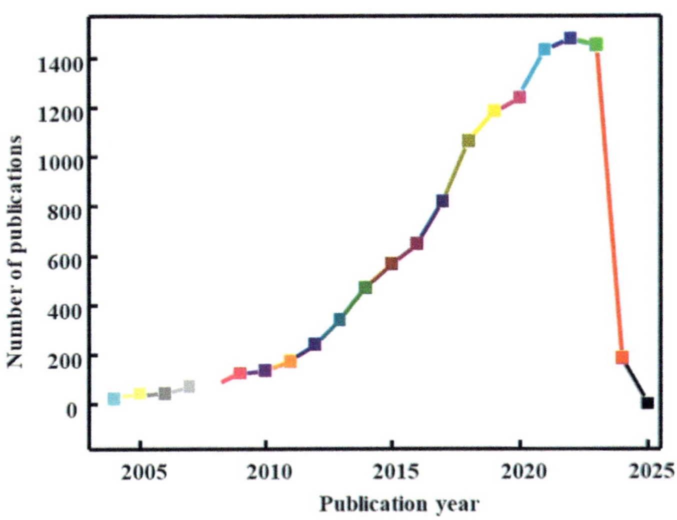

FIGURE 5.5 Number of publications with the keyword "photocatalytic water splitting" from 2003 to 2025. *Source: From Web of Science.*

photocatalytic water splitting in recent years, particularly regarding bandgap tuning, crystal facet engineering, and the development of low-cost, high-performance cocatalysts. The primary focus of these efforts encompasses several aspects of the photocatalytic water-splitting process.

5.4.1 Selection, design, and development of efficient photocatalytic materials

Domen's group has been at the forefront of advancing photocatalytic water splitting toward industrial application for over 40 years (Takata and Domen, 2009). The team has concentrated on enhancing the quantum efficiency of $SrTiO_3$-based photocatalysts to facilitate industrial-scale hydrogen production through photocatalytic water splitting. Reports indicate that Al-doped $SrTiO_3$ can achieve an apparent quantum efficiency of 30% under UV-A light irradiation (Ham et al., 2016). The group focused on improving spatial charge separation in crystals featuring various facets, successfully creating Al-doped $SrTiO_3$ decorated with Rh/Cr_2O_3 and CoOOH cocatalysts through selective photodeposition. The anisotropic charge transport across the different facets of the Al-doped $SrTiO_3$ in this photocatalyst was crucial for attaining a remarkable 96% external quantum yield. Similarly, Liu et al. reported an apparent quantum efficiency of 56% for 1.5% single-atom copper-loaded TiO_2. This study highlighted that even 380 days postsynthesis, the catalyst exhibited consistent activity compared to initial observations. Notably, the work provided a straightforward and rapid synthesis approach for an affordable and highly stable solar photocatalyst utilizing TiO_2 (Zhang et al., 2022). In another study, Yang et al. reported the development of an effective photocatalytic system through the implementation of a donor–acceptor structure composed of β-ketene-cyano groups integrated into nanosheets of a covalent organic framework. This research notably surpassed all prior reports regarding the efficiency of polymeric materials designed for HER, achieving an apparent quantum efficiency of 82.5% at 450 nm light (Li et al., 2022). Additionally, comparable studies involving CdTe/V-In_2S_3 and InGaN/GaN nanowires decorated with $Rh/Cr_2O_3/Co_3O_4$ have emerged as potential candidates for large-scale photocatalytic water-splitting applications, demonstrating high throughput (Pinaud et al., 2013; Zhou et al., 2023).

5.4.2 Industrialization of hydrogen generation from photocatalytic water splitting

The exploration and advancement of industrial-scale hydrogen production are primarily conducted by the group led by Li et al. (Li et al., 2014, 2013; Zhu et al., 2015). With over 20 years of experience in photocatalytic water splitting utilizing suspended photocatalysts, the team successfully developed a sustainable solar hydrogen production unit that achieved a quantum efficiency of 71% in 2020. This study was part of the hydrogen farm project (HFP) initiated by the team (Zhao et al., 2020). The project has led to the successful implementation of a 1 m^2 photocatalytic panel for solar energy storage, representing a groundbreaking achievement as it effectively addresses the technical challenges that have historically hindered the practical application of large-scale hydrogen production through solar water splitting. The experiments utilized $BiVO_4$ as the photocatalyst, incorporating facet-selective loading of dual cocatalysts to enhance the spatial separation of photoinduced charge carriers. The comprehensive study conducted over two decades has enabled the group to gain critical insights into the importance of charge separation mechanisms and their role in the practicality and cost-effectiveness of the hydrogen production process.

5.5 Case study

5.5.1 Title

In situ synthesis of $AgNbO_3/g\text{-}C_3N_4$ photocatalyst for efficient photocatalytic H_2 generation using microwave heating (Chen et al., 2019).

5.5.2 Objective

To improve the photocatalytic hydrogen evolution performance of $g\text{-}C_3N_4$ through modification with $AgNbO_3$ nanocubes using a microwave heating method.

5.5.3 Scope

This study was conducted under laboratory conditions with a focus on developing a novel photocatalyst for hydrogen generation under visible light. The project used microwave heating to reduce synthesis time and achieve close contact between the two semiconductors. The primary limitation was that the project was focused on fundamental research, without immediate plans for industrial-scale application.

5.5.4 Audience

This case study is intended for researchers in the fields of materials science, photocatalysis, and sustainable energy. It may also be useful for engineers and scientists interested in hydrogen production, semiconductor materials, and green chemistry.

5.5.5 Rationale

With the depletion of fossil fuels and the need for sustainable energy sources, hydrogen is considered a promising alternative fuel. However, the efficiency of existing photocatalysts for hydrogen generation under visible light is limited. This study aimed to address this issue by developing a new $AgNbO_3/g\text{-}C_3N_4$ photocatalyst that offers enhanced photocatalytic performance and greater efficiency in hydrogen production.

5.5.6 Expected results and deliverables

The study aimed to deliver a new photocatalyst that shows significantly higher photocatalytic hydrogen evolution rates under visible light compared to existing materials. The deliverables included experimental data demonstrating improved photocatalytic performance, a detailed characterization of the synthesized materials, and insights into the mechanism of charge separation and migration in the composite.

5.5.7 Safety considerations

Key safety considerations included handling chemicals like Ag_2O, Nb_2O_5, NH_4HF_2, and melamine, which require appropriate personal protective equipment (PPE) such as gloves, goggles, and lab coats. Microwave heating equipment also required careful handling to prevent burns or exposure to microwave radiation.

5.5.8 Actions taken/workflow/tools used/simulations and analyses

Synthesis: $AgNbO_3$ was prepared using a hydrothermal method with Ag_2O, Nb_2O_5, and NH_4HF_2 as precursors. g-C_3N_4 was in situ synthesized on the $AgNbO_3$ surface using microwave heating.

Characterization: The composite was characterized using X-ray diffraction (XRD), Fourier transform-infrared (FT-IR) spectroscopy, X-ray photoelectron spectroscopy (XPS), photoluminescence (PL) spectroscopy, and transient photocurrent response measurements.

Photocatalytic testing: The hydrogen evolution rate under visible light illumination was measured to assess photocatalytic performance.

Data analysis: The results were analyzed to understand the impact of material composition, structure, and synthesis method on the photocatalytic activity.

5.5.9 Challenges and solutions

Challenge: High-temperature thermal treatment could decompose g-C_3N_4.

Solution: The microwave heating method allowed for a rapid synthesis process that minimized decomposition risks by reducing the treatment time to 40 minutes.

Challenge: Efficient charge separation and migration were required for high photocatalytic activity.

Solution: The in situ synthesis method ensured intimate contact between $AgNbO_3$ and g-C_3N_4, reducing interface energy and improving electron transfer.

5.5.10 Results

The $AgNbO_3$/g-C_3N_4 composite synthesized via microwave heating showed a hydrogen evolution rate of 88 $\mu mol \cdot g^{-1} \cdot h^{-1}$ under visible light, which was 2.0 times higher than that of g-C_3N_4 alone. The composite also demonstrated good stability over multiple recycling runs.

5.5.11 Learning and knowledge outcomes

The microwave heating method is effective for synthesizing g-C_3N_4-based composites quickly and efficiently.

The intimate contact between $AgNbO_3$ and g-C_3N_4 enhances photocatalytic performance by improving charge separation and reducing recombination rates.

Future studies could explore the use of different microwave settings, alternative semiconductors, and scaling up the synthesis for practical applications.

5.6 Constraints in photocatalytic water splitting

While there are reports on photocatalytic materials demonstrating remarkable activity for overall water splitting, most catalytic materials, particularly those sensitive to visible light, still fall short of the criteria required for commercial energy production. A variety of elements lead to this inadequate performance, such as excessive charge recombination, sluggish surface kinetics, and disproportionate back reactions. Inadequate alignment of band structures and a lack of sufficient overpotentials further exacerbate these issues. Mechanisms like Z or S schemes address these challenges to some degree; however, the charge transfer mechanism and the rate of charge separation in these modified schemes present additional obstacles for practical applications. The highest hydrogen production achieved via photocatalytic water splitting is approximately 0.7% for Al-doped $SrTiO_3$ and around 1% for $SrTiO_3$:La, as well as for Rh/Au/$BiVO_4$:Mo Z-scheme photocatalysts. The production values currently observed fall significantly short of the 5% threshold recommended by the United States Department of Energy for viable applications. The limitations in this area drive the advancement of photocatalytic water splitting.

Table 5.1 provides a comprehensive summary of the findings from previous studies, detailing the outcomes of various water-splitting reactions conducted in photoelectrochemical and photochemical cells. The table specifically highlights the performance of different semiconductor-based photocatalytic materials, with a particular emphasis on the rates of H_2 and O_2 evolution.

5.7 Conclusions

In conclusion, photocatalytic water splitting marks a significant advancement in the pursuit of sustainable energy solutions. By harnessing sunlight to split water into hydrogen and oxygen, it offers a clean and renewable hydrogen fuel source. This chapter has reviewed the fundamental principles of photocatalysis, key factors influencing photocatalytic water-splitting efficiency, and recent advancements in photocatalyst design. Despite substantial progress, many challenges remain, particularly regarding efficiency, stability, and scalability for industrial applications. Improving photocatalytic performance while ensuring affordability and minimizing environmental and health risks requires exploring enhancement strategies for all photocatalyst types. Achieving greater efficiency in photocatalytic water splitting involves addressing current limitations through bandgap engineering. Various approaches—such as nonmetal and morphology engineering, metal doping, surface modifications, metal ion implantation, dye sensitization, noble metal deposition, and heterojunction construction—have been employed to increase light absorption within the visible spectrum and prevent rapid charge recombination, thereby enabling charge carriers

TABLE 5.1 Photocatalytic water splitting via photochemical cell reactions employing various photocatalysts.

Photocatalysts	Light source	Reaction solution	Rate of evolution (μmol/h)		References
			H_2	O_2	
Pt/$SrTiO_3$:Rh, $BiVO_4$	300 W Xe lamp	5 mM $FeCl_2/FeCl_3$	0.8	0.4	Yu et al. (2011)
NiO/$Sr_2Nb_2O_7$	400 W Hg lamp	Distilled water	110	36	Kudo et al. (2000)
NiO/$Sr_2Ta_2O_7$	400 W Hg lamp	Distilled water	1000	480	Kudo et al. (2000)
Pt/TiO_2	400 W Hg lamp	2.17 M Na_2CO_3	568	287	Sayama and Arakawa (1997)
Pt/$SrTiO_3$:Rh, $BiVO_4$	300 W Xe lamp with filters: $\lambda > 420$ nm	2 mM $FeCl_3$	15	7.2	Kato et al. (2004)
Pt/$SrTiO_3$:Rh, $BiVO_4$	300 W Xe lamp with filters: $\lambda > 420$ nm	2 mM $FeCl_3$	19	8.9	Kato et al. (2004)
Pt/$SrTiO_3$:Rh, WO_3	300 W Xe lamp with filters: $\lambda > 420$ nm	2 mM $FeCl_3$	7.8	4.0	Kato et al. (2004)
ZrO_2	400 W Hg lamp	Distilled water	72	36	Sayama and Arakawa (1993)
ZrO_2	400 W Hg lamp	1.09 M Na_2CO_3	142	75	Sayama and Arakawa (1993)
Pt/ZrO_2	400 W Hg lamp	0.94 M $NaHCO_3$	120	61	Sayama and Arakawa (1993)
Ru_2O/ZrO_2	400 W Hg lamp	Distilled water	11	5	Sayama and Arakawa (1993)
Cu/ZrO_2	400 W Hg lamp	Distilled water	14	6	Sayama and Arakawa (1993)
Pt/g-C_3N_4 nanosheets/TiO_2	300 W Xe lamp (> 420 nm)	Triethanolamine aqueous solution	40	–	Zhang et al. (2017)
Pt/P@P-doped g-C_3N_4	300 W Xe lamp (> 420 nm)	Triethanolamine aqueous solution	94.18	–	Feng et al. (2018)
$AgNbO_3$/g-C_3N_4	Visible light	–	88	–	Chen et al. (2019)

to participate effectively in hydrogen production reactions. Future developments in material science are expected to yield photocatalysts with improved stability, optimized bandgaps, and well-aligned conduction and valence band edges, ultimately boosting hydrogen production rates and efficiencies. Deepening our understanding of the complex mechanisms underlying photocatalysis, especially the role of defects in photocatalytic performance, will be crucial for refining water-splitting systems. While research in this area is still limited, further exploration could unlock new insights for improvement. With advancements in materials science, nanotechnology, and solar energy conversion, the potential of photocatalytic water splitting appears highly promising. Addressing the remaining challenges and enhancing these systems could pave the way toward a more sustainable and energy-efficient future.

References

Aldosari, O.F., Hussain, I., 2024. Unlocking the potential of TiO_2-based photocatalysts for green hydrogen energy through water-splitting: recent advances, future perspectives and techno feasibility assessment. International Journal of Hydrogen Energy 59, 958–981.

Ali, M., Pervaiz, E., Noor, T., Rabi, O., Zahra, R., Yang, M., 2021. Recent advancements in MOF-based catalysts for applications in electrochemical and photoelectrochemical water splitting: a review. International Journal of Energy Research 45, 1190–1226.

Arifin, K., Yunus, R.M., Minggu, L.J., Kassim, M.B., 2021. Improvement of TiO_2 nanotubes for photoelectrochemical water splitting: review. International Journal of Hydrogen Energy 46, 4998–5024.

Cesar, I., Sivula, K., Kay, A., Zboril, R., Grätzel, M., 2009. Influence of feature size, film thickness, and silicon doping on the performance of nanostructured hematite photoanodes for solar water splitting. Journal of Physical Chemistry C 113, 772–782.

Chen, X., Li, C., Grätzel, M., Kostecki, R., Mao, S.S., 2012. Nanomaterials for renewable energy production and storage. Chemical Society Reviews 41, 7909.

Chen, P., Xing, P., Chen, Z., Hu, X., Lin, H., Zhao, L., et al., 2019. In-situ synthesis of $AgNbO_3/g\text{-}C_3N_4$ photocatalyst via microwave heating method for efficiently photocatalytic H_2 generation. Journal of Colloid and Interface Science 534, 163–171.

Costa, M.B., de Araújo, M.A., Tinoco, M.V., de, L., Brito, J.F., de, M., et al., 2022. Current trending and beyond for solar-driven water splitting reaction on WO_3 photoanodes. Journal of Energy Chemistry 73, 88–113.

Finke, E., Leandri, C., F., Tshijik Karumb, H., Zheng, E., Hoffmann, D., R., et al., 2021. Economically advantageous pathways for reducing greenhouse gas emissions from industrial hydrogen under common, current economic conditions. Energy and Environmental Science 14, 1517–1529.

Feng, J., Zhang, D., Zhou, H., Pi, M., Wang, X., Chen, S., 2018. Coupling P nanostructures with P-doped $g\text{-}C_3N_4$ as efficient visible light photocatalysts for H_2 evolution and RhB degradation. ACS Sustainable Chemistry and Engineering 6, 6342–6349.

Fu, C., Wu, X., Yang, J., 2018. Material design for photocatalytic water splitting from a theoretical perspective. Advanced Materials 30, 1802106.

Fu, B., Wu, Z., Cao, S., Guo, K., Piao, L., 2020. Effect of aspect ratios of rutile TiO_2 nanorods on overall photocatalytic water splitting performance. Nanoscale 12, 4895–4902.

Fujishima, A., Honda, K., 1972. Electrochemical photolysis of water at a semiconductor electrode. Nature 238, 37–38.

Fujiwara, T., Sasahara, A., Happo, N., Kimura, K., Hayashi, K., Onishi, H., 2020. Single-crystal model of highly efficient water-splitting photocatalysts: a $KTaO_3$ wafer doped with calcium cations. Chemistry of Materials 32, 1439–1447.

Galińska, A., Walendziewski, J., 2005. Photocatalytic water splitting over $Pt\text{–}TiO_2$ in the presence of sacrificial reagents. Energy and Fuels 19, 1143–1147.

Ham, Y., Hisatomi, T., Goto, Y., Moriya, Y., Sakata, Y., Yamakata, A., et al., 2016. Flux-mediated doping of $SrTiO_3$ photocatalysts for efficient overall water splitting. Journal of Materials Chemistry A 4, 3027–3033.

Han, N., Liu, P., Jiang, J., Ai, L., Shao, Z., Liu, S., 2018. Recent advances in nanostructured metal nitrides for water splitting. Journal of Materials Chemistry A 6, 19912–19933.

Hasija, V., Kumar, A., Sudhaik, A., Raizada, P., Singh, P., Van Le, Q., et al., 2021. Step-scheme heterojunction photocatalysts for solar energy, water splitting, CO_2 conversion, and bacterial inactivation: a review. Environmental Chemistry Letters 19, 2941–2966.

Hisatomi, T., Kubota, J., Domen, K., 2014. Recent advances in semiconductors for photocatalytic and photoelectrochemical water splitting. Chemical Society Reviews 43, 7520–7535.

Huang, Y., Meng, L., Xu, W., Li, L., 2023. The positive versus negative effects of defect engineering for solar water splitting: a review. Advanced Functional Materials 33, 2305940.

Ishaq, T., Yousaf, M., Bhatti, I.A., Batool, A., Asghar, M.A., Mohsin, M., et al., 2021. A perspective on possible amendments in semiconductors for enhanced photocatalytic hydrogen generation by water splitting. International Journal of Hydrogen Energy 46, 39036–39057.

Ismael, M., 2020. A review and recent advances in solar-to-hydrogen energy conversion based on photocatalytic water splitting over doped-TiO_2 nanoparticles. Solar Energy 211, 522–546.

Jeffry, L., Ong, M.Y., Nomanbhay, S., Mofijur, M., Mubashir, M., Show, P.L., 2021. Greenhouse gases utilization: a review. Fuel 301, 121017.

Jiang, C., Moniz, S.J.A., Wang, A., Zhang, T., Tang, J., 2017. Photoelectrochemical devices for solar water splitting – materials and challenges. Chemical Society Reviews 46, 4645–4660.

Jin, W., Shin, C., Lim, S., Lee, K., Yu, J.M., Seo, K., et al., 2023. Natural leaf-inspired solar water splitting system. Applied Catalysis B 322, 122086.

Kato, H., Hori, M., Konta, R., Shimodaira, Y., Kudo, A., 2004. Construction of Z-scheme type heterogeneous photocatalysis systems for water splitting into H_2 and O_2 under visible light irradiation. Chemistry Letters 33, 1348–1349.

Kudo, A., Kato, H., Nakagawa, S., 2000. Water splitting into H_2 and O_2 on new $Sr_2M_2O_7$ (M=Nb and Ta) photocatalysts with layered perovskite structures: factors affecting the photocatalytic activity. The Journal of Physical Chemistry B 104, 571–575.

Kudo, A., Kato, H., Tsuji, I., 2004. Strategies for the development of visible-light-driven photocatalysts for water splitting. Chemistry Letters 33, 1534–1539.

Kundu, S., Bramhaiah, K., Bhattacharyya, S., 2020. Carbon-based nanomaterials: in the quest of alternative metal-free photocatalysts for solar water splitting. Nanoscale Advances 2, 5130–5151.

Li, C., Liu, J., Li, H., Wu, K., Wang, J., Yang, Q., 2022. Covalent organic frameworks with high quantum efficiency in sacrificial photocatalytic hydrogen evolution. Nature Communications 13, 2357.

Li, R., Han, H., Zhang, F., Wang, D., Li, C., 2014. Highly efficient photocatalysts constructed by rational assembly of dual-cocatalysts separately on different facets of $BiVO_4$. Energy and Environmental Science 7, 1369–1376.

Li, R., Zhang, F., Wang, D., Yang, J., Li, M., Zhu, J., et al., 2013. Spatial separation of photogenerated electrons and holes among {010} and {110} crystal facets of $BiVO_4$. Nature Communications 4, 1432.

Lin, L., Hisatomi, T., Chen, S., Takata, T., Domen, K., 2020. Visible-light-driven photocatalytic water splitting: recent progress and challenges. Trends in Chemistry 2, 813–824.

Lin, S., Huang, H., Ma, T., Zhang, Y., 2021. Photocatalytic oxygen evolution from water splitting. Advanced Science 8, 2002458.

Liu, G., Sheng, Y., Ager, J.W., Kraft, M., Xu, R., 2019. Research advances towards large-scale solar hydrogen production from water. EnergyChem 1, 100014.

Ma, X., Shi, Y., Liu, J., Li, X., Cui, X., Tan, S., et al., 2022. Hydrogen-bond network promotes water splitting on the TiO_2 surface. Journal of the American Chemical Society 144, 13565–13573.

Maeda, K., 2011. Photocatalytic water splitting using semiconductor particles: history and recent developments. Journal of Photochemistry and Photobiology C 12, 237–268.

Moridon, S.N.F., Arifin, K., Yunus, R.M., Minggu, L.J., Kassim, M.B., 2022. Photocatalytic water splitting performance of TiO_2 sensitized by metal chalcogenides: a review. Ceramics International 48, 5892–5907.

Mu, L., Zhao, Y., Li, A., Wang, S., Wang, Z., Yang, J., et al., 2016. Enhancing charge separation on high symmetry $SrTiO_3$ exposed with anisotropic facets for photocatalytic water splitting. Energy and Environmental Science 9, 2463–2469.

Ni, M., Leung, M.K.H., Leung, D.Y.C., Sumathy, K., 2007. A review and recent developments in photocatalytic water-splitting using TiO_2 for hydrogen production. Renewable and Sustainable Energy Reviews 11, 401–425.

Nishioka, S., Osterloh, F.E., Wang, X., Mallouk, T.E., Maeda, K., 2023. Photocatalytic water splitting. Nature Reviews Methods Primers 3, 42.

Ohno, T., Bai, L., Hisatomi, T., Maeda, K., Domen, K., 2012. Photocatalytic water splitting using modified GaN:ZnO solid solution under visible light: long-time operation and regeneration of activity. Journal of the American Chemical Society 134, 8254–8259.

Osterloh, F.E., 2017. Photocatalysis versus photosynthesis: a sensitivity analysis of devices for solar energy conversion and chemical transformations. ACS Energy Letters 2, 445–453.

Peiris, S., de Silva, H.B., Ranasinghe, K.N., Bandara, S.V., Perera, I.R., 2021. Recent development and future prospects of TiO_2 photocatalysis. Journal of the Chinese Chemical Society 68, 738–769.

Pinaud, B.A., Benck, J.D., Seitz, L.C., Forman, A.J., Chen, Z., Deutsch, T.G., et al., 2013. Technical and economic feasibility of centralized facilities for solar hydrogen production via photocatalysis and photoelectrochemistry. Energy and Environmental Science 6, 1983–2002.

Rahman, M., Tian, H., Edvinsson, T., 2020. Revisiting the limiting factors for overall water-splitting on organic photocatalysts. Angewandte Chemie 132, 16418–16433.

Rather, R.A., Mehta, A., Lu, Y., Valant, M., Fang, M., Liu, W., 2021. Influence of exposed facets, morphology and hetero-interfaces of $BiVO_4$ on photocatalytic water oxidation: a review. International Journal of Hydrogen Energy 46, 21866–21888.

Rosman, N.N., Yunus, R.M., Shah, N.R.A.M., Shah, R.M., Arifin, K., Minggu, L.J., et al., 2022. An overview of co-catalysts on metal oxides for photocatalytic water splitting. International Journal of Energy Research 46, 11596–11619.

Sahani, S., Malika Tripathi, K., Il Lee, T., Dubal, D.P., Wong, C.-P., Chandra Sharma, Y., et al., 2022. Recent advances in photocatalytic carbon-based materials for enhanced water splitting under visible-light irradiation. Energy Conversion and Management 252, 115133.

Salomao, A.C., dos Santos Araujo, M., dos Santos, H.L.S., Medina, M., Mascaro, L.H., Andrade Junior, M.A.S., 2021. Towards highly efficient chalcopyrite photocathodes for water splitting: the use of cocatalysts beyond Pt. ChemSusChem 14, 4671–4679.

Sasaki, Y., Nemoto, H., Saito, K., Kudo, A., 2009. Solar water splitting using powdered photocatalysts driven by Z-schematic interparticle electron transfer without an electron mediator. Journal of Physical Chemistry C 113, 17536–17542.

Sayama, K., Arakawa, H., 1993. Photocatalytic decomposition of water and photocatalytic reduction of carbon dioxide over zirconia catalyst. The Journal of Physical Chemistry 97, 531–533.

Sayama, K., Arakawa, H., 1997. Effect of carbonate salt addition on the photocatalytic decomposition of liquid water over Pt–TiO_2 catalyst. Journal of the Chemical Society, Faraday Transactions 93, 1647–1654.

Shangguan, W., 2007. Hydrogen evolution from water splitting on nanocomposite photocatalysts. Science and Technology of Advanced Materials 8, 76–81.

Sheng Chua, C., Ansovini, D., Jun Lee, C.J., Ting Teng, Y., Ting Ong, L., Chi, D., et al., 2016. The effect of crystallinity on photocatalytic performance of Co_3O_4 water-splitting cocatalysts. Physical Chemistry Chemical Physics 18, 5172–5178.

Shi, M., Li, G., Li, J., Jin, X., Tao, X., Zeng, B., et al., 2020. Intrinsic facet-dependent reactivity of well-defined BiOBr nanosheets on photocatalytic water splitting. Angewandte Chemie 132, 6652–6657.

Song, X., Wei, G., Sun, J., Peng, C., Yin, J., Zhang, X., et al., 2020. Overall photocatalytic water splitting by an organolead iodide crystalline material. Nature Catalysis 3, 1027–1033.

Tachibana, Y., Vayssieres, L., Durrant, J.R., 2012. Artificial photosynthesis for solar water-splitting. Nature Photonics 6, 511–518.

Takata, T., Domen, K., 2009. Defect engineering of photocatalysts by doping of aliovalent metal cations for efficient water splitting. Journal of Physical Chemistry C 113, 19386–19388.

Takata, T., Jiang, J., Sakata, Y., Nakabayashi, M., Shibata, N., Nandal, V., et al., 2020. Photocatalytic water splitting with a quantum efficiency of almost unity. Nature 581, 411–414.

Tsai, F.-T., Deng, Y.-T., Pao, C.-W., Chen, J.-L., Lee, J.-F., Lai, K.-T., et al., 2020. The HER/OER mechanistic study of an FeCoNi based electrocatalyst for alkaline water splitting. Journal of Materials Chemistry A 8, 9939–9950.

Walter, M.G., Warren, E.L., McKone, J.R., Boettcher, S.W., Mi, Q., Santori, E.A., et al., 2010. Solar water splitting cells. Chemical Reviews 110, 6446–6473.

Wang, M., Zhen, W., Tian, B., Ma, J., Lu, G., 2018. The inhibition of hydrogen and oxygen recombination reaction by halogen atoms on over-all water splitting over Pt-TiO_2 photocatalyst. Applied Catalysis B: Environmental 236, 240–252.

Weber, M.F., Dignam, M.J., 1986. Splitting water with semiconducting photoelectrodes—efficiency considerations. International Journal of Hydrogen Energy 11, 225–232.

Yaseen, S., Tahir, M.B., Wattoo, A.G., 2022. Photocorrosion inhibition of sulphide-based nanomaterials for energy production through photocatalytic water splitting. International Journal of Energy Research 46, 634–666.

Ye, S., Shi, W., Liu, Y., Li, D., Yin, H., Chi, H., et al., 2021. Unassisted photoelectrochemical cell with multimediator modulation for solar water splitting exceeding 4% solar-to-hydrogen efficiency. Journal of the American Chemical Society 143, 12499–12508.

Yoshinaga, T., Saruyama, M., Xiong, A., Ham, Y., Kuang, Y., Niishiro, R., et al., 2018. Boosting photocatalytic overall water splitting by Co doping into Mn_3O_4 nanoparticles as oxygen evolution cocatalysts. Nanoscale 10, 10420–10427.

Yu, S.-C., Huang, C.-W., Liao, C.-H., Wu, J.C.S., Chang, S.-T., Chen, K.-H., 2011. A novel membrane reactor for separating hydrogen and oxygen in photocatalytic water splitting. Journal of Membrane Science 382, 291–299.

Yuan, L., Han, C., Yang, M.-Q., Xu, Y.-J., 2016. Photocatalytic water splitting for solar hydrogen generation: fundamentals and recent advancements. International Reviews in Physical Chemistry 35, 1–36.

Zeng, Y., Zhao, M., Huang, Z., Zhu, W., Zheng, J., Jiang, Q., et al., 2022. Surface reconstruction of water splitting electrocatalysts. Advanced Energy Materials 12, 2201713.

Zhang, D., He, W., Ye, J., Gao, X., Wang, D., Song, J., 2021. Polymeric carbon nitride-derived photocatalysts for water splitting and nitrogen fixation. Small 17, 2005149.

Zhang, G., Wang, X., 2019. Oxysulfide semiconductors for photocatalytic overall water splitting with visible light. Angewandte Chemie International Edition 58, 15580–15582.

Zhang, H., Liu, F., Wu, H., Cao, X., Sun, J., Lei, W., 2017. In situ synthesis of g-C_3N_4/TiO_2 heterostructures with enhanced photocatalytic hydrogen evolution under visible light. RSC Advances 7, 40327–40333.

Zhang, J., Ling, B., He, Y., Zhu, Y., Wang, Z., 2022. Life cycle assessment of three types of hydrogen production methods using solar energy. International Journal of Hydrogen Energy 47, 14158–14168.

Zhang, L., Mohamed, H.H., Dillert, R., Bahnemann, D., 2012. Kinetics and mechanisms of charge transfer processes in photocatalytic systems: a review. Journal of Photochemistry and Photobiology C 13, 263–276.

Zhang, Y., Zhao, J., Wang, H., Xiao, B., Zhang, W., Zhao, X., et al., 2022. Single-atom Cu anchored catalysts for photocatalytic renewable H_2 production with a quantum efficiency of 56%. Nature Communications 13, 58.

Zhao, Y., Ding, C., Zhu, J., Qin, W., Tao, X., Fan, F., et al., 2020. A hydrogen farm strategy for scalable solar hydrogen production with particulate photocatalysts. Angewandte Chemie International Edition 59, 9653–9658.

Zhou, P., Navid, I.A., Ma, Y., Xiao, Y., Wang, P., Ye, Z., et al., 2023. Solar-to-hydrogen efficiency of more than 9% in photocatalytic water splitting. Nature 613, 66–70.

Zhu, J., Fan, F., Chen, R., An, H., Feng, Z., Li, C., 2015. Direct imaging of highly anisotropic photogenerated charge separations on different facets of a single $BiVO_4$ photocatalyst. Angewandte Chemie International Edition 54, 9111–9114.

CHAPTER

6

Photoelectrochemical water splitting

Seema Rani, Soumyadip Sharangi, Subhabrata Das, Mansi Pahuja, and Kaushik Ghosh

Quantum Materials and Devices Unit, Institute of Nano Science and Technology, Mohali, Punjab, India

6.1 Introduction

Sustainable sources of energy, for example, wind energy, hydropower, and solar power are some of the environmentally friendly and economically viable alternatives to coal and petroleum. Among these, solar energy, due to its abundance and inexhaustibility, holds the promise of a sustainable energy-fulfilled world. Every hour, the earth receives as much energy from the sun as will be required by human civilization for an entire year. In recent decades, significant research considerable research endeavors have been focused on converting solar energy into electricity through photovoltaic means. Owing to ongoing enhancements in solar cell efficiencies, the prospect of the conversion of solar energy into chemical fuel is gaining momentum. Solar water splitting, mimicking nature's process of photosynthesis, holds the promise of storing solar energy as chemical fuels, a concept refined over billions of years through natural evolution.

Solar-assisted production of hydrogen through water-splitting techniques can be broadly categorized into three distinctive methods: the photoelectrochemical system (PEC), the photocatalyst system (PC), and the integrated photovoltaic-electrochemical system (PV-EC). Among these, the photocatalytic system (PC) stands out as the simplest and most cost-effective approach with potential scalability for water splitting. However, despite its simplicity and cost-effectiveness, the photocatalytic system has inherent drawbacks that constrain its extensive utilization in water splitting. The main challenge in the photocatalytic system (PC) for hydrogen production is the necessity for immediate separation of generated hydrogen and oxygen, consuming extra energy and reducing overall efficiency. Additionally, the large-scale practical implementation of PC systems is impeded by significant implementation challenges (Alfaifi et al., 2018). In contrast, photovoltaic-photoelectrochemical hydrogen production suffers from reduced efficiency due to the additional wiring and connections required. Consequently, PEC hydrogen production emerges as the preferred choice, thanks to its ability to operate under less severe process conditions. In the PEC system, oxygen and hydrogen evolution reactions (OER and HER) occur at two different electrodes. This division

Photochemical Splitting of Water
DOI: https://doi.org/10.1016/B978-0-443-29064-0.00018-8

enables the easy separation of hydrogen and oxygen, offering a feasible method for commercial hydrogen production. Because the PEC system consists of fewer device components, it demonstrates superior cost-effectiveness compared to complex PV-EC devices.

PEC employs both types of semiconductors, n-type and p-type, as photoelectrode materials. Utilizing an n-type photoelectrode involves the collection of electrons in the photoanode, followed by their transport to the counter electrode to initiate OER. Conversely, with a p-type semiconductor, HER occurs at the semiconductor/liquid interface, transmitting electrons toward the electrolyte. This elegant arrangement enhances the efficiency and practicality of hydrogen production through PEC means.

The essential component of a PEC system is semiconducting photoelectrode, and the ideal material for this photoelectrode must meet multiple criteria simultaneously. These criteria include proficient absorption of light, effective charge separation and transport, and the promotion of oxygen and hydrogen evolution at the surface. In the early exploration of PEC water splitting, first reported in 1968, a semiconducting TiO_2 electrode was illuminated in an aqueous solution, generating electron–hole pairs (Boddy 1968). At TiO_2 surface, water oxidation takes place by the holes, leading to the evolution of oxygen, while the electrons migrate toward the Pt counter electrode. At the counter electrode, these electrons help in water reduction producing hydrogen. Fujishima and Honda (1972) later reported hydrogen production by PEC water splitting via a TiO_2 photocatalyst. Following this discovery, various n-type (ZnO, TiO_2, Fe_2O_3, WO_3) and p-type (NiO, Cu_2O, $ZnRh_2O_4$) semiconductors have been thoroughly researched for their applicability in PEC water splitting (Grimes, Varghese and Ranjan 2008; Augustyński, Alexander and Solarska 2011; Yang and Moon 2019). However, finding a perfect material has proven challenging due to its requirement of being both economical and abundantly available, while also delivering high performance, including chemical stability. Some materials, like $SrTiO_3$, WO_3, $BaTiO_3$, and SnO_2, are stable in aqueous electrolytes but exhibit lower photocatalytic activity due to large band gap structures. Conversely, GaAs, CdSe, and CuO, with smaller band gaps, corrode rapidly when immersed in electrolyte solutions. Intermediate band gap semiconductor materials, such as Fe_2O_3, show low hole mobility, resulting in less quantum efficiency. Overcoming these deficiencies involves various modification methods, including doping, bilayered systems, and nanostructured formation, to achieve the desired properties (Walter et al. 2010; Dias and Mendes 2017; Chang, Kumar and Shen 2021).

This chapter offers an in-depth exploration of PEC water splitting, covering fundamental principles related to hydrogen and oxygen generation through PEC. First, the mechanism and procedures involved in PEC water splitting are introduced, followed by an exploration of semiconductor materials suitable for PEC reactions. Subsequently, factors potentially influencing PEC water splitting have been discussed, and its prospective implications are evaluated.

6.2 Photoelectrochemical water splitting: working principles

6.2.1 Basis of photoelectrochemical water splitting

The fundamental operational mechanism of PEC water splitting depends on transforming light energy into chemical energy in an electrochemical cell involving two (or three) electrodes, with at least one being a semiconductor and having the ability to produce electron–hole pairs upon light absorption. Photons are initially converted into electrical energy upon absorption, which is then used for water electrolysis to separate water into its fundamental components (Dias and Mendes 2017). Hence, the initial stage involves the creation of electron–hole pairs through light absorption:

$$4hv \rightarrow 4e^- + 4h^+; \tag{6.1}$$

In a PEC cell featuring an acidic electrolyte abundant in H^+ ions, the reactions for hydrogen and oxygen evolution can be expressed as follows:

$$4H^+ + 4e^- \rightarrow 2H_2;$$

$$E^o_{H^+/H_2} = +0.00\ V_{NHE} \tag{6.2}$$

$$2H_2O + 4h^+ \rightarrow 4H^+ + O_2;$$

$$E^o_{H_2O/O_2} = -1.23\ V_{NHE} \tag{6.3}$$

In an alkaline environment, the dissociation of water generates hydroxyl anions, leading to the following expressions for the hydrogen and oxygen evolution reactions:

$$4H_2O + 4e^- \rightarrow 2H_2 + OH^-;$$

$$E^o_{H_2O/OH^-} = -0.828\ V_{NHE} \tag{6.4}$$

$$4OH^- + 4h^+ \rightarrow 2H_2O + O_2;$$

$$E^o_{O_2/OH^-} = +0.401\ V_{NHE} \tag{6.5}$$

Hence, the entire reaction of the process can be formulated as follows:

$$2H_2O + 4hv \rightarrow 2H_2 + O_2 \tag{6.6}$$

Therefore the generation of one molecule of H_2 requires the creation of two electron–hole pairs, while oxygen evolution is relatively more complex, as each O_2 molecule necessitates the production of four electron–hole pairs. Hence, the OER, involving a four-electron transfer process combined with the extraction of four protons from water molecules leading to the creation of O-O bond represents a particularly challenging stage in the comprehensive process of water splitting (Dong, Yan and Bi 2023). Under typical circumstances, the Gibbs free energy alteration when a water molecule converts into hydrogen and oxygen equals 237.2 kJ/mol (Ager et al. 2015). The fact that the standard electrodynamic potential has a negative value suggests that the process of water splitting is thermodynamically unfavorable, requiring additional energy as the reaction proceeds uphill. The Nernst equation indicates that a minimum of 1.23 eV of energy is needed for the electrochemical process. Hence, the semiconductor needs to absorb energy exceeding 1.23 eV, to facilitate conversion of water into its corresponding constituents, hydrogen and oxygen. However, practical cells often require a potential exceeding 1.6 V due to additional energy losses, such as overpotential at electrodes and ionic conductivity loss in electrolytes (Smith 2016). Hence, materials with larger band gaps (1.6–2.4 eV) are practical for water splitting, considering the significant kinetic overpotentials required for both half-reactions. Moreover, the band

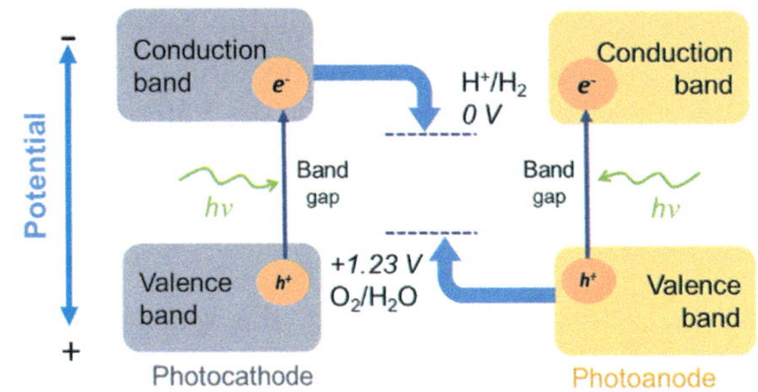

FIGURE 6.1 The fundamental mechanism of water splitting using two semiconductors.

positions of valence band (VB) and conduction band (CB) with respect to redox potentials of water are vital (Chang et al. 2021). The maxima of the VB must lie at a more positive potential than the potential required for water oxidation (1.23 V vs normal hydrogen electrode [NHE]) for a photoanode, while the CB bottom of a photocathode has to be negative enough than the potential required for the reduction of H^+ to H_2 (Fig. 6.1). Therefore the semiconductors utilized for water splitting must have an appropriate band gap for absorbing photons with energy exceeding 1.23 eV, a well-aligned band structure to facilitate OER and HER, and ensure effective separation and transfer of photoexcited carriers for optimal efficiency.

6.2.2 Working mechanism of a photoelectrochemical cell

The most basic configuration of a PEC cell involves an electrolyte solution containing a semiconductor electrode which is photoactive and a counter, where the photoactive electrode may either be a p-type or an n-type semiconductor. Since the photoelectrode and electrolyte have different Fermi levels, charge transfer takes place to achieve thermal equilibrium after submerging of the semiconducting electrode in the electrolyte solution. At the semiconductor—electrolyte interface, band bending occurs to equalize Fermi levels, as the majority charge carriers migrate toward the electrolyte from the semiconductor. Consequently, a space-charge region emerges in semiconductor near interface, depleting it of mobile charge carriers. This region generates an electric field, hindering the further movement of charge carriers. For example, for an n-type semiconductor electrode, electrons transfer from the electrode into the solution because the Fermi level usually resides above the electrolyte's redox potential. The ionized dopants in the semiconductor will provide the electrode with an additional positive charge as a result of equilibration. Conversely, there will be an extra negative charge in the solution (Fig. 6.2A). In contrast, in a p-type semiconductor, a positive charge layer occurs in the electrolyte, and a negative space charge layer forms within the semiconductor when holes migrate into the electrolyte since the semiconductor's initial Fermi level is lower than the electrolyte's. The positive charge disperses across the depletion layer, characterized by a width denoted as "W," whereas the negative charge is confined to a significantly narrower region known as the Helmholtz layer, situated in middle of electrode and electrolyte. This Helmholtz layer, with a width of just a few angstroms, is composed of ion species adsorbed from the electrolyte (Lopes et al. 2014).

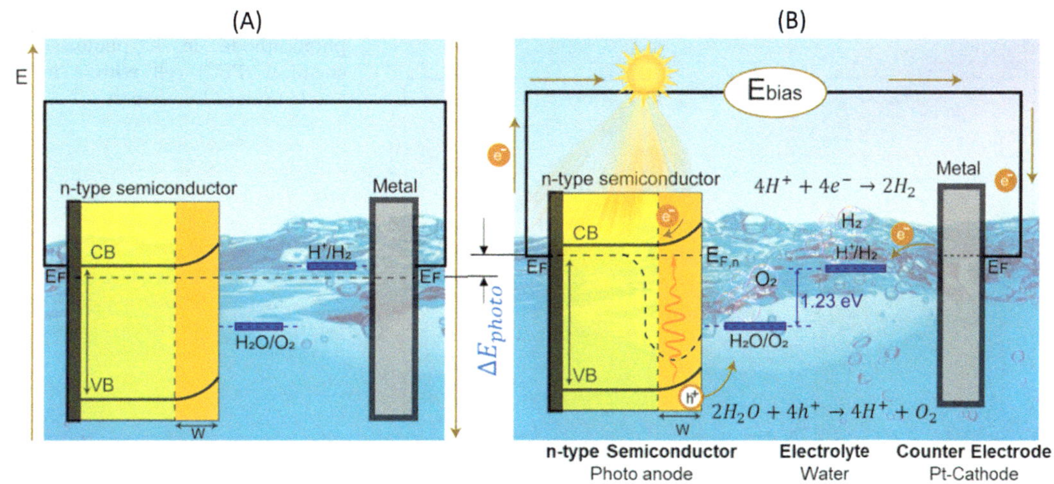

FIGURE 6.2 Band diagram of a photoanode in a photoelectrochemical (PEC) cell with a metal cathode (A) under dark, and (B) under illumination.

When exposed to light, the semiconducting photoelectrode absorbs photons with energies more than its band gap, thus triggering the electron–hole pairs generation as described by Eq. (6.1). The built-in electric field then causes these electrons and holes to move toward or away from the electrode–electrolyte interface. Specifically, for n-type semiconductors (as illustrated in Fig. 6.2B), photogenerated electrons within the space charge region move away from semiconductor/electrolyte interface and holes migrate into the electrolyte. This results in the establishment of a nonequilibrium electron and hole population under illumination, characterized by a quasi-Fermi level. Near the surface of semiconductor, an electric field is generated by the change in quasi-Fermi level, leading to the generation of current and voltage. The experimentally measurable voltage, termed "photovoltage" or "open-circuit voltage (V_{OC})," is evaluated by assessing the potential variance between quasi-Fermi levels of holes and electrons in no current flow conditions (Jiang et al. 2017). At the counter electrode, the electrons aid in the reduction of water to produce hydrogen after leaving semiconducting material's back contact and traveling through the external circuit—either according to Eqs. (6.2) or (6.4), based on the electrolyte medium. In this process, water can either react with the surface holes of the semiconductor (Eq. 6.3), or the holes at the surface of the photoelectrode may react with OH^- ions when they diffuse back (Eq. 6.5), leading to the oxidation of holes and the production of oxygen. An alternative configuration involves a p-type semiconductor acting as a photocathode, as illustrated by Fig. 6.3.

6.2.3 Photoelectrochemical water splitting efficiency analysis

The efficiency of energy conversion is a critical performance measure for PEC cells. The process of PEC water splitting requires four main stages: (1) semiconductor absorption of solar energy, (2) photogeneration of charge carriers, (3) transport of generated carriers to

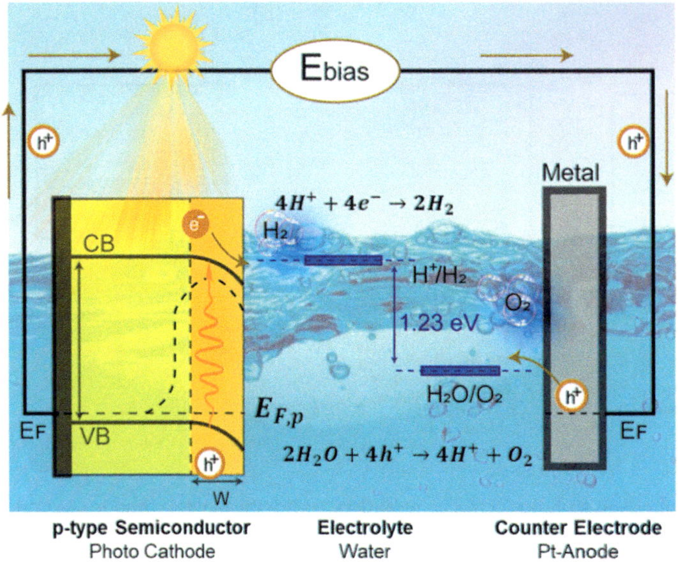

FIGURE 6.3 Band diagram of a photocathode in a photoelectrochemical (PEC) cell with a metal anode under illumination.

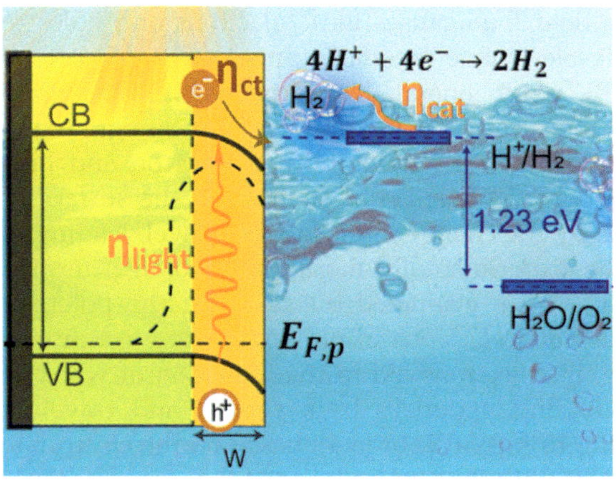

FIGURE 6.4 Charge transfer process in photoelectrochemical (PEC) cell using p-type semiconductor, indicating different efficiencies.

electrode surface, (iv) involvement of charge carriers in catalytic reduction/oxidation reactions (see Fig. 6.4). The ratio of photogenerated charge carriers to incident photons is referred to as light harvesting efficiency (η_{light}). These carriers undergo separation and transport via two mechanisms: drift, influenced by the inherent electric field, and diffusion, due to gradients in the charge carrier population. Nevertheless, not every photogenerated charge carrier manages to reach electrode surface and fraction of collected carriers at the interface is termed charge transfer efficiency (η_{ct}). Furthermore, not all charge carriers involved in catalytic reactions make it to the electrode surface. The ratio of charge carriers participating in catalytic reactions to the number of charge carriers collected at the electrode surface is termed as quantum efficiency of charge transfer from the catalyst to the redox species within electrolyte (η_{cat}).

Hence, external quantum efficiency of PEC system, commonly referred to as solar-to-hydrogen (STH) conversion efficiency, is expressed as:

$$\mathrm{EQE} = \eta_{STH} = \frac{\text{Total Energy Generated}}{\text{Total Input Energy}} = \frac{\text{Number of } H_2 \text{ molecules produced}}{\text{Number of incident photons}} = \eta_{light} \times \eta_{ct} \times \eta_{cat} \tag{6.7}$$

Another way to express the STH efficiency can be formulated as follows:

$$\eta_{STH} = \frac{\text{Chemical Energy Generated}}{\text{Input Solar Energy}} = \frac{\phi_{H2} \times \Delta G^0}{P_{light} \times A} \tag{6.8}$$

where ϕ_{H_2} represents the rate at which hydrogen is produced (measured in mmol $H_2 \cdot s^{-1}$) and ΔG^o denotes the Gibbs free energy change per mole of hydrogen ($\Delta G^o =$ 237 kJ/mol). The parameters P_{light} and A stand for incident light power and surface area of the system, respectively.

The optimal approach to confirm that the generated photocurrent stems from water splitting, as opposed to occurrences like electrode photocorrosion or other side reactions, involves calculating the Faradaic efficiency, given by the formula:(6.9)

$$\begin{aligned}\text{Faradaic Efficiency} &= \frac{\text{Actual Gas Evolved}}{\text{Theoretical Gas Evolved}} \\ &= \frac{\text{Gas Measured}}{\text{Gas Evolved based on the photocurrent}} \\ &= \frac{\text{Gas Measured}}{\left(\frac{J_{ph} \times A \times T}{e} / n\right) / \mathrm{NA}} \times 100\%\end{aligned} \tag{6.9}$$

In this context, J_{ph} represents the photocurrent density (obtained from the photocurrent–voltage characteristics in A/cm^2) produced over a period T (in seconds). Additionally, A represents the area of the electrode which receives light, N_A is Avogadro's constant, e is the charge of an electron, and n represents electrons participated in reaction (i.e., two for HER, four for OER). Actual gas evolved can be measured through techniques such as gas collection or water displacement, where the volume of the collected gas is measured and converted to moles. Alternatively, analytical methods like gas chromatography or specialized sensors such as oxygen electrodes can be employed for accurate quantification of evolved gases. The Faradaic efficiency is usually expressed as a percentage. If the Faradaic efficiency is 100%, it indicates that all the observed gas is due to the desired electrochemical reaction, while values below 100% suggest the presence of side reactions or undesired processes. Since Eq. (6.8) requires measuring the rate of gas production, especially in miniature devices, can pose challenges, whereas assessing the STH efficiency is comparatively straightforward by presuming a Faradaic efficiency of 100%, as follows:

$$\eta_{STH} = \frac{V_{redox} \times J_{ph}}{P_{light}} \tag{6.10}$$

Here, V_{redox} represents the redox potential of water splitting.

Further, when an external bias (V_{bias}) is introduced to the photoelectrode, then externally provided electrical energy needs to be deducted. Consequently, another term is commonly employed called the applied bias photon-to-current efficiency (ABPE), formulated by the expression:

$$\text{ABPE} = \frac{P_{light} - P_{in}}{P_{light}} = \frac{J_{ph}(V_{redox} - V_{bias})}{P_{light}} \tag{6.11}$$

Quantum efficiencies, like incident photon-to-current efficiency (IPCE), are frequently evaluated at particular incident frequencies/wavelengths to understand how well the photoelectrode responds to light at different energy levels. It is characterized as number of charge carriers that contribute to photocurrent for each incident photon, as described by the following equation:

$$\text{IPCE}(\lambda) = \frac{\text{Total energy of converted electrons}}{\text{Total energy of incident photons}} = \frac{\left(\frac{J_{ph}(\lambda)}{e}\right) \times \left(\frac{hc}{\lambda}\right)}{P(\lambda)} \times 100\% \tag{6.12}$$

where $J_{ph}(\lambda)$ represents the photocurrent density resulting from incident light at a specific wavelength; c represents the speed of light; h denotes Planck's constant; $P(\lambda)$ denotes the incident light power density with λ being the incident wavelength.

Optical losses, for example, reflection and transmission, which can significantly affect the precision of IPCE, are frequently disregarded by IPCE evaluations. To remedy this limitation and enhance the precision of quantum efficiency determination, absorbed photon-to-current conversion efficiency (APCE) comes into play. It is employed to correct optical losses, providing a more refined measure of the internal quantum efficiency. The mathematical expression for APCE is as follows:

$$\text{APCE}(\lambda) = \frac{\text{IPCE}(\lambda)}{A(\lambda)} = \frac{\text{IPCE}(\lambda)}{1 - R(\lambda) - T(\lambda)} \times 100\% \tag{6.13}$$

where $R(\lambda)$, $A(\lambda)$, and $T(\lambda)$ denote optical reflection, absorption, and transmission, respectively.

6.3 Materials for photoelectrochemical water splitting

The suitable components for PEC water splitting should satisfy some basic requirements. These requirements include enhanced absorption of visible wavelengths and extended stability in aqueous environment, reduced kinetic overpotentials for redox reactions, and suitable band edge positions for facilitating the oxidation and reduction of water. The semiconductor's effectiveness is further enhanced by inexpensive, earth-abundant materials and efficient charge conduction.

A key parameter shaping the semiconductor's performance is its band gap energy, determining the spectral region in which it absorbs light, with a minimum requirement of 1.23 eV for water splitting. Under ideal conditions and no losses, a semiconductor with a 1.23 eV band gap could achieve a maximum STH conversion efficiency of 47.4% (Weber and Dignam 1984). However, practical considerations introduce energy losses in the solar conversion process due to photon absorption exceeding the band gap of the semiconductor, resulting in quick recombination of electron–hole pairs and subsequent energy release as heat or photons. Additional losses occur during electron transport, resulting in an overall energy loss exceeding 0.8 eV in PEC water-splitting systems, implying that the optimal band gap for the semiconductor should be between 2.0 and 2.4 eV (Grimes 2007; Krol et al. 2008). Higher band gaps contribute to improved chemical stability and resistance against corrosion; however, there is a trade-off as they diminish the absorption of visible light. Thus achieving an optimal balance among band gap, absorption, and stability becomes crucial when designing efficient photoelectrodes for PEC water-splitting systems. Fig. 6.5 demonstrates the band gaps and positions of

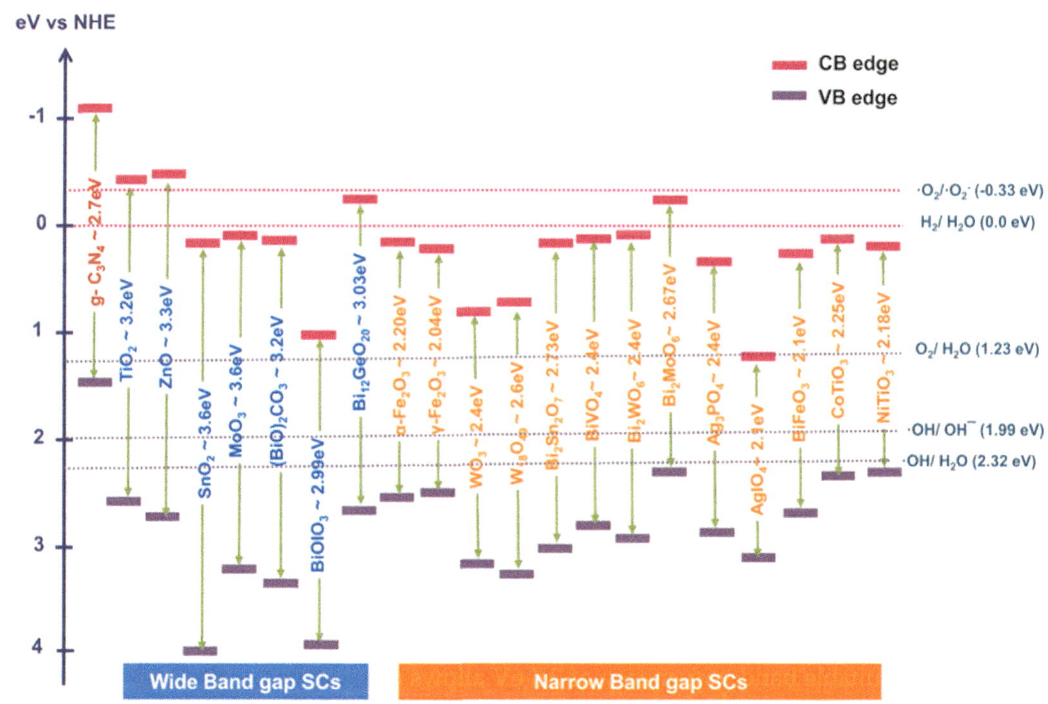

FIGURE 6.5 Band gaps and band alignments of various semiconductors drawn with the redox potentials for water splitting at zero pH (Das et al., 2022). Source: *Adapted with permission from Das, S., Deka, T., Ningthoukhangjam, P., Chowdhury, A., Nair, R.G., 2022. A critical review on prospects and challenges of metal-oxide embedded g-C_3N_4-based direct Z-scheme photocatalysts for water splitting and environmental remediation. Applied Surface Science Advances 11, 100273.*

different semiconductors in relation to the redox potentials of H_2 and O_2 at pH zero. It is important to note that the electrolyte's pH can affect the band locations of oxide semiconductors. This variability occurs because the valence band potential is sensitive to pH, shifting by 0.059 V per pH unit in relation to the electrolyte's redox potential (Jiang et al. 2017). Furthermore, it is essential for the semiconductor to enable effective separation and transportation of charge carriers to mitigate rapid charge recombination, a common cause of low STH efficiencies. Ensuring catalytic activity and stability is vital to prevent concerns such as photocorrosion, especially in materials like metal sulfides. Additionally, the material should be economically viable and comprised of elements abundant in the earth to ensure practical and scalable applications.

Because of the strict requirements mentioned earlier, not one single semiconductor material has been identified as fulfilling the criteria of unbiased water splitting, achieving high STH conversion efficiency, and ensuring long-term stability. The success of PEC water-splitting technology relies on finding a solution to this challenge. In recent decades, several metal oxides (like ZnO, TiO_2, α-Fe_2O_3, Cu_2O, $BiVO_4$, WO_3) and nonoxides semiconductors (including silicon, CdS, GaP, TaON, InP, Ta_3N_5, GaAs) have been widely investigated (Grimes et al. 2008; Osterloh 2008). A common choice is metal oxides because they are stable in water and not too expensive. However, they typically exhibit suboptimal semiconducting characteristics, like carrier mobility, in comparison to other conventional semiconductors, such as III–V semiconductors and silicon. Overcoming these limitations while leveraging the beneficial properties of metal oxides remains a significant challenge.

6.3.1 Photocathodes

In PEC, hydrogen evolution occurs at the cathode, necessitating a higher cathodic current for the conversion of protons to hydrogen. Photocathodes are basically p-type semiconductors having holes as majority charge carriers. They should be photoelectrochemically stable for a very long time during the whole reaction where water is converted into H_2 and O_2. While single-crystal phosphides are commonly examined, polycrystalline materials are often more practical for large-scale applications, given their enhanced stability, as seen in polycrystalline Si. Although polycrystalline Si demonstrates high stability in an acidic environment, its durability is constrained by surface oxidation. To counteract oxide formation, methyl groups are typically employed as a coating material on the electrode surface. SiC has also emerged as an intriguing photocathode with p-type semiconductor behavior, exhibiting the unique ability to store hydrogen gas within the solid material without compromising its lattice structure, attributed to its exceptional chemical stability (Dorp et al. 2009).

Among extensively studied metal oxides, Cu_2O has found great utility as both a photocathode and photoanode, particularly standing out as an appealing p-type oxide for PEC HER. Its suitable band gap energy of 2.4 eV allows efficient harvesting of visible light. With a well-aligned conduction band (−0.7 V vs reversible hydrogen electrode [RHE]) and valence band (1.4 V vs RHE) for water splitting, it can achieve a high photocurrent density and an STH efficiency of approximately 18%. This material, derived from abundant, cost-effective sources, is environmentally friendly. However, its stability is compromised by easy reduction to Cu in an aqueous electrolyte by the presence of Cu+ ions within its band gap, owing to their redox potential. To address this issue, protective layers are often deposited on the Cu_2O material (Jang and Lee 2019). Calcium ferrite ($CaFe_2O_4$) stands out among p-type oxides for its impressive PEC stability, requiring no additional protective layers. With a band gap energy of 1.9–2.0 eV, its conduction and valence band positions are strategically aligned to facilitate efficient water splitting when exposed to visible light (Wheeler and Choi 2018). Recent studies focus on $CuFeO_2$ and $CuRhO_2$ as photocathode materials, with $CuFeO_2$ preferred for its widely available constituents, excellent absorption properties, high conductivity, and lower band gap (1.5 eV) (Dias and Mendes 2017). Lanthanum iron oxide ($LaFeO_3$), which has a perovskite-type structure is also a p-type semiconducting material having a lower band gap energy (2.1 eV), shows promise as a photocathode, capable of generating both H_2 and O_2 under visible light (Wheeler and Choi 2017). Further, copper bismuth oxide ($CuBi_2O_4$), featuring a spinel molecule with AB_2O_4 moiety, presents another compelling photocathode material with a band gap of 1.6–1.8 eV. Its p-type semiconductor behavior, driven by Cu vacancies, leads to high internal photovoltage and moderate to high stability in an aqueous medium (Yang and Moon 2019; Seo et al. 2021).

It is worth mentioning that all these semiconductors are low-cost, but none fulfill all ideal requirements for PEC water splitting. Regarding this, Sb_2Se_3 emerges as a nearly perfect semiconductor for PEC water splitting because of its narrow band gap, cost-effectiveness, and outstanding optoelectronic properties. The Sb-Se system's thermodynamic properties prevent undesirable secondary phases, a common issue in other multivalent compounds. Moreover, Sb_2Se_3 exhibits intrinsic stability against photocorrosion in strong acidic media. Although possessing these advantages, the photocurrent density of Sb_2Se_3 photocathodes, while steadily improving, still falls below the theoretical maximum. Enhancements in photovoltage and fill factor are crucial for realizing unassisted water splitting (Yang et al. 2020).

6.3.2 Photoanodes

In the realm of PEC, photoanodes fundamentally embody n-type semiconductors. Within the conduction band of these semiconductors, a higher concentration of electrons (e^-) is present. The photoanode facilitates the evolution of O_2. The junction between the semiconductor and the electrolyte is where the oxidation of water to O_2 occurs. An optimal photoanode material should possess a relatively lower band gap energy, enabling efficient harnessing of a substantial fraction of visible light. Additionally, it should feature a positive valence band edge, providing the necessary overpotential for facilitating the water oxidation process (Bhatt and Lee 2015).

Ever since its identification by Fujishima and Honda (1972), titanium dioxide (TiO_2) has garnered interest for its effective photocatalytic water splitting and robust photooxidation capabilities. However, because of its notably high band gap energy (3.2 eV), it can exclusively absorb light within the UV region, resulting in a low overall STH production efficiency (2.2%) and suboptimal PEC performance. To address this limitation, the addition of certain cations, such as Ba^{2+} and Sr^{2+}, into the TiO_2 crystal is employed to shift the conduction band energy to more negative values (Walter et al. 2010). Bismuth vanadate ($BiVO_4$) is another well-known material for photoanodes due to its elevated theoretical photocurrent density and an expanded range of photo response wavelengths. With a band gap of 2.4 eV, it is among some of the exceptional visible light absorbers. However, its efficiency is hindered by limited carrier separation and transfer, impacting the oxygen evolution process and overall PEC cell activity. To enhance its water-splitting activity, modified $BiVO_4$ variants are developed through doping, cocatalyst loading, substrate modification, etc. These processes play a pivotal role in enhancing carrier separation, transfer efficiency, and stability by mitigating photocorrosion, thereby elevating the overall PEC activity (Shi et al. 2022).

Tungsten trioxide (WO_3) is another material which presents itself as a compelling material with a relatively lower band gap energy (2.6–2.8 eV), allowing absorption of both, near UV region light as well as a portion of the visible light spectrum, particularly in the blue range. Exhibiting stability in acidic environments ($pH < 3$) over an extended duration, its STH efficiency stands at approximately 4.8%. To enhance both photocurrent density and stability, nano-crystallized WO_3 is employed. Due to its highly crystalline structure, it can inhibit carrier recombination by minimizing charge trapping. Consequently, using a WO_3 thin film as a photoanode attains a noteworthy photocurrent density (Dias and Mendes 2017). α-Fe_2O_3 has also appeared as a valuable material as a photoanode for PEC applications due to its relatively lower band gap energy of approximately 2.0–2.2 eV. To add to this, it also possesses qualities of being nontoxic, cost-effective, environmentally friendly, and readily available. However, its conduction band is positioned at a potential that is overly positive concerning the range where hydrogen evolution occurs. To address this issue, the addition of certain tetravalent or pentavalent metal ions, such as Ti^{4+}, Si^{4+}, Ta^{5+}, and Nb^{5+}, is employed. This augmentation enhances the number of charge carriers in the material, subsequently increasing conductivity and achieving high photocurrent density with STH efficiency being around 14%–17% (Augustyński et al. 2011).

Copper vanadate ($Cu_3V_2O_8$), a ternary oxide of the n-type with a band gap energy of approximately 2.0 eV, proves to be a suitable material for photoanode. It exhibits stability in a neutral to slightly alkaline solution ($pH < 9.2$) and the positioning of the valence band maxima is conducive to water oxidation. The introduction of Mo-doping in $Cu_3V_2O_8$ allows for the tuning of its electronic properties, leading to increased O_2 production under photolytic conditions and an overall enhancement in photocatalytic activity (Seabold and Neale 2015). Zinc oxide (ZnO) also presents a multitude of appealing characteristics for serving as a photoanode material, including exceptional stability, elevated electron mobility, and a substantial binding energy (approximately 60 meV), at room temperature, and an environmentally benign nature (Guo et al. 2018).

Tables 6.1 and 6.2 provide a performance comparison of the various photocathode and photoanode materials.

6.4 Role of electrolyte

In PEC systems, the electrolyte assumes a versatile role, shaping the dynamics of the process at various levels. It serves as a vital conduction medium, establishing a pathway for ions to traverse between the photoanode and photocathode. This flow of ions is crucial, as it helps complete the electrical circuit and enables the coordinated movement of electrons and ions during the redox reactions involved in water splitting. Beyond its role as a mere

TABLE 6.1 A comparative analysis of performances of various photocathode materials.

Photocathode material	Band gap	Electrolyte	Performance	Stability	Advantages and challenges	References
$Si/In_2O_3/CuO/ZnO$	1.1–2.5 eV	0.5 M H_2SO_4	Photocurrent = 120 mA/cm^2 @ 0 V_{RHE}	Stable	Facile charge transfer; integration with $BiVO_4$	Abdullah et al. (2024)
$InAlN/Cu_2O$	1.6–2.1 eV	1.0 M Na_2SO_4	Photocurrent = 2.54 mA/cm^2 @ 0 V_{RHE}	Good stability	Enhanced charge separation	Zeng et al. (2024)
$SrTiO_3/Si$	3.2 eV	0.1 M KOH	Overpotential of 0.326 V_{RHE} @ −0.1 mA/cm^2	Enhanced with STO layer	Epitaxial protection; corrosion resistance	Ho, Smiljanić, et al. (2023)
$Rh/BaTiO_3$	3.2 eV (modified)	NaOH	0.1% STH	100 h	Visible light absorption; onset potential	Shi et al. (2023)
$CuBi_2O_4$	1.5 eV	0.5 M Na_2SO_4	Photocurrent = 3.0 mA/cm^2 @ 0 V_{RHE}	Moderate	High onset potential; nanostructure control	Meena et al. (2023)
$g\text{-}C_3N_4/Cu_2O$	2.0 eV	1.0 M Na_2SO_4	Photocurrent = 2.33 mA/cm^2@ 0 V_{RHE}	Moderate	Improved light harvesting; carrier transport	Bae et al. (2024)
$Cu{:}NiO_x/Sb_2Se_3/TiO_2/Pt$	1.4–1.7 eV	1.0 M H_2SO_4	Photocurrent = 0.56 mA/cm^2 @ 0 V_{RHE}	High	Good interfacial charge transfer	Zhan, Chen and Long (2024)
$SrTiO_3/rGO/Si$	3.2 eV	0.5 M H_2SO_4	Onset potential = 0.326 V_{RHE} @ −0.1 mA/cm^2	Long-term stability	Easy charge separation; passivation layer	Ho, Smiljanic, et al. (2023)
$CuInS_2$	1.45 eV	0.5 M Na_2SO_4	Photocurrent = 18 mA/cm^2	Good	Effective hydrogen production; efficiency boost	Gunawan et al. (2023)
$MoSe_2/SiNWs$	1.5 eV	1.0 M KOH	Photocurrent = 14.56 mA/cm^2 @ −1.2 V_{RHE}	High	Improved charge transfer	(Sharma et al., 2023)
$Bi_2S_3\text{-}Cu_3BiS_3$	1.2–1.5 eV	0.1 M H_2SO_4	Photocurrent = 7.8 mA/cm^2 @ 0 V_{RHE}	21 h	Enhanced light scattering; stability issues	Moon et al. (2023)
Graphene/p-Si	1.1 eV	1.0 M Na_2SO_4	Photocurrent density = 3.9×achieved without graphene	Enhanced with graphene	Decreased interfacial barrier	Zhu et al. (2023)
$Ag_2S/Au/Al_2O_3$	1.1 eV	0.1 M KOH	Photocurrent = 2.50 mA/cm^2 @ 0.95 V Ag/AgCl	Improved	Plasmonic enhancement; structural challenges	Yadav and Singh (2024)

TABLE 6.2 A comparative analysis of performances of various photoanode materials.

Photoanode material	Band gap	Electrolyte	Performance	Stability	Advantage and challenges	References
$CoTiO_3$@g-C_3N_4	2.2 eV	0.1 M Na_2SO_4	Photocurrent = 55.51 mA/cm^2 @ 1.23 V_{RHE}	Good	Visible light driven, charge separation; needs better charge transport	Sundararaj et al. (2024)
CdS/FST	2.42 eV	0.5 M Na_2SO_4	Photocurrent = 24.5 mA/cm^2@ 1.23 V_{RHE}	Stable	High photocurrent, good charge separation; needs better long-term stability	Sawal et al. (2024)
Pyridine-anchored TiO_2	3.2 eV	0.1 M $HClO_4$	Photocurrent = 200 μA/cm^2	Moderate	Enhanced electron transfer; moderate stability	Tang et al. (2024)
Hematite	2.1 eV	0.5 M NaOH	Photocurrent = 5.1 mA/cm^2 @ 1.23 V_{RHE}	Enhanced by doping	Nanostructure for charge separation; short hole diffusion length	Park et al. (2024)
NiFeO/$BiVO_4$	2.4 eV	0.1 M Na_2SO_4	Photocurrent = 5.87 mA/cm^2 @ 1.23 V_{RHE}	High stability	Efficient water oxidation; requires better interface design	Wang et al. (2024)
Co-MOF/MXene/$BiVO_4$	2.4 eV	0.5 M NaOH	Photocurrent = 4.26 mA/cm^2 @ 1.23 V_{RHE}	Improved with MXene	Improved charge separation with MXene; recombination issues	Zhong et al. (2024)
GaN-nanowire	3.4 eV	0.1 M KOH	Photocurrent = 2.2 mA/cm^2 @ 0.6 V_{RHE}	High stability	High photocurrent density; needs better carrier transport	Noh et al. (2024)
Molecular TiO_2	3.2 eV	0.5 M H_2SO_4	94% Faradaic efficiency	Long-term stability	Stable anchoring of molecular components; loss of photoactivity over time	Yan, Sakai and Ozawa (2023)
$BiVO_4$/Mo-$BiVO_4$/CoPi	2.4 eV	0.5 M KOH	Photocurrent = 2.7 mA/cm^2 @ 1.23 V_{RHE}	Improved	Enhanced light absorption; needs better electron–hole recombination control	Wang et al. (2024) Wang et al. (2023)
$SrTiO_3$-TiO_2	3.2 eV	0.1 M $HClO_4$	Photocurrent = 1.95 mA/cm^2 @ 1.5 V Ag/AgCl	88.5 h	Durable structure; surface roughness increase over time	Kolodziejak et al. (2023)
N,S-GQDs/$MoSe_2$	1.8 eV	0.1 M NaOH	–	High photostability	Enhanced carrier separation; photodegradation issues	Zhang et al. (2024)
Sn-Fe_2O_3	2.0 eV	0.5 M NaOH	Photocurrent = 1.40 mA/cm^2@ 1.23 V_{RHE}	Improved with CoOx	Enhanced roughness and charge transfer; electron–hole recombination	Huang, Zhao and Li (2024)
Gamma-ray enhanced TiO_2	3.18 eV	0.1 M Na_2SO_4	Photocurrent = 100.12 μA/cm^2 @ 1.23 V_{NHE}	Enhanced by gamma irradiation	Increased crystallinity and stability; needs better charge transport	Aida Mohamed et al. (2024)

conductor, the electrolyte actively participates in enhancing charge transfer kinetics. A thoughtful selection of the electrolyte contributes to improved efficiency by minimizing losses associated with undesired phenomena like recombination. The electrolyte acts as a critical determinant in preventing surface recombination and trapping of charge carriers, thereby enhancing stability of the PEC system over time. At its core, the key factor is choosing the right redox pairs in the electrolyte. These pairs, engaging in reduction and oxidation reactions, make it easy for charge carriers to move between the photoelectrode and the electrolyte. The careful selection of redox pairs significantly affects how well the PEC system works and how long it lasts. The pH compatibility emerges as another crucial aspect, with the pH of the electrolyte closely linked to optimal functioning of photoelectrode material, as it is crucial in influencing redox potentials and reaction rates, emphasizing the importance of matching it with the unique traits of the photoelectrode. Additionally, the concentration of the electrolyte solution is important. Higher concentrations, usually above 0.5 M, are essential in reducing ohmic potential losses, thereby decreasing resistance and improving the effectiveness of the PEC system. Stability is a fundamental aspect of sustainable PEC performance and is further supported by the electrolyte. The electrolyte plays a crucial role in preventing problems such as photocorrosion, ensuring the long-lasting functionality of the PEC cell over extended periods (Dias and Mendes 2017).

6.5 Advancements in photoelectrochemical water splitting

6.5.1 Strategies to enhance photoelectrochemical efficiency

6.5.1.1 Band structure engineering

The modification of a photoelectrode's band structure via the introduction of elements is a prevalent approach to enhance PEC efficiency. Doping semiconductor materials involve introducing foreign ions (cations or anions) to modify the band structure, simultaneously enhancing visible light absorption and charge carrier mobility. For instance, when dealing with a photoelectrode characterized by a wide band gap, like TiO_2, adjusting its band structure via element doping is frequently employed to reduce the band gap. This redshifts the absorption edge, leading to an elevation in PEC efficiency. On the other hand, visible spectrum absorption coefficients in broad band gap photoelectrodes that have undergone element doping are frequently low. In addition, the introduction of elements can lead to deep-level defects and crystal defects, which could increase the chance of photoinduced electron–hole pair recombination and decrease IPCE in the UV spectrum. Consequently, despite the redshift in the action spectra due to band gap narrowing, only marginal enhancements in STH conversion might be attained for wide band gap photoelectrodes.

Wide band gap photoelectrodes can occasionally have their charge balanced by codoping with two nonisovalent elements in order to overcome these difficulties. An example of this may be seen in the synergistic effect that is obtained by the cotreatment of TiO_2 nanowire arrays with hydrogenation and nitridation (codoping with H^+ and N^{3-}), which boosts water oxidation under visible light. To significantly enhance the STH efficiency of a PEC cell, it is crucial to select photoelectrode materials with low intrinsic band gaps. To enhance the concentration of carriers and the diffusion length of minority carriers, photoelectrodes with small band gaps, such as Fe_2O_3 and $BiVO_4$, are frequently doped with nonisovalent elements to modify their band structures. Notably, reducing the effective mass of minority carriers proves beneficial in enhancing the diffusion length of minority carriers, thereby improving photocurrent density (Li et al. 2013).

6.5.1.2 *Surface treatment of the semiconductor*

Surface treatment is emphasized as a productive way to increase PEC efficiency. It serves a number of purposes, such as protecting against photocorrosion and creating surface and interface layers to passivate, enhancing charge collection with textured conductive substrates, adjusting band edge positions for better photovoltage, and selectively removing carriers with higher catalytic activity. By addressing segregation close to the surface or surface states that serve as hubs for photoinduced carrier recombination, surface passivation is helpful. PEC characteristics may be impacted by segregation, which is the enrichment of a material ingredient. For instance, electrochemical surface treatment effectively removes indium-rich segregation in an $In_{0.20}Ga_{0.80}N$ photoelectrode, resulting in a threefold increase in IPCE at 400 nm and improved stability under long-time illumination for practical solar hydrogen production (Li et al., 2013). By applying an analogous technique, the IPCE of Mo-doped $BiVO_4$ photoanodes is increased through electrochemical surface preparation, which significantly increases photocurrent under visible light illumination (Pattengale and Huang 2016). Enhancing photoelectrode stability is another criterion that involves adding protective overlayers that function as physical separators between the electrolyte and the semiconductor. These overlayers should be sufficiently thick, semitransparent, and conductive to avert constraints on light absorption efficiency or charge transfer. The third criterion aims to increase charge carrier collection efficiency by adding conductive scaffolds to the photoelectrode structure, such as nano-textured transparent conducting oxide and carbonaceous materials (Dias and Mendes 2017).

6.5.1.3 *Morphology control of the photoelectrode*

Morphology control, particularly in porous or nanostructured photoelectrodes like nanorods, nanotubes, and nanowires, is essential. Nanostructured photoelectrodes offer advantages such as reduced transfer distances for minority carriers and increment of the surface reaction sites, aiding in lowering carrier recombination and improving PEC performance. Promising prospects for designing high-performance photoelectrodes lie in three-dimensional nanowire networks, particularly those characterized by extended optical paths, one-dimensional conducting channels, and substantial surface areas. Additionally, for a highly efficient photoelectrode, the film must be of sufficient thickness for increased light absorption, and photogenerated charge carriers generated deep inside the electrode should reach the surface for effective collection. The collection of charge carriers depends upon many factors such as. the penetration depth of light, the absorption coefficient, minority carrier diffusion length, and the extent of the space-charge region (Li et al. 2013).

6.5.1.4 Alternative approaches

Other techniques involve the use of heterojunction photoelectrodes, typically composed of semiconductors with a small band gap covered by one with a larger band gap. These structures aim to enhance not just the IPCE through improved light absorption but also improve photochemical stability by preventing direct interaction between narrow band gap semiconductors and electrolytes. The attributes of heterojunction photoelectrodes are chiefly established by the outer layer. For instance, the deposition of an In_2O_3 overlayer on $LaTiO_2N$ photoanodes can significantly increase photocurrents by two to three times, credited to enhanced separation of charges, reduced instances of charge recombination, and/or improved extraction of charges facilitated by a surface dipole (Leroy et al. 2011). Heterojunction photoelectrodes, involving p- and n-type hematite semiconductors, have also been reported favoring PEC water-splitting reactions.

6.5.2 Silicon-based electrodes for PEC water splitting: a case study

Silicon (Si) holds great promise for revolutionizing PEC water splitting, a process crucial for sustainable hydrogen production and solar energy conversion. Despite its inherent advantages such as excellent light-harvesting properties and superior charge-transport capabilities, several challenges have impeded its widespread application in PEC systems. In response, researchers have proposed and developed various innovative approaches to improve the performance of Si-based electrodes. Here, we delve into some of these comprehensive advancements.

The advancement of Si photocathodes for PEC water splitting confronts several key hurdles: limited light absorption by untreated Si substrates, instability in acidic environments, and insufficient photocatalytic efficiency. To tackle these challenges, some innovative approaches have been proposed. One of these involves the creation of black Si through reactive ion etching, followed by the deposition of TiO_2 onto black Si using atomic layer deposition, and the subsequent growth of ReS_2 on the TiO_2 layer via chemical vapor deposition. The resulting structure demonstrates significantly enhanced PEC performance (Cheng et al. 2022). Amorphous silicon (a-Si) is widely utilized in the fabrication of solar cells because of its outstanding ability to capture light effectively and its straightforward and inexpensive preparation process. Yet, the application of a-Si as photoelectrodes for water splitting has remained unexplored because of its insufficient photovoltage. In an investigation, a group of scientists developed single-junction photocathodes and photoanodes based on amorphous silicon, followed by the construction of dual-photoelectrode setups for unaided solar water electrolysis. By employing p/i/n and n/i/p junction a–Si as photoabsorbers, significant advancements were observed. The assembled biphotocathode system demonstrated an STH efficiency of 0.61%, highlighting the potential of a-Si in facilitating unaided overall water splitting (Zhang et al. 2020). Further, the shallow valence band edge restricts the PEC performance of silicon. Usually, the metal oxides are combined with silicon to enhance the overall efficiency. However, this frequently impedes carrier transport within the electrode because of deep valence band edge of metal oxides. As an alternative approach, researchers integrated a ferroelectric poly(vinylidenefluoride-trifluoroethylene) layer with silicon increasing both the saturated current density and photovoltage.

Leveraging distinctive ferroelectric properties of the employed material, they could finely adjust the Schottky barrier between silicon and electrolyte by controlling the orientation of the ferroelectric domains. This integration led to a remarkable improvement, raising the photovoltage, while simultaneously boosting the current density (Cui et al. 2015). In another study, a simple technique was used to create a nanoporous 6H-silicon carbide photoelectrode coated with Ni-FeOOH nanorods acting as a cocatalyst for water oxidation. The porous nanostructure of the photoanode demonstrates significantly increased photocurrent density with a zero turn-on potential (Li et al. 2020). Moreover, another team established a PEC water-splitting system based on Si NW arrays with directly deposited 2D-NbS_2 as the photocathode system. This resulting 2D-NbS_2-Si NWs HER catalyst exhibits enhanced kinetics at the Si-NbS_2 interfaces, leading to an improved onset potential. These findings highlight the synergistic advantages of various electrocatalysts and nanowire architectures in improving the surface kinetics of Si for PEC water splitting (Gnanasekar et al. 2019). Thus a semiconductor photoelectrode with surface nanostructures shows great potential for PEC due to its extensive active surface area, efficient light absorption, and minimized charge transport length.

Additionally, to address the stability issues, transition metal/metal oxides are also employed to passivate the silicon surface and improve the catalytic properties. A study addressed the challenge of fabricating stable photoelectrodes for solar water electrolysis devices based on silicon by advancing the creation of a photoanode based on CuO/silicon with a remarkably lower onset potential for water oxidation. The photoanode utilizes a CuO/Cu layer, serving as a protective and hole-conducting layer, as well as an electrocatalyst to reduce overpotential. The CuO-based coating offers a cost-effective and low-toxicity solution for enhancing the performance of unstable materials in solar energy conversion (Shi et al. 2016). In another study, electrodeposition, a cost-effective method, is utilized to fabricate effective catalysts based on Co for water oxidation on silicon substrates. Various morphologies of Co-based catalysts are attained through adjustments in precursor solution additives. The performance of resultant silicon photoanodes, featuring either amorphous CoO_x nanowalls or Co nanoparticles, varies with catalyst thickness and coverage. The completely coated CoO_x NWs/n-Si photoanode displays superior stability compared to partially coated Co NPs/n-Si. These findings indicate that transition-metal-based catalysts are a promising strategy for developing high-performance Si-based photoanodes through surface passivation (Lee et al. 2020).

6.6 Case Study

The following section presents a case study focused on developing a cost-effective, metal-free, and environmentally friendly photocathode for efficient solar-to-hydrogen conversion, with improved PEC-HER kinetics. The objective was to address the challenges of traditional Si/metal-based systems, including high production costs, inadequate light-trapping efficiency, and instability under harsh conditions. A comprehensive report of the case study is provided below; however, more details are available in a previous publication (Riyajuddin et al., 2021).

6.6.1 Title

Development of a Scalable, Metal-Free Photocathode for Efficient PEC-HER-Based Solar-to-Hydrogen Conversion

6.6.2 Objective

The objective of this case study was to develop a cost-effective, metal-free, and environmentally friendly photocathode for efficient solar-to-hydrogen conversion through enhanced PEC-HER kinetics. Specifically, the study aimed to address the limitations of traditional silicon- and metal-based systems, such as high costs, poor light-trapping capabilities, and instability in harsh conditions, by designing a more stable, scalable, and efficient alternative.

6.6.3 Scope

This project sought to establish a low-cost and scalable method for fabricating a durable photocathode. The study primarily focused on improving the PEC-HER performance of Si photocathodes by utilizing passivation layers and metal-free catalysts. Limitations included access to specific materials like tantalum pentoxide (Ta_2O_5) and N-doped graphene quantum dots (NGQD), budget constraints for scalability, and time limitations for long-term durability testing.

6.6.4 Audience

This case study will be useful for researchers and engineers working in renewable energy, particularly in solar-to-hydrogen conversion technologies, as well as those focused on material science and photoelectrochemical systems. It can also inform policymakers and investors interested in the commercialization of green energy technologies. The primary audience includes academics, industry professionals, and students engaged in PEC systems and hydrogen fuel technologies.

6.6.5 Rationale

The study aimed to overcome key challenges faced by conventional Si/metal systems in PEC-HER applications, such as instability, high production costs, and sluggish kinetics. The primary issue being addressed was the creation of an effective, scalable, and durable photocathode that could enhance the overall efficiency of solar-to-hydrogen conversion. Silicon's dominance in the photovoltaic industry provided a foundation, but the study explored an alternate pathway to expand its applicability to PEC-HER, focusing on a metal-free solution to reduce costs and improve environmental sustainability.

6.6.6 Expected Results and Deliverables

From this case study, readers will learn how to design and fabricate a high-performance, metal-free photocathode for PEC-HER, with a specific focus on enhancing stability and catalytic efficiency. The key deliverables include data on the performance metrics of the photocathode, such as ABPE, overpotential, Tafel slope, and durability under operational conditions.

6.6.7 Safety Considerations

Key safety considerations include handling of hazardous chemicals like strong acids, ensuring proper laboratory protocols for high-voltage electrochemical systems, and maintaining appropriate disposal methods for materials used during fabrication. Readers must also take care when working with semiconductor materials and catalysts, following proper handling and storage procedures.

6.6.8 Actions Taken/Workflow/Tools Used/Simulations and Analyses

The workflow for this case study involved several key steps:

1. Fabrication of vertically aligned p-Silicon Nanowires (p-SiNWs).
2. Conformal coating of the nanowires with a passivating layer of Ta_2O_5 to reduce surface instability.
3. Integration of NGQD as a metal-free electrocatalyst on the surface.
4. Testing the photoelectrochemical properties of the p-SiNWs-Ta_2O_5-NGQD photocathode, including its ABPE, overpotential, and Tafel slope measurements.
5. Analyzing the charge transfer process, surface passivation, and durability of the photocathode under operational conditions.

6.6.9 Challenges and Solutions

Key challenges included the instability of bare Si in PEC-HER, high reflectance loss, and surface oxidation. The introduction of a Ta_2O_5 passivation layer solved the instability issue by reducing lattice mismatch and improving charge transfer from Si to the electrolyte. The NGQD layer, with its high density of active sites and non-corrosive nature, addressed the need for a durable and efficient catalyst, significantly enhancing the overall performance of the system.

6.6.10 Results

The p-SiNWs-Ta_2O_5-NGQD photocathode achieved an ABPE of ~21.1%, which is 6 times higher than the p-SiNWs-NGQD matrix, with a low overpotential of -449 mV at 5 mA/cm^2 and a Tafel slope of 78 mV/dec. The photocathode demonstrated long-term durability, maintaining performance for at least 10 hours at 10 mA/cm^2 with minimal potential loss, confirming the effectiveness of the design.

6.6.11 Learning and Knowledge Outcomes

The study demonstrated that passivating the Si surface with Ta_2O_5 and using NGQD as a metal-free catalyst can significantly improve PEC-HER efficiency and durability. Future investigations should focus on optimizing the fabrication process for larger-scale applications and exploring other metal-free catalysts. The results confirmed the potential for scalable, environmentally friendly photocathodes in hydrogen fuel production, offering a viable pathway toward the commercialization of Si-based, metal-free PEC systems.

6.6 Conclusion

PEC water splitting holds immense promise as an environmentally conscious and sustainable method for the production of H_2, leveraging solar energy to facilitate the electrolysis of water. Despite existing challenges and limitations, ongoing research efforts continue to advance the efficiency, stability, and scalability of PEC systems. While metal oxides exhibit proficiency as cathode materials due to their stability and resistance to photocorrosion, they face challenges such as a large band gap, limited absorption coefficient, and inadequate electrical conductivity. These issues can be addressed through nanostructuring techniques, like the creation of nanowires, nanotubes, or nanopores, which can effectively increase carrier mobility, offering potential enhancements in efficiency and stability. Photocathodes, however, currently exhibit low photocurrent performance, prompting the need for the development of photocathodes with improved photocurrent performance at positive potentials to create highly efficient electrochemical cells.

6.7 Future outlook

The PEC field looks forward to the discovery of novel materials, with current studies predominantly focusing on materials containing one to two metal cations. This exploration opens avenues for identifying new materials, composites, and methodologies to achieve efficient and stable water splitting. Despite significant progress, there is considerable room for investigation, offering the potential for transformative technologies independent of current photovoltaic market influences. Creating a system that strikes a compromise between performance, cost, and complexity within a self-contained apparatus is the main objective of PEC research. Achieving commercial viability in PEC water splitting necessitates meeting ambitious technological and economic milestones which include achieving an STH conversion efficiency threshold of ~10%, ensuring a lifespan of at least 10 years, and maintaining a hydrogen threshold cost between $2 and $4 per kilogram. The semiconductor material's role in harvesting solar light is pivotal for determining the electrochemical cell's efficiency. Recent research has identified three parallel pathways for future exploration: enhancing the stability of efficient materials, increasing the efficiency of stable materials, and exploring materials with high efficiency and stability. Future research can progress in the direction of synthesizing materials characterized by their high stability and efficiency, enhancing the stability of already discovered efficient semiconductor materials, and increasing the efficiency of already reported highly stable materials.

References

Abdullah, H., Lin, J.-H., Kuo, D.-H., Shuwanto, H., Tsai, M.-L., Chiang, C.-H., 2024. Unbiased overall photoelectrocatalytic water-splitting over micropyramid Si/In_2O_3/CuO/ZnO photocathode and Ni/$BiVO_4$ photoanode under simulated solar-light illumination. International Journal of Hydrogen Energy 72, 521–536.

Ager, J.W., Shaner, M.R., Walczak, K.A., Sharp, I.D., Ardo, S., 2015. Experimental demonstrations of spontaneous, solar-driven photoelectrochemical water splitting. Energy and Environmental Science 8 (10), 2811–2824.

Aida Mohamed, N., Sieh Kiong, T., Fazli Ismail, A., Asri Mat Teridi, M., 2024. Novel gamma-ray enhanced TiO_2 nanoparticles photoanode for efficient photoelectrochemical (PEC) water splitting. Applied Surface Science 642, 158602.

Alfaifi, B.Y., Ullah, H., Alfaifi, S., Tahir, A.A., Mallick, T.K., 2018. Photoelectrochemical solar water splitting: from basic principles to advanced devices. Veruscript Functional Nanomaterials 2 bdjoc3.

Augustyński, J., Alexander, B.D., Solarska, R., 2011. Metal oxide photoanodes for water splitting. Topics in Current Chemistry 303, 1–38.

Bae, H., Bhamu, K.C., Mane, P., Burungale, V., Kumar, N., Lee, S.H., et al., 2024. Rationally designed core-shell of 2D-g-C_3N_4/Cu_2O nanowires heterojunction photocathode for efficient photoelectrochemical water splitting. Materials Today Energy 40, 101484.

Bhatt, M.D., Lee, J.S., 2015. Recent theoretical progress in the development of photoanode materials for solar water splitting photoelectrochemical cells. Journal of Materials Chemistry A 3 (20), 10632–10659.

Boddy, P.J., 1968. Oxygen evolution on semiconducting TiO_2. Journal of The Electrochemical Society 115 (2), 199.

Chang, J.-H., Kumar, M., Shen, S.-Y., 2021. Fundamentals of photoelectrochemical water splitting. Nanostructured Materials for Photoelectrochemical Water Splitting. IOP Publishing.

Cheng, P., Liu, Y., Ziegler, M., Klingenhof, M., Wang, D., Zhang, Z., et al., 2022. Improving silicon photocathode performance for water reduction through dual interface engineering and integrating ReS_2 photocatalyst. ACS Applied Energy Materials 5 (7), 8222–8231.

Cui, W., Xia, Z., Wu, S., Chen, F., Li, Y., Sun, B., 2015. Controllably interfacing with ferroelectric layer: a strategy for enhancing water oxidation on silicon by surface polarization. ACS Applied Materials and Interfaces 7 (46), 25601–25607.

Das, S., Deka, T., Ningthoukhangjam, P., Chowdhury, A., Nair, R.G., 2022. A critical review on prospects and challenges of metal-oxide embedded g-C_3N_4-based direct Z-scheme photocatalysts for water splitting and environmental remediation. Applied Surface Science Advances 11, 100273.

Dias, P., Mendes, A., 2017. Hydrogen production from photoelectrochemical water splitting. In: Meyers, R.A., (Ed.), Encyclopedia of Sustainability Science and Technology. Springer, New York, NY, pp. 1–52.

Dong, G., Yan, L., Bi, Y., 2023. Advanced oxygen evolution reaction catalysts for solar-driven photoelectrochemical water splitting. Journal of Materials Chemistry A 11 (8), 3888–3903.

Dorp, D.H., Van Hijnen, N., Di Vece, M., Kelly, J.J., 2009. SiC: a photocathode for water splitting and hydrogen storage. Angewandte Chemie International Edition 48 (33), 6085–6088.

Fujishima, A., Honda, K., 1972. Electrochemical photolysis of water at a semiconductor electrode. Nature 238 (5358), 37–38.

Gnanasekar, P., Periyanagounder, D., Varadhan, P., He, J.-H., Kulandaivel, J., 2019. Highly efficient and stable photoelectrochemical hydrogen evolution with 2D-NbS_2/Si nanowire heterojunction. ACS Applied Materials and Interfaces 11 (47), 44179–44185.

Grimes, C.A., 2007. Synthesis and application of highly ordered arrays of TiO_2 nanotubes. Journal of Materials Chemistry 17 (15), 1451–1457.

Grimes, C.A., Varghese, O.K., Ranjan, S., 2008. Oxide semiconducting materials as photoanodes. In: Grimes, C.A., Varghese, O.K., Ranjan, S., (Eds.), Light, Water, Hydrogen: The Solar Generation of Hydrogen by Water Photoelectrolysis. Springer US, Boston, MA, pp. 191–255.

Gunawan, Adi, Wijaya, R., Suseno, A., Lusiana, R.A., Septina, W., Harada, T., 2023. Synthesis of $CuInS_2$ thin film photocathode with variation of sulfurization sources and Pt-In_2S_3 modification for photoelectrochemical water splitting. Journal of Electroanalytical Chemistry 945, 117683.

Guo, S., Zhao, X., Zhang, W., Wang, W., 2018. Optimization of electrolyte to significantly improve photoelectrochemical water splitting performance of ZnO nanowire arrays. Materials Science and Engineering B 227, 129–135.

Ho, H.-C., Smiljanić, M., Jovanović, Z., Čekada, M., Kovač, J., Koster, G., et al., 2023. Robust $SrTiO_3$ passivation of silicon photocathode by reduced graphene oxide for solar water splitting. ACS Applied Materials and Interfaces 15 (37), 44482–44492.

Ho, H.-C., Smiljanic, M., Jovanović, Z., Jaklič, B., Kovač, J., Čekada, M., et al., 2023. $SrTiO_3$ epitaxially protected Si-photocathode for photoelectrochemical water splitting. ECS Meeting Abstracts MA2023-02 (47), 2293.

Huang, Q., Zhao, Y., Li, Y., 2024. Improved photoelectrochemical water splitting performance of Sn-doped hematite photoanode with an amorphous cobalt oxide layer. International Journal of Hydrogen Energy 51, 1176–1183.

Jang, Y.J., Lee, J.S., 2019. Photoelectrochemical water splitting with p-type metal oxide semiconductor photocathodes. ChemSusChem 12 (9), 1835–1845.

Jiang, C., Moniz, S.J.A., Wang, A., Zhang, T., Tang, J., 2017. Photoelectrochemical devices for solar water splitting – materials and challenge. Chemical Society Reviews 46 (15), 4645–4660.

Kolodziejak, K., Sar, J., Wysmulek, K., Orlinski, K., Piotrowski, P., Gajewski, M., et al., 2023. Durability of $SrTiO_3$-TiO_2 eutectic composite as a photoanode for photoelectrochemical water splitting. RSC Advances 13 (50), 35422–35428.

Krol, R., Van De Liang, Y., Schoonman, J., 2008. Solar hydrogen production with nanostructured metal oxides. Journal of Materials Chemistry 18 (20), 2311–2320.

Lee, S.A., Lee, T.H., Kim, C., Choi, M.-J., Park, H., Choi, S., et al., 2020. Amorphous cobalt oxide nanowalls as catalyst and protection layers on n-type silicon for efficient photoelectrochemical water oxidation. ACS Catalysis 10 (1), 420–429.

Leroy, C.M., Maegli, A.E., Sivula, K., Hisatomi, T., Xanthopoulos, N., Otal, E.H., et al., 2011. $LaTiO_2N/In_2O_3$ photoanodes with improved performance for solar water splitting. Chemical Communications 48 (6), 820–822.

Li, B., Jian, J., Chen, J., Yu, X., Sun, J., 2020. Nanoporous 6H-SiC photoanodes with a conformal coating of Ni-FeOOH nanorods for zero-onset-potential water splitting. ACS Applied Materials and Interfaces 12 (6), 7038–7046.

Li, M., Luo, W., Liu, Q., Zhuang, Z., Li, Z., Liu, B., et al., 2013. An efficient In0.30Ga0.70N photoelectrode by decreasing the surface recombination centres in a H_2SO_4 aqueous solution. Journal of Physics D 46 (34), 345103.

Li, Z., Luo, W., Zhang, M., Feng, J., Zou, Z., 2013. Photoelectrochemical cells for solar hydrogen production: current state of promising photoelectrodes, methods to improve their properties, and outlook. Energy and Environmental Science 6 (2), 347–370.

Lopes, T., Andrade, L., Formal, F.L., Gratzel, M., Sivula, K., Mendes, A., 2014. Hematite photoelectrodes for water splitting: evaluation of the role of film thickness by impedance spectroscopy. Physical Chemistry Chemical Physics 16 (31), 16515–16523.

Meena, B., Kumar, M., Hocking, R.K., Juodkazis, S., Biju, V., Subramanyam, P., et al., 2023. Exploring $CuBi_2O_4$ as a promising photocathode material for PEC water splitting. Energy and Fuels 37 (18), 14280–14289.

Moon, S., Park, J., Lee, H., Yang, J.W., Yun, J., Park, Y.S., et al., 2023. Bi_2S_3-Cu_3BiS_3 mixed phase interlayer for high-performance Cu_3BiS_3-photocathode for 2.33% unassisted solar water splitting efficiency. Advanced Science 10 (6), 2206286.

Noh, S., Shin, J., Lee, J., Oh, H., Yu, Y., Kim, J., 2024. Improvement in Photoelectrochemical water splitting performance of GaN-nanowire photoanode using MXene. ACS Applied Materials and Interfaces 16.

Osterloh, F.E., 2008. Inorganic materials as catalysts for photochemical splitting of water. Chemistry of Materials 20 (1), 35–54.

Park, J., Yoon, K.-Y., Ghule, B.G., Kim, H., Jang, J.-H., 2024. Morphology-engineered hematite photoanode for photoelectrochemical water splitting. ACS Energy Letters.

Pattengale, B., Huang, J., 2016. The effect of Mo doping on the charge separation dynamics and photocurrent performance of $BiVO_4$ photoanodes. Physical Chemistry Chemical Physics 18 (48), 32820–32825.

Sawal, M.H., Jalil, A.A., Abdullah, T.A.T., Hassan, N.S., Bahari, M.B., Izzudin, N.M., et al., 2024. n-n Heterojunction CdS/FST photoanode for enhanced photoelectrochemical water splitting. International Journal of Hydrogen Energy.

Seabold, J.A., Neale, N.R., 2015. All first row transition metal oxide photoanode for water splitting based on $Cu_3V_2O_8$. Chemistry of Materials 27 (3), 1005–1013.

Seo, G., Kim, B., Hwang, S.W., Shin, S.S., Cho, I.S., 2021. High-performance bulky crystalline copper bismuthate photocathode for enhanced solar water splitting. Nano Energy 80, 105568.

Shi, H., Guo, H., Wang, S., Zhang, G., Hu, Y., Jiang, W., et al., 2022. Visible light photoanode material for photoelectrochemical water splitting: a review of bismuth vanadate. Energy and Fuels 36 (19), 11404–11427.

Shi, K., Zhang, B., Liu, K., Zhang, J., Ma, G., 2023. Rhodium-doped barium titanate perovskite as a stable p-type photocathode in solar water splitting. ACS Applied Materials and Interfaces 15 (40), 47754–47763.

Sharma, R., Devi, S.P., Arora, S.K., 2023. $MoSe_2$ nanosheets/SiNWs heterojunction-based photocathode for efficient photoelectrochemical water splitting applications. Nanotechnology 34 (40), 405705.

Shi, Y., Gimbert-Suriñach, C., Han, T., Berardi, S., Lanza, M., Llobet, A., 2016. CuO-functionalized silicon photoanodes for photoelectrochemical water splitting devices. ACS Applied Materials and Interfaces 8 (1), 696–702.

Smith, W.A., 2016. Photoelectrochemical cell design, efficiency, definitions, standards, and protocols. In: Giménez, S., Bisquert, J., (Eds.), Photoelectrochemical Solar Fuel Production: From Basic Principles to Advanced Devices. Springer, pp. 163–197.

Sundararaj, S.B., Amir, H., Viswanathan, C., Thangavelu, S., 2024. Photoelectrochemical water splitting: a visible-light-driven $CoTiO_3$@g-C_3N_4-based photoanode interface follows the type II heterojunction scheme. Langmuir 40 (31), 16582–16594.

Tang, K., Shao, J.-Y., Yan, Y., Zhong, Y.-W., 2024. Photoelectrochemical cells with a pyridine-anchored bilayer photoanode for water splitting. Langmuir 40 (12), 6244–6252.

Walter, M.G., Warren, E.L., McKone, J.R., Boettcher, S.W., Mi, Q., Santori, E.A., et al., 2010. Solar water splitting cells. Chemical Reviews 110 (11), 6446–6473.

Wang, L., Wang, H., Bu, Q., Mei, Q., Zhong, J., Bai, B., et al., 2024. Al-O bridged NiFeOx/$BiVO_4$ photoanode for exceptional photoelectrochemical water splitting. Chinese Chemical Letters 110139.

Wang, L., Zhang, Y., Li, W., Wang, L., 2023. Recent advances in elaborate interface regulation of $BiVO_4$ photoanode for photoelectrochemical water splitting. Materials Reports 3 (4), 100232.

Wang, N., Zhang, Y., Xu, X., An, M., Gu, J., Song, W., et al., 2024. Nanoplate structured $BiVO_4$ homojunction photoanode for boosting photoelectrochemical water splitting. Journal of Alloys and Compounds 988, 174274.

Weber, M.F., Dignam, M.J., 1984. Efficiency of splitting water with semiconducting photoelectrodes. Journal of The Electrochemical Society 131 (6), 1258.

Wheeler, G.P., Choi, K.-S., 2017. Photoelectrochemical properties and stability of nanoporous p-type $LaFeO_3$ photoelectrodes prepared by electrodeposition. ACS Energy Letters 2 (10), 2378–2382.

Wheeler, G.P., Choi, K.-S., 2018. Investigation of p-type $Ca_2Fe_2O_5$ as a photocathode for use in a water splitting photoelectrochemical cell. ACS Applied Energy Materials 1 (9), 4917–4923.

Yadav, J., Singh, J.P., 2024. Surface plasmonic hot hole driven $Ag_2S/Au/Al_2O_3$ photocathode for enhanced photoelectrochemical water splitting performance. Renewable Energy 228, 120615.

Yan, X., Sakai, K., Ozawa, H., 2023. Highly efficient and stable molecular-based TiO_2 photoanodes for photoelectrochemical water splitting achieved by pyridyl anchoring technique. ACS Catalysis 13 (20), 13456–13465.

Yang, W., Kim, J.H., Hutter, O.S., Phillips, L.J., Tan, J., Park, J., et al., 2020. Benchmark performance of low-cost Sb_2Se_3 photocathodes for unassisted solar overall water splitting. Nature Communications 11 (1), 861.

Yang, W., Moon, J., 2019. Recent advances in earth-abundant photocathodes for photoelectrochemical water splitting. ChemSusChem 12 (9), 1889–1899.

Zeng, H., Chang, J.-C., Qu, Y., Wang, W., Birch, J., Hsiao, C.-L., et al., 2024. Interface-engineered InAlN/Cu_2O photocathode with accelerated charge separation for boosting photoelectrochemical water splitting. Solar RRL 8 (10), 2400094.

Zhan, Y., Chen, Y.-C., Long, X., 2024. Interfacial engineering enabling solution-processed Cu:NiOx/Sb_2Se_3/TiO_2/Pt photocathodes for highly efficient photoelectrochemical water-splitting. Sustainable Energy and Fuels 8 (16), 3670–3679.

Zhang, D., Cao, Y., Karuturi, S.K., Du, M., Liu, M., Xue, C., et al., 2020. Enabling unassisted solar water splitting by single-junction amorphous silicon photoelectrodes. ACS Applied Energy Materials 3 (5), 4629–4637.

Zhang, L., Sun, J., Zhao, M., Wei, Y., Luo, T., Zhao, Z., et al., 2024. Quantum dots mediated heterojunction coupling $MoSe_2$ photoanode for photoelectrochemical water splitting. Molecules 29 (5), 1070.

Zhong, S., Kang, B., Cheng, X., Chen, P., Fang, B., 2024. Construction of a Co-MOF/MXene/$BiVO_4$ composite photoanode for efficient photoelectrochemical water splitting. ACS Sustainable Chemistry and Engineering 12 (3), 1233–1246.

Zhu, X., Yu, X., Li, J., Li, H., Wang, S., Xu, Y., et al., 2023. Graphene enhanced photo-electrochemical water splitting by diminishing Pt/p-Si photocathode interfacial barrier. ChemPhotoChem 7 (7) e202200272.

CHAPTER

7

Selection of materials and cell design components for photochemical splitting of water

Sapna Balayan[1], *Mamta Shandilya*[2], *Itika Kainthla*[2], *and Gun Anit Kaur*[2]

[1]Amity Institute of Nanotechnology (AINT), Amity University Uttar Pradesh (AUUP), Noida, Uttar Pradesh, India [2]School of Physics and Materials Science, Shoolini University of Biotechnology and Management Sciences, Solan, Himachal Pradesh, India

7.1 Introduction

The rapid expansion of the world's population and industrial output is leading to a substantial rise in worldwide energy usage (Jiao et al., 2016; Wang et al., 2019). The majority of this energy comes from conventional fossil fuels like natural gas, coal, and oil. Yet, the excessive use of these fossil fuels results in high carbon emissions, causing significant damage to the environment (Shih et al., 2018). This environmental degradation necessitates exploring and developing renewable energy sources (including solar, ocean, wave, and tidal, wind) (Zhao et al., 2019; Chu and Majumdar, 2012). Solar energy, in particular, is abundant, renewable, and clean, making it suitable for a wide range of applications, such as the photoelectric process, desalination, and solar evaporation (Ni et al., 2016; Elimelech and Phillip, 2011; Boddy, 1968). Industries face a major obstacle in creating systems that can function using renewable energy sources instead of depending on fossil fuels. Hydrogen (H_2), for instance, is a potential recyclable energy carrier for large-scale use. However, its production typically requires significant amounts of natural gas, resulting in CO_2 emissions. To address this, photo-catalytically generated H_2 through water-splitting offers a cleaner alternative (Fig. 7.1; Bolton, 1996; Domen et al., 2000; Abe, 2010; Kudo and Miseki, 2009; Kitano and Hara, 2010). This

Photochemical Splitting of Water
DOI: https://doi.org/10.1016/B978-0-443-29064-0.00006-1

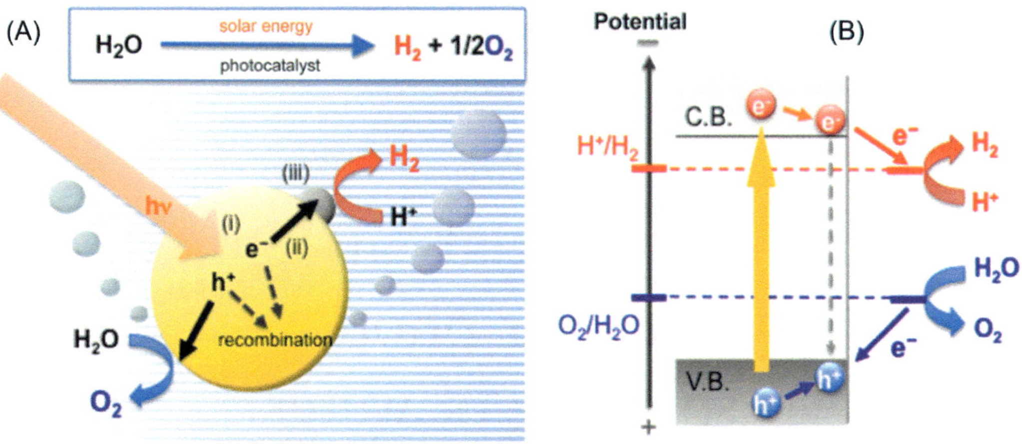

FIGURE 7.1 (A) Illustrates water-splitting over a semiconductor photocatalyst. The process combines surface chemical reactions among compounds and carriers, migration, separation of charges, and photon absorption. Moreover, the interaction of holes and e^- can occur without participating in reactions. (B) It is important for heterogeneous semiconductor photocatalyst-assisted water-splitting to have an excess negative charge at the lower end of the conduction band, as compared to the reduction potential required for H_2 generation. Conversely, there should be an excess of positive charge at the upper part of the valence band, in comparison to the oxidation potential needed for O_2 production. Source: *Reproduced from Abe, R., 2010. Recent progress on photocatalytic and photoelectrochemical water-splitting under visible light irradiation. Journal of Photochemistry and Photobiology C: Photochemistry Reviews 11, 179–209. Copyright 2020, Elsevier.*

process involves converting light energy into chemical energy, which enhances the water-splitting reaction as follows:

$$H_2O + h\upsilon \rightarrow H_2 + \frac{1}{2}O_2 \Delta G = 238 \text{ kJ/mol} \quad (7.1)$$

In a groundbreaking development, Fujishima and Honda introduced the concept of photochemical water-splitting in 1972 paving the way for a new era of renewable energy research and technology. As we strive to address the pressing global energy and environmental challenges, their pioneering work continues to inspire and guide innovative endeavors in sustainable energy production (Zhao et al., 2022). They demonstrated a photoelectrochemical system comprising platinum (Pt) electrodes and a titanium dioxide (TiO_2) photoelectrode, laying the foundation for the use of solar energy in driving heterogeneous catalytic reactions (Fujishima and Honda, 1972). Since then, solar energy has increasingly replaced traditional energy sources in these applications (Zhao et al., 2019; Sastre et al., 2014). Despite significant advancements, semiconductor photocatalysis faces two major limitations that hinder its large-scale application: low apparent quantum efficiency and poor sunlight absorption in the infrared and visible regions (Zhu et al., 2018; Ghoussoub et al., 2019). These issues result in energy loss and sluggish reaction kinetics. To overcome these challenges, the incorporation of photothermal catalytic processes into water-splitting has shown promise. This approach utilizes single-atom silver incorporated g-C_3N_4 (SAAg-g-C_3N_4), significantly increasing the photocatalytic H_2 evolution reaction rate and enhancing its stability and activity. Photothermal catalysis combines the benefits of photocatalysis and thermal activation, thereby improving solar energy conversion efficiency. In the case of SAAg-g-C_3N_4, the single-atom silver acts as a highly effective catalytic site, promoting efficient charge separation and transfer, which are critical for the hydrogen evolution reaction (HER). This enhancement addresses two primary limitations of semiconductor photocatalysis by improving quantum efficiency and expanding the light absorption range. Research in this area is ongoing, with scientists exploring various materials and methods to optimize the photothermal catalytic process. The goal is to achieve a sustainable and efficient system for H_2 production that relies entirely on renewable energy sources. The development has the potential to greatly lessen our reliance on fossil fuels, which would help to minimize environmental harm and support a greener, more enduring energy future. In conclusion, the shift from conventional fossil fuels to renewable sources is essential to address the environmental challenges of high carbon emissions. With its abundance and cleanliness, solar energy plays a crucial role in this transition. Developing efficient systems for renewable energy utilization, particularly in the context of H_2 production through H_2O splitting, represents a significant step towards achieving sustainable energy solutions. Advances in photothermal catalysis, exemplified by the use of SAAg-g-C_3N_4, offer promising avenues for enhancing the efficiency and applicability of these renewable energy systems. The possibility of achieving large-scale, clean energy production becomes more viable as research advances, setting the stage for a more environmentally friendly future (Li et al., 2021). Over a decade, a range of green energy systems has been developed to efficiently produce H_2. These include 2-electrode water electrolysis, water-splitting driven by a photo-electrode device, solar cells, thermoelectric devices, triboelectric nanogenerator, pyroelectric and water-gas shift reaction (Refer to Fig. 7.2). All these green energy systems are capable of efficiently driving water-dissociation for H_2 production (Li et al., 2020).

The section delves into the complex procedure of utilizing energy to facilitate the water-splitting reaction, an essential approach for eco-friendly H_2 generation. It discusses various semiconductor materials, their properties, and their roles in enhancing photocatalytic efficiency, emphasizing the importance of catalytic activity, charge separation, and light absorption. Additionally, the chapter elaborates on innovative cell designs that optimize the interaction between light and photocatalysts, thereby maximizing H_2 yield. By integrating recent advancements in material science and engineering, this chapter aims to guide researchers and engineers in developing more efficient and robust systems for clean energy generation through photochemical water-splitting.

7.2 Background of water-splitting

7.2.1 Fundamentals of photochemical reactions

A. Fujishima and K. Honda's 1972 established a theory for water dissipation using a metal-oxide semiconductor (TiO_2) suggested that visible light can decompose water molecules into O and H without needing external voltage. The photochemical reaction comprises the collection of sunlight, separation of charge, and its transmission, as well as oxidation and catalytic reaction (Yang et al., 2013). The light needs to have a wavelength of

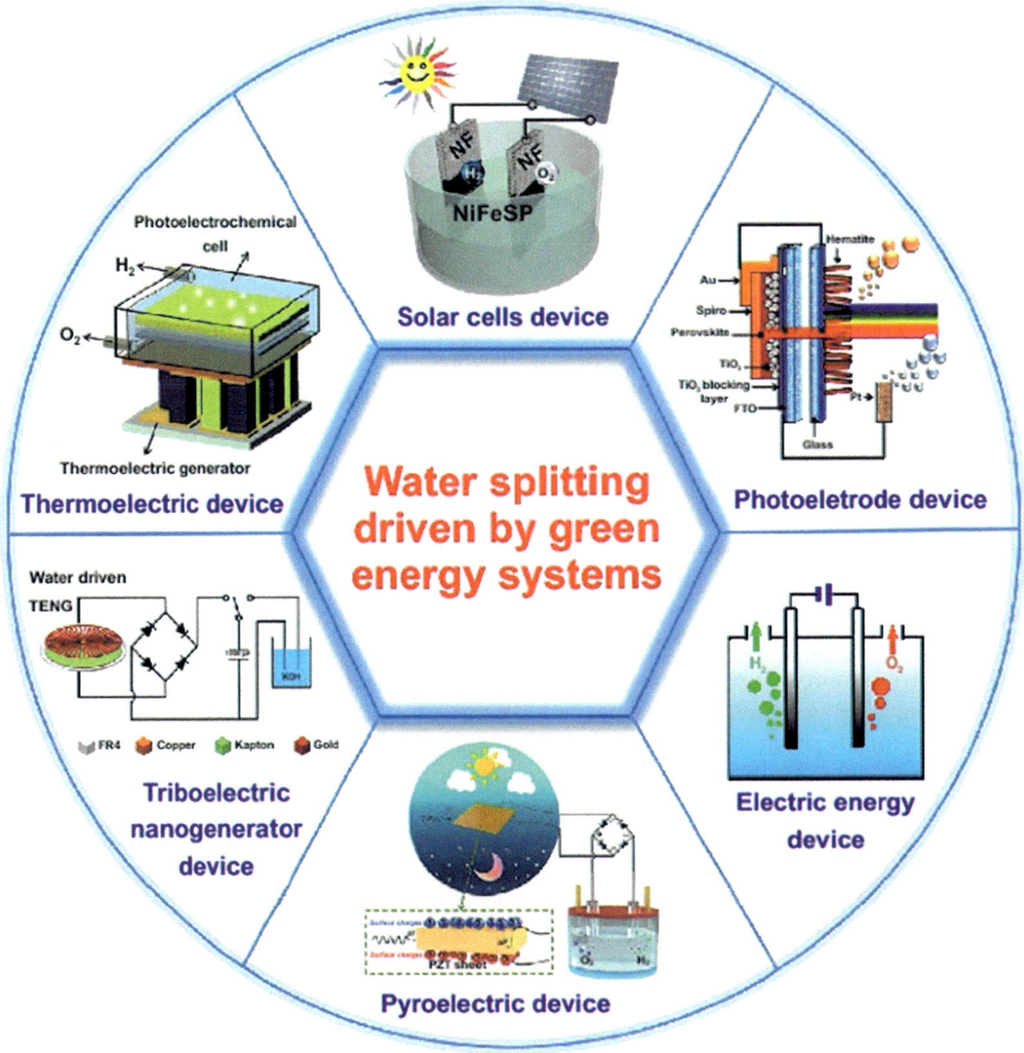

FIGURE 7.2 The splitting of water-splitting can be fueled by different environmentally friendly energy systems. Source: *Reproduced from Li, X., Zhao, L., Yu, J., Liu, X., Zhang, X., Liu, H., Zhou, W., 2020. Water splitting: from electrode to green energy system. Nano-Micro Letters 12, 1–29. Copyright 2020, Nano-Micro Lett.*

less than 1000 nm in order to produce the necessary energy for water-splitting, additionally, a 1.5G air-to-mass ratio is required (Fig. 7.3; Klotz et al., 2017; Fujishima and Honda, 1972; Jha et al., 2024). In the presence of sunlight, the photo-electrode initiates photocarriers, causing them to interact with water and thereby produce H_2 and oxygen (Chen et al., 2021). This process demands photoelectrodes with specific qualities such as high carrier mobility, wide absorption range, environmental friendliness, high stability, and extended carrier life (Chen et al., 2021; Zhou et al., 2013; Chouhan et al., 2017).

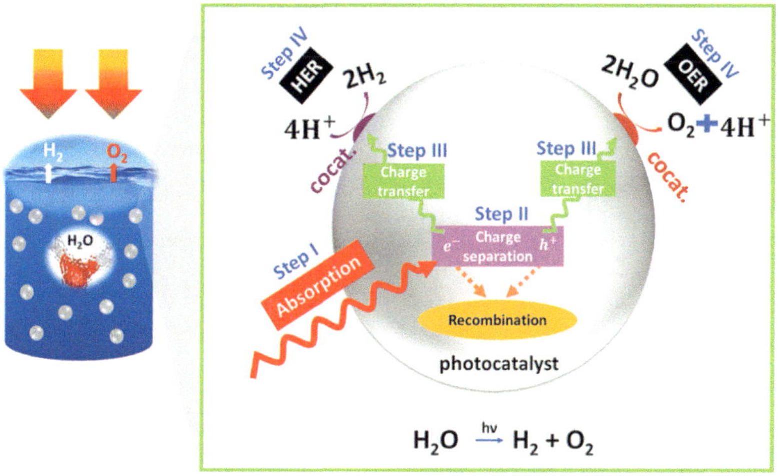

FIGURE 7.3 The principle of photoelectrochemical water-splitting is illustrated, showing the critical water-splitting reactions in the final stage. *HER*, hydrogen-evolution reaction; *OER*, oxygen-evolution reaction. Source: *Reproduced from Jha, B.K., Chaule, S., Jang, J.-H., 2024. Enhancing photocatalytic efficiency with hematite photoanodes: principles, properties, and strategies for surface, bulk, and interface charge transfer improvement. Materials Chemistry Frontiers 8, 2197–2226. Copyright 2024, RSC.*

Water is dissociated into molecular H_2 and oxygen due to photoelectrochemical water-splitting, where the photon conversions lead to the alteration in the semiconductor surface into electrochemical energy and result in the elevation of photon energy exceeding the semiconductor bandgap (Fig. 7.4). Minority charge carriers are an important factor in successfully completion of the process following the electrochemical reaction and migration on the interface. The main mechanics occurring on the interface of a photoelectrochemical cell are electrochemical reactions, charge generation, transportation, separation, and light absorption. In semiconductors, electrons move to the conduction band, leaving behind gaps in the valence band. The movement of photogenerated charge carriers to the electrolyte interface in the depletion region is facilitated by the electric field, initiating important chemical reactions.

The electric field is created by the variance between the redox potential of water and the semiconductor's Fermi level. In a photoelectrochemical setup with a photoanode, the generated electrons move to the external circuit towards a counter-electrode, leading to water reduction for the generation of H_2 through the HER. Furthermore, the migration of holes to the surface for water oxidation and production of molecular oxygen (oxygen-evolution reaction [OER]) and the leftover charge carriers recombine, and the released energy limits the photo-electrode activity (Hamdani and Bhaskarwar, 2021; Hoffmann et al., 1995; Kochuveedu, 2016; Linsebigler et al., 1995; Steier and Holliday, 2018).

7.2.1.1 Photoelectrochemical process

Step I: Photon absorption: During early stage of the photoelectrochemical process, a semiconductor electrode, as a photoactive material, absorbs photons, causing the excitation of e^- from the valence band to the conduction band, thus creating an e^- and hole pair. It is

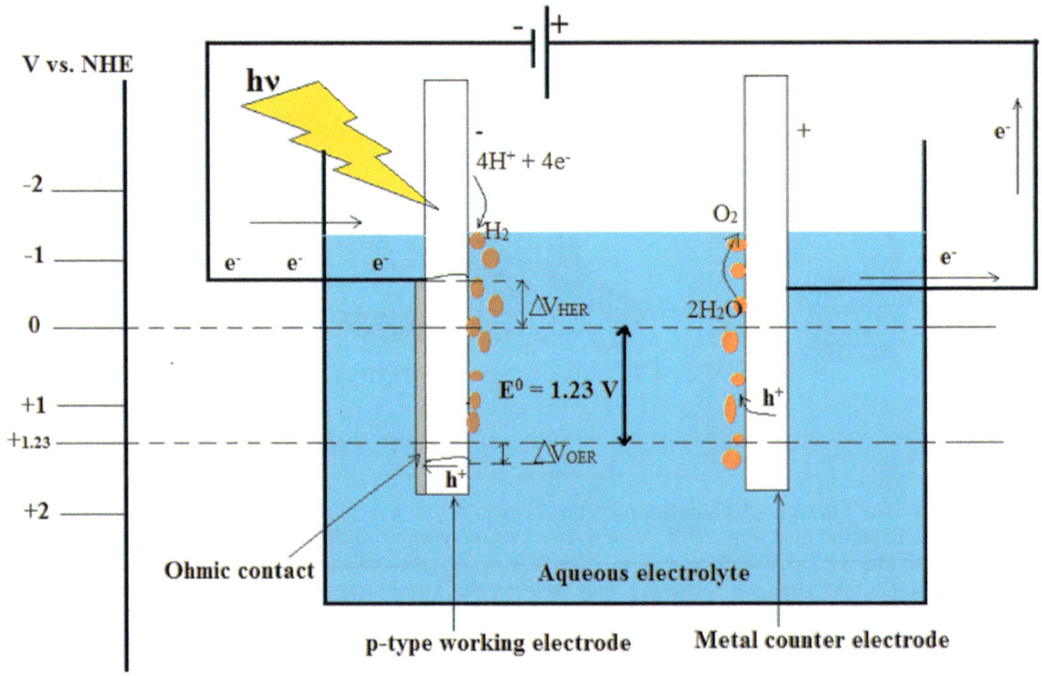

FIGURE 7.4 In p-type semiconductors, +ve charged holes serve as the primary charge carriers, while in n-type semiconductors, −ve charged e⁻ fulfill this role. Following the introduction of these charge carriers, there is a subsequent migration of the charges, leading to establishing an equilibrium state. *Source: Reproduced from Hamdani, I., Bhaskarwar, A., 2021. Recent progress in material selection and device designs for photoelectrochemical water-splitting. Renewable and Sustainable Energy Reviews 138, 110503. Copyright 2024, Elsevier.*

crucial for the semiconductor's band positions to meet the bandgap requirement. The orientation of the conduction band is an important factor in the water-dissociation. It should be considered that the -ve charge should be more at the conduction band, considering the potential for water reduction; moreover, the valence band should be highly +vely charged compared to the oxidation potential in the system. These band positions should be placed precisely to facilitate critical reactions (Jha et al., 2024).

Step II: Charge separation: In a photovoltaic energy converter (PEC), efficient exciton separation is crucial to avoid recombination. The condition can be fulfilled with the potential generation in the semiconductor or by accommodating hole and e⁻ acceptors. There are distinct factors such as refractive index, optical penetration length, absorption coefficient, band nature, gap, and position that affect photon absorption (Jha et al., 2024).

Step III: Charge transfer: The semiconductor material exhibits the diffusion of electrons and holes following exciton dissociation. This aids in the transportation of e⁻ charge carriers from areas of higher to lower concentration and then to the electrolyte interface. The charge carriers are driven or drifted towards the specific electrode by the applied electric field or built-in potential of the semiconductor, facilitating their transportation and focusing on mitigating recombination and efficient charge collection. Various attributes of semiconductors, such as surface states, effective carrier mass, depletion width, conductivity, resistivity, flat-band potential, dielectric constant, carrier mobility, diffusion length, exciton binding energy, space charge layer, and carrier lifetime, exert a considerable impact on the distribution of charge within materials. Effective separation of charges is a critical factor in evaluating the overall efficiency of photoelectrochemical systems because it directly impacts catalytic activity. The recombination of charge carriers in semiconductor materials is crucial for enhancing the efficiency of the PEC system. These materials exhibit distinct properties at interfaces like electrode-electrolyte, substrate-electrode, and electrode-catalyst. It is essential to study and address these recombination processes while enhancing semiconductor properties to minimize their impact on PEC systems. In particular, a thorough investigation of these recombination phenomena is necessary to develop strategies that effectively mitigate their influence and improve the performance of PEC systems (semiconductor-based) (Jha et al., 2024).

Step IV: Catalytic process: The catalytic process is obtained at the junction of electrolyte and electrode; the reduction of redox species in the electrolyte is carried out by the electrons excited by light, whereas the holes play a role in the oxidation of other species. These catalytic reactions are fundamentals for performing experiments such as water-dissociation or CO_2 reduction. The speed of the catalytic reactions is influenced by the characteristics of the electrolyte, including the size of effective ions, viscosity, activity coefficient, ohmic drop (iR drop), and pH. Effective catalysis can be obtained with charge transfer resistance, exchange current density, and electrocatalytic nature. Moreover, the semiconductor-electrolyte interface can enable efficient chemical reactions and charge transfer. These interfaces must possess various parameters for photoelectrochemical cells, such as specific morphology, energy levels, high electrical conductivity, suitable chemical composition, efficient charge transfer, and large surface area. Proper and careful designing of this interface can result in higher efficiency and low energy loss of the photoelectrochemical system (Jha et al., 2024). The O-H bond in the water molecule is very difficult to cleave directly (Ruscic et al., 2002), making it one of the toughest X-H bonds (Chandra and Uchimaru, 2000). Transition-metal complexes offer a hopeful resolution to this challenge by activating water. This potential arose from noting a substantial reduction of nearly 60 kcal/mol to break the O-H bond leading to the generation of titanocene(III) aqua-complex 1 due to the reaction among Cp_2TiCl and H_2O (Barrero et al., 2002; Cuerva et al., 2006; Paradas et al., 2010; Gansäuer et al., 2012; Martínez et al., 2020, 2023) (Fig. 7.5). Active Cp_2TiCl is produced by reducing commercially available Cp_2TiCl_2 and Mn or Zn (Nugent and Rajanbabu, 1988).

The C radicals obtained during the reaction among functional groups and Cp_2TiCl can be eliminated with aqua titanium(III) complex 1. Subsequently, it was studied to explore processes of water-dissociation that include a photochemical reaction as a crucial stage. An early method of splitting water used a ruthenium pincer complex and light. The H_2 and O_2 were efficiently produced from water by the system in a sequential manner. H_2 was created through a thermal process, while O_2 was generated when the water was illuminated. When dearomatized Ru(II) pincer hydride complex 2 reacts with water, it forms a monomeric aromatic Ru(II) hydrido-hydroxo complex 3 (refer to Fig. 7.6). It's important to note that there is no change in the ruthenium's oxidation state during this reaction (Martínez and Rodríguez-García, 2024).

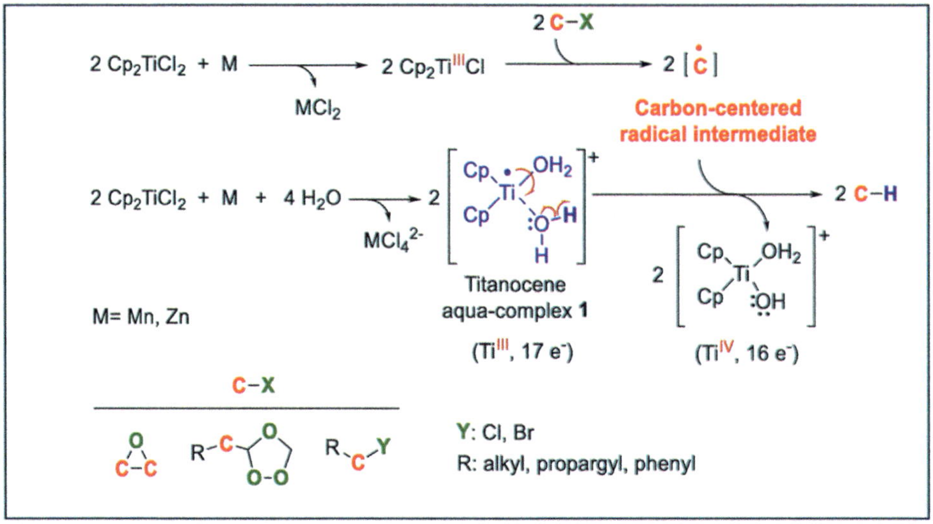

FIGURE 7.5 Cp_2TiCl forms a proper aqua-complex 1, and it undergoes the reduction of a C-centered radical intermediate. Source: *Reproduced from Martínez, A.R., Rodríguez-García, I., 2024. Photochemical water-splitting via transition-metal complexes. Journal of Catalysis 432, 115415. Copyright 2024, Elsevier.*

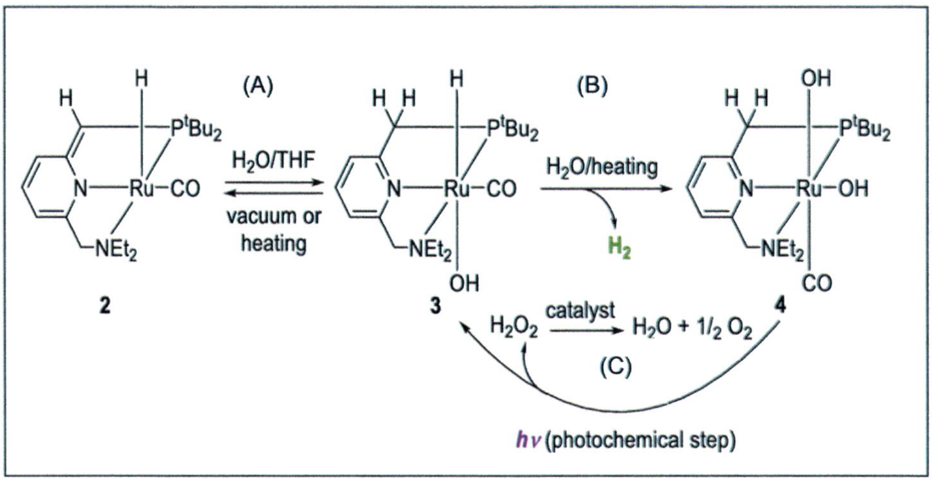

FIGURE 7.6 Photochemical water-dissociation occurs via ruthenium transition-metal complexes. (A) Water reacts with the Ru(II) pincer complex 2. (B) Complex 3 is thermally activated with water, leading to H_2 generation. (C) O_2 is formed after photolysis of complex 4. Source: *Reproduced from Martínez, A.R., Rodríguez-García, I., 2024. Photochemical water-splitting via transition-metal complexes. Journal of Catalysis 432, 115415. Copyright 2024, Elsevier.*

7.2.1.2 Thermodynamics of water-splitting

Thermodynamics of water-splitting can be explained as the energetics underlying the dissociation of water molecules into H_2 and O_2. The laws of thermodynamics control the process and are essential in renewable energy technologies such as electrolysis and photoelectrochemical cells. The principle states that during the water-splitting the O-H bonds can be dissociated with Gibbs free energy barrier and result in fresh bonds. Provided the standard conditions the reaction must proceed spontaneously and ΔG must be negative. This explains that sufficient energy was provided to drive the reaction, where the electrical signals or sunlight acts as the input energy (Ismail and Bahnemann, 2014; Swierk and Mallouk, 2013; Abe, 2010; Singh, 2021).

For the entirety of the water-splitting process, the alteration in Gibbs free energy (ΔG) can be computed by summing the modifications in Gibbs free energy of the two half-reactions:

$$\Delta G = \Delta G(\mathrm{OER}) + \Delta G(\mathrm{HER})$$

Calculation of Gibbs free energy change for each half-reaction:

$$\Delta G = \Delta H - T\Delta S$$

7.2.1.3 *Kinetics of photochemical processes*

The kinetic study of the photochemical water-dissociation process explain the reaction rate, mechanism, and pathway of photo conversion to chemical energy. The molecules absorb the photons, leading to electronic excitations and subsequent transformations. The "Rate laws" facilitate studying the relation between the concentration of analytes, the rate of reaction, and parameters such as light absorption, absorption cross-section reaction, and other photochemical reactions. The complexity of photochemical kinetics can be obtained through the energy and electron transfer process, as well as excited state lifetimes that facilitate diverse reaction pathways and transient species. Moreover, the higher sensitivity of the system can be obtained by engineering solvent environment, temperature, and catalyst. While programming the kinetics of photochemical processes, the light energy can be incorporated for diverse applications (Haussener et al., 2012).

7.3 Material selection for photochemical splitting of water

7.3.1 Criteria for material selection

Materials intended for use as photoelectrodes in a photoelectrochemical reactor must meet two essential criteria (a) optional functionality-to achieve optimal absorption within the visible range of the solar spectrum. and (b) catalytic functionality - to facilitate the required HER and OER.

Various studies report the preparation of materials that have characteristics that can facilitate the water-splitting reactions, such as particle morphology, environmental and economic considerations, absorption coefficients, crystallographic, charge carrier dynamics, stability, durability, catalytic activity, band edge position, tuneability, compatibility with substrates and Interfaces and phase purity (Maeda and Domen, 2010). Researchers have coined a flowchart to aid the selection of suitable materials as photo-electrode in PEC cells, as depicted in Fig. 7.7. However, it is challenging to identify the important factors that

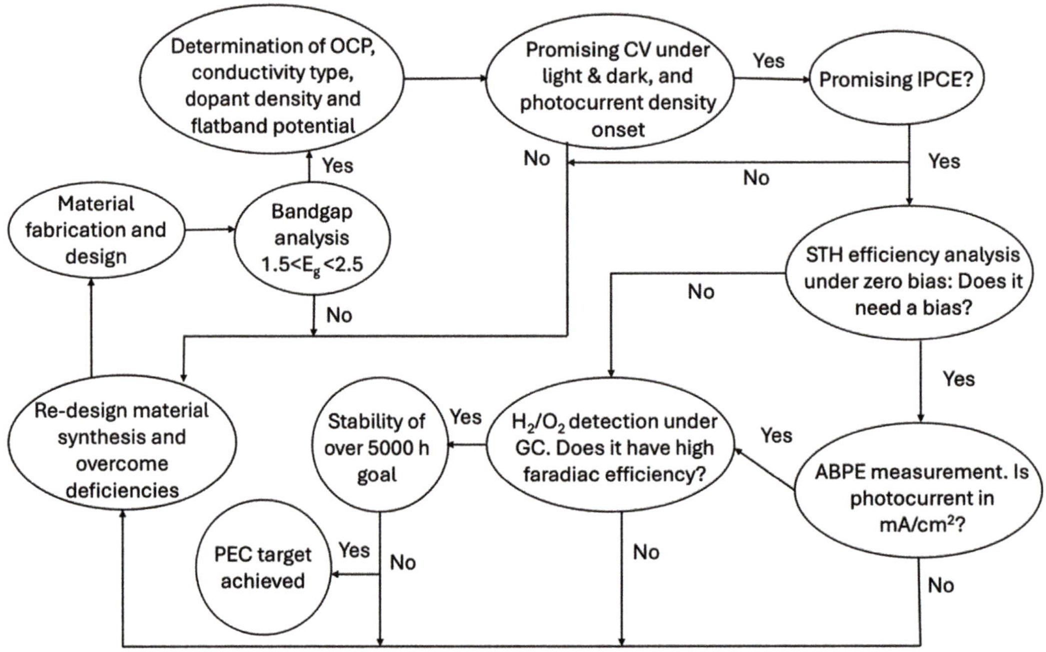

FIGURE 7.7 Flowchart displays single photo-electrode-based photoelectrochemical device. *IPCE*, Incident photon-to-current efficiency; *GC*, gas chromatography; *OCP*, open circuit potential; *CV*, cyclic voltammetry; *STH*, solar-to-hydrogen; *ABPE*, applied bias photon-to-current efficiency; *PEC*, photovoltaic energy converter. *Source: Reproduced from Hamdani, I., Bhaskarwar, A., 2021. Recent progress in material selection and device designs for photoelectrochemical water-splitting. Renewable and Sustainable Energy Reviews 138, 110503. Copyright 2024, Elsevier.*

dominate the activity of the photochemical cell as the process is carried out in a complicated sequence and with multiple steps.

- **First step:** The first step is designing and creating material for PEC use. After material fabrication, the bandgap of the material is analyzed. The ideal bandgap for efficient water-splitting is between 1.5 eV and 2.5 eV. If the bandgap falls within this range, the process moves forward. If not, the material needs to be redesigned to overcome this deficiency, looping back to the material synthesis step. Further, several important characteristics of the material are evaluated, such as open circuit potential, to measure its potential under illumination and no current. Conductivity Type: whether the material is n-type or p-type. Dopant Density and Flat-band Potential: which influence charge separation and transport. If these are promising, the next step is to measure the current-voltage characteristics.
- **Promising cyclic voltammetry under light and dark, and photocurrent density onset:** A cyclic voltammetry test is conducted under both light and dark conditions to check the material's response. If the response is promising, the next step is to check for incident photon-to-current efficiency.

- **Promising incident photon-to-current efficiency?:** The **IPCE** measures how efficiently the material converts photons into current. If the IPCE is promising, the next step is to measure the Solar-to-Hydrogen (STH) efficiency under zero bias.
- **Solar-to-hydrogen efficiency analysis under zero bias:** At this stage, the material's STH efficiency is evaluated without applying external bias. If the material requires a bias to achieve better results, the next step is to perform an applied bias photon-to-current efficiency measurement to check the photocurrent density.
- **Applied bias photon-to-current efficiency measurement (photocurrent in mA/cm²):** If the photocurrent density is promising (i.e., high enough), the next step is to measure the stability.
- **H_2/O_2 detection under gas chromatography – high faradaic efficiency?:** Gas chromatography is used to detect the production of H_2 and oxygen. The Faradaic efficiency, which measures how much charge goes into generating the desired products, is evaluated. If the material shows high Faradaic efficiency, the next step is to check its long-term stability.
- **Stability of over 5000 hours:** The material must show stability for at least 5000 hours, which is critical for long-term use in PEC applications. If the material achieves this goal, the PEC target is reached.
- **Looping back:** If at any point the material fails to meet the criteria (e.g., nonpromising IPCE, nonideal photocurrent, or instability), the material must be redesigned to overcome its deficiencies, and the process begins again from the fabrication and design step.
- **Final goal: PEC target achieved:** After successfully passing through all these steps, the PEC material is deemed ready for practical application, having met the targets for efficiency, stability, and performance.

7.3.2 Photoactive materials

When exposed to light, these materials can exhibit significant changes in their properties or behavior. They can be incorporated in a variety of uses, including optoelectronics, photonic devices, photovoltaics, sensors and photocatalysis.

7.3.2.1 Metal oxides

The size of the particles and the degree of crystallinity are crucial factors in determining the ability of conventional metal-oxide photocatalysts to catalyze the process of splitting water through photocatalysis (Tong et al., 2012; Ikeda et al., 2013; Méndez-Ramos et al., 2013). Over the last 4 decades, numerous materials for photocatalysts have been reported for water-splitting (H_2 and O_2) in the presence of visible and ultraviolet light. This facilitates H_2 production on a large-scale with a clean and recyclable process. The researchers have already proposed various photocatalysts and reported high efficiency in the presence of UV illumination. The efficient photocatalytic materials are composed of d-Block Elements +ve ions with a d0 electronic config. such as Mo^{6+}, Zr^{4+}, Ta^{5+}, Nb^{5+}, Ta^{5+}, Ti^{4+}, and W^{6+}, or d10 electronic config. such as Sb^{5+}, Ge^{4+}, Sn^{4+}, In^{3+}, and Ga^{3+} as principal +ve ions, and the conduction bands constitute empty d or sp orbitals (Law et al., 2005; Chen et al., 2010; Moriya et al., 2013; Bard and Fox, 1995).

7.3.2.2 *Metal sulfides*

It is essential to advance the development of functional bio-inspired catalysts to integrate large-scale sustainable H_2 production. Although nature contains nitrogenase and hydrogenase capable of mediating the HER, it is improbable that enzyme-based devices can facilitate high-level H_2 generation (Mcpherson and Vincent, 2014; Eckenhoff et al., 2013; Kaur-Ghumaan and Stein, 2014; Hallenbeck and Benemann, 2002). The sophisticated biocatalysts exhibit exceptional performance in natural settings with remarkable catalytic selectivity. However, they quickly lose their effectiveness in harsh conditions, such as highly acidic or alkaline environments. Drawing inspiration from the structure and composition of nitrogenase and hydrogenase, scientists have created a range of metal sulfides that serve as highly effective electrocatalysts for the HER. This marks a significant advancement in the area of noble metal-free HER electrocatalysts. Recently, WS_2 has been in demand due to its unique electronic and structural properties, which can be integrated into HER when their properties are programmed with specific requirements. Recent research in this field has produced several successful WS_2 nanomaterials, with strained chemically exfoliated WS_2 nanosheets being a notable example (Zou and Zhang, 2015). WS_2 has garnered considerable attention lately because of its structural and electronic resemblances to MoS_2. It is crucial to develop precision-engineered nanostructured WS_2 with adjustable properties to fully utilize its potential in HER applications. Recent research in this area has produced several successful instances of such WS_2 nanomaterials, with strained chemically exfoliated WS_2 nanosheets serving as a prominent example.

Metal sulfides play an important role in photochemical water-splitting as they possess various advantageous characteristics, such as high absorption coefficients, band gaps, and considerable valence and conduction band orientation for O_2 and H_2 reactions. The reactions occur when exposed to light energy, causing the breakdown of H_2O and potential production of H_2 fuel. Numerous metal sulfides have been documented to be effective in the process of photochemical water-splitting. Molybdenum Sulfide (MoS_2), Cadmium Sulfide (CdS), Cobalt Sulfide (CoS), Iron Sulfide (FeS, FeS_2), Tungsten Disulfide (WS_2), Zinc Sulfide (ZnS), and Copper Sulfides (CuS, Cu_2S). These materials are focused on overcoming various challenges, such as stability and toxicity. The materials are optimized through doping or combining with material to enhance catalytic activity, charge separation, and light absorption. CdS has been reported as the best metal for the photocatalyst as it possesses a narrow bandgap (2.4 eV) and is capable of absorbing < 510 nm wavelength (Kamat, 2007). The available flat-band potential of CdS, which is 0.87 V, is high and facilitates the H_2O reduction (Meissner et al., 1988; Reber and Rusek, 1986; Arora et al., 1998; Darwent and Porter, 1981).

Hydrogenases are a highly efficient group of bio-enzymes that catalyze p^+ and e^- to produce molecular H_2 with minimal overpotential, approaching the thermodynamic potential closely. Their catalytic efficiency exceeds that of Pt, which is widely recognized as the best catalyst for the HER. It's worth noting that regardless of the type of hydrogenase, the active sites of these bio-enzymes exclusively contain iron (Fe) and/or nickel (Ni) atoms. Specifically, the active sites of the extensively studied bio-enzymes, [NiFe] and [FeFe] hydrogenases, consist of metal-sulfur clusters deeply embedded in proteins. CoS_2 materials, especially those with well-defined nanostructures, have emerged as strong competitors among catalysts for the HER in absence of noble metals. Jin's group has effectively manufactured metallic CoS_2 material with three distinct shapes: film, microwire, and nanowire (Faber et al., 2014). An extensive examination of these materials' compositions, behaviors, and durabilities was performed, leading to the establishment of a clear relationship between structure and performance. Their research yielded two crucial findings: (1) Enhancing the electrode's catalytic activity for the HER was achieved by manipulating CoS_2 at micro- and nano scales, significantly increasing its effective surface area. This emphasizes the crucial impact of morphology on CoS_2 overall catalytic effectiveness. (2) The operational stability of CoS_2 was improved by observing the restructuring at micro- and nano levels, allowing the release of gas bubbles from the interface. In a different investigation, Peng et al. effectively created a self-supporting electrode using a highly efficient ternary composite material consisting of CoS_2 nanosheets, graphene, and carbon nanotubes for the HER (Peng et al., 2014). The production of this substance was a two-step process: first, graphene was used to grow CoS_2 nanosheets, which were then combined with carbon nanotubes through vacuum filtration. Because of its significant specific surface area, outstanding electrical conductivity, and a nano-porous structure within a unified material system, this nanohybrid film has proven to be a highly efficient catalyst for H_2 evolution despite not containing precious metals (Zou and Zhang, 2015).

7.3.2.3 *Other semiconductors*

The promising approach for H_2 generation by photochemical water-dissociation can be found in semiconductors where solar energy is harnessed to supply renewable and sustainable energy carrier. The semiconductors absorb sunlight, and the generated e^- hole pairs are used to process the reaction for H_2 production. The system's productivity can be obtained with the characteristics of semiconductors, such as their stability, mobility, band edge positions, and bandgap. The semiconductors have been classified into four classes based on their electronic config. (1): D0 metal (Mo^{6+}, Ta^{5+}, Zr^{4+}, W^{6+}, Nb^{5+}, and Ti^{4+}); (2) d10 metal (Sb^{5+}, Sn^{4+}, Ga^{3+}, In^{3+}, and Ge^{4+}), (3) f0 metal (Ce^{4+}), and (4) nonoxide photocatalysts (Ismail and Bahnemann, 2014). Bismuth vanadate ($BiVO_4$, BVO) is commonly utilized as a material for the positive electrode in solar water-splitting. However, it faces difficulties in effectively separating charges due to its limited ability to transport charges. Hematite has several advantages, such as being widely available in nature, environmental friendly, exhibiting higher photochemical stability, having a narrow bandgap, and having the potential to obtain higher STH efficiency. Nonetheless, its limited light absorption capacity, short lifetime of excited states, slow kinetics of the oxygen-evolution reaction, short distance for holes to travel, and low electrical conductivity lead to the recombination of e^- and holes to recombine within the bulk material's, interfaces, and surfaces restricting the activity of hematite electrodes for converting light to electricity (Jha et al., 2024).

7.4 Cell Design

Cell design for water-splitting combines different components to capture solar energy and break H_2O into H_2 and O_2. The semiconductor-based photo-electrode is a crucial component of the system as it captures solar energy and creates e^- hole pairs. The charge carriers are responsible for the electrolyte and electrode reaction. These devices need considerable engineering for their optimized performance some of the key factors are discussed below:

7.4.1 Photoanode and photocathode design

Electrode configuration: A photochemical cell system is a composition of the counter electrolyte, working, and electrodes. It is necessary to consider the electrode material before designing, as it can influence charge transfer efficiency and overall cell performance. Electrodes such as indium tin oxide/fluorine-doped tin oxide electrode materials can enhance charge transport.

7.4.2 Light absorption and management

Optical design: The system should be designed with light-trapping structures, antireflection coatings, and optical filters to enhance performance and reduce and minimize reflection or transmission losses. The photochemical cell operation includes four stages: (1) light absorption, charge carrier, (2) generation, (3) separation, and (4) transportation to semiconductor-liquid junction) for the reaction on the surface. The semiconductors with appropriate bandgap absorb the light, which excites electron-hole pairs. These electrons separate, creating a hole and moving toward the photocathode (Kumar et al., 2022).

7.4.3 Material and electrolyte selection

Photoactive material: It is the most critical part of designing a photochemical cell; the material used in the cell facilitates the efficiency and spectral response of the system. Recently, semiconductors such as TiO_2, CdS, and perovskite materials have been studied they are capable of absorbing subsequent amounts of light and generating charge carriers. The material used for cell development must have suitable bandgap energy to absorb photons in the desired wavelength range. Semiconductors (photoactive materials) are crucial in photoelectrochemical water-splitting. The charge carriers must provide a required overpotential for reaction at the electrodes. During the experimentation, the semiconductors are capable of producing approx. 400 mV of photovoltage below its bandgap (Polman and Atwater, 2012). Semiconductors absorb solar radiation to enable the proper driving of OER and HER, allowing them to function within the bandgap range of 1.6–2.6 eV (Hisatomi et al., 2014; Yang et al., 2019). Various n-type semiconductors, such as $g\text{-}C3N_4$, ZrO_2, CdS, and TiO_2, have an appropriate band edge for OER. The wide bandgap of these materials results in low PEC efficiency in relevance to semiconductors that possess a narrow bandgap (Subramanyam et al., 2022). Another p-type semiconductor (p-Si, Sb_2Se_3, $CuInS_2$, Cu_2S, and Cu_2O) possesses a narrow bandgap and CB edges that make it optimal for the absorption of light these semiconductors are used for the development of photocathodes (Jang and Lee, 2019).

7.4.4 Integration of components

The photochemical cells devices are capable of converting light energy into electrical energy by undergoing photochemical reactions. Substrate Engineering: The surface between different cell components must be engineered properly for efficient charge separation, transport, and collection. Engineering techniques, such as surface modification, passivation layers, and heterojunction formation, can enhance the performance of photochemical cells. Electrolyte Selection: The electrolyte plays a critical role in charge transfer kinetics, redox reactions, and stability of the photochemical cell. The electrolytes can be liquid, solid-state, or gel-based, and they may contain redox-active species or sacrificial agents to facilitate charge transfer and minimize charge recombination. The PEC setup combines two electrode systems in a cell containing an aqueous electrolyte; one electrode or both can be photoactive (Bak et al., 2002). The cell is occupied with an optical window that acts as a photo-electrode that allows light interaction.

7.5 Future perspective and recommendations

As global energy demands rise and the impacts of fossil fuels on climate change become more pronounced, the quest for sustainable and clean energy sources intensifies. Photochemical water-splitting, a process that uses solar energy to generate H_2 from H_2O, is a promising technology for producing renewable energy. This chapter delves into the intricacies of optimizing materials and design strategies to enhance the efficiency and practicality of this technology. To significantly enhance PEC efficiency, the following parameters are absolutely crucial: (1) efficiency of photon absorption, (2) separation of charges within the bulk of the semiconductor, and (3) efficiency of charge injection into the surface. Looking to the future, several key areas are poised for significant advancement and impact.

7.5.1 Advanced material innovations

One of the primary drivers for progress in photochemical water-dissociation will be the development of novel materials. Future research is likely to focus on discovering and synthesizing new semiconductor materials with improved light absorption, charge separation, and catalytic properties. Emerging materials such as perovskites, transition-metal dichalcogenides, and organic-inorganic hybrids are anticipated to have a significant impact. Furthermore, adjusting the bandgap of these materials to better align with the solar spectrum will be a crucial area for investigation. Innovations in nanostructuring and surface modification techniques will also enhance the efficiency of these materials by increasing their surface area and catalytic activity.

7.5.2 Integration of nanotechnology

Nanotechnology will be at the forefront of enhancing photochemical water-splitting. Integrating nanoparticles, nanowires, and quantum dots can significantly improve light absorption and charge carrier dynamics. Future research will likely focus on optimizing these nanomaterials' size, shape, and composition to maximize their efficiency. Furthermore, developing plasmonic materials that can concentrate light at the nanoscale will offer new avenues for boosting photocatalytic activity. These advancements will improve the efficiency and lower the cost of photochemical H_2O-dissociation systems.

7.5.3 Artificial photosynthesis

Drawing inspiration from natural photosynthesis, artificial photosynthetic systems are poised to revolutionize photochemical water-splitting. Future developments will likely involve creating more efficient and stable biomimetic systems under various environmental conditions. Incorporating bio-inspired catalysts and mimicking the complex electron transfer processes found in nature will enhance these systems' overall efficiency and durability. Additionally, advancements in synthetic biology may allow for the integration of living organisms or their components into photochemical cells, leading to hybrid systems that combine biological and artificial elements.

7.5.4 System integration and scalability

Significant progress in system integration and scalability is essential for photochemical water-splitting to become a viable commercial technology. Future research will focus on designing photochemical cells that can be easily scaled up without compromising efficiency. This includes developing robust and cost-effective cell architectures, optimizing light management strategies, and enhancing the stability of materials for the longer term. Integration with other renewable energy technologies, such as solar panels and wind turbines, will also be crucial for creating hybrid systems that provide a continuous and reliable energy supply.

7.5.5 Economic and environmental considerations

The future of photochemical water-splitting will also depend on addressing economic and environmental challenges. Research efforts will be directed towards reducing the cost of materials and manufacturing processes to make the technology economically competitive with traditional energy sources. Additionally, lifecycle assessments will be crucial in minimizing the environmental impact of photochemical water-splitting. Sustainable sourcing of raw materials, recycling of components, and minimizing energy consumption during production will be key factors in achieving this goal.

The future of photochemical water-splitting is bright, with numerous opportunities for innovation and improvement. By focusing on advanced material development, integrating nanotechnology, drawing inspiration from natural processes, and addressing scalability and economic challenges, researchers can unlock this technology's full potential. As these advancements unfold, photochemical water-splitting could become a cornerstone of the global renewable energy landscape, contributing significantly towards a future of sustainable and renewable energy.

7.6 Conclusion

The exploration of materials and cell design for the photochemical splitting of water is a multidisciplinary endeavor that intersects chemistry, physics, materials science, and engineering. This section has given a thorough summary of the important factors and factors to consider in enhancing the materials and the structure of photochemical cells for the effective splitting of water. The selection of materials and cell design for photochemical splitting of water represents a dynamic and rapidly evolving field. By concentrating on enhancing material characteristics, improving cell structures, and resolving real-world implementation obstacles, scientists and engineers can unleash the complete potential of this promising technology. Progress in this area will make a substantial contribution to the growth of clean and renewable energy sources, opening doors to a more sustainable and environmentally friendly future. Through continued innovation and interdisciplinary collaboration, photochemical water-splitting can become a cornerstone technology in the global transition towards renewable energy. As the demand for renewable energy is increasing rapidly, it is required that systems be developed that can increase energy production without maximum carbon emission. Photochemical water-splitting is one of the systems that can produce a large amount of H_2 with less amount of carbon emission here, the H_2O is broken down into H_2 and oxygen using solar energy. This system can help protect the environment, as compared with the traditional methods of fuel production. In this chapter, the photochemical for water-splitting is discussed, and also, the chapter provides a brief about the parameters or conditions that need to be considered while designing the photochemical cells.

References

Abe, R., 2010. Recent progress on photocatalytic and photoelectrochemical water splitting under visible light irradiation. Journal of Photochemistry Photobiology C: Photochemistry Reviews 11, 179–209.

Arora, M.K., Sinha, A., Upadhyay, S., 1998. Active cadmium sulfide photocatalysts for hydrogen production from water. Industrial & Engineering Chemistry Research 37, 3950–3955.

Bak, T., Nowotny, J., Rekas, M., Sorrell, C., 2002. Photo-electrochemical hydrogen generation from water using solar energy. Materials-related aspects. International Journal of Hydrogen Energy 27, 991–1022.

Bard, A.J., Fox, M.A., 1995. Artificial photosynthesis: solar splitting of water to hydrogen and oxygen. Accounts of Chemical Research 28, 141–145.

Barrero, A.F., Oltra, J.E., Cuerva, J.M., Rosales, A., 2002. Effects of solvents and water in Ti (III)-mediated radical cyclizations of epoxygermacrolides. Straightforward synthesis and absolute stereochemistry of (+)-3α-hydroxyreynosin and related eudesmanolides. The Journal of Organic Chemistry 67, 2566–2571.

Boddy, P., 1968. Oxygen evolution on semiconducting TiO_2. Journal of the Electrochemical Society 115, 199.

Bolton, J.R., 1996. Solar photoproduction of hydrogen: a review. Solar Energy 57, 37–50.

Chandra, A.K., Uchimaru, T., 2000. A DFT study on the C–H bond dissociation enthalpies of haloalkanes: correlation between the bond dissociation enthalpies and activation energies for hydrogen abstraction. The Journal of Physical Chemistry A 104, 9244–9249.

Chen, X., Shen, S., Guo, L., Mao, S.S., 2010. Semiconductor-based photocatalytic hydrogen generation. Chemical Reviews 110, 6503–6570.

Chen, X., Xu, W., Shi, Z., Ji, Y., Lyu, J., Pan, G., et al., 2021. Plasmonic gold nanorods decorated Ti_3C_2 MXene quantum dots-interspersed nanosheet for full spectrum photoelectrochemical water splitting. Chemical Engineering Journal 426, 130818.

Chouhan, N., Liu, R.-S., Zhang, J., 2017. Photochemical Water Splitting: Materials and Applications. CRC Press.

Chu, S., Majumdar, A., 2012. Opportunities and challenges for a sustainable energy future. Nature 488, 294–303.

Cuerva, J.M., Campaña, A.G., Justicia, J., Rosales, A., Oller-López, J.L., Robles, R., et al., 2006. Water: the ideal hydrogen-atom source in free-radical chemistry mediated by Ti^{III} and other single-electron-transfer metals? Angewandte Chemie International Edition 45, 5522.

Darwent, J.R., Porter, G., 1981. Photochemical hydrogen production using cadmium sulphide suspensions in aerated water. Journal of the Chemical Society, Chemical Communications 145–146.

Domen, K., Kondo, J.N., Hara, M., Takata, T., 2000. Photo-and mechano-catalytic overall water splitting reactions to form hydrogen and oxygen on heterogeneous catalysts. Bulletin of the Chemical Society of Japan 73, 1307–1331.

Eckenhoff, W.T., Mcnamara, W.R., Du, P., Eisenberg, R., 2013. Cobalt complexes as artificial hydrogenases for the reductive side of water splitting. Biochimica et Biophysica Acta (BBA) - Bioenergetics 1827, 958–973.

Elimelech, M., Phillip, W.A., 2011. The future of seawater desalination: energy, technology, and the environment. Science 333, 712–717.

Faber, M.S., Dziedzic, R., Lukowski, M.A., Kaiser, N.S., Ding, Q., Jin, S., 2014. High-performance electrocatalysis using metallic cobalt pyrite (CoS_2) micro-and nanostructures. Journal of the American Chemical Society 136, 10053–10061.

Fujishima, A., Honda, K., 1972. Electrochemical photolysis of water at a semiconductor electrode. Nature 238, 37–38.

Gansäuer, A., Behlendorf, M., Cangönül, A., Kube, C., Cuerva, J.M., Friedrich, J., et al., 2012. H_2O activation for hydrogen-atom transfer: correct structures and revised mechanisms. Angewandte Chemie International Edition 13, 3266–3270.

Ghoussoub, M., Xia, M., Duchesne, P.N., Segal, D., Ozin, G., 2019. Principles of photothermal gas-phase heterogeneous CO_2 catalysis. Energy & Environmental Science 12, 1122–1142.

Hallenbeck, P.C., Benemann, J.R., 2002. Biological hydrogen production; fundamentals and limiting processes. International Journal of Hydrogen Energy 27, 1185–1193.

Hamdani, I., Bhaskarwar, A., 2021. Recent progress in material selection and device designs for photoelectrochemical water-splitting. Renewable and Sustainable Energy Reviews 138, 110503.

Haussener, S., Xiang, C., Spurgeon, J.M., Ardo, S., Lewis, N.S., Weber, A.Z., 2012. Modeling, simulation, and design criteria for photoelectrochemical water-splitting systems. Energy & Environmental Science 5, 9922–9935.

Hisatomi, T., Kubota, J., Domen, K., 2014. Recent advances in semiconductors for photocatalytic and photoelectrochemical water splitting. Chemical Society Reviews 43, 7520–7535.

Hoffmann, M.R., Martin, S.T., Choi, W., Bahnemann, D.W., 1995. Environmental applications of semiconductor photocatalysis. Chemical Reviews 95, 69–96.

Ikeda, T., Xiong, A., Yoshinaga, T., Maeda, K., Domen, K., Teranishi, T., 2013. Polyol synthesis of size-controlled Rh nanoparticles and their application to photocatalytic overall water splitting under visible light. The Journal of Physical Chemistry C 117, 2467–2473.

Ismail, A.A., Bahnemann, D.W., 2014. Photochemical splitting of water for hydrogen production by photocatalysis: a review. Solar Energy Materials and Solar Cells 128, 85–101.

Jang, Y.J., Lee, J.S., 2019. Photoelectrochemical water splitting with p-type metal oxide semiconductor photocathodes. ChemSusChem 12, 1835–1845.

Jha, B.K., Chaule, S., Jang, J.-H., 2024. Enhancing photocatalytic efficiency with hematite photoanodes: principles, properties, and strategies for surface, bulk, and interface charge transfer improvement. Materials Chemistry Frontiers 8, 2197–2226.

Jiao, F., Li, J., Pan, X., Xiao, J., Li, H., Ma, H., et al., 2016. Selective conversion of syngas to light olefins. Science 351, 1065–1068.

Kamat, P.V., 2007. Meeting the clean energy demand: nanostructure architectures for solar energy conversion. The Journal of Physical Chemistry C 111, 2834–2860.

Kaur-Ghumaan, S., Stein, M., 2014. [NiFe] hydrogenases: how close do structural and functional mimics approach the active site? Dalton Transactions 43, 9392–9405.

Kitano, M., Hara, M., 2010. Heterogeneous photocatalytic cleavage of water. Journal of Materials Chemistry 20, 627–641.

Klotz, D., Grave, D.A., Rothschild, A., 2017. Accurate determination of the charge transfer efficiency of photoanodes for solar water splitting. Physical Chemistry Chemical Physics 19, 20383–20392.

Kochuveedu, S.T., 2016. Photocatalytic and photoelectrochemical water splitting on TiO_2 via photosensitization. Journal of Nanomaterials 2016, 4073142.

Kudo, A., Miseki, Y., 2009. Heterogeneous photocatalyst materials for water splitting. Chemical Society Reviews 38, 253–278.

Kumar, M., Meena, B., Subramanyam, P., Suryakala, D., Subrahmanyam, C., 2022. Recent trends in photoelectrochemical water splitting: The role of cocatalysts. NPG Asia Materials 14, 88.

Law, M., Greene, L.E., Johnson, J.C., Saykally, R., Yang, P., 2005. Nanowire dye-sensitized solar cells. Nature Materials 4, 455–459.

Li, X., Zhao, L., Yu, J., Liu, X., Zhang, X., Liu, H., et al., 2020. Water splitting: from electrode to green energy system. Nano-Micro Letters 12, 1–29.

Li, X., Zhao, S., Duan, X., Zhang, H., Yang, S.-Z., Zhang, P., et al., 2021. Coupling hydrothermal and photothermal single-atom catalysis toward excellent water splitting to hydrogen. Applied Catalysis B: Environment and Energy 283, 119660.

Linsebigler, A.L., Lu, G., Yates, J.R., J.T., 1995. Photocatalysis on TiO_2 surfaces: principles, mechanisms, and selected results. Chemical Reviews 95, 735–758.

Maeda, K., Domen, K., 2010. Photocatalytic water splitting: recent progress and future challenges. The Journal of Physical Chemistry Letters 1, 2655–2661.

Martínez, A.R., Enríquez, L., Jaraíz, M., Thiessen, T., Sidhu, J., Mcindoe, J.S., et al., 2023. Understanding the color change of the solutions of Cp_2TiCl upon addition of water. Applied Organometallic Chemistry 37, e6979.

Martínez, A.R., Morales, L.P., Ojeda, E.D., Rodríguez, M.C., Rodríguez-García, I., 2020. The proven versatility of Cp_2TiCl. The Journal of Organic Chemistry 86, 1311–1329.

Martínez, A.R., Rodríguez-García, I., 2024. Photochemical water splitting via transition metal complexes. Journal of Catalysis 432, 115415.

Mcpherson, I.J., Vincent, K.A., 2014. Electrocatalysis by hydrogenases: lessons for building bio-inspired device. Journal of the Brazilian Chemical Society 25, 427–441.

Meissner, D., Memming, R., Kastening, B., 1988. Photoelectrochemistry of cadmium sulfide. 1. Reanalysis of photocorrosion and flat-band potential. The Journal of Physical Chemistry 92, 3476–3483.

Méndez-Ramos, J., Ruiz-Morales, J., Acosta-Mora, P., Del-Castillo, J., Yanes, A., 2013. Rare-earth doped nano-glass-ceramics for extending spectral response of water-splitting semiconductor electrodes by high intense UV-blue up-conversion: turning the sun into blue. Journal of Power Sources 238, 313–317.

Moriya, Y., Takata, T., Domen, K., 2013. Recent progress in the development of (oxy) nitride photocatalysts for water splitting under visible-light irradiation. Coordination Chemistry Reviews 257, 1957–1969.

Ni, G., Li, G., Boriskina, S.V., Li, H., Yang, W., Zhang, T., et al., 2016. Steam generation under one sun enabled by a floating structure with thermal concentration. Nature Energy 1, 1–7.

Nugent, W.A., Rajanbabu, T., 1988. Transition-metal-centered radicals in organic synthesis. Titanium (III)-induced cyclization of epoxy olefins. Journal of the American Chemical Society. 110, 8561–8562.

Paradas, M., Campana, A.G., Jiménez, T., Robles, R., Oltra, J.E., Bunuel, E., et al., 2010. Understanding the exceptional hydrogen-atom donor characteristics of water in Ti^{III}-mediated free-radical chemistry. Journal of the American Chemical Society 132, 12748–12756.

Peng, S., Li, L., Han, X., Sun, W., Srinivasan, M., Mhaisalkar, S.G., et al., 2014. Cobalt sulfide nanosheet/graphene/carbon nanotube nanocomposites as flexible electrodes for hydrogen evolution. Angewandte Chemie 126, 12802–12807.

Polman, A., Atwater, H.A., 2012. Photonic design principles for ultrahigh-efficiency photovoltaics. Nature Materials 11, 174–177.

Reber, J.F., Rusek, M., 1986. Photochemical hydrogen production with platinized suspensions of cadmium sulfide and cadmium zinc sulfide modified by silver sulfide. The Journal of Physical Chemistry 90, 824–834.

Ruscic, B., Wagner, A.F., Harding, L.B., Asher, R.L., Feller, D., Dixon, D.A., et al., 2002. On the enthalpy of formation of hydroxyl radical and gas-phase bond dissociation energies of water and hydroxyl. The Journal of Physical Chemistry A 106, 2727–2747.

Sastre, F., Puga, A.V., Liu, L., Corma, A., Garcia, H., 2014. Complete photocatalytic reduction of CO_2 to methane by H_2 under solar light irradiation. Journal of the American Chemical Society 136, 6798–6801.

Shih, C.F., Zhang, T., Li, J., Bai, C., 2018. Powering the future with liquid sunshine. Joule 2, 1925–1949.

Singh, S.B., 2021. Nanomaterials for water splitting: a greener approach to generate hydrogen. Handbook of Nanomaterials and Nanocomposites for Energy and Environmental Applications, pp. 1201–1220

Steier, L., Holliday, S., 2018. A bright outlook on organic photoelectrochemical cells for water splitting. Journal of Materials Chemistry A 6, 21809–21826.

Subramanyam, P., Meena, B., Biju, V., Misawa, H., Challapalli, S., 2022. Emerging materials for plasmon-assisted photoelectrochemical water splitting. Journal of Photochemistry Photobiology C: Photochemistry Reviews 51, 100472.

Swierk, J.R., Mallouk, T.E., 2013. Design and development of photoanodes for water-splitting dye-sensitized photoelectrochemical cells. Chemical Society Reviews 42, 2357–2387.

Tong, H., Ouyang, S., Bi, Y., Umezawa, N., Oshikiri, M., Ye, J., 2012. Nano-photocatalytic materials: possibilities and challenges. Advanced Materials 24, 229–251.

Wang, J., Kattel, S., Hawxhurst, C.J., Lee, J.H., Tackett, B.M., Chang, K., et al., 2019. Enhancing activity and reducing cost for electrochemical reduction of CO_2 by supporting palladium on metal carbides. Angewandte Chemie International Edition 58, 6271–6275.

Yang, J., Wang, D., Han, H., Li, C., 2013. Roles of cocatalysts in photocatalysis and photoelectrocatalysis. Accounts of Chemical Research 46, 1900–1909.

Yang, W., Prabhakar, R.R., Tan, J., Tilley, S.D., Moon, J., 2019. Strategies for enhancing the photocurrent, photovoltage, and stability of photoelectrodes for photoelectrochemical water splitting. Chemical Society Reviews 48, 4979–5015.

Zhao, H., Jian, L., Gong, M., Jing, M., Li, H., Mao, Q., et al., 2022. Transition-metal-based cocatalysts for photocatalytic water splitting. Small Structures 3, 2100229.

Zhao, Y., Gao, W., Li, S., Williams, G.R., Mahadi, A.H., Ma, D., 2019. Solar-versus thermal-driven catalysis for energy conversion. Joule 3, 920–937.

Zhou, M., Lou, X.W.D., Xie, Y., 2013. Two-dimensional nanosheets for photoelectrochemical water splitting: possibilities and opportunities. Nano Today 8, 598–618.

Zhu, L., Gao, M., Peh, C.K.N., Ho, G.W., 2018. Solar-driven photothermal nanostructured materials designs and prerequisites for evaporation and catalysis applications. Materials Horizons 5, 323–343.

Zou, X., Zhang, Y., 2015. Noble metal-free hydrogen evolution catalysts for water splitting. Chemical Society Reviews 44, 5148–5180.

CHAPTER

8

Noble metal-based catalysts for water splitting

Lekshmi Krishna[1], *Paulose Thomas*[1,2], *Angel Susan Cherian*[1], *and Anju K. Nair*[1]

[1]Department of Physics, Mar Thoma College, Tiruvalla, Kerala, India [2]Department of Physics, Mar Thoma College for Women, Perumbavoor, Kerala, India

8.1 Introduction

By decreasing the world's dependency on burning fossil fuels for energy, the use of hydrogen as a fuel source has the potential to slow down global warming. Electrocatalysts are important in this economy because they can increase the speed, efficiency, and selectivity of the hydrogen evolution reaction (HER), which is the main mechanism for producing $H_2(g)$ (Dresselhaus and Thomas, 2001). Unfortunately, because HER needs a costly electrocatalyst—mostly platinum—to produce measurable amounts of energy, its applicability for commercial applications is highly constrained. One of the main challenges facing the hydrogen economy is coming up with completely new, less expensive alternatives or developing HER electrocatalysts with lower Pt needs (Yang and Saidi, 2017, Yang et al., 2020).

The process of water splitting, which breaks down water (H_2O) into its component elements of hydrogen (H_2) and oxygen (O_2), has attracted a lot of interest from scientists because it may be a viable way to produce hydrogen. When utilized in fuel cells or internal combustion engines, hydrogen is regarded as a clean fuel that only produces water as a byproduct, which makes it a desirable substitute for fossil fuels (Turner, 2004). Water splitting is significant because it can help solve global energy issues in addition to having the potential to generate clean energy. There is a pressing need to discover sustainable and ecologically suitable alternative energy sources due to the depletion of fossil fuel supplies and the negative effects of their use on the environment. A workable solution to this issue is water splitting, which is fueled by renewable energy sources like solar or wind (Fujishima and Honda, 1972).

The process of water splitting can be achieved through various methods, including electrolysis, photocatalysis, and thermochemical processes. The most developed and most applied of these is electrolysis, which involves passing an electric current through water to split it into hydrogen and oxygen (Pinaud et al., 2013). While thermochemical water splitting

Photochemical Splitting of Water
DOI: https://doi.org/10.1016/B978-0-443-29064-0.00010-3

includes high-temperature chemical processes powered by concentrated solar power or nuclear energy, photocatalytic water splitting employs sunlight to drive the reaction, mimicking natural photosynthesis (Gray, 2009).

Water splitting is important for reasons other than only energy generation. It affects climate change mitigation as well as environmental sustainability. Water splitting could lessen the need for carbon-intensive energy sources and help cut greenhouse gas emissions by offering a clean, renewable source of hydrogen (Lewis and Nocera, 2006). The oxygen that is created as a byproduct of water splitting can also be used in other industrial processes, which increases the technology's overall sustainability and efficiency.

The hydrogen generated has the potential to serve as a clean fuel and present a viable substitute for fossil fuels, which are accountable for substantial greenhouse gas emissions. However, the catalysts utilized to speed up the process have a significant impact on how well water splits. Because of their extraordinary catalytic qualities, noble metals like platinum (Pt), palladium (Pd), and gold (Au) have become important actors in this field. Noble metals are distinguished by their strong resistance to oxidation and corrosion, which is essential for catalyst stability over the long term in challenging conditions. They are able to efficiently assist oxygen evolution reaction (OER) and HER due to their distinct electronic structures. For example, because of its high exchange current density and low overpotential, platinum is considered to be one of the most effective catalysts for HER (Zhang et al., 2019). Similarly, because they effectively mediate the multielectron transfer stages necessary in oxygen synthesis, ruthenium and iridium oxides are among the greatest catalysts for OER (Chen et al., 2020).

Noble metals have great catalytic activity, but their high cost and restricted availability prevent them from being used widely in water splitting. Due to this, a lot of research has been done to either find other, earth-abundant minerals that can complement or replace these metals, or to reduce the amount of noble metal that is needed. Techniques that have showed promise in increasing catalytic activity while reducing the consumption of precious metals include alloying noble metals with less expensive elements, creating core-shell structures, and supporting noble metals on conductive substrates (Wang et al., 2021). Hybrid catalysts that combine the best qualities of both kinds of metals—noble and nonnoble—have been developed as a result, offering enhanced durability and performance (Jiang et al., 2022).

The goal of the chapter is to present a thorough analysis of the part these metals play in the development of water electrolysis science and technology. The process of splitting H_2O into O_2 and H_2 using an external energy source—typically electricity—is essential for producing hydrogen, a potential renewable energy carrier. The chapter's scope covers a number of important topics. First, it explores the basic ideas of water splitting, covering the kinetics and thermodynamics of the procedure. The difficulties in water splitting are discussed in this section, with special attention to the high energy barriers that must be overcome and the need of selecting the right catalyst materials for effective operation.

Because of their remarkable electrochemical qualities, noble metals including palladium (Pd), ruthenium (Ru), iridium (Ir), and platinum (Pt) are presented as excellent possibilities for catalytic materials. The special properties of these metals—such as their high catalytic activity, superior conductivity, and corrosion resistance—that make them appropriate for HER are covered in this chapter. These are the two half-reactions that cause water to split, and when it comes to catalyzing these reactions, noble metals have demonstrated better performance than other substances.

However, the chapter also discusses the drawbacks of using noble metals, chief among them being their high price and rarity. The economic feasibility and scalability of noble metal-based catalysts for industrial applications are greatly impacted by these considerations. As a result, the chapter examines a number of approaches to overcoming these difficulties, such as the creation of alloy catalysts, nanostructuring, and the application of composites based on noble metals. These methods seek to preserve or even improve catalytic performance while requiring less noble metal. The chapter's focus is further broadened to include new developments in noble metal substitutes as well as the contribution of theoretical and computational research to the comprehension and development of more effective catalysts. It looks at how nonnoble metals, transition metal dichalcogenides, and perovskite materials can be used as low-cost substitutes in hybrid systems that are supported by noble metals. The goal is to give a fair assessment of the current status of the field's research as well as its potential future paths.

The chapter examines the wider ramifications of using noble metals in water splitting, such as sustainability, environmental impact, and the possibility of incorporating noble metal catalysts into renewable energy systems, in addition to the scientific and technical elements. This comprehensive approach guarantees that readers will not only comprehend the materials science at hand in great detail, but will also recognize the larger context in which these technologies are being created and implemented. By the time this chapter ends, readers will have a solid grasp of the role that noble metals play in water splitting, as well as the opportunities and problems that come with using them. They will also have a better idea of what lies ahead for innovation in this important field of clean energy research.

8.2 Fundamentals of water splitting

Water splitting is a chemical process in which H_2O is broken down into its constituent elements, O_2 and H_2. This procedure is essential for many uses, such as the generation of hydrogen for fuel cells, the storage of renewable energy, and environmental control. Despite major developments in technology and materials science, the fundamentals of electrochemistry and basic chemistry still underpin the concepts of water splitting.

8.2.1 Thermodynamics of water splitting

Thermodynamics is the study of energy transitions in chemical reactions. The thermodynamic parameters that are most important for water splitting are enthalpy (ΔH) and Gibbs free energy (ΔG). For the electrolysis of water, the usual Gibbs free energy change is roughly 237.13 kJ/mol. The electrochemical process yields this value:

$$2H_2O(1) \rightarrow 2H_2(g) + O_2(g) \tag{8.1}$$

The Gibbs free energy change (ΔG) is given by:

$$\Delta G = \Delta H - T\Delta S \tag{8.2}$$

where T is the temperature, ΔS is the entropy change, and ΔH is the enthalpy change. At 298 K (standard circumstances), ΔH for water splitting is roughly 285.83 kJ/mol and ΔS is 163.3 J/(mol•K) (Noggle and Smith, 1998). The overpotential, or excess voltage needed above and beyond the theoretical value to drive the reaction at a practical rate, affects the electrolysis process' efficiency. Many factors, including as resistive losses and electrode dynamics, can lead to overpotential.

8.2.2 Kinetics of water splitting

The study of reaction rates and processes is part of the kinetics of water splitting. The surface characteristics and electrode materials have an impact on the electrolysis reaction rate. The following half-reactions at the anode and cathode can be used to characterize the water-splitting reaction:

Anode reaction (OER):

$$2H_2O(1) \rightarrow O_2(g) + 4H + (aq) + 4e^- \tag{8.3}$$

Cathode reaction (HER):

$$4H + (aq) + 4e^- \rightarrow 2H_2(g) \tag{8.4}$$

The overall reaction is:

$$2H_2O(1) \rightarrow 2H_2(g) + O_2(g) \tag{8.5}$$

The electrode material, the presence of catalysts, and the operating parameters (such as temperature and pressure) all affect how quickly these reactions occur. For instance, it is well known that catalysts such as iridium oxide and platinum can dramatically increase reaction speeds by reducing the activation energy (Fujishima and Honda, 1972).

8.3 Catalysis in water splitting

A key component in raising the effectiveness of water splitting is catalysts. They speed up the reaction by lowering the activation energy needed for the process. Metal oxides and noble metals are frequently employed as catalysts in electrolysis. Research is still being done to create catalysts that are more affordable and effective, including compounds based on nonnoble metals (Koper, 2013). Water splitting, which uses a variety of techniques to split water into hydrogen and oxygen, is an essential step in the production of renewable energy. Three main methods of splitting water are discussed in this paper: photocatalysis, electrolysis, and photoelectrochemical (PEC) processes.

8.3.1 Electrolysis

The most well-known technique for splitting water into hydrogen and oxygen is electrolysis, which uses an electric current to fuel the chemical reaction. An electrolyzer, which has two electrodes—a cathode and an anode—separated by an electrolyte, is where the process takes place.

In electrolysis, water is decomposed into hydrogen and oxygen gas. The overall reaction can be represented by:

$$2H_2O(1) \rightarrow 2H_2(g) + O_2(g) \tag{8.6}$$

The anode reaction involves the oxidation of water:

$$2H_2O(1) \rightarrow O_2(g) + 4H^+ + 4e^- \tag{8.7}$$

The cathode reaction involves the reduction of protons to form hydrogen:

$$4H^+ + 4e^- \rightarrow 2H_2(g) \tag{8.8}$$

Different types of electrolyzers are used, including:

- Proton exchange membrane (PEM) electrolyzers: Use a solid polymer electrolyte and offer high efficiency and quick start-up times (Barbir, 2005).
- Alkaline electrolyzers: Employ a liquid alkaline electrolyte and are more mature but less efficient compared to PEM electrolyzers (Srinivasan, 2007).
- Solid oxide electrolyzers: Operate at high temperatures and are efficient but require significant energy input (Zimmerman et al., 2015).

8.3.2 Photocatalysis

With photocatalysis, light is used to accelerate the water-splitting process; semiconductor materials are usually used as catalysts. A photocatalyst absorbs photons during the process, producing electron–hole pairs that aid in the water-splitting reaction.

The photocatalytic water-splitting reaction can be expressed as:

$$H_2O(1) + hv \rightarrow 2H_2(g) + O_2(g) \tag{8.9}$$

where hv represents the energy from photons absorbed by the photocatalyst. Common photocatalysts include titanium dioxide (TiO_2) and hematite (Fe_2O_3).

8.3.2.1 Materials and mechanisms

- Titanium dioxide (TiO_2): Widely used due to its stability and efficiency under UV light. However, it has limited activity under visible light (Fujishima and Honda, 1972).
- Hematite (Fe_2O_3): Promising for visible light photocatalysis but requires efficient charge separation (Kumar et al., 2010).

8.3.3 Photoelectrochemical processes

Electrolysis and photocatalysis are combined in PEC processes. PEC cells work by converting light into electrical energy, which powers the PEC cell's water-splitting mechanism. A semiconductor material in a PEC cell absorbs light to produce electron–hole pairs. The water-splitting processes are then propelled by these charge carriers:

At the photoanode: Water oxidation occurs:

$$2H_2O(1) \rightarrow O_2(g) + 4H^+ + 4e^- \quad (8.10)$$

At the photocathode: Hydrogen is generated:

$$4H^+ + 4e^- \rightarrow 2H_2(g) \quad (8.11)$$

8.3.3.1 Materials and systems

- Cadmium telluride (CdTe) and copper indium gallium selenide (CIGS): Used for their high efficiency in converting sunlight into electrical energy (Chen et al., 2015).
- Tandem cells: Combine multiple semiconductor layers to improve efficiency (Tong et al., 2017).

8.4 Noble metals: properties and characteristics

Noble metal catalysts are unique materials that use precious metals like platinum, gold, or palladium to speed up chemical reactions. They are elements renowned for their resistance to corrosion and oxidation. This inherent stability is due to their electronic configuration, making them less reactive than other metals. These metals are very resistant to rust and other damage and have unique properties that make them incredibly good at helping different chemical processes happen (Zeng et al., 2020).

8.4.1 Types of noble metals

8.4.1.1 Platinum

Platinum (Pt) is a noble metal that has garnered immense attention in various fields due to its exceptional chemical and physical properties. It belongs to the platinum group metals (PGMs) and is distinguished by its exceptional conductivity, strong corrosion resistance, and notable catalytic activity, particularly in redox reactions. Because of these characteristics, platinum is an essential element for many industrial uses, such as electronics, fuel cells, vehicle catalytic converters, and petroleum refining. Platinum also has outstanding catalytic activity and stability (Shao et al., 2016).

8.4.1.2 Palladium

PGMs include the noble metal palladium (Pd), which is prized for its exceptional corrosion resistance, chemical stability, and catalytic abilities. Since its discovery by William Hyde Wollaston in 1803, palladium has grown to be a crucial component of many industrial processes, most notably those involving electronics, hydrogen storage, and catalysis. Palladium finds widespread application in catalysis, particularly in automotive catalytic converters, where it catalyzes the oxidation of hydrocarbons and carbon monoxide, as well as the reduction of nitrogen oxides into less toxic compounds, thereby reducing harmful emissions (Schlapbach and Züttel, 2001).

8.4.1.3 *Rhodium*

One of the rarest and most precious transition metals in PGMs is rhodium (Rh). Its remarkable corrosion resistance, high reflectivity, and catalytic capabilities define it. Owing to its special qualities, rhodium is widely used in many industrial applications. Catalytic converters, in particular, are crucial for lowering dangerous emissions from automobile exhaust systems (Buchwalter et al., 2015).

8.4.1.4 *Ruthenium*

Ruthenium (Ru) is a rare transition metal. It is well known for having exceptional chemical and physical qualities that make it useful in many different fields, especially materials science, electronics, and catalysis. It works especially well in reactions involving hydrogenation, oxidation, and metathesis. Ruthenium nanoparticles are being researched in nanotechnology for their potential in antibacterial agents, cancer therapy, and catalysis, among other uses (Yu et al., 2013).

8.4.1.5 *Iridium*

Iridium is well known for having a high melting point—roughly 2446°C—and for having exceptional oxidation resistance, even at elevated temperatures. Because of these characteristics, it is perfect for usage in high-temperature settings, like crucibles and molds used in the manufacturing of single crystals, spark plugs, and other aerospace components. Iridium is important in catalysis because it can help with hydrogenation, dehydrogenation, and other chemical reactions. It is especially useful in the development of organic synthesis and automotive catalytic converters.

8.4.1.6 *Gold*

Gold (Au) is a noble metal with a long history and a variety of uses because of its special chemical and physical characteristics. It is well known for having outstanding oxidation and corrosion resistance, great electrical conductivity, and malleability, which make it indispensable in a variety of industries, including electronics and medical. Because gold is chemically inert and reluctant to produce oxides, it has earned its noble reputation. Due to its resistance to tarnish and corrosion in both air and water, gold objects from ancient civilizations frequently maintain their original condition. Gold is the perfect material for applications needing long-term durability because of this feature as well. Gold nanoparticles have special optical and electrical characteristics because of their huge surface area-to-volume ratio and quantum size effects (Haruta, 2003).

8.4.1.7 *Osmium*

Within the platinum group, osmium (Os) is a rare, hard, dense transition metal with exceptional chemical and physical properties. Osmium is one of the elements that is least common in the crust of the Earth and has a bluish-silver sheen. Osmium is a rather innocuous element chemically, but it can combine with other elements to generate volatile, highly poisonous compounds with strong oxidizing properties, as osmium tetroxide (OsO_4) (Cotton and Wilkinson, 1988).

8.4.2 Properties of noble metals

Noble metals, including gold (Au), silver (Ag), platinum (Pt), palladium (Pd), rhodium (Rh), ruthenium (Ru), iridium (Ir), and osmium (Os), are a group of metals characterized by their exceptional resistance to corrosion and oxidation, as well as their excellent catalytic and electrical properties. These metals are highly valued in various industrial, technological, and medical applications due to their unique physical and chemical attributes.

8.4.2.1 Electronic configuration of noble metals

The unique electrical configurations of noble metals, which usually include gold (Au), silver (Ag), and PGMs including platinum (Pt), palladium (Pd), rhodium (Rh), ruthenium (Ru), osmium (Os), and iridium (Ir), are what set them apart. Their chemical and physical properties are largely determined by these arrangements. One factor contributing to the chemical inertness of noble metals is that their electronic configuration usually ends in a partially or fully filled d-subshell. d-Orbitals and vhemical inertness: Because filled d-orbitals are not easily involved in chemical bonding, they are a contributing factor to the low reactivity of noble metals. Noble metals have exceptional resistance to corrosion, in large part because of this stability (Greenwood and Earnshaw, 1997).

8.4.2.2 High surface area and thermal stability

The great surface area and thermal stability of noble metals are essential for their usage in high-temperature applications and catalysis. More active sites for catalytic reactions are provided by the increased surface area of noble metals, particularly when they are employed as nanoparticles. Because of this, they function as extremely efficient catalysts in reactions involving hydrogenation, oxidation, and fuel cells.

Noble metals are able to withstand high temperatures without losing their structure or capacity to function. This characteristic is very crucial for high-temperature sensors and catalytic converters, among other uses (Haruta and Date, 2001).

8.4.2.3 Resistance to corrosion

Since they do not readily lose electrons to form oxides or other compounds, noble metals have an exceptional resistance to corrosion and oxidation, which is one of their most well-known characteristics. This property is especially valuable in harsh chemical environments where other metals would quickly degrade. Noble metals have an exceptional resistance to corrosion and oxidation because of their unique electronic structure and low reactivity.

Noble metals are intrinsically resistant to oxidation, even in the absence of a protective film, in contrast to other metals that depend on the formation of a protective oxide layer (passivation) to prevent corrosion. This makes them perfect for usage in electronic parts, jewelry, and medical equipment. Few established research works related to properties of noble metals are as follows. Somorjai and Park (2008) examine the contribution of surface atoms to the overall catalytic processes and the function of surface area in the catalytic activity of noble metals. The authors stress that noble metals' surface atoms have a special electronic structure that makes them incredibly active in catalytic processes. They also showed that the higher proportion of surface atoms in smaller noble metal particles results in better catalytic activities and more effective catalytic processes. Furche and Perdew (2006) examine the electrical structure of noble metals, focusing on the role that electron affinity plays in the chemical inertness of these metals. It offers a thorough examination of the elements that contribute to noble metal stability and discovered that noble metals' low electron affinity, which contributes to their chemical inertness and corrosion resistance, is what causes them to prevent the formation of negative ions. Li et al. (2015) examine the structural integrity and high-temperature catalytic performance of noble metal nanoparticles in order to determine their thermal stability. They proved that noble metal nanoparticles are appropriate for high-temperature industrial applications because they maintain their catalytic activity even at high temperatures. Wang and Chen (2013) discuss the noble metals' ability to withstand corrosion and emphasize how useful they are in situations when other materials tend to fail. And established that noble metals are perfect for use in hostile chemical conditions because they retain their structural integrity in extremely corrosive environments. Erlebacher and Sieradzki (2001) introduce the concept of nanoporous noble metals and their high surface area, which enhances their catalytic and sensing capabilities. The authors demonstrated that nanoporous noble metals have significantly increased surface areas, leading to enhanced catalytic activity and sensitivity in sensor applications. The research demonstrated that unique electrical and optical properties resulting from quantum size effects in noble metal nanoclusters are not observed in bulk materials, providing new opportunities for their use in nanoelectronics and photonics.

8.5 Noble metals in hydrogen evolution reaction

8.5.1 Mechanism of hydrogen evolution reaction

A key mechanism in electrochemical and PEC systems for the production of hydrogen gas is HER. Optimizing the efficiency of hydrogen production requires a thorough understanding of the HER mechanism. At the cathode of electrochemical cells, the HER involves the reduction of protons (H3 ions) to create molecular hydrogen (H_2). The general response to HER is:

$$2H^+ + 2e^- \rightarrow H_2 \tag{8.12}$$

This reaction can be divided into several steps, including adsorption of protons, electron transfer, and desorption of hydrogen molecules.

8.5.2 Mechanistic pathways

The Volmer–Heyrovsky mechanism and the Volmer–Tafel mechanism are the two main processes via which the HER can proceed. These routes outline various electrochemical processes.

8.5.2.1 Volmer–Heyrovsky mechanism

- Volmer step (hydrogen adsorption): Protons from the solution are adsorbed onto the electrode surface, where they are reduced by electrons to form adsorbed hydrogen atoms:

$$(H^+ + e^- \rightarrow H^*)$$

- Heyrovsky step (hydrogen evolution): Two adsorbed hydrogen atoms combine to form a hydrogen molecule, which then desorbs from the electrode surface: $H^{*}+H^{*} \rightarrow H_2$

8.5.2.2 Volmer–Tafel mechanism

- Volmer step (hydrogen adsorption): Similar to the first step in the Volmer–Heyrovsky mechanism, protons are adsorbed and reduced to form adsorbed hydrogen atoms: ($H^{+}+e^{-} \rightarrow H^{*}$)
- Tafel step (hydrogen evolution): Two adsorbed hydrogen atoms combine to form a hydrogen molecule directly, which then desorbs: ($H^{*}+H^{*} \rightarrow$H2)

8.5.3 Factors affecting hydrogen evolution reaction

Several factors influence the HER efficiency, including:

8.5.3.1 Electrode material

The performance of the HER is greatly impacted by the electrode material selection. Because of their high activity and stability, platinum and platinum-based alloys are regarded as good catalysts (Greeley et al., 2006). But in search of more affordable options, nonprecious elements including nickel, cobalt, and their compounds are also being researched (Zhou et al., 2018).

8.5.3.2 pH of the solution

The pH affects the HER system. The Volmer–Heyrovsky mechanism is primarily followed by the reaction in acidic environments (low pH). The Volmer–Tafel mechanism is typically followed by the reaction in alkaline conditions (high pH) (Koper, 2013).

8.5.3.3 Overpotential

The operating circumstances and catalyst selection affect the needed overpotential to drive the HER. More effective catalysts are indicated by lower overpotentials (Jaramillo et al., 2007). This mechanism is typically favored at lower overpotentials.

8.5.4 Platinum metal-based catalysts for hydrogen evolution reaction

HER requires platinum (Pt) as a highly efficient catalyst in order to produce hydrogen through water splitting. Its distinct electrical characteristics and surface chemistry, which promote the development of hydrogen, are the sources of its efficiency. One of the most preferred active HER catalysts for producing H_2 is platinum (Pt). Nanostructured materials have been the focus of recent developments in platinum-based catalysts because they perform better than bulk platinum. High surface area-to-volume ratios provided by platinum nanoparticles improve catalytic activity. Faster electron and proton transport is also made possible by the small size of nanoparticles (Jiang et al., 2012). Different aspects of platinum, such as its nanowires and nanosheets, can have a big impact on HER efficiency. For example, because of their superior electrical characteristics and larger surface area, nanowires can demonstrate greater catalytic activity (Yuan et al., 2024).

The largest barrier to platinum's industrial use in hydrogen evolution processes (HER) is its high cost and poor stability in corrosive electrolytes. To address this, researchers have focused on ways to lower the required amount of platinum while maintaining its high efficiency (Ruqia and Choi, 2018). Developing carbon-based composites supported by platinum nanoparticles (Pt NPs) is a workable approach. These carbon-based materials can be engineered to have the necessary nanostructure and high specific surface area, which will give the Pt NPs plenty of active sites and the perfect matrix. The enhanced electrical conductivity of the carbon substrate accelerates electron transport and improves HER kinetics, which benefits Pt-supported carbon-based hybrids as well.

Chen et al. conducted a thorough investigation on the dissolving behavior of platinum (Pt) used as a counter electrode during the electrochemical measurement of HER activity. Their findings indicate that Pt is dissolved chemically and electrochemically in acidic or alkaline electrolytes, resulting in Pt redeposition on the working electrode, which significantly boosts HER activity. As illustrated in Fig. 8.1A, the higher potential limit of the Pt counter electrode increases with cyclic voltammetry (CV) scans that stretch to bigger negative values at the lower potential limit. The increase in the higher voltage limit on the Pt counter electrode varies in different catalyst systems (Fig. 8.1B, D, F). The maximum current of the carbon fabric is much greater than that of MoS_2 (Fig. 8.1A, C). This implies that a little potential differential on the Pt counter electrode results from a large change in the circuit's current (Fig. 8.1B). Conversely, the MoS_2 current hits a plateau in the oxygen adsorption region of the Pt CV curve, indicating that a little variation in current causes a discernible shift in the potential of the Pt counter electrode (Fig. 8.1D). The maximum current shown by the Pt counter electrode's CV curve (Fig. 8.1E) is even higher than that of the carbon cloth, suggesting a progressive rise in maximum potential with increasing current. Interestingly, the upper potential limit on the working electrode remains at 0 V versus Ag/AgCl, despite the lower potential limit on the Pt counter electrode decreasing as the potential scan progresses toward more negative values (Fig. 8.1B, D, F). This drop can be attributed to the higher current observed at position W3, Fig. 8.1A, C, E, during the backward scan. In materials like carbon cloth, a larger negative potential results in higher polarization (Fig. 8.1A). A lower minimum potential on the Pt counter electrode is the result of this higher current during the backward scan with greater negative scanning (Fig. 8.1B) (Chen et al., 2017).

The best electrocatalysts for HER in water splitting are thought to be platinum-based catalysts. Platinum's exceptional catalytic activity and stability make it a popular choice for the HER in acidic conditions. Even at high reaction rates, it also needs tiny overpotentials. Platinum's commercial potential is limited by its high cost and scarcity. In the past few decades, scientists have worked to create Pt-based catalysts that will increase platinum's catalytic efficiency and utilization. Among these attempts are methods to engineer the catalysts' shape, composition, and crystal phase. For instance, it has been demonstrated that PtP_2@PNC is more selective and has better oxygen reduction reaction (ORR) activity than commercial Pt/C. It is also resistant to CO and methanol poisoning (Yu et al., 2020). Transition metals such as iron (Fe), cobalt (Co), or nickel (Ni) can be alloyed with Pt to enhance its performance in alkaline electrolytes. This can improve the catalyst's cost-effectiveness; increase Pt dispersion; and reduce the amount of Pt needed.

Pt single atom with NiFe alloy may reach a current density of $-10\ mA/cm^2$, has a near-zero onset potential, and a reduced overpotential. Pt-NiFe alloy performs exceptionally

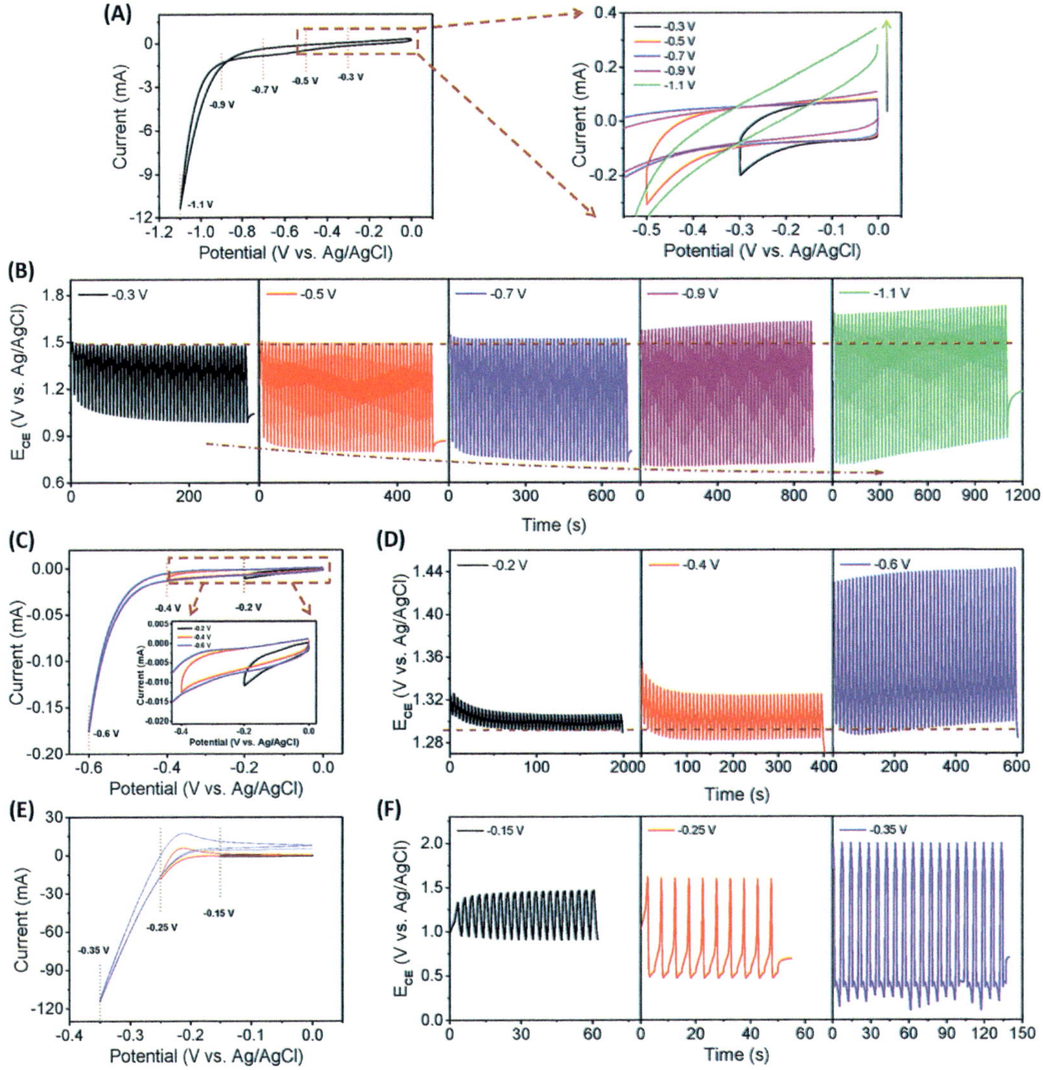

FIGURE 8.1 (A, C, E) CV curves with a fixed upper potential limit at 0 V versus Ag/AgCl and a different lower potential limit for the carbon cloth (2 cm^2), MoS_2 (0.196 cm^2), and the Pt plate (2 cm^2), respectively. (B, D, F) The corresponding potential on the Pt counter electrode. Scan rate: 100 mV/s. Source: *Copyright reserved to the American Chemical Society (Chen et al., 2017).*

well in alkaline environments. Following a long-term stability test, Pt-NiFe is very stable. In a study, murexide and cetyltrimethylammonium chloride (CTAC) were used as the costructure-directing agents to create flower-like platinum-cobalt-ruthenium alloy nanoassemblies (PtCoRu NAs). This method was found to be generic and easy to use. With a low overpotential (η, 22 mV) to drive 10 mA/cm^2, a small Tafel slope (46 mV/dec), and a high exchange current density (j0, 3.30 mA/cm^2) during the long-term electrolysis, the as-prepared PtCoRu NAs demonstrated remarkably enhanced electrocatalytic performance for the HER in 1.0 mM KOH. The Pt/CoNi alloy exhibits exceptional stability over an extended period of use, with a mass activity of 32.4 A mgPt at an overpotential of 20 mV and a turnover frequency of up to 32.86 s at 20 mV.

8.5.5 Palladium metal-based catalysts

In HERs, palladium (Pd) can be utilized in place of platinum (Pt). Nanoparticles of palladium (Pd) have been proposed as a viable substitute for catalysts based on platinum in the HER. Pd nanoparticles, which have similar catalytic capabilities but are more abundant and less expensive, are being thoroughly investigated for their potential to facilitate efficient and economical hydrogen synthesis by water splitting. Using a silica hard template (SBA-15), Pd nanoparticles (Pd–NP) can be produced and utilized as an electrocatalyst in electrochemical methanol reformation (ECMR) procedures. Pd–NPs can also be anchored onto graphene to form a nanocomposite that exhibits enhanced catalytic activity for the hydrogen evolution reaction (HER).

Like platinum, palladium is a member of the noble metal family, which is renowned for its superior catalytic qualities. Pd is useful for HER because of its potent hydrogen adsorption and desorption properties. Palladium's catalytic effectiveness is ascribed to its facile formation of hydrides, a property that is pivotal for the development of hydrogen (Tian et al., 2016). Furthermore, palladium's strong activity during the HER process is attributed to its effective electron transport during the d-band electronic structure (Marković et al., 1997). Palladium nanoparticles (Pd NPs) have replaced bulk Pd as the material of choice for HER, greatly enhancing its catalytic performance. High surface area-to-volume ratios provided by Pd NPs enhance the quantity of active sites accessible for the reaction (Zhao et al., 2017). Pd NPs' catalytic activity is significantly influenced by their size and shape. Because they have a bigger surface area and more active sites for HER, smaller nanoparticles often show stronger catalytic activity (Kou et al., 2018). Furthermore, distinct geometries, such cubes, rods, and spheres, reveal distinct crystallographic facets that might modify the hydrogen atom-binding strength and, consequently, the HER efficiency (Chen et al., 2014). When assessing HER activity, Pd NPs' surface structure is a key factor. The electronic characteristics of Pd can be adjusted to improve catalytic efficiency by adding strain, flaws, or alloying with other metals to the surface (Xiao et al., 2017). Alloying Pd with elements such as Ni, Co, or Au has been shown to improve stability and reduce the overpotential required for HER (Zheng et al., 2015).

Pd-based alloys and composites have been investigated by researchers in an effort to improve Pd NPs' performance for HER even more. Pd's electrical environment can be changed by alloying it with other transition metals, which will increase catalytic stability and efficiency. It has been demonstrated that Pd alloys containing nickel (Ni) or cobalt (Co) have better HER activity than Pd that is pure. The electronic structure of Pd is altered by the presence of Ni or Co, which lowers the energy barrier for the adsorption and desorption of hydrogen (Liu et al., 2018). Additionally, these alloys show enhanced resistance to corrosion and deactivation in both alkaline and acidic conditions. To improve the performance of Pd NPs, carbon-based supports like graphene and carbon nanotubes are frequently employed. Stronger contact between Pd NPs and the carbon support reduces agglomeration and improves stability, whereas high conductivity of carbon materials enables faster electron transport during HER (Wang et al., 2015). For instance, Pd NPs scattered across graphene have demonstrated exceptional HER activity with low overpotential and great durability (Cheng et al., 2016).

When employed as an electrocatalyst for the hydrogen evolution process (HER), palladium nanoparticles contained within hollow graphitized carbon nanofibers demonstrate a remarkable rise in activity after 30,000 reaction cycles, even outperforming the performance of platinum on carbon. The PdNP@GNF hybrid was created by immersing GNF in a $Pd_2(dba)_3$ (tris(dibenzylideneacetone)dipalladium(0)) chloroform solution. This caused Pd nanoparticles to develop spontaneously, mainly at the graphitic step-edges within the GNF. The production of evenly dispersed Pd nanoparticles with an average diameter of 4.37 ± 1.03 nm was verified by high-resolution transmission electron microscopy (HRTEM) pictures (Fig. 8.2A Additionally, HRTEM confirmed that the Pd nanoparticles in the PdNP@GNF were metallic, with a lattice fringe spacing of 0.22 nm that corresponds to the (111) planes of the palladium face-centered cubic lattice (Aygün et al., 2021). The existence of metallic Pd in the PdNP@GNF hybrid material was further verified by comparing the powder XRD patterns for PdNP@GNF, GNF, and a palladium reference (Fig. 8.2C).

8.5.6 Iridium-based catalysts for hydrogen evolution reaction

Iridium (Ir) nanoparticles have garnered significant attention as catalysts for the HER, particularly as researchers seek alternatives to platinum (Pt) due to its high cost and limited availability. Iridium is a good option for HER applications because of its specific qualities, which include high stability, resistance to corrosion, and effective catalytic activity, especially in acidic settings. As a metal belonging to the platinum group, iridium shares many of the same catalytic characteristics as platinum, including superior stability and resistance to corrosion in acidic environments. Iridium is particularly appealing for long-term HER applications because, in contrast to platinum, it is less likely to degrade under severe electrochemical conditions (McCrory et al., 2013). For HER, iridium nanoparticles exhibit strong catalytic activity, requiring a little overpotential to initiate the reaction. Protons are adsorbed onto the iridium surface as part of the catalytic process of HER on iridium, which is then followed by electron transfer and the production of hydrogen gas (H_2) (Greeley et al., 2009). Optimizing HER performance requires striking an equilibrium between adsorption strength and desorption kinetics, and the surface chemistry of iridium makes this possible efficiently, enabling both stages of the reaction (Kou et al., 2018).

The remarkable stability of iridium nanoparticles in abrasive environments—especially in acidic electrolytes, which are frequently employed in proton exchange membrane (PEM) electrolyzers—is one of its main benefits. Because of its excellent oxidation resistance, iridium is less likely to dissolve and degrade than many other catalysts in comparable circumstances (Seh et al., 2017). Iridium nanoparticle production has been tuned to produce particular nanostructures that improve catalytic activity. To create iridium nanoparticles with regulated size, shape, and surface morphology, techniques such as chemical reduction, thermal decomposition, and electrochemical deposition are frequently used (Zhang et al., 2019). Iridium nanoparticles' size can be adjusted to enhance their surface area and active

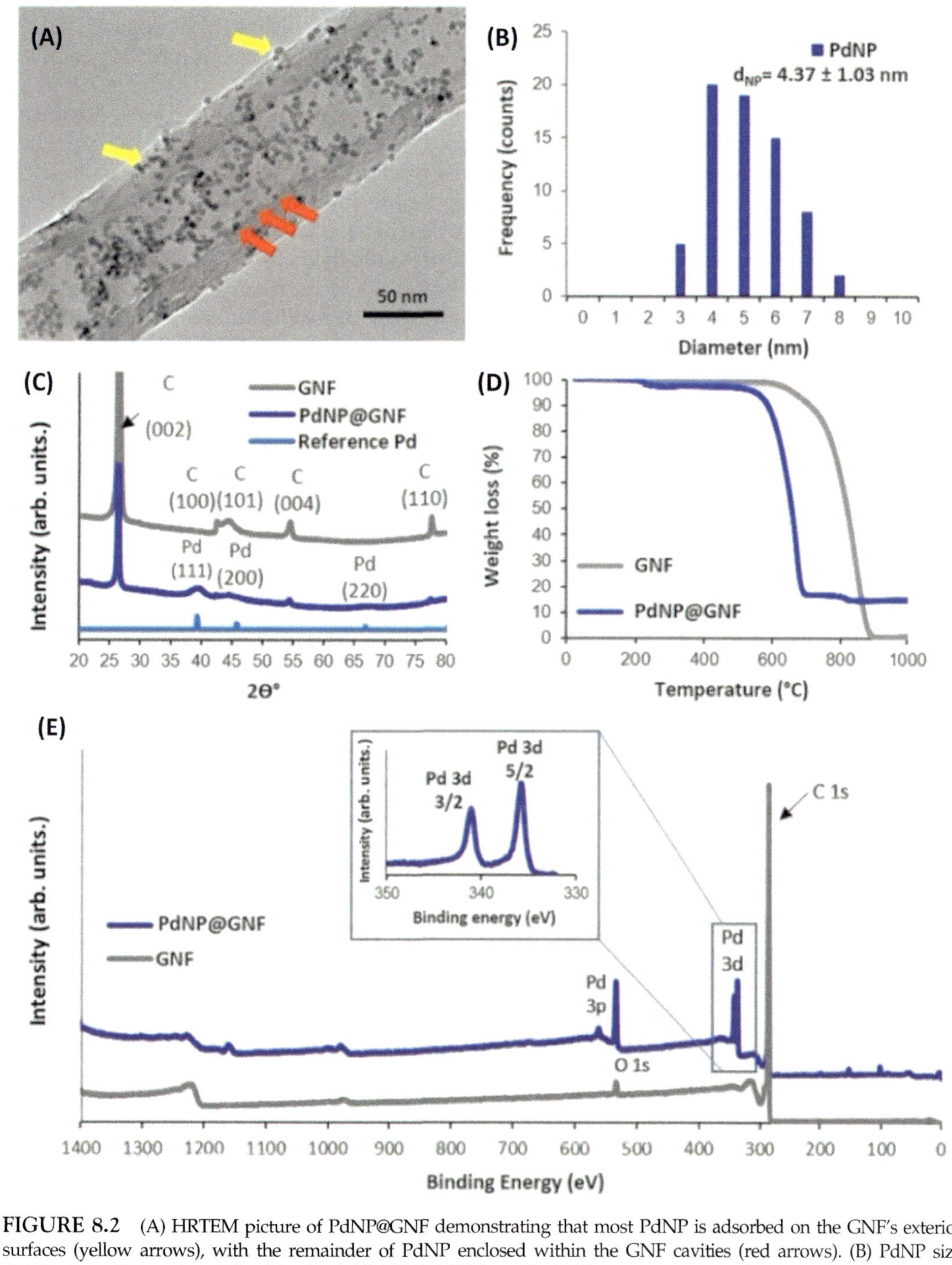

FIGURE 8.2 (A) HRTEM picture of PdNP@GNF demonstrating that most PdNP is adsorbed on the GNF's exterior surfaces (yellow arrows), with the remainder of PdNP enclosed within the GNF cavities (red arrows). (B) PdNP size distribution within and outside of the GNF. (C) PdNP@GNF powder XRD patterns compared to empty GNF and Pd as a metal reference. (D) Airborne TGA measurements for GNF (gray) and PdNP@GNF (blue). (E) High-resolution XPS spectra of PdNP@GNF in the Pd 3d region, as well as wide-scan XPS spectra of GNF and PdNP@GNF. Source: *Copyright reserved to the Creative Commons Attribution 4.0 Unported License, Aygün, M., Guillen-Soler, M., Vila-Fungueiriño, J.M., Kurtoglu, A., Chamberlain, T.W., Khlobystov, A.N., et al., 2021. Palladium nanoparticles hardwired in carbon nanoreactors enable continually increasing electrocatalytic activity during the hydrogen evolution reaction. ChemsusChem, 14 (22), 4973–4984.*

sites, both of which have a direct impact on HER activity. According to Jiang et al. (2019), smaller nanoparticles usually provide more active sites per unit mass, improving catalytic effectiveness.

In order to further enhance the HER performance of iridium nanoparticles, advanced nanostructuring methods have been investigated, such as the formation of core-shell structures or alloying with other metals. These methods can alter iridium's electrical structure, improving its catalytic qualities and possibly lowering the quantity of iridium needed (Fabbri et al., 2017). Iridium nanoparticles have outstanding HER activity and stability, but, similar to platinum, their high cost prevents them from being widely used in commercial applications. The creation of alloys based on iridium and the incorporation of iridium as a constituent of composite materials are two approaches taken to lower iridium loading while preserving good catalytic performance (Song and Hu, 2018). In some applications, iridium's resistance to deterioration may provide longer-lasting performance than platinum's, thereby offsetting the expense by lowering the frequency of catalyst replacement.

8.5.7 Ruthenium-based catalysts for hydrogen evolution reaction

Since ruthenium (Ru) exhibits superior catalytic qualities than those of platinum (Pt) at a lower cost, it has become a very attractive catalyst for HER. Ru-based catalysts, especially in acidic and alkaline media, have garnered a lot of interest due to their potential to replace or lessen the dependency on Pt in electrochemical water splitting. Because of its electronic structure, which enables efficient hydrogen intermediate adsorption and desorption, ruthenium exhibits high catalytic efficiency in HERs. Because Ru and Pt have comparable hydrogen-binding energies, Ru can catalyze the HER at low overpotentials. Ru is also more affordable and plentiful than Pt, which makes it a more economical choice for extensive uses (Cui et al., 2018). Nanostructuring, which raises the active surface area and facilitates better accessibility to active areas, can greatly improve Ru's catalytic performance.

Ru nanoparticles' huge surface area and improved electron transport capabilities contribute to their strong catalytic activity for HER. To increase their stability and catalytic efficacy, these nanoparticles can be spread across a variety of substrates, such as materials based on carbon (Zheng et al., 2015). Excellent HER activity has been shown by ultrathin Ru nanosheets, especially in alkaline environments. These nanosheets' two-dimensional shape makes more active sites visible, which increases catalytic effectiveness (Liu et al., 2017).

Researchers have looked into alloying Ru with other metals and creating Ru-based composites in an effort to improve the HER performance and stability of Ru-based catalysts even further. It has been demonstrated that alloying Ru with other transition metals, such as cobalt (Co) or nickel (Ni), increases HER activity and stability. According to Cao et al. (2019), these alloys have synergistic effects that improve hydrogen adsorption and reduce the overpotential needed for HER. Ru supported on carbon-based materials, like graphene or carbon nanotubes, performs better in HER because the carbon supports high electrical conductivity and wide surface area. According to Zhao et al. (2016), these composites give Ru nanoparticles a stable substrate, increasing their endurance and lowering the total catalyst loading.

Maintaining the long-term stability of Ru-based HER catalysts is a major challenge, especially in corrosive conditions. According to recent research, Ru-based catalysts can continue to have strong catalytic activity for extended periods of time, even in challenging circumstances (Jiang et al., 2018). To completely comprehend the breakdown mechanisms and create plans to increase these catalysts' endurance, more research is necessary. Even though Ru-based catalysts have a lot of potential for HER, current research is concentrated on creating extremely effective Ru-based alloys and composites in order to further reduce the Ru content. Nonnoble metal substitutes that can complement Ru and improve HER performance at the lowest possible cost are also being investigated.

8.6 Noble metal-based photocatalysts

Using light energy to speed up a chemical reaction when a catalyst is present is known as photocatalysis. Water splitting is the most often used application for producing hydrogen and breaking down organic contaminants. Chemical processes are propelled by electron–hole pairs created by photocatalysts' absorption of photons. By promoting charge separation and lengthening the lifetime of charge carriers, noble metals can greatly improve these photocatalytic activities when paired with semiconductor materials (Schrauben et al., 2012). Noble metal-based photocatalysts have attracted a lot of interest in the photocatalysis community because of their special qualities, which include good conductivity, strong electron affinity, and the capacity to sustain localized surface plasmon resonances (LSPRs). Due to these characteristics, noble metals such as platinum (Pt), silver (Ag), and gold (Au) are very useful for increasing the productivity of photocatalytic processes, which include the creation of hydrogen, the reduction of CO_2 and the degradation of environmental pollutants.

8.6.1 Types of noble metal-based photocatalysts

8.6.1.1 Gold-based photocatalysts

Since gold nanoparticles (AuNPs) have significant plasmonic characteristics, they are widely used in photocatalysis. Gold nanoparticles are very useful for visible light-driven photocatalysis because they show strong LSPR in the visible spectrum. Water splitting, CO_2 reduction, and the degradation of organic pollutants are just a few of the uses for Au-based photocatalysts (Xiao et al., 2014).

When paired with wide-bandgap semiconductors such as TiO_2, AuNPs can improve visible light absorption. The formation of electron–hole pairs is enhanced by the LSPR effect in AuNPs and subsequently transmitted to the semiconductor for catalytic processes. For instance, Au/TiO_2 composites have demonstrated improved efficacy in visible light pollutant degradation and water splitting (Tong et al., 2012). The capacity of AuNPs to sustain visible light-excited LSPRs is well known. Gold can efficiently capture light thanks to its plasmonic characteristics and produce "hot electrons," which are energizing electrons that can propel chemical reactions like water splitting (Zheng et al., 2015). According to Christopher et al. (2011), AuNPs' distinct ability to absorb and scatter light at particular wavelengths is dependent on their size, shape, and surroundings.

Hot electrons and holes are produced when AuNPs are exposed to light, which excites the LSPRs. Then, these active charge carriers can take part in the redox processes required for the splitting of water (Ingram and Linic, 2011). The produced hot electrons are either delivered straight to the reaction site, where they decrease protons to form hydrogen, or they are moved to the conduction band of a semiconductor (such as TiO_2). The holes have the ability to oxidize water simultaneously, producing oxygen (Mukherjee et al., 2013). According to Thomann et al. (2011), the creation of a Schottky junction between AuNPs and a semiconductor material (such as TiO_2 or ZnO) improves charge separation and lowers recombination losses, which raises photocatalytic efficiency.

Effective photocatalysis depends on the absorption of a wide range of light, including visible and near-infrared wavelengths, which is made possible by AuNPs' plasmonic effect (Linic et al., 2011). Effective water splitting depends on faster electron transfer and decreased recombination, both of which are made possible by the presence of AuNPs in the charge carrier dynamics (Zhou et al., 2014). Au-based photocatalysts are stable in severe reaction conditions because gold has a high resistance to oxidation and corrosion (Sivula et al., 2011).

By maximizing light absorption and charge separation, combining AuNPs with other semiconductor materials such as TiO_2, ZnO, or g-C_3N_4 has been demonstrated to increase photocatalytic activity (Tang et al., 2014). Gold's plasmonic capabilities can now be more effectively utilized thanks to the development of Au-core/shell structures with other metals or oxides, which have also increased photocatalyst stability and efficiency (Xiao et al., 2015). The electrical and plasmonic characteristics of AuNPs can be adjusted by adding additional metals to create alloys or doped structures, which will increase the water splitting efficiency even more (Liu et al., 2015).

8.6.1.2 Silver-based photocatalysts

Gold nanoparticles can be adjusted to the near-ultraviolet and visible ranges, but silver nanoparticles have a greater LSPR frequency. The promise of silver (Ag)-based photocatalysts in water-splitting applications has drawn a lot of attention recently. This is mainly because of their high electrical conductivity, special plasmonic characteristics, and capacity to improve visible light absorption. Ag-based photocatalysts work especially well in applications that need UV light with a high energy content (Li et al., 2015). AgNPs can increase semiconductors' photocatalytic activity by causing charge separation and transfer through plasmons. Well-researched systems, Ag/AgCl and Ag/ZnO composites, exhibit strong photocatalytic effectiveness in the degradation of organic dyes and water splitting when exposed to UV and visible light (Zhang et al., 2013; Christopher and Moskovits, 2017). Ag nanoparticles are adaptable for improving photocatalytic processes because their LSPR impact may be modified by varying their size, shape, and the surrounding dielectric environment (Zhang et al., 2018).

By adding Ag nanoparticles to TiO_2, the LSPR effect increases the material's ability to absorb visible light, increasing the water-splitting process' overall efficiency (Chen et al., 2010). Ag nanoparticles can also act as electron sinks, which can improve charge separation by lowering electron–hole pair recombination (Li et al., 2015). Composites of silver and zinc oxide (Ag-ZnO) are another well-known system. According to Zheng et al. (2014), the use of Ag nanoparticles increases the photocatalytic water-splitting efficiency by extending ZnO's photocatalytic activity into the visible range. Ag has also been used with materials based on graphene to create efficient photocatalysts. Ag-graphene composites take advantage of graphene's large surface area and superior electron mobility in addition to Ag nanoparticles' plasmonic enhancement. High-performance water splitting depends on effective charge separation and transfer, which is achieved by this combination (Liu et al., 2011). Additionally, Ag nanoparticle stabilization, aggregation prevention, and long-term catalytic activity maintenance are all made possible by the graphene matrix.

8.6.1.3 Platinum-based photocatalysts

Platinum is a very good cocatalyst because of its ability to separate and transmit charges efficiently, while being less plasmonic than Au or Ag. To increase photocatalytic activity, Pt nanoparticles are frequently mixed with other plasmonic materials (Zhao et al., 2016). PtNPs act as electron traps to help protons reduce to hydrogen when they are put onto semiconductors such as TiO_2. For HER in water splitting, Pt-based photocatalysts are therefore very effective. Further enhancing photocatalytic effectiveness is the strong contact between Pt and the semiconductor, which can enhance charge separation and lower recombination rates (Fujishima et al., 2000).

The Pt-based photocatalyst's capacity to absorb light, produce charge carriers, and effectively transfer those carriers to the HER's surface determines how effective this process will be (Zhao et al., 2018). By optimizing the electrical characteristics and increasing the surface area, Pt NPs with regulated sizes and shapes enhance catalytic activity. For instance, because of improved hydrogen adsorption and desorption, Pt NPs with a high surface-to-volume ratio show higher HER activity (Jiang et al., 2012). The efficiency of the photocatalyst is increased when Pt is combined with carbon-based substances like graphene or carbon nanotubes. According to Zhang et al. (2016), these composites have excellent electrical conductivity, which promotes effective charge transfer and raises the HER efficiency as a whole. Plasmonic effects can further improve the efficacy of HER in Pt-based photocatalysts by stimulating surface plasmons, which can result in increased light absorption and localized heating (Ma et al., 2019). High catalytic activity can be sustained over time by stabilizing Pt NPs and preventing their aggregation through the incorporation of supports such as TiO_2 (Wang et al., 2015).

8.7 Noble metal-free alternatives

Noble metals, such as platinum and iridium, have traditionally been the preferred catalysts for water splitting, they excel in catalyzing HER and OER, respectively, the two half-reactions involved in water splitting. They exhibit exceptional catalytic activity due to their unique electronic properties and strong adsorption of water molecules. However, their widespread commercialization is hindered by several factors: Noble metals, such as platinum and iridium, are among Earth's most precious and expensive elements. Their scarcity and high demand drive their prices, making them a significant cost factor in water-splitting systems. The use of noble metals can substantially increase the overall cost of these systems, limiting their commercial viability.

The limited availability of noble metals raises concerns about their long-term sustainability. As demand for these metals increases, particularly in applications like water splitting and fuel cells, supply chain disruptions and price fluctuations are risks. This uncertainty can hinder the development and deployment of large-scale water-splitting systems. While noble metals are generally known for their durability, they can still be susceptible to degradation under certain operating conditions. High temperatures, corrosive environments, and exposure to contaminants can lead to the deterioration of noble metal catalysts, reducing their performance and lifespan. This can result in increased maintenance costs and potential system downtime.

8.7.1 Development of cost-effective alternatives

To address the limitations of noble metals in water splitting, researchers have explored a wide range of noble metal-free materials as potential catalysts. These materials encompass various classes, including:

8.7.1.1 Transition metal-based catalysts

Transition metals, such as nickel, cobalt, and iron, have been extensively investigated as alternatives to noble metals. Combining transition metals with other elements can modify their electronic properties and improve catalytic activity. For instance, nickel-iron alloys have shown promising results for OER. Li et al. (2016) demonstrated the effectiveness of a nickel-iron–layered double hydroxide catalyst for OER. Li et al. (2017) further explored the potential of NiFe-LDH nanosheets with controlled orientation and exposed active sites for efficient OER. Designing catalysts with hierarchical structures can increase the surface area available for reaction. For example, core-shell structures or nanostructured materials can enhance catalytic performance. Wang et al. (2016) reported a core-shell structured $NiCo_2O_4$@$NiMoO_4$ nanorod array for efficient oxygen evolution. Liu et al. (2017) presented a hierarchical nanostructured $NiCo_2O_4/NiFe_2O_4$ composite for OER.

Supporting transition metal catalysts on conductive and stable materials can enhance their performance and durability. Ge et al. (2017) demonstrated the effectiveness of nitrogen-doped graphene as a metal-free catalyst for oxygen reduction reaction. Wang et al. (2017) reported a covalent organic framework-supported cobalt nanoparticles as a highly efficient and stable catalyst for OER.

Metal Organic Frame work (MOFs) are highly porous materials with customizable structures. They can be tailored to create active sites for water splitting by incorporating metal ions or functional groups. Wang et al. (2017) developed a metal-organic framework with embedded cobalt single atoms as a high-performance catalyst for OER. Li et al. (2017) developed a highly active and stable perovskite oxide catalyst for OER. These noble metal-free materials offer a promising path toward developing more affordable and sustainable water-splitting technologies. Continued research and development in this area are essential to overcome the challenges associated with these materials and realize their full potential for clean energy production.

8.8 Comparative analysis of noble metals versus noble metal-free catalysts

While noble metal-free catalysts have made significant strides in recent years, they still face challenges in catalytic activity, stability, and selectivity compared to noble metals. Table 8.1 provides a comparative analysis of the critical properties of noble metals and noble metal-free catalysts for water splitting.

TABLE 8.1 Comparison of noble metals and noble metal-free catalysts for water splitting.

Property	Noble metals	Noble metal-free catalysts
Catalytic activity	High	Generally lower, but can be improved through design and optimization
Stability	Good	It can be variable, depending on the specific material and operating conditions
Selectivity	High	Can be less selective, leading to competing side reactions
Cost	High	Low
Availability	Limited	Abundant

Noble metals, particularly platinum and iridium, exhibit exceptional catalytic activity for HER and OER. Their unique electronic properties and strong adsorption of water molecules contribute to their superior performance. While noble metal-free catalysts have significantly improved, they often exhibit lower catalytic activity than noble metals. However, through careful design and optimization, their performance can be enhanced.

8.8.1 Stability

Noble metals are generally known for their strength and durability, making them suitable for long-term operation in water-splitting systems. On the other hand, noble metal-free catalysts can exhibit varying degrees of stability. The specific material and operating conditions can influence their stability. Some noble metal-free catalysts may be more prone to degradation or deactivation over time, limiting their lifespan.

8.8.2 Selectivity

Noble metals typically exhibit high selectivity for the desired water-splitting reactions, minimizing competing side reactions. Noble metal-free catalysts may be less selective, leading to the formation of undesired products or byproducts. This can reduce the overall efficiency of the water-splitting process and increase energy consumption.

8.8.3 Cost

Noble metals are significantly more expensive than noble metal-free catalysts. This high cost is a significant barrier to widespread commercializing water-splitting systems based on noble metals. On the other hand, noble metal-free catalysts offer a more cost-effective alternative, potentially making water-splitting technology more accessible and economically viable.

8.8.4 Availability

Noble metals are relatively scarce, which can limit their availability and potentially lead to supply chain disruptions. Noble metal-free catalysts derived from more abundant elements are generally more readily available. This can ensure a more stable and reliable supply for water-splitting applications.

Despite the challenges, developing noble metal-free catalysts is crucial to realizing a sustainable and scalable water-splitting technology. Continued research and innovation are essential to further improve their performance and enable widespread adoption in various applications, such as renewable energy production and hydrogen fuel generation.

8.9 Challenges and limitations

While noble metal-free catalysts have shown significant promise as alternatives to noble metals in water splitting, they still face several challenges that hinder their widespread commercialization. This chapter will delve into the key challenges and limitations associated with cost and resource scarcity, scalability and practical implementation, and stability and long-term performance issues.

8.9.1 Cost and resource scarcity

Despite the development of cost-effective alternatives, noble metal-free catalysts often require specialized synthesis techniques or the use of rare earth elements, which can increase their production costs. For instance, the synthesis of certain transition metal-based catalysts may involve complex chemical reactions or the use of specialized equipment, leading to higher manufacturing costs. Additionally, the availability and price of rare earth elements can fluctuate, impacting the overall cost of noble metal-free catalysts (Li et al., 2016).

For example, the synthesis of perovskite oxide catalysts, which are a promising class of noble metal-free materials, often involves the use of rare earth elements such as lanthanum or cerium. These elements can be relatively expensive and their availability may be subject to supply chain disruptions. As highlighted by Suntivich et al. (2011), the choice of rare earth elements can significantly influence the catalytic activity and stability of perovskite oxide catalysts. Furthermore, scaling up the production of these catalysts to meet the demands of large-scale water-splitting systems can be challenging. Ensuring consistent quality, reproducibility, and cost-effectiveness at a larger scale requires careful optimization of synthesis processes and materials selection. This may involve developing new manufacturing techniques or investing in advanced equipment, which can increase capital expenditures.

8.9.2 Scalability and practical implementation

Integrating noble metal-free catalysts into practical water-splitting devices poses several challenges. Electrode design, electrolyte compatibility, and system efficiency are key factors to consider. Ensuring that the catalyst is well-dispersed and adheres effectively to the electrode surface is crucial for optimal performance. The choice of electrolyte can also significantly impact the catalyst's activity and stability (Wang et al., 2016).

8.9.3 Electrode design and catalyst dispersion

The design of electrodes plays a crucial role in the performance of water-splitting systems. The electrode material, surface area, and morphology can influence the mass transfer of reactants and products, as well as the distribution of the catalyst. Optimizing electrode design can enhance the efficiency and durability of water-splitting devices.

Additionally, ensuring that the catalyst is well-dispersed and adheres effectively to the electrode surface is essential for maximizing its catalytic activity. Poor catalyst dispersion or detachment can lead to reduced performance and increased energy consumption. Various techniques, such as electrodeposition, spray coating, or drop casting, can be used to achieve uniform catalyst distribution on the electrode surface (Liu et al., 2017).

8.9.4 Electrolyte selection

The choice of electrolyte can significantly impact the reaction kinetics, selectivity, and stability of the catalyst. The electrolyte's pH, conductivity, and compatibility with the electrode material and catalyst are important factors to consider. For example, alkaline electrolytes are often preferred for the oxygen evolution reaction due to their higher conductivity and lower overpotential. However, the choice of electrolyte may also influence the stability and durability of the catalyst. Liu et al. (2017) demonstrated the importance of electrode design and electrolyte selection in achieving efficient OER using a hierarchical nanostructured $NiCo_2O_4/NiFe_2O_4$ composite. They found that a combination of a porous carbon electrode and an alkaline electrolyte provided optimal performance.

8.9.5 Long-term performance and durability

The long-term performance of noble metal-free catalysts in real-world applications may be influenced by factors such as temperature fluctuations, pH variations, and exposure to contaminants. These factors can affect the catalyst's stability, activity, and selectivity over time. Exposure to contaminants, such as impurities in the water source or corrosion products from the electrode material, can also have a detrimental effect on the catalyst's performance (Ge et al., 2017). Strategies such as catalyst encapsulation, protective coatings, and periodic maintenance can help to improve the longevity and performance of these catalysts. Integrating noble metal-free catalysts into practical water-splitting devices requires careful consideration of electrode design, electrolyte selection, and long-term performance. By optimizing these factors, it is possible to enhance the efficiency, durability, and commercial viability of water-splitting systems based on noble metal-free catalysts.

8.9.6 Stability and long-term performance issues

Noble metal-free catalysts may exhibit varying degrees of stability and durability under operating conditions. Factors such as temperature, pH, and exposure to contaminants can influence their performance and lifespan. Ensuring the long-term stability and reliability of these catalysts is crucial for the commercial viability of water-splitting technologies.

Temperature fluctuations can have a significant impact on the stability and performance of noble metal-free catalysts. High temperatures can accelerate degradation processes, such as sintering or phase transformation, leading to a decrease in catalytic activity and surface area. On the other hand, low temperatures can slow down reaction kinetics, reducing the overall efficiency of the water-splitting process.

The pH of the electrolyte can also influence the stability and activity of noble metal-free catalysts. For example, some transition metal-based catalysts may be more stable in alkaline environments, while others may be more susceptible to corrosion or leaching in acidic conditions. The pH can also affect the adsorption of water molecules and the formation of intermediate species during the water-splitting reaction. Exposure to contaminants can have a detrimental effect on the performance of noble metal-free catalysts. Impurities in the water source, corrosion products from the electrode material, or other contaminants can poison the catalyst's active sites, reducing its catalytic activity. Additionally, contaminants can accelerate degradation processes or promote side reactions, leading to decreased efficiency and selectivity.

8.10 Recent trends and future directions

Noble metals, particularly platinum and iridium, have long been the preferred catalysts for water splitting due to their exceptional catalytic activity. However, their high cost and scarcity have driven researchers to explore alternative materials. This chapter will discuss recent innovations in noble metal catalyst design, the development of hybrid and composite materials incorporating noble metals, and the future prospects for noble metals in water-splitting technologies. Recent advancements in nanotechnology and materials science have led to innovative designs for noble metal catalysts. These designs aim to maximize catalytic efficiency while minimizing the use of precious metals.

Fabrication of noble metal nanoparticles with controlled size, shape, and morphology has been a significant area of research. These nanostructures can exhibit enhanced catalytic activity due to their increased surface area and unique electronic properties. For example, platinum nanoparticles with specific facets or hollow structures have shown improved performance for HER. Chen et al. (2016) synthesized shape-controlled Pt nanocubes with exposed {100} facets, which demonstrated enhanced electrocatalytic activity for HER. Zhang et al. (2017) fabricated hollow Pt nanostructures that exhibited superior performance for HER compared to solid Pt nanoparticles.

Alloying noble metals with nonnoble metals can modify their electronic properties and improve their catalytic activity. For instance, alloying platinum with copper or cobalt can enhance its activity for OER. Greeley et al. (2007) used computational high-throughput screening to identify promising platinum-based alloy catalysts for OER. The choice of support material can significantly influence the performance of noble metal catalysts. Highly conductive and stable supports, such as carbon nanotubes or metal oxides, can enhance the dispersion and utilization of noble metals. Wang et al. (2016) demonstrated the effectiveness of graphene-supported NiFe-layered double hydroxide nanosheets for efficient oxygen evolution.

Single-atom catalysts, where individual noble metal atoms are dispersed on a support material, have emerged as a promising approach for maximizing noble metal utilization.

These catalysts can exhibit high catalytic activity and selectivity due to the unique electronic properties of isolated metal atoms. Qiao et al. (2011) pioneered the concept of single-atom catalysts and highlighted their potential for various catalytic reactions.

By incorporating these innovative design strategies, researchers have been able to develop noble metal catalysts with improved performance and reduced noble metal usage, making them more sustainable and economically viable for water-splitting applications.

Hybrid and composite materials combining noble metals with other materials have emerged as promising strategies for enhancing the catalytic performance of water-splitting systems. These materials can offer synergistic effects, such as improved activity, stability, and selectivity. This section will discuss various hybrid and composite materials incorporating noble metals and their potential applications in water splitting. Combining noble metals with metal oxides can create synergistic effects that enhance catalytic activity and stability. The metal oxide component can provide a high surface area, support for the noble metal, and modify the electronic properties of the catalyst. Platinum-cerium oxide composites have been extensively studied for OER. The cerium oxide component can provide oxygen vacancies and enhance the charge transfer kinetics, improving catalytic activity. Suntivich et al. (2011) investigated perovskite oxides, a class of metal oxide materials, for oxygen evolution catalysis and identified trends and design principles.

Noble metals supported on carbon-based materials, such as graphene or carbon nanotubes, can provide high conductivity, stability, and a large surface area for the dispersion of noble metal nanoparticles. Noble metals supported on graphene can exhibit enhanced catalytic activity and durability. Wang et al. (2016) used graphene as a support material for NiFe-layered double hydroxide nanosheets to enhance their catalytic activity for oxygen evolution. Carbon nanotubes can provide a high surface area and a conductive pathway for electron transfer.

Core-shell structures, where a noble metal core is surrounded by a protective shell, can improve the durability and selectivity of catalysts. The shell can protect the noble metal core from degradation and prevent agglomeration. Platinum-based core-shell structures have been widely studied for HER. The protective shell can be made of a metal oxide, carbon, or another suitable material. Wang et al. (2018) reported a highly efficient and stable Pt-Co alloy electrocatalyst for OER, which was fabricated using a core-shell structure.

8.11 Future prospects for noble metals in water-splitting technologies

Despite developing noble metal-free alternatives, noble metals are still expected to play a crucial role in water-splitting technologies, especially in high-performance and demanding applications. Future research will focus on:

8.11.1 Novel designs

Developing innovative catalyst designs, such as hierarchical structures, core-shell architectures, and single-atom catalysts, can enhance catalytic activity and selectivity.

8.11.2 Synthesis methods

Exploring new synthesis methods, including atomic layer deposition, colloidal synthesis, and electroplating, can improve the control over particle size, shape, and dispersion.

8.11.3 Support materials

Investigating novel support materials with tailored properties, such as high conductivity, stability, and compatibility with noble metals, can enhance catalytic performance.

8.11.4 Synergistic effects

Investigating new combinations of noble metals with other materials, such as metal oxides, carbon-based materials, or other catalysts, to achieve synergistic effects and improve catalytic performance.

8.11.5 Tailored properties

Designing hybrid and composite materials with specific properties, such as enhanced stability, selectivity, or activity, for water-splitting applications.

8.12 Conclusion

In conclusion, noble metals play a pivotal role in advancing water-splitting technologies due to their exceptional catalytic properties. The unique electronic structures of noble metals such as platinum, palladium, gold, silver, and iridium, enable efficient HER and OER by providing optimal active sites, promoting charge transfer, and enhancing reaction kinetics.

Despite their outstanding performance, the high cost and scarcity of noble metals pose significant challenges for their widespread application. Research efforts have increasingly focused on optimizing noble metal utilization through the development of alloy catalysts, core-shell structures, and the integration of noble metals with earth-abundant materials. Additionally, strategies like plasmonic enhancement, Schottky junction formation, and nanoscale engineering have shown promise in improving catalytic efficiency while reducing noble metal content.

Future research should continue exploring novel approaches to further minimize noble metal usage without compromising catalytic activity. This includes the development of advanced materials and hybrid systems that combine noble metals with other catalytic components to achieve synergistic effects. Ultimately, these advancements will contribute to more sustainable and economically viable water-splitting technologies, paving the way for efficient hydrogen production and a cleaner energy future.

References

Aygün, M., Guillen-Soler, M., Vila-Fungueiriño, J.M., Kurtoglu, A., Chamberlain, T.W., Khlobystov, A.N., et al., 2021. Palladium nanoparticles hardwired in carbon nanoreactors enable continually increasing electrocatalytic activity during the hydrogen evolution reaction. ChemsusChem 14 (22), 4973–4984.

Barbir, F., 2005. PEM Fuel Cells: Theory and Practice. Elsevier.

Buchwalter, P., Rosé, J., Braunstein, P., 2015. Multimetallic catalysis based on rhodium complexes: opportunities and challenges. Chemical Reviews 115 (1), 28–126.

Cao, X., et al., 2019. Ruthenium-based catalysts for hydrogen evolution reaction: recent advances and future perspectives. ChemElectroChem 6 (1), 98–105.

Chen, A., Chen, X., Gao, Y., 2014. Shape-controlled palladium nanocrystals for hydrogen evolution reaction. Chemical Society Reviews 43 (5), 1387–1398.

Chen, R., Yang, C., Cai, W., Wang, H.-Y., Miao, J., Zhang, L., et al., 2017. Use of platinum as the counter electrode to study the activity of nonprecious metal catalysts for the hydrogen evolution reaction. ACS Energy Letters 2 (5), 1070–1075.

Chen, X., Liu, L., Huang, F., 2010. Visible light photocatalytic hydrogen production activity of TiO_2 and Ag-TiO_2 nanocomposites. Applied Catalysis B 95 (1-2), 93–99.

Chen, X., et al., 2015. Photoelectrochemical water splitting using CdTe-based systems. Chemical Society Reviews 44 (8), 2243–2263.

Chen, Y., et al., 2016. Shape-controlled synthesis of Pt nanocubes with exposed {100} facets for enhanced electrocatalytic hydrogen evolution. Journal of the American Chemical Society 138 (40), 13308–13311.

Chen, Z., Duan, X., Wei, W., Wang, S., Ni, B.-J., Amal, R., 2020. Photothermal enhanced water electrolysis for efficient hydrogen production: a critical review. Journal of Materials Chemistry A 8 (39), 20483–20498.

Cheng, N., Stambula, S., Wang, D., 2016. Platinum single-atom and cluster catalysis of the hydrogen evolution reaction. Nature Communications 7, 13638.

Christopher, P., Moskovits, M., 2017. Hot charge carrier transmission from plasmonic nanostructures. Annual Review of Physical Chemistry 68, 379–398.

Christopher, P., Xin, H., Linic, S., 2011. Visible-light-enhanced catalytic oxidation reactions on plasmonic silver nanostructures. Nature Chemistry 3 (6), 467–472.

Cotton, F.A., Wilkinson, G., 1988. Advanced Inorganic Chemistry, fifth ed. John Wiley & Sons, New York.

Cui, W., et al., 2018. Ruthenium catalysts for hydrogen evolution reaction: theoretical understanding and recent progress. Journal of Catalysis 367, 50–62.

Dresselhaus, M.S., Thomas, I.L., 2001. Alternative energy technologies. Nature 414, 332.

Erlebacher, J., Sieradzki, K., 2001. Nanoporous noble metals. Nature 410, 450–453. Available at https://doi.org/10.1038/35068529.

Fabbri, E., Habereder, A., Waltar, K., Kötz, R., Schmidt, T.J., 2017. Developments and perspectives of oxide-based catalysts for the oxygen evolution reaction. Catalysis Science and Technology 4 (11), 3800–3821.

Fujishima, A., Honda, K., 1972. Electrochemical photolysis of water at a semiconductor electrode. Nature 238 (5358), 37–38.

Fujishima, A., Rao, T.N., Tryk, D.A., 2000. Titanium dioxide photocatalysis. Journal of Photochemistry and Photobiology C 1 (1), 1–21.

Furche, F., Perdew, J.P., 2006. The electron affinity of noble metals. The Journal of Physical Chemistry B 110, 20957–20964. Available at https://doi.org/10.1021/jp0656443.

Ge, X., et al., 2017. Nitrogen-doped graphene as a highly efficient metal-free catalyst for oxygen reduction reaction. Energy and Environmental Science 10 (2), 370–378.

Gray, H.B., 2009. Powering the planet with solar fuel. Nature Chemistry 1 (1), 7–8.

Greeley, J., et al., 2006. Computational study of hydrogen evolution on transition-metal surfaces. Journal of the American Chemical Society 128 (1), 188–190. Available at https://doi.org/10.1021/ja056525j.

Greeley, J., et al., 2007. Computational high-throughput screening of alloy catalysts for the oxygen reduction reaction. Angewandte Chemie International Edition 46 (35), 6157–6160.

Greeley, J., Jaramillo, T.F., Bonde, J., Chorkendorff, I., Nørskov, J.K., 2009. Computational high-throughput screening of electrocatalytic materials for hydrogen evolution. Nature Materials 8 (2), 120–125.

Greenwood, N.N., Earnshaw, A., 1997. Chemistry of the Elements, second ed. Butterworth-Heinemann.

Haruta, M., 2003. Gold as a novel catalyst in the 21st century: preparation, working mechanism and applications. Chemical Record 3 (2), 75–87.

Haruta, M., Date, M., 2001. Advances in the catalysis of Au nanoparticles. Applied Catalysis A 222 (1–2), 427–437.

Ingram, D.B., Linic, S., 2011. Water splitting on composite plasmonic-metal/semiconductor photoelectrodes: evidence for selective plasmon-induced formation of charge carriers near the semiconductor surface. Journal of the American Chemical Society 133 (14), 5202–5205.

Jaramillo, T.F., et al., 2007. Identification of active edge sites for electrochemical H2 evolution from a MoS_2 monolayer. Science 317 (5834), 100–102. Available at https://doi.org/10.1126/science.1146310.

Jiang, B., Li, Z., Zhen, W., Huang, Y., Ding, S., 2022. Noble metal/transition metal hybrids for efficient water splitting: from fundamentals to applications. Advanced Functional Materials 32 (8), 2108535.

Jiang, H., Zhang, J., Liu, X., 2012. Nanostructured platinum as catalysts for hydrogen evolution. Journal of the American Chemical Society 134 (32), 13271–13274.

Jiang, N., Shibata, N., Yokoyama, T., 2019. Synthesis of highly dispersed iridium nanoparticles on nitrogen-doped graphene as an efficient electrocatalyst for the hydrogen evolution reaction. Journal of Materials Chemistry A 7 (19), 11934–11943.

Jiang, Q., et al., 2018. Ruthenium-based catalysts for hydrogen evolution: toward a sustainable hydrogen economy. Nano Today 21, 65–82.

Koper, M.T.M., 2013. Catalysis of the hydrogen evolution and oxidation reactions: an electrochemical perspective. Journal of Electroanalytical Chemistry 662, 259–269. Available at https://doi.org/10.1016/j.jelechem.2010.06.010.

Kou, T., Zhu, W., Liu, J., 2018. The size effect of Pd nanoparticles on catalytic activity of hydrogen evolution reaction. Nano Energy 47, 24–31.

Kumar, P.R., et al., 2010. Hematite-based photocatalysts for solar water splitting. Journal of Materials Chemistry 20 (27), 5703–5713.

Lewis, N.S., Nocera, D.G., 2006. Powering the planet: chemical challenges in solar energy utilization. Proceedings of the National Academy of Sciences 103 (43), 15729–15735.

Li, H., Bian, Z., Zhu, J., Huo, Y., Li, H., Lu, Y., 2015. Mesoporous Au/TiO_2 nanocomposites with enhanced photocatalytic activity. Journal of the American Chemical Society 129 (14), 4538–4539.

Li, X., et al., 2016. A highly active and stable nickel-iron layered double hydroxide catalyst for oxygen evolution reaction. Journal of the American Chemical Society 138 (40), 13232–13235.

Li, X., Yu, J., Jaroniec, M., Chen, X., 2015. Cocatalysts for selective photoreduction of CO_2 into solar fuels. Chemical Reviews 115 (19), 9858–9878.

Li, Y., et al., 2017. NiFe-LDH nanosheets with controlled orientation and exposed active sites for efficient oxygen evolution reaction. Advanced Materials 29 (28), 1701329.

Linic, S., Christopher, P., Ingram, D.B., 2011. Plasmonic-metal nanostructures for efficient conversion of solar to chemical energy. Nature Materials 10 (12), 911–921.

Liu, W., et al., 2017. Ultrathin ruthenium nanosheets with enhanced activity for the hydrogen evolution reaction in alkaline media. Nature Communications 8, 15438.

Liu, X., Zhang, N., Xu, Y.-J., 2015. Metal nanoparticles supported on 2D graphene-based sheets for heterogeneous catalysis. ChemCatChem 7 (20), 2765–2787.

Liu, Y., et al., 2017. A hierarchical nanostructured $NiCo_2O_4$/$NiFe_2O_4$ composite for efficient oxygen evolution reaction. Journal of Materials Chemistry A 5 (18), 8997–9003.

Liu, Y., et al., 2017. Carbon nanotube-supported $CoMoO_4$ nanorods for efficient oxygen evolution reaction. Nanoscale 9 (12), 4528–4534.

Liu, Y., Jiang, Y., Zhang, H., 2018. Pd-Ni alloy nanoparticles as highly active and stable catalysts for the hydrogen evolution reaction. Journal of Materials Chemistry A 6 (9), 4234–4241.

Liu, Y., Yu, L., Hu, Y., Zhang, J., Xia, H., 2011. Graphene oxide-silver nanocomposites with ion-exchange and antibacterial properties. ACS Nano 5 (9), 7169–7177.

Ma, H., Liu, S., Zhang, X., 2019. Plasmonic enhancement in platinum-based photocatalysts for efficient hydrogen production. Nano Energy 60, 754–762.

Marković, N.M., Schmidt, T.J., Ross, P.N., 1997. Oxygen reduction reaction on palladium and platinum single-crystal electrodes. Journal of Physical Chemistry B 101 (27), 5405–5413.

McCrory, C.C.L., Jung, S., Peters, J.C., Jaramillo, T.F., 2013. Benchmarking heterogeneous electrocatalysts for the oxygen evolution reaction. Journal of the American Chemical Society 135 (45), 16977–16987.

Mukherjee, S., Libisch, F., Large, N., et al., 2013. Hot electrons do the impossible: plasmon-induced dissociation of H_2 on Au. Nano Letters 13 (1), 240–247.

Noggle, J.H., Smith, J.M., 1998. Principles of Chemical Thermodynamics. Prentice Hall.

Pinaud, B.A., Benck, J.D., Seitz, L.C., Forman, A.J., Chen, Z., Deutsch, T.G., et al., 2013. Technical and economic challenges of solid-oxide photoelectrochemical cells. Energy and Environmental Science 6 (1), 198–204.

Qiao, B., et al., 2011. Single-atom catalysts: a new frontier in heterogeneous catalysis. Chemical Society Reviews 40 (12), 6347–6376.

Ruqia, B., Choi, S.-I., 2018. Pt and Pt-$Ni(OH)_2$ electrodes for the hydrogen evolution reaction in alkaline electrolytes and their nanoscaled electrocatalysts. ChemSusChem 11, 2643–2653.

Schrauben, J.N., Sellers, K., Valdez, C.N., Barlow, S., Marder, S.R., Mayer, J.M., 2012. Nitrogen reduction by photogenerated electrons in CdS quantum dots. Journal of the American Chemical Society 134 (14), 9447–9450.

Seh, Z.W., Kibsgaard, J., Dickens, C.F., Chorkendorff, I., Nørskov, J.K., Jaramillo, T.F., 2017. Combining theory and experiment in electrocatalysis: insights into materials design. Science 355 (6321), eaad4998.

Shao, M., et al., 2016. Platinum-based nanostructures for electrocatalytic oxygen reduction. Chemical Reviews 116 (6), 3594–3657.

Sivula, K., Le Formal, F., Grätzel, M., 2011. Solar water splitting: progress using hematite (α-Fe_2O_3) photoelectrodes. ChemSusChem 4 (4), 432–449.

Somorjai, G.A., Park, J.Y., 2008. Surface chemistry of catalysis by metals. The Journal of Chemical Physics 128, 182504. Available at https://doi.org/10.1063/1.2913328.

Song, F., Hu, X., 2018. Exfoliation of layered double hydroxides for enhanced oxygen evolution catalysis. Nature Communications 5, 4477.

Srinivasan, S., 2007. Electrochemical Hydrogen Production. Springer.

Suntivich, J., et al., 2011. Perovskite oxides for oxygen evolution catalysis: activity trends and design principles. Science 334 (6053), 1383–1385.

Schlapbach, T., Züttel, A., 2001. Hydrogen-storage materials for mobile applications. Nature 414, 353–358.

Tang, J., Durrant, J.R., Klug, D.R., 2014. Mechanism of O_2 production from water splitting: nature of charge carriers and catalytic activity. Advanced Functional Materials 24 (26), 4446–4454.

Thomann, I., Pinaud, B.A., Chen, Z., et al., 2011. Plasmon enhanced solar-to-fuel energy conversion. Nano Letters 11 (8), 3440–3446.

Tian, H., Zhang, H., Liu, Z., 2016. Palladium-based electrocatalysts for hydrogen evolution reaction: from water splitting to water electrolysis. Journal of Materials Chemistry A 4 (31), 11711–11729.

Tong, H., et al., 2017. Tandem photoelectrochemical cells for enhanced water splitting. Advanced Energy Materials 7 (3), 1601394.

Tong, H., Ouyang, S., Bi, Y., Umezawa, N., Oshikiri, M., Ye, J., 2012. Nano-photocatalytic materials: possibilities and challenges. Advanced Materials 24 (2), 229–251.

Turner, J.A., 2004. Sustainable hydrogen production. Science 305 (5686), 972–974.

Wang, H., et al., 2016. A core-shell structured $NiCo_2O_4$@$NiMoO_4$ nanorod array for efficient oxygen evolution. Advanced Materials 28 (38), 8673–8681.

Wang, H., et al., 2016. Graphene-supported NiFe layered double hydroxide nanosheets for efficient oxygen evolution. Advanced Energy Materials 6 (10), 1600218.

Wang, J., et al., 2017. A metal-organic framework with embedded cobalt single atoms as a high-performance catalyst for oxygen evolution reaction. Journal of the American Chemical Society 139 (45), 16751–16754.

Wang, X., et al., 2017. Covalent organic framework supported cobalt nanoparticles as a highly efficient and stable catalyst for oxygen evolution reaction. Journal of the American Chemical Society 139 (34), 11836–11843.

Wang, X., Xu, Y., Liu, S., 2015. Titanium dioxide-supported platinum nanoparticles for enhanced photocatalytic hydrogen production. Journal of Catalysis 329, 24–31.

Wang, X., Chen, Y., 2013. Corrosion resistance of noble metals. Electrochimica Acta 109, 161–168. Available at https://doi.org/10.1016/j.electacta.2013.07.012.

Wang, Y., et al., 2017. A metal-organic framework with embedded nickel single atoms as a high-performance catalyst for hydrogen evolution reaction. Nature Communications 8, 15664.

Wang, Y., et al., 2018. A highly efficient and stable Pt-Co alloy electrocatalyst for the oxygen evolution reaction. Nature Communications 9, 5183.

Wang, Y., Vogel, A., Sachs, M., Sprick, R.S., Wilbraham, L., Moniz, S.J.A., et al., 2021. Current understanding and challenges of solar-driven hydrogen generation using polymeric photocatalysts. Nature Energy 6 (4), 450–460.

Xiao, M., Wang, Z., Lyu, M., et al., 2015. Plasmon-enhanced chemical reactions. Journal of the American Chemical Society 137 (9), 3044–3051.

Xiao, M., Wang, Z., Lyu, M., Luo, B., Wang, S., Liu, G., et al., 2014. Tuning the electron localization of Au cocatalyst for enhanced photocatalytic hydrogen evolution. Advanced Functional Materials 24 (18), 2770–2776.

Xiao, P., Chen, W., Wang, X., 2017. Tuning the electronic properties of palladium for enhanced hydrogen evolution reaction. Journal of Catalysis 354, 1–9.

Yu, F.Y., Lang, Z.L., Yin, L.Y., et al., 2020. Pt-O bond as an active site superior to Pt° in hydrogen evolution reaction. Nature Communications 11, 490. Available at https://doi.org/10.1038/s41467-019-14274-z.

Yang, T.T., Tan, T.L., Saidi, W.A., 2020. High activity toward the hydrogen evolution reaction on the edges of MoS2-supported platinum nanoclusters using cluster expansion and electrochemical modeling. Chemistry of Materials 32, 1315.

Yang, T.T., Saidi, W.A., 2017. Tuning the hydrogen evolution activity of β-Mo_2C nanoparticles via control of their growth conditions. Nanoscale 9, 3252.

Yu, Q., et al., 2013. Ruthenium nanoparticles: synthesis, properties, and applications in catalysis. Chemical Reviews.

Yuan, D., Hu, Z., Chen, Z., Liu, J., Sun, J., Song, Y., et al., 2024. Atomic-level tailoring of the electronic metal–support interaction between Pt-Co_3O_4 interfaces for high hydrogen evolution performance. The Journal of Physical Chemistry Letters 15 (13), 3486–3492. Available at https://doi.org/10.1021/acs.jpclett.4c00199.

Zeng, Z., et al., 2020. Recent advances in platinum-based catalysts for the oxygen reduction reaction: synthesis and electrochemical mechanisms. Advanced Materials 32 (25), 1902333.

Zhang, H., et al., 2019. Understanding the reaction mechanism of electrochemical water splitting by Ir-based electrocatalysts. Advanced Materials 31 (16), 1806926.

Zhang, J., et al., 2017. Hollow Pt nanostructures with enhanced electrocatalytic activity for hydrogen evolution reaction. Nano Letters 17 (1), 453–459.

Zhang, N., Zhang, Y., Xu, Y.J., 2013. Recent progress on advanced design for photocatalytic materials and applications. Nanoscale 5 (5), 1949–1961.

Zhang, Q., Li, Y., Ackerman, M.M., Yu, Y., Chen, X., 2018. Ag nanoparticle based plasmonic photocatalysis: mechanism, design and application. Chemical Reviews 118 (6), 2927–2954.

Zhang, Y., Zhang, L., Li, Z., 2016. Enhanced hydrogen evolution reaction performance using platinum/graphene composites. Nano Energy 27, 302–310.

Zhao, X., Yang, Y., Zhang, L., 2018. Recent developments in platinum-based photocatalysts for hydrogen evolution. Energy and Environmental Science 11 (5), 1207–1220.

Zhao, Y., et al., 2016. Ruthenium nanoparticles supported on nitrogen-doped graphene for efficient hydrogen evolution in both acidic and alkaline solutions. Journal of the American Chemical Society 138 (5), 1673–1681.

Zhao, Z., Yang, Y., Jiang, L., 2017. Palladium nanoparticles with high-index facets as high-performance electrocatalysts for hydrogen evolution reaction. ACS Catalysis 7 (8), 4796–4802.

Zheng, Y., et al., 2015. Toward design of synergistically active carbon-supported ruthenium catalysts for electrocatalytic hydrogen evolution. ACS Nano 9 (3), 2395–2403.

Zheng, Y., Jiao, Y., Jaroniec, M., Qiao, S.Z., 2014. Advances in heterogeneous photocatalysts for hydrogen evolution. Angewandte Chemie International Edition 54 (1), 52–65.

Zhou, H., et al., 2018. Non-precious metal catalysts for hydrogen evolution reaction. Nature Reviews Materials 3 (4), 17058. Available at https://doi.org/10.1038/natrevmats.2017.58.

Zhou, Y., Zhang, Y., Lin, Z., 2014. Plasmon-enhanced photoelectrochemical water splitting. Chemical Reviews 114 (19), 9543–9572.

Zimmerman, J.B., et al., 2015. Solid oxide electrolysis: the key to high-efficiency hydrogen production. Energy Conversion and Management 91, 1–16.

CHAPTER

9

Noble metal-free catalysts – oxides, chalcogenides, nitrides, oxynitrides for water splitting

Swapnali Walake[1], *Pragati Thakur*[2], *and Yogesh Jadhav*[1,3]

[1]Symbiosis Centre for Nanoscience and Nanotechnology (SCNN), Symbiosis International University, Pune, Maharashtra, India [2]Department of Chemistry, Savitribai Phule Pune University, Pune, Maharashtra, India [3]BacPlexTechnologies Private Limited, C/O AIC IISER Pune SEED Foundation IISER Pune Campus, Pashan, Pune, Maharashtra, India

9.1 Introduction

The transition from unsustainable and conventional fossil fuels to renewable sources of energy that are abundant, low-cost, and less harmful is a better choice to accomplish net-zero emissions in the world by 2050 (Rogelj et al., 2021; Wadhai et al., 2021). Therefore the need for technologies where clean and green energy sources are generated, stored, and transported is being explored. Different types of clean energy sources originated including solar energy, geothermal, wind energy, hydropower, bioenergy, and H_2 energy contributed to the world's environmental solutions over the consumption of fossil fuels like petrol, diesel, coal, natural gas, etc. Among all mentioned renewable energy sources, hydrogen has become a highly appreciated green energy source and also contributing to the world's economy. 140 MJ/kg of energy is the gravimetric energy density of H_2 gas that can be generated after its consumption with the evolution of water vapors but not harmful gases.

Production of clean air is another solution for air pollution that requires a high amount of O_2 in the environment diminishing carbon dioxide (CO_2) to fix increasing global warming. Therefore the production of O_2 through chemical reactions and industrial processes will help to resolve the issue. Since fossil fuels are limited as compared to renewable energy sources the use of simple methods to produce H_2 and a clean environment via O_2 is a challenge to all the researchers in the world. Herein, water splitting takes advantage of simultaneously producing gaseous H_2 and O_2 using external energy such as heat, potential, solar energy, etc. Depending on these, water splitting can be performed via photocatalytic,

Photochemical Splitting of Water
DOI: https://doi.org/10.1016/B978-0-443-29064-0.00012-7

EC, PEC, and thermochemical methods that are low-cost, bench-top, easily transportable set-ups than other industrial set-ups. These methods include catalysts that undergo several chemical reactions and interactions with water-based electrolytes to perform Homogeneous and Heterogeneous catalysis. Heterogeneous catalysis typically takes place in two different types of systems (solid and liquid) takes place during water splitting reactions called HER and OER. More importantly, the development of appropriate catalysts is a key factor in these types of reactions.

Based on chemical structures, catalysts are classified as transition metal oxides (TMOs), chalcogenides, nitrides, and oxynitrides. This class of catalysts is highly explored by the researchers identifying the literature gaps in their synthesis methodologies (Lu and Zhang, 2022; Suen et al., 2017), experimental parameters during HER and OER (Raveendran et al., 2023), stability (Zhao et al., 2022), cost-effectiveness, and reusable catalysts (Walake et al., 2023). The present chapter consists of synthesis methodologies of catalysts, fundamental parameters of experiments to validate HER, insights into HER activity for TMOs, Chalcogenides, Nitrides, Oxynitrides catalysts, and future challenges associated with it. The chapter also provides helpful insights for developing better catalysts of HER using magnetic catalysts that are recoverable and reusable without harm to the environment. Moreover, transition metals are most frequently used as catalysts in HER and OER. The use of noble-metal-free catalysts like TMOs, Chalcogenides, Nitrides and Oxynitrides to avoid toxicity, high-cost and thereby increasing the efficiency of water splitting will open new doors for their future use and advancements.

9.2 Electrochemical and photo-electrochemical water splitting

In EC and PEC water splitting, a suitable electrocatalyst is used which is used to increase the charge transfer in the reactions, that is, reduction and oxidation of water. It does not directly consume in the reactions but helps to adsorption of water molecules on the surface of it and initiate transfer of charge among reactant(water) and electrode. It involves the main electrolytic parameters, that is, Overpotential (η), Tafel slope (b), Exchange current density (i_0), turnover frequency (TOF) which evaluate performance of the electrocatalyst (Suen et al., 2017; Anantharaj et al., 2021). Therefore before proceeding further, we are giving detailed information about these parameters below.

9.2.1 Overpotential (η)

Overpotential (η) is an additional driving force essential by a catalyst that start water splitting reaction from its equilibrium where well-known Nernst equation can be derived,

$$\eta = E - E_0 \tag{9.1}$$

The overpotential (η) is a potential that needs to be applied to attain a current density (10 mA/cm^2) that is equal to the 12.3% power conversion efficiency (PCE) of the water splitting reaction. Thermodynamic potential for water splitting is 1.23 V. Indeed, in practice, water splitting cell requires over/higher potential to be applied where overall water splitting potential can be given as,

$$V_{OP} = 1.23V + \eta a + \eta c + \eta \Omega \quad (9.2)$$

where intrinsic activation barriers of the cathode and anode are overcome by overpotentials ηc and ηa respectively. This can be decreased by using a highly active electrocatalyst. The lesser the overpotential, the better active is electrocatalyst. However, an overpotential determines the apparent activity of the electrocatalyst but not an intrinsic activity because the EC surface area does not possess good uniformity and planar nature where the current density is calculated by normalizing electric current (j) by geometrical area.

9.2.2 Exchange current density (i_0)

Exchange current density (i_0) is an important factor and determines the performance of the electrocatalyst at equilibrium. At equilibrium, when cathodic (jc) and anodic (ja) are equal, there is zero net current. From experimental parameters, i_0 can be derived from curve overpotential vs $\log_{10}$ (i). Platinum (Pt) has the highest current density of (η) vs logarithmic current density $\log_{10}$ (i). Exchange current density (i_0) is determined after extrapolating the curve to $\eta = 0$. The higher the value of i_0, the more active is the electrocatalyst.

9.2.3 Tafel slope (b)

Tafel slope is simply the relation between the overpotential and exchange current density. As already known, the Butler–Volmer equation describes reaction kinetics of HER and OER where the Tafel slope is derived using the slope of the curve η vs $\log_{10}$(i). The magnitude of the slope, that is, Tafel slope describes how fast the current increases with overpotential. Lower value of overpotential and rapidly increased current gives smaller value of Tafel slope.

9.2.4 Turn over frequency

Since above mentioned kinetic parameters are usually reported by the researchers for every novel electrocatalyst for its apparent activity TOF is very crucial and less reported. The amount of H_2 produced is not reported by EC and PEC techniques for water splitting. Moreover, TOF is another parameter that can determine the catalyst's intrinsic activity of as a function of overpotential. Most commonly, TOF can be calculated as,

$$TOF = j \times \frac{N_A}{F \times n \times \Gamma} \quad (9.3)$$

where, j = current density,

N_A = Avogadro constant

F = Faraday constant

N = number of electrons transferred to generate one molecule of the product

Γ = surface concentration or the exact number of active sites catalyzing the reaction.

9.3 Photocatalytic water splitting

Along with EC and PEC water splitting, photocatalytic water splitting has improved efficiency and cost, being a key solution for sustainable H_2 energy generation and a clean environment in the form of O_2. Therefore nanomaterials including transition and nontransition elements are considered photocatalysts that require light energy to cross an activation barrier of water splitting. Photocatalytic HER is one of the promising technologies that doesn't require external potential and efficiently works in the overall pH range. The light energy in this type of water splitting is solar power which is eco-friendly and low-cost. Solar energy of specific power is incident on the solution containing photocatalyst placed in a closed reactor in determined concentration for several hours for estimation of gases evolved using gas chromatography as shown in Fig 9.1. Herein, the stability of photocatalyst plays a very important role. Different strategies are developed for material synthesis that is suitable for solution and thin film-based photocatalysis. The pH of a solution, temperature of the system, concentration of a solution, and morphology of the photocatalysts affect photocatalytic activity. According to the literature, film-based photocatalytic measurements give good H_2 evolution efficiency but contain complex procedures to develop an electrode. In the photocatalytic H_2 production process, the light photons having energy equal to the band gap of the semiconductor incident on the surface of semiconductor photocatalysts. It excites valence band electrons to the conduction band (C.B) and creates an empty hole in the valence band. This excited electron in the C.B is accepted by dissociated water molecule and produces hydrogen via reduction at cathode. This interesting phenomenon continues until the overall consumption of photocatalyst in the reactor and evolved gaseous

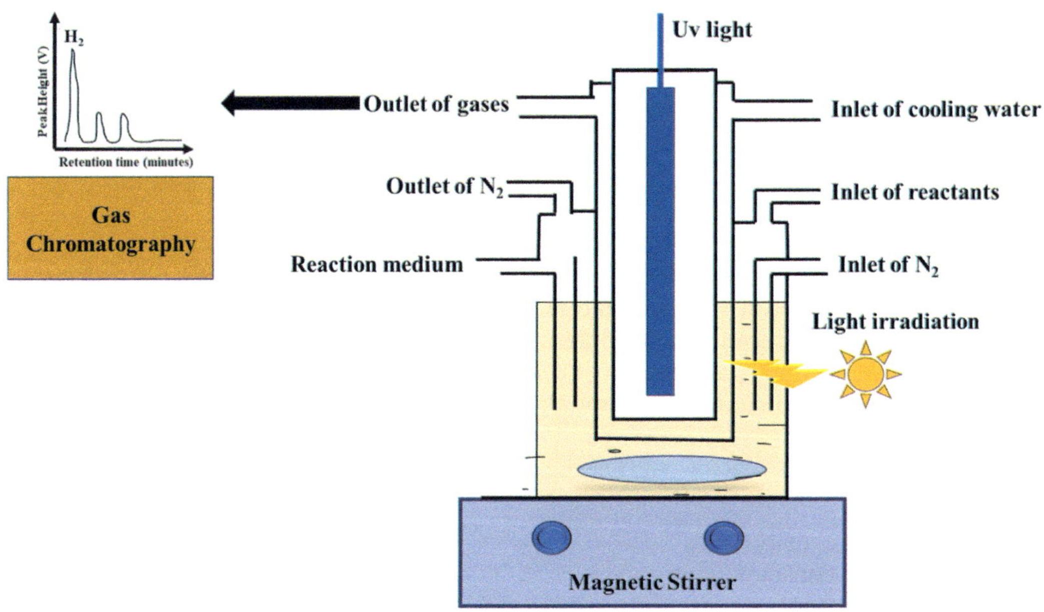

FIGURE 9.1 Showing block diagram of reactor used in photocatalytic hydrogen evolution reactions and oxygen evolution reactions

H_2 is quantified using a Gas chromatography instrument. Moreover, an aqueous solution containing photocatalysts requires high solubility of photocatalysts for greater efficiency. The care should be taken to avoid recombinatin losses for efficient energy generation. For the quantitative estimation of H_2 and O_2 evolution, the apparent quantum efficiency is considered and calculated at specific wavelength at each monochromatic wavelength using following formula.

$$\textit{Apparent quantum efficiency (AQE)} = \frac{n \times \text{moles of gas production per hour moles of photon flux per hour}}{\text{moles of photon flux per hour}} \times 100 \quad (9.4)$$

where $n = 2$ for H_2 evolution and 4 for O_2 evolution.

9.4 Transition metal oxides, chalcogenides, nitrides, oxynitrides catalysts

9.4.1 Synthesis methodologies

In the 21st century, nanotechnology has become a core of the material science research field due to the development of several nanomaterials having a high surface area to volume ratio and smaller size. The development of earth-abundant, stable, low-cost, and active nanostructured catalysts is a prime need for diminishing the use of fossil fuels for green energy generation. Therefore different methods of synthesis and functionalization have been employed for the synthesis of TMOs, chalcogenides, nitrides, and oxynitrides in recent research work. Research is narrowed down on the developing TMOs and non-TMOs-based nanostructures for enhanced performance of water splitting (Santra et al., 2022; Shit et al., 2024). Fig 9.2 illustrates the different types of catalysts for water splitting.

9.5 Mechanism of water splitting

The water splitting generally takes place via redox reactions including two half-cell reactions at the cathode and anode. The HER reactions proceed on the cathode via the reduction in alkaline media whereas OER reactions proceed on the anode via oxidation of water molecules in acidic media and vice-versa. There is no such involvement of catalysts in the reactions, but they play an important role in the adsorption of water molecules (Walake et al., 2023), dissociation, and formation of reaction intermediates.

In acidic media for HER reactions,

$$\text{Anode}(+)\text{: } H_2O_{(l)} \rightarrow \frac{1}{2}O_{2(g)} + 2H^+ + 2e^-. \quad (9.5)$$

$$\text{Cathode}(-)\text{: } 2H^+ + 2e^- \rightarrow H_{2(g)}. \quad (9.6)$$

$$\text{Net reaction: } H_2O_{(l)} \rightarrow H_{2(g)} + \frac{1}{2}O_{2(g)}. \quad (9.7)$$

In alkaline media for HER reactions,

$$\text{Anode}(+)\text{: } 2OH^- \rightarrow H_2O_{(l)} + \frac{1}{2}O_2 + 2e^-. \tag{9.8}$$

$$\text{Cathode}(-)\text{: } 2H_2O + 2e^- \rightarrow H_{2(g)} + 2OH^-. \tag{9.9}$$

$$\text{Net Reaction: } H_2O_{(l)} \rightarrow H_{2(g)} + \frac{1}{2}O_{2(g)}. \tag{9.10}$$

Furthermore, in overall water splitting three main types of reaction mechanisms are involved. The first step is the Volmer step includes dissociation of adsorbed water molecules. The next step is the rate-determining step (RDS) including Heyrovsky and Tafel reactions and depends on the overall activity of the catalysts involved. Generally, noble metals-based catalysts follow the Tafel reaction whereas noble-metal-free catalysts follow the Heyrovsky reaction and evolve H_2. It is necessary that Gibb's free energy of Hydrogen atom adsorption on the catalyst surface (ΔG_{H^*}) is minimal to proceed with the HER reaction (Yin et al., 2020).

1. Volmer step
   In acidic media,

$$H_3O^+ + e^- \rightarrow H^* \tag{9.11}$$

   In alkaline media.

$$H_2O + e^- \rightarrow H^* \tag{9.12}$$

2. Heyrovsky step
   In acidic media,

$$H_3O^+ + e^- + H^* \rightarrow H_2 \tag{9.13}$$

   In alkaline media,

$$H_2O + e^- + H_* \rightarrow H_2 \tag{9.14}$$

The OER process is a slow and complex process that includes the transfer of four electrons at anode surface. There is a possibility of the formation of reaction intermediates, that is, OH^-, O^-, and HOO^- and the binding energies of these intermediate describes the RDS.

In acidic media,

$$H_2O \rightarrow HO^* \tag{9.15}$$

$$H_2O + HO^* \rightarrow O^* \tag{9.16}$$

$$H_2O + O^* \rightarrow HOO^* \tag{9.17}$$

$$HOO^* \rightarrow O_2 \tag{9.18}$$

In alkaline media,

$$OH^- \rightarrow HO^* \tag{9.19}$$

$$OH^- + HO^* \rightarrow O^* \tag{9.20}$$

$$HO^- + O^* \rightarrow HOO^* \tag{9.21}$$

$$HOO^* + OH^* \rightarrow O_2 \tag{9.22}$$

9.5.1 Metal oxides

The metal oxide nanomaterials also called TMOs typically known as semiconductors that are primarily composed of transition metal atoms are explored due to the presence of d-orbital that is favorable in easy electron transfer required for water reduction and oxidation in HER and OER respectively. More effectively, their PEC stability and photo absorption ability, favorable band edge positions, and wide band gap have led to their outstanding performance in overall water splitting. Metal oxide semiconductor nanomaterials have one valence band (V.B) edge that prefers OER and one C.B edge that prefers HER in water splitting where C.B edge minima should be close to zero and V.B edge maxima should be far from zero for water splitting. The trap states between the V.B and C.B create difficulties in water splitting as they allow recombination losses of charge carriers. Therefore to avoid recombination, efforts have been made via surface passivation of catalysts (Liu et al., 2014). The high density of surface states and slow reaction kinetics in metal oxide decrease the water splitting performance. Multinary metal oxides are reported in the literature for water splitting as tabulated in Tables 9.1 and 9.2.

Synthesis routes such as physical, chemical, and biological are the most common ways of lab-level synthesis of catalysts where top-down and bottom-up approaches are involved. For an efficient process of water splitting, uniform particle size, narrow size distribution, hierarchical architectures, and high dispersion stability are desirable and can be achieved via chemical synthesis routes whereas physical synthesis routes can develop high-quality thin films having good adhesion. Biological synthesis routes make the process and products more biocompatible, nontoxic, environment-friendly, and simple with the disadvantage of lower scaling of products required for industry. Several metal oxide nanomaterials including noble-metal free binary metal oxides, and ternary metal oxides called bimetallic metal oxides are reported as good catalysts for HER and OER processes as illustrated in Tables 9.1 and 9.2.

9.5.2 Chalcogenides

Chalcogenides referred to as Transition metal chalcogenides (TMCs) based nanomaterials consist of sulfides, selenides, and tellurides-based nanomaterials. Moreover, they are again classified as binary, ternary, and quaternary chalcogenides as shown in Fig 9.2. Examples of binary chalcogenides are CdS, Cdse, MnSe, etc. For ternary are Cesium dinickel selenide ($CsNi_2Se_2$) and quaternary chalcogenides are Cu_2ZnSnS_4 (CZTS), Cu_2MnSnS_4 (CMTS), etc. Metal chalcogenides are getting remarkable attention as electrocatalysts due to their greater

TABLE 9.1 Oxide-based electro/photo electrocatalysts for HER and OER processes.

S. no.	Catalyst	Synthesis route	Type of water splitting	Electrolyte/ sacrificial agent	Morphology	Overpotential (mV)		Tafel Slope (mV/dec)	References
						HER	OER		
1.	WO_3 Thin film on Ni foam	DC magnetron sputtering	PEC	0.5 M H_2SO_4	Porous nanowires	322	–	45–110	Bayrak Pehlivan et al. (2021)
2.	WO_3/C@ CoO	Hydrothermal process and immersion	EC	1 M KOH	Octopus tentacles	55	–	115	Lv et al. (2018)
3.	ZnO	Chemical Bath Deposition	PEC	1 M Na_2SO_4	Nanorods	–	–	–	Deshmukh et al. (2017)
4.	ZnO	Electrodeposition	PEC	0.5 M Na_2SO_4	Nanorods	–	–	–	Rokade et al. (2017)
5.	Fe_2O_3	Hydrothermal & annealing	EC	1 M KOH	Nanospheres	321	–	140	Ali et al. (2018)
6.	CoO_x @N doped carbon hydrides	Precursors calcination	EC	1 M KOH	Flake-like morphology	85	–	115	Jin et al. (2015)
7.	WO_3	Liquid exfoliation	EC	0.5 M H_2SO_4	Nanosheets	37	–	38	Zheng et al. (2017)
8.	Co/CoO/ $Co_2Mo_3O_8$	Anodization	EC	–	Porous nanoflowers	78	–	–	Bian et al. (2023)
9.	$TiO_{1.23}$	Anodization	EC	0.5 M H_2SO_4	Porous Nanoparticles	198	–	88	Swaminathan et al. (2016)
10.	$WO_3.H_2O$	Precipitation	EC	1 M Na_2SO_4	Nanoflowers	290	–	131	Li et al. (2019)
11.	Co_3O_4/CeO_2	Electrodeposition	EC	0.5 M H_2SO_4	Nanosheets	423	–	–	Huang et al. (2021)
12.	Co_3O_4 {111}	Hydrothermal followed by thermal decomposition	EC	1 M KOH	Nanocubes	–	285	49	Liu et al. (2017)
	Co_3O_4 {110}					–	362	119	
	Co_3O_4 {112}					–	341	101	
	Co_3O_4 {001}					–	312	73	

OER, Oxygen evolution reactions; *HER*, hydrogen evolution reactions.

TABLE 9.2 Oxide photocatalysts for photocatalytic HER.

S. no.	Photocatalyst	Synthesis method	Sacrificial agent	Light irradiation	Morphology	H_2 evolved (μmol/h/g)	References
1.	V-Doped TiO_2	Hydrothermal	Methanol	Visible	Nanoparticles	120	Agegnehu et al. (2016)
2.	Fe, Ni-doped TiO_2	Hydrothermal	Ethanol + water	Visible	Nanoparticles	361.64	Sun et al. (2012)
3.	S-doped TiO_2	Furnace heating	Water	Visible	Nanofibers	9610	Sun et al. (2017)
4.	CeO_2	Chemical precipitation	Ethanol + water	Visible	Nanoparticles	140	Fiorenza et al. (2021)
5.	TiO_2 doped with ransition metals	Impregnation	Ethanol +water		Nanoparticles		Díaz et al. (2021)
	1. TiO_2			Visible		16	
	2. TiO_2			Uv		3	
	3. TiO_2 -Fe			Uv		2100	
	4. TiO_2-Co			Uv		2100	
	5. TiO_2-Ni			Uv		2100	
	6. TiO_2-Cu			Visible		220	
	7. TiO_2-Cu			Uv		5000	
	8. TiO_2-Zn			–		–	

HER, Hydrogen evolution reactions.

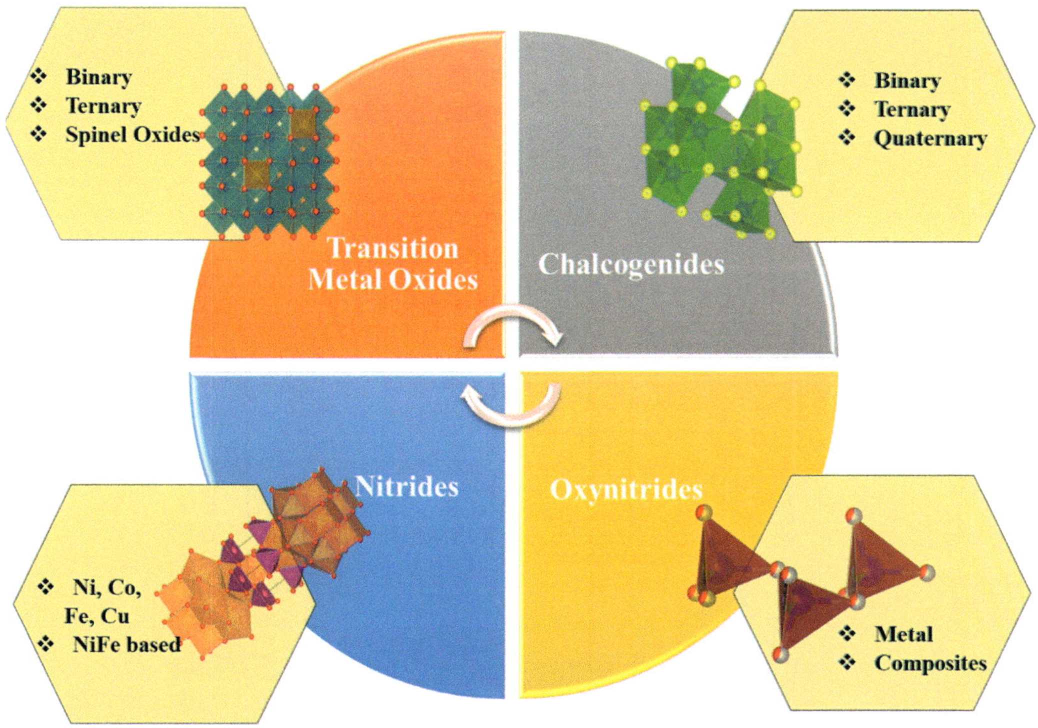

FIGURE 9.2 Illustration of types of catalysts with their types.

conductivity and layered-nonlayered structures. The efficiency of water splitting depends on the different physicochemical properties of a metal chalcogenide (Majhi and Yadav, 2023). The synthesis methods for their development are generally microwave-assisted, sonochemical methods, sono-EC method, low-temperature solid-state synthesis, solvothermal synthesis, templated synthesis, microemulsion methods, sol-gel method as well as thermal decomposition using organic solvents (Ridhova et al., 2021). Most of these methods fall under chemical methods of synthesis and are explored significantly in catalyst development. After synthesis, the morphology obtained for TMCs is generally 2D layered sheets. Typically for nonlayered chalcogenide materials, the bottom-up synthesis method is preferred while for layered crystal structures, top-down approaches are implied. Atoms are bonded with strong covalent linkages in different crystalline planes with weak van der Wall forces among stacked layers. Nonlayered TMCs do not possess parent layered structures and have sizes at an atomic level and small unit cells of 3D crystals with a high surface area to volume ratio. Porous nanoflowers of $MoO_x/Fe_{1-x}S$ as nonlayered TMC also exhibited lowered HER overpotential in alkaline media (Liu et al., 2022). Indispensable capping agents or templates are used to stop the unwanted growth of nonlayered TMCs. On the other hand, MoS_2 is in the form of 2D layered sheets and electrocatalytically inactive for HER reactions, but the size and the morphological advancements played an important role in giving active sites to Mo edge (1010) as reported by Zhang and Liang (2018). TMCs also contain low-coordinated surface atoms with some dangling bonds that create defects and are responsible for large no. of active sites required for EC, PEC, and photocatalytic water splitting (Wang et al., 2021). Furthermore, the transition of 2D TMCs to 1D nanostructures has gained attention via ion intercalation beneficial for active sites at the edges of 1D nanostructures. The conductivity of TMCs is higher than that of TMOs and lower than that of nitrides and oxynitrides. Therefore efforts are taken to increase the conductivity of TMCs using different approaches. TMCs can absorb a small wavelengths of the solar spectrum and possess higher recombination losses after illumination with light with separation and transportation of charge carriers (Shahid et al., 2024). Along with sulfides, tellurides, and selenides are gaining importance as prominent electrocatalysts due to their higher conductivity than sulfides (Peng et al., 2020). Moreover, strategies have to be developed for improving their performances in EC, PEC as well as photocatalytic water splitting and avoid sluggish reaction kinetics and structure stability. Intercalation of metals and nonmetals, polymers, and organic compounds rather than doping is preferred in this case as reported in (Wang et al., 2021; Jung et al., 2016; Wang et al., 2018a,b). EC, PEC and photocatalytic performance of TMCs based catalysts are illustrated in Tables 9.3 and 9.4.

9.5.3 Nitrides

Paramount importance has been given to the low-cost, earth-abundant, and efficient H_2 generation from water splitting. There is a significant importance of N atom doping in metal oxide but their industrial scale is limited due to the lower energy of formation of nitrides than oxide, which is the reason for the scarcity of nitrides in electrocatalysis (Jamil et al., 2021). Different transition metals (Ni, CO, Fe, Mo, NiFe, etc.) based nitrides (TMNs) have emerged for the same purpose. They are developed by the insertion of an electronegative N atom into the parent compound. Ternary nitrides such as Co_2Mo_3N and Ni_2Mo_3N are elucidated as good HER catalysts for overpotentials 149 and 158 mV respectively (Sun et al., 2023).

They are also used for heterostructure formation in HER catalysts. TMNs possess good conductivity, enhanced d-electron density, and low corrosion due to N atoms making suitable catalysts for HER and OER with the same electronic structure as that of Pt.

Another important category of nitrides is known as graphitic carbon nitride (g-C_3N_4) emerged as a novel catalyst in water splitting since it consists of large no. of N atoms replaced in benzene ring or organic compounds such as urea, melamine, etc. Furthermore, due to the N atom doping, it is highly conductive and resembles a layered structure like graphene. g-C_3N_4 individually shows good photocatalytic activity and behaves as a bifunctional catalyst showing H_2 evolution of 42.6 μmol/h/g and O_2 evolution of 18.7 μmol/h/g. Furthermore, in photocatalytic water splitting, combining g-C_3N_4 with metal oxides has increased the adsorption-desorption reactions between reaction intermediates and catalysts as studied in (Zhou et al., 2021). Thakur et al. in 2021 reported the decreased conduction band edge of TiO_2- g-C_3N_4 nanocomposite that led to H_2 evolution of 2531 μmol/h/g (Wadhai et al., 2021). Furthermore, heterojunction based on $NiCo_2O_4$ @ g-C_3N_4 was reported by Babu et al. and obtained enhanced photocatalytic activity with H_2 evolution of 3,533 μmol/h/g larger than pristine g-C_3N_4(2056 μmol/h/g) and $NiCo_2O_4$(2334 μmol/h/g). Reactive facets of NiMoN ternary nitride nanowires were reported as bifunctional electrocatalysts with HER overpotential of 38 mV and OER overpotential of 54 mV (Chang et al., 2018). Apart from these excellent properties, research has been focused on the structural stability of TMNs (Tang et al., 2022). Wang et al. in 2016

TABLE 9.3 TMCs-based electrocatalysts for HER and OER processes.

S. no.	Catalyst	Synthesis route	Type of water splitting	Electrolyte/ sacrificial agent	Morphology	Overpotential (mV) at 10 mA/cm^2		Tafel Slope (mV/dec)	References
						HER	OER		
1.	Ni_3OH_2 @ MoS_2	Hydrothermal	EC	1 M KOH	Nanosheets	73	–	114	Zhang and Liang (2018)
2.	$CuCo_2S_4$	Hydrothermal	EC	1 M KOH	Nanowires	65	–	42	Czioska et al. (2018)
3.	$Fe_{4.5}Ni_{4.5}S_8$	High-temperature heating	EC	0.5 M H_2SO_4	–	190	–	72	Piontek et al. (2019)
4.	Sb_2S_3	Reflux	EC	1 M KOH	Pellet drums	72	–	193	Maiti and Srivastava (2020)
5.	Ni_3S_2	Hydrothermal	EC	1 M KOH	Porous Nanoparticles	32	–	61	Liu et al. (2020)
6.	$Ni_{0.9}Fe_{0.1}S_3$	Solid-state synthesis	EC	1 M KOH	Nanosheets	72	–	73	Song et al. (2017)
7.	NiTe-NiSe	Hydrothermal	EC	1 M KOH	Nanosheets	76	164	–	Wang et al. (2022a)
8.	$Co_xNi_{1-x}Te$	Inward diffusion	EC	0.1 M KOH	Microfibers		289	117	Bhol et al. (2023)

OER, Oxygen evolution reactions; *HER*, hydrogen evolution reactions.

TABLE 9.4 TMCs-based photocatalysts for HER.

S. no.	Photocatalyst	Synthesis method	Sacrificial agent	Light irradiation	Morphology	H_2 evolved (μmol/h/g)	References
1.	$CdSe/CuInS_2$	Solvothermal	Ethanol + water	Visible	Micro flowers	10,610.37	Wang et al. (2022b)
2.	$NiS/Zn_xCd_{1-x}S$	Solid solution	Na_2S and Na_2SO_3	Visible	Microrods	16,780	Wang et al. (2022b)
3.	MoS_2/CdS	One pot organic synthesis	Lactic acid	Visible	Nanosheets	1472	Chen et al. (2015a)
4.	MoS_2/TiO_2	Hydrothermal	Methanol	Visible	Nanobelts	600	Li et al. (2016)

HER, Hydrogen evolution reactions.

studied long-term stability over 2000 linear sweep voltammetry cycles of Ni_3FeN nanosheets as bifunctional catalysts for EC HER and OER and reported overpotential as 223 mV (Wang et al., 2016). Therefore nitride-based electrocatalysts as well as photocatalysts are enhancing HER and OER performance as given in Tables 9.5 and 9.6.

9.5.4 Oxynitrides

Metal oxynitrides possess small band gap and hence can absorb the light of visible range where they are significantly used for photocatalytic H_2 evolution. The conventional synthesis route of metal oxynitrides comprises thermal ammonolysis of oxides under N_2 gas flow. Some N atoms in nitrides are replaced by O_2, hence combining the properties of both oxides and nitrides, they possess good water splitting activity (Ahmed and Xinxin, 2016). Perovskite oxynitrides that are sensitive to visible light are explored for PEC and photocatalytic applications. A recent study reported by Ahmed et al. showed that Cobalt-based oxynitrides play an important role in EC water splitting with an OER overpotential of 350 at alkaline pH and 425 mV at neutral pH for synthetic seawater (Badreldin et al., 2023). Most of the reports available for oxynitrides support OER in water splitting (Gebauer et al., 2016; Dai et al., 2019; Chen et al., 2015b; Wang et al., 2018) reviewed the doping of different metal atoms in oxynitrides beneficial for the anion-regulated activity of OER (Wang et al., 2018a). Anion-regulated OER activity involves more active sites created at metal centers and showed improved OER performance. The literature values for EC, PEC and photocatalytic HER and OER performance are given in Tables 9.7 and 9.8.

9.6 Strategies to improve water splitting

Since the noble-metal-free catalysts - oxides, chalcogenides, nitrides, and oxynitrides for HER and OER are getting attention some pitfalls in their performance arise depending on their intrinsic and extrinsic properties that can be modified using different approaches. Some of these are listed below and overviewed based on the literature. Intrinsic catalytic activity can be modified using O_2-vacancy control, phase-structure engineering, heteroatom doping, etc. whereas extrinsic catalytic activity can be modified using morphology, temperature, pH, sacrificial agent, the concentration of catalyst, etc. as illustrated in Fig 9.3. Furthermore, we have illustrated factors affecting intrinsic catalytic activity because the extrinsic activity of the catalyst is apparent and discussed in many reports.

9.6.1 Oxygen-vacancy control

It is one of the most important approaches to change/engineer an O_2-vacancy in the metal oxides to make it an effective catalyst in EC, PEC, and photocatalytic water splitting. The electronic band structure of metal oxides containing O_2-vacancy or metal oxide creates trap states or defects and may differ from pure metal oxide nanomaterials in terms of band gap, crystalline structure, and conductivity. As stated in a recent review, easy charge transport is feasible using this strategy so that recombination will be avoided. O_2-vacancy defects can be

TABLE 9.5 Nitride electrocatalysts for HER and OER processes.

S. no.	Catalyst	Synthesis route	Type of water splitting	Electrolyte/ sacrificial agent	Morphology	Overpotential (mV)		Tafel Slope (mV/dec)	References
						HER	OER		
1.	Ni_3N	Nitridation in NH_3	EC	1 M KOH	Nanorods	121	–	109	Xing et al. (2016)
2.	TiO_xN_y	Thermal nitridation	EC	0.5 M H_2SO_4	Porous channels	–	300	–	Gebauer et al. (2016)
3.	Ni_3N	Calcination	EC	1 M KOH	Nanosheets	–	350	45	Xu et al. (2015)
4.	NiFeN@ NiMoN	Hydrothermal	EC	Seawater + 1M KOH	Nanorods	180	348	58	Yu et al. (2019)
5.	CoS_2-$gC_3N_4CoS_2$-BN	Hydrothermal	EC	0.1 M NaOH	Spheres on Nanosheets	108 123	417 376	120 121	Borthakur et al. (2021)

OER, Oxygen evolution reactions; *HER*, hydrogen evolution reactions; *EC*, electrochemical.

TABLE 9.6 Nitride-based photocatalysts for HER reactions.

S. no.	Photocatalyst	Synthesis method	Sacrificial agent	Light irradiation	Morphology	H_2 evolved (μmol/h/g)	References
1.	Fe-gC_3N_4	Solvothermal	Ethanol + water	Visible	Microspheres	16,200	Chen et al. (2023)
2.	g-CN_x	Ultrasonication	Triethanolamine	Visible	Nanosheets	9600	Sha et al. (2023)
3.	gC_3N_4-TiO_2 P doped gC_3N_4-TiO_2	Furnace heating	Methanol	Visible	Nanosheets	9202531	Wadhai et al. (2021)

HER, Hydrogen evolution reactions.

TABLE 9.7 Oxynitride electrocatalysts for HER and OER processes.

S. no.	Catalyst	Synthesis route	Type of water splitting	Electrolyte/ sacrificial agent	Morphology	Overpotential (mV)		Tafel Slope (mV/dec)	References
						HER	OER		
1.	TiO_xN_y	Thermal nitridation	EC	0.5 M H_2SO_4	Porous channels	–	300	–	Gebauer et al. (2016)

OER, Oxygen evolution reactions; *HER*, hydrogen evolution reactions.

TABLE 9.8 Oxynitride-based photocatalysts for HER reactions.

S. no.	Photocatalyst	Synthesis method	Sacrificial agent	Light irradiation	Morphology	H_2 and O_2 evolved (μmol/h/g)		References
						H_2	O_2	
1.	$LaTa_{0.9}Zr_{0.1}O_1 + yN_2y$	High temperature Polymerization	0.05 M $AgNO_3$	Visible light	Particles with irregular shapes	360	–	Wang et al. (2019)
2.	$BaNbO_2N$	Solid-state reaction method	30 mM $AgNO_3$	Visible light	–	–	250	Seo et al. (2019)
3.	TaON/CdS	Precipitation	–	Natural sunlightXe lamp	–	7.8 11.08	– –	Kalia et al. (2023)
4.	TiO_xN_y	Thermal nitridation	0.5 M H_2SO_4	Visible light		1546	–	Ravikumar et al. (2023)
5.	TaON	Nitridation	Water + Methanol	Visible light	–	15	660	Hitoki et al. (2002)

HER, Hydrogen evolution reactions.

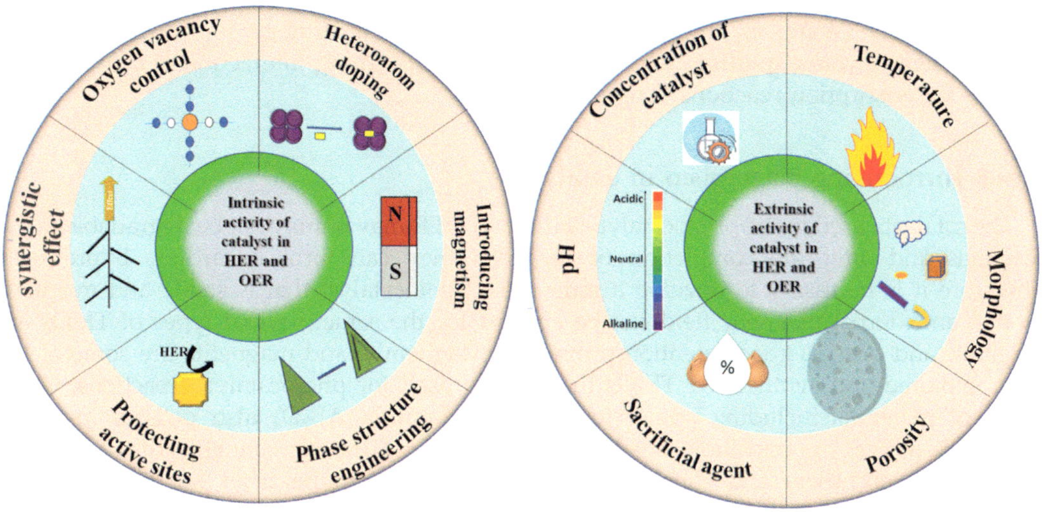

FIGURE 9.3 Illustration of parameters that affect (A) intrinsic and (B) extrinsic catalytic activity in hydrogen evolution reactions.

made in the material using high-temperature calcination, reduction using chemicals, and EC treatment (Ouyang et al., 2023). Sacrificial agent, that is, sodium sulfite was used by Gong et al. resulting in enhanced photocurrent density in PEC activity of $BiVO_4$ (Feng et al., 2020). However, the way of introducing O_2 vacancies must be strategically planned because large no. of surface O_2 species can decrease in performance. Therefore engineering the surface of metal oxide via the introduction of O_2 species or removal of O_2 species is an open challenge to all researchers.

9.6.2 Phase-structure engineering

Phase structure and crystal facet engineering can improve water splitting performance thereby enhancing intrinsic properties of metal oxide nanomaterials. The proper orientation of the crystal facet gives a large no. of active sites for adsorption of water molecules in water splitting (Wang et al., 2023a). Changing the orientation of the crystal facet will add anisotropy in the metal oxide and therefore in the direction of external light, nanomaterial enhances the PEC and photocatalytic activity. Furthermore, the phase of nanomaterial imparts different behavior of nanomaterial as seen in TiO_2.TiO_2 in its anatase phase shows the best water splitting activity whereas Rutile and mixed phase TiO_2 does not suit (Eidsvag et al., 2021).

9.6.3 Heteroatom doping

Heteroatom doping is necessary for optimizing active centers via electronic and structural changes. The electronic band is the most important parameter in photocatalytic HER and OER and to decrease the band gap for making suitable catalysts, different metal, and nonmetal atoms are doped in the parent compound. It has been getting a remarkable interest in recent years (Wang et al., 2023a,b). Moreover, it is a strategy for intrinsic activity enhancement of water splitting, which contributes to the strong interaction of central metal to OH^- for adsorption reactions (Yuan et al., 2023).

9.6.4 Introducing magnetism in catalysts

The catalysts used in the photocatalytic HER and OER may or may not be degradable in the reactions and after the completion of reactions they can form hazardous byproducts. Therefore it is necessary to develop a removable photocatalyst that is stable and magnetic so that an external magnetic field can be applied for the same. Several types of TMOs are magnetic and possess good retention properties. Maghemite and magnetite are some of the most explored and synthesized TMOs that can be used for photocatalytic reactions. Other types of materials including spinel ferrites with formula AB_2O_4 also possess magnetic properties where A is transition metal and B is iron (Fe). However, spinel ferrites are stable, good photo absorbers and tunable photocatalysts and can be used for heterojunction formation and commonly used for thermochemical water splitting (Goikoetxea et al., 2019). Walake et al. recently reported a review on photocatalytic HER using spinel ferrites where strategies for efficient H_2 evolution are overviewed (Walake et al., 2023). In the year 2016, Ren et al. reviewed the importance of ferrites as photocatalysts wherein very less no. of ferrites as photoanodes were employed (Ren et al., 2016). Furthermore, magnetic nanocatalysts such as Fe_3O_4 are used with IrO_2 for the reduction of bubbling and increasing rate of OER reactions (Kaya et al., 2021). By using magnetic catalysts, the external magnetic field can be applied to the reaction to drive charged species toward respective electrodes (Wei and Xu, 2022).

9.7 Conclusion and future prospects

In summary, three types of water splitting reactions with the catalysts based on TMOs, chalcogenides, nitrides, and oxynitrides are discussed in the present chapter. The synthesis strategies that will affect intrinsic and extrinsic catalytic activity are summarized. Most of the noble-metal-free catalysts show enhanced HER activity than OER. The dissociation ability and stability of most of these catalysts are less explored and can be studied in the future. One of the less explored aspects of these types of catalysts is their performance without sacrificial agents for photocatalytic HER and OER which can be explored further. For EC and PEC, heterojunction formation is suitable and can enhance catalytic activity wherein improvements in catalysts based on their long-term stability in water can be studied. Furthermore, the reuse and removal of catalysts and proper disposal are important to avoid environmental damage in the coming days, therefore introduction of magnetic materials will solve the present problem. Accordingly, the development of a core-shell noble-metal-free catalyst with synergistic effects of light absorption, lower charge recombination, and intrinsic magnetism can give a novel way for efficient HER and OER activity. The morphological advancements such as high porosity, and layered structure for more active sites can be initiated. Furthermore, the industrialization of noble-metal-free catalysts remains a challenge that can be overcome with the proper understanding of industrial requirements and cost analysis.

References

Agegnehu, A.K., Pan, C.-J., Tsai, M.-C., Rick, J., Su, W.-N., Lee, J.-F., et al., 2016. Visible light responsive noble metal-free nanocomposite of V-doped TiO_2 nanorod with highly reduced graphene oxide for enhanced solar H_2 production. International Journal of Hydrogen Energy 41, 6752–6762.

Ahmed, M., Xinxin, G., 2016. A review of metal oxynitrides for photocatalysis. Inorganic Chemistry Frontiers 3, 578–590.

Ali, R.N., Naz, H., Zhu, X., Xiang, J., Hu, G., Xiang, B., 2018. Synthesis, characterization and applications of ph controlled Fe_2O_3 nanoparticles for electrocatalytic hydrogen evolution reaction. Materials Research Express 6.

Anantharaj, S., Karthik, P.E., Noda, S., 2021. The significance of properly reporting turnover frequency in electrocatalysis research. Angewandte Chemie (International Edition in English) 60, 23051–23067.

Badreldin, A., Abed, J., Hassan, N., El-Ghenymy, A., Suwaileh, W., Wubulikasimu, Y., et al., 2023. Sulfide interlayered cobalt-based oxynitrides for efficient oxygen evolution reaction in neutral pH water and seawater. Applied Catalysis B: Environment and Energy 330, 122599.

Bayrak Pehlivan, I., Atak, G., Niklasson, G.A., Stolt, L., Edoff, M., Edvinsson, T., 2021. Electrochromic solar water splitting using a cathodic WO_3 electrocatalyst. Nano Energy 81, 105620.

Bhol, P., Patil, S., Barman, N., Siddharthan, E., Thapa, R., Saxena, M., et al., 2023. Design and fabrication of cobalt x nickel (1-x) telluride microfibers on nickel foam for battery-type supercapacitor and oxygen evolution reaction study. Materials Today Chemistry 30, 101557.

Bian, H., Chen, Z., Chen, T., Humayun, M., Zhou, B., Liao, W., et al., 2023. Anodically designing of refreshable bi-metallic oxides for highly-efficient hydrogen evolution. Chemical Engineering Journal 466, 143045.

Borthakur, P., Boruah, P.K., Das, M.R., Ibrahim, M.M., Altalhi, T., El-sheshtawy, H.S., et al., 2021. Cos_2 Nanoparticles supported on rGO, g-C_3N_4, BCN, MoS_2, and WS_2 two-dimensional nanosheets with excellent electrocatalytic performance for overall water splitting: electrochemical studies and DFT calculations. ACS Applied Energy Materials 4, 1269–1285.

Chang, B., Yang, J., Shao, Y., Zhang, L., Fan, W., Huang, B., et al., 2018. Bimetallic nimon nanowires with a preferential reactive facet: an ultraefficient bifunctional electrocatalyst for overall water splitting. Chemsuschem 11, 3198–3207.

Chen, J., Wu, X.J., Yin, L., Li, B., Hong, X., Fan, Z., et al., 2015a. One-pot synthesis of cds nanocrystals hybridized with single-layer transition-metal dichalcogenide nanosheets for efficient photocatalytic hydrogen evolution. Angewandte Chemie (International Edition in English) 54, 1210–1214.

Chen, P., Xu, K., Fang, Z., Tong, Y., Wu, J., Lu, X., et al., 2015b. Metallic CO_4N porous nanowire arrays activated by surface oxidation as electrocatalysts for the oxygen evolution reaction. Angewandte Chemie (International Edition in English) 54, 14710–14714.

Chen, L., Zhang, W., Wang, J., Li, X., Li, Y., Hu, X., et al., 2023. High piezo/photocatalytic efficiency of Ag/Bi_5O_7I nanocomposite using mechanical and solar energy for N_2 fixation and methyl orange degradation. Green Energy & Environment 8, 283–295.

Czioska, S., Wang, J., Teng, X., Chen, Z., 2018. Hierarchically structured $CuCo_2S_4$ nanowire arrays as efficient bifunctional electrocatalyst for overall water splitting. ACS Sustainable Chemistry & Engineering 6, 11877–11883.

Dai, J., Zhang, W., Li, T., Bai, L., He, W., Ji, L., et al., 2019. The advanced multi-functional electrocatalyst efficiently built from multi-integrated sites for overall water splitting and rechargeable zinc-air battery. Electroanalysis 31, 1381–1389.

Deshmukh, P.R., Sohn, Y., Shin, W.G., 2017. Chemical synthesis of ZnO nanorods: Investigations of electrochemical performance and photo-electrochemical water splitting applications. Journal of Alloys and Compounds 711, 573–580.

Díaz, L., Rodríguez, V.D., González-rodríguez, M., Rodríguez-castellón, E., Algarra, M., Nunez, P., et al., 2021. M/Tio_2 (M=Fe, Co, Ni, Cu, Zn) catalysts for photocatalytic hydrogen production under UV and visible light irradiation. Inorganic Chemistry Frontiers 8, 3491–3500.

Eidsvag, H., Bentouba, S., Vajeeston, P., Yohi, S., Velauthapillai, D., 2021. TiO_2 as a photocatalyst for water splitting-an experimental and theoretical review. Molecules 26, 1687.

Feng, S., Wang, T., Liu, B., Hu, C., Li, L., Zhao, Z.J., et al., 2020. Enriched surface oxygen vacancies of photoanodes by photoetching with enhanced charge separation. Angewandte Chemie (International Edition in English) 59, 2044–2048.

Fiorenza, R., Balsamo, S.A., Condorelli, M., D'urso, L., Compagnini, G., Scirè, D., 2021. Solar photocatalytic H_2 production over CeO_2-based catalysts: influence of chemical and structural modifications. Catalysis Today 380, 187–198.

Gebauer, C., Fischer, P., Wassner, M., Diemant, T., Jusys, Z., Hüsing, N., et al., 2016. Performance of titanium oxynitrides in the electrocatalytic oxygen evolution reaction. Nano Energy 29, 136–148.

Goikoetxea, N.B., Gómez-Mancebo, M.B., Fernández-Saavedra, R., Borlaf, F., García-Pérez, F., Jiménez, J.A., et al., 2019. Understanding water-splitting thermochemical cycles based on nickel and cobalt ferrites for hydrogen production. International Journal of Hydrogen Energy 44, 17578–17585.

Hitoki, G.O., Takata, T., Kondo, J.N., Hara, M., Kobayashi, H., Domen, K., 2002. (Oxy) nitrides as new photocatalysts for water splitting under visible light irradiation. Electrochemistry 70 (6), 463–465.

Huang, J., Sheng, H., Ross, R.D., Han, J., Wang, X., Song, B., et al., 2021. Modifying redox properties and local bonding of Co_3O_4 by CeO_2 enhances oxygen evolution catalysis in acid. Nature Communications 12, 3036.

Jamil, R., Ali, R., Loomba, S., Xian, J., Yousaf, M., Khan, K., et al., 2021. The role of nitrogen in transition-metal nitrides in electrochemical water splitting. Chem Catalysis 1, 802–854.

Jin, H., Wang, J., Su, D., Wei, Z., Pang, Z.A., Wang, Y., 2015. In situ cobalt–cobalt oxide/N-doped carbon hybrids as superior bifunctional electrocatalysts for hydrogen and oxygen evolution. Journal of the American Chemical Society 137, 2688–2694.

Jung, Y., Zhou, Y., Cha, J.J., 2016. Intercalation in two-dimensional transition metal chalcogenides. Inorganic Chemistry Frontiers 3, 452–463.

Kalia, R., Pirzada, B.M., Kunchala, R.K., Naidu, B.S., 2023. Noble metal free efficient photocatalytic hydrogen generation by taon/cds semiconductor nanocomposites under natural sunlight, 48, 16246–16258.

Kaya, M.F., Demir, N., Rees, N.V., El-Kharouf, A., 2021. Magnetically modified electrocatalysts for oxygen evolution reaction in proton exchange membrane (PEM) water electrolyzers. International Journal of Hydrogen Energy 46, 20825–20834.

Li, H., Wang, Y., Chen, G., Sang, Y., Jiang, H., He, J., et al., 2016. Few-layered MoS_2 nanosheets wrapped ultrafine TiO_2 nanobelts with enhanced photocatalytic property. Nanoscale 8, 6101–6109.

Li, Y., Yu, Z.G., Wang, L., Weng, Y., Tang, C.S., Yin, X., et al., 2019. Electronic-reconstruction-enhanced hydrogen evolution catalysis in oxide polymorphs. Nature Communications 10, 3149.

Liu, L., Jiang, Z., Fang, L., Xu, H., Zhang, H., Gu, X., et al., 2017. Probing the crystal plane effect of co3o4 for enhanced electrocatalytic performance toward efficient overall water splitting. ACS Applied Materials & Interfaces 9, 27736–27744.

Liu, R., Zheng, Z., Spurgeon, J., Yang, X., 2014. Enhanced photoelectrochemical water-splitting performance of semiconductors by surface passivation layers. Energy & Environmental Science 7, 2504–2517.

Liu, T., Diao, P., Lin, Z., Wang, H., 2020. Sulfur and selenium doped nickel chalcogenides as efficient and stable electrocatalysts for hydrogen evolution reaction: the importance of the dopant atoms in and beneath the surface. Nano Energy 74, 104787.

Liu, Y., Gu, X., Jiang, W., Li, H., Ma, Y., Liu, C., et al., 2022. In situ synthesis of morphology-controlled MoOx/Fe_{1-x} S bifunctional catalysts for high-efficiency and stable alkaline water splitting. Dalton Transactions 51, 9486–9494.

Lu, S., Zhang, T.J.C.C., 2022. Strategies for designing efficient electrocatalytic her catalysts at the atomic scale. Chem Catalysis 2, 1505–1509.

Lv, Y., Liu, Y., Chen, C., Wang, T., Zhang, M., 2018. Octopus tentacles-like WO_3/C@CoO as high property and long life-time electrocatalyst for hydrogen evolution reaction. Electrochimica Acta 281, 1–8.

Maiti, A., Srivastava, S.K., 2020. N, Ru co-doped pellet drum bundle-like Sb(2)S(3): An efficient hydrogen evolution reaction and hydrogen oxidation reaction electrocatalyst in alkaline medium. ACS Applied Materials & Interfaces 12, 7057–7070.

Majhi, K.C., Yadav, M., 2023. Transition metal-based chalcogenides as electrocatalysts for overall water splitting. ACS Engineering Au 3 (5), 278–284. In this issue. https://doi.org/10.1021/acsengineeringau.3c00014.

Ouyang, J., Lu, Q.C., Shen, S., Yin, S.F., 2023. Surface oxygen species in metal oxide photoanodes for solar energy conversion. Nanomaterials 13.

Peng, X., Yan, Y., Jin, X., Huang, C., Jin, W., Gao, B., et al., 2020. Recent advance and prospectives of electrocatalysts based on transition metal selenides for efficient water splitting. Nano Energy 78.

Piontek, S., Junge Puring, K., Siegmund, D., Smialkowski, M., Sinev, I., Tetzlaff, D., et al., 2019. Bio-inspired design: bulk iron-nickel sulfide allows for efficient solvent-dependent CO(2) reduction. Chemical Science 10, 1075–1081.

Raveendran, A., Chandran, M., Dhanusuraman, R., 2023. A comprehensive review on the electrochemical parameters and recent material development of electrochemical water splitting electrocatalysts. RSC Advances 13, 3843–3876.

Ravikumar, M.P., Quach, T.-A., Urupalli, B., Murikinati, M.K., Venkatakrishnan, S.M., Do, T.-O., et al., 2023. Observation of inherited plasmonic properties of tin in titanium oxynitride (TiO_xN_y) for solar-drive photocatalytic applications. Environmental Research 229, 115961.

Ren, B., Huang, Y., Han, C., Nadagouda, M.N., Dionysiou, D.D., 2016. Ferrites as photocatalysts for water splitting and degradation of contaminants. Ferrites and Ferrates: Chemistry and Applications in Sustainable Energy and Environmental Remediation.

Ridhova, A., Puspasari, V., Amal, M.I., 2021. Synthesis methods for chalcogenides and chalcogenides-based nanomaterials for photocatalysis. Chalcogenide-Based Nanomaterials as Photocatalysts. Elsevier.

Rogelj, J., Geden, O., Cowie, A., Reisinger, A., 2021. Three ways to improve net-zero emissions targets. Nature 591 (7850), 365–368.

Rokade, A., Rondiya, S., Date, A., Sharma, V., Prasad, M., Pathan, H., et al., 2017. Electrochemical synthesis of core-shell ZnO/CdS nanostructure for photocatalytic water splitting application. Energy Procedia 110, 121–127.

Santra, S., Streibel, V., Sharp, I.D., 2022. Emerging noble metal-free mo-based bifunctional catalysts for electrochemical energy conversion. Nano Research 15, 10234–10267.

Seo, J., Hisatomi, T., Domen, K., 2019. Visible-light-driven water splitting using perovskite-type oxynitride photoanodes. Electrochemical Society Meeting Abstracts 235. pp. 1603.

Sha, P., You, Y., Wen, D., Wu, Z., Wang, Q., Bu, D., et al., 2023. Two-dimensional ultrathin graphic carbon nitrides with extended π-conjugation as extraordinary efficient hydrogen evolution photocatalyst. Small 19, 2205834.

Shahid, M.U., Najam, T., Helal, H.H., Hossain, I., El-bahy, S.M., El-bahy, Z.M., et al., 2024. Transition metal chalcogenides and phosphides for photocatalytic H_2 generation via water splitting: a critical review. International Journal of Hydrogen Energy 62, 1113–1138.

Shit, S., Kuila, T., Srivastava, S.K., 2024. Noble-metal-free bifunctional electrocatalysts for overall water splitting in alkaline medium. Advances in catalysts research. Springer, pp. 279–337.

Song, B., Li, K., Yin, Y., Wu, T., Dang, L., Cabán-Acevedo, M., et al., 2017. Tuning mixed nickel iron phosphosulfide nanosheet electrocatalysts for enhanced hydrogen and oxygen evolution. ACS Catalysis 7, 8549–8557.

Suen, N.T., Hung, S.F., Quan, Q., Zhang, N., Xu, Y.J., Chen, H.M., 2017. Electrocatalysis for the oxygen evolution reaction: recent development and future perspectives. Chemical Society Reviews 46, 337–365.

Sun, S., Zhang, J., Gao, P., Wang, Y., Li, X., Wu, et al., 2017. Full visible-light absorption of TiO_2 nanotubes induced by anionic S_2^{2-} doping and their greatly enhanced photocatalytic hydrogen production abilities. Applied Catalysis B: Environmental 206, 168–174.

Sun, T., Fan, J., Liu, E., Liu, L., Wang, Y., Dai, H., et al., 2012. Fe and Ni co-doped TiO_2 nanoparticles prepared by alcohol-thermal method: application in hydrogen evolution by water splitting under visible light irradiation. Powder Technology 228, 210–218.

Sun, Y., Sviridova, E., Kamp, M., Zhang, J., Kienle, L., Moran, D.A.J., et al., 2023. Elucidating catalytic sites governing the performance toward the hydrogen evolution reaction in ternary nitride electrocatalysts. ACS Applied Energy Materials 6, 1265–1273.

Swaminathan, J., Subbiah, R., Singaram, V., 2016. Defect-rich metallic Titania (TiO1.23) an efficient hydrogen evolution catalyst for electrochemical water splitting. ACS Catalysis 6, 2222–2229.

Tang, S., Zhang, Z., Xiang, J., Yang, X., Shen, X., Song, F., 2022. Recent advances in transition metal nitrides for hydrogen electrocatalysis in alkaline media: from catalyst design to application. Frontiers in Chemistry 10, 1073175.

Wadhai, S., Jadhav, Y., Thakur, P., 2021. Synthesis of metal-free phosphorus doped graphitic carbon nitride-P25 (TiO_2) composite: characterization, cyclic voltammetry and photocatalytic hydrogen evolution. Solar Energy Materials and Solar Cells 223, 110958.

Walake, S., Jadhav, Y., Kulkarni, A., 2023. Novel spinel nanomaterials for photocatalytic hydrogen evolution reactions: an overview. Energies 16, 1407.

Wang, C., Zhang, Q., Yan, B., You, B., Zheng, J., Feng, L., et al., 2023a. Facet engineering of advanced electrocatalysts toward hydrogen/oxygen evolution reactions. Nanomicro Letters 15, 52.

Wang, Q., He, R., Yang, F., Tian, X., Sui, H., Feng, L., 2023b. An overview of heteroatom doped cobalt phosphide for efficient electrochemical water splitting. Chemical Engineering Journal 456, 141056.

Wang, H.-F., Tang, C., Li, B.-q, Zhang, Q., 2018a. A review of anion-regulated multi-anion transition metal compounds for oxygen evolution electrocatalysis. Inorganic Chemistry Frontiers 5, 521–534.

Wang, M., Williams, D., Lahti, G., Teshima, S., Dominguez Aguilar, D., Perry, R., et al., 2018b. Chemical intercalation of heavy metal, semimetal, and semiconductor atoms into 2d layered chalcogenides. 2D Materials 5, 045005.

Wang, J., Huang, J., Chen, G., Chen, W., Li, T., Meng, A., et al., 2022a. In-situ engineered heterostructured nickel tellur-selenide nanosheets for robust overall water splitting. Chemical Engineering Journal 446, 137297.

Wang, Y., Peng, J., Xu, Y., Zhao, R., Han, J., Wang, L., 2022b. Facile fabrication of CdSe/CuInS2 microflowers with efficient photocatalytic hydrogen production activity. International Journal of Hydrogen Energy 47, 8294–8302.

Wang, Y., Jin, S., Pan, G., Li, Z., Chen, L., Liu, G., et al., 2019. Zr doped mesoporous $LaTaON_2$ for efficient photocatalytic water splitting. Journal of Materials Chemistry A 7, 5702–5711.

Wang, Y., Ren, B., Zhen Ou, J., Xu, K., Yang, C., Li, Y., et al., 2021. Engineering two-dimensional metal oxides and chalcogenides for enhanced electro- and photocatalysis. Science Bulletin 66, 1228–1252.

Wang, Y., Xie, C., Liu, D., Huang, X., Huo, J., Wang, S., 2016. Nanoparticle-stacked porous nickel-iron nitride nanosheet: a highly efficient bifunctional electrocatalyst for overall water splitting. ACS Applied Materials & Interfaces 8, 18652–18657.

Wei, C., Xu, Z.J.J.C.J.O.C., 2022. The possible implications of magnetic field effect on understanding the reactant of water splitting. Chinese Journal of Catalysis 43, 148–157.

Xing, Z., Li, Q., Wang, D., Yang, X., Sun, X., 2016. Self-supported nickel nitride as an efficient high-performance three-dimensional cathode for the alkaline hydrogen evolution reaction. Electrochimica Acta 191, 841–845.

Xu, K., Chen, P., Li, X., Tong, Y., Ding, H., Wu, X., et al., 2015. Metallic nickel nitride nanosheets realizing enhanced electrochemical water oxidation. Journal of the American Chemical Society 137, 4119–4125.

Yin, J., Jin, J., Lin, H., Yin, Z., Li, J., Lu, M., et al., 2020. Optimized metal chalcogenides for boosting water splitting. Advanced Science 7, 1903070.

Yu, L., Zhu, Q., Song, S., Mcelhenny, B., Wang, D., Wu, C., et al., 2019. Non-noble metal-nitride based electrocatalysts for high-performance alkaline seawater electrolysis. Nature Communications 10, 5106.

Yuan, L.-J., Sui, X.-L., Pan, H., Wang, Z.-b, 2023. Strategies and mechanism for enhancing intrinsic activity of metal–nitrogen–carbon catalysts in electrocatalytic reactions. Renewables 1, 514–540.

Zhang, X., Liang, Y., 2018. Nickel hydr(oxy)oxide nanoparticles on metallic Mos(2) nanosheets: a synergistic electrocatalyst for hydrogen evolution reaction. Advanced Science 5, 1700644.

Zhao, S., Yang, Y., Tang, Z., 2022. Insight into structural evolution, active sites, and stability of heterogeneous electrocatalysts. Angewandte Chemie International Edition 61, e202110186.

Zheng, T., Sang, W., He, Z., Wei, Q., Chen, B., Li, H., et al., 2017. Conductive tungsten oxide nanosheets for highly efficient hydrogen evolution. Nano Letters 17, 7968–7973.

Zhou, P., Zhai, G., Lv, X., Liu, Y., Wang, Z., Wang, P., et al., 2021. Boosting the electrocatalytic HER performance of Ni_3N-V_2O_3 via the interface coupling effect. Applied Catalysis B: Environmental 283, 119590.

CHAPTER

10

Metal ion incorporated metal oxides for water splitting

Anju John[1], *Jain Maria Thomas*[2], *and Jesty Thomas*[1]

[1]Research Department of Chemistry, Kuriakose Elias College, Kottayam, Kerala, India

[2]Department of Chemistry, St. George's College, Aruvithura, Kottayam, Kerala, India

10.1 Introduction

In an era defined by the pressing need for sustainable energy sources and the mitigation of environmental concerns, the quest for efficient and clean energy solutions has become paramount. Among the diverse array of renewable energy technologies, hydrogen has developed as an ideal applicant for the clean energy landscape of the future. Its versatility, high energy content, and negligible greenhouse gas emissions when utilized as a fuel make it an attractive choice for addressing the confronts of climatic variations and green energy. However, the generation of hydrogen from conventional approaches, such as steam methane reforming or coal gasification, currently relies heavily on fossil fuels, thus compromising the overall environmental benefits of hydrogen as an energy carrier (Li et al., 2020). In this context, the advancement of alternative and sustainable methods like electrochemical water splitting, water splitting driven by thermoelectric devices, pyroelectric devices, solar light-driven water splitting devices, including photoelectrochemical, photovoltaic, dye-activated solar converters (Ahmad et al., 2015) for hydrogen production has garnered significant attention from researchers and policymakers alike.

Solar energy holds tremendous economic and societal significance due to its abundant, clean, and free source of energy that can be utilized to meet our power demands for generations to come. This elegant and environmentally friendly approach, which mirrors the natural photosynthesis has the potential to revolutionize the hydrogen production landscape and usher in a new era of sustainable energy, making it an integral component of the emerging hydrogen economy (Gopinath and Marimuthu, 2022). The notion of photochemical water splitting relies on the ability of certain materials, semiconductor based photocatalysts, to exploit the sunlight and drive the transformation of aqueous molecules into H_2 and O_2. Metal oxides are widely used due to their favorable characteristics like low toxicity, easy production, high crystalline structure, adaptable optoelectronic properties,

Photochemical Splitting of Water
DOI: https://doi.org/10.1016/B978-0-443-29064-0.00009-7

diverse forms, availability, and easy affordability (Wang et al., 2018). They act as photocatalysts and photoelectrodes, facilitating the hydrogen and oxygen evolution process. Metallic oxides like Fe_2O_3 and Cu_2O are given in Table 10.1 and are crucial as their capability to absorb sunlight and create charge transferors, driving the water-splitting reactions. Their stability, tunable bandgap, and corrosion resistance make them ideal for long-term operation in aqueous systems, and engineering these materials enhances proficiency, making them central to advancing clean energy technologies.

The pivotal semiconducting light-activated catalysts, like TiO_2, ZnO, and WO_3 (Maeda and Domen, 2013), garnered an immense deal of interest as photocatalysts in photochemical splitting of water for their appropriate band edge voltage, excellent chemical persistence, and affordability. Nevertheless, pure metal oxides face performance limitations, such as lack of active spots, increased recoupling of light-induced charge transferors, and inadequate conductivity of electric charge. To address these issues, various methods, including doping, morphological changes, surface modifications, synthesis of composite materials, and nanoscale design, are employed to enhance the attributes of metal oxides and their effectiveness in specific applications (Martha et al., 2015). This chapter embarks on a thorough investigation of the photochemical water-splitting process to elucidate the usage of metal oxide semiconducting light-stimulated catalysts and delve into their essential parameters, benefits, obstacles, and methods of enhancement to boost their catalytic performance. The chapter aims to explicate how metal doping enhances photocatalytic activity by examining changes in electronic structure, charge carrier dynamics, and surface properties. It investigates the physical and chemical peculiarities of doped metallic oxides and identifies optimal doping strategies to maximize efficiency. Comparative analysis helps to determine promising candidates based on hydrogen evolution rates, quantum efficiency, and photocurrent density. The chapter also examines surface and interface characteristics and their impact on performance, bandgap influence on light absorption, and charge separation.

TABLE 10.1 Metal oxides and their properties.

Metal oxide	Bandgap (eV)	Conduction band edge potential (V w.r.t NHE)	Valence band edge potential (V w.r.t NHE)	Stability (pH range)	Photocatalytic efficiency
TiO_2	3.0–3.2	−0.1 to −0.3	2.7–2.9	0–14	Moderate
ZnO	3.2–3.3	−0.3	3.0	6–12	Moderate
SnO_2	3.6–4.0	−0.1	3.5	0–14	Low
Fe_2O_3	2.1–2.2	0.2	2.3	2–12	High
Cu_2O	2.0–2.2	−0.7	1.3	5–9	High
WO_3	2.6–2.8	0.4	3.0	1–9	Moderate
$BiVO_4$	2.4–2.5	0.0	2.4	2–12	High
$SrTiO_3$	3.2–3.4	−0.3	3.0	0–14	Moderate
$BaTiO_3$	3.2–3.4	−0.1	3.1	2–12	Low

10.2 Mechanism of metal oxide–assisted photochemical water splitting

Metal oxides are wide bandgap-valued semiconductors whose electronic configuration is a decisive factor in photocatalytic reaction. Metal oxide semiconductors contain valence band (VB) and conduction band (CB); disparity in energy among these two bands is referred to as the band gap (Eg). As semiconductors are exposed to photons (solar irradiation) having energy, E ≥ Eg, electrons in VB absorb energy from this incident photonic particle and get excited to CB (Fig. 10.1).

Photoinduced excitons that transit to the surface of the catalyst undergo redox reactions with the reactants adhered on the semiconductor surface, forms the fundamental processes behind photocatalytic hydrogen generation and the deterioration of toxins. In the case of hydrogen production, it is essential for the CB level to have more negatively charged than hydrogen liberation level ($E_{H2/H2O}$) and for the VB level to have more positively charged than oxygen liberation level ($E_{O2/H2O}$) to achieve effectual oxygen generation through photocatalysis. Moreover, usage of so-called sacrificial reagents can boost the competence of the photolysis of water. The primary responsibility of these sacrificial agents is to split the photoinduced charge carriers and make accessible for various reactions and act as electron donors or acceptor foragers (Galińska and Walendziewski, 2005).

The smallest possible photon energy expected for the light-induced reaction is 1.23 eV, analogous to a wavelength around 1000 nm in the near-IR region. Fig. 10.2 depicts the bandgap energies and band edge positions among various semiconductor photocatalysts. Semiconductors such as silicon carbide (SiC), titanium dioxide (TiO_2), tin dioxide (SnO_2), and zinc oxide (ZnO) exhibit wide bandgaps, ranging from 3 to 3.8 electron volts (eV), and rely on high-energy ultraviolet radiation for their photocatalytic processes. In contrast, materials like gallium arsenide (GaAs), cadmium selenide (CdSe), iron (III) oxide (Fe_2O_3), gallium phosphide (GaP), cadmium sulfide (CdS), and tungsten trioxide (WO_3) possess

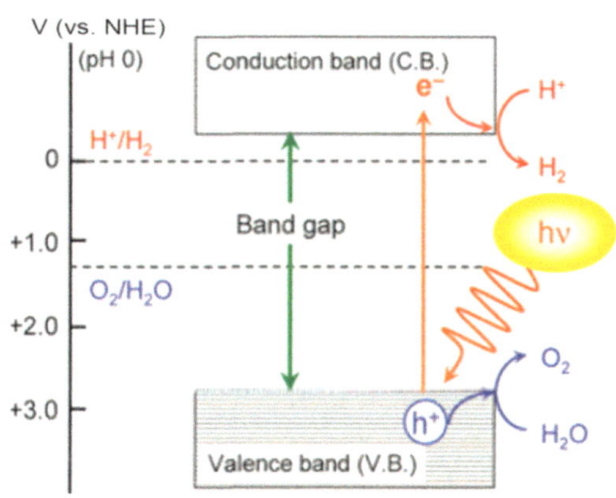

FIGURE 10.1 Fundamental principle of photolysis of water on a light-activated catalyst (Maeda and Domen, 2013). *Source: Replicated with permission from Maeda, K., Domen, K., 2013. 7.22 - Photochemical water splitting using nanostructured metal oxides. In: Comprehensive Inorganic Chemistry II, Second Edition), Second Edition Elsevier, pp. 587–614. https://doi.org/10.1016/B978-0-08-097774-4.00721-X. Copyright 2007 American Chemical Society.*

narrow bandgaps, spanning from 1.4 to 2.6 eV and utilize visible light to drive their photocatalytic reactions. Both wide and narrow bandgap semiconductors encounter drawbacks, including swift coalescing and limited production of light-driven charge transferors, which hampers their catalytic performance under light. The requirement of harmful UV light for wide bandgap semiconductors and their tendency for rapid recombination create significant challenges for using sunlight as a practical energy source in photocatalysis. In practice, relying on UV light largely precludes the exploitation of daylight as a viable power supplier for photocatalysis.

The bandgap of metal oxides can widen or narrow depending on the kind of metal dopant used. Furthermore, doping changes the band-edge regions and increases carrier density. The transition metal ions combine with oxygen to create localized areas on the surface of doped metal oxides. These specific regions function as sites that trap and transport electrons when subjected to bandgap excitation (Kalanur et al., 2020). When the host material absorbs photons, it generates pairs of electrons and holes. The excited electron resulting from this process moves to the CB of the doped metallic oxides. The VB of the base material is employed in the process of oxidizing H_2O and the reducing H^+ ions take place at integrated metal situated on the base material's surface (Hameed et al., 2004). This chapter explores contemporary research and advancements in the field of semiconducting metallic oxides, such as TiO_2, ZnO, and WO_3, by incorporating suitable metals, focusing on their role in using solar light to facilitate the splitting of water.

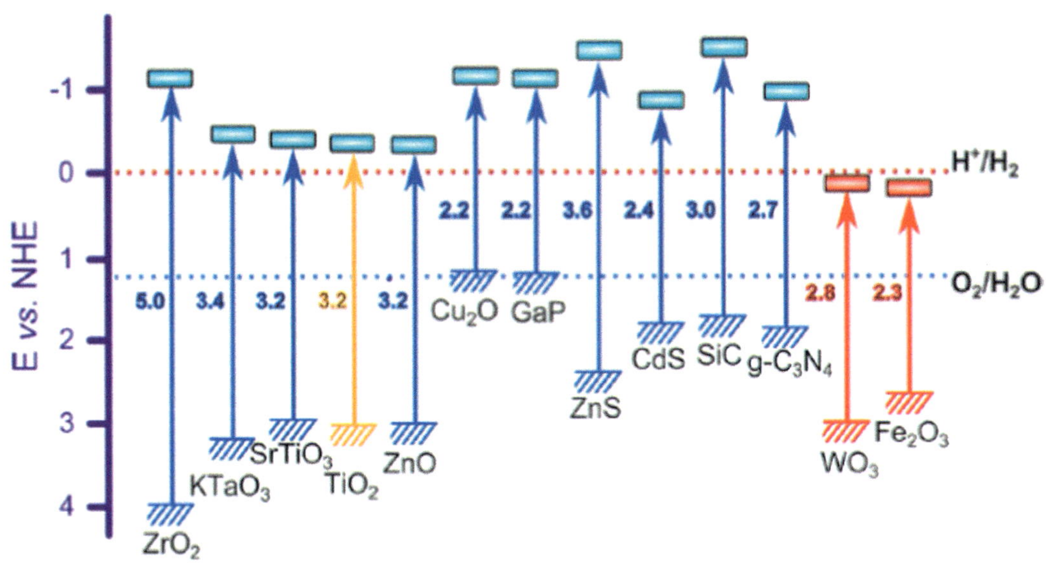

FIGURE 10.2 The bandgap and band edge of different semiconductor photocatalysts (Christoforidis and Fornasiero, 2017). *Source: Replicated from Christoforidis, K.C., Fornasiero, P., 2017. Photocatalytic hydrogen production: a rift into the future energy supply. ChemCatChem 9, 1523–1544. https://doi.org/10.1002/cctc.201601659 with permission of the Wiley Publisher.*

10.3 TiO_2-based photocatalysts

Titanium dioxide is a versatile light-activated catalyst, renowned for its unique electronic structure and reactivity. It stands out as a universal photocatalyst due to its exceptional attributes, including photostability, well-suited band edge positions, and chemical inertness. However, its broad bandgap limits visible light absorption. Doping TiO_2 with foreign elements such as metals is a promising way to stimulate TiO_2 under sunlight and restrain the reunification of charge transferors (Kumaravel et al., 2019). Metal dopants enhance the spectral response and charge carrier separation in TiO_2 through several mechanisms; create electron traps to suppress the recombination rate, leverage surface plasmon resonance, introduce new intermediate energy levels in between the bandgap and enable visible light captivation, and hybridize with TiO_2 orbitals to modify the band edifice and improve charge transferor's motion. The mechanism of metal dopants incorporated TiO_2-mediated photochemical water splitting is given in Fig. 10.3. Table 10.2 comprises the studies on photocatalytic hydrogen generation by metal-doped TiO_2. Inert elements, including Pd, Pt, Ag, and Au, are often utilized as cocatalysts and can enhance photogenerated electron and hole separation, extend their lifetime, and prevent inverse reactions among O_2 and H_2, thereby improving photocatalytic activity and water-splitting ability of TiO_2 (Rafique et al., 2023). For instance, Pd nanoparticle loaded TiO_2 (0.18 at.% Pd/TiO_2) nano layers with a huge fraction of exposed (001) planes, produced via chemical reduction exhibited superior water-splitting capability by an H_2 production rate of 3096 μmol/g/h, with CH_3OH as the sacrificial agent (Wu et al., 2016). Also, 0.3 wt.% Pd-doped TiO_2 nanoparticles (NPs),

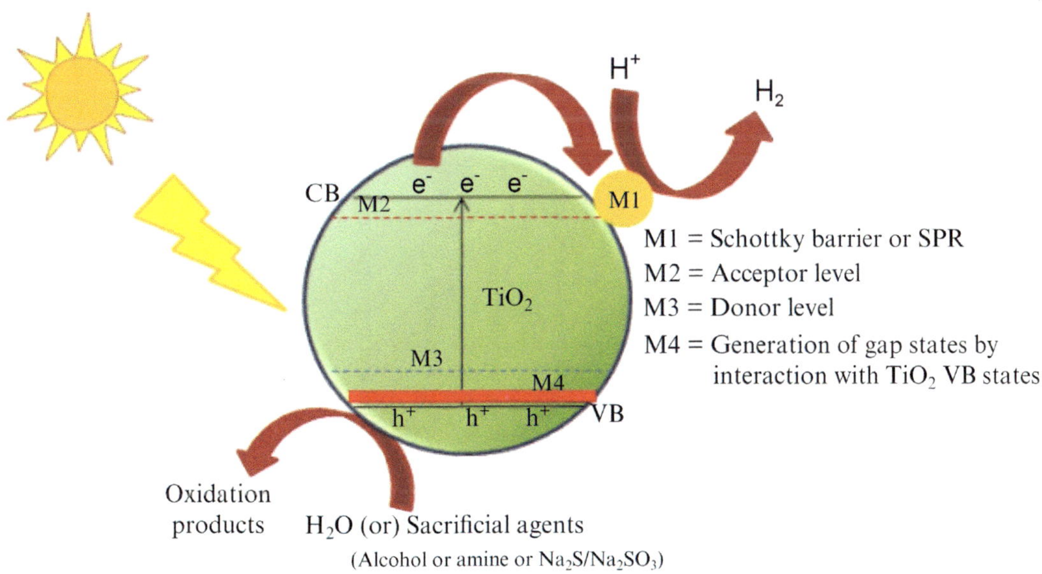

FIGURE 10.3 The possible mechanism of metal dopants incorporated TiO_2 mediated photochemical water splitting. *Source: Replicated from Kumaravel, V., Mathew, S., Bartlett, J., Pillai, S.C., 2019. Photocatalytic hydrogen production using metal doped TiO_2: a review of recent advances. Applied Catalysis B 1021–1064. https://doi.org/10.1016/j.apcatb.2018.11.080 with the permission of Elsevier publisher.*

TABLE 10.2 Summary of studies on hydrogen generation by metal-doped TiO_2.

Catalyst	Synthetic method	Source of light	H_2 evolution	Quantum efficiency	References
Pd-TiO_2	Ultrasonic dispersion	Xe lamp (300 W)	3096 μmol/g/h	3.4%	Wu et al. (2016)
Pd-TiO_2	Ionic liquid-assisted hydrothermal treatment	Xe lamp (300 W)	1250 μmol/g	–	Ravishankar et al. (2018)
Pd/TiO_2/ZSM-5	Impregnation	7 W UVC lamp	1148 μmol/g/h	–	Enzweiler et al. (2018)
Pt/TiO_2	In situ photo deposition	Xe lamp (300 W)	560 μmol/g/h	–	Yang et al. (2020)
Pt/TiO_2	Polygonal barrel-sputtering	UV light	107.8 μmol/h	–	Matsubara et al. (2019)
Pt/TiO_2	Photo deposition method	UV light	22.0 μmol/h	–	Matsubara et al. (2019)
Ag/TiO_2	Reductive reaction	UV lamp	23.5 mmol/g/h	19%	Gogoi et al. (2020)
Ag/TiO_2	Sol-gel and impregnation method	Liquid phase plasma (UV and visible)	1.6 mmol/h/g	–	Park et al. (2020)
Fe-impregnated MIL-125(Ti)(Fe/TiO_2)	Calcination and pyrolysis	500 W Xe/Hg lamp	230 μmol/g/h	–	Valero-Romero et al. (2019)
Fe -Ni/TiO_2	Alcohol-thermal method	Hg and Xe lamp	361.64 μmol/h/g	–	Sun et al. (2012)
Ti^{3+}/Ag/TiO_2	Microwave-induced reduction and photochemical deposition	Xenon arc lamp (350 W)	37.1 μmol/h/g	–	Lu et al. (2020)

Fe /TiO_2/g-C_3N_4	Impregnation method solid-state Dispersion method	Solar radiation (average light intensity 1360 W/m^2)	5200 μmol/h/g	–	Sadanandam et al. (2018)
Fe/TiO_2 Cr/TiO_2	Radio-frequency magnetron sputtering	Tungsten halogen lamp (250 W)	(15.5 μmol/h) (5.3 μmol/h)	–	Dholam et al. (2009)
Fe-Ni/Ag/TiO_2	Solvothermal method	Xenon lamp (500 W)	793.86 μmol/h/g	–	Sun et al. (2015)
Ag/Ce/TiO_2	Electrochemical anodic oxidation and microwave aided reduction	Hg lamp (UV) Xenon lamp (visible light)	1.34 μmol/(cm^2·h)	7.7%	Fan et al. (2015)
W/TiO_2/Au	Precipitation and annelation	Xenon arc lamp (300 W)	24 mmol/g/h	–	Liu et al. (2017)
NiO_x-modified Au/TiO_2	Photo deposition and impregnation and CPH method	Xenon lamp (300 W)	5.5 μmol/h	–	Tanaka et al. (2017)
Cu/TiO_2	Sol–gel and annealing	Xenon lamp	16.6 mmol/g h	45.5%	Lee et al. (2019)
CuSA-TiO_2	Precipitation and annealing	Xenon lamp	101.7 mmol/ g/h	56%	Zhang et al. (2022)
GO:Ni:TiO_2	Precipitation and annealing	Solar irradiation	4.97 mmol/g/h	3.11%	Atrees et al. (2021)
Cu/TiO_2	Impregnation method	UV lamp (300 W) and visible lamp (halogen, 500 W)	5000 μmol/h/ g (uv) 220 μmol/h/g (visible)	–	Díaz et al. (2021)
Ni/Zn/TiO_2	Deposition-precipitation method	Xenon light source (300 W)	17 mmol/h/g	24%	Neațu et al. (2020)
Cu/Ni/TiO_2	Wet impregnation and coimpregnation	Solar irradiation	35.4 mmol/g/h	–	Naresh et al. (2021)

synthesized through ionic liquid-assisted hydrothermal treatment, gave 1250 μmol/g hydrogen generation under light irradiation in a H_2O/ethanol system as a scavenger (Ravishankar et al., 2018). Studies reported that Pd/TiO_2/ZSM-5 photocatalyst displayed an improved hydrogen generation rate (1148 μmol/g/h) under UV radiation which can be credited to the combined effects of surface area per unit and porous nature of ZSM-5 zeolite, fine dispersal of catalytic activator and interaction among palladium and TiO_2 (Enzweiler et al., 2018) nanoparticles. Similarly, Pt incorporation also supports the water-splitting ability; Maeda (2014) reported that rutile TiO_2 integrated with 0.1 wt.% platinum nanoparticles produced about 14 and 7 μmol/h of H_2 and O_2, correspondingly, whereas bare TiO_2 did not show any activity. Nevertheless, the photocatalytic efficacy of Pt/TiO_2 is significantly decreased in acidic and alkaline environments due to a photo-erosion impact and the absence of an adequate water reduction mechanism. Galińska and Walendziewski (2005) focused on the impact of sacrificial reagents (CH_3OH, Na_2S, ethylenediammine tetraacetic acid (EDTA), I^- and IO_3^- ions) on photoinduced water cleavage proficiency of Pt/TiO_2 catalyst under UV radiance and testified that the usage of methanol resulted in enhanced hydrogen evolution than EDTA and Na_2S as sacrificial reagents. Studies reveal that 2.50 wt.% Pt incorporated anatase TiO_2 with a great surface area displayed enhanced photoactivity in the generation of hydrogen and exhibited a quantum yield of 22.6% (Guayaquil-Sosa et al., 2017). The polygonal deposition method resulted in smaller Pt nanoparticles which are finely distributed on the TiO_2 support, showing that the H_2 and O_2 gases with evolution rates (H_2: 107.8 μmol/h, O_2: 51.7 μmol/h) is 5–10 times superior to Pt/TiO_2 prepared by conventional photodeposition method and achieved stoichiometric water decomposition where molar fractions of H_2 versus O_2 were near to 2 (Matsubara et al., 2019). As is known, in lack of a sacrificial agent, O_2^- will devour adjacent H_2 to form H_2O. Yang et al. (2020) suggested a novel approach for the photo dissociation of water by regulating water imbibition on the Pt/TiO_2 surface, which stops H_2 from interacting with O_2^- to renew water. The effectiveness of the approach could attain 560 μmol/g/h and the total sunlight energy transduction proficiency was established to be 0.46%.

Well-distributed Ag ions on TiO_2 lead to an effective charge migration from TiO_2 to silver via Ti-Ag-O bonds (Gogoi et al., 2020) and reported a higher hydrogen evolution rate. The plasmonic resonance of Ag (Lu et al., 2020) has increased TiO_2 absorption capacity, lengthened the light spectrum response, and curbed reunification of photo-generated excitons of TiO_2. Studies report that the hydrogen generation rate of Ag-doped TiO_2 nanotubes (1.34 μmol/(cm^2·h)) produced with 0.002 M $AgNO_3$ solution was found to be 3.3 times as that of pristine TiO_2 (Wu et al., 2013). Park et al. (2020) described that liquid-phase plasma (LPP) can produce hydrogen through photocatalytic water splitting over Ag-doped TiO_2 photocatalysts. This method offers greater plasma density and wider spatial coverage, with additives enhancing optical properties and methanol increasing hydrogen evolution in water. Bimetallic catalysts demonstrate enhanced catalytic efficacy in comparison to their monometallic counterparts. Efficient photocatalysis using visible light was achieved by incorporating Ce and Ag nanoparticles in anatase Ag–Ce/TiO_2 nanotubes, enhancing absorption capacity due to surface plasmon (Fan et al., 2015). Lanthanide ions, especially Ce^{4+}/Ce^{3+}, act as Lewis acids and efficiently fence photo-excited CB electrons, transmitting them to absorbed O_2, thereby impedes the recoupling of photogenerated e^-– h^+ pairs.

The efficacy of Au/TiO_2 was ascribed to the incorporation of deformity, while suboptimal visibility of light uptake in the TiO_2 was enhanced by the gold plasmonic waves (Tanaka et al., 2017). The study demonstrates that a Ni-modified Au/TiO_2 catalyst with a NiOx cocatalyst can significantly increase the proportion of H_2 and O_2 evolution, indicating that loading NiOx cocatalyst separately without alloying with Au particles can enhance the overall water-splitting activity. Tungsten metal ion doped with Au/TiO_2 preserved its anatase structure, narrowed the bandgap, and acted as a competent capping reagent to alleviate AuNPs, thereby preventing crystal growth and allowing the uniform distribution on the TiO_2 surface (Liu et al., 2017).

Transition metals like Cu, Ni, Co, and Fe, which are copious in the earth's crust are being studied as substitutes for noble metals in photocatalysis. Transition metal loading, such as Fe-doping (Ismael 2020), has been explored as a tactic to expand the excitation energies of semiconducting material into the optical spectrum, enabling better consumption of sunlight energy. Valero-Romero et al. (2019) synthesized Fe-doped titania photocatalysts by depositing Fe^{3+} on the surface of Ti-containing metal organic framework (MOF) MIL-125. At low Fe concentrations, Fe-doped titania catalysts exhibited a red shift in photosensitivity toward visible light. The rate of hydrogen generation rose from 76 to 230 µmol/g/h for 0.5 wt.% Fe/TiO_2 compared to pristine TiO_2, showing a threefold increase in hydrogen generation upon UV irradiation. Another research investigated the synthesis of Cr or Fe incorporated TiO_2 thin films via RF deposition and colloidal gel technique (Dholam et al., 2009). The results demonstrate that metal-blended TiO_2 achieved via deposition is more effective than that attained through the sol-gel method in generating a red edge shift in visible spectrum. Xiao et al. (2020) adopted the high temperature salt technique (MSM) to prepare Ni cocatalysts on TiO_2 nanoparticles, ensuring strong Ni-O bonds and maintaining semiconductor integrity. During the MSM process, Ni/TiO_2 samples maintain their crystal structure, morphology, and light-harvesting performance, while Ni atoms enhance oxygen vacancy formation on TiO_2. The Ni-doped TiO_2 catalyst exhibited moderate H_2 production (5.5 mol/h), which was further improved by a thermal reduction treatment (573 K, 5% H_2/N_2 flow, 0.5 h). Ni/TiO_2 catalyst when supported with graphene oxide (Atrees et al., 2021), demonstrated 3.11% quantum efficacy for hydrogen generation with an H_2 production rate of 4.97 mmol/h/g. The dual doping of Fe and Ni in TiO_2 enhanced light-induced H_2 generation through the photodissociation of water (Sun et al., 2012). The crystallite size of the codoped TiO_2 was about 14 nm and had surface area per unit of 98.35 m^2/g. Another study observed that Ag ions incorporated Fe and Ni-doped TiO_2 catalysts showed the highest H_2 generation rate and attained 793.86 µmol/h/g, which was significantly superior to pure TiO_2 (Sun et al., 2015). The fine dispersion of Fe^{3+} and carbon nitride expanded titania's photo response, allowing efficient electron migration. The Fe-integrated TiO_2 with graphitic carbon nitride catalyst (Sadanandam et al., 2018) showed elevated photoactivity and enhanced hydrogen production, with a 17 times higher hydrogen evolution rate.

Recently, single-atom catalysts (SACs) (Li et al., 2023, Hu et al., 2021, Lee et al., 2019) have been emphasized for their ability to maximize reaction active sites, resembling homogeneous catalysis, by utilizing isolated metal atoms. Cu nanoparticles with readily changeable valence states (Xiao et al., 2023) have shown promise for competent charge parting and transport, directing to heightened catalytic properties, and present themselves as a cost-effective alternative to Pt/TiO_2 for hydrogen production. Zhang et al. (2022) achieved a highly dispersed copper single-atoms about 1.5 wt.% on TiO_2(CuSA/TiO_2), exhibited a higher H_2 generation rate of 101.7 mmol/g/h compared to photocatalyst PtSA-TiO_2 (95.3 mmol/g/h), along with excellent stability even after storing for 380 days. This catalyst CuSA-TiO_2, which demonstrates a photonic efficiency of 56% at λ 365 nm, represents a noteworthy advancement in the field of photo-assisted hydrogen synthesis. The photo-deposited 1% Cu on TiO_2 nanoparticles resulted in visibly colored titania with distinct visible absorbance properties and exhibited an H_2 generation rate of about 8500 μmol/h/g. The poorer activity in photo-deposited metal TiO_2 samples with Co or Ni is due to variations in the quantity of metal species and disparities in the redox activity of metal ions (Díaz et al., 2021). By introducing $Na_2S_2O_3$ or methanol as hole scavengers, Neațu et al. (2020) could achieve 17 mmol/h/g of H_2 production and an obvious quantum yield (AQY) of 24% with Ni-Zn/TiO_2 ternary photocatalyst. Another ternary Cu-Ni-TiO_2 catalyst (Naresh et al., 2021) could attain an approximate H_2 output of 0.20 mmol/g/h in lack of a sacrificial reagent, while it reached a value of about 35.4 mmol/g/h when mixture of H_2O and CH_3OH was employed.

10.4 ZnO-based photocatalysts

Zinc oxide (ZnO) is another extensively studied semiconductor photocatalyst with a unique crystalline structure, wide bandgap, and high electron mobility. Its remarkable unique defect chemistry imparts it with exceptional photocatalytic properties. The framework of ZnO presents a crucial role in its capability to harness light energy and initiate photochemical reactions. The pursuit of appropriate photocatalysts comprised of transition metal oxides to facilitate the water oxidation reaction (Cao et al., 2022). ZnO has materialized as a prominent contender due to its limited toxicity, abundance, and inexpensive and effortless fabrication.

However, the widespread utilization of ZnO is limited by two primary constraints including the limited absorption of sunlight and quick recoupling rate of charge carriers. Photon uptake competence of ZnO is primarily confined to the UV zone due to its widened optical bandgap (3.3 eV) and elevated exciton binding energy of 60 meV (Bharat et al., 2019). Additionally, the small surface area of ZnO significantly diminishes its photocatalytic activity. Moreover, ZnO exhibits photo instability in aqueous solutions, specifically due to its photocorrosion while subjected to ultraviolet radiation. Hence, proficient utilization of ZnO for the photolysis of water necessitates enhanced design and synthesis strategies to surmount the limitations inherent to ZnO. Studies reported that the activity of pure ZnO nanowires (NWs) in water splitting (442 μmol/h/g of H_2) surpasses that of ZnO nanoparticles (86 μmol/hg of H_2) and this enhancement potentially attributed to surface imperfections and oxygen vacancies (Machín et al., 2018). Guo et al. (2015) reported that microporous 3D ZnO prepared from the precursors, zinc nitrate and NaOH, having considerable surface area of 288 m^2/g, showed excellent undiminished hydrogen generation activity upon simulated solar illumination for 24 hours. Extensive investigations have been carried out to enhance the effectiveness of ZnO photocatalysts, revealing that the incorporation of ionic metals (Samadi et al., 2016) into the crystal framework of ZnO is an efficacious method for tailoring its suitability for various photocatalytic applications, provided that the dopant metal possesses a lower Fermi level than the host material and studies on metal-doped ZnO with enhanced photocatalytic hydrogen evolution efficiency is summarized in Table 10.3.

TABLE 10.3 Summary of studies on photocatalytic hydrogen evolution by metal-doped ZnO.

Catalyst	Synthetic method	Source of light	H_2 evolution	Quantum efficiency	References
Ag/ZnO	Rf sputtering	UV and visible light	~38 (UV) and ~24 (visible) μmol/h	–	Trang et al. (2020)
Au/ZnO	Dispersion and sonication	UV-visible light	853 μmol/h/g	–	Machín et al. (2018)
Ag/AgCl/ZnO	Conventional single pot refluxing method and microwave-assisted method	Xe light source	6082 (one pot) and 6782 (microwave) μmol/h/g	–	Mandawat et al. (2019)
Ca/ZnO	Sol-gel self-ignition method	Xe lamp (300 W)	819.75 μmol/h	–	Ahmad et al. (2020)
Ce/ZnO	Solution-free mechanochemical method	Xenon arc lamp (300 W)	43 μmol/h/g	–	Patil et al. (2021)
Cu/ZnO	Precipitation method	LEDS (nominal power: 10 W; light intensity: 32 mW/cm^2)	4770 μmol/L	–	Vaiano and Iervoli (2019)
Sr-ZnO/CNT	Sol–gel method	Dielectric barrier discharge (DBD)	2760 μmol/h/g	–	Irfan et al. (2023)
Eu/ZnO/GO	Hydrothermal method	Xe lamp (300 W)	229 μmol/h/g	–	Gurbani and Neelu (2023)

Noble metal gold (Au), which possesses a lower Fermi level than ZnO, has been found to effectively address the limitations of ZnO (Kunthakudee et al., 2022; Ahmad et al., 2023). Usage of Au (10 wt.%)@ZnO nanowire composites has demonstrated higher hydrogen production activity compared to Au (10 wt.%)@ZnO nanoparticles, achieving hydrogen generation of 853 μmol/h/g (Machín et al., 2018). Another study reported that Au nanoparticles have been utilized to sensitize ZnO core, which displays promising characteristics leading to significantly enhanced current density, conversion efficacy, and boosted photo-electrochemical dissociation of water (Zhang et al., 2015a).

When silver metal is employed as a dopant, its various characteristics including resistance to photo-corrosion, anchoring on surfaces, and the existence of surface plasmon bands within the visible range have significantly increased the water-splitting ability of ZnO (Ahmad et al., 2019). Utilization of RF-sputtering approach allows excellent silver dispersion into ZnO matrices as well as simultaneous control of Ag NP size via solitary sputtering time adjustments. Heterostructure Ag/ZnO (AZO) nanocomposites synthesized on an indium tin oxide via RF deposition displayed favorable catalytic properties owing to the interactivities between Ag and ZnO, exhibited remarkable performance in H_2 evolution with noticeable photonic efficiency of 4.9% and 2.7%, separately under UV and visible radiance (Trang et al., 2020). Mandawat et al. (2019) conveyed that Ag@AgCl/ZnO nanocomposite synthesized via microwave method is found to have higher hydrogen generation rates (6782.32 μmol/h/g) than the nanocomposite obtained through single pot refluxing method with 20% methanol under AM 1.5 G irradiation. Nevertheless, expensive and weak catalytic competence of inert metals cause a significant challenge in H_2 evolution which can be conquered by single-atom catalysts. For example, Liu et al. (2021) observed that loading of a single Pt atom catalyst on ZnO-ZnS heterostructure having surplus amount of Zn vacancies has notably boosted the activity of the catalyst.

Recently, alkali earth metal, Sr-doped ZnO was found to have high catalytic activity for hydrogen synthesis using an argon plasma discharge light source (Irfan et al., 2023). The author found that when they modified 25 wt.% Sr-doped ZnO with carbon nanotubes (2 wt.%) ($S_{25}ZC_2$), H_2 production increased by 2760 μmol/h/g. However, the catalytic performance of $S_{25}ZC_2$ declined vaguely over time and the inclusion of Sr into the ZnO/CNT photocatalyst serves to modulate the surface alkalinity of the catalyst, thereby augmenting the rate of hydrogen generation during the process of water splitting. Ahmad et al. (2020) reported that when 3 mol% Ca is doped with ZnO (Ca/ZnO) particles, it showed an increased hydrogen generation rate (819.75 μmol/h), which was twofold of earlier stated optimum ZnO catalysts. This increased performance can be attributed to several factors, including smaller grain size, heightened light harvest, diminished interfacial confrontation, a narrowed bandgap, and a higher Ca/ZnO photocatalyst surface area. The inclusion of transition metals like Cu in the ZnO lattice led to a substantial upgrade in the catalytic action for producing H_2 from aqueous glycerol. 1.08 mol% of Cu-loaded ZnO achieved hydrogen production about 2600 μmol/L with 5 wt.% aqueous glycerol and concentration of catalyst was 1.5 g/L at pH 6. By raising the initial glycerol dosage to 10 wt.%, the hydrogen generation rate escalated to around 4770 μmol/L (Vaiano and Iervoli, 2019). When 15% Fe metal ion was incorporated with the ZnO photoanode, the catalyst presented a significant visible light-assisted water-dissociation performance with a photoconversion competence of 2.4% (Khan et al., 2020). Metal lanthanides like Ce, Eu, Sm, and Tb-doped ZnO catalysts exhibited remarkable activity in the photolysis of water. Incorporation of Ce^{4+} ions in ZnO led to lattice expansion, lowered the crystal size, and improved the surface of Ce/ZnO (Patil et al., 2021). The mechanistic pathway ensued by the Ce/ZnO for hydrogen production is depicted in Fig. 10.4. The boosted catalytic action of Eu-integrated ZnO@GO is owed to the development of a p–n interface among the materials, that is, GO and Eu/ZnO, where they function as p-type and n-type regions, respectively, facilitating electron transfer and minimizing charge recombination (Gurbani and Neelu, 2023). Ahmad et al. (2022) synthesized terbium and samarium dual-doped ZnO/carbon nanotube (CNT) composites, with CNTs serving as a supportive material and found that dual-doped TS-ZnO/CNTs led to higher activity in hydrogen evolution compared to the Sm-doped and Tb-integrated ZnO/CNTs catalysts. By employing triethanolamine as a sacrificial reagent, TS-ZnO/CNTs catalyst displayed notable hydrogen production of 2683 mol/h/g and showed remarkable constancy over five sequential hydrogen evolution runs.

10.5 WO_3-based photocatalysts

WO_3 has garnered considerable interest as a nontoxic n-type semiconductor photocatalyst, owing to its robust visible light captivation capabilities and low bandgap energy of 2.7–3.1 eV (Murillo-Sierra et al., 2021). Hodes et al. (1976) used WO_3 for solar water splitting in 1976. It is relatively stable in aqueous conditions throughout a large range of pH (Murillo-Sierra et al., 2021). Additionally, it exhibits biocompatibility, affordability, good hole mobility, excellent bandgap flexibility, resilience to photocorrosion, and electron transport qualities. Theoretically, when employed as an anode in a photoelectrochemical (PEC) water photolysis system, WO_3 can convert solar energy into H_2 with a maximum efficiency of 4.8% (Liu et al., 2012). A narrow position of CB edge and rapid charge carrier recoupling are the main drawbacks of WO_3 photocatalyst which limits its

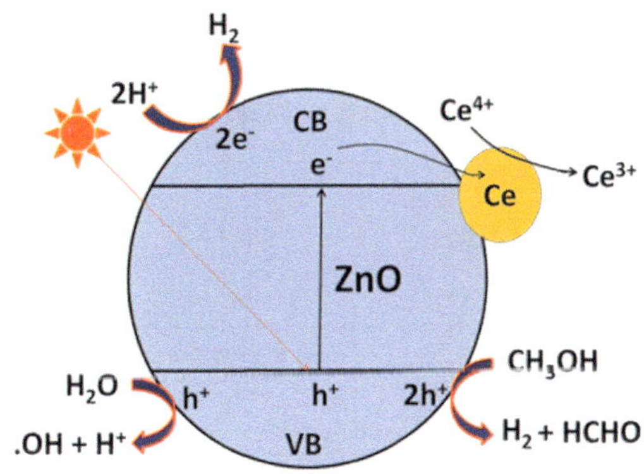

FIGURE 10.4 The mechanism of photochemical hydrogen production by Ce-doped Z. *Replicated from Patil, A.B., Jadhav, B.D., Bhoir, P.V., 2021. Efficient photocatalytic hydrogen production over Ce/ZnO from aqueous methanol solution. Materials for Renewable and Sustainable Energy 10, 1–9. https://doi.org/10.1007/s40243-021-00199-5 with the permission of Springer publisher.*

applications (Shandilya et al., 2022). CB minimum of bulk should be 0.4 eV less than the redox potential of hydrogen to make it appropriate for solar light absorption (Wang et al., 2012). Since its conduction band lacks the energy necessary for H_2O reduction, it cannot effectively split water to produce H_2 on its own (Camposeco et al., 2018). Numerous techniques, including doping with metals or nonmetals, altering the structure, and creating heterojunctions, might increase its effectiveness. Among these, modification of WO_3 via inclusion of metal ions is a viable approach to upgrade its efficiency by tailoring the electronic structure and properties (Hwang et al., 2002). Near the conduction band, new energy levels are formed as a result of the integration of metal ions (Fig. 10.5). These new energy levels help in the modification of the photocatalyst's bandgap thereby enhancing its efficiency (Shandilya et al., 2022). A highly stable crystal configuration is formed because of the substitution of the transition and posttransition metal ions that are present in the lattice locations of WO_3. But alkali and alkaline earth metals occupy at the interstitial sites (Kalanur et al., 2019b).

It was found that when magnesium is doped, the CB edge shifts upward, rising well over the hydrogen reduction potential (Hwang et al., 2002). Yin et al. (2020) incorporated WO_3 films with alkali ions (Li, Na, and K) in the presence of Na_2SO_4 electrolyte by doctor blade method. In comparison to the other alkali metal dopants, Li/WO_3 films had the highest PEC performance of 0.43 mA/cm^2 at 0.65 V Ag/AgCl. This was due to the smallest radius of Li^+ ion which facilitates the insertion into WO_3 lattice. The current density of the bare WO_3 was around 0.4 mA/cm^{-2} at 1.23 V versus Reversible Hydrogen Electrode (RHE), and for 0.11 at% Ba/WO_3, it improved to 1.18 mA/cm^{-2} (Kalanur et al., 2019b). Barium intercalation into WO_3 led to a reduction of bandgap, an upsurge in carrier density, and an increase in incident photon to current efficiency levels and photocurrent. Li et al. (2015) performed a hydrothermal approach to generate WO_3 films with plate-like structure doped with aluminum from an aluminum-tungsten layer sputtered by a radio frequency magnetron on tin oxide doped with fluorine (FTO) glass. A photocurrent density of 1.14 mA/cm^2 at 1.2 V versus Ag/AgCl was observed upon

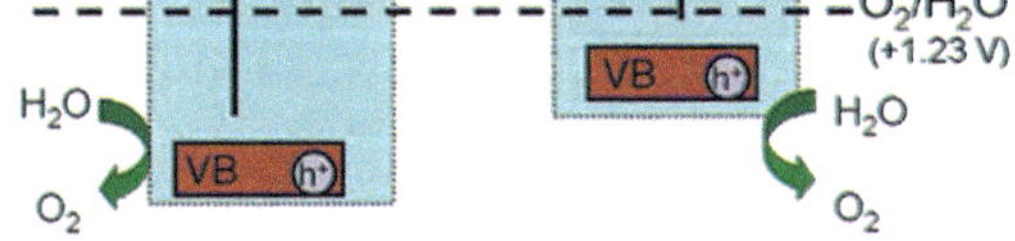

FIGURE 10.5 Schematic representation of enhancement in the photocatalytic hydrogen generation of WO_3 after metal doping. *Source: Replicated with the permission of Wang, F., Di Valentin, C., Pacchioni, G., 2012. Doping of WO_3 for photocatalytic water splitting: hints from density functional theory. The Journal of Physical Chemistry C 116 (16), 8901–8909. https://doi.org/10.1021/jp300867j. Copyright 2012 American Chemical Society.*

doping with 0.77 at% Al, which was 1.3 times greater than on comparison with pristine WO_3 photoanodes (0.87 mA/cm^2). The greatest current density of 2.18 mA/cm^2 was generated by the In/WO_3 photoanode, which was synthesized using a hydrothermal process with 3 wt.% indium inclusion at 1.23 V versus RHE. This was around three times greater than the pure monoclinic WO_3 photoanode (Kumar et al., 2020). As studied by Zou et al. (2011), doping WO_3 with Bi increases its photocatalytic water-splitting activity. Bi^{3+} doping increases the H_2 evolution under UV light, by approximately 167 μmol/h. The valence band of Bi/WO_3 consists mainly of O 2p and Bi 6p orbitals, with minor Bi 6s involvement, while the conduction band is primarily formed by W 5d orbitals, with a small contribution from Bi 6p orbitals. The hybridization of O 2p and Bi 6s levels leads to a more dispersed valence band, enhancing photoinduced hole mobility and oxidation reactions (Fu et al., 2006; Zhang et al., 2014; Cao et al., 2014; Sun and Wang, 2014). But Kalanur et al. (2018) synthesized Bi-incorporated WO_3 thin films on FTO glass substrate via a hydrothermal route. The VB maximum, CB minimum, and bandgap of WO_3 are altered by the introduction of Bi into the lattice. PEC experiments showed a marked increase in photocurrent, from 0.401 mA/cm^2 for WO_3 to 1.511 mA/cm^2 for optimal Bi/WO_3 at 1.23 V versus the reversible hydrogen electrode. Doping of Sn hydrothermally increases the photocurrent to 0.427 mA/cm^2 in contrast to the bare WO_3 (0.253 mA/cm^2) (Kalanur, 2019). Band edge positions in WO_3 shift downward toward higher potentials as a result of posttransition metal atom doping.

Fe/WO_3 nanoflakes were vertically grown on fluorine-doped tin oxide using a hydrothermal method for PEC water splitting (Zhang et al., 2015b). Performance of WO_3 is significantly enhanced by the Fe integration due to reduced bandgap and improve in the density of charge transferors. Development of additional band states through doping is probably the reason for the wider absorption of light observed in doped films. The photocurrent density of the 2 mol% Fe/WO_3 was 0.88 mA/cm^2, approximately 30% higher than WO_3 at 1.23 V against RHE. An investigation using X-ray photoelectron spectroscopy verified that the produced films had Fe^{3+} substituted at the W^{6+} position. An analysis of the effects of different d-block metals, on the photoinduced water splitting of WO_3 under UV laser light, revealed that Ni doping produced the highest efficiency (Hameed et al., 2004). The d^8 configuration of 10% Ni, having suitable CB edge of −0.5 V, caused an increase of about 19% upon doping (Xiao et al., 2017). The photocatalytic water splitting shows an H_2 production of 12.06 mmol/g for 1% Ag/WO_3 nanoplates under 8 hours of visible light irradiation (Naaz and Ahmad, 2023). Kalanur et al. (2020) describe a methodical process for merging WO_3 films with first-row transition elements (V to Zn). The purpose of this work was to modify the band edge location and lower the bandgap energy via band edge engineering WO_3 films. The dopant type directly influenced the crystalline morphology, charge carrier density, and oxygen vacancies in WO_3 films. All of the other dopants, except V and Cr, particularly the Co-WO_3 films, showed an improvement in photoelectroactivity, which reduced PEC performance. The films achieved a maximum anodic photocurrent density of 0.636 mA/cm^2 at 1.23 V versus RHE and a 62.9% incident photon-to-current efficiency at 300 nm. Also, the band edge positions shifted higher, making location more ideal for oxidizing water. The onset potential was moved to less positive values leading to the reduction of overpotential necessary for water oxidation. In addition, the bandgap energy decreased to 2.52 eV from 2.62 eV in comparison to the pristine WO_3 film. Copper doping has been found to be beneficial for increasing the photocurrent density (at 1.23 V RHE, 2.8 mA/cm^2) of WO_3 photoanodes for water dissociation in PEC cells (Ma and Hu, 2021). A notable increase in the incoming efficiency to current conversion was caused by the enhancement of parting of charges and charge transference effectiveness, as well as the reduction of Gibbs energy for the oxygen evolution reaction due to Cu^{2+} inclusion. Ta-doped WO_3 is another system synthesized via the hydrothermal method (Kalanur and Seo, 2019a). This system produced a change in the band edge positions and a drop in the bandgap energy. On comparison with WO_3, it showed higher efficacy in conversion of photon into current at 1.23 V versus RHE. Photocurrent was observed to raise to approximately 0.6 mA/cm^2 at 1.23 V versus RHE after 1.88 at% of Ta doping. Nb-doped WO_3 is another system being studied (Kalanur et al., 2019). The photocurrent values for Nb-doped and undoped WO_3 produced were 0.253 and 0.372 mA/cm^2, respectively. The photocurrent values acquired from the solar simulator are found to be compatible with these values. The above studies reveal that metal-doped WO_3 is a potential semiconductor photoanode material that can be employed as a renewable energy source and comprised in Table 10.4.

10.6 Case studies

10.6.1 Objectives

To investigate the role of metal ions in metal oxides to enhance their photocatalytic efficiency in water splitting. The specific aims include:

- Understanding synthesis techniques
- Characterizing metal ion–incorporated metal oxides
- Evaluating photocatalytic properties
- Comparing photochemical water splitting performance with pure metal oxides
- Identifying challenges and opportunities
- Demonstrating practical applications

10.6.2 Scope

Comparative Analysis:

- Comparison of the performance of metal ion–incorporated metal oxides with their pure counterparts.
- Identification of the specific advantages provided by different metal ions in improving photocatalytic efficiency.

Challenges and Solutions:

- Identification of the key challenges in the synthesis, stability, and reproducibility of metal ion–incorporated metal oxides materials.
- Exploration of potential solutions and future research directions to address these challenges.

TABLE 10.4 Comparative studies on metal-doped WO_3.

Catalyst	Synthetic method	Source of light	H_2 evolution	Photocurrent density	Quantum efficiency	References
Mg/WO_3	Impregnation method	Hg lamb 450 W	3.0 μmol/g/ h	–	–	Hwang et al. (2002)
Ba/WO_3	Hydrothermal method	Sun 2000 solar simulator	–	1.18 mA/ cm^2	–	Kalanur and, Seo (2019b)
Bi/WO_3	Microwave synthesis	Hg lamp (500 W)	167 μmol/h (UV) 40 μmol/h (visible)	–	–	Zou et al. (2011)
Bi/WO_3	Hydrothermal method	Sun 2000 solar simulator	–	1.511 mA/ cm^2	–	Kalanur et al. (2018)
Al/WO_3	Hydrothermal method	150 W Xe lamp	–	1.14 mA/cm^2	–	Li et al. (2015)
Sn/WO_3	Hydrothermal method	Sun 2000 solar simulator	–	0.427 mA/ cm^2	–	Kalanur and Shankara (2019)
Fe/WO_3	Hydrothermal method	450 W Xe lamp	–	0.88 mA/ cm^2		Zhang et al. (2015b)
Ag/WO_3	Hydrothermal method	200 W Mercury-Xenon Lamp	1.51 mmol/g/h	–	14.11%	Naaz and Ahmad (2023)
In/WO_3	Hydrothermal method	100 mW/cm^2 solar simulator	–	2.18 mA/cm^2	–	Kumar et al. (2020)
Ta/WO_3	Hydrothermal method	Sun 2000 solar simulator	–	0.6 mA/cm^2	–	Kalanur and, Seo (2019a)
Nb/WO_3	Hydrothermal method	Sun 2000 solar simulator	–	0.414 mA/ cm^2	–	Kalanur et al. (2019)

Applications and Implications:

- Discussion of practical applications of MIMO materials in renewable energy, particularly, in hydrogen production.
- Consideration of the broader implications for sustainable energy technologies and environmental impacts.

10.6.3 Audience

The intended audiences of the study are the following:

- Researchers and scientists
- Industry professionals
- Academics and educators
- Environmentalists and policymakers
- Investors and entrepreneurs

By targeting this diverse audience, the case study aims to foster a deep understanding of the potential of metal ion–incorporated metal oxides for water splitting and to promote collaboration across different sectors to advance renewable energy technologies.

10.6.4 Rationale

- To find alternative renewable sources of energy to reduce the dependency on fossil fuels and mitigate the environmental impacts.
- To contribute to the efficient and scalable technologies for hydrogen production through photochemical water splitting.

10.6.5 Expected results and deliverables

- Identification of specific metal ions that significantly improve the photocatalytic performance of metal oxides.
- Quantitative improvements in hydrogen and oxygen production rates under various light conditions.
- Development of optimized synthesis methods and standardized procedures that ensure reproducibility of metal ions–doped metal oxides.

10.6.6 Safety Considerations

- Proper handling of toxic materials, requiring appropriate protective equipment and ventilation to prevent exposure.
- Ensuring proper waste disposal and spill management procedures is essential to avoid environmental contamination and adhere to regulatory standards.

10.6.7 Actions Taken/Workflow/Tools Used/Simulations and Analyses

- **Material Preparation:** Selection and storage of metal ions and metal oxides.
- **Synthesis:** Mix the metal ions and oxides in a solvent, stir and heat the mixture to incorporate metal ions, dry the mixture, and perform calcination to achieve the desired crystalline structure.
- **Characterization:** Use X-ray diffraction (XRD), scanning electron microscopy (SEM), and energy-dispersive X-ray spectroscopy (EDX) to confirm the material's structure and composition.
- **Water Splitting Setup:** Prepare the apparatus, coat electrodes with the synthesized material, and prepare the electrolyte solution, ensuring all connections are secure and calibrated.
- **Conducting the Experiment:** Initiate the water splitting reaction by applying voltage, monitor reaction parameters (voltage, current, gas evolution), and systematically collect data.
- **Postexperiment Analysis:** Analyze data to evaluate efficiency and stability, ensure reproducibility with multiple trials, and follow proper disposal and decontamination procedures to maintain safety.

10.6.8 Challenges and Solutions

The challenges are the following:

- **Low Efficiency and Yield**: Photochemical water splitting often has low quantum efficiency and hydrogen yield due to limited light absorption and rapid recombination of electron-hole pairs.
- **Control of Reaction Conditions**: In laboratory setups, maintaining optimal pH, temperature, and light intensity can be difficult.
- **Scalability of Synthesis:** Producing high-quality metal-doped metal oxides in sufficient quantities for experiments is challenging.
- **Reactor Design:** Designing efficient photoreactors ensures uniform light distribution and maximizes interaction between light, catalyst, and water.
- **Safety Concerns in Storage**: Hydrogen is highly flammable, posing significant safety risks, especially at high pressures.
- **Low Storage Density**: Storing hydrogen in gaseous form requires large volumes or high pressures.
- **Material Degradation**: Repeated hydrogen absorption and desorption can degrade storage materials over time.

These challenges or issues can be solved by

- **Enhanced Catalysts**: Focus on developing and optimizing highly active and stable metal-doped metal oxide catalysts to increase the efficiency and yield of hydrogen production.
- **Automated Systems**: Implementing automated systems to precisely monitor and control reaction conditions.
- **Scalable Synthesis Methods**: Optimizing scalable methods like hydrothermal synthesis and sol-gel processes.

- **Advanced Reactor Designs**: Developing photoreactors with rotating catalyst beds, reflective surfaces, and transparent materials to enhance light utilization.
- **Safe Storage Materials**: Using metal hydrides, carbon nanotubes, or MOFs for safe, low-pressure hydrogen storage.
- **Integrated Systems**: Creating integrated systems for photochemical water splitting and hydrogen storage.
- **Robust Storage Materials**: Research and implement advanced hydrogen storage technologies that offer high capacity, safety, and stability, such as solid-state storage materials and novel sorbent materials, and withstand multiple absorption-desorption cycles without degradation.

10.6.9 Results

- The outcome of compiling studies on photochemical water splitting and hydrogen generation showed a variety of materials and approaches, including metal oxides, semiconductors, and molecular catalysts.
- Comparative analysis highlighted differences in efficiency, stability, and scalability among different methods, offering insights into their relative performance.
- Overall, the comparative study provided a comprehensive overview of current advancements and guided future research directions in photochemical water-splitting technologies.

10.6.10 Learning and Knowledge Outcomes

10.6.10.1 What was learned?

The synthesis and application of metal ion–incorporated metal oxides require precise control over material composition and synthesis conditions to optimize catalytic performance.

10.6.10.2 What could have been done differently?

Exploring additional synthesis methods or varying catalyst compositions could have provided insights into enhancing performance metrics such as activity and stability.

What must not be changed?

Consistent adherence to robust characterization techniques ensures the reliability and credibility of experimental outcomes.

10.6.10.3 What recommendations do you have for future investigation?

Future studies should focus on scaling up synthesis processes and evaluating the practical applicability of these materials in real-world water splitting applications. Exploring novel catalyst designs and exploring synergistic effects with other materials could further enhance efficiency and durability.

10.7 Conclusion and future perspective

Despite the emergence of numerous alternative types of photocatalysts throughout the years, metal oxide photocatalysts continue to maintain their dominance owing to their facile synthesis and distinctive physicochemical properties. In this study, we summarize the important prospects of metal ion-integrated metallic oxide for photodissociation of water. The selective addition of transition metal ions like Fe, Cr, Cu to metal oxide matrices like TiO_2, ZnO, and WO_3 modifies their surface characteristics, electron transport kinetics, and active site density, thereby improving the charge-carrier parting and red shifting absorption maxima to the visible range. Inert elements like Au, Ag, Pt, and Pd are found to be efficient in augmenting light absorption through plasmonic effects and reducing charge recombination by trapping photoexcited electrons. However, high cost restricts their applications. Study reveals that rare earth metals like Ce, La, Nd, and Eu improve optical properties by introducing f-orbital states, leading to better visible light absorption, enhancement of thermal and chemical stability of host metal oxides. But the effects of rare earth doping are complex and highly dependent on the specific rare earth element and its concentration, making optimization more challenging.

Several limitations, including preparation technique, security, cost, commercial use, and efficiency, continue to restrict the use of metal ion-modified metal oxide for water photolysis. Also, conversion proficiency of photoactivated metal oxide, with respect to solar-to-hydrogen conversion, remains exceedingly feeble ($<2\%$). Recently, single-atom catalysts supported metal oxides have emerged as one of the finest methods to maximize active reaction sites thereby enhancing the photocatalytic functioning of metal oxide. Besides, several strategies, encompassing the regulation of oxygen defects, the establishment of interfacial junctions, and the engineering of phase-facet-morphology along with the amalgamation of nanosized metal particles, have embraced to boost the performance of these metal oxide photocatalysts. Promising prospects exist for metal ion-incorporated metal oxides in water splitting in the future, include control at the atomic level, nanostructuring, sustainability, scalability, and integration with renewable energy producers. Also, it is necessary to clarify the basic problems associated with photocatalytic water-splitting events using theoretical predictions. By thoughtfully employing these tactics in future investigations, the development of productive energy materials for sustainable energy solutions can be facilitated. Advancement of extremely competent, stable, and inexpensive catalysts, essential for the growth of clean hydrogen production technologies will be made possible by ongoing research and innovation in these fields.

References

Ahmad, I., Ahmed, E., Ahmad, M., 2019. The excellent photocatalytic performances of silver doped ZnO nanoparticles for hydrogen evolution. SN Applied Sciences 1, 1–12. Available at https://doi.org/10.1007/S42452-019-0331-9.

Ahmad, I., Ahmed, E., Ahmad, M., Akhtar, M.S., Basharat, M.A., Khan, W.Q., et al., 2020. The investigation of hydrogen evolution using Ca doped ZnO catalysts under visible light illumination. Materials Science in Semiconductor Processing 105, 104748. Available at https://doi.org/10.1016/j.mssp.2019.104748.

Ahmad, H., Kamarudin, S.K., Minggu, L.J., Kassim, M., 2015. Hydrogen from photo-catalytic water splitting process: a review. Renewable and Sustainable Energy Reviews 43, 599–610. Available at https://doi.org/10.1016/j.rser.2014.10.101.

Ahmad, I., Shukrullah, S., Naz, M.Y., Ahmed, E., Ahmad, M., 2022. Rare earth metals co-doped ZnO/CNTs composite as high performance photocatalyst for hydrogen production from water_triethanolmine mixture. International Journal of Hydrogen Energy 47 (15), 9283–9294. Available at https://doi.org/10.1016/j.ijhydene.2022.01.006.

Ahmad, I., Shukrullah, S., Naz, M.Y., Bhatti, H.N., Khalid, N.R., Ullah, S., 2023. Rational design of ZnO–CuO–Au S-scheme heterojunctions for photocatalytic hydrogen production under visible light. International Journal of Hydrogen Energy 48 (34), 12683–12698. Available at https://doi.org/10.1016/j.ijhydene.2022.11.289.

Atrees, M.S., Ebraheim, E.E., Ali, M.E.M., Khawassek, Y.M., Mahmoud, M.S., Almutairi, M.M., 2021. Synergetic effect of metal-doped GO and TiO_2 on enhancing visible-light-driven photocatalytic hydrogen production from water splitting. Energy Sources, Part A 43 (4), 484–494. Available at https://doi.org/10.1080/15567036.2019.1629130.

Bharat, T.C., Mondal, S., Gupta, H.S., Singh, P.K., Das, A.K., 2019. Synthesis of doped zinc oxide nanoparticles: a review. Materials Today 11, 767–775. Available at https://doi.org/10.1016/j.matpr.2019.03.041.

Camposeco, R., Castillo, S., Rodriguez-González, V., Hinojosa-Reyes, M., Medina-Álvares, M.I., Mejía-Cente, I., 2018. Promotional effect of Rh nanoparticles on WO_3/TiO_2 titanate nanotube photocatalysts for boosted hydrogen production. Journal of Photochemistry and Photobiology A 353, 114–121. Available at https://doi.org/10.1016/j.jphotochem.2017.11.014.

Cao, C., Zhang, B., Lin, S., 2022. p-type ZnO for photocatalytic water splitting. APL Materials 10 (3). Available at https://doi.org/10.1063/5.0083753.

Cao, S., Zhou, P., Yu, J., 2014. Recent advances in visible light Bi-based photocatalysts. Chinese Journal of Catalysis 35 (7), 989–1007. Available at https://doi.org/10.1016/S1872-2067(14)60075-9.

Christoforidis, K.C., Fornasiero, P., 2017. Photocatalytic hydrogen production: a rift into the future energy supply. ChemCatChem 9 (9), 1523–1544. Available at https://doi.org/10.1002/cctc.201601659.

Dholam, R., Patel, N., Marco, A., Antonio, M., 2009. Hydrogen production by photocatalytic water-splitting using Cr-or Fe-doped TiO_2 composite thin films photocatalyst. International Journal of Hydrogen Energy 34 (13), 5337–5346. Available at https://doi.org/10.1021/cs500865y.

Díaz, L., Rodríguez, V.D., González-Rodríguez, M., Rodríguez-Castellón, E., Algarra, M., Núñez, P., et al., 2021. M/TiO_2 (M=Fe, Co, Ni, Cu, Zn) catalysts for photocatalytic hydrogen production under UV and visible light irradiation. Inorganic Chemistry Frontiers 8 (14), 3491–3500. Available at https://doi.org/10.1039/D0QI01311K.

Enzweiler, H., Yassue-Cordeiro, P.H., Schwaab, M., Barbosa-Coutinho, E., Olsen Scaliante, M.H.N., Fernandes, N.R.C., 2018. Evaluation of Pd-TiO_2/ZSM-5 catalysts composition effects on hydrogen production by photocatalytic water splitting. International Journal of Hydrogen Energy 43 (13), 6515–6525. Available at https://doi.org/10.1016/j.ijhydene.2018.02.077.

Fan, X., Wan, J., Liu, E., Sun, L., Hu, Y., Li, H., et al., 2015. High efficiency photoelectrocatalytic hydrogen generation enabled by Ag deposited and Ce doped TiO_2 nanotube arrays. Ceramics International 41 (3), 5107–5116. Available at https://doi.org/10.1016/j.ceramint.2014.12.083.

Fu, H., Zhang, L., Yao, W., Zhu, Y., 2006. Photocatalytic properties of nanosized Bi_2WO_6 catalysts synthesized via a hydrothermal process. Applied Catalysis B 66 (1-2), 100–110. Available at https://doi.org/10.1016/j.apcatb.2006.02.022.

Galińska, A., Walendziewski, J., 2005. Photocatalytic water splitting over Pt–TiO_2 in the presence of sacrificial reagents. Energy and Fuels 19 (3), 1143–1147. Available at https://doi.org/10.1021/ef0400619.

Gogoi, D., Namdeo, A., Golder, A.K., Peela, N.R., 2020. Ag-doped TiO_2 photocatalysts with effective charge transfer for highly efficient hydrogen production through water splitting. International Journal of Hydrogen Energy 45 (4), 2729–2744. Available at https://doi.org/10.1016/j.ijhydene.2019.11.127.

Gopinath, M., Marimuthu, R., 2022. A review on solar energy-based indirect water-splitting methods for hydrogen generation. International Journal of Hydrogen Energy 47 (89), 37742–37759. Available at https://doi.org/10.1016/j.ijhydene.2022.08.297.

Guayaquil-Sosa, J.F., Benito, S.-R., Valadés-Pelayo, P.J., Lasa, H.de, 2017. Photocatalytic hydrogen production using mesoporous TiO_2 doped with Pt. Applied Catalysis B 211, 337–348.

Guo, S.-y., Zhao, T.-j., Jin, Z.-q., Wan, X.-m., Wang, P.-g., Shang, J., et al., 2015. Self-assembly synthesis of precious-metal-free 3D ZnO nano/micro spheres with excellent photocatalytic hydrogen production from solar water splitting. Journal of Power Sources 293, 17–22. Available at https://doi.org/10.1016/j.apcatb.2017.04.029.

Gurbani, N., Neelu, C., 2023. P–N heterojunction system Eu-doped ZnO@ GO for photocatalytic water splitting. Global Challenges 7 (4), 2200106. Available at https://doi.org/10.1002/gch2.202200106.

Hameed, A., Gondal, M.A., Yamani, Z.H., 2004. Effect of transition metal doping on photocatalytic activity of WO_3 for water splitting under laser illumination: role of 3d-orbitals. Catalysis Communications 5 (11), 715–719. Available at https://doi.org/10.1016/j.catcom.2004.09.002.

Hodes, G., Cahen, D., Manassen, J., 1976. Tungsten trioxide as a photoanode for a photoelectrochemical cell (PEC). Nature 260 (5549), 312–313. Available at https://doi.org/10.1038/260312a0.

Hu, X., Song, J., Luo, J., Zhang, H., Sun, Z., Li, C., et al., 2021. Single-atomic Pt sites anchored on defective TiO_2 nanosheets as a superior photocatalyst for hydrogen evolution. Journal of Energy Chemistry 62, 1–10. Available at https://doi.org/10.1016/j.jechem.2021.03.003.

Hwang, D.W., Kim, J., Park, T.J., Lee, J.S., 2002. Mg-doped WO_3 as a novel photocatalyst for visible light-induced water splitting. Catalysis Letters 80, 53–57. Available at https://doi.org/10.1023/A:1015322625989.

Irfan, M., Afzal, S., Hussain, M., Naz, M.Y., Shukrullah, S., Rahman, S., et al., 2023. Testing of Sr-doped ZnO/CNT photocatalysts for hydrogen evolution from water splitting under atmospheric dielectric barrier plasma exposure. ACS Omega 8 (21), 18891–18900. Available at https://doi.org/10.1021/acsomega.3c01262.

Ismael, M., 2020. Enhanced photocatalytic hydrogen production and degradation of organic pollutants from Fe (III) doped TiO_2 nanoparticles. Journal of Environmental Chemical Engineering 8 (2), 103676. Available at https://doi.org/10.1016/j.jece.2020.103676.

Kalanur, S.S., 2019. Structural, optical, band edge and enhanced photoelectrochemical water splitting properties of tin doped WO_3. Catalysts 9 (5), 456. Available at https://doi.org/10.3390/catal9050456.

Kalanur, S.S., Noh, Y.-G., Seo, H., 2020. Engineering band edge properties of WO_3 with respect to photoelectrochemical water splitting potentials via a generalized doping protocol of first-row transition metal ions. Applied Surface Science 509, 145253. Available at https://doi.org/10.1016/j.apsusc.2020.145253.

Kalanur, S.S., Seo, H., 2019a. Aligned nanotriangles of tantalum doped tungsten oxide for improved photoelectrochemical water splitting. Journal of Alloys and Compounds 785, 1097–1105. Available at https://doi.org/10.1016/j.jallcom.2019.01.226.

Kalanur, S.S., Seo, H., 2019b. Intercalation of barium into monoclinic tungsten oxide nanoplates for enhanced photoelectrochemical water splitting. Chemical Engineering Journal 355, 784–796. Available at https://doi.org/10.1016/j.cej.2018.08.210.

Kalanur, S.S., Yoo, I.-H., Cho, I.S., Seo, H., 2019c. Niobium incorporated WO_3 nanotriangles: band edge insights and improved photoelectrochemical water splitting activity. Ceramics International 45 (7), 8157–8165. Available at https://doi.org/10.1016/j.ceramint.2019.01.117.

Kalanur, S.S., Yoo, I.-H., Eom, K., Seo, H., 2018. Enhancement of photoelectrochemical water splitting response of WO_3 by means of Bi doping. Journal of Catalysis 357, 127–137. Available at https://doi.org/10.1016/j.jcat.2017.11.012.

Khan, H.R., Aamir, M., Akram, B., Tahir, A.A., Malik, M.A., Choudhary, M.A., et al., 2020. Superior visible-light assisted water splitting performance by Fe incorporated ZnO photoanodes. Materials Research Bulletin 122, 110627. Available at https://doi.org/10.1016/j.materresbull.2019.110627.

Kumaravel, V., Mathew, S., Bartlett, J., Pillai, S.C., 2019. Photocatalytic hydrogen production using metal doped TiO_2: a review of recent advances. Applied Catalysis B 1021–1064. Available at https://doi.org/10.1016/j.apcatb.2018.11.080.

Kumar, M., M., Kanta Sahu, T., Alam, S., Qureshi, M., 2020. Tuning the electronic structure of monoclinic tungsten oxide nanoblocks by indium doping for boosted photoelectrochemical performance. Chemistry 15 (22), 3886–3896. Available at https://doi.org/10.1002/asia.202000787.

Kunthakudee, N., Puangpetch, T., Ramakul, P., Serivalsatit, K., Hunsom, M., 2022. Light-assisted synthesis of Au/TiO_2 nanoparticles for H_2 production by photocatalytic water splitting. International Journal of Hydrogen Energy 47 (56), 23570–23582. Available at https://doi.org/10.1016/j.ijhydene.2022.05.150.

Lee, B.-H., Park, S., Kim, M., Sinha, A.K., Lee, S.C., Jung, E., et al., 2019. Reversible and cooperative photoactivation of single-atom Cu/TiO_2 photocatalysts. Nature Materials 18 (6), 620–626. Available at https://doi.org/10.1038/s41563-019-0344-1.

Liu, B., Su, S., Zhou, W., Wang, Y., Wei, D., Yao, L., et al., 2017. Photo-reduction assisted synthesis of W-doped TiO_2 coupled with Au nanoparticles for highly efficient photocatalytic hydrogen evolution. CrystEngComm 19 (4), 675–683. Available at https://doi.org/10.1039/C6CE02417C.

Liu, X., Wang, F., Wang, Q., 2012. Nanostructure-based WO_3 photoanodes for photoelectrochemical water splitting. Physical Chemistry Chemical Physics 14 (22), 7894–7911. Available at https://doi.org/10.1039/C2CP40976C.

Liu, Y., Zhu, Q., Tayyab, M., Zhou, L., Lei, J., Zhang, J., 2021. Single-atom Pt loaded zinc vacancies ZnO–ZnS induced type-V electron transport for efficiency photocatalytic H_2 evolution. Solar RRL 5 (11), 2100536. Available at https://doi.org/10.1002/solr.202100536.

Li, C.-f., Wei-guo, P., Zhang, Z.-r., Wu, T., Guo, R.-t., 2023. Recent progress of single-atom photocatalysts applied in energy conversion and environmental protection. Small 2300460. Available at https://doi.org/10.1002/smll.202300460.

Li, W., Zhan, F., Li, J., Liu, C., Yang, Y., Li, Y., et al., 2015. Enhancing photoelectrochemical water splitting by aluminum-doped plate-like WO_3 electrodes. Electrochimica Acta 160, 57–63. Available at https://doi.org/10.1016/j.electacta.2015.01.095.

Li, X., Zhao, L., Yu, J., Liu, X., Zhang, X., Liu, H., et al., 2020. Water splitting: from electrode to green energy system. Nano-Micro Letters 12, 1–29. Available at https://doi.org/10.1007/s40820-020-00469-3.

Lu, L., Wang, G., Xiong, Z., Hu, Z., Liao, Y., Wang, J., et al., 2020. Enhanced photocatalytic activity under visible light by the synergistic effects of plasmonic and Ti^{3+-}doping at the Ag/TiO2-x heterojunction. Ceramics International 46 (8), 10667–10677. Available at https://doi.org/10.1016/j.ceramint.2020.01.073.

Machín, A., Cotto, M., Duconge, J., Arango, J.C., Morant, C., Pinilla, S., et al., 2018. Hydrogen production via water splitting using different Au@ ZnO catalysts under UV–vis irradiation. Journal of Photochemistry and Photobiology A 353, 385–394. Available at https://doi.org/10.1016/j.jphotochem.2017.11.050.

Maeda, K., 2014. Photocatalytic properties of rutile TiO_2 powder for overall water splitting. Catalysis Science and Technology 4 (7), 1949–1953. Available at https://doi.org/10.1039/C4CY00251B.

Maeda, K., Domen, K., 2013. 7.22 - Photochemical water splitting using nanostructured metal oxides. In: Comprehensive Inorganic Chemistry II, Second Edition), Second Edition. Elsevier, pp. 587–614. Available at https://doi.org/10.1016/B978-0-08-097774-4.00721-X.

Mandawat, N.K., Sharma, H.S., Chouhan, N., 2019. Nanocomposite Ag@ AgCl/ZnO for efficient hydrogen generation through water splitting. SN Applied Sciences 1, 1–12. Available at https://doi.org/10.1007/s42452-019-0578-1.

Martha, S., Chandra Sahoo, P., Parida, K.M., 2015. An overview on visible light responsive metal oxide based photocatalysts for hydrogen energy production. RSC Advances 5 (76), 61535–61553. Available at https://doi.org/10.1039/C5RA11682A.

Matsubara, K., Inoue, M., Hagiwara, H., Abe, T., 2019. Photocatalytic water splitting over Pt-loaded TiO_2 (Pt/TiO_2) catalysts prepared by the polygonal barrel-sputtering method. Applied Catalysis B 254, 7–14. Available at https://doi.org/10.1016/j.apcatb.2019.04.075.

Ma, Y., Hu, Y.H., 2021. Surface-copper-doped WO_3 photoanode for photoelectrochemical water splitting. Applied Physics Letters 118 (22). Available at https://doi.org/10.1063/5.0050614.

Murillo-Sierra, J.C., Hernández-Ramírez, A., Hinojosa-Reyes, L., Guzmán-Mar, J.L., 2021. A review on the development of visible light-responsive WO_3-based photocatalysts for environmental applications. Chemical Engineering Journal Advances 5, 100070. Available at https://doi.org/10.1016/j.ceja.2020.100070.

Naaz, F., Ahmad, T., 2023. Ag-doped WO_3 nanoplates as heterogenous multifunctional catalyst for glycerol acetylation, electrocatalytic and enhanced photocatalytic hydrogen production. Langmuir 39 (27), 9300–9314. Available at https://doi.org/10.1021/acs.langmuir.3c00386.

Naresh, G., Vijay Kumar, B., Sasikumar, Venugopal, A., 2021. Improved H_2 yields over Cu-Ni-TiO_2 under solar light irradiation: behaviour of alloy nano particles on photocatalytic H_2O splitting. Applied Catalysis B 299, 120654. Available at https://doi.org/10.1016/j.apcatb.2021.120654.

Neațu, F., Laura, E.A., Trandafir, M.M., Negrea, R.F., Florea, M., Teodorescu, C.M., et al., 2020. Degenerated TiO_2 semiconductor modified with Ni and Zn as efficient photocatalysts for the water splitting reaction. ChemCatChem 12 (18), 4642–4651. Available at https://doi.org/10.1002/cctc.202000691C.

Park, Y.-K., Kim, B.-J., Jeong, S., Jeon, K.-J., Chung, K.-H., Jung, S.-C., 2020. Characteristics of hydrogen production by photocatalytic water splitting using liquid phase plasma over Ag-doped TiO_2 photocatalysts. Environmental Research 188, 109630. Available at https://doi.org/10.1016/j.envres.2020.109630.

Patil, A.B., Jadhav, B.D., Bhoir, P.V., 2021. Efficient photocatalytic hydrogen production over Ce/ZnO from aqueous methanol solution. Materials for Renewable and Sustainable Energy 10, 1–9. Available at https://doi.org/10.1007/s40243-021-00199-5.

Rafique, M., Hajra, S., Irshad, M., Usman, M., Imran, M., Assiri, M.A., et al., 2023. Hydrogen production using tio2 based photocatalysts: a comprehensive review. ACS Omega 8 (29), 25640–25648. Available at https://doi.org/10.1021/acsomega.3c00963.

Ravishankar, T.N., Vaz, M.de O., Ramakrishnappa, T., Teixeira, S.R., Dupont, J., Pai, R.K., et al., 2018. The heterojunction effect of Pd on TiO_2 for visible light photocatalytic hydrogen generation via water splitting reaction and photodecolorization of trypan blue dye. Journal of Materials Science 29, 11132–11143. Available at https://doi.org/10.1007/s10854-018-9197-7.

Sadanandam, G., Zhang, L., Scurrell, M.S., 2018. Enhanced photocatalytic hydrogen formation over Fe-loaded TiO_2 and g-C_3N_4 composites from mixed glycerol and water by solar irradiation. Journal of Renewable and Sustainable Energy 10 (3). Available at https://doi.org/10.1063/1.5021350.

Samadi, M., Zirak, M., Naseri, A., Khorashadizade, E., Moshfegh, A.Z., 2016. Recent progress on doped ZnO nanostructures for visible-light photocatalysis. Thin Solid Films 605, 2–19. Available at https://doi.org/10.1016/j.tsf.2015.12.064.

Shandilya, P., Sambyal, S., Sharma, R., Mandyal, P., Fang, B., 2022. Properties, optimized morphologies, and advanced strategies for photocatalytic applications of WO_3 based photocatalysts. Journal of Hazardous Materials 428, 128218. Available at https://doi.org/10.1016/j.jhazmat.2022.128218.

Sun, T., Fan, J., Liu, E., Liu, L., Wang, Y., Dai, H., et al., 2012. Fe and Ni co-doped TiO_2 nanoparticles prepared by alcohol-thermal method: application in hydrogen evolution by water splitting under visible light irradiation. Powder Technology 228, 210–218. Available at https://doi.org/10.1016/j.powtec.2012.05.018.

Sun, T., Liu, E., Liang, X., Hu, X., Fan, J., 2015. Enhanced hydrogen evolution from water splitting using Fe-Ni codoped and Ag deposited anatase TiO_2 synthesized by solvothermal method. Applied Surface Science 347, 696–705. Available at https://doi.org/10.1016/j.apsusc.2015.04.162.

Sun, S., Wang, W., 2014. Advanced chemical compositions and nanoarchitectures of bismuth based complex oxides for solar photocatalytic application. RSC Advances 4 (88), 47136–47152. Available at https://doi.org/10.1039/C4RA06419D.

Tanaka, A., Teramura, K., Hosokawa, S., Kominami, H., Tanaka, T., 2017. Visible light-induced water splitting in an aqueous suspension of a plasmonic Au/TiO_2 photocatalyst with metal co-catalysts. Chemical Science 8 (4), 2574–2580. Available at https://doi.org/10.1039/C6SC05135A.

Trang, T.N.Q., Phan, T.B., Nam, N.D., Thu, V.T.H., 2020. In situ charge transfer at the Ag@ ZnO photoelectrochemical interface toward the high photocatalytic performance of H_2 evolution and RhB degradation. ACS Applied Materials and Interfaces 12 (10), 12195–12206. Available at https://doi.org/10.1021/acsami.9b15578.

Vaiano, V., Iervoli, G., 2019. Photocatalytic hydrogen production from glycerol aqueous solution using Cu-doped ZnO under visible light irradiation. Applied Sciences 9 (13), 2741. Available at https://doi.org/10.3390/app9132741.

Valero-Romero, M.J., Santaclara, J.G., Oar-Arteta, L., Van Koppen, L., Osadchii, D.Y., Jorge, G., et al., 2019. Photocatalytic properties of TiO_2 and Fe-doped TiO_2 prepared by metal organic framework-mediated synthesis. Chemical Engineering Journal 360, 75–88. Available at https://doi.org/10.1016/j.cej.2018.11.13.

Wang, F., Di Valentin, C., Pacchioni, G., 2012. Doping of WO_3 for photocatalytic water splitting: hints from density functional theory. The Journal of Physical Chemistry C 116 (16), 8901–8909. Available at https://doi.org/10.1021/jp300867j.

Wang, J., van Ree, T., Wu, Y., Zhang, P., Gao, L., 2018. Metal oxide semiconductors for solar water splitting. Metal Oxides in Energy Technologies. Elsevier, pp. 205–249. Available at https://doi.org/10.1016/B978-0-12-811167-3.00008-0.

Wu, F., Hu, X., Fan, J., Liu, E., Sun, T., Kang, L., et al., 2013. Photocatalytic activity of Ag/TiO_2 nanotube arrays enhanced by surface plasmon resonance and application in hydrogen evolution by water splitting. Plasmonics 8, 501–508. Available at https://doi.org/10.1007/s11468-012-9418-5.

Wu, J., Lu, S., Ge, D., Zhang, L., Chen, W., Gu, H., 2016. Photocatalytic properties of Pd/TiO_2 nanosheets for hydrogen evolution from water splitting. RSC Advances 6 (72), 67502–67508. Available at https://doi.org/10.1039/C6RA10408H.

Xiao, M., Lei, Z., Bin, L., Miaoqiang, L., Zhiliang, W., Hengming, H., et al., 2020. Molten-salt-mediated synthesis of an atomic nickel Co-catalyst on TiO_2 for improved photocatalytic H_2 evolution. Angewandte Chemie 132 (18), 7297–7301. Available at https://doi.org/10.1002/ange.202001148.

Xiao, B., Shen, C., Luo, Z., Li, D., Kuang, X., Wang, D., et al., 2023. Cu surface doped TiO_2: constructing Cu single-atoms active sites and broadening the photo-response range for efficient photocatalytic hydrogen production. Chemical Engineering Journal 468, 143650. Available at https://doi.org/10.1016/j.cej.2023.143650.

Xiao, Y.-H., Xu, C.-Q., Zhang, W.-D., 2017. Facile synthesis of Ni-doped WO_3 nanoplate arrays for effective photoelectrochemical water splitting. Journal of Solid-State Electrochemistry 21, 3355–3364. Available at https://doi.org/10.1007/s10008-017-3680-6.

Yang, L., Feng, N., Wang, Q., Chu, Y., Xu, J., Deng, F., 2020. Surface water loading on titanium dioxide modulates photocatalytic water splitting. Cell Reports Physical Science 1 (2), 100013. Available at https://doi.org/10.1016/j.xcrp.2019.100013.

Yin, X., Qiu, W., Li, W., Wang, K., Yang, X., Du, L., et al., 2020. Effects of alkali ion on boosting WO_3 photoelectrochemical performance by electrochemical doping. International Journal of Hydrogen Energy 45 (38), 19257–19266. Available at https://doi.org/10.1016/j.ijhydene.2020.05.057.

Zhang, N., Ciriminna, R., Pagliaro, M., Xu, Y.-J., 2014. Nanochemistry-derived Bi_2 WO_6 nanostructures: towards production of sustainable chemicals and fuels induced by visible light. Chemical Society Reviews 43 (15), 5276–5287. Available at https://doi.org/10.1039/C4CS00056K.

Zhang, Y., Jianhong, Z., Hui, W., Bin, X., Wen, Z., Xinbo, Z., et al., 2022. Single-atom Cu anchored catalysts for photocatalytic renewable H_2 production with a quantum efficiency of 56%. Nature Communications 13 (1), 58. Available at https://doi.org/10.1038/s41467-021-27698-3.

Zhang, C., Shao, M., Ning, F., Xu, S., Li, Z., Wei, M., et al., 2015a. Au nanoparticles sensitized ZnO nanorod@ nanoplatelet core–shell arrays for enhanced photoelectrochemical water splitting. Nano Energy 12, 231–239. Available at https://doi.org/10.1016/j.nanoen.2014.12.037.

Zhang, T., Zhu, Z., Chen, H., Bai, Y., Xiao, S., Zheng, X., et al., 2015b. Iron-doping-enhanced photoelectrochemical water splitting performance of nanostructured WO_3: a combined experimental and theoretical study. Nanoscale 7 (7), 2933–2940. Available at https://doi.org/10.1039/C4NR07024K.

Zou, L., Liu, Q., Xu, Q., Mi, G., 2011. Microwave synthesis of mesoporous WO_3 doping with bismuth and photocatalytic oxidation of water to H_2. Korean Journal of Chemical Engineering 28, 1299–1303. Available at https://doi.org/10.1007/s11814-010-0500-4.

CHAPTER

11

Plasmonic material and quantum dot sensitized photocatalysts for water splitting

Nikita Chaudhary, Jyoti Jyoti, and Kaushik Ghosh

Quantum Materials and Devices Unit, Institute of Nano Science and Technology, Mohali, Punjab, India

11.1 Introduction

A typical two-electrode photoelectrochemical (PEC) system consists of a metal counter electrode and a semiconductor photoelectrode. Here, the semiconductor acts as a light absorber, which produces electron–hole pair for carrying oxygen and hydrogen evolution, respectively (Jin et al., 2021). However, poor light harvesting, charge carrier recombination, and chemical stability pose a potent challenge for improving PEC efficiency. In this context, plasmonic materials as well as quantum dot integration in the photoelectrode material strategically improve the overall efficiency. In this aspect, plasmonic materials help to augment light-harvesting capability and reduce the charge carrier recombinations. They provide two routes for decay of surface plasmons after photoexcitation by hot electron generation or by the radiative scattering of resonant photons (Kamat and Hartland, 2018; Subramanyam et al., 2022). Combining metal nanoparticles with semiconductors is the traditional way of creating nanocomposites. These metal nanoparticles localize the charge carriers and transmits to the semiconductors within 10 nm range. Moreover, an enhanced light absorption is observed within semiconductor because of the increased path length of light after multiple light scattering from metal nanoparticles. The inherently high refractive index of semiconductors also leads to reflection losses which can be reduced nearly to zero by light scattering from various metal nanoparticles (Warren and Thimsen, 2012). Finally, there is a high loss of less energetic photons in large band gap semiconductors. This can be addressed by the plasmonic nanoparticles which can inject low energy electrons in the conduction band improving the overall photo absorption (Warren and Thimsen, 2012). The integration of quantum dots (QDs) in semiconductors enhances the light absorption in a similar manner by utilizing low energy photons and reduces the electron–hole recombination

Photochemical Splitting of Water
DOI: https://doi.org/10.1016/B978-0-443-29064-0.00002-4

in the photoelectrode. Various semiconductors serve as scaffolds for QDs depending on semiconductor surface morphology along with the shape, surface, and size of QDs (Selopal et al., 2020). This chapter aims at discussing the various aspects of the application of plasmonic materials and quantum dots for sensitizing photoelectrode materials.

11.2 Plasmonic materials for PEC

The plasmonic materials are known to transform light energy into plasmonic energy; however, their intraband transitions are short-lived creating a challenge for achieving high solar energy conversion. An alternate strategy for this problem is to form plasmon-metal-semiconductor heterojunction. Here, metal nanoparticles facilitate energy transfer from plasmons to semiconductor by serving as light-scattering centers. Thus the semiconductor can utilize the low and high band gap energy. Additionally, a broadband light absorption is achieved in the semiconductors by the localized surface plasmon resonance (LSPR) phenomena.

11.3 Surface plasmon resonance

The change in the electromagnetic field of the incident light creates an impact on the freely available electrons. They experience a time dependent opposite to this variable field giving rise to electronic oscillations (Fig. 11.1A). The oscillation frequency is a function of the effective electron mass, electron density as well as the permittivity of metal nanoparticles. The oscillations confined to the surface of the conduction band and exhibiting strong interactions with incident light are known as surface plasmons (Warren and Thimsen, 2012; Kamat and Hartland, 2018). The dielectric function consists of a real part (**R**) and an imaginary part, where the real part determines the light interaction with the metal nanoparticles. There are three possible cases for the real part of the dielectric equation:

1. **R** > 0: When the real part is positive, that is, plasmon frequency is less than light frequency, the electrons do not oscillate leading to light transmission or absorption depending on the interband transitions (Fig. 11.1A and C).
2. **R** < 0: When the real part is negative, that is, plasmon frequency is greater than light frequency, the light gets reflected from the surface (Fig. 11.1A and C).
3. **R**=0: When the real part becomes zero, there is an occurrence of resonance between plasmon frequency and light frequency leading to intense high amplitude oscillations of the electrons in metal nanoparticles creating high electron density. This phenomena is referred to as surface plasmon resonance (SPR) (Jin et al., 2021).

The SPR effect can be further categorized into two parts known as LSPR and propagating surface plasmon polaritons (SPP) (Brown et al., 2016). The occurrence of LSPR depends on the fulfillment of two conditions. First, the metal nanoparticle size should be less than the wavelength of light and second, oscillations should occur on the interface of metal and dielectric. Further, the geometrical aspects, like shape and dimensions, of the metal

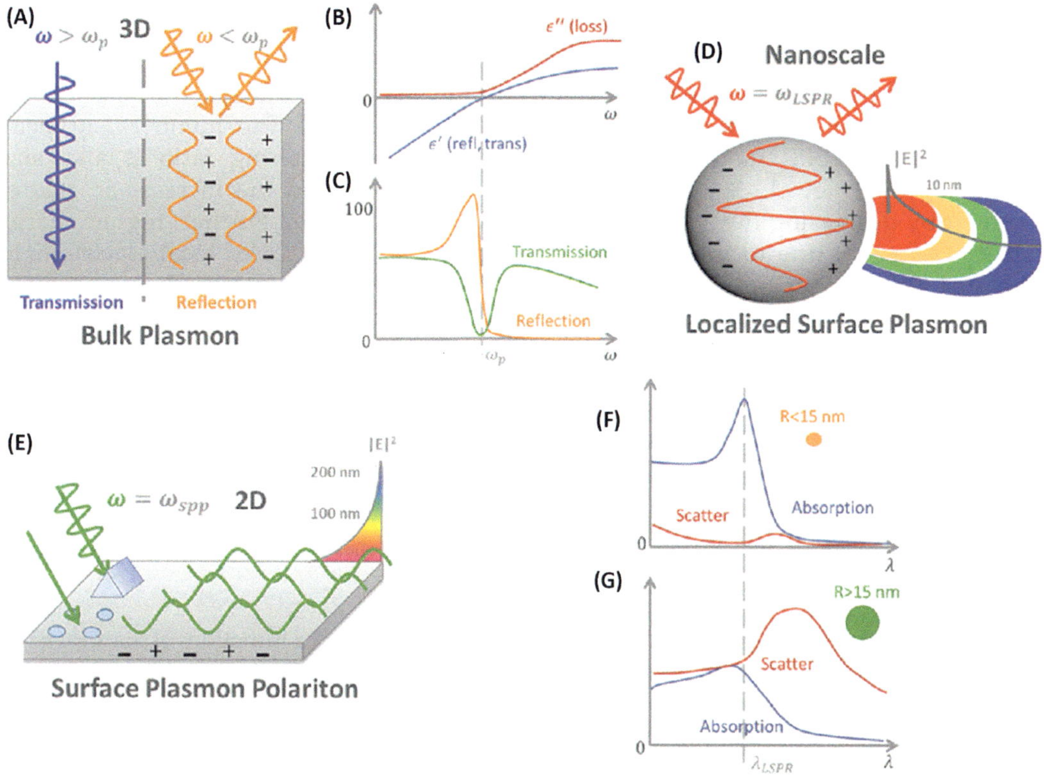

FIGURE 11.1 Light wave interaction (A) in bulk, (B) when real part of dielectric function is positive, (C) when real part of dielectric function is positive, (D) LSPR effect, (E) SPP mechanism, SPP based on particle size, (F) < 15 nm, and (G) > 15 nm. Source: *Reprinted from Subramanyam, P., et al., 2022. Emerging materials for plasmon-assisted photoelectrochemical water splitting. Journal of Photochemistry and Photobiology C 51, 100472. https://doi.org/10.1016/j.jphotochemrev.2021.100472. Copyright 2022, Elsevier.*

nanoparticle along with the refractive index and dielectric properties of the surroundings also impact the extent of LSPR (Lieberman et al., 2008; Ghobadi et al., 2018). Absorption or scattering of light is dependent on the size of the metal nanoparticle. When metal nanoparticle size is smaller than 15 nm, LSPR energy is converted into heat due to electron–electron scattering, thus absorption takes place (Fig. 11.1F). However, when size is larger than 15 nm, a large scattering cross-section is created by the re-radiation of plasmonic energy (Fig. 11.1G). The SPPs, on the other hand, are created due to the perpendicular propagation of electromagnetic excitations along the metal–dielectric junction (Fig. 11.1E) (Li et al., 2015). This phenomena is dependent on the geometry of the metal nanoparticle, however, at least one dimension of the structure should be of the order of excitation wavelength. The occurrence of radiative decays and scattering leads to a lower lifetime of LSPR as compared to SPP. The next section discusses the plasmonic properties of metallic nanoparticles and their dependence on the geometrical aspects as well as the chemical composition of the material.

11.3.1 Size

The careful tuning of dimensions of nanoparticles is the easiest way of engineering plasmonic properties. The LSPR frequency and the shape of LSPR spectral lines can be tuned by adjusting the dimensions of metal nanoparticles. The number and peak position of the LSPR mode as well as absorption to scattering ratio is also size dependent (Kelly et al., 2003). Similarly, Kim et al. reported the increase in photo-response by decreasing the Au nanodots size from 83 to 50 nm. Further, higher charge density is created due to the enhancement of the quality factor of local electric field when the size of Au nanoparticles is reduced. Thus the overall plasmon enhancement increases the PEC performance of the photoelectrode (Zhu et al., 2020). Another study reported the variation of light scattering by Au nanoparticles with size 20–80 nm. The 80 nm-sized particles exhibit maximum light scattering whereas, the 40 nm-sized particles provide highest adsorption cross section, comparable to various conventional dyes (Gomes Silva et al., 2011). Following the same trend, various reports indicate the increase in PEC performance with decrease in nanoparticle size to an optimum level (Lee et al., 2017).

11.3.2 Shape

The tunability of the resonance wavelength is found to be a function of nanoparticle shape. Recent studies show the dependence of SPR behavior on metal nanoparticle shape. Zhang et al. have demonstrated the enhancement of PEC performance with the application of Au nanoparticles and nanoprisms. The absorption rate is higher in nanoprisms due to the enhanced LSPR effect (Zhang et al., 2014). Similarly, the performance of TiO_2 electrode varied when sensitized with Au nanoparticles (NPs) and Au nanorods (NRs). A reduction in the the surface trap states as well as electron–hole recombination is observed with the application of AuNPs yielding higher photocurrent. The degree of surface passivation is higher in case of NPs as compared to NRs due to the presence of surface ligands in nanorod architecture (Pu et al., 2013). The use of Pd NRs and Pd NPs showcased similar results when combined with $BiVO_4$ electrode (Nishijima et al., 2010).

11.3.3 Mechanisms

The two basic decay pathways in LSPR include nonradiative and radiative decay (Fig. 11.2). Radiative decay leads to photon emission (Brown et al., 2016) and further energy transfer to the surroundings, whereas, nonradiative decays create electron–electron scattering leading to the formattion of hot carriers (Ghobadi et al., 2018). These hot carriers often lead to intra- and interband excitation of electrons within the conduction band. They can also travel to the nearest semiconductor with appropriate band alignment. Depending on the above-mentioned pathways, there are four mechanisms for LSPR-assisted solar to hydrogen conversion namely: (1) light absorption and scattering, (2) plasmon-induced near-field energy transfer, (3) hot carrier injection, and (4) local enhancement of electric field (LEF) (Linic, Christopher and Ingram, 2011).

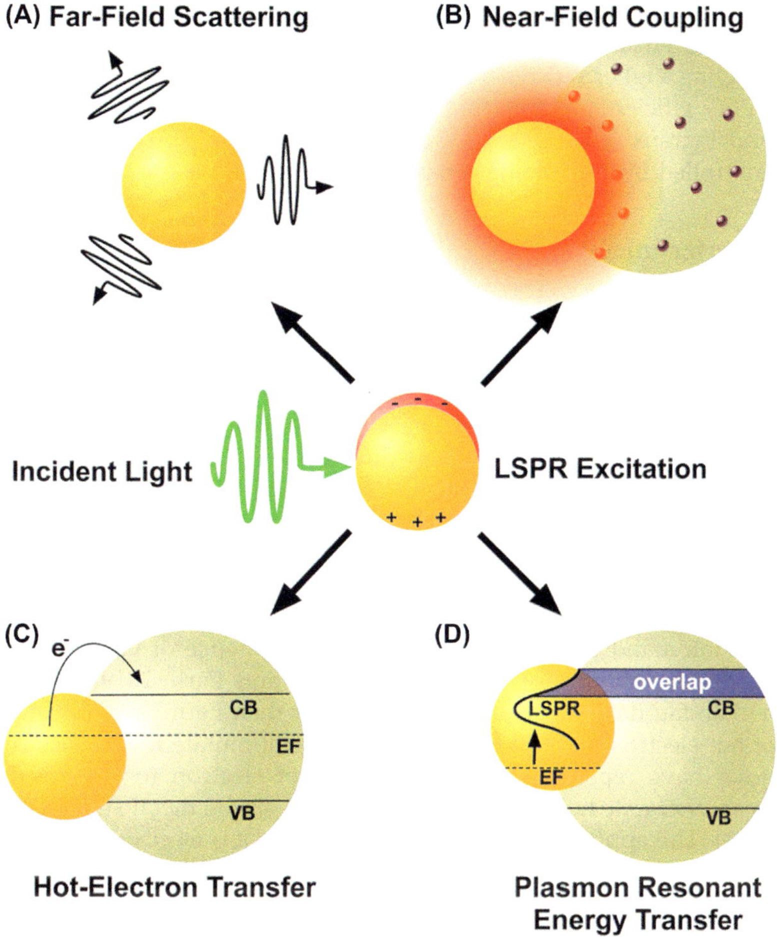

FIGURE 11.2 Schematic illustration of plasmon enhancement mechanisms of radiative effects (A) far-field scattering and (B) near-field coupling; and nonradiative effects: (C) hot electron transfer, and (D) plasmon resonant energy transfer (Erwin et al., 2016). Source: *Reproduced with permissions from Erwin, W.R., et al., 2016. Light trapping in mesoporous solar cells with plasmonic nanostructures. Energy and Environmental Science 9 (5), 1577–1601. https://doi.org/10.1039/C5EE03847B. Copyright 2016, Royal Society of Chemistry.*

11.3.4 Light absorption and scattering

Light scattering takes place when a particle smaller than the wavelength of incident light is illuminated. The emergence of SPR phenomena creates a large cross-section area, transforming the oscillating electrons into secondary energy source. The large scattering cross-section area linked to SPR is the origin of this process. The light scattering by metallic nanoparticles leads to local amplification of electromagnetic field on the surface or within the photoactive material leading to increase in optical path length and absorption cross-section of the semiconductor (Lieberman et al., 2008). Further, fast-varying dipoles are induced in the semiconductor due to plasmonic oscillations leading to formation of polarized molecules. The increased concentration of polarized molecules acts as catalytically active centers giving higher efficiency. Various metallic nanoparticles have been found to improve the PEC efficiency depending on their size, concentration, and shape (Linic, Christopher and Ingram, 2011; Valenti et al., 2016). Therefore absorption and scattering of light can be controlled by optimizing the plasmonic metals.

11.3.5 Hot carrier injection

The interband and intraband electronic transitions from other bands to the unfilled conduction band leads to various nonradiative decays resulting in the formation of hot electrons (Zhang et al., 2018). The presence of hot carriers over a Schottky barrier leads to collection of low energy photons, thereby, increasing light-harvesting capability (Zhong et al., 2015). Various theoretical studies suggest that interband transitions in noble metals Au and Ag occur upon visible light excitation due to their p-band positions leading to hot carrier formations. The intraband transitions can lead to the formation of carriers having energies upto E_f+hv (Kang et al., 2019). These intraband transitions can be maximized by terminating the interband transitions. The use of materials having large interband energy thresholds can maximize the intraband excitation spectrum. For instance, the incidence of visible light can trigger only intraband transitions in case of interband energy level of 3.7 eV. However, if the interband energy gap is 2.4 eV, the incidence of visible light can lead to interband transitions creating hot carriers (Lee, Park and Park, 2019). Further, these hot carriers follow Landau-damping decay process rather than Fermi–Dirac distribution. Also, the lifetime of hot electrons should be long enough to stimulate the surface reactions.

Recent studies have reported the capability of hot electron to drive water reduction process, whereas, energetic holes could carry out OER on the metal surface (Primo, Corma and García, 2011). The application of AgNPs-sensitized TiO_2 electrode has demonstrated a higher photocurrent due to presence of hot electrons as well as improved stability than TiO_2. Chen et al. reported similar results where the increase in photocurrent has been correlated to the presence of Au nanospheres on ZnO electrode. Further, Fowler's law has been used for photocurrent versus wavelength curve fitting in order to confirm that hot electrons are responsible for this current enhancement (Chen et al., 2012).

11.3.6 Plasmon-induced resonant energy transfer

This mechanism governs the nonradiative transfer mechanism between metal and semiconductor prior to energy dissipation. In plasmon-induced resonant energy transfer (PIRET) process, excited plasmonic metals dissipate via electron–electron interactions creating strong dipoles (Subramanyam et al., 2022). The dipole–dipole interactions induce electron–hole pairs within the semiconductor leading to red- or blue-shifted energy transfer, that is, without photon emission, enhancing the performance in the visible region. The basic criteria for PIRET is the overlap of plasmonic and semiconductor resonance wavelengths. However, a longer plasmonic lifespan with respect to semiconductor's dipole lifetime is necessary in order to achieve a large charge carrier population. Several reports including α-Fe_2O_3 (Naldoni et al., 2016) and nitrogen (N)-doped TiO_2 (Ingram et al., 2011) sensitized with Au nanoparticles have reported an enhanced PEC performance where the absorption of the material overlapped with the plasmonic absorption. In addition, PIRET is favorable in case of N-doped TiO_2 as compared to TiO_2.

The proximity between the acceptor (semiconductor) and the donor (metal) also determines the favorability of PIRET mechanism. A smaller distance will lead to larger resonance overlap giving higher PIRET efficiency. Recently, an Au@SiO_2@Cu_2O heterojunction has shown distance-dependent PIRET enhancement. The dipole–dipole interactions are able to create more number of charge carriers showcasing the increased optical extraction by the semiconductor due to broadband absorption as compared to mere interfacial charge transfer (Subramanyam et al., 2022). Thus the PIRET process improves the solar spectrum utilization, thereby, increasing the photocurrent of the system.

The overall performance enhancement can be attributed to other mechanisms including quantum mechanical tunneling and catalytic properties of metal nanostructures. The composition and light-absorbing properties of the metal determine the overall photoelectrochemical performance. Two parallel process of PIRET and direct electron transfer (DET) or either one of them could be the possible reason for performance enhancement. The presence of direct contact of the metal surface with the semiconductor material is the prime requirement for DET to take place. Further, the energy of hot electrons can be harvested through the PEC system by forming a direct contact between metal nanostructures and semiconductor material. Usually, Landau damping of surface plasmons is thought to be the reason behind plasmon-induced hot electron injection (PHET) process, where hot electron pass across metal–semiconductor interface (Tan et al., 2017). Therefore dephasing of metal center and the dielectric substrate as well as their interface determines the method of hot electron generation (Jiang et al., 2019). The electron-phonon scattering and ultrafast thermalization of hot electrons, however, pose challenge to the effectiveness of PHET (Wu et al., 2015). To overcome this issue, the hot electron transfer in CdSe-Au NRs explained the concept of plasmon-induced charge-transfer transition (PICTT). In this process, electrons move to the CB of the semiconductor after excitation by plasmons, which is in compliance with the momentum and energy conservation postulates and excluding the competing process of hot electron relaxation (Song et al., 2021). This mechanism represents a new pathway circumventing the loss of energy by metallic nanoparticles.

In addition to above process, thermoplasmic effect comes in picture for lowering the endothermic enthalpy as well as the potential barrier for harnessing Vis/NIR region for hydrogen production. The simultaneous hot electron injection reduces the activation energy threshold of the reaction and facilitates interfacial charge transfer as well (Agarwal et al., 2017). Therefore the Faradaic efficiency and overall kinetics of semiconductor materials enhance. Another issue that has been observed over the past few years is the migration of hot carriers. The hot holes are better energy convertors than hot electrons; however, their short mean free path and lifetime create a barrier for their utilization. Further, hot holes and hot electrons possess a nonsymmetric behavior of the energy distribution curve with respect to the Fermi level. Therefore the efficient capture of hot holes is necessary for PEC enhancement. In this direction, amorphous TiO_2 covered with Au NPs demonstrated the efficient defect mediated hole extraction capability. Here, TiO_2 served as hole collecting as well as a protective layer while Au act as the primary absorber material. The electrons near or above Fermi level are transferred to the counter electrode and the external circuit to carry out HER while the TiO_2 shell delivers oxidation reaction (Vahidzadeh et al., 2021).

11.4 Design

The major considerations while designing SPR-based photoelectrodes include (Thimsen et al., 2011):

1. Semiconductor should not absorb enough light to get effective plasmon enhancement.
2. Effective resonant coupling between metal and semiconductor.
3. Formation of trap state and Fermi level pinning at the metal–semiconductor interface.
4. Schottky barrier formation at the interface for maximum photovoltage.

Metal nanoparticles are used to transfer energy to semiconductors with low light absorption capability and thus, generate electron–hole pairs. With appropriate energy coupling, the decay of localized surface plasmon can be avoided which is a function of distance. This distance dependence can be specific to the material, however, highest enhancement of near field in semiconductors can be achieved near to the interface (Standridge, Schatz and Hupp, 2009). Further, the presence of trap state at the interface serve as pathways for recombination and affect the PEC performance. Finally, a Schottky barrier similar to the macroscopic contact is formed with the use of nanoparticles less than 10 nm size. Usually, the metal-semiconductor Schottky barrier is smaller than the semiconductor–electrolyte interface due to more surface state at this interface leading to Fermi level pinning. Therefore electrode configuration with enhanced metal-semiconductor barrier height gives higher performance (Pala et al., 2009).

11.5 PEC system based on quantum dots

In a PEC system, QDs are employed as photoactive materials that absorb photons from sunlight and generate electron–hole pairs. There are various methods for depositing QDs onto the substrate, including spin-coating, dip-coating, drop-casting, or spray-coating. They can get adsorbed on the nanotube and nanorod architecture, they can also sit inside the mesoporous film depending on their shapes and sizes (Fig. 11.3). Charge separation becomes effective due to band shifting between QD and host semiconductor. QDs extend the absorbance toward the visible region by narrowing the band gap of host material. In turn, QDs' narrower bandgap allows photons to generate charge carriers injected into host semiconductors.

The following four pathways are used to classify these unwanted charge recombination events, which occurs within QD, at the interfaces of QDs and electrolyte, and interface of QDs and host material. These four pathways are as follows:

1. Charge recombination caused due to defects in QDs.
2. At interface of QDs and electrolyte-oxidized ions from electrolyte and electrons from conduction band of QDs can recombine.
3. At the interface of host material and electrolyte, electrons from CB of host material can react with oxidized ions from the electrolyte.
4. At interface of host material and QDs, holes from VB of QD can combine with CB of host material.

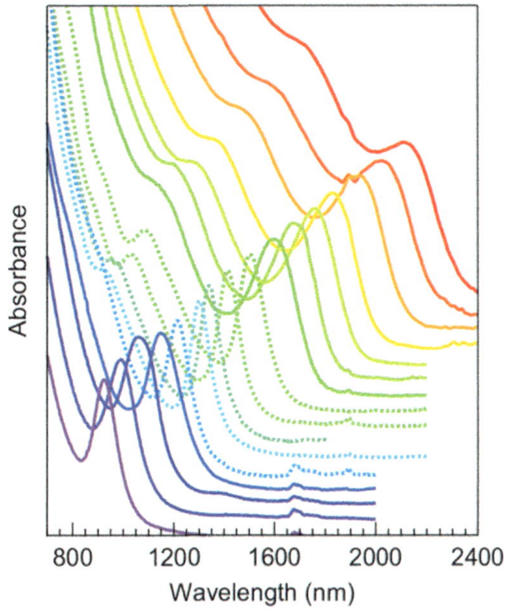

FIGURE 11.3 Absorbance spectra of PbS QDs with change in size from 3 to 10 nm (Moreels et al., 2011). Source: *Reproduced with permission from Moreels, I., et al., 2011. Size-tunable, bright, and stable PbS quantum dots: a surface chemistry study. ACS Nano 5 (3), 2004–2012. https://doi.org/10.1021/nn103050w. Copyright 2011, American Chemical Society.*

11.6 Photoanode based on quantum dots

For hydrogen generation, photoanode acts as the site for electron generation which are later relocated to the counter electrode through an external circuit. In this section, an overview of critical approaches for engineering electron acceptor materials, adjusting photoanode configurations, and tuning QDs parameters such as size, shape, composition, and surface is given.

11.6.1 Engineering materials for electron acceptors

The n-type semiconductors with narrower band gaps, such as CdTe, GaAs, and CdS, achieve extended lifetimes, regulated carrier concentrations, strong carrier mobility, and visible spectrum absorption but their H_2 production is hampered by charge-carrier recombination. Also, various semiconductors react with water and form oxide layers. In this aspect, metal oxides become potential candidates with their low cost, high stability, and suitable band gap. Light is effectively absorbed by narrow band gap metal oxides like WO_3 (Li et al., 2020; Zheng et al., 2020) and hematite, but their high resistivity reduces the efficiency of charge carrier transmission. SnO_2 and other posttransition metal oxide (Li et al., 2016) exhibit superior photogeneration, but at a lower efficiency. Some of the large band gap metallic oxides including TiO_2, ZnO, and Ga_2O_3 are explored due to their properties like higher electron mobility, environmental friendliness, and economic feasibility, however, they absorb only in the UV region.

In this background, doping facilitates the hole transfer process, thereby, a broader spectrum range gets absorbed by the material. These host-semiconductor charge acceptors offer a surface for photocatalytic processes as well as conducting channels for charge transfer. Mesoporous electron acceptor materials often suffer from significant charge recombination, despite much research. A faster electron transfer from the semiconductor to the solid/liquid interface can be facilitated by the amalgamation of supporting semiconductors with low resistive materials like metallic oxides, which can enhance the photocurrent. Examples of such combinations are TiO_2 combined with a metal–organic framework (MOF) (Han et al., 2019) 1D graphene nanoribbon networks, multiwalled carbon nanotubes (MWCNTs), and Au nanoparticle-decorated MWCNTs.

11.6.2 Adjusting the light harvesting

Extending the photoelectrodes' light absorption edge is crucial for raising the STH (solar to hydrogen) efficiency of PEC systems. Because of their attractive band structure and ease of synthesis, visible-active chalcogenide QDs become the subject of ongoing research (Chen et al., 2010). Because of their easy production, tunable optical/electronic properties (Fig. 11.4), and

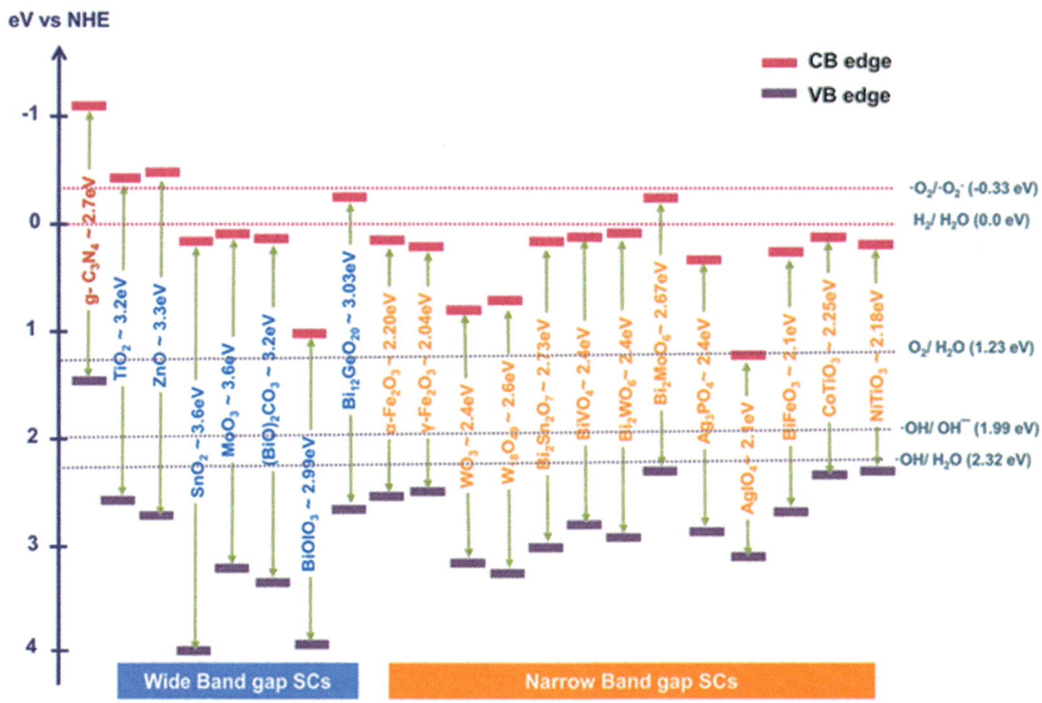

FIGURE 11.4 Band edge positions of some common wide and narrow band gap semiconductors with respect to eV versus NHE. (Das et al., 2022).

effective absorption of visible light, cadmium chalcogenide QDs make for appealing sensitizers. When QDs were deposited, light absorption was expanded into the visible spectrum and shift in the absorption band edge can be adjusted by tuning the dimensions of the QDs. Only 5% of the sun spectrum is made up of UV, with the visible and near-infrared regions having the strongest distribution intensities.

Narrow-bandgap QDs, such as lead chalcogenide QDs, are utilized as a way to expand the light absorption edge of the photoelectrode and thus, increase the solar absorption efficiency. For example, PbS QDs can be grown in TiO_2 films with mercaptoacetic acid (MAA)-surface modified photodeposition, achieving 1.15% STH efficiency in PEC systems. However, this extends light-harvesting range but reduces electron injection force. In order to address this problem, multiple types of QDs with varying light absorption ranges were assembled to create a QDs cosensitized photoanode, which improved photon collection (Jin-nouchi et al., 2010). Combining tandem setups with QDs-sensitization is one possible remedy. Some examples are double-sided ZnO nanowire arrays cosensitized with QDs grown over indium tin oxide (ITO) substrate. Such an assembly of CdS–ZnO (CBD) layer at the front side collects light, which is then absorbed by the CdSe–ZnO (CBD) layer at back side. Alternative design involves successively placing QD sensitizers on same host semiconductor. For example, CdSe QDs on SILAR-coated ZnO nanowires increase photocurrent density due to higher absorbance and stepwise band alignment, as seen in CdS/CdSe cosensitized TiO_2, CdS/PbS/ZnS cosensitized TiO_2, and PbS/CdS cosensitized TiO_2 (Wang et al., 2010).

Although covering QDs with a shell improves light harvesting, the in situ fabrication process (SILAR) is laborious and fraught with problems such as excessive charge recombination because of interfacial defects/traps and control of QD coverage, which compromise photoanode stability. Due to the shortcomings of in situ SILAR, thick shell (1.5 nm) colloidal QDs were developed and these considerably enhanced absorption cross-section in comparison to plane QDs (Qian et al., 2017).

11.6.3 Customizing quantum dot stability and charge dynamics

By adjusting their size, QDs' bandgap and electron-hole wavefunctions can be readily tuned. However, because of electron-hole wavefunction dependence on band gap of the underlying material and confinement energy of electrons and holes, size engineering naturally has restrictions when it comes to adjusting the band gap (Alivisatos, 1996; Smith and Nie, 2010). Broader QDs can collect solar light over a broader area even though they are often less successful at injecting charge because of the slight pushing force produced by relative positions of conduction band of host material and QDs. Because QDs are sensitive to the chemical environment on their surface, surface defects can serve as charge traps. The increase in number of recombination sites upon sun exposure and weak chemical/photostability results in poor device performance. To overcome this issue, core-shell architecture of QDs using wide band gap semiconductor materials as passivation layers to enhance stability as well as charge dynamics were designed. PbS QDs were covered with a CdS coating layer on a TiO_2 mesoporous film using a SILAR technique, which stabilized the material and improved visible response.

Depending on the relative alignment of conduction (CB) and valence band (VB) edges, there are four primary types of core-shell architectures of QDs (shown in Fig. 11.5). Similar

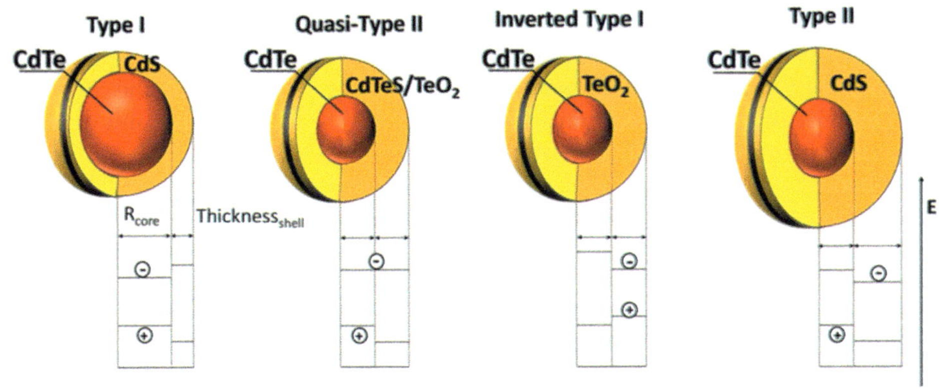

FIGURE 11.5 Different excitonic localization regimes in core/shell QDs (Vale et al., 2019). +, Holes; −, electrons. Source: *Reproduced with permission from Vale, B.R.C., et al., 2019. Ligand induced switching of the band alignment in aqueous synthesized CdTe/CdS core/shell nanocrystals. Scientific Reports 9 (1), 8332. https://doi.org/10.1038/s41598-019-44787-y. Copyright 2019, Springer.*

to CdTe/ZnS, the CB and VB of type-I core/shell QDs are both situated in the shell band gap. This robust shell development passivates the surface atoms by confining excitons with the nanocrystal core along with the confinement of core semiconductor within the potential energy well. Furthermore, this configuration creates a separation from the surrounding environment. CdTe/TeO_2 represents inverted type-I QDs, where the band edges are present in the core band gap leading partial or complete delocalization of electrons and hole in shell area. This makes it possible to improve the charge injection rate and retrieve charges efficiently. Increasing the thickness of the smaller band gap shell can also broaden the absorption spectrum to include longer wavelengths.

Type-II core-shell configuration consists of a staggered CB and VB alignment, where one carrier resides in the shell and other in the core creating a spatial distribution of photoinduced charge carriers (e.g., CdSe/ZnTe, CdTe/CdS). By varying the shell's thickness and core size, one may manage the band energy offset between the two, which determines the effective band gap of QDs. A quasi-Type-II configuration (e.g., CdTe/CdTeS/TeO_2) is formed when a low band energy offset is present which can contain either holes or electrons and the other carriers get delocalized in the overall core-shell area (Chen et al., 2013). This spatial separation of charge carriers reduces coulombic interaction as well as the possibility wave function overlap and thus, decreases charge recombination. These charge transfer characteristics make Type-II and quasi-Type-II configurations of QDs suitable for PEC systems (García-Santamaría et al., 2011). Additionaly, surface passivation and less number of trap states in thick shell QDs enhance the chemical and photostability in comparison with pure, thin QDs (García-Santamaría et al., 2009; Navarro-Pardo et al., 2020).

11.6.4 Modifying quantum dot shape

QDs' electrical structure, optical qualities, and wavelength are all greatly influenced by their shape. In spherical QDs, electron and holes separate in all direction creating wave function overlap and thus, faster charge carrier recombination. Advanced colloidal nanofabrication techniques have made possible the development of nonspherical QDs, which can be tuned by adjusting synthesis parameters like core structure, temperature, and ligands. Because of their longer PL lifetime, these nonspherical QDs increase the likelihood of exciton extraction and enhance the electron hole wavefunction spatial separation. For instance, CdS shell thickness in the dot-in-rod CdSe/CdS QD (Rainò et al., 2011) is less than 1 nm, yet this does not compromise the effective hole transmission, whereas, the dot-in-rod architecture of QDs can create quasi-Type-II band structure where electrons can transfer into the shell of rod, yielding faster electron transfer.

11.6.5 Building of heavy metal-free quantum dots

Use of heavy metals (such as Cd, Hg, and Pb) or including these elements in production of QDs is another obstacle that requires attention in the commercialization of QDs-based photoelectrodes. Because single-element carbon dots have advantages over toxic metal-free QD-based photosensitizers, including earth abundance, chemical stability, affordability, and broad optical absorption, they have been developed as an eco-friendly substitute. These QDs have the ability to increase light absorption range of photoelectrodes while maintaining good stability; nevertheless, because of significant charge recombination, their current densities are still limited. Synthesis of carbon-based QDs that absorb in broad UV-NIR region and tunable band gap that can complement wide band gap host semiconductors may be main focus of future research. Graphitic carbon nitride QDs (g-C_3N_4) are propitious carbon-based materials that can work in synergy with TiO_2 nanowires to boost light absorption.

For PEC H_2 evolution, V-VI group QDs such as SnS, SnSe, and SnTe have been employed. With heavy metal-free NIR QDs, SnSe nanocrystal-sensitized TiO_2 photoanodes demonstrated a record-breaking saturation current density of roughly 8.2 mA/cm^2. However, harmful heavy metals are present in CdSe QDs, which are used in the manufacture of SnSe nanocrystals. In order to remove hazardous components and reactive precursors, further research should concentrate on direct synthesis techniques. $CuInS_2$, $CuInSe_xS_{2-x}$, and $CuInSe_2$, which are ternary materials based on I–III–VI semiconductors, are being investigated for water reduction because of their composition-dependent optoelectronic capabilities, size, and cost-effective manufacture. But modest photocurrent densities are claimed in most investigations. Although noncolloidal NIR QDs can be produced at higher current densities, the process requires expensive and complex molecular beam epitaxy.

In the core-shell structure of ZnS, usual Type-I band alignment slows down charge transfer or separation. To address this problem, quasi-Type-II colloidal $CuInSe_2$/ $CuInS_2$core/thick shell QDs was developed using absorption spanning UV–vis–NIR ranges. The quasi-Type-II colloidal $CuInSe_2$/$CuInS_2$ core/thick shell QDs resolves the issue of slower charge transfer present in ZnS core-shell QDs with Type-I band alignment due to their broad absorption range of UV to NIR region. The original CdSe/CdS QDs template can be finely tuned to get exact control over the spherical form and shell thickness. Outstanding saturation photocurrent density of up to 3 mA/cm^{-2} was demonstrated by the QDs-sensitized TiO_2 PEC photoanodes, which is double the current density observed in pure CuInSe. When a graded type $CuInSe_xS_{1-x}$ interfacial layer was added to $CuInSe_2$/$CuInS_2$

QDs, the saturated photocurrent density increased by 70%. Future research should concentrate on creating a straightforward synthetic method to create $CuInSe_2/CuInS_2$ and on modifying the surface and interface of I–III–VI group QDs.

11.7 Surface customization of quantum dots

11.7.1 Introduction of passivating layer

In order to minimize energy loss due to electron-holes recombination, wide band gap material coatings on QDs-based photoelectrodes can efficiently reduce direct contact of electrolyte to photoanode surface. Most widely utilized layer is ZnS coating because it greatly improves photoelectrode performance. ZnS has a lower energy level than QD, which modifies band alignment, inhibits charge recombination, and encourages electron flow to electron acceptors in TiO_2. $ZnFe_2O_4$ (band gap 1.9 eV) overlayers were recently developed; they function as hole receptor for CdS and as a passivating layer to decrease CdS photocorrosion. This facilitates hole transfer because of the stair-like band arrangement between ZnO, CdS, and $ZnFe_2O_4$.

11.7.2 Cocatalysts

Cocatalysts like Mo_2S, IrO_x, and Co-based compounds are widely recognized for their practical applications in reducing reaction energy barriers, alleviating hole accumulation, and increasing charge transfer of photoanode and electrolyte interfaces. To speed up the water oxidation reaction, for example, a Co-Pi/CdS/TiO_2 nanowire array (Ai et al., 2015) is used as hole transfer network.

11.7.3 The choice of surface ligands

Colloidal QDs are commonly coated with insulating ligands that can be some organic molecules, inorganic ions, polymers, or molecular chalcogenide complexes. QDs are utilized in photoexcited electron microscopy (PEC) photoanodes. These ligands can adjust the phase, shape, and optical characteristics of synthesized QDs in addition to guaranteeing solution processability and modulating precursor reactivity and nucleation/growth rate control. Without photogenerated charges tunneling through the ligand layer, redox processes cannot proceed. Oleylamine, oleic acid, dodecanethiol, and trioctylphosphine oxide are examples of surfactant ligands that can modify trap states present on surface and function as insulating source, preventing charge carrier movement during PEC. Wapping out initial ligands entirely or partially for single atoms or short chain surfactants is a workable alternative. Examples of capped PbS/CdS materials include those with cetyltrimethylammonium bromide, hybrid inorganic ligands, InP/ZnS QDs, CdSe/CdS QDs, and PbS/CdS capped PdS/CdS QDs. Thus we can conclude that QDs can be engineered to change their structures, energy levels, forms, and surfaces in order to influence electron-hole wave function and absorption processes, which in turn affects the dynamics of excitons and charge carriers.

This can enhance PEC performance together with modifying electron/hole acceptor and photoelectrode combinations. Future works must focus on moderating QDs, controlling the architecture of PEC systems, and developing supporting semiconductors in order to get high photon capture, better charge separation, and photon-to-electron conversion efficiency in an aqueous environment.

11.8 Photocathode based on quantum dot

Till now QDs sensitizers that are being used include graphene QDs, PbS QDs (Raissi et al., 2018), CdSe QDs (H. Li et al., 2018), CdSeTe QDs, carbon QDs (Wang et al., 2017), and so on. An effective bias-free tandem PEC system may not be achieved if the photoanode and photocathode are not properly matched in terms of photocurrent density and catalytic efficiency (Cesar et al., 2006). In order to do this, it is important to create photocathode that has excellent light absorption capacity, good conductivity, and is stable at the same time.

11.8.1 Modifying hole transfer process

PEC performance and charge separation is strongly impacted by electron hole transfer processes. In photocathodes, holes are taken out and moved to host materials. Therefore the engineering and selection of these materials is essential to PEC performance. Because of its high electrical mobility, wide absorption spectrum, and earth abundance, porous p-Si is a promising photocathode material. Its electrochemically etched surface area makes it possible to load QDs that have band gaps lying from 1.8 to 2.2 eV, which improves light absorption in visible part of solar spectrum. Nevertheless, issues like corrosion and expensive material prices arise in real-world applications. Metal oxides like Cu_2O (Pan et al., 2018), $CaFe_2O_4$ (Díez-García and Gómez, 2016), $CuFeO_2$ (Prévot, Guijarro and Sivula, 2015), and $CuBiO_4$ are being explored in place of hole acceptors because of their easy preparation and lower cost. Copper-based chalcogenides, such as Cu_2S, $CuIn_xGa_{1-x}Se_2$, $CuInS_2$, and $CuGaSe_2$, exhibit an excellent visible-light photo-response but are often unstable and suffer from charge recombination.

More effective ligands that satisfy energy needs of consecutive hole transfer process and interact with QDs for ultrafast transfer can be added to improve hole transfer. A two-purpose connector known as MAA was created to secure colloidal CdSe QDs onto NiO films. At the integrated CdSe/MAA/NiO photocathode interface, hole transfer from VB of CdSe to VB of NiO is facilitated by HOMO level of MAA which lies between VB of both CdSe and NiO (Liu et al., 2015).

To fabricate CdSe/PTZ/NiO photocathode, heterocyclic PTZ ligand is used which is electron-rich structure. In comparision to direct charge transfer from CdSe QDs to MAA, hole transfer from excited CdSe to surface bound PTZ is faster. In comparison to CdSe/MAA/NiO, slower charge recombination results in an increased photocurrent density. It is also possible to add a layer of hole transfer material to extract and transfer photogenerated holes.

11.8.2 Electron transfer engineering

Cocatalysts are added to QD-based PEC photocathodes to greatly improve the processes of proton reduction and electron capture. The activation energy required to produce H_2 on semiconductor surface can be decreased by catalysts like MoS_2/CdSe/NiO (Guo et al., 2021), NiS/CdTe/NiO, Co(dmgBF$_2$)(H$_2$O)$_2$/CdSeTe/NiO, and $Fe_2S_2(CO)_6$/CdSe/NiO. To create CoP/CdSe/NiO, a molecular cobaloxime catalyst, CoP was covalently bonded to CdSe/NiO. The reduction potentials of Co_{III}/Co_{II} and Co_{II}/Co_{I} in the cobaloxime complex are more negative than CB of CdSe QDs, although VB of CdSe QD is more positive than that of NiO. Because of this, there is less chance of recombination with NiO holes when electrons are transferred favorably from CdSe QDs to cobalt catalyst. Thiol groups or sulfide ions from the capping agent accounted for most of the holes in CdSe QDs. Sulfide ions from capping agents is reason for most of holes in CdSe QDs. To address this problem, thiol-free capping agents must be taken into consideration. In summary, we can say, for the electron transfer process, selecting the right cocatalyst is essential. Additionally, a cocatalyst can simultaneously increase the stability and efficiency of the photoelectrode by choosing the right type of QD and carefully selecting its morphology.

Selecting the appropriate techniques for anchoring the cocatalysts is another essential step. For instance, a CdS/NiO photocathode's photocurrent increased 1.42 times upon the addition of a cobaloxime catalyst. However, a discernible decrease in photocurrent occurred in less than 600 seconds due to inadequate contact of catalysts and electrode surface.

11.9 Quantum dots assembly

Thanks to their size-dependent band gap, QDs absorption can be tuned and ordered arrangement of QDs can be fabricated of selected size. Photocurrent can be efficiently improved by introducing random size QDs but it can lead to charge recombination (Nie et al., 2018).

Taking inspiration from QDs rainbow solar cells (Kongkanand et al., 2008), photocathode was made by spin coating colloidal CdSe QDs of different sizes layer by layer on NiO/ITO substrate. There was a minor shift in the CB and VB edges of CdSe QDs as their diameter changes from 2.1 to 4.2 nm. The positioning of various QDs on the photoelectrode's surface to boost PEC efficiency is made possible by this approach, which provides insights for future photoelectrode design. In order to efficiently enhance light harvesting, QD-based upconversion nanocrystal assemblies were also created in place of photosensitizers which help to convert low energy photons to high energy photons, that is, transform IR to visible light.

Subsequent investigations may concentrate on enhancing the surface passivation of QD assemblies, substituting environment friendly $CuInSe_2$–$CuInS_2$ QDs for heavy metal-based ones, incorporating core-shell structure QDs into photocathode architecture, and refining PEC design through manipulation of QD surface, composition, and structure.

11.10 Tandem quantum dot-based PEC architecture

PEC cells are only able to function when an applied bias is present, as demonstrated by Eq. (11.1), which also limits the STH efficiency. An external voltage bias supply is positioned between counter electrode and photoelectrode to compensate for any potential deficit and achieve efficient charge separation (Lin et al., 2013). During two electrode tests, when an external voltage is supplied, efficiency η of photon-to-current is defined as

$$\eta = \frac{|j\,(\mathrm{mA/cm^2})| \times \left(V_{\mathrm{th}} - V_{\mathrm{app}}\right)}{P_{\mathrm{Sun}}(\mathrm{mW/cm^2})} \tag{11.1}$$

where j is the photocurrent density under an applied voltage bias V_{app} (V_{app} is real potential difference between working electrode and counter electrode). Faradaic efficiency is considered as unity for both reactions and it is assumed that there is no corrosion at photoelectrode.

Due to the complicated and expensive electrochemical workstations required for the process, solar energy is currently being used more and more in PEC systems to drive reactions. Common methods involve linking photoelectrodes to photovoltaic cells (PV-PEC systems) and arranging photoanodes and photocathodes both in series. When dye-sensitized solar cells (DSSCs) were used as voltage source, no significant modification in PEC system was observed, suggesting that solar cells might be used as a backup power source. The PEC device and a PV cell arranged in series are integrated in the PV-PEC system, and for PV device to gather light, the front PEC electrode needs to have enough emission capacity. The photogenerated electrons in the CdS/TiO_2 recombine with the holes in the solar cell upon passing through front CdS-sensitized TiO_2 and reaching rear PV, whereupon they contribute to the creation of O_2. These PV-based tandem systems, however, have limited system viability because to their complex setups and high output voltage that is dependent on illumination intensity. Alternatively, a tandem unaided PEC cell that combines the photoanode and photocathode might possibly harvest a larger amount of solar energy by employing two separate light absorbers.

The PEC system can be self-powered and perform efficient water splitting using only solar power because water splitting is carried out directly at interface of semiconductor and electrolyte via photogenerated minority carriers while majority carriers recombine at ohmic contact between photoelectrodes. Light generated holes in CdS/TiO_2 photoanode and electrons in CdSe/NiO photocathode in a photoanode-photocathode tandem device separated water molecules; at the external circuit, the electrons in the TiO_2 nanorod and the holes in the NiO nanosheet recombined. Comparable to natural photosynthesis, this device achieved average gas evolution rates of 2.24 (H_2) and 1.07 (O_2) $\mu mol/h/cm^2$ at zero bias condition. There have also been reports of ternary QD-based tandem systems, utilizing Cu_3SbS_4/NiO nanosheet and $CuSbS_2/TiO_2$ nanorod as the photocathode and photoanode, respectively.

11.11 Case study for TiO_2-based photoanode

Titanium dioxide (TiO_2) has been widely explored for PEC application owing to its properties like high chemical stability, large band gap, and efficient light-harvesting capability. However, practical large scale application requires further efficiency enhancements. In this regard, various researchers have investigated the efficiency enhancement by the application of different metallic nanoparticles. Recently, Li et al. showed a higher PEC performance by the application of alumium (Al) nanoparticles on TiO_2 nanocavity arrays. The enhancement showed a size dependence where 20 nm-sized nanoparticles exhibited highest performance. Further, the presence of a thin oxide layer of Al_2O_3 prevented further oxidation of metallic nanoparticles and reduced the surface charge recombination (Li et al., 2018). Similarly, Celebi et al. designed a one-pot synthesis method for Au-sensitized TiO_2 photoelectrode demonstrating SPR mediated faster charge separation. A 97% efficiency enhancement was achieved with a very low loading percentage of Au (0.154 wt.%) (Celebi et al., 2021). Peerakiatkhajohn et al. investigated performance enhancement in heterojunction system via Au-sensitized TiO_2 layer on Al_2O_3@Cu_2O-based photoelectrode exhibiting a Z-scheme mechanism. This bifunctional system generated hydrogen in self biasing conditions along with the degradation of humic acid (Peerakiatkhajohn et al., 2021). QD sensitization has also been used as an alternative strategy for PEC enhancement of TiO_2 nanotubes electrode. Abdelkarim et al. utilized CdS/CdSe QDs p-n heterojucntion which improved the current density by 51 times as compared to bare TiO_2 heterostructure. Further, the application of CuSe coating formed a p-n heterojunction improving the charge carrier separation (Abdelkarim et al., 2022). Similarly, the application of black phosphorus QDs by Zheng et al. increased the current density by two times as compared to bare TiO_2 nanotube arrays. This was achieved due to the efficiency charge separation and an increased surface area (Zheng et al., 2020). Finally, Chaulagain et al. demonstrated the combination of surface plasmon resonance and QDs where they utilized Au nanoparticles along with g-C_3N_4 QDs. The spectral overlap of g-C_3N_4 emission band and plasmon band of Au lead to the maximum PEC performance enhancement via exciton to plasmon energy transfer. Thus both plasmonic materials and QDs highlight enormous potential in increasing the overall PEC performance (Chaulagain et al., 2022).

11.12 Conclusion

Two main obstacles to the PEC system's efficiency are low light absorption and charge recombination. Plasmonic materials (in Table 11.1) and QDs (in Table 11.2) have the potential to mitigate the stability and efficiency concerns associated with hydrogen production. By creating plasmon-sensitized photoelectrodes, one can lessen charge recombination and make use of a larger portion of the solar spectrum. Although the use of noble metals, such as palladium, silver, and gold, has been shown to improve overall performance, cost-effectiveness is a major concern. The application of low cost metal and nonmetal nanostructures is a better alternative to tune the PEC efficiency. In addition, various metal-semiconductor nanostructures namely, core-shell, sandwich type, and plasmonic crystals

TABLE 11.1 Comparison of plasmonic materials-based catalysts.

S. no.	Catalyst	Yield	Advantage	References
1.	g-C_3N_4/Ag@SiO_2	11.4 μmol/h	High yield	Chen et al. (2015)
2.	AuNP/TiO_2	16.577 mmol/g/h	By-product in biodiesel production	Zhang et al. (2023)
3.	2D-like TiB_2 flakes		Operation in alkaline and acidic conditions	Zabelina et al. (2023)
4.	1D/2D Co_3O_4/$ZnIn_2S_4$	3844.12 μmol/h/g	High yield. No additional cocatalyst	Zhang et al. (2023)
5.	Ag/$MgCO_3$	867 mmol/h/g	H_2 production in natural solution condition	Wang et al. (2023)
6.	Au decorated $LaFeO_3$-g-C_3N_4-$BiFeO_3$	698.4 μmol/h/g	Natural elements	Arif et al. (2023)
7.	3D/3D ReS_2/$ZnIn_2S_4$-S_v	1.08 mmol/g/h	Furfural generation rates	Hu et al. (2023)
8.	MoSx homojunction	3.43 mmol/h/g	Low cost	Li et al. (2023)
9.	CdS/Bi/TiO_2	673.81 μmol/h/cm^2	Cyclic H_2 production and dye degradation	Wang et al. (2022)
10.	Ag_2O/Bi_2WO_6	3.06 mmol/g/h	97% reusability five times	El-Shishtawy et al. (2023)

TABLE 11.2 Comparison of quantum dots-based catalysts.

S. no.	Catalyst	Photocurrent density (mA/ cm^2)	Solar energy conversion efficiency (e.g., H_2 generation rate)	References
1.	Red carbon QDs/TiO_2-glucose-Pt/CNFs	4	36 mL/cm^2/day	Wang et al. (2021)
2.	Carbon QDs/TiO_2 NRs	3	ABPE of 1.29% at 0.615 V versus RHE	Liang et al. (2019)
3.	NiOOH/FeOOH/carbon QDs /$BiVO_4$	5.99	ABPE of 2.29% at 0.6 V versus RHE	Ye et al. (2017)
4.	Carbon QDs/TiO_2 NTs	1	14.1 μmol/h	Zhang et al. (2013)
5.	$CuZnInS_3$ QDs/TiO_2	4.4	34 μmol/cm^2/h	Liu et al. (2021)
6.	$AgInS_2$ QDs/ZnS passivation/TiO_2	5.7	33.5 μmol/cm^2/h	Pan et al. (2018)

(Continued)

TABLE 11.2 (Continued)

S. no.	Catalyst	Photocurrent density (mA/ cm^2)	Solar energy conversion efficiency (e.g., H_2 generation rate)	References
7.	CuGaS2/ZnSe core/shell QDs/TiO_2	3.5	23 μmol/cm^2/h	You et al. (2022)
8.	SnSe/ZnSe core/shell QDs/TiO_2	7	60 mL/cm^2/day	Ren et al. (2021)
9.	Zn-Cu$InSe_2$/CuInS_2 core/shell QDs/ TiO_2	3.1	26.5 mL/cm^2/day	Tong et al. (2018)
10.	Cu$InSe_2$/(CuInSexS1−x)5/CuInS_2 core/multishell QDs/TiO_2	4.5	43.2 mL/cm^2/day	Zhao et al. (2023)

are being explored as well. Further, the use of QDs provide the benefits of band gap tenability, broad absorption spectrum, less surface trap states along with high durability and low cost. However, emphasis is currently on shifting toward development of green QDs to replace toxic elements like Pb, Hg, and Cd. Few strategies include developing core-shell structures and the use of cocatalysts also. Therefore the integration of plasmonic materials and QD could help to further enhance the efficiency of hydrogen production.

References

Abdelkarim, O., et al., 2022. Constructing quantum dots sensitized TiO_2 nanotube p-n heterojunction for photoelectrochemical hydrogen generation. Chemical Engineering Journal 446, 137312. Available at https://doi.org/10.1016/j.cej.2022.137312.

Agarwal, D., et al., 2017. Engineering localized surface plasmon interactions in gold by silicon nanowire for enhanced heating and photocatalysis. Nano Letters 17 (3), 1839–1845. Available at https://doi.org/10.1021/acs.nanolett.6b05147.

Ai, G., et al., 2015. Solar water splitting by TiO_2/CdS/Co–Pi nanowire array photoanode enhanced with Co–Pi as hole transfer relay and CdS as light absorber. Advanced Functional Materials 25 (35), 5706–5713. Available at https://doi.org/10.1002/adfm.201502461.

Alivisatos, A.P., 1996. Semiconductor clusters, nanocrystals, and quantum dots. Science 271 (5251), 933–937. Available at https://doi.org/10.1126/science.271.5251.933.

Arif, N., et al., 2023. Enhanced charge separation in dual Z-scheme Au decorated $LaFeO_3$-g-C_3N_4-$BiFeO_3$ system for efficient H_2 production. Fuel 336, 126832. Available at https://doi.org/10.1016/j.fuel.2022.126832.

Brown, A.M., et al., 2016. Nonradiative plasmon decay and hot carrier dynamics: effects of phonons, surfaces, and geometry. ACS Nano 10 (1), 957–966. Available at https://doi.org/10.1021/acsnano.5b06199.

Celebi, N., et al., 2021. Ligand-free fabrication of Au/TiO_2 nanostructures for plasmonic hot-electron-driven photocatalysis: photoelectrochemical water splitting and organic-dye degredation. Journal of Alloys and Compounds 860, 157908. Available at https://doi.org/10.1016/j.jallcom.2020.157908.

Cesar, I., et al., 2006. Translucent thin film Fe_2O_3 photoanodes for efficient water splitting by sunlight: nanostructure-directing effect of Si-doping. Journal of the American Chemical Society 128 (14), 4582–4583. Available at https://doi.org/10.1021/ja060292p.

Chaulagain, N., et al., 2022. Synergistic enhancement of the photoelectrochemical performance of TiO_2 nanorod arrays through embedded plasmon and surface carbon nitride Co-sensitization. ACS Applied Materials and Interfaces 14 (21), 24309–24320. Available at https://doi.org/10.1021/acsami.2c02649.

Chen, H.M., et al., 2010. Quantum dot monolayer sensitized ZnO nanowire-array photoelectrodes: true efficiency for water splitting. Angewandte Chemie International Edition 49 (34), 5966–5969. Available at https://doi.org/10.1002/anie.201001827.

Chen, H.M., et al., 2012. Plasmon inducing effects for enhanced photoelectrochemical water splitting: X-ray absorption approach to electronic structures. ACS Nano 6 (8), 7362–7372. Available at https://doi.org/10.1021/nn3024877.

Chen, O., et al., 2013. Compact high-quality CdSe–CdS core–shell nanocrystals with narrow emission linewidths and suppressed blinking. Nature Materials 12 (5), 445–451. Available at https://doi.org/10.1038/nmat3539.

Chen, J., et al., 2015. Nanogap engineered plasmon-enhancement in photocatalytic solar hydrogen conversion. Advanced Materials Interfaces 2 (14), 1500280. Available at https://doi.org/10.1002/admi.201500280.

Das, S., et al., 2022. A critical review on prospects and challenges of metal-oxide embedded g-C_3N_4-based direct Z-scheme photocatalysts for water splitting and environmental remediation. Applied Surface Science Advances 11, 100273. Available at https://doi.org/10.1016/j.apsadv.2022.100273.

Díez-García, M.I., Gómez, R., 2016. Investigating water splitting with $CaFe_2O_4$ photocathodes by electrochemical impedance spectroscopy. ACS Applied Materials and Interfaces 8 (33), 21387–21397. Available at https://doi.org/10.1021/acsami.6b07465.

El-Shishtawy, R.M., et al., 2023. Efficient visible-light-driven H_2 evolution over sol-gel processed Bi_2WO_6 nanocrystals anchored with Ag_2O support. Ceramics International 49 (9), 14274–14280. Available at https://doi.org/10.1016/j.ceramint.2023.01.014.

Erwin, W.R., et al., 2016. Light trapping in mesoporous solar cells with plasmonic nanostructures. Energy and Environmental Science 9 (5), 1577–1601. Available at https://doi.org/10.1039/C5EE03847B.

García-Santamaría, F., et al., 2009. Suppressed Auger recombination in "giant" nanocrystals boosts optical gain performance. Nano Letters 9 (10), 3482–3488. Available at https://doi.org/10.1021/nl901681d.

García-Santamaría, F., et al., 2011. 'Breakdown of volume scaling in auger recombination in CdSe/CdS heteronanocrystals: the role of the core–shell interface. Nano Letters 11 (2), 687–693. Available at https://doi.org/10.1021/nl103801e.

Ghobadi, T.G.U., et al., 2018. Strategies for Plasmonic hot-electron-driven photoelectrochemical water splitting. ChemPhotoChem 2 (3), 161–182. Available at https://doi.org/10.1002/cptc.201700165.

Gomes Silva, C., et al., 2011. Influence of excitation wavelength (UV or visible light) on the photocatalytic activity of titania containing gold nanoparticles for the generation of hydrogen or oxygen from water. Journal of the American Chemical Society 133 (3), 595–602. Available at https://doi.org/10.1021/ja1086358.

Guo, L., et al., 2021. Silver-catalyzed direct regioselective C3 phosphonation of 4 *H* -pyrido[1,2- *a*]pyrimidin-4-ones with *H* -phosphites. Asian Journal of Organic Chemistry 10 (7), 1660–1664. Available at https://doi.org/10.1002/ajoc.202100235.

Han, H., et al., 2019. Highly ordered N-doped carbon dots photosensitizer on metal–organic framework-decorated ZnO nanotubes for improved photoelectrochemical water splitting. Small 15 (40), 1902771. Available at https://doi.org/10.1002/smll.201902771.

Hu, J., et al., 2023. Bifunctional honeycomb hierarchical structured 3D/3D $ReS_2/ZnIn_2S_4$-Sv heterojunction for efficient photocatalytic H_2-evolution integrated with biomass oxidation. Chemical Engineering Journal 453, 139957. Available at https://doi.org/10.1016/j.cej.2022.139957.

Ingram, D.B., et al., 2011. Predictive model for the design of plasmonic metal/semiconductor composite photocatalysts. ACS Catalysis 1 (10), 1441–1447. Available at https://doi.org/10.1021/cs200320h.

Jiang, K., et al., 2019. Single platinum atoms embedded in nanoporous cobalt selenide as electrocatalyst for accelerating hydrogen evolution reaction. Nature Communications 10 (1), 1743. Available at https://doi.org/10.1038/s41467-019-09765-y.

Jin, L., et al., 2021. Quantum dots-based photoelectrochemical hydrogen evolution from water splitting. Advanced Energy Materials 11 (12), 2003233. Available at https://doi.org/10.1002/aenm.202003233.

Jin-nouchi, Y., et al., 2010. PbS quantum dot-sensitized photoelectrochemical cell for hydrogen production from water under illumination of simulated sunlight. ChemPhysChem 11 (17), 3592–3595. Available at https://doi.org/10.1002/cphc.201000593.

Kamat, P.V., Hartland, G.V., 2018. Plasmons for energy conversion. ACS Energy Letters 3 (6), 1467–1469. Available at https://doi.org/10.1021/acsenergylett.8b00721.

Kang, H., et al., 2019. Stabilization of silver and gold nanoparticles: preservation and improvement of plasmonic functionalities. Chemical Reviews 119 (1), 664–699. Available at https://doi.org/10.1021/acs.chemrev.8b00341.

Kelly, K.L., et al., 2003. The optical properties of metal nanoparticles: the influence of size, shape, and dielectric environment. The Journal of Physical Chemistry B 107 (3), 668–677. Available at https://doi.org/10.1021/jp026731y.

Kongkanand, A., et al., 2008. Quantum dot solar cells. Tuning photoresponse through size and shape control of CdSe–TiO_2 architecture. Journal of the American Chemical Society 130 (12), 4007–4015. Available at https://doi.org/10.1021/ja0782706.

Lee, C., Park, Y., Park, J.Y., 2019. Hot electrons generated by intraband and interband transition detected using a plasmonic Cu/TiO_2 nanodiode. RSC Advances 9 (32), 18371–18376. Available at https://doi.org/10.1039/C9RA02601K.

Lee, M.G., et al., 2017. Dominance of plasmonic resonant energy transfer over direct electron transfer in substantially enhanced water oxidation activity of $BiVO_4$ by shape-controlled Au nanoparticles. Small 13 (37), 1701644. Available at https://doi.org/10.1002/smll.201701644.

Li, H., et al., 2018. Interface engineering of colloidal CdSe quantum dot thin films as acid-stable photocathodes for solar-driven hydrogen evolution. ACS Applied Materials and Interfaces 10 (20), 17129–17139. Available at https://doi.org/10.1021/acsami.7b19229.

Li, L., et al., 2023. Amorphization-crystallization synergism on MoSx homojunction for boosting photocatalytic H_2 production of TiO_2 in alkaline medium. Applied Surface Science 608, 155173. Available at https://doi.org/10.1016/j.apsusc.2022.155173.

Li, M., et al., 2016. Carbon quantum dots decorated Cu_2S nanowire arrays for enhanced photoelectrochemical performance. Nanoscale 8 (16), 8559–8567. Available at https://doi.org/10.1039/C5NR06908D.

Li, M., Cushing, S.K., Wu, N., 2015. Plasmon-enhanced optical sensors: a review. The Analyst 140 (2), 386–406. Available at https://doi.org/10.1039/C4AN01079E.

Li, Y., et al., 2020. An effective strategy of constructing a multi-junction structure by integrating a heterojunction and a homojunction to promote the charge separation and transfer efficiency of WO_3. Journal of Materials Chemistry A 8 (13), 6256–6267. Available at https://doi.org/10.1039/D0TA00452A.

Li, Z., et al., 2018. Drastic enhancement of photoelectrochemical water splitting performance over plasmonic Al@TiO_2 heterostructured nanocavity arrays. Nano Energy 51, 400–407. Available at https://doi.org/10.1016/j.nanoen.2018.06.083.

Liang, Z., et al., 2019. Hydrogenated TiO_2 nanorod arrays decorated with carbon quantum dots toward efficient photoelectrochemical water splitting. ACS Applied Materials and Interfaces 11 (21), 19167–19175. Available at https://doi.org/10.1021/acsami.9b04059.

Lieberman, I., et al., 2008. Plasmon-resonance-enhanced absorption and circular dichroism. Angewandte Chemie 120 (26), 4933–4935. Available at https://doi.org/10.1002/ange.200800231.

Lin, Y., et al., 2013. Amorphous Si thin film based photocathodes with high photovoltage for efficient hydrogen production. Nano Letters 13 (11), 5615–5618. Available at https://doi.org/10.1021/nl403265k.

Linic, S., Christopher, P., Ingram, D.B., 2011. Plasmonic-metal nanostructures for efficient conversion of solar to chemical energy. Nature Materials 10 (12), 911–921. Available at https://doi.org/10.1038/nmat3151.

Liu, B., et al., 2015. A solution-processed, mercaptoacetic acid-engineered CdSe quantum dot photocathode for efficient hydrogen production under visible light irradiation. Energy and Environmental Science 8 (5), 1443–1449. Available at https://doi.org/10.1039/C5EE00331H.

Liu, C., et al., 2021. Tuning the composition of heavy metal-free quaternary quantum dots for improved photoelectrochemical performance. Journal of Materials Chemistry A 9 (9), 5825–5832. Available at https://doi.org/10.1039/D0TA11481B.

Moreels, I., et al., 2011. Size-tunable, bright, and stable PbS quantum dots: a surface chemistry study. ACS Nano 5 (3), 2004–2012. Available at https://doi.org/10.1021/nn103050w.

Naldoni, A., et al., 2016. Solar-powered plasmon-enhanced heterogeneous catalysis. Nanophotonics 5 (1), 112–133. Available at https://doi.org/10.1515/nanoph-2016-0018.

Navarro-Pardo, F., et al., 2020. 1D/2D cobalt-based nanohybrids as electrocatalysts for hydrogen generation. Advanced Functional Materials 30 (14), 1908467. Available at https://doi.org/10.1002/adfm.201908467.

Nie, R., et al., 2018. Artificial photosynthesis of methanol by Mn:CdS and CdSeTe quantum dot cosensitized titania photocathode in imine-based ionic liquid aqueous solution. ChemCatChem 10 (15), 3342–3350. Available at https://doi.org/10.1002/cctc.201800190.

Nishijima, Y., et al., 2010. Plasmon-assisted photocurrent generation from visible to near-infrared wavelength using a Au-nanorods/TiO_2 electrode. The Journal of Physical Chemistry Letters 1 (13), 2031–2036. Available at https://doi.org/10.1021/jz1006675.

Pala, R.A., et al., 2009. Design of plasmonic thin-film solar cells with broadband absorption enhancements. Advanced Materials 21 (34), 3504–3509. Available at https://doi.org/10.1002/adma.200900331.

Pan, L., et al., 2018. Boosting the performance of Cu_2O photocathodes for unassisted solar water splitting devices. Nature Catalysis 1 (6), 412–420. Available at https://doi.org/10.1038/s41929-018-0077-6.

Peerakiatkhajohn, P., et al., 2021. Bifunctional photoelectrochemical process for humic acid degradation and hydrogen production using multi-layered p-type Cu_2O photoelectrodes with plasmonic Au@TiO_2. Journal of Hazardous Materials 402, 123533. Available at https://doi.org/10.1016/j.jhazmat.2020.123533.

Prévot, M.S., Guijarro, N., Sivula, K., 2015. Enhancing the performance of a robust sol–gel-processed p-type delafossite $CuFeO_2$ photocathode for solar water reduction. ChemSusChem 8 (8), 1359–1367. Available at https://doi.org/10.1002/cssc.201403146.

Primo, A., Corma, A., García, H., 2011. Titania supported gold nanoparticles as photocatalyst. Physical Chemistry Chemical Physics 13 (3), 886–910. Available at https://doi.org/10.1039/C0CP00917B.

Pu, Y.-C., et al., 2013. Au nanostructure-decorated TiO_2 nanowires exhibiting photoactivity across entire UV-visible region for photoelectrochemical water splitting. Nano Letters 13 (8), 3817–3823. Available at https://doi.org/10.1021/nl4018385.

Qian, H.-L., et al., 2017. High-crystallinity covalent organic framework with dual fluorescence emissions and its ratiometric sensing application. ACS Applied Materials and Interfaces 9 (29), 24999–25005. Available at https://doi.org/10.1021/acsami.7b08060.

Rainò, G., et al., 2011. Probing the wave function delocalization in CdSe/CdS dot-in-rod nanocrystals by time- and temperature-resolved spectroscopy. ACS Nano 5 (5), 4031–4036. Available at https://doi.org/10.1021/nn2005969.

Raissi, M., et al., 2018. Improved efficiency of PbS quantum dot sensitized NiO photocathodes with naphthalene diimide electron acceptor bound to the surface of the nanocrystals. Solar Energy Materials and Solar Cells 181, 71–76. Available at https://doi.org/10.1016/j.solmat.2017.12.029.

Ren, S., et al., 2021. Near-infrared heavy-metal-free SnSe/ZnSe quantum dots for efficient photoelectrochemical hydrogen generation. Nanoscale 13 (6), 3519–3527. Available at https://doi.org/10.1039/D0NR09154E.

Selopal, G.S., et al., 2020. Core/shell quantum dots solar cells. Advanced Functional Materials 30 (13), 1908762. Available at https://doi.org/10.1002/adfm.201908762.

Smith, A.M., Nie, S., 2010. Semiconductor nanocrystals: structure, properties, and band gap engineering. Accounts of Chemical Research 43 (2), 190–200. Available at https://doi.org/10.1021/ar9001069.

Song, J., et al., 2021. Highly efficient plasmon induced hot-electron transfer at Ag/TiO_2 interface. ACS Photonics 8 (5), 1497–1504. Available at https://doi.org/10.1021/acsphotonics.1c00321.

Standridge, S.D., Schatz, G.C., Hupp, J.T., 2009. Distance dependence of plasmon-enhanced photocurrent in dye-sensitized solar cells. Journal of the American Chemical Society 131 (24), 8407–8409. Available at https://doi.org/10.1021/ja9022072.

Subramanyam, P., et al., 2022. Emerging materials for plasmon-assisted photoelectrochemical water splitting. Journal of Photochemistry and Photobiology C 51, 100472. Available at https://doi.org/10.1016/j.jphotochemrev.2021.100472.

Tan, S., et al., 2017. Plasmonic coupling at a metal/semiconductor interface. Nature Photonics 11 (12), 806–812. Available at https://doi.org/10.1038/s41566-017-0049-4.

Thimsen, E., et al., 2011. Influence of plasmonic Au Nanoparticles On The Photoactivity of Fe_2O_3 electrodes for water splitting. Nano Letters 11 (1), 35–43. Available at https://doi.org/10.1021/nl1022354.

Tong, X., et al., 2018. Near-infrared, heavy metal-free colloidal "giant" core/shell quantum dots. Advanced Energy Materials 8 (2), 1701432. Available at https://doi.org/10.1002/aenm.201701432.

Vahidzadeh, E., et al., 2021. Harvesting hot holes in plasmon-coupled ultrathin photoanodes for high-performance photoelectrochemical water splitting. ACS Applied Materials and Interfaces 13 (36), 42741–42752. Available at https://doi.org/10.1021/acsami.1c10698.

Vale, B.R.C., et al., 2019. Ligand induced switching of the band alignment in aqueous synthesized CdTe/CdS core/shell nanocrystals. Scientific Reports 9 (1), 8332. Available at https://doi.org/10.1038/s41598-019-44787-y.

Valenti, M., et al., 2016. Plasmonic nanoparticle-semiconductor composites for efficient solar water splitting. Journal of Materials Chemistry A 4 (46), 17891–17912. Available at https://doi.org/10.1039/C6TA06405A.

Wang, F., et al., 2017. In situ decoration of CuSCN nanorod arrays with carbon quantum dots for highly efficient photoelectrochemical performance. Carbon 125, 344–351. Available at https://doi.org/10.1016/j.carbon.2017.09.073.

Wang, G., et al., 2010. Double-sided CdS and CdSe quantum dot Co-sensitized ZnO nanowire arrays for photoelectrochemical hydrogen generation. Nano Letters 10 (3), 1088–1092. Available at https://doi.org/10.1021/nl100250z.

Wang, Q., et al., 2022. Construction of Bi-assisted modified CdS/TiO_2 nanotube arrays with ternary S-scheme heterojunction for photocatalytic wastewater treatment and hydrogen production. Fuel 322, 124163. Available at https://doi.org/10.1016/j.fuel.2022.124163.

Wang, Q., et al., 2023. Metal-carbonate interface promoted activity of Ag/$MgCO_3$ catalyst for aqueous-phase formaldehyde reforming into hydrogen. Fuel 337, 126897. Available at https://doi.org/10.1016/j.fuel.2022.126897.

Wang, X., et al., 2021. Colloidal carbon quantum dots as light absorber for efficient and stable ecofriendly photoelectrochemical hydrogen generation. Nano Energy 86, 106122. Available at https://doi.org/10.1016/j.nanoen.2021.106122.

Warren, S.C., Thimsen, E., 2012. Plasmonic solar water splitting. Energy and Environmental Science 5 (1), 5133–5146. Available at https://doi.org/10.1039/C1EE02875H.

Wu, K., et al., 2015. Efficient hot-electron transfer by a plasmon-induced interfacial charge-transfer transition. Science 349 (6248), 632–635. Available at https://doi.org/10.1126/science.aac5443.

Ye, K.-H., et al., 2017. Correction: carbon quantum dots as a visible light sensitizer to significantly increase the solar water splitting performance of bismuth vanadate photoanodes. Energy and Environmental Science 10 (2), 642. Available at https://doi.org/10.1039/C7EE90006F 642. Available at.

You, Y., et al., 2022. Tailoring the optoelectronic properties of eco-friendly $CUGAS_2$/ZNSE core/shell quantum dots for boosted photoelectrochemical solar hydrogen production. EcoMat 4 (5), e12206. Available at https://doi.org/10.1002/eom2.12206.

Zabelina, A., et al., 2023. Plasmon coupling inside 2D-like TiB_2 flakes for water splitting half reactions enhancement in acidic and alkaline conditions. Chemical Engineering Journal 454, 140441. Available at https://doi.org/10.1016/j.cej.2022.140441.

Zhang, J., et al., 2023. Insight into synergistic enhancement of photothermal catalytic hydrogen production by plasmonic Au-NP/TiO_2 in the presence of glycerol. Energy Conversion and Management 277, 116626. Available at https://doi.org/10.1016/j.enconman.2022.116626.

Zhang, X., et al., 2013. Carbon quantum dot sensitized TiO_2 nanotube arrays for photoelectrochemical hydrogen generation under visible light. Nanoscale 5 (6), 2274. Available at https://doi.org/10.1039/c3nr34142a.

Zhang, X., et al., 2014. Shape-dependent localized surface plasmon enhanced photocatalytic effect of ZnO nanorods decorated with Ag. International Journal of Hydrogen Energy 39 (16), 8238–8245. Available at https://doi.org/10.1016/j.ijhydene.2014.03.153.

Zhang, Y., et al., 2018. Surface-plasmon-driven hot electron photochemistry. Chemical Reviews 118 (6), 2927–2954. Available at https://doi.org/10.1021/acs.chemrev.7b00430.

Zhang, Y., et al., 2023. Fabricating 1D/2D Co_3O_4/$ZnIn_2S_4$ core–shell heterostructures with boosted charge transfer for photocatalytic hydrogen production. Applied Surface Science 610, 155272. Available at https://doi.org/10.1016/j.apsusc.2022.155272.

Zhao, H., et al., 2023. Engineering the interfacial structure of heavy metal-free colloidal heterostructured quantum dots for high-efficiency photoelectrochemical water oxidation without co-catalyst. Advanced Energy and Sustainability Research 4 (2), 2200142. Available at https://doi.org/10.1002/aesr.202200142.

Zheng, G., et al., 2020. WO_3/Cu_2O heterojunction for the efficient photoelectrochemical property without external bias. Applied Catalysis B 265, 118561. Available at https://doi.org/10.1016/j.apcatb.2019.118561.

Zheng, L., et al., 2020. Black phosphorus quantum dot-sensitized TiO_2 nanotube arrays with enriched oxygen vacancies for efficient photoelectrochemical water splitting. ACS Sustainable Chemistry and Engineering 8 (42), 15906–15914. Available at https://doi.org/10.1021/acssuschemeng.0c04819.

Zhong, Y., et al., 2015. Cocatalyst effects on hydrogen evolution in a plasmon-induced water-splitting system. The Journal of Physical Chemistry C 119 (16), 8889–8897. Available at https://doi.org/10.1021/acs.jpcc.5b01720.

Zhu, H., et al., 2020. Mapping hot electron response of individual gold nanocrystals on a TiO_2 photoanode. Nano Letters 20 (4), 2423–2431. Available at https://doi.org/10.1021/acs.nanolett.9b05125.

CHAPTER

12

Carbon-based materials for water-splitting

Mansi Pahuja, Mohd Afshan, Daya Rani, E.M. Harini, and Kaushik Ghosh

Quantum Materials and Devices Unit, Institute of Nano Science and Technology, Mohali, Punjab, India

12.1 Introduction

Hydrogen is increasingly acknowledged as a promising and clean source of energy with zero carbon content capable of mitigating emission of greenhouse gases and other pollutants (Sundmacher et al., 2008; Turner, 2004). Environmentally benign water-splitting technique is employed for the generation of green hydrogen fuel stepping towards renewable and sustainable pathways in the ground of automobile industries (Yu et al., 2020; Pahuja et al., 2022). Water-splitting can be achieved through electrochemical, photo-electrochemical and photocatalytic processes (Bonke et al., 2015; You and Sun, 2018; Yu et al., 2020). All of these processes require catalysts to expedite the reactions and attain a high efficiency of hydrogen production, paving the way for commercialization. Considerable efforts have already been expended in the fabrication of a stable, cost-efficient, highly active, and environmentally friendly catalyst to replace noble metals, which are scarce and costly (Deng et al., 2014; Su et al., 2014; Zeng and Li, 2015). In recent years, diverse strategies have been employed to modify the morphology, expose crystalline facets, and enhance the surface area of the catalyst, proven to be a notable enhancement in the efficiency of hydrogen generation. However, these are not enough for the practical feasibility due to the following reasons, (1) catalysts are prone to oxidation in the long-term contact of the electrolyte under applied potential; (2) the advance and costly synthetic procedure of the catalyst hinders them for commercialization; (3) using binders to develop an electrode which blocks the active surface area of the catalyst. Similarly in photocatalysis, although huge efforts have been made in synthesizing metal oxide, sulfide, and nitride catalysts, yet the efficiencies of green fuel generation processes still fall short of practical commercialization standards. The inefficiencies of photocatalysts can be ascribed to the recombining of charge carriers, predominantly taking place at crystalline defects and the grain boundaries dispersed throughout the

Photochemical Splitting of Water
DOI: https://doi.org/10.1016/B978-0-443-29064-0.00004-8

photocatalyst. In these regions, the diffusion pathway of photogenerated charge carriers is substantial, and the active sites density is insufficient.

In such scenario, carbon-based materials are a hot topic of research in electrochemical as well as photocatalytic water-splitting that includes graphene, graphene oxide (GO)/reduced graphene oxide (rGO), carbon nanotubes (CNTs), carbon quantum dots, fullerenes and other carbon materials such as carbon fibers, activated carbon, and carbon black. This significant interest owes to their distinctive characteristics, including a tunable band diagram, chemically inert, extremely large surface area with customizable crystalline structure exhibiting active crystallographic facets. The porous and organized configuration of carbon-based materials promotes the addition of functional groups, leading to a marked improvement in their photocatalytic properties. Carbon materials exhibit diverse applications, serving as photosensitizers, support materials, cocatalysts, and semiconductors, and the capability to alter the band gap and accept and transfer electrons. The extensive surface area of carbon materials not only increases the number of adsorption sites for more efficient H_2 generation but also provides stability to the decorated photoactive material by strongly adhering to it. Carbon nanomaterials are characterized by exceptional thermal conductivity and a work function that exceeds that of typical semiconductors. This ability allows for the efficient extraction of electrons from the conduction band, lowering the recombination rate. Moreover, the electronic band structure as well as electrical conductivity can be fine-tuned by introducing heteroatoms onto carbonaceous materials. These modifications affect their properties which include the transportation of charge carriers and broad-band absorption, ultimately enhancing the photocatalytic activity.

A remarkable benchmark has been made by carbon-based materials not only in the field of photocatalysis but also in electrocatalysis where they are increasingly recognized as a promising alternative to Pt-based catalysts, holding substantial potential to address critical challenges in the green hydrogen fuel generation. The significant research focus on carbon-based catalysts for electrocatalytic and photocatalytic water-splitting has prompted a need for a comprehensive review of this research domain. This chapter highlights the latest progress in the use of carbon-based catalysts for water-splitting. In addition, the various approaches for augmenting the hydrogen generation efficiency of carbon-based materials are also discussed. The insights gained from these achievements can contribute to the future development of innovative, cost-effective catalysts with high activity and prolonged lifetimes for practical applications.

12.2 Carbon nanomaterials

The most multipurpose element in the whole periodic table is "Carbon" due to its ability to form a wide variety of bonds, both with itself and with numerous other elements, characterized by different types and strengths. The existence of various allotropes of carbon is due to the capability of carbon to hybridize in various configurations like sp, sp^2 and sp^3. There are two types of carbon allotropes, naturally available such as amorphous carbon, diamond and graphite and synthetically made which are carbon nanohorns, graphene, CNTs, fullerenes, and nanodiamonds.

Generally, materials are known to be nanomaterials if one of the dimension of particles is in between 1 and 100 nm in size. Nanomaterials consisting of carbon atoms are collectively known as carbon-based or carbon nanomaterials. They are most commonly classified based on their geometrical structure. They exhibit particles with various shapes, such as tubes, horns, spheres, and ellipsoids. Nanoparticles shaped like tubes are known as CNTs, those with a horn shape are called nanohorns, and spherical or ellipsoidal particles fall into the category of fullerenes. Carbon nanomaterials have a wide range of applications such as in catalysis, sensors, gas storage, electronics, composites, displays, antifouling paints, textiles, durable batteries, gas biosensors, and more. Some of the carbon-based materials have been explored in this chapter, particularly in water-splitting applications.

Carbon atoms (Georgakilas et al., 2015), due to their diverse hybridization states, can form various allotropes, including diamond and graphite, as primary examples, alongside an array of structures ranging from simple molecules to extended chains. The distinct arrangement of carbon atoms in the crystalline lattices of graphite and diamond leads to notable differences in their physical characteristics. In graphite, carbon atoms are organized in stacked layers with a honeycomb arrangement, held together by weak van der Waals forces. This weak interlayer interaction allows π electrons to move freely, imparting graphite its semimetallic nature. The overlap of valence and conduction bands at a single point results in a zero optical band gap, giving graphite a dark appearance and high light absorption.

Conversely, diamonds exhibit sp^3 hybridization, forming strong covalent bonds in a tetrahedral geometry and a face-centered Bravais lattice. This structure lacks electron mobility, making diamond an excellent electrical insulator with an optical band gap of approximately 5.5 eV. Diamond's high transparency spans a wide optical range beyond the visible spectrum, although absorption bands due to impurities such as nitrogen, boron, hydrogen, and nickel may be observed. Its transmission remains flat for wavelengths above ~225 nm, with moderate absorption between 2.6 and 6.2 μm caused by multiphonon interactions (Melchakova et al., 2024).

Beyond these crystalline forms, carbon can also exist in amorphous configurations. Amorphous carbon (aC) incorporates both sp^2 and sp^3 hybridized bonds (Robertson, 1986), with properties resembling either graphite or diamond, depending on the dominant hybridization. Graphite-like aC tends to have low mechanical strength and high extinction coefficients, while diamond-like aC exhibits exceptional hardness and optical transparency, with an optical band gap reaching up to 4.5 eV. The introduction of hydrogen can further modify aC properties, enabling both hard and polymer-like forms suitable for applications such as biomaterials. Amorphous carbon can also appear as nanoparticles or carbon nanostructures, offering unique optical, electrical, and mechanical properties for various applications.

12.2.1 Graphene

In 2D carbonaceous materials, the conversation is incomplete without graphene, developed by Novoselov et al. who efficaciously make an innovation by synthesizing stable 2-Dimensional carbon layer under ambient conditions, commonly known as graphene nanosheets and this became a scientific and technological revolution of October 2004. Graphene, an allotrope of carbon, is a single isolated monolayer sheet where atoms are densely packed in a sp^2 honeycomb lattice with hybridized C—C bonds and a π- π conjugated electron cloud. Owing to its exceptional thermal, optical, mechanical, and electrical properties, graphene quickly emerged as a compelling component in the fabrication of photonic devices, as well as in applications related to fuel conversion, fuel storage, environmental technologies, sensing, and catalysis. It became a magical material in the domain of materials science due to its astoundingly high surface area (2630 m^2/g), outstanding electron mobility (200,000 $cm^2/V/s$), high mechanical strength (1060 GPa), high thermal conductivity (3000 W m/K), 97.7% optical transmittance, and < 0.1% reflectance (monolayer graphene nanosheet) which proves to be an outstanding option for light harvesting and the manufacturing of energy conversion devices. However, it has a zero band gap, resulting from the interaction between antibonding π^* orbitals and π orbitals at the Brillouin zone. Also, inherent hydrophobicity of graphene constrains its direct use as a photocatalyst in water-splitting technique. Thanks to its exceptional electron mobility and large surface area, graphene can efficiently serve as an electron mediator and sink. Additionally, it provides a consistent surface for the uniform coverage of other photocatalysts. The broadening of the energy band gap via introducing heteroatoms or functional groups, metal oxides, and composites, can increase the catalytic activity of graphene via modification of the electronic density within the graphene structure. The electroneutrality of graphene is disrupted leading to the formation of regions with unbalanced charges which become as active sites upon doping with heteroatoms. Hydrogenation alters the sp^2 hybridization of graphene, transitioning some points to sp^3 hybridization and consequently increasing the band gap which enhances the photocatalytic activity considerably. In addition, due to its electron-accepting and conductive characteristics, graphene can serve as a cocatalyst alongside other catalysts.

Bellamkonda et al. synthesized a novel composite material that eliminates the need for noble metals, composed of multiwalled carbon nanotubes (MWCNTs), graphene, and titanium dioxide (CNT-GR-TiO_2) for the photocatalytic water-splitting under light irradiation. The CNT-GR-TiO_2 nanocomposite achieves a remarkable hydrogen production rate of 29 mmol/h/g under full-spectrum light exposure, which is eight times greater than that of commercial titanium dioxide (Degussa P25). Additionally, the solar energy conversion efficiency reaches 14.6%. Spectroscopic and photocatalytic analyses reveal that graphene acts as an electron reservoir, promoting efficient charge transfer at the interface during the water-splitting process (Bellamkonda et al., 2019). Similarly, Daulbayev et al. developed a Bio-waste-derived few-layered graphene (FLG)/$SrTiO_3$/PAN system using an electrospinning technique. A FLG fabricated from walnut shell and rice husk is used as a cocatalyst in a composite system. The results revealed that the presence of graphene not only contributed in decreasing the band gap of the system but also enhanced the effective separation of charge carriers. Furthermore, these composite photocatalytic systems exhibited a higher rate of hydrogen release compared to pure $SrTiO_3$ during the water-splitting reaction. Although the characterization data reveal the existence of defects in the synthesized FLGs, their optical and electrical characteristics are similar to those of pure graphene. This similarity allows for the enhancement of the hydrogen evolution efficiency by incorporating them into the c-$SrTiO_3$/PAN photocatalyst (Daulbayev et al., 2021), shown in Fig. 12.1.

Likewise, Mateo et al. utilized biomass wastes to synthesize N-doped defective graphene. The addition of 1 wt% Al-doped $SrTiO_3$ along with $RhCrO_x$ nanoparticles (RhCrOx/STO:Al) as a cocatalyst to N-doped defective graphene resulted in a photocatalytic hydrogen production rate of 153 mmol H_2 per gram of photocatalyst and 74 mmol O_2 per gram after

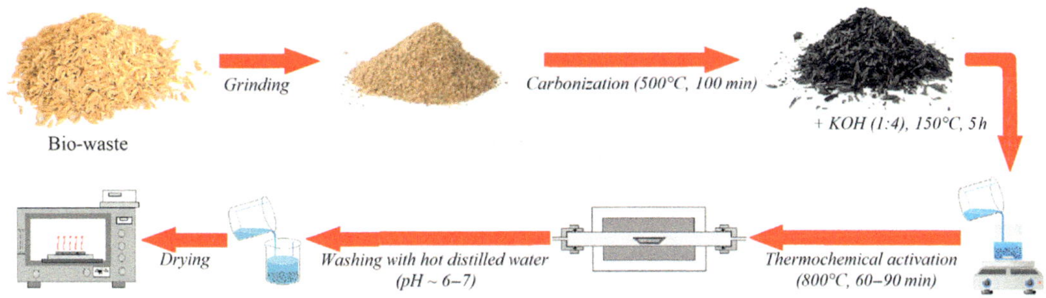

FIGURE 12.1 Diagram showing the process of synthesizing few-layered graphene from bio-waste (Daulbayev et al., 2021). Source: *Reprinted with permission from Daulbayev, C., Sultanov, Fail, et al., 2021. Bio-waste-derived few-layered graphene/SrTiO$_3$/PAN as efficient photocatalytic system for water splitting. Applied Surface Science 549, 149176. https://doi.org/10.1016/j.apsusc.2021.149176, Copyright 2021, Elsevier.*

24 hours of UV-Vis light exposure. Notably, this hybrid N-doped defective graphene-RhCrOx/STO:Al photocatalyst exhibited stable performance across four consecutive photocatalytic cycles. These findings suggest that incorporating small amounts of RhCrO$_x$/STO:Al as cocatalysts on N-doped graphene significantly enhances its photocatalytic efficiency, making it a promising candidate for water-splitting applications (Mateo et al., 2019). Qiao et al. also fabricated a core-shell structure, Co@CoO, on N-doped graphene, showcasing a synergistic effect with graphene. This synergy proves advantageous for light absorption and the efficient separation of photo-generated charges, enhancing the photocatalytic activity (Qiao et al., 2020). Machín et al. reported the presence of ZnO particles on graphene sheets serves to capture light energy and function as electron mediators. The Au@ZnO NPs-graphene catalysts exhibited increased hydrogen production, reaching 759 μmol/hg at 400 nm, particularly noteworthy with the 10% Au@ZnO NPs-graphene. This improvement in hydrogen generation is 8.8 times greater than that of ZnO NPs catalyst (Machín et al., 2020).

12.2.2 Graphene oxide

Graphene with zero band gap has high electronic conductivity which restricts its application in photocatalysis. By heteroatom doping or introducing lattice defects band opening of graphene can be done. GO and rGO are obtained by doping of graphene by especially oxygenated functional groups, these oxygenated groups transform some of the conducting sp^2-carbon atoms of graphene to nonconducting sp^3 carbon as a result band opening takes place. The band gap of GO shows a particular range of 2.4–4.3 eV, depending on oxygen content. Increasing the oxygen level results in a widening of the band gap, achieved by shifting the valence band maximum from the graphene π orbital to the oxygen 2p orbital while keeping the graphene π* orbital as the conduction band minimum. Carbonyl, hydroxyl, and epoxy groups are present on the basal plane of GO, while carboxylic groups are present near the edges as a result GO is hydrophilic, has high surface area, facile mass production, mechanical strength, ease of functionalization, compatibility with various electrolytes, low manufacturing cost, and has tunable properties which makes it suitable for water-splitting (electro-, photo- and photoelectrochemical catalytic).

GO in its bare form exhibits suboptimal electrocatalytic performance for overall water-splitting. To enhance its efficacy, improvements can be achieved through strategies such as doping with external atoms, or by creating composite or heterostructured materials. Narwade et al. reported that Rhodium-graphene oxide (Rh-GO) demonstrates remarkable enhancements in electrochemical activity for both hydrogen evolution reaction (HER) and oxygen evolution reaction (OER). Specifically, Rh-GO exhibits an exceptionally low over-potential of merely 2 mV for the HER @10 mA/cm^2. Furthermore, it displayed reduced Tafel slope of 10 mV/dec indicative of improved kinetics in the hydrogen evolution process. In the case of the OER, Rh-GO demonstrated an overpotential of 230 V at 10 mA/cm^2, accompanied by a smaller Tafel slope of 27 mV/dec. These findings underscore the outstanding electrocatalytic capabilities of Rh-GO, emphasizing its capability as a proficient catalyst for water-dissociation reactions (Narwade et al., 2020).

Yeh et al. observed various GO photocatalysts with different oxygen content and assessed their photocatalytic activity for water-splitting. Their findings revealed that increasing oxygen content led to a decrease in the GO valence band maximum as shown in Fig. 12.2. Only the GO with the highest oxidation potential demonstrated the ability to produce O_2, indicating its suitability for water oxidation, while all GO variants showed H_2 evolution (Yeh et al., 2011). Same author, Yeh et al. also produced a significant amount of H_2 at a rate of 2833 μmol/h, illustrated the photocatalytic potential of GO in water-splitting using a sacrificial agent. The H_2 production rate was decreased to 47 μmol/h when sacrificial agents were not present (Yeh et al., 2010). The photocatalytic properties of metal selenides, oxides, etc., can be enhanced by preparing the composite or heterostructure with GO by facile electrons and holes separation due to its π conjugation structure of GO which enhances its electron mobility. In composite GO can form a bond with other atoms of the compound via π electrons and can tune the light absorption window. Zhang et al. synthesized GO-TiO_2 nanocomposites using a sol-gel process and then calcined, showed an amazing 1.9-fold increase in hydrogen evolution activity when compared to pure TiO_2. Under Xe-lamp illumination, this improved performance was seen in an aqueous polysulfide solution (Zhang et al., 2010).

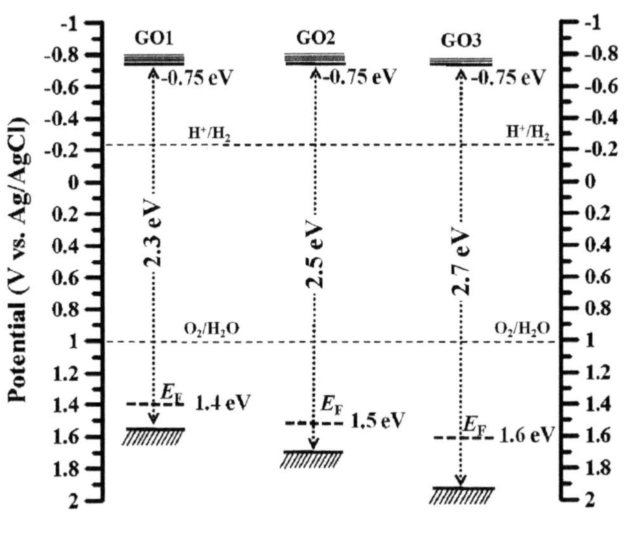

FIGURE 12.2 Energy diagram of GO samples in comparison with potentials for water reduction and oxidation (Yeh et al., 2011). *GO*, Graphene oxide. *Source: Reprinted with permission from Yeh, T.-F., et al., 2011. Graphite oxide with different oxygenated levels for hydrogen and oxygen production from water under illumination: the band positions of graphite oxide. The Journal of Physical Chemistry C 115 (45), 22587–22597. https://doi.org/10.1021/jp204856c, Copyright 2011, American Chemical Society.*

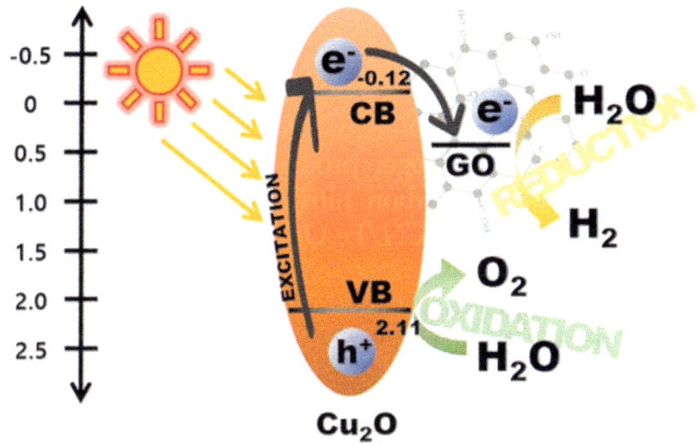

FIGURE 12.3 Diagrammatic representation of the charge transfer process within the GO-Cu_2O photo-cathode (Heo et al., 2023). *GO*, Graphene oxide. Source: *Reprinted with permission from Heo, J., et al., 2023. Surface engineering of Cu_2O photocathodes via facile graphene oxide decoration for improved photoelectrochemical water splitting. ACS Omega 8 (36), 32794–32803. https://doi.org/10.1021/acsomega.3c03585, Copyright 2023, American Chemical Society.*

Furthermore, GO can also be used for photoelectrochemical overall water-splitting. Heo et al. synthesized GO-Cu_2O composite as a photocathode demonstrated photo-current density ranging from −0.39 to −1.20 mA/cm^2 showcasing a threefold increase compared to pristine Cu_2O photocathode (Heo et al., 2023). There was an improvement in the charge transport and light absorption characteristics of the composite owing to the introduction of GO in Cu_2O explained in Fig. 12.3.

12.2.3 Reduced graphene oxide

RGO or rGO is obtained after chemical or photo-thermal reduction of GO which results in regeneration of the sp^2-carbon network and improved electrical conductivity compared to GO. The enhanced conductivity, increased active sites, improved electron transfer kinetics, high surface area, lower overpotential, better stability, tunable properties, and versatility make RGO a preferable choice over GO as a catalyst in various electrochemical reactions. RGO can be easily integrated into composite materials with other catalytic components, creating synergistic effects that further enhance its electrocatalytic, photo-catalytic, and photoelectrochemical performance.

The CoP/RGO nanocomposite, as reported by Jiao et al., displays low overpotentials of 105 mV in acidic conditions and 150 mV in basic conditions at a current density of 10 mA/cm^2. This composite demonstrates excellent HER activity across a wide pH range from 0 to 14, outperforming both pure CoP and RGO. Additionally, in basic medium, the nanocomposite demonstrated exceptional OER activity, with an overpotential of 340 mV at 10 mA/cm^2 of current density owing to high electrical conductivity and stability of composite with RGO incorporation (Jiao et al., 2016).

The band gap of RGO exhibits a range of approximately 1.00–1.69 eV, contingent upon the extent of reduction. Achieving the energy requirements for photocatalytic water-splitting can be challenging due to this variability. Therefore RGO is often integrated with other photocatalysts in nanocomposites or heterostructures to improve the performance of photocatalytic water-splitting processes. Wang et al. revealed that RGO/Pt-TiO_2 nanocomposites with a 2 wt.% RGO doping exhibited a remarkable solar-driven 1075.68 μmol/h/g hydrogen production rate. This rate surpassed that of bare TiO_2 by 81 times and Pt/TiO_2 samples by five times, highlighting the superior efficiency of the nanocomposite in facilitating hydrogen production. This enhanced performance was attributed to the two-dimensional π-conjugation structure inherent in RGO with exceptional electron-accepting and transport properties (Wang et al., 2017). Panet et al. observed that hydrogen production rate 43.6 μmol/h and oxygen production rate 21.2 μmol h for polymeric carbon nitride (PCN)/RGO/Fe_2O_3 composite which was high as compared to bare PCN and PCN/Fe_2O_3. The presence of RGO allowed for easy charge transfer between the PCN and Fe_2O_3 interface, which was responsible for increased photocatalytic performance of the composite (Pan et al., 2019).

Furthermore, Aragaw designed rGO-GO nano-hybrid material for photoelectrochemical overall water-splitting exhibited maximum photocurrent - 61.35 μA/cm^2 for rGO 1 wt% in GO which was high as compared to pristine GO (−42.80 μA/cm^2) owing to the strong – stacking (π - π) interactions among rGO and GO, as seen in Fig. 12.4, which result in better electron acquiring and transportation by rGO and consequently lower recombination of charge carriers, the photoelectrochemical (PEC) performance of the rGO/GO nano-hybrid photoelectrode was enhanced (Aragaw, 2020). Elbakkay et al. syntehsized S-TiO_2/S-RGO photoanode reaching 3.36 mA/cm^2 photocurrent density at 1 V versus Ag/AgCl. This achievement marked a threefold improvement compared to the photocurrent density of the

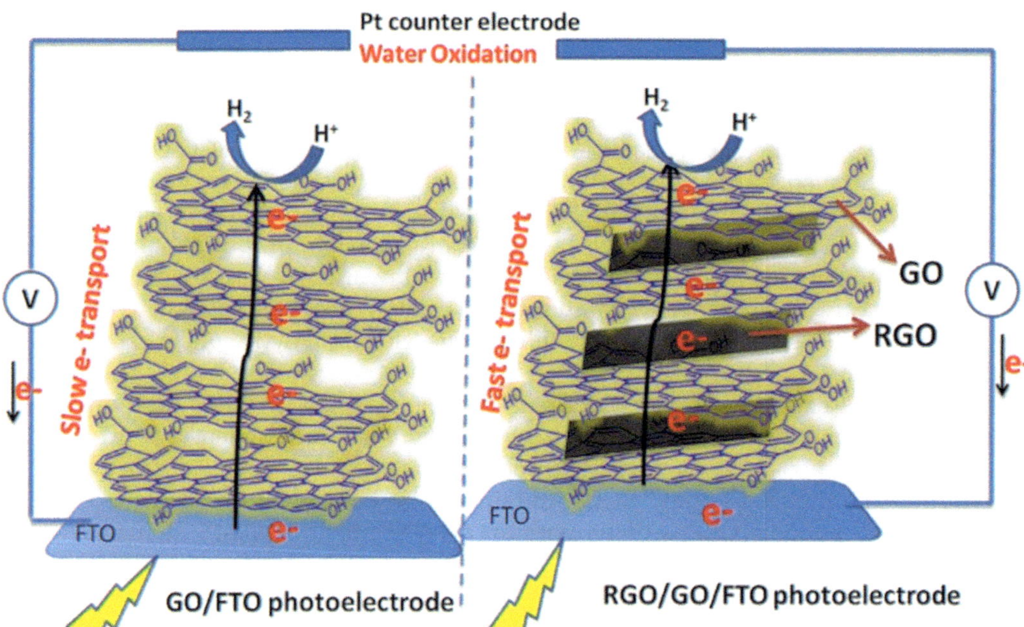

FIGURE 12.4 Illustration depicting the proposed electron transport process in the GO and RGO/GO nano-hybrid photoelectrodes in the process of photoelectrochemical water reduction (Aragaw, 2020). *rGO*, Reduced graphene oxide; *GO*, graphene oxide. Source: *Reprinted with permission from Aragaw, B.A., 2020. Reduced graphene oxide-intercalated graphene oxide nano-hybrid for enhanced photoelectrochemical water reduction. Journal of Nanostructure in Chemistry 10 (1), 9–18. https://doi.org/10.1007/s40097-019-00324-x, Copyright 2020, Springer.*

bare synthesized TiO_2 which ascribed to the enhanced photo-generated holes separation rate and electrons as well as the utilization of an efficient amount of visible light in the presence of RGO (Elbakkay et al., 2018). Tables 12.1 and 12.2 represent different materials heterostructure and composites with GO and rGO used for overall water-splitting (electrochemical, photocatalysis, and photoelectrochemical).

12.2.4 Carbon nanotubes

Iijima's pioneering identification of CNTs in 1991 was a significant breakthrough (Pawar et al., 2015). The sp^2 hybridized structure of CNTs consists of concentric arrangements of carbon hexagons. Recognized among diverse carbon nanomaterials due to their hollow cylindrical structure and an impressive diameter-to-length ratio of 1:132,000,000, CNTs exhibit unique properties. Conjugated p-electrons within CNTs enhance the speed of electron transfer, and van der Waals forces bind them together akin to ropes. CNTs exist in single-walled and multiwalled forms, each demonstrating distinct properties arising from structural differences. Furthermore, unique structures like fullerites, nanoknots, toroids, and double-walled CNTs add to the variety of this fascinating nanomaterial (Khare and Bose, 2005). Among carbon's various allotropes, CNTs stand out as highly promising and extensively researched, characterized by sp^2 hybridization. Recognized for their unique attributes, CNTs are known for high electric conductivity, chemical stability, and their prominent role as catalytic support materials (Kingston et al., 2014; Ma et al., 2017). They are categorized as single-walled/double-walled or multiwalled depending on the graphitic layers number (Rahmandoust and Ayatollahi, 2016). Various manufacturing techniques, such as chemical vapor deposition (CVD), laser ablation and arc discharge, have been

TABLE 12.1 Different materials with GO and rGO involved in photocatalytic water-splitting.

S. no.	Material	H_2 production rate	O_2 production rate	References
1.	Ni-doped GO/TiO_2	3 mmol/g.min	–	Atrees et al. (2021)
2.	GO/Rh-$SrTiO_3$	55.83μmol/g/h	23.26μmol/g/h	Zhang et al. (2023)
3.	Eu-doped ZnO@GO	255.8μmol h/g	–	Gurbani and Chouhan (2023)
4.	rGO-CdS-Sodium Alginate	5001μmol/g/h	–	Sarkar et al. (2023)
5.	Ag:$ZnIn_2S_4$/RGO	6343.86μmol/g/h	–	Gao et al. (2020)
6.	Fe/MgO-rGO	432μmol/g/h	–	Sharmin et al. (2021)
7.	TiO_2/$In_{0.5}WO_3$-rGO	304μmol/g/h	–	Shaheer et al. (2021)
8.	PANI-TiO_2/rGO	806μmol/g/h	–	Ma et al. (2020)
9.	rGO-Er-TiO_2	115.37μmol/g/h	–	Cho et al. (2021)
10.	$LaFeO_3$/RGO	82 mmol/g/h	–	Samajdar et al. (2023)

rGO, reduced graphene oxide; *GO*, graphene oxide.

TABLE 12.2 Different materials with GO and rGO are involved in electrochemical catalysis water-splitting.

Electrocatalyst	Electrolyte	HER overpotential (mV) Tafel slope (mV/dec)	OER overpotential (mV) and Tafel slope (mV/dec)	Cell potential	References
rGO/Co_2P	1 M KOH	134 and 58	378 and 77	1.7	Zhao et al. (2020)
Fe-Co-CN/rGO	1 M KOH	215 and 54	308 and 38	1.5	Fang et al. (2021)
Comox/NiFeOx/N-RGO	1 M KOH	–	372 and 92	1.6	Jia et al. (2020)
NiFe-N-CNt-rGO	1 M KOH	230 and 84	270 and 42	1.5	Wan et al. (2019)
PrGO/NiCoP	1 M KOH	106 and 58.3	281.3 and 60.1	1.5	Dong et al. (2019)
NGO/Ni_7S_6	0.1 M KOH		380 and 45	–	Jayaramulu et al. (2017)
V-$Ni_{0.06}Fe_{0.06}$ MOF/GO	1 M KOH	90 and 150	210 and 97	–	Gopi et al. (2022)
rGO/NF	1 M KOH	169 and 108.23	400 and 104	–	Kumar et al. (2023)

rGO, Reduced graphene oxide; *HER*, hydrogen evolution reaction; *OER*, oxygen evolution reaction; *GO*, graphene oxide.

developed, where arc discharge being one of the earliest techniques for developing CNTs (Khare and Bose, 2005). Arc discharge involves the thermal dispersion of graphite electrodes in plasma, while CVD utilizes hydrocarbon gases to produce single- and MWCNTs (Kharissova and Kharisov, 2014). Functioning as promising catalytic materials, CNTs serve diverse roles as catalysts, additives, or supports. Numerous metals and their oxides have been identified as excellent catalysts for processes like hydrogen generation, synthetic fuel production, and the synthesis of intermediates for polymers and artificial fibers (Chen et al., 2013; Chernyak et al., 2016; Mierczynski et al., 2016). The challenge of poor dispersion of CNTs in solvents, attributed to low chemical compatibility, weak van der Waals interactions, and hydrophobicity, has been mitigated by leveraging their extensive functionalization capabilities, rendering them more soluble in a broad range of solvents (Gopannagari and Chaturvedi, 2015; Kemnade et al., 2015).

Photocatalytic water-splitting, employing nanomaterials modified with CNTs, emerges as a promising technology for cost-effective and stable hydrogen production. Integrating CNTs with other photocatalysts further enhances conductivity and promotes effective charge transfer during the reaction. To optimize future performance, there is a need for increased efforts in the in-situ synthesis of CNT-based composites to reinforce the synergistic interactions between CNTs and other nanostructures. The details performance comparison mainly the H_2 and O_2 production rate of the different composites with CNT and different metal nonmetal doping with CNT has been discussed in Table 12.3.

In photocatalysts, CNTs are distinguished as attractive carbon semiconductors, showcasing unique electrical traits such as high conductivity and electroluminescence, along with a substantial surface area and strong tensile strength (Pawar et al., 2015). Enhancing their photocatalytic activity is achievable through the synergistic coupling of CNTs with other catalytic materials. For example, a CNTs/CdS composite was produced through the solvothermal decomposition of a single-source molecular precursor in the presence of ethylene glycol (Pawar et al., 2016). While pure CdS can generate hydrogen at a rate of 14.5 $\mu mol \cdot h^{-1} g^{-1}$ under a xenon lamp with 0.1 M Na_2S and 0.05 M Na_2SO_3 as sacrificial agents, the CdS/CNTs composite demonstrates improved hydrogen photocatalytic activity, yielding 52 μmol/h/g under identical conditions (Ye et al., 2012). Iron oxide nanoparticles, recognized for their exceptional adsorption, electrocatalytic activity and biocompatibility can be combined with MWCNTs to form efficient photocatalysts for hydrogen synthesis (Nate et al., 2020). The photoelectrochemical efficacy of MWCNTs/Fe_2O_3 nanocomposites reaches 85 Lh/m^2. Nanocomposites comprising α-Fe_2O_3 and MWCNTs exhibit a noteworthy photoelectrochemical H_2 production rate of 85 Lh/m^2, showcasing the ability of MWCNTs to facilitate charge separation and transportation (Rai et al., 2015). Several other catalyst systems have demonstrated significant H_2 evolution rates: TiO_2/functionalized CNTs (Reddy et al., 2018), $Zn_{0.5}Cd_{0.5}S$-TiO_2/MWCNTs (Rout, 2011), ascorbic acid-functionalized group-CNT/CdS (Gopannagari et al., 2018), and CNTs/$ZnFe_2O_4$ (Dang et al., 2016) showcase rates of 2134 μmol/h/g, 21.9 μmol/h/g, 120.1 mmol/h/g, and 18 μmol/h/g, respectively. These improvements result from enhanced light absorption, improved charge separation, accelerated charge transfer, and more even distribution of the catalyst. Both $ZnFe_2O_4$ and CdS suffer from high recombination rates of charge carriers, which can be alleviated by incorporating CNTs. The photocatalytic activity of $ZnFe_2O_4$, diminished due to its high recombination rate for hydrogen and oxygen, is enhanced by doping the CNT/

TABLE 12.3 Materials with CNTs showing H_2 and O_2 production rate using sacrificial agent.

S. no.	Material	Light source	Sacrificial agent	Activity of H_2 production rate	Activity of O_2 production rate	References
1.	$CNT/ZnFe_2O_4$	Mercury lamp	Methanol	18.0 μmol/g/h	–	Dang et al. (2016)
2.	Co/NCNTs-800	Visible light	–	14.7 mmol/h/g	51%	Meng et al. (2018)
3.	ZnO/CNTs	UV-visible	Water-methanol (1:4)	1 μmol/h/g	–	Kemnade et al. (2015)
4.	$Zn_{0.5}Cd_{0.5}STiO_2$/Multiwalled CNTs	Solar light	–	21.9 mmol/h/g	–	Ng et al. (2017)
5.	CNT-GR-TiO_2	Sunlight irradiation (AM 1.5 G, 135 mW/cm^2)	–	29 mmol/h/g	–	Bellamkonda et al. (2019)
6.	Ascorbic acid/CNTs/CdS	Sunlight	–	120.1 mmol/h/g	–	Gopannagari et al. (2018)
7.	CNTs/CdS	Xe-lamp	Na_2S and Na_2SO_3	52 μmol/h/g	–	Ye et al. (2012)
8.	CuO/NiO/CNTs	Visible	Triethanolamine	1 mmol/h/g	–	Kang et al. (2012)
9.	MWCNTs oxide	Visible light		423.7 mol/h/g	–	Yang et al. (2015)
10.	TiO_2/FCNTs	Visible light	Glycerol	2134 mmol/h/g	–	M et al. (2015)
11.	Co-NCNT-800	Visible light	–	14.7 mmol/h/g	51%	Meng et al. (2018)

CNTs, Carbon nanotubes.

$ZnFe_2O_4$ surface with Pt. This improvement arises from the enhanced hydrogen production, driven by the reduction of H^+ ions on both the CNT and Pt-doped $ZnFe_2O_4$ surfaces (Dang et al., 2016). Fig. 12.5 presents the energy level diagram for the CNT/$ZnFe_2O_4$ composite.

Additionally, TiO_2, which has a restrained ability to absorb solar UV radiation, underwent modification with carbonaceous materials to improve solar energy absorption and mitigate electron-hole recombination. The incorporation of functionalized CNTs with titanium dioxide (TiO_2/FCNTs) resulted in a notable surge in the hydrogen production rate, reaching 2134 mmol/h/g under solar irradiation and employing glycerol as a sacrificial agent (Reddy et al., 2018), shown in Fig. 12.6.

Nevertheless, CuO/NiO photocatalysts synthesized hydrothermally with CNTs as structural support demonstrate active reaction sites and achieve H_2 production of ~1.0 mmol/h/g under visible light (Kang et al., 2012). CNT-cadmium sulfide (SF/CNTs/CdS) composites with surface functionalization show significantly improved photocatalytic hydrogen production, reaching a rate of 6.1 mmol/h/g under solar irradiation with platinum as a cocatalyst, surpassing the performance of pure CdS nanorods (Gopannagari et al., 2018). Moreover, ternary photocatalysts comprising cobalt (Co) coupled with nitrogen-doped carbon nanotubes (N-CNTs) were synthesized through the coheating of cobalt acetate and melamine at 800°C (Meng et al., 2018). The resulting composite, denoted as Co/N-CNTs, demonstrates a notable hydrogen production rate of 14.7 mmol/h/g. In addition, Meng et al. developed Co nanoparticles supported on nitrogen-doped carbon nanotubes (Co-NCNT-T, where T represents the pyrolysis temperature) using a simple pyrolysis method at different temperatures. This study (Meng et al., 2019) highlights the significant performance of Co-NCNT-800, demonstrating a notable HER rate of 14.7 mmol/h/g and an oxygen yield of 51%. Optimized conditions resulted in a turnover number of 25.8 and a turnover frequency of 0.86 s for the water oxidation process. This exceptional activity stems from efficient electron transfer between the catalyst and photosensitizer, as evidenced by photoluminescence (PL) spectra and transient photocurrent response experiments. Yuan et al., introduced Ni@NCNTs/NF-L, serving as a dual-function layer for

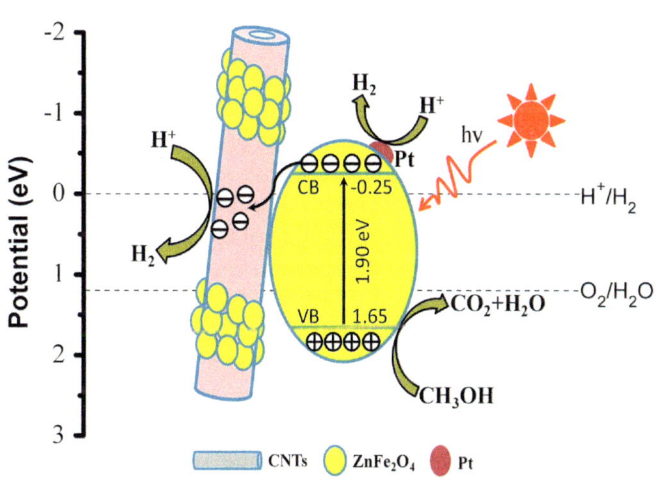

FIGURE 12.5 Diagram illustrating the energy levels of the CNTs/$ZnFe_2O_4$ hybrid (Dang et al., 2016). Source: *Reprinted with permission from Dang, H., et al., 2016. Hydrothermal preparation and characterization of nanostructured CNTs/$ZnFe_2O_4$ composites for solar water splitting application. Ceramics International 42 (8), 10520–10525. https://doi.org/10.1016/j.ceramint.2016.03.019, Copyright 2016, Elsevier.*

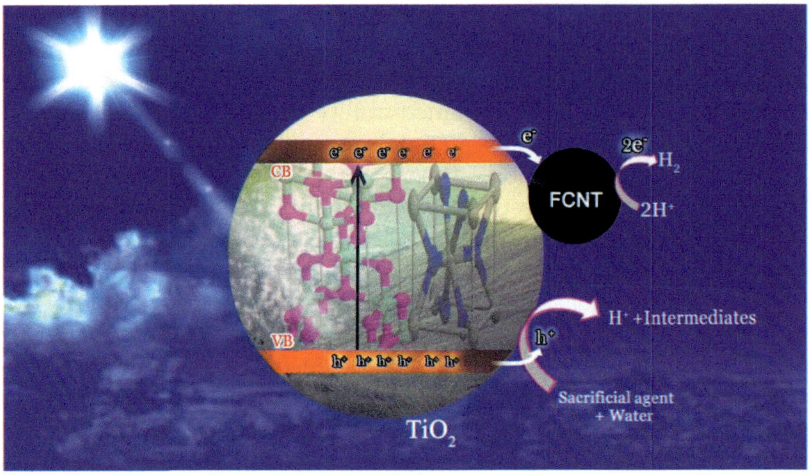

FIGURE 12.6 Illustrates the possible mechanism of hydrogen production in FCNTs/ TiO_2 nanocomposite photocatalysts (Reddy et al., 2018). Source: *Reprinted with permission from Reddy, N.R., et al., 2018. Multiwalled carbon nanotubes in titania based nanocomposite as trap for photoexcitons for enhanced photocatalytic hydrogen production under solar light irradiation, Materials Research Bulletin 106. https://doi.org/10.1016/j.materresbull.2018.06.009., Copyright 2018, Elsevier.*

photothermal conversion and efficient HER electrocatalysis, enabling integrated water-splitting. Fabricated via laser processing and calcination, Ni@NCNTs/NF-L incorporates Ni nanoparticles into NCNT. Additionally, NiFe-L, produced by laser ablation (Yuan et al., 2021), facilitates the OER electrode. Integration with a thermoelectric generator lowers the water-splitting voltage from 1.947 V to 1.213 V at a current density of 50 mA/cm^2 under AM 1.5 G illumination, enabling large-scale solar-powered water-splitting.

12.2.5 C_{60} or Fullerenes-based materials

C_{60}, also referred to as buckminsterfullerene or a buckyball, is a carbon allotrope that features a spherical structure with 60 carbon atoms arranged in nonplanar hexagonal and pentagonal rings. Each pentagon is surrounded by five hexagons and the hexagons are surrounded alternatively by three, five and six-membered rings possessing a curved structure like a soccer ball. C_{60} has found a potential spot as a catalyst and cocatalyst in water-splitting especially in electrocatalysis and photocatalysis due to its unique structural and electronic properties. Numerous reports have discussed the use of C_{60} for its structural benefits, to mention Fernandez-Delgado et al. studied the electrocatalytic properties of C60 with different dimensions by solvent engineering to produce 0D C_{60} molecules,1D C_{60} nanotubes and 2D C_{60} nanosheets (Fernandez-Delgado et al., 2020). HER measurements of C_{60} structures with different morphologies were carried out in acidic environments and was discovered that C_{60} nanotubes perform better for HER with an onset potential of −0.13 V compared to C_{60} nanosheets and C_{60} molecules with an onset potential of −0.21 V and −0.54 V, respectively, concluding that the mass transport properties of metal-free carbon catalysts depend directly on their dimensionality. The curved surfaces of these materials possess intense localized electric fields producing the tip effect thus lowering the Gibbs-free energy for hydrogen absorption at these sites. However, for OER the dimensionality didn't play a significant role due to the lack of tip effect for the absorption of oxygen molecules. The OER properties of C_{60} can be improved by composite creation. Gao et al. studied the electrocatalytic properties of C_{60}-SWCNT as a trifunctional catalyst for hydrogen evolution reaction (HER), oxygen evolution reaction (OER) and oxygen reduction reaction (ORR) and observed that C60-SWCNT$_{15}$ presented a lower Tafel slope of 46.7 mV/dec for OER compared to pristine C_{60} 50.5 mV/dec and SWCNTs 79.7 mV/dec attributed to the facile charge transfer from C_{60} to SWCNT enhancing their electrocatalytic activity (Gao et al., 2019), shown in Fig. 12.7.

Improving the photocatalytic properties of C_{60} has been achieved through dimensional reduction. Jones and Peng conducted a theoretical analysis of the photocatalytic behavior of one-dimensional fullerene polymers and two-dimensional fullerene networks using first-principles calculations. They found that carbon atoms in the one-dimensional polymer chains have twice as many energetically favorable adsorption sites as those in the two-dimensional networks, making them highly effective for catalyzing the HER (Jones and Peng, 2023). In another study, Song et al. studied the photocatalytic effects of C_{60}/graphene/g-C_3N_4 composite under visible light radiation (Song et al., 2017). Hydrogen evolution activities tested for various samples revealed that CGG-10 composite produces 5449.5 μmol/g hydrogen in 10 hours which is much higher compared to g-C_3N_4. Incorporating C_{60} and graphene-enhanced the photo-catalytic performance of CGG-10, attributed to C_{60} strong electron-withdrawing ability and graphene's efficient charge transport capabilities, which extended the lifespan of photogenerated charge carriers by reducing recombination.

C_{60} has not been single-handedly explored for photoelectrochemical water-splitting so far however, it is used for composite creation with other nanomaterials to improve their stability

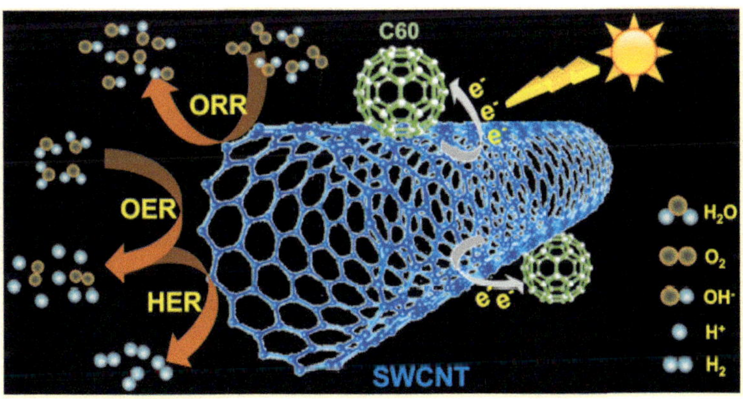

FIGURE 12.7 Diagrammatic representation of the charge transfer mechanism and ORR/OER/HER activities on C_{60}-SWCNTs (Gao et al., 2019). *HER*, Hydrogen evolution reaction; *OER*, oxygen evolution reaction. Source: *Reprinted with permission from Gao, R., et al., 2019. C_{60}-adsorbed single-walled carbon nanotubes as metal-free, pH-universal, and multifunctional catalysts for oxygen reduction, oxygen evolution, and hydrogen evolution. Journal of the American Chemical Society 141 (29), 11658–11666. https://doi.org/10.1021/jacs.9b05006, Copyright 2019, American Chemical Society.*

behaving as a photoelectron transfer promoter. Tan et al. used a layer of fullerene on top of the photo-absorbing layer of Sb_2Se_3 to improve the stability of the photocathode by preventing the electron accumulation over TiO_2 by rapidly transferring them to the Pt cocatalyst facilitating HER, improving the stability as shown in Fig. 12.7. Stability analysis carried out in strong acidic environments (pH 1) for $Pt/C_{60}/TiO_2/Sb_2Se_3$($PC_{60}TS$) and $Pt/TiO_2/Sb_2Se_3$(PTS) emphasized that $PC_{60}TS$ withstanding 98% of its initial photocurrent after 10 hours of testing compared to PTS that started degrading in performance after 1 hour of operation (Tan et al., 2019).

Similarly, in another work Wang et al. used a layer of C_{60} between the photo-absorbing layer of GeSe and TiO_2 protection layer to improve the stability of Pt/TiO_2 ball/GeSe photocathodes (Wang et al., 2021). Pt/C60 pat.-TiO_2 ball/GeSe was exceptionally stable for a period of 60 hours retaining 82.8% of its initial photocurrent density compared to Pt/TiO_2 ball/GeSe that retained 48.7% after 3.5 hours. This exceptional performance (Fig. 12.8) was attributed to the redistribution of the accumulated photoelectrons over the TiO_2 surface, by the C_{60} layer, through a fast electron transfer to the Pt cocatalyst layer thus reducing TiO_2 self-reduction. Chai et al. studied the photocatalytic properties of C_{60} decorated SWCNTs and its composite with TiO_2 where the composite C_{60}-d-CNTs/TiO_2 presented a higher H_2 evolution of 651 μmol/h arising from the efficient charge transfer properties of C_{60} and a stable photocurrent value of about 10 and 2.8 times greater than bare TiO_2 and SWCNTs (Chai et al., 2013). The details performance comparison of the different composites with C_{60} has been discussed in Table 12.4.

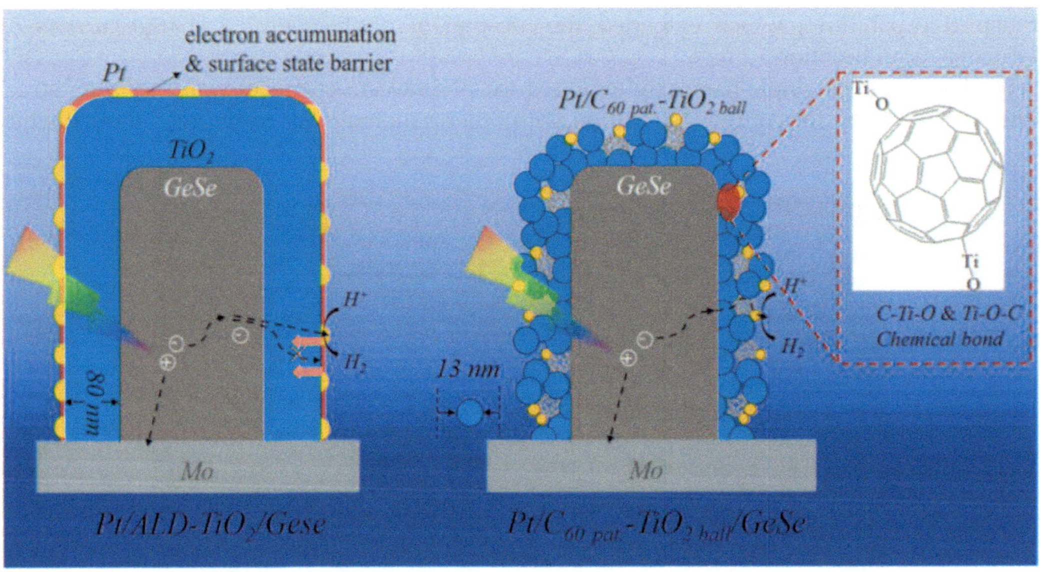

FIGURE 12.8 A schematic representation highlights the limited electron transfer efficiency of the Pt/ALD-TiO_2/GeSe electrode, contrasted with the enhanced carrier transport mechanisms observed in the Pt/C_{60}-patched TiO_2 ball/GeSe system (Wang et al., 2021). Source: *Reprinted with permission from Wang, K., et al., 2021. Efficient carrier transfer route via the bridge of C_{60} particle to TiO_2 nanoball based coverage layer enables stable and efficient cadmium free GeSe photocathode for solar hydrogen evolution. Applied Catalysis B: Environmental 297, 120437. https://doi.org/10.1016/j.apcatb.2021.120437, Copyright 2021, Elsevier.*

TABLE 12.4 Heterostructure formation with C_{60} involved in photocatalytic water-splitting.

S. no.	Sample	Sacrificial agent	H_2 production rate	References
1.	Few - layer C_{60}	No	91 μmol/g/h	Wang et al. (2023)
2.	C_{60}/g-C_3N_4 nanowire	TEOA	8.73 μmol/h	Long et al. (2020)
3.	Fe_3O_4@C_{60}@Pt	TEOA	29,157 μmol/g in 3 h	Zhang et al. (2022)
4.	H_2O_2/C_{60}–TiO_2	TEOA	41.6 mmol/ g/h	Wang et al. (2022b)
5.	10C C/P25	TEOA	597.2 mmol/g/h	Hou et al. (2020)
6.	C_{60}/CdS	Lactic acid	1.73 mmol/g/h	Cai et al. (2017)
7.	MoS_2-C_{60}	TEOA	6.89 μmol/g/h	Guan et al. (2018)
8.	g-C_3N_4/C_{60}	TEOA	266 μmol/g/h	Chen et al. (2017)
9.	C_{60}-d-CNTs/TiO_2	TEOA	651 μmol/h	Chai et al. (2013)
10.	CdS-F/G	–	127.2 μmol/h/g	Wang et al. (2022a)

12.2.6 Other carbon-based materials

Water-splitting properties of other carbon-based materials like onion-like carbons (OLCs), carbon black, activated carbon, carbon nanoplatelets and carbon quantum dots have also been studied due to their ability to act as semiconductors, photosensitisers, electron sponges and supporting materials. Mostly carbon-based materials were prevalently used to prevent charge carrier recombination by their effective separation and fast transfer during the photocatalytic water-splitting process. Wang et al. utilized carbon quantum dots implanted in graphitic carbon nitride nanotubes for photo-catalytic water-splitting with an H_2 production rate of 3538.3 mmol/h/g which is 2.4 times higher than pristine g-C_3N_4 (Wang et al., 2018). In another study, Han et al. explored the use of graphitic carbon nitride nanoplatelets for photocatalytic hydrogen production. The catalyst synthesized through the ball milling process, named Ultrathin graphitic carbon nitride nanoplatelets (UGCNPs) (Han et al., 2015), possessed a superior H_2 evolution rate of 1365 μmol/h/g, attributing to a 13.7-fold improvement in H_2 evolution rates, compared to bulk graphitic carbon nitride. This performance enhancement is attributed to the enhanced surface area, reduced thickness and a negative conduction band shift that is significant with exfoliation of bulk graphitic carbon nitride into Ultrathin graphitic carbon nitride nanoplatelets. Also, Sayed et al. investigated photocatalytic H_2 production in a hybrid of graphene nanoplates and self-doped TiO_2-Pd. In this work, the sample with 3 wt.% of graphene shows improved activity for hydrogen production using photons energy with a hydrogen evolution rate of 288 μmol/h/g which is attributed to improved light absorption promoted by graphene and Ti^{3+} dopant, enhanced electron trapping due to graphene that improves the charge carriers lifetime and increased number of active sites provided by Pd (Sayed et al., 2014). The scope of carbon-based materials for water-splitting is not only limited to the above-reported materials.

Sekar et al. studied the electrocatalytic properties of biomass-derived activated carbon on WO_3 Nanoflakes producing an overpotential of 320 mV at 10 mA/cm^2 for OER and 360 mV at 10 mA/cm^2 in 1 M KOH owing to the synergistic effect from both activated carbon with superior electrical conductivity and high surface area of WO_3 nanoflake (Sekar et al., 2020). In another work, Lv et al. studied the electrocatalytic properties of nitrogen-doped graphene quantum dots and Ni_3S_2 nanocomposite (Lv et al., 2017). The catalyst grown on Ni foam delivers an overpotential of 218 mV at 10 mA/cm^2 for HER and 216 mV at 10 mA/cm^2 for oxygen evolution reaction in alkaline media serving as a bi-functional catalyst. Through theoretical studies, the reason for the outstanding performance of the catalyst was found to be attributed to the introduction of nitrogen-doped graphene quantum dots (NGQDs) to the Ni_3S_2 resulting in a change in the electronic structure of Ni_3S_2 giving rise to active edge sites and improving the electrical conductivity and catalytic performance.

Shi et al. studied the photo electrocatalytic properties of OLC with controllable hybridization of carbon atom incorporated $BiVO_4$ where the OLCs were used to improve the charge carrier separation (Shi et al., 2023). BiVO4/OLC photoanode with a 40% of sp^2-carbon content exhibits a photocurrent density of 4.8 mA/cm^2 with a bulk charge separation efficiency of 91% compared to pristine $BiVO_4$ with 2.848 mA/cm^2 (1.23 V vs reversible hydrogen electrode (RHE)) current density with 56% efficiency implying that charge carrier recombination was suppressed by the addition of OLCs. Carbon-based catalysts have substantially gained humungous importance as catalysts/cocatalysts for water-splitting because of their low toxicity, high electrical conductivity, and facile structural tunability.

12.3 Case study

The study focuses on elucidating the photocurrent degradation mechanism of TiO_2 coated Sb_2Se_3 photocathode and simultaneously proposes an interface modification methodology using C_{60} as a photoelectron transfer promoter to enhance the stability. Photo-corrosion is one of the major problems hampering the long-term stability of photo-absorbing semiconductor in aqueous environments. Numerous strategies have been proposed to reduce the photo-corrosion of the photocathode and some of them include the use of a TiO_2 passivating layer to physically separate the semiconducting cathode from the electrolyte, also the use of film-type RuO_x or MoS_x cocatalysts on top of the TiO_2 layer significantly enhances the overall stability of photocathodes. However, these previous approaches only focused on physical separation to prevent the absorbing layer from being exposed to the electrolyte, whereas nearly no attention was paid to the deterioration of the TiO_2 protecting layer itself when in direct contact with corrosive electrolytes. Hence in this study C_{60} has been used as an electron transfer promotor to improve the electron transfer from TiO_2 to Pt preventing its dissolution and improving the stability.

12.3.1 Title

Fullerene as a Photoelectron Transfer Promoter Enabling Stable TiO_2-Protected Sb_2Se_3 Photocathodes for Photo-Electrochemical Water-Splitting

12.3.2 Objective

The objective of this case study is to eliminate the photo-corrosion of photocathodes in the presence of aqueous environments for improving the stability of the photocathode which is prime to prevent the degradation of performance efficiency over time. Specifically, it aims in elucidating the mechanism of TiO_2 degradation in the presence of light and proposing strategies to prevent its degradation which includes the use of a C_{60} photoelectron transfer promotor increasing the electron transfer efficiency from TiO_2 to Pt suggesting that interface engineering plays a pivotal role in supressing the degradation of photocathodes directly reflecting in the stability.

12.3.3 Scope

This case study aims to reduce the photodegradation of TiO_2 using interface engineering using C_{60} to reduce the electron accumulation over the TiO_2 surface preventing its reduction. However, the project has limitation with industrial scalability due to lab scale experiments and use of toxic materials like Se hampering the scaling. The workflow involves fabrication $Pt/TiO_2/Sb_2Se_3$ and $Pt/C_{60}/TiO_2/Sb_2Se_3$ followed by routine and electrochemical characterization of the two configurations to understand the photodegradation mechanism of TiO_2.

12.3.4 Audience

This case study explores the photodegradation mechanism of the TiO_2 passivation layer on photocathodes used in photoelectrochemical water-splitting, offering important insights into interface engineering and the detailed processes of passivation layer degradation in aqueous electrolytes. The findings are valuable to researchers in materials science and nanotechnology who are focused on water-splitting technologies for H_2 production. It also provides essential guidance for engineers and scientists developing photocathodic systems and passivation strategies for hydrogen production. Graduate students and postdoctoral researchers relying on photo electrocatalysis will benefit from the methods described for minimizing TiO_2 dissolution and enhancing photocathode stability. Beyond academia, industry professionals in the renewable energy sector focused on hydrogen production can apply these insights to design more cost-effective and reliable hydrogen generation systems.

12.3.5 Rationale

This research focuses on improving the stability of photocathodes under aqueous electrolytic conditions. Proposed solutions include the application of TiO_2 passivation layers combined with thin film cocatalysts such as RuOx or MoSx to enhance durability. Additionally, the study explores the use of electron transfer promoters to prevent the accumulation of photogenerated electrons on the TiO_2 surface, thereby mitigating self-reduction and ensuring long-term stability.

12.3.6 Expected results and deliverables

This case study includes strategies to prevent photocurrent degradation of TiO_2 over the photo-absorbing material for its stable operation in harsh aqueous electrolytic environments. The goal is to develop stable and reliable photocathodes for H_2 production using photo-electrochemical water-splitting approach which is crucial for addressing the environmental challenges and renewable energy deployment.

12.3.7 Safety considerations

To ensure optimal operational conditions, it is critical to handle nanomaterials with care, following strict safety protocols. Additionally, personnel involved in the synthesis and handling of nanomaterials should be thoroughly trained and informed about the properties, hazards, and environmental impacts of the precursor materials being used. This includes understanding the reactivity, toxicity, and potential risks associated with the chemicals to ensure safe handling and proper waste disposal. Special attention must also be given to the scalability of the developed cathode material for practical applications. Factors such as the material's abundance, environmental impact, and toxicity play a significant role in determining its feasibility for large-scale adoption.

12.3.8 Actions taken/workflow/tools used/simulations and analyses

This study investigates the photodegradation mechanism of the passivation layer on photocathodes. The process begins with the synthesis of an Sb_2Se_3 light-absorbing cathode, followed by the immediate deposition of a TiO_2 layer using atomic layer deposition. This is followed by C_{60} spin coating and a platinum (Pt) cocatalyst is then sputtered onto the fabricated $C_{60}/TiO_2/Sb_2Se_3$ electrode using an Auto Sputter Coater. Subsequently, electrical contact with the electrode is established using a copper wire and silver paste. To protect the nonessential regions of the electrode, an epoxy resin is applied as a sealant. Various tools are used to study the electron transfer kinetics at the interface and subsequently interpreting the results to understand the mechanism of photocurrent degradation.

12.3.9 Challenges and solutions

A significant challenge in scalability lies in achieving long-term stability without severe photocorrosion of the light-absorbing semiconductor under electrochemical reduction conditions. An effective approach to mitigate this issue is the use of a TiO_2 protection strategy, which prevents photocathode corrosion by isolating the semiconductor from the electrolyte. While TiO_2 protection enhances the stability of photocathodes compared to their unprotected counterparts, gradual photocurrent degradation remains unavoidable. The incorporation of film-type RuO_x or MoS_x cocatalysts on the TiO_2 layer has been shown to significantly improve the overall stability of photocathodes. Additionally, a novel interface modification methodology introduces fullerene (C_{60}) as a photoelectron transfer promoter in TiO_2-protected photocathode systems, such as $Pt/C_{60}/TiO_2/Sb_2Se_3$, further enhancing their performance and stability.

12.3.10 Results

This study presents a novel interface-modification methodology that significantly enhances the photo-electrochemical stability of TiO_2-protected Sb_2Se_3 photocathodes by introducing a C_{60} photoelectron transfer promoter at the Pt/TiO_2 interface. The findings reveal that the accumulation of photoelectrons on the TiO_2 surface induces reductive dissolution and the formation of trap states, which ultimately result in photo-current degradation. This research provides a comprehensive understanding of the degradation mechanisms affecting TiO_2-protected photocathodes and highlights the potential of C_{60} as a promising and versatile photoelectron transfer promoter. Additionally, it offers valuable insights into developing durable photocathodes for photo-electrochemical water-splitting.

12.3.11 Learning and knowledge outcomes

Interface engineering and the incorporation of various materials at the interface effectively mitigate charge carrier accumulation over the passivation layer, preventing photocurrent degradation. Future research should focus on developing novel light-absorbing materials, passivation layers, and electron transfer promoters that are both abundant and nontoxic, ensuring compatibility with industrial scalability. Adopting a systematic approach to integrate large-scale fabrication techniques with existing technologies is essential for advancing practical applications.

12.3.12 Conclusion (best propeies of carbon materials)

To address the continuously growing global energy demand and counteract the escalating concentrations of greenhouse gases, photocatalytic green fuel generation using earth-abundant, affordable, and highly active materials is adopted that renders the usage of fossil fuels. A family of carbon-based catalysts is fabricated as an alternative to noble metals. This chapter provides an overview of the advancements made in recent years in the research and development of carbon-based photocatalysts in water-splitting due to their enthralling properties such as huge surface area, variable geometrical configuration due to wide variety of allotropes, porous, inert, non-toxic, tuning of band structure, and excellent mobility of charge carriers. Several carbon materials such as Graphene, GOGraphene oxide, CNTs, Fullerenes, and other carbon materials have been investigated in this chapter and highlighted their contribution to narrowing down the band gap of the photocatalysts, their proficiency in accepting and transporting electrons and harnessing the maximum energy within the solar spectrum, ultimately serve as an effective co-catalyst in photo-assisted water-splitting. Additionally, carbon materials' performance is enhanced through doping, the incorporation of functional groups, the creation of hybrids with other semiconductors, and interface engineering. These approaches aim to achieve higher quantum yields in the production of green hydrogen. Hence, carbon-based materials play a crucial role in actively engaging in photocatalysis to enhance hydrogen efficiency, meeting the practical standards required for commercialization.

References

Aragaw, B.A., 2020. Reduced graphene oxide-intercalated graphene oxide nano-hybrid for enhanced photoelectrochemical water reduction. Journal of Nanostructure in Chemistry 10 (1), 9–18. Available at https://doi.org/10.1007/s40097-019-00324-x.

Atrees, M.S., et al., 2021. Synergetic effect of metal-doped GO and TiO_2 on enhancing visible-light-driven photocatalytic hydrogen production from water splitting. Energy Sources, Part A: Recovery, Utilization, and Environmental Effects 43 (4), 484–494. Available at https://doi.org/10.1080/15567036.2019.1629130.

Bellamkonda, S., et al., 2019. Highly active and stable multi-walled carbon nanotubes-graphene-TiO_2 nanohybrid: an efficient non-noble metal photocatalyst for water splitting. Catalysis Today 321–322, 120–127. Available at https://doi.org/10.1016/j.cattod.2017.10.023.

Bonke, S.A., et al., 2015. Renewable fuels from concentrated solar power: towards practical artificial photosynthesis. Energy & Environmental Science 8 (9), 2791–2796. Available at https://doi.org/10.1039/C5EE02214B.

Cai, Q., et al., 2017. Fullerene (C_{60})/CdS nanocomposite with enhanced photocatalytic activity and stability. Applied Surface Science 403, 151–158. Available at https://doi.org/10.1016/j.apsusc.2017.01.135.

Chai, B., et al., 2013. Synthesis of C_{60} -decorated SWCNTs (C_{60}-d-CNTs) and its TiO_2 -based nanocomposite with enhanced photocatalytic activity for hydrogen production. Dalton Transactions 42 (10), 3402–3409. Available at https://doi.org/10.1039/C2DT32458J 2003.

Chen, C., et al., 2013. Revealing the enhanced catalytic activity of nitrogen-doped carbon nanotubes for oxidative dehydrogenation of propane. Chemical Communications 49 (74), 8151–8153. Available at https://doi.org/10.1039/C3CC41500G.

Chen, X., et al., 2017. A facile mechanochemical route to a covalently bonded graphitic carbon nitride (g-C_3N_4) and fullerene hybrid toward enhanced visible light photocatalytic hydrogen production. Nanoscale 9 (17), 5615–5623. Available at https://doi.org/10.1039/C7NR01237C.

Chernyak, S.A., et al., 2016. Co catalysts supported on oxidized CNTs: evolution of structure during preparation, reduction and catalytic test in Fischer-Tropsch synthesis. Applied Catalysis A: General 523, 221–229. Available at https://doi.org/10.1016/j.apcata.2016.06.012.

Cho, H., et al., 2021. Enhanced photocatalytic activity of TiO_2 nanotubes decorated with erbium and reduced graphene oxide. Applied Surface Science 565, 150459. Available at https://doi.org/10.1016/j.apsusc.2021.150459.

Dang, H., et al., 2016. Hydrothermal preparation and characterization of nanostructured CNTs/$ZnFe_2O_4$ composites for solar water splitting application. Ceramics International 42 (8), 10520–10525. Available at https://doi.org/10.1016/j.ceramint.2016.03.019.

Daulbayev, C., Sultanov, F., et al., 2021. Bio-waste-derived few-layered graphene/$SrTiO_3$/PAN as efficient photocatalytic system for water splitting. Applied Surface Science 549, 149176. Available at https://doi.org/10.1016/j.apsusc.2021.149176.

Deng, J., et al., 2014. Highly active and durable non-precious-metal catalysts encapsulated in carbon nanotubes for hydrogen evolution reaction. 'Energy & Environmental Science 7 (6), 1919–1923. Available at https://doi.org/10.1039/C4EE00370E.

Dong, T., et al., 2019. Hierarchical nickel-cobalt phosphide hollow spheres embedded in P-doped reduced graphene oxide towards superior electrochemistry activity. Carbon 149, 222–233. Available at https://doi.org/10.1016/j.carbon.2019.04.050.

Elbakkay, M.H., et al., 2018. S-TiO_2/S-reduced graphene oxide for enhanced photoelectrochemical water splitting. Applied Surface Science 439, 1088–1102. Available at https://doi.org/10.1016/j.apsusc.2018.01.070.

Fang, W., et al., 2021. Metal-organic framework derived Fe-Co-CN/reduced graphene oxide for efficient HER and OER. Electrochimica Acta 365, 137384. Available at https://doi.org/10.1016/j.electacta.2020.137384.

Fernandez-Delgado, O., et al., 2020. Facile synthesis of C_{60} -nano materials and their application in high-performance water splitting electrocatalysis. Sustainable Energy & Fuels 4 (6), 2900–2906. Available at https://doi.org/10.1039/D0SE00399A.

Gao, R., et al., 2019. C_{60}-adsorbed single-walled carbon nanotubes as metal-free, pH-universal, and multifunctional catalysts for oxygen reduction, oxygen evolution, and hydrogen evolution. Journal of the American Chemical Society 141 (29), 11658–11666. Available at https://doi.org/10.1021/jacs.9b05006.

Gao, Y., et al., 2020. Atomic insights for Ag interstitial/substitutional doping into $ZnIn_2S_4$ nanoplates and intimate coupling with reduced graphene oxide for enhanced photocatalytic hydrogen production by water splitting. Applied Catalysis B: Environmental 279, 119403. Available at https://doi.org/10.1016/j.apcatb.2020.119403.

Georgakilas, V., et al., 2015. Broad family of carbon nanoallotropes: classification, chemistry, and applications of fullerenes, carbon dots, nanotubes, graphene, nanodiamonds, and combined superstructures. Chemical Reviews 115 (11), 4744–4822. Available at https://doi.org/10.1021/cr500304f.

Gopannagari, M., Chaturvedi, H., 2015. Light induced aggregation of specific single walled carbon nanotubes. Nanoscale 7 (40), 16590–16596. Available at https://doi.org/10.1039/C5NR03091A.

Gopannagari, M., et al., 2018. Influence of surface-functionalized multi-walled carbon nanotubes on CdS nanohybrids for effective photocatalytic hydrogen production. Applied Catalysis B: Environmental 236, 294–303. Available at https://doi.org/10.1016/j.apcatb.2018.05.009.

Gopi, S., et al., 2022. Bifunctional electrocatalysts for water splitting from a bimetallic (V doped-NixFey) metal–organic framework MOF@graphene oxide composite. International Journal of Hydrogen Energy 47 (100), 42122–42135. Available at https://doi.org/10.1016/j.ijhydene.2021.05.028.

Guan, J., et al., 2018. Hybridizing MoS_2 and C_{60} via a van der Waals heterostructure toward synergistically enhanced visible light photocatalytic hydrogen production activity. International Journal of Hydrogen Energy 43 (18), 8698–8706. Available at https://doi.org/10.1016/j.ijhydene.2018.03.148.

Gurbani, N., Chouhan, N., 2023. P–N heterojunction system Eu-doped ZnO@GO for photocatalytic water splitting. Global Challenges 7 (4), 2200106. Available at https://doi.org/10.1002/gch2.202200106.

Han, Q., et al., 2015. Facile production of ultrathin graphitic carbon nitride nanoplatelets for efficient visible-light water splitting. Nano Research 8 (5), 1718–1728. Available at https://doi.org/10.1007/s12274-014-0675-9.

Heo, J., et al., 2023. Surface engineering of Cu_2O photocathodes via facile graphene oxide decoration for improved photoelectrochemical water splitting. ACS Omega 8 (36), 32794–32803. Available at https://doi.org/10.1021/acsomega.3c03585.

Hou, J., et al., 2020. A mild and simple method to fabricate commercial TiO_2 (P25) and C_{60} composite for highly enhancing H_2 generation. International Journal of Hydrogen Energy 45 (4), 2852–2861. Available at https://doi.org/10.1016/j.ijhydene.2019.11.180.

Jayaramulu, K., et al., 2017. Nanoporous nitrogen-doped graphene oxide/nickel sulfide composite sheets derived from a metal-organic framework as an efficient electrocatalyst for hydrogen and oxygen evolution. Advanced Functional Materials 27 (33), 1700451. Available at https://doi.org/10.1002/adfm.201700451.

Jia, X., et al., 2020. CoNx/NiFeOx/nitrogen-doping reduced graphene oxide nanocomposite derived from layered double hydroxide precursor as an efficient bifunctional electrocatalyst for oxygen electrocatalytic reactions. Ionics 26 (4), 1885–1894. Available at https://doi.org/10.1007/s11581-020-03465-0.

Jiao, L., Zhou, Y.-X., Jiang, H.-L., 2016. Metal–organic framework-based CoP/reduced graphene oxide: high-performance bifunctional electrocatalyst for overall water splitting. Chemical Science 7 (3), 1690–1695. Available at https://doi.org/10.1039/C5SC04425A.

Jones, C., Peng, B., 2023. Boosting photocatalytic water splitting of polymeric C_{60} by reduced dimensionality from two-dimensional monolayer to one-dimensional chain. The Journal of Physical Chemistry Letters 14 (51), 11768–11773. Available at https://doi.org/10.1021/acs.jpclett.3c02578.

Kang, S.-Z., et al., 2012. Composite photocatalyst containing Eosin Y and multiwalled carbon nanotubes loaded with CuO/NiO: mixed metal oxide as an active center of H_2 evolution from water. Applied Surface Science 258 (16), 6029–6033. Available at https://doi.org/10.1016/j.apsusc.2012.02.118.

Kemnade, N., et al., 2015. Non-destructive functionalisation for atomic layer deposition of metal oxides on carbon nanotubes: effect of linking agents and defects. Nanoscale 7 (7), 3028–3034. Available at https://doi.org/10.1039/C4NR04615C.

Khare, R., Bose, S., 2005. Carbon nanotube based composites a review. Journal of Minerals and Materials Characterization and Engineering 04. Available at https://doi.org/10.4236/jmmce.2005.41004

Kharissova, O.V., Kharisov, B.I., 2014. Variations of interlayer spacing in carbon nanotubes. RSC Advances 4 (58), 30807–30815. Available at https://doi.org/10.1039/C4RA04201H.

Kingston, C., et al., 2014. Release characteristics of selected carbon nanotube polymer composites. Carbon 68, 33–57. Available at https://doi.org/10.1016/j.carbon.2013.11.042.

Kumar, C., et al., 2023. Petroleum coke derived reduced graphene oxide as an electrocatalyst for water splitting. Diamond and Related Materials 140, 110433. Available at https://doi.org/10.1016/j.diamond.2023.110433.

Long, D., et al., 2020. Synergetic effect of C_{60}/g-C_3 N_4 nanowire composites for enhanced photocatalytic H_2 evolution under visible light irradiation. ChemCatChem 12 (7), 2022–2031. Available at https://doi.org/10.1002/cctc.201901958.

Lv, J.-J., et al., 2017. Incorporating nitrogen-doped graphene quantum dots and Ni_3S_2 nanosheets: a synergistic electrocatalyst with highly enhanced activity for overall water splitting. Small 13 (24), 1700264. Available at https://doi.org/10.1002/smll.201700264.

M, M., et al., 2015. Nanohybrid of titania/carbon nanotubes-nanohorns: a promising photocatalyst for enhanced hydrogen production under solar irradiation. International Journal of Hydrogen Energy 40, 1665–1674. Available at https://doi.org/10.1016/j.ijhydene.2014.11.117.

Ma, L., et al., 2017. Fabrication and water treatment application of carbon nanotubes (CNTs)-based composite membranes: a review. Membranes 7 (1), 16. Available at https://doi.org/10.3390/membranes7010016.

Ma, J., et al., 2020. Fabrication of PANI-TiO_2/rGO hybrid composites for enhanced photocatalysis of pollutant removal and hydrogen production. Renewable Energy 156, 1008–1018. Available at https://doi.org/10.1016/j.renene.2020.04.104.

Machín, A., et al., 2020. Biomimetic catalysts based on Au@ZnO–graphene composites for the generation of hydrogen by water splitting. Biomimetics 5 (3), 39. Available at https://doi.org/10.3390/biomimetics5030039.

Mateo, D., et al., 2019. N-doped defective graphene decorated by strontium titanate as efficient photocatalyst for overall water splitting. Applied Catalysis B: Environmental 252, 111–119. Available at https://doi.org/10.1016/j.apcatb.2019.04.011.

Melchakova, I.A., et al., 2024. Pentagonal symmetry in aperiodic cyclic multiply twinned nano- and mesodiamonds. The Journal of Physical Chemistry A 128 (40), 8591–8614. Available at https://doi.org/10.1021/acs.jpca.4c02792.

Meng, X., et al., 2018. Co nanoparticles decorated with nitrogen doped carbon nanotubes for boosting photocatalytic water splitting. ACS Sustainable Chemistry & Engineering 7. Available at https://doi.org/10.1021/acssuschemeng.8b05632.

Meng, X., et al., 2019. Co nanoparticles decorated with nitrogen doped carbon nanotubes for boosting photocatalytic water splitting. ACS Sustainable Chemistry & Engineering 7 (1), 1753–1759. Available at https://doi.org/10.1021/acssuschemeng.8b05632.

Mierczynski, P., et al., 2016. MWCNTs as a catalyst in oxy-steam reforming of methanol. RSC Advances 6 (84), 81408–81413. Available at https://doi.org/10.1039/C6RA15618E.

Narwade, S.S., et al., 2020. Graphene oxide decorated with Rh nanospheres for electrocatalytic water splitting. ACS Applied Nano Materials 3 (12), 12288–12296. Available at https://doi.org/10.1021/acsanm.0c02762.

Nate, Z., et al., 2020. A simple in-situ flame synthesis of nanocomposite (MWCNTs-Fe_2O_3) for electrochemical sensing of proguanil in pharmaceutical formulation. Diamond and Related Materials 111. Available at https://doi.org/10.1016/j.diamond.2020.108178.

Ng, B.-J., et al., 2017. All-solid-state Z-scheme photocatalyst with carbon nanotubes as an electron mediator for hydrogen evolution under simulated solar light. Chemical Engineering Journal 316, 41–49. Available at https://doi.org/10.1016/j.cej.2017.01.054.

Pahuja, M., et al., 2022. Se-incorporated porous carbon/Ni_5P_4 nanostructures for electrocatalytic hydrogen evolution reaction with waste heat management. ACS Applied Nano Materials 5 (1), 1385–1396. Available at https://doi.org/10.1021/acsanm.1c03943.

Pan, Z., Zhang, G., Wang, X., 2019. Polymeric carbon nitride/reduced graphene oxide/Fe_2O_3: All-solid-state Z-scheme system for photocatalytic overall water splitting. Angewandte Chemie International Edition 58 (21), 7102–7106. Available at https://doi.org/10.1002/anie.201902634.

Pawar, R.C., et al., 2015. Gold nanoparticle modified graphitic carbon nitride/multi-walled carbon nanotube (g-C_3N_4/CNTs/Au) hybrid photocatalysts for effective water splitting and degradation. RSC Advances 5 (31), 24281–24292. Available at https://doi.org/10.1039/C4RA15560B.

Pawar, A.S., Garje, S.S., Revaprasadu, N., 2016. Synthesis and characterization of CdS nanocrystallites and OMWCNT-supported cadmium sulfide composite and their photocatalytic activity under visible light irradiation. Materials Chemistry and Physics 183, 366–374. Available at https://doi.org/10.1016/j.matchemphys.2016.08.040.

Qiao, S., et al., 2020. Core-shell cobalt particles Co@CoO loaded on nitrogen-doped graphene for photocatalytic water-splitting. International Journal of Hydrogen Energy 45 (3), 1629–1639. Available at https://doi.org/10.1016/j.ijhydene.2019.10.157.

Rahmandoust, M., Ayatollahi, M.R., 2016. Characterization of Carbon Nanotube Based Composites under Consideration of Defects 39. Available at https://doi.org/10.1007/978-3-319-00251-4.

Rai, S., et al., 2015. Photoactivity of MWCNTs modified α-Fe_2O_3 photoelectrode towards efficient solar water splitting. Renewable Energy 83. Available at https://doi.org/10.1016/j.renene.2015.04.053.

Reddy, N.R., et al., 2018. Multiwalled carbon nanotubes in titania based nanocomposite as trap for photoexcitons for enhanced photocatalytic hydrogen production under solar light irradiation. Materials Research Bulletin 106. Available at https://doi.org/10.1016/j.materresbull.2018.06.009.

Robertson, J., 1986. Amorphous carbon. Advances in Physics 35 (4), 317–374. Available at https://doi.org/10.1080/00018738600101911.

Rout, U.K., 2011. Prospects of India's energy and emissions for a long time frame. Energy Policy 39 (9), 5647–5663.

Samajdar, S., et al., 2023. Exploration of 1D-2D $LaFeO_3$/RGO S-scheme heterojunction for photocatalytic water splitting. International Journal of Hydrogen Energy 48 (47), 17838–17851. Available at https://doi.org/10.1016/j.ijhydene.2023.01.271.

Sarkar, A., et al., 2023. Enhanced photocatalytic Hydrogen generation by splitting water using sodium alginate decorated rGO-CdS hybrid photo-catalyst. Materials Today: Proceedings 76, A1–A8. Available at https://doi.org/10.1016/j.matpr.2023.02.095.

Sayed, F.N., et al., 2014. Photocatalytic hydrogen generation from water using a hybrid of graphene nanoplatelets and self doped TiO_2–Pd. RSC Advances 4 (26), 13469–13476. Available at https://doi.org/10.1039/C3RA47974A.

Sekar, S., et al., 2020. Enhanced water splitting performance of biomass activated carbon-anchored WO_3 nanoflakes. Applied Surface Science 508, 145127. Available at https://doi.org/10.1016/j.apsusc.2019.145127.

Shaheer, A.R.M., et al., 2021. Reduced graphene oxide as a solid-state mediator in $TiO_2/In_{0.5}WO_3$ S-scheme photocatalyst for hydrogen production. Solar Energy 213, 260–270. Available at https://doi.org/10.1016/j.solener.2020.11.030.

Sharmin, F., Chandra Roy, D., Basith, M.A., 2021. Photocatalytic water splitting ability of Fe/MgO-rGO nanocomposites towards hydrogen evolution. International Journal of Hydrogen Energy 46 (77), 38232–38246. Available at https://doi.org/10.1016/j.ijhydene.2021.09.072.

Shi, Y., et al., 2023. Onion-like carbon with controllable carbon atom hybridization embedded in $BiVO_4$ photoanode for enhanced photoelectrochemical water splitting. Applied Surface Science 614, 156120. Available at https://doi.org/10.1016/j.apsusc.2022.156120.

Song, L., et al., 2017. C_{60}/graphene/g-C_3N_4 composite photocatalyst and mutually- reinforcing synergy to improve hydrogen production in splitting water under visible light radiation. Ceramics International 43 (10), 7901–7907. Available at https://doi.org/10.1016/j.ceramint.2017.03.115.

Su, Y., et al., 2014. Cobalt nanoparticles embedded in N-doped carbon as an efficient bifunctional electrocatalyst for oxygen reduction and evolution reactions. Nanoscale 6 (24), 15080–15089. Available at https://doi.org/10.1039/C4NR04357J.

Sundmacher, K., et al., 2008. Beyond Oil and Gas: The Methanol Economy.

Tan, J., et al., 2019. Fullerene as a photoelectron transfer promoter enabling stable TiO_2-protected Sb_2Se_3 photocathodes for photo-electrochemical water splitting. Advanced Energy Materials 9 (16), 1900179. Available at https://doi.org/10.1002/aenm.201900179.

Turner, J.A., 2004. Sustainable hydrogen production. Science 305 (5686), 972–974. Available at https://doi.org/10.1126/science.1103197.

Wan, W., et al., 2019. Transition metal electrocatalysts encapsulated into N-doped carbon nanotubes on reduced graphene oxide nanosheets: efficient water splitting through synergistic effects. Journal of Materials Chemistry A 7 (25), 15145–15155. Available at https://doi.org/10.1039/C9TA03213D.

Wang, P., et al., 2017. The fundamental role and mechanism of reduced graphene oxide in rGO/Pt-TiO_2 nanocomposite for high-performance photocatalytic water splitting. Applied Catalysis B: Environmental 207, 335–346. Available at https://doi.org/10.1016/j.apcatb.2017.02.031.

Wang, Y., et al., 2018. Carbon quantum dot implanted graphite carbon nitride nanotubes: excellent charge separation and enhanced photocatalytic hydrogen evolution. Angewandte Chemie 130 (20), 5867–5873. Available at https://doi.org/10.1002/ange.201802014.

Wang, K., et al., 2021. Efficient carrier transfer route via the bridge of C_{60} particle to TiO_2 nanoball based coverage layer enables stable and efficient cadmium free GeSe photocathode for solar hydrogen evolution. Applied Catalysis B: Environmental 297, 120437. Available at https://doi.org/10.1016/j.apcatb.2021.120437.

Wang, W., et al., 2022a. Fullerene–graphene acceptor drives ultrafast carrier dynamics for sustainable CdS photocatalytic hydrogen evolution. Advanced Functional Materials 32 (23), 2201357. Available at https://doi.org/10.1002/adfm.202201357.

Wang, Y.-Q., et al., 2022b. H_2O_2-assisted preparation of microspherical C60–TiO_2 photocatalyst and its hydrogen evolution mechanism. International Journal of Hydrogen Energy 47 (28), 13503–13514. Available at https://doi.org/10.1016/j.ijhydene.2022.02.101.

Wang, T., et al., 2023. Few-layer fullerene network for photocatalytic pure water splitting into H_2 and H_2O_2. Angewandte Chemie International Edition 62 (40), e202311352. Available at https://doi.org/10.1002/anie.202311352.

Yang, P., et al., 2015. Pure carbon nanodots for excellent photocatalytic hydrogen generation. RSC Advances 5 (27), 21332–21335. Available at https://doi.org/10.1039/C5RA01924A.

Ye, A., et al., 2012. CdS–graphene and CdS–CNT nanocomposites as visible-light photocatalysts for hydrogen evolution and organic dye degradation. Catalysis Science & Technology 2 (5), 969–978. Available at https://doi.org/10.1039/C2CY20027A.

Yeh, T., et al., 2010. Graphite oxide as a photocatalyst for hydrogen production from water. Advanced Functional Materials 20 (14), 2255–2262. Available at https://doi.org/10.1002/adfm.201000274.

Yeh, T.-F., et al., 2011. Graphite oxide with different oxygenated levels for hydrogen and oxygen production from water under illumination: the band positions of graphite oxide. The Journal of Physical Chemistry C 115 (45), 22587–22597. Available at https://doi.org/10.1021/jp204856c.

You, B., Sun, Y., 2018. Innovative strategies for electrocatalytic water splitting. Accounts of Chemical Research 51 (7), 1571–1580. Available at https://doi.org/10.1021/acs.accounts.8b00002.

Yu, J.M., et al., 2020. High-performance and stable photoelectrochemical water splitting cell with organic-photoactive-layer-based photoanode. Nature Communications 11 (1), 5509. Available at https://doi.org/10.1038/s41467-020-19329-0.

Yuan, H., et al., 2021. Laser patterned and bifunctional Ni@*N*-doped carbon nanotubes as electrocatalyst and photothermal conversion layer for water splitting driven by thermoelectric device. Applied Catalysis B: Environmental 283, 119647. Available at https://doi.org/10.1016/j.apcatb.2020.119647.

Zeng, M., Li, Y., 2015. Recent advances in heterogeneous electrocatalysts for the hydrogen evolution reaction. Journal of Materials Chemistry A 3 (29), 14942–14962. Available at https://doi.org/10.1039/C5TA02974K.

Zhang, X.-Y., et al., 2010. Graphene/TiO_2 nanocomposites: synthesis, characterization and application in hydrogen evolution from water photocatalytic splitting. Journal of Materials Chemistry 20 (14), 2801. Available at https://doi.org/10.1039/b917240h.

Zhang, W., et al., 2022. Enhancing photocatalytic activity of C_{60}-based photocatalyst for visible-light-driven hydrogen evolution with the coupling of Fe_3O_4 microbead to modulate charge separation efficiency. Catalysis Communications 171, 106514. Available at https://doi.org/10.1016/j.catcom.2022.106514.

Zhang, S., et al., 2023. Visible-light-driven GO/Rh-$SrTiO_3$ photocatalyst for efficient overall water splitting. Catalysts 13 (5), 851. Available at https://doi.org/10.3390/catal13050851.

Zhao, X., et al., 2020. Cobalt phosphide-embedded reduced graphene oxide as a bifunctional catalyst for overall water splitting. ACS Omega 5 (12), 6516–6522. Available at https://doi.org/10.1021/acsomega.9b04143.

CHAPTER

13

Metal organic frameworks water splitting

Olugbenga Akande[1], *and Jude. A. Okolie*[2]

[1]Department of Computer Science and Electrical Engineering, Handong Global University, Pohang, Gyeongsangbukdo, Republic of Korea [2]Engineering Pathways, Gallogly College of Engineering, University of Oklahoma, Norman, OK, United States

13.1 Introduction

Heavy metals are extensively utilized across industries, producing pollutants containing these substances, which present significant environmental hazards. For instance, manganese, lead, chromium, copper, nickel, cadmium, and mercury are particularly concerning due to their carcinogenic and mutagenic properties (Balali-Mood et al., 2021). These metals have widespread ecological implications, adversely affecting plants, animals, and human communities, thus highlighting pressing contemporary issues.

Metals like mercury, lead, nickel, cadmium, copper, and chromium, prominently listed on the Substance Priority List managed by the United States Agency for Toxic Substances and Disease Registry, are highlighted by Parker et al. (2023). The existence of several metallic elements, like chromium, copper, nickel, and lead, in water sources stems from the residues generated by diverse industries encompassing tanneries, smelters, metal plating, alloy production, mining operations, lead accumulators, and chemical manufacturing (Hedau et al., 2024). Regrettably, these waste byproducts are often discharged into rivers and tributaries without adequate treatment, resulting in profound harm to natural ecosystems.

The cumulative repercussions of these practices have inflicted substantial damage on ecological systems, underscoring the urgency of addressing and rectifying these harmful behaviors.

The utilization of heavy metals in diverse industrial applications offers significant advantages attributable to their unique properties. These materials, characterized by excellent resistance to corrosion, a low melting point, favorable malleability, and exceptional thermal and electrical conductivity, have led to their extensive global utilization and trade (Prasad et al., 2022).

The US Geological Survey reports that cadmium is used in many different applications: 83% of Ni-Cd batteries, 8% of pigments, 7% of varnishes and overlays, 1.2% of plastic stabilizers, and 0.8% of various other uses, including nonferrous alloys, electronic components, and solar equipment. Industrial fertilizers, waste burning, mining activities,

Photochemical Splitting of Water
DOI: https://doi.org/10.1016/B978-0-443-29064-0.00008-5

industrial effluent, and coal/oil combustion increase environmental pollution (Soltani et al., 2021; Akande et al., 2023; Okolie et al., 2023). Creating effective treatment methods is imperative to remove heavy metallic element ions from the effluent before they are released into water bodies nearby.

Adsorption, ion exchange, membrane separation, chemical precipitation, and flotation are techniques to eliminate heavy metal pollutants in wastewater. Nonetheless, these methods generally result in labor-intensive processes, sludge, and high operational expenses (Razzak et al., 2022; Fei and Hu, 2023). Cost and simplicity of execution are two factors that make choosing the best remediation technique difficult.

Among these approaches, adsorption is known to be a highly efficient and economical solution for extracting heavy metals from water (Gupta et al., 2021). Its effectiveness is attributed to its straightforwardness, simplicity, affordability, and efficiency. The adsorbent's functional surface areas and pore structure make it more effective physically and chemically interacting with heavy metal ions.

Additionally, adsorption allows for the recovery and reuse of specific adsorbents through appropriate desorption techniques. Extensive scientific research has demonstrated the capability of various adsorbent materials to absorb heavy metals (Soltani et al., 2021). Nevertheless, practical limitations related to costs and time restrict the widespread use of many adsorbents, driving ongoing research to find economical and valuable materials for heavy metal extraction from wastewater.

Cadmium materials are gaining prominence due to the development of compounds featuring high levels of sulfur, nitrogen, and oxygen, forming strong and durable interactions with heavy metal ions. Consequently, materials have been designed to emphasize their adsorption capacity rather than their processability (Hedau et al., 2024).

This chapter examines metal-organic frameworks (MOFs) for water-splitting, scrutinizing their effectiveness and potential. It delves into various water-splitting methods, including synthesis techniques and photocatalytic degradation; additionally, it explores the practical usage of MOFs in water-splitting processes, emphasizing their role and significance in this crucial area of research while also focusing on water sources containing heavy metals (Fig. 13.1) the chapter examines heavy metal extraction techniques and methodologies, highlighting the utilization of MOFs in this context.

13.2 Overview of different metal organic frameworks used in water splitting

The breakdown of water into hydrogen and oxygen via an appropriate catalyst is an up-and-coming method of sustainable hydrogen production. Due to their tunable porosities, vast surface area, and diverse metal centers, MOFs have recently attracted interest as novel materials for water-splitting. The following section discusses several types of MOFs developed for water-splitting applications with appropriate categorization while presenting their mechanisms and properties.

MOFs represent a crystalline material of metal ions synchronized using organic ligands toward establishing 1D, 2D, or 3D structures. The great versatility in their design extends to properties by designing MOFs for specific properties suitable for catalytic applications, among which is water-splitting.

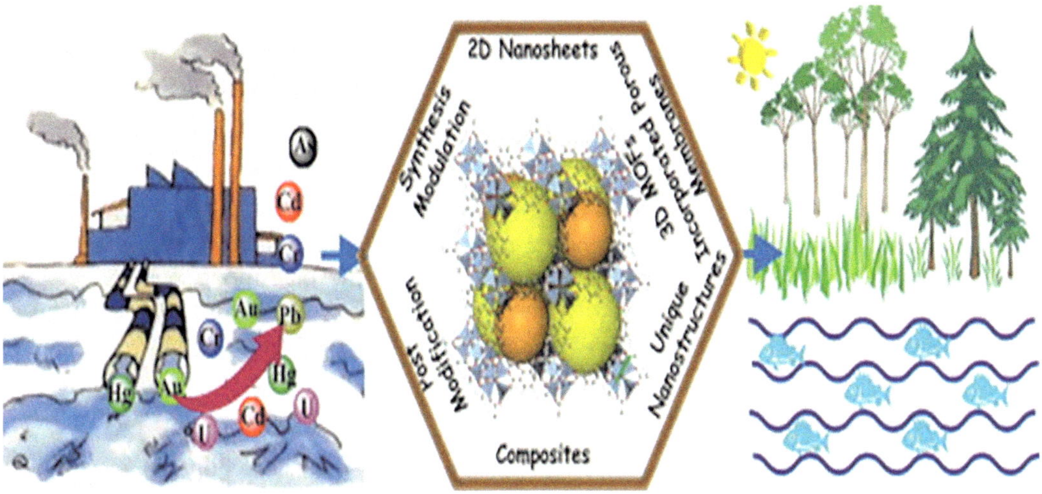

FIGURE 13.1 MOF water-splitting to extract heavy metals (Xu et al., 2021). *MOF*, Metal-organic framework.

13.2.1 Metal-organic framework as photocatalysts

A photocatalytic MOF utilizes light energy to promote the water-splitting reaction. Typically, they have organic linkers that facilitate charge transfer and metal cores that neutralize visible light. On light irradiation, the metal centers in the Photons are absorbed by MOF, resulting in pairs of electron-holes. The next step is for these charge carriers to reach the surface and participate in water oxidation and reduction reactions. For instance, the efficiency for water-splitting by photocatalysis using MIL-101(Cr) is very high, considering the substantial pore size and extensive surface area it offers to absorb more light and increase the charge separation efficiency (Luo et al., 2020).

13.2.2 Two-dimensional MOF (2D metal-organic frameworks)

The unique feature of this class of MOFs is their layered structures that lead to a large surface area and exceedingly fast charge transfer pathway, thus lending applications in photoelectrocatalysis. The hierarchically layered structure allows an efficient separation of charge carriers with their transportation after being generated due to light absorption. Substantial surface areas with many functional sites are used for the water-splitting reaction. A typical example is a 2D MOF-based on the Zn-TCPP tetrakis (4-carboxyphenyl) porphyrin with high photoelectrocatalytic activity, which is assigned to an effective charge separation and transfer ability at molecular interfaces (Ebrahimi and Krivosudský, 2022).

13.2.3 Nanoporous metal/covalent organic framework-based electrocatalysts

This makes nanoporous frameworks ideal candidates, which could have both high stability and tunable porosity for the electrocatalysis of water-splitting. The fabrication of many active sites in proximity, along with improved mass movement, is assisted by nanopores and contributes to increased catalytic activity in the water-splitting reaction. In this nanoporous structure, nickel MOF-74 exhibits outstanding catalytic activity in the HER; thus it harbors enormous potential in the electrocatalyst field (Cui et al., 2024).

13.2.4 Nickel phosphonate metal-organic framework

The nickel-based centers bear high catalytic activity, possessing stability in reactions of splitting water. These active sites are supposed to be the centers of nickel, whose effectiveness and conductivity of the MOF are enhanced with phosphonate groups. Among them, Ni-MOF-74 stands out due to its impressive reactivity and longevity in environments with a high pH, positioning it as a strong contender for practical water-splitting purposes (Zhang et al., 2020a,b).

13.2.5 Cobalt-based metal-organic framework

The cobalt-based MOFs show high activity in the O_2 evolution reaction (OER), a crucial step in the water-splitting process. However, these Co centers inside a MOF material generally provide the associated OER activity via the active sites for water adsorption and its breakdown. Co-MOF-71 has been well studied for its activity regarding the OER with high efficiency and stability under different conditions (Duan et al., 2022).

13.2.6 Metal doping in cerium metal-organic framework

Doping of other metals into cerium-based MOFs is one way to increase catalytic performance by introducing extra active sites while modifying the electronic characteristics. Thus the modification by metal doping enhances light absorption and charge separation, improving photocatalytic and electrocatalytic activities.

The synergistic effect of transition metals such as Fe or Mn doped into Ce-MOF has significantly enhanced the photocatalytic performance in water-splitting (Li et al., 2022).

13.2.7 Mixed metal-metal-organic framework

The MM-MOFs, featuring multiple types of metal centers in one framework, therefore show more synergistic catalytic effects that enhance water-splitting performance. More than one catalytic pathway can be facilitated due to the different metal centers in one system, thereby improving the efficiency of the water decomposition process.

Fe/Ni-MOF is a bimetallic example with superior activity in HER and OER due to the cooperative interaction of the iron and nickel centers (Chen et al., 2020). A summary of various MOFs used in water-splitting is shown in Table 13.1.

13.3 Advances in metal-organic framework photocatalysts for efficient water-splitting

One possible method for sustainable hydrogen production is to develop MOF-based photocatalysts for water-splitting. MOFs show good promise in this respect, as they exhibit a

TABLE 13.1 Classification and analysis of MOFs for water-splitting.

MOF type	Structure	Properties	Mechanism	References
Metal Organic frameworks as photocatalysts	3D structures	large surface area, visible light active	Photocatalytic water-splitting	Luo et al. (2020)
Two dimensional MOF (2D MOFs)	Layered structures	High surface area, fast charge transfer	Photoelectrocatalytic	Ebrahimi and Krivosudský (2022)
Nanoporous metal/covalent organic framework-based electrocatalysts	Nanoporous structures	High stability, tunable porosity	Electrocatalytic	Cui et al. (2024)
Nickel phosphonate MOF	3D structures with Ni centers	High catalytic activity, robust	Electrocatalytic	Zhang et al. (2020a),
Cobalt-based MOFs	3D structures with Co centers	High O_2 evolution reaction (OER) activity	Electrocatalytic	Duan et al. (2022)
Metal doping in cerium MOFs	Doped 3D structures	Enhanced catalytic activity	Photocatalytic/electrocatalytic	Li et al. (2022)
Mixed metal-MOF (MM-MOFs)	Mixed metal centers in 3D structures	Synergistic catalytic effects	Photocatalytic/electrocatalytic	Chen et al. (2020)

MOFs, Metal-organic frameworks.

variety of dimensionalities: two-dimensional (2D), nanoporous MM-MOFs, and so on. The building blocks of MOFs are organic ligands coupled with metal ions, which are then packed into a crystalline structure with pores.

This tunability allows the engineering of electronic and structural properties at will, which is crucial for efficient water-splitting. MOFs can absorb a broad light spectrum and afford charge separation and transfer—two critical parameters for decomposition (Luo et al., 2020).

The 2D characteristics of MOF configurations involve many unique aspects: very high surface area, ease of access, little thickness, absorption of light, and charge mobility. For example, a study of 2D cobalt MOF nanosheets shows a dramatic improvement in the efficiency of hydrogen evolution, specifically, an apparent quantum efficiency (AQE) of 10.2% under irradiation by visible light, in comparison with ordinary 3D MOFs that have an AQE of approximately 5%–7% (Mao et al., 2021).

Meanwhile, high surface areas and porosity in nanoporous MOFs and COFs enhance catalytic performance to offer more access to water molecules and higher diffusion rates. For example, the water-splitting efficiency of nanoporous nickel-based MOFs could reach approximately 12% when integrated into the configuration of a photoelectrochemical cell. This is a leap from nonporous MOF counterparts, generally in the 5%–8% efficiencies (Shah et al., 2024).

Nickel phosphonate MOFs are stable and exhibit good electrical conductivity. In water-splitting applications, nickel-based MOFs have presented an efficiency increase of 14%, attributed to the favorable binding with water molecules and intermediate species. Significantly higher efficiencies, around 8%–10%, are reported for traditional nickel catalysts (Taffa et al., 2022).

The cobalt-based MOFs serve as efficient photocatalysts for both the HER and OER. Optimized cobalt MOFs have reached a water-splitting efficiency of about 15%, whereas standard cobalt catalysts usually get the range of 9%–12%. This has been attributed to the interaction between cobalt clusters and organic ligands in MOF-based systems (Fan and Tahir, 2022).

Doping of cerium MOFs with different metals increases their photocatalytic properties. For instance, doping them with iron increases the efficiency to 16%, while undoped cerium MOFs usually achieve 10%–12%. Metal doping creates new active sites of the material, enhancing the light absorption process, which is very important in water-splitting (Chen et al., 2023).

MM-MOFs incorporate multiple kinds of metal ions into a single framework and take the synergistic effect from different metals to have optimal catalytic sites to achieve improved overall catalytic activity. MM-MOFs have shown water-splitting efficiencies of up to 18%, which is remarkably higher than their single-metal-MOF counterparts that generally attain only 10%–13% (Kharissova et al., 2024).

Water-splitting efficiency via MOFs as photocatalysts is significantly enhanced by their structural versatility, tunability, and high surface area. 2D, nanoporous, and MM-MOFs offer unique advantages that improve light absorption, charge separation, and catalytic activity. Specific examples like nickel phosphonate MOFs, cobalt-based MOFs, and metal-doped cerium MOFs demonstrate the potential of these materials to revolutionize hydrogen production through efficient and sustainable water-splitting. Table 13.2 summarizes the relative water-splitting efficiency of several MOFs. The comparative efficiencies highlight

TABLE 13.2 Comparative efficiencies of MOFs in water-splitting.

Type of MOF	Efficiency	Citation
2D MOFs	Up to 10.2%	Mao et al. (2021)
Nanoporous MOFs	Around 12%	Shah et al. (2024)
Nickel phosphonate MOFs	Approximately 14%	Taffa et al. (2022)
Cobalt-based MOFs	About 15%	Fan and Tahir (2022)
Metal-doped cerium MOFs	Up to 16%	Chen et al. (2023)
Mixed metal-MOF (MM-MOFs)	Up to 18%	Kharissova et al. (2024)

MOFs, Metal-organic frameworks.

the promising future of MOFs in achieving higher water-splitting performance and advancing renewable energy technologies.

13.3.1 Exploring the catalytic potential and synthesis strategies of metal-organic framework

Extensive scientific research has been dedicated to MOFs because of their diverse uses in catalysis and water-splitting (Li et al., 2023). The crucial process of water-splitting, which releases hydrogen and oxygen gasses from water molecules, has potential applications in producing sustainable fuels (Engge et al., 2021). Their unique permeability characterizes MOFs due to large surface areas. They comprise metal ions or aggregates joined by organic ligands, particularly water-splitting, which are highly beneficial for catalytic applications (Verma et al., 2023).

Modifying MOFs by incorporating catalytically active sites, such as metal centers, into their structure introduces catalysts essential for water-splitting. The porous structure of MOFs allows reactants easy access to these catalytic sites, speeding up the water-splitting reaction. Also, the large area of MOFs' surface offers additional effective sites for catalytic activities, boosting efficiency (Soltani et al., 2021).

Some MOFs can absorb and then exploit light in the visible or ultraviolet range, making them appropriate for photocatalytic water-splitting (Wang et al., 2017). This mechanism drives water-splitting by producing a pair of electrons and holes, which participate in the redox processes due to absorbed light energy. MOFs must be stable in water-splitting environments, which involve contact with water, high temperatures, and harsh chemical conditions (Samia et al., 2024).

Effective water-splitting requires the simultaneous transfer of electrons and protons. MOFs can be engineered to enhance electron and proton transfer activities, vital for overall redox processes (Zhu et al., 2018). Specific modifications can optimize the H_2 Evolution Reaction (HER) or the OER. For example, adjustments to specific metal centers within MOFs can efficiently catalyze proton conversion in HER. In contrast, other metal centers may be optimized for the OER's conversion of water into oxygen. These modification techniques are highly effective (Yaqoob et al., 2021). Fig. 13.2 shows a Nickel Cobalt Organic Framework with high activity.

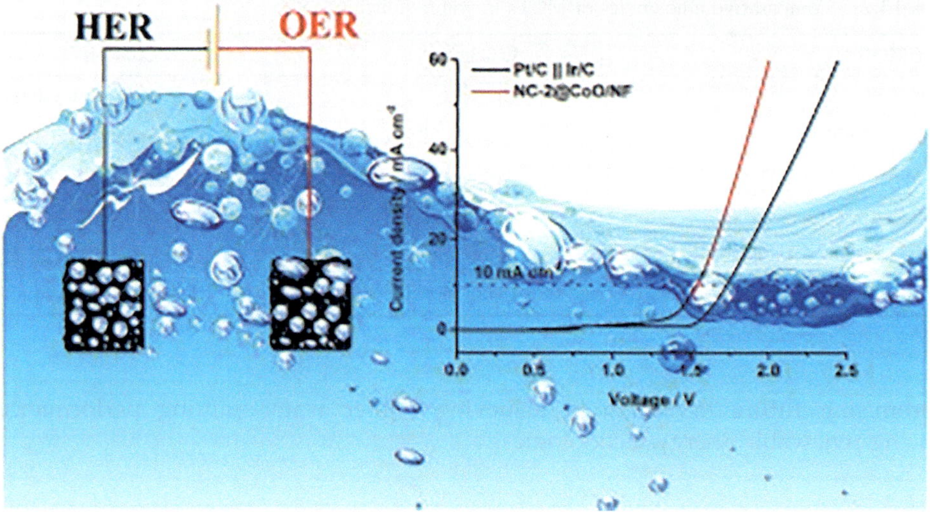

FIGURE 13.2 Highly active nickel cobalt organic framework (Van Phuc et al., 2022). *OER*, O_2 evolution reaction; *HER*, H_2 evolution reaction.

Researchers are now concentrating on creating and producing MOFs with increased catalytic capability, stability, and efficiency for water-splitting. MOFs, or unique hybrid crystalline nanoporous materials, have garnered much attention in several scientific and technological domains during the last ten years (Wang and Sun, 2021). Through reticular synthesis, MOFs may be carefully changed to improve their structural characteristics, leading to distinctive structures, many surface groups with functional properties, and enhanced textural qualities (Yaghi, 2016).

A detailed examination of the physicochemical properties of MOFs enables these modifications. Recent studies have demonstrated that MOFs are excellent surfaces for absorbing contaminants, especially heavy metals. With well-ordered crystalline structures, MOFs consist of networks of metal ions (such as Cu^{2+}, Mg^{2+}, Zr^{2+}, Zn^{2+}, Co^{2+}, Fe^{3+}, Cd^{2+}, Cr^{2+}, and Al^{3+}) or groups of metal atoms serving as fundamental building blocks (Soltani et al., 2021). The metal ions within MOFs can form diverse geometric structures, increasing the material's versatility. These metals also bind to various sites within organic groups, including amines, carboxylates, tetrazoles, and sulfonates (Xie et al., 2020).

MOFs come in various geometric forms, including cubic, trigonal bipyramidal, square planar, octahedral, tetrahedral, T-/Y-shaped, and linear. MOFs are perfect for applications that include medication administration, water-splitting, chemical detection, gas storage and separation, and extraction of heavy metals. Due to their extensive surface areas, they offer significant advantages, distinct architectures, porosity, and variable functionalities (Rocío-Bautista et al., 2019). Common acronyms for MOFs include HKUST-1 (Hong Kong University of Science and Technology), Iso Reticular Metal-Organic Frameworks, Materials of Institute Lavoisier, Zeolitic Imidazolate Framework, and UiO MOF (University of Oslo).

The synthesis of MOFs can be influenced by several factors, including temperature, solvent composition, final size and structure, organic ligands, metal ions, counterions, and crystallization kinetics (Zhu et al., 2018). These elements have an impact on crystal growth and nucleation. Usually, liquid phase solutions comprising metal salts and ligands are combined to create MOFs. Selecting the appropriate solvent is crucial, significantly impacting the reaction's equilibrium conditions and activation energy. This choice requires careful consideration of the solvent's reactivity, solubility, and redox potential. Solid-state synthetic methods are sometimes used despite the difficulty of creating single crystals. Typically, the gradual evaporation of the reaction solution is how MOF crystals are produced (Wang and Sun, 2021).

The primary method for producing MOFs involves solvothermal techniques at high pressures and temperatures, although conventional synthesis methods are also widely used. Recent advances in material synthesis have introduced new mechanochemical, electrochemical, microwave, and sonochemical methods renowned for their exceptional efficiency, cost-effectiveness, and ability to produce exceptionally pure products (Fareza et al., 2022). Detailed methodologies for MOF synthesis are well-documented in numerous articles and reviews, so they will not be discussed further here (Raptopoulou, 2021; Zhang et al., 2023).

MOFs are widely used because of their exceptional efficacy in absorbing heavy metals, ease of application, and effectiveness in extracting heavy metal ions. To boost the competitiveness of MOF powder in wastewater treatment, scientists are continually working to diversify the materials used in the adsorption process of MOF powder. This section explores the functions of MOFs and MOF membranes in extracting heavy metal ions from water (Ibrahim et al., 2021). Table 13.3 enumerates various MOF synthesis methods tailored to specific reaction conditions.

13.3.2 Extracting heavy metals via metal-organic framework-assisted water-splitting

Since their surface areas are large and have adaptable pore sizes, MOFs are highly effective for the adsorption of heavy metal ions. Interactions between MOFs and heavy metal ions, facilitated by metal nodes or clusters, significantly enhance their adsorption capacity (Jeoung et al., 2020). The selectivity of MOFs is modified by altering their organic ligands or metal centers to target specific heavy metal ions.

Functionalizing MOFs with specific groups enhances their affinity for heavy metal ions. This process, achieved by adding extra binding sites or improving the MOF's stability, can significantly boost its performance in water (An et al., 2023). By modifying organic ligands, scientists can tailor the functionality of MOFs to capture specific heavy metal ions preferentially without interference from others.

MOFs may be recycled and utilized again for many heavy metal adsorption cycles. This involves removing the accumulated metal ions from MOF, allowing the material to be recycled without a considerable reduction in its adsorption capability (Wang et al., 2023).

Certain MOFs also possess photocatalytic capabilities, which, when exposed to light, can be harnessed to either convert or breakdown heavy metal ions. Combining light-initiated reactions with the adsorption process could enhance the removal or transformation of heavy metals, potentially complementing other water treatment methods like membrane filtration or precipitation for more effective heavy metal removal (Wang et al., 2017).

While MOFs show potential for removing heavy metals, addressing water stability and scalability concerns for practical applications is critical. Fig. 13.3 presents a conceptual

TABLE 13.3 Categorization of MOF synthesis methods by reaction conditions (Ma et al., 2023).

MOFs categories	Synthesis method	Mechanism	Adsorption capacity	Target heavy metal	Condition	Regeneration	References
UiO-66-NH_2	Solvothermal	Photoreduction	198.7 mg/g	Pb(II)	STP	4 cycles	Huang et al. (2022a)
Cu-BTC	Hydrothermal	Photoreduction	167.2 mg/g	Hg(II)	STP	5 cycles	Vishnyakov et al. (2003)
Cd-MOFs	Microwave-assisted	Photoreduction	71.4 mg/g	Cr(VI)	STP	3 cycles	Li et al. (2015)
MIL-53	Ultrasonication	Photo-oxidation	92.6 mg/g	Methyl orange dye	25°C	6 cycles	Cui et al. (2012)
Zn-MOFs	Solvothermal	Photo-oxidation	248.7 mg/g	Rhodamine B dye	STP	4 cycles	Nguyen (2021)
NH_2-MIL-125	Solvothermal	Photoreduction	175.4 mg/g	Ag(I)	30°C	5 cycles	Furukawa et al. (2013)
ZIF-8	STP synthesis	Photo-oxidation	104.7 mg/g	Methylene blue dye	25°C	3 cycles	Gatou et al. (2023)
Cu-BTC/G.O.	Hydrothermal	Photoreduction	152.6 mg/g	Cd(II)	STP	4 cycles	Lan et al. (2019)
Fe-MIL-101	Solvothermal	Photoreduction	198.4 mg/g	Cr(VI)	25°C	3 cycles	Mondloch et al. (2014)
UiO-66	Microwave-assisted	Photo-oxidation	167.9 mg/g	Orange II dye	STP	5 cycles	Wang et al. (2011)

MOFs, Metal-organic frameworks.

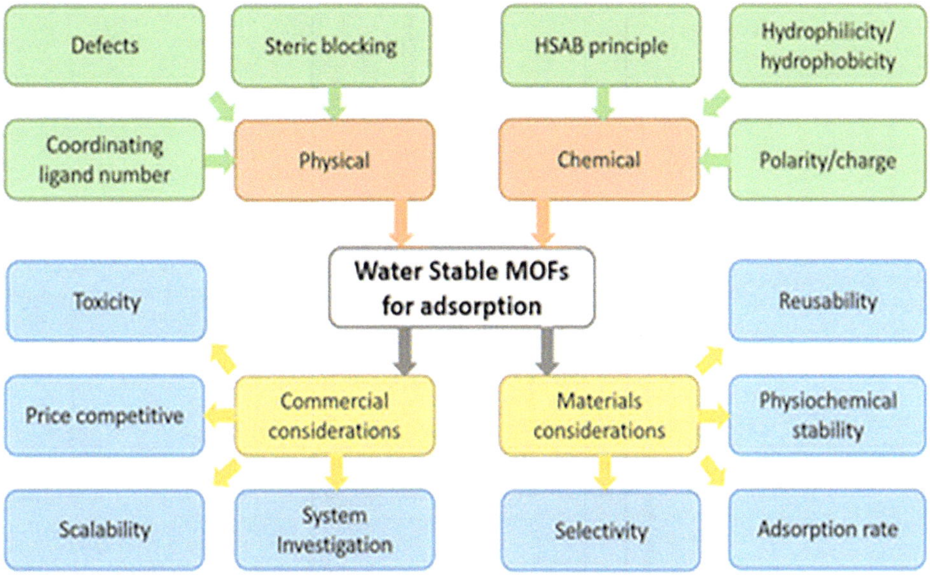

FIGURE 13.3 An illustration of the theoretical foundations of water-stable MOFs for metal adsorption in aqueous systems (Zadehahmadi et al., 2023).

framework outlining the design principles for achieving water-stable MOFs specifically tailored for metal adsorption from water. Current research is dedicated to developing new MOF materials, improving their characteristics, and devising efficient processes for treating large volumes of water. Additionally, considerations for the proper disposal or regeneration of used MOFs and the potential release of captured heavy metals are essential components of a comprehensive approach to water treatment (Xu et al., 2021). Table 13.4 presents the photocatalytic degradation of heavy metals through MOF-based materials.

Khalid et al. (2023) detailed the production of an aerosolized graphene oxide coating on NiCo-MOF framework on a Nimesh, significantly boosting its HER's electrocatalytic activity in alkaline conditions by enhancing the active surface area. This composite design facilitated efficient electron transfer within the material, improving catalytic efficiency. Additionally, nanoflakes coated with rGOAS further optimized HER performance, illustrating the synergistic effect between rGOAS and NiCo-MOF in enhancing electroconductivity and mass transfer to active sites.

In a complementary analysis, Jahan et al. (2013) utilized a solvothermal approach incorporating graphene oxide to develop a catalyst based on Cu-MOF for HER. Testing under acidic conditions demonstrated its superior performance, exceeding standards and reaching a potential of −0.209 V versus reference half cell electrode (RHE) at a current density of 30 mA cm^2, accompanied by a Tafel slope of 84 mV/dec

Wu et al. (2017) studied cobalt-based MOFs (CTGU-5, CTGU-6) synthesized through solvothermal methods, highlighting their stable frameworks and distinctive coordination patterns with H_2O molecules, which affect their catalytic properties for HER. In a separate investigation, Duan et al. (2017) examined FeNi-MOF nanosheet arrays in alkaline

TABLE 13.4 Degradation of heavy metals via photocatalysis with MOF-based materials (Ma et al., 2023).

MOFs/Material	Metals	Kinetic model	Rate expression	References
NH_2-MIL-125(Ti)	Cr(VI)	Langmuir–Hinshelwood	$r = k\theta C/(1 + KC)$	Largitte and Pasquier (2016)
Ag/ZIF-8	Hg(II)	Pseudo-first-order	$\ln(C0/C) = k't$	Fan et al. (2018)
UiO-66-NH_2	Pb(II)	Pseudo-second-order	$t/Qt = 1/k'Qe2 + t/Qe$	Wang et al. (2022)
MIL-53(Fe)	Cd(II)	Elovich	$qt = (1/\beta)\ln(\alpha\beta) + (1/\beta)\ln t$	Jin et al. (2023)
Zr-fum	Ni(II)	Intraparticle diffusion	$qt = k'\sqrt{t}$	Largitte and Pasquier (2016), Huang et al. (2022b)
Cysteine-MIL-101	Hg(II)	Chick–Watson	$\ln(N/N0) = -k'C't$	Largitte and Pasquier (2016), Huang et al. (2022b)
Cd-MOFs	*Escherichia coli*	Hom	$\ln(N/N0) = -k'C'ntm - 1$	Che et al. (2022)

MOFs, Metal-organic frameworks.

conditions, showing that iron-induced structural vacancies contributed to their sustained electrocatalytic activity and improved reactivity. Additionally, Chen et al. (2020) presented tungsten-modified MOF catalysts that exhibited low overpotential, rivaling Pt/C catalysts in alkaline environments, thereby emphasizing their potential for HER.

Zhu et al. (2018) explored MIL-101 as a chromium-based MOF enhanced with AuNi alloy nanoparticles for HER via chemocatalysis despite challenges with precursor mobility in its narrow pores. Concurrently, Cheng et al. (2018) investigated AuPdCo alloy nanoparticles on amine-modified MIL-101, showcasing impressive turnover frequencies in formic acid hydrolysis and highlighting the interaction between trimetallic nanoparticles and the MOF framework.

These studies utilized diverse analytical methods to characterize material properties and electrocatalytic behaviors, including Raman spectroscopy, microscopy, X-ray diffraction (XRD), and thermal analysis. Electrochemical assessments using rotating disk electrode (RDE) and glassy carbon electrode (GCE) setups provided insights into HER rates, validating the efficacy of modified MOF catalysts in various electrolytic environments. Subsequent applications of modified electrocatalyst inks processed through ultrasonic dispersion facilitated consistent electrode coverage and prepared surfaces for detailed electrochemical evaluations (Zadehahmadi et al., 2023). Fig. 13.4 illustrates the synthetic and postsynthetic modification processes crucial for optimizing MOF catalysts in electrocatalytic applications.

The HER experiment involved loading the Pd catalyst at a precise density of 0.0199 mg/cm^2. A newly developed 0.1 M potassium hydroxide (KOH) electrolyte was used under controlled conditions maintained by a water-bath thermostat. The experimental setup included room temperature rotation at 1600 rpm. A comprehensive examination of HER kinetics was performed using an overpotential of 0.2 V relative to the reversible RHE, with a frequency range from 100 kHz to 0.1 Hz. Electrochemical impedance spectroscopy provided further insights into the process. To ensure consistent potential measurements, the RHE was utilized, converting all potentials to the standard $E_{Ag|AgCl} + 0.197 + 0.059$ pH at a pH level of 12, as outlined by Ipadeola and Ozoemena (2020). Hence, this meticulous setup and choice of parameters aimed to precisely investigate the kinetics of the HER under controlled conditions, allowing for a comprehensive analysis of the electrochemical processes involved.

13.4 Types of water-splitting methods for extraction of heavy metals

The significant environmental risk posed by heavy metal pollutants in water necessitates their efficient removal, as these contaminants have the potential to adversely impact ecosystems and pose threats to human health (Carolin et al., 2023). Heavy metal pollutants in water remain a critical environmental concern, prompting the exploration of diverse methodologies for their elimination.

13.4.1 Chemical precipitation

Chemical precipitation involves introducing specific chemicals into the water, forming insoluble precipitates containing heavy metal ions. Lime, comprised of calcium hydroxide,

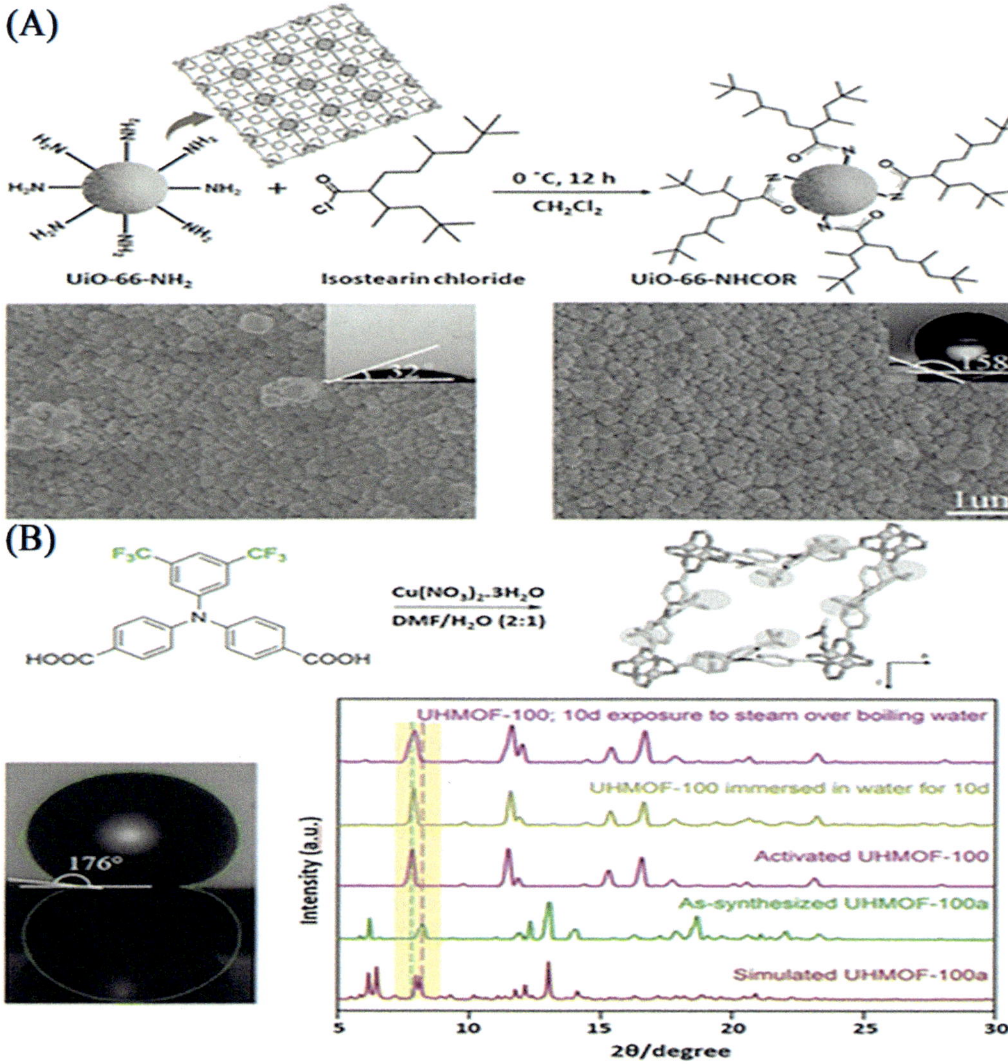

FIGURE 13.4 Synthetic and postsynthetic modification of MOF. (A) Isostearin chloride postsynthetic alteration of UiO-66-NH_2 to decrease hydrophilicity. (B) Producing an extremely hydrophobic MOF, UHMOF-100, by attaching trifluoromethyl groups to the ligand (Zadehahmadi et al., 2023). *MOF*, metal-organic framework.

sodium hydroxide, and sulfide-containing chemicals, is an example of a substance capable of inducing precipitation (Carolin et al., 2023).

13.4.2 Ion exchange

This process involves replacing heavy metal ions in water with ions attached to a solid exchange material. Zeolites, activated carbon, and certain resins are effective ion exchange materials that help extract heavy metal ions from water (Carolin et al., 2023).

13.4.3 Adsorption

Adsorption entails binding heavy metal ions onto a solid substance's surface, enabling the subsequent isolation of the adsorbent. Efficient adsorbents like activated carbon, clay minerals, and modified polymers commonly extract heavy metals from the environment (Carolin et al., 2023).

13.4.4 Membrane filtration

This method uses semipermeable membranes to isolate heavy metal ions based on size, charge, or other properties. Processes like reverse osmosis, ultrafiltration, and nanofiltration are extensively employed to extract heavy metals from water systems (Carolin et al., 2023).

13.4.5 Electrochemical methods

These methods employ an electric current to drive and concentrate heavy metal ions onto electrodes. Electrocoagulation, electroflocculation, and electrochemical precipitation remove heavy metals from water (Fareza et al., 2022).

13.4.6 Biological methods

This method utilizes bacteria, algae, and plants to absorb or accumulate heavy metals. Examples include phytoremediation and microbial remediation, where plants extract metals. Additionally, chelating agents such as ethylenediaminetetraacetic acid (EDTA) and citric acid form stable compounds, facilitating the removal of heavy metal ions (Fareza et al., 2022).

The specific heavy metals at play are one consideration when deciding on a procedure, as well as their concentrations and water quality standards (Barakat, 2011). Environmental concerns and proper management of recovered heavy metals are essential to prevent further pollution. Current research focuses on developing new, more efficient methods for heavy metal extraction from water (Ibrahim et al., 2021).

13.5 Extraction of heavy metals using hydrogen evolution reaction

Employing MOFs as catalysts for the HER process represents a promising avenue. These materials exhibit unique physicochemical characteristics that make them suitable for various applications, given their resilience in acidic and alkaline settings and the potential to enhance their effectiveness through simple modification methods (Chen and Xu, 2019). Nonetheless, despite these strengths, several practical challenges need to be overcome. Initially, catalysts derived from MOFs encountered stability issues when introduced into different working environments (Li et al., 2023), highlighting a critical barrier in their extensive use. To effectively leverage MOFs as catalysts, reliable electrocatalysts must be established, particularly those exhibiting stability in acidic and alkaline conditions.

Overcoming these challenges and establishing dependable electrocatalysts based on MOFs is crucial for effective utilization.

Once these difficulties are surmounted, MOFs can considerably improve the water-splitting process, thus enabling more sustainable hydrogen production; however, presently, hydrogen is produced mainly from fossil fuels (Akande and Lee, 2022; Smith et al., 2023; Tabat et al., 2023). The desirable characteristics of MOFs, including their unique physicochemical properties and capacity for modification, make them viable candidates for creating composites with significant surface areas and carbon support structures, essential requisites for effective catalysts in the HER process. Implementing these strategies could substantially improve hydrogen generation efficiency, thereby contributing to advancing environmentally friendly energy production.

13.6 Analyzing chemical bonding via raman spectroscopy

Using Raman spectroscopy, researchers have analyzed the vibrational characteristics of chemical bonds in various compounds such as MOF-derived carbon (MOFDC), Pd/MOFDC, Co/MOFDC, and Co-MOF (Ipadeola and Ozoemena, 2020). The existence of Co-O bonds in the spectra shows the successful synthesis of Co-MOF. Subsequent thermal gravimetric analysis demonstrated that Co-MOF undergoes decomposition and carbonization at higher temperatures, forming Co/MOFDC (Todaro et al., 2016; Akande et al., 2024).

Co/MOFDC displayed distinct vibrational modes associated with Co metal and D and G bands typical of MOFDC. Most Co nanoparticles were eliminated through chemical etching while adding Pd nanoparticles caused a slight rearrangement of the D and G bands. This change resulted in a higher ID/IG ratio, suggesting an increase in the presence of structural defects (Mofokeng et al., 2020). Understanding these changes at catalytic sites during the catalysis process necessitates comprehensive investigation.

Considering the overpotential required for the HER and other evaluation responses, finding a suitable electrocatalyst is crucial (Appel and Helm, 2014). Developing a functionalized carbon-based catalyst with enhanced resilience to acidic and alkaline environments, numerous active sites, and specific structural morphologies is recommended (Basyooni-M. Kabatas, 2023).

Recognizing solar energy as a sustainable resource, various types of solar cells are proposed to be combined with these HER catalysts distinctively to facilitate environmentally friendly hydrogen production. Investigating the factors influencing reaction rates and kinetics is crucial in advancing an effective electrocatalyst (Song et al., 2022). By aligning these catalysts with solar technologies, we can enhance their utility in renewable energy systems, fostering more efficient hydrogen production.

13.7 Case study

Rapid, Selective Heavy Metal Removal from Water via a MOF/Polydopamine Composite: A Case Study.

13.7.1 Objective

The case study aimed to assess the effectiveness of a MOF combined with polydopamine (Fe-BTC/PDA composite) in rapidly and selectively removing heavy metal contaminants, specifically lead (Pb^{2+}) and mercury (Hg^{2+}), from water. The goal was to create a cost-effective, water-stable material that can remove trace levels of these metals, providing an efficient solution to water contamination problems in real-world and experimental settings (Sun et al., 2018).

13.7.2 Scope

The project faced limitations in terms of scalability and resource availability. While the composite material was tested successfully in laboratory settings, real-world applications were confined to small-scale water samples. Additionally, the study was conducted within the budget and material constraints typical of academic research, limiting extensive field testing or commercialization (Sun et al., 2018).

13.7.3 Audience

This case study is intended for environmental engineers, materials scientists, chemists, policymakers, and researchers focused on water treatment and environmental remediation technologies. The collaborative research team involved scientists from multiple disciplines, including chemistry, materials science, and environmental engineering. It will also be useful to decision-makers in the water treatment industry (Sun et al., 2018).

13.7.4 Rationale

Heavy metal contamination, particularly lead and mercury, poses severe environmental and health risks. Traditional water purification methods struggle to remove these contaminants quickly and efficiently. This study addresses the need for rapid, selective, cost-effective heavy metal remediation techniques. The Fe-BTC/PDA composite was designed to offer an improved solution by leveraging MOFs' high adsorption capacity and catalytic properties (Sun et al., 2018).

13.7.5 Expected results and deliverables

The case study expected to demonstrate the high efficiency of the Fe-BTC/PDA composite in removing Pb^{2+} and Hg^{2+} from contaminated water, achieving over 99.8% removal within seconds. Deliverables included experimental data on removal rates, material characterization, and guidelines for the potential scale-up of this technology for field applications (Sun et al., 2018).

13.7.6 Safety considerations

The safe synthesis and handling of the Fe-BTC/PDA composite required careful consideration of chemical safety. Polydopamine synthesis and heavy metal exposure posed risks requiring proper waste disposal, ventilation, and personal protective equipment use. The material was tested to ensure stability, and no significant leaching of harmful components occurred during its use in water (Sun et al., 2018).

13.7.7 Actions taken/workflow/tools used/simulations and analyses

Synthesis of Fe-BTC/PDA composite: The MOF (Fe-BTC) was treated with dopamine to induce spontaneous polymerization, forming polydopamine (PDA) inside the MOF's pores. This combination resulted in a highly porous composite with increased heavy metal scavenging capabilities.

Characterization: Various tools were employed to analyze the structure and composition of the composite, such as XRD, to confirm the crystallinity of the composite. Scanning Electron Microscopy: To study surface morphology. X-ray Photoelectron Spectroscopy: To verify polymerization and interaction between PDA and Fe-BTC (Sun et al., 2018).

Adsorption and removal studies: The composite was tested in water samples containing Pb^{2+} and Hg^{2+}, achieving rapid removal. Concentration levels were measured using Inductively Coupled Plasma methods to confirm the material's effectiveness (Sun et al., 2018).

Simulation and modeling: Density functional theory was used to model the interaction between PDA and metal ions, confirming the solid binding affinities contributing to high removal rates (Sun et al., 2018).

13.7.8 Challenges and solutions

Challenges: Optimization of Polymer Loading; Different amounts of PDA were incorporated into the MOF to balance porosity and removal capacity. Excessive PDA could reduce porosity, compromising efficiency (Sun et al., 2018).

Scalability issues: While laboratory tests showed high effectiveness, scaling the technology for real-world applications posed challenges due to cost and material availability.

Solutions: Optimizing PDA loading resulted in a mass composite with 19% PDA, offering the best balance between porosity and heavy metal removal efficiency (Sun et al., 2018).

Collaboration with industrial partners is suggested for future scaling, exploring cost-effective production methods (Sun et al., 2018).

Results: The study showed that the Fe-BTC/PDA composite removed over 99.8% of Pb^{2+} and Hg^{2+} from contaminated water within seconds, reducing concentrations to safe drinking levels. The composite demonstrated remarkable selectivity, even in the presence of other ions (e.g., Na^{+}) at concentrations 14,000 times higher than Pb^{2+}. The material proved fully regenerable, maintaining efficiency over multiple cycles (Sun et al., 2018).

13.7.9 Learning and knowledge outcomes

The study provided several key insights: Material Design; The interaction between the PDA and the MOF was critical to the material's performance, offering a template for future MOF-polymer composites.

Selective removal: The composite showed excellent selectivity for Pb^{2+} and Hg^{2+} over other metal ions, making it highly suitable for real-world water treatment applications.

Scalability considerations: Future work should make the composite more cost-effective and scalable, including field trials and long-term stability studies.

13.7.10 Recommendations for future investigation

Investigate the long-term stability and cost-effectiveness of Fe-BTC/PDA in larger-scale applications.

Explore other MOF-polymer combinations to address different types of water contaminants. Collaborate with industrial stakeholders to scale-up the composite production for commercial use.

13.8 Conclusion

In summary, MOFs can be excellent multifunctional materials to catalyze water-splitting and, simultaneously, for environmental issues like separating heavy metals from aqueous environments. Their special structural flexibility, large surface area, and tunable characteristics promote them as suitable candidates for improving the HER proficiency and supporting the extraction of heavy metal pollutants through adsorption and photocatalysis processes. This chapter highlighted several ways MOFs may benefit sustainable energy production and environmental cleanup. These properties highlight their potential for industrial applications, including the capacity to mediate rapid electron and proton transfer and stability in various pH environments. Furthermore, the progress in analytical techniques provides insight into MOF structures, and their dynamic changes can be revealed rigorously during catalysis. The knowledge learned from these studies could guide MOF enhancement strategies to achieve higher performance and durability.

Despite these advancements, several challenges remain to be addressed, including MOF materials stability, the scalability of their synthesis methods, and the need for cost-effective production. Overcoming these challenges with new synthesis methods, functionalization strategies, and coupling with sustainable energy sources will be critical in making the most of MOFs at a technical level. Moving forward, future research will need to be centered on developing MOF catalysts with optimized activity, selectivity, and robustness in complex environmental matrices. Finally, it will be necessary to create MOF-based technologies that can be scaled for industrial water treatment and H_2 production if broad adoption and impact are to occur.

MOFs stand at the forefront of the material science world and promise game-changing solutions toward several SDGs related to clean energy and environmental well-being. We argue here that such progress will require further interdisciplinary research and collaboration to explore the potential of MOFs for meeting societal needs in water purification, energy sustainability, and beyond.

References

Akande, O., Lee, B., 2022. Plasma steam methane reforming (PSMR) using a microwave torch for commercial-scale distributed hydrogen production. International Journal of Hydrogen Energy 47, 2874–2884. Available at https://doi.org/10.1016/j.ijhydene.2021.10.258.

Akande, O., Lee, B., Okolie, J.A., Museba, H.N., 2023. Hydrogen-rich syngas generation through microwave plasma reforming of greenhouse gases. International Journal of Hydrogen Energy 48, 34649–34658. Available at https://doi.org/10.1016/j.ijhydene.2023.05.262.

Akande, O., Jimoh, T., Okoye, P.U., Okolie, J.A., 2024. Analytical Methods, Modeling Approaches and Challenges of Nanomaterial-Based Hydrogen Storage. CRC Press eBooks, pp. 166–181. Available at https://doi.org/10.1201/9781003371007-9.

An, Y., Lv, X., Jiang, W., Wang, L., Shi, Y., Hang, X., 2023. The stability of MOFs in aqueous solutions—research progress and prospects. Green Chemical Engineering. Available at https://doi.org/10.1016/j.gce.2023.07.004.

Appel, A.M., Helm, M.L., 2014. Determining the overpotential for a molecular electrocatalyst. ACS Catalysis 4, 630–633. Available at https://doi.org/10.1021/cs401013v.

Balali-Mood, M., Naseri, K., Tahergorabi, Z., Khazdair, M.R., Sadeghi, M., 2021. Toxic mechanisms of five heavy metals: mercury, lead, chromium, cadmium, and arsenic. Frontiers in Pharmacology 12, 643972. Available at https://doi.org/10.3389/fphar.2021.643972 -643972.

Barakat, M.A., 2011. New trends in removing heavy metals from industrial wastewater. Arabian Journal of Chemistry 4, 361–377. Available at https://doi.org/10.1016/j.arabjc.2010.07.019.

Carolin, C.F., Kamalesh, T., Kumar, P.S., Rangasamy, G., 2023. A critical review on the sustainable approaches for the removal of toxic heavy metals from water systems. Industrial & Engineering Chemistry Research 62, 8575–8601. Available at https://doi.org/10.1021/acs.iecr.3c00709.

Che, J., Chen, K., Song, J., Tu, Y., Reymick, O.O., Chen, X., 2022. Fabrication of γ-cyclodextrin-based metal-organic frameworks as a carrier of cinnamaldehyde and its application in fresh-cut cantaloupes. Current Research in Food Science 5, 2114–2124. Available at https://doi.org/10.1016/j.crfs.2022.10.025.

Chen, L., Xu, Q., 2019. Metal-organic framework composites for catalysis. Matter 1, 57–89. Available at https://doi.org/10.1016/j.matt.2019.05.018.

Chen, L., Wang, H.-F., Li, C., Xu, Q., 2020. Bimetallic metal–organic frameworks and their derivatives. Chemical Science 11, 5369–5403. Available at https://doi.org/10.1039/d0sc01432j.

Chen, J., Gao, H., Tao, Z., Wang, L., Li, R., Wang, G., 2023. Recent progress in mixed rare earth metal-organic frameworks: from synthesis to application. Coordination Chemistry Reviews 485, 215121. Available at https://doi.org/10.1016/j.ccr.2023.215121.

Cheng, M., Lai, C., Liu, Y., Zeng, G., Huang, D., Zhang, C., 2018. Metal-organic frameworks for highly efficient heterogeneous Fenton-like catalysis. Coordination Chemistry Reviews 368, 80–92. Available at https://doi.org/10.1016/j.ccr.2018.04.012.

Cui, Y., Xu, H., Yue, Y., Guo, Z., Yu, J., Chen, Z., 2012. A luminescent mixed-lanthanide metal-organic framework thermometer. Journal of the American Chemical Society 134, 3979–3982. Available at https://doi.org/10.1021/ja2108036.

Cui, M., Xu, B., Ding, Y., Sun, W., Liu, H., Dou, S., 2024. Deriving multi-metal nanomaterials on metal-organic framework platforms for oxygen electrocatalysis. Energy & Environmental Science. Available at https://doi.org/10.1039/d3ee04513g.

Duan, J., Chen, S., Zhao, C., 2017. Ultrathin metal-organic framework array for efficient electrocatalytic water splitting. Nature Communications 8, 15341. Available at https://doi.org/10.1038/ncomms15341 -15341.

Duan, D., Feng, J., Guo, D., Gao, J., Liu, S., Wang, Y., et al., 2022. MOF-derived cobalt manganese phosphide as highly efficient electrocatalysts for hydrogen evolution reaction. International Journal of Hydrogen Energy 47, 12927–12936. Available at https://doi.org/10.1016/j.ijhydene.2022.02.037.

Ebrahimi, A., Krivosudský, L., 2022. Metalloporphyrin metal-organic frameworks: eminent synthetic strategies and recent practical exploitations. Molecules/Molecules Online/Molecules Annual 27, 4917. Available at https://doi.org/10.3390/molecules27154917.

Engge, Y., Maulana, F., Nurhuda, M., Istiroyah, 2021. Dissociation of water into hydrogen and oxygen through a combination of electrolysis and photocatalyst. IOP Conference Series: Earth and Environmental Science 743. Available at https://doi.org/10.1088/1755-1315/743/1/012054.

Fan, G., Zheng, X., Luo, J., Peng, H., Lin, H., Bao, M., 2018. Rapid synthesis of Ag/AgCl@ZIF-8 as a highly efficient photocatalyst for degradation of acetaminophen under visible light. Chemical Engineering Journal 351, 782–790. Available at https://doi.org/10.1016/j.cej.2018.06.119.

Fan, W.K., Tahir, M., 2022. Recent advances on cobalt metal-organic frameworks (MOFs) for photocatalytic CO_2 reduction to renewable energy and fuels: a review on current progress and future directions. Energy Conversion and Management 253, 115180. Available at https://doi.org/10.1016/j.enconman.2021.115180.

Fareza, A.R., Nugroho, F.A.A., Abdi, F.F., Fauzia, V., 2022. Nanoscale metal oxides–2D materials heterostructures for photoelectrochemical water splitting—a review. Journal of Materials Chemistry A 10, 8656–8686. Available at https://doi.org/10.1039/d1ta10203f.

Fei, Y., Hu, Y.H., 2023. Recent progress in removal of heavy metals from wastewater: a comprehensive review. Chemosphere 335. Available at https://doi.org/10.1016/j.chemosphere.2023.139077.

Furukawa, H., Cordova, K.E., O'Keeffe, M., Yaghi, O.M., 2013. The chemistry and applications of metal-organic frameworks. Science 341. Available at https://doi.org/10.1126/science.1230444.

Gatou, M.-A., Vagena, I.-A., Lagopati, N., Pippa, N., Gazouli, M., Pavlatou, E.A., 2023. Functional MOF-based materials for environmental and biomedical applications: a critical review. Nanomaterials 13. Available at https://doi.org/10.3390/nano13152224.

Gupta, A., Sharma, V., Sharma, K., Kumar, V., Choudhary, S., Mankotia, P., 2021. A review of adsorbents for heavy metal decontamination: growing approach to wastewater treatment. Materials 14. Available at https://doi.org/10.3390/ma14164702.

Hedau, B., Park, S.-J., Kang, B.-C., Ha, T.-J., 2024. Bifunctional carbon quantum dot-embedded metal-organic framework nanohybrid as a highly efficient electrocatalyst for water splitting and CO_2 reduction. Carbon 216. Available at https://doi.org/10.1016/j.carbon.2023.118527.

Huang, L., Cao, H., Ma, J., Wang, X., 2022a. Efficient removal of Pb(II) by UiO-66-NH_2: a combined experimental and spectroscopic studies. Environmental Nanotechnology, Monitoring & Management 18. Available at https://doi.org/10.1016/j.enmm.2022.100741.

Huang, X., Huang, L., Babu Arulmani, Y., SR, J., L.Q., T.J., 2022b. Research progress of metal-organic frameworks and their derivatives for adsorption of anions in water: a review. Environmental Research 204. Available at https://doi.org/10.1016/j.envres.2021.112381.

Ibrahim, A.O., Adegoke, K.A., Adegoke, R.O., AbdulWahab, Y.A., Oyelami, V.B., Adesina, M.O., 2021. Adsorptive removal of different pollutants using metal-organic framework adsorbents. Journal of Molecular Liquids 333. Available at https://doi.org/10.1016/j.molliq.2021.115593.

Ipadeola, A.K., Ozoemena, K.I., 2020. Alkaline water-splitting reactions over Pd/Co-MOF-derived carbon obtained via microwave-assisted synthesis. RSC Advances 10, 17359–17368. Available at https://doi.org/10.1039/d0ra02307h.

Jahan, M., Liu, Z., Loh, K.P., 2013. A graphene oxide and copper-centered metal-organic framework composite as a tri-functional catalyst for HER, OER, and ORR. Advanced Functional Materials 23, 5363–5372. Available at https://doi.org/10.1002/adfm.201300510.

Jeoung, S., Kim, S., Kim, M., Moon, H.R., 2020. Pore engineering of metal-organic frameworks with coordinating functionalities. Coordination Chemistry Reviews 420. Available at https://doi.org/10.1016/j.ccr.2020.213377.

Jin, Y., Li, Y., Du, Q., Chen, B., Chen, K., Zhang, Y., 2023. Efficient adsorption of Congo red by MIL-53(Fe)/chitosan composite hydrogel spheres. Microporous and Mesoporous Materials 348. Available at https://doi.org/10.1016/j.micromeso.2022.112404.

Khalid, W., Abdul Karim, M.R., Atif, M., Shehzad, W., Marwat, M.A., Yaqoob, K., 2023. Ultrasounds-assisted solvothermal synthesis of Ni-Co-Mn MOFs/PANI-CNTs nanocomposites with enhanced electrochemical energy storage performance. Energy & Environment. Available at https://doi.org/10.1177/0958305x231196126.

Kharissova, O.V., González, L.T., Aguilar, D.A.C., Hernández, M.G., Uflyand, I.E., Kulkarni, N.V., et al., 2024. Bi- and Trimetallic MOFs and their MOF-Derived Nanocarbons in Electrocatalytic Water Splitting Processes, in Elsevier eBooks. pp. 147–171. doi:10.1016/b978-0-323-96125-7.00009-5

Lan, G., Ni, K., Lin, W., 2019. Nanoscale metal-organic frameworks for phototherapy of cancer. Coordination Chemistry Reviews 379, 65–81. Available at https://doi.org/10.1016/j.ccr.2017.09.007.

Largitte, L., Pasquier, R., 2016. A review of the kinetics adsorption models and their application to the adsorption of lead by an activated carbon. Chemical Engineering Research and Design 109, 495–504. Available at https://doi.org/10.1016/j.cherd.2016.02.006.

Li, T., Zhang, L., Li, X., Wang, X., Jin, Z., 2022. Design and synthesis of phosphating bimetallic CeCo-MOF for substantially improved photocatalytic hydrogen evolution. Journal of Materials Chemistry C 10, 8750–8761. Available at https://doi.org/10.1039/d2tc00922f.

Li, X., Gao, X., Ai, L., Jiang, J., 2015. Mechanistic insight into the interaction and adsorption of Cr(VI) with zeolitic imidazolate framework-67 microcrystals from aqueous solution. Chemical Engineering Journal 274, 238–246. Available at https://doi.org/10.1016/j.cej.2015.03.127.

Li, C., Zhang, H., Liu, M., Lang, F.-F., Pang, J., Bu, X.-H., 2023. Recent progress in metal-organic frameworks (MOFs) for electrocatalysis. Industrial Chemistry & Materials 1, 9–38. Available at https://doi.org/10.1039/d2im00063f.

Luo, H., Zeng, Z., Zeng, G., Zhang, C., Xiao, R., Huang, D., et al., 2020. Recent progress on metal-organic frameworks based- and derived-photocatalysts for water splitting. Chemical Engineering Journal 383, 123196. Available at https://doi.org/10.1016/j.cej.2019.123196.

Ma, Q., Li, Y., Tan, Y., Xu, B., Cai, J., Zhang, Y., 2023. Recent advances in metal-organic framework (MOF)-based photocatalysts: design strategies and applications in heavy metal control. Molecules 28. Available at https://doi.org/10.3390/molecules28186681.

Mao, S., Shi, J.-W.S., Zhang, G., Ji, Y., Lv, X., Wang, Y., Xu, B., et al., 2021. Cu (II) decorated thiol-functionalized MOF as an efficient transfer medium of charge carriers promoting photocatalytic hydrogen evolution. Chemical Engineering Journal 404, 126533. Available at https://doi.org/10.1016/j.cej.2020.126533.

Mofokeng, T.P., Ipadeola, A.K., Tetana, Z.N., Ozoemena, K.I., 2020. Defect-engineered nanostructured Ni/MOF-derived carbons for an efficient aqueous battery-type energy storage device. ACS Omega 5, 20461–20472. Available at https://doi.org/10.1021/acsomega.0c02563.

Mondloch, J.E., Katz, M.J., Planas, N., Semrouni, D., Gagliardi, L., Hupp, J.T., 2014. Are Zr_6-based MOFs water stable? Linker hydrolysis vs. capillary-force-driven channel collapse. Chemical Communications 50. Available at https://doi.org/10.1039/c4cc02401j.

Nguyen, H.L., 2021. Metal-organic frameworks for photocatalytic water splitting. Solar RRL 5. Available at https://doi.org/10.1002/solr.202100198.

Okolie, J.A., Jimoh, T., Akande, O., Okoye, P.U., Ogbaga, C.C., Adeleke, A.A., 2023. Pathways for the valorization of animal and human waste to biofuels, sustainable materials, and value-added chemicals. Environments 10. Available at https://doi.org/10.3390/environments10030046.

Parker, K.H., Bishop, J.M., Serieys, L.E.K., Mateo, R., Camarero, P.R., Leighton, G.R.M., 2023. A heavy burden: metal exposure across the land-ocean continuum in an adaptable carnivore. Environmental Pollution 327. Available at https://doi.org/10.1016/j.envpol.2023.121585.

Prasad, S.V.S., Prasad, S.B., Verma, K., Mishra, R.K., Kumar, V., Singh, S., 2022. The role and significance of magnesium in modern day research-a review. Journal of Magnesium and Alloys 10, 1–61. Available at https://doi.org/10.1016/j.jma.2021.05.012.

Raptopoulou, C.P., 2021. Metal-organic frameworks: synthetic methods and potential applications. Materials 14. Available at https://doi.org/10.3390/ma14020310.

Razzak, S.A., Faruque, M.O., Alsheikh, Z., Alsheikhmohamad, L., Alkuroud, D., Alfayez, A., 2022. A comprehensive review on conventional and biological-driven heavy metals removal from industrial wastewater. Environmental Advances 7. Available at https://doi.org/10.1016/j.envadv.2022.100168.

Rocío-Bautista, P., Taima-Mancera, I., Pasán, J., Pino, V., 2019. Metal-organic frameworks in green analytical chemistry. Separations 6. Available at https://doi.org/10.3390/separations6030033.

Samia, Saeed, F., Jia, L., Arain, M., Ahmed, A., Yikai, F., Zhenda, C., Hussain, I., Ashraf, G.A., Ahmed, S.B., Dai, H., 2024. Emerging trends in metal-organic framework (MOFs) photocatalysts for hydrogen energy using water splitting: a state-of-the-art review. Journal of Industrial and Engineering Chemistry 131, 54–135. Available at https://doi.org/10.1016/j.jiec.2023.10.055.

Shah, S.H.A., Shah, A., Iftikhar, F.J., 2024. Nanoporous metal/covalent organic framework-based electrocatalysts for water splitting. ACS Applied Nano Materials. Available at https://doi.org/10.1021/acsanm.4c00397.

Smith, E.K., Barakat, S.M., Akande, O., Ogbaga, C.C., Okoye, P.U., Okolie, J.A., 2023. Subsurface combustion and gasification for hydrogen production: reaction mechanism, techno-economic and lifecycle assessment. Chemical Engineering Journal 480, 148095. Available at https://doi.org/10.1016/j.cej.2023.148095.

Soltani, R., Pelalak, R., Pishnamazi, M., Marjani, A., Shirazian, S., 2021. A water-stable functionalized NiCo-LDH/MOF nanocomposite: green synthesis, characterization, and its environmental application for heavy metals adsorption. Arabian Journal of Chemistry 14. Available at https://doi.org/10.1016/j.arabjc.2021.103052.

Song, H., Luo, S., Huang, H., Deng, B., Ye, J., 2022. Solar-driven hydrogen production: recent advances, challenges, and future perspectives. ACS Energy Letters 7, 1043–1065. Available at https://doi.org/10.1021/acsenergylett.1c02591.

Sun, D.T., Peng, L., Reeder, W.S., Moosavi, S.M., Tiana, D., Britt, D.K., 2018. Rapid, selective heavy metal removal from water by a metal-organic framework/polydopamine composite. ACS Central Science 4, 349–356. Available at https://doi.org/10.1021/acscentsci.7b00605.

Tabat, M.E., Omoarukhe, F.O., Güleç, F., Adeniyi, D.E., Mukherjee, A., Okoye, P.U., et al., 2023. Process design, exergy, and economic assessment of a conceptual mobile autothermal methane pyrolysis unit for onsite hydrogen production. Energy Conversion and Management 278, 116707. Available at https://doi.org/10.1016/j.enconman.2023.116707.

Taffa, D.H., Balkenhohl, D., Amiri, M., Wark, M., 2022. Minireview: Ni–Fe and Ni–Co metal-organic frameworks for electrocatalytic water-splitting reactions. Small Structures 4. Available at https://doi.org/10.1002/sstr.202200263.

Todaro, M., Alessi, A., Sciortino, L., Agnello, S., Cannas, M., Gelardi, F.M., 2016. Investigation by Raman spectroscopy of the decomposition process of HKUST-1 upon exposure to air. Journal of Spectroscopy 2016, 1–7. Available at https://doi.org/10.1155/2016/8074297.

Van Phuc, T., Jana, J., Ravi, N., Kang, S.G., Chung, J.S., Choi, W.M., et al., 2022. Highly active Ni/Co-metal organic framework bifunctional electrocatalyst for water splitting reaction. International Journal of Hydrogen Energy 47, 22787–22795. Available at https://doi.org/10.1016/j.ijhydene.2022.05.097.

Verma, C., Rasheed, T., Anwar, M.T., Quraishi, M.A., 2023. From metal-organic frameworks (MOFs) to metal-doped MOFs (MDMOFs): current and future scenarios in environmental catalysis and remediation applications. Microchemical Journal 192. Available at https://doi.org/10.1016/j.microc.2023.108954.

Vishnyakov, A., Ravikovitch, P.I., Neimark, A.V., Bülow, M., Wang, Q.M., 2003. Nanopore structure and sorption properties of Cu–BTC metal–organic framework. Nano Letters 3, 713–718. Available at https://doi.org/10.1021/nl0341281.

Wang, C., Xie, Z.K.E., Lin, W., 2011. Doping metal-organic frameworks for water oxidation, carbon dioxide reduction, and organic photocatalysis. Journal of the American Chemical Society 133, 13445–13454. Available at https://doi.org/10.1021/ja203564w.

Wang, C., Liu, X., Yang, T., Sridhar, D., Algadi, H., Bin Xu, B., 2023. An overview of metal-organic frameworks and their magnetic composites for the removal of pollutants. Separation and Purification Technology 320. Available at https://doi.org/10.1016/j.seppur.2023.124144.

Wang, H., Shuai, W., Shixing, W., Tang, J., Chen, Y., Zhang, L., 2022. Adenosine-functionalized UiO-66-NH_2 to efficiently remove Pb(II) and Cr(VI) from aqueous solution: thermodynamics, kinetics and isothermal adsorption. Journal of Hazardous Materials 425. Available at https://doi.org/10.1016/j.jhazmat.2021.127771.

Wang, W., Xu, X., Zhou, W., Shao, Z., 2017. Recent progress in metal-organic frameworks for applications in electrocatalytic and photocatalytic water splitting. Advanced Science 4, 1600371. Available at https://doi.org/10.1002/advs.201600371 -1600371.

Wang, X., Sun, M., 2021. Metal-organic framework (MOF) based materials for electrochemical hydrogen production: a mini review. International Journal of Electrochemical Science 16. Available at https://doi.org/10.20964/2021.05.42.

Wu, Y.P., Zhou, W., Zhao, J., Dong, W.W., Lan, Y.Q., Li, D.S., et al., 2017. Surfactant-assisted phase-selective synthesis of new cobalt MOFs and their efficient electrocatalytic hydrogen evolution reaction. Angewandte Chemie (International Ed. in English) 56 (42), 13001–13005. Available at https://doi.org/10.1002/anie.201707238.

Xie, L.S., Skorupskii, G., Dincă, M., 2020. Electrically conductive metal-organic frameworks. Chemical Reviews 120, 8536–8580. Available at https://doi.org/10.1021/acs.chemrev.9b00766.

Xu, G.-R., An, Z.-H., Xu, K., Liu, Q., Das, R., Zhao, H.-L., 2021. Metal organic framework (MOF)-based micro/nanoscaled materials for heavy metal ions removal: the cutting-edge study on designs, synthesis, and applications. Coordination Chemistry Reviews 427. Available at https://doi.org/10.1016/j.ccr.2020.213554.

Yaghi, O.M., 2016. Reticular chemistry—construction, properties, and precision reactions of frameworks. Journal of the American Chemical Society 138, 15507–15509. Available at https://doi.org/10.1021/jacs.6b11821.

Yaqoob, L., Noor, T., Iqbal, N., Nasir, H., Zaman, N., Talha, K., 2021. Electrochemical synergies of Fe–Ni bimetallic MOF CNTs catalyst for OER in water splitting. Journal of Alloys and Compounds 850. Available at https://doi.org/10.1016/j.jallcom.2020.156583.

Zadehahmadi, F., Eden, N.T., Mahdavi, H., Konstas, K., Mardel, J.I., Shaibani, M., 2023. Removal of metals from water using MOF-based composite adsorbents. Environmental Science: Water Research & Technology 9, 1305–1330. Available at https://doi.org/10.1039/d2ew00941b.

Zhang, J., Bai, T., Huang, H., Yu, M., Fan, X., Chang, Z., 2020a. Metal–metal-organic-framework-based photocatalysts optimized by spatially separated cocatalysts for overall water splitting. Advanced Materials 32. Available at https://doi.org/10.1002/adma.202004747.

Zhang, M., Xu, W., Li, T., Zhu, H., Zheng, Y., 2020b. In situ growth of tetrametallic FeCoMnNi-MOF-74 on nickel foam as efficient bifunctional electrocatalysts for the evolution reaction of oxygen and hydrogen. Inorganic Chemistry 59, 15467–15477. Available at https://doi.org/10.1021/acs.inorgchem.0c02504.

Zhang, Q., Yan, S., Yan, X., Lv, Y., 2023. Recent advances in metal-organic frameworks: synthesis, application and toxicity. Science of the Total Environment 902. Available at https://doi.org/10.1016/j.scitotenv.2023.165944.

Zhu, B., Zou, R., Xu, Q., 2018. Metal-organic framework based catalysts for hydrogen evolution. Advanced Energy Materials 8. Available at https://doi.org/10.1002/aenm.201801193.

CHAPTER

14

2D materials for water splitting

Misbah Mirza[1], *Muhammad Safdar*[2], *Nazia Yasmin*[1], *and Sadia Yasmin*[3]

[1]Department of Physics, The Women University Multan, Multan, Pakistan [2]Department of Basic Sciences and Humanities, Khawaja Fareed University of Engineering and Information Technology, Rahim Yar Khan, Pakistan [3]National Transmission and Despatch Company, Multan, Pakistan

14.1 Introduction

The world's power crisis and environmental degradation are among the most pressing problems that need to be solved. It is expected that by 2050, the world's energy needs would grow between 30 and 50 TW. Still accounts for roughly around 83% of the energy produced in the world today comes from fossil fuels like natural gas, coal, and oil. Meeting this demand under the current quo is concerning (Qureshy et al., 2019). CO_2 is the main byproduct of burning fossil fuels and accounts for over 76% of the annual emissions of greenhouse gases (Du Plessis and Du Plessis, 2019). These emissions have increased drastically over the past few decades, posing a serious threat to the environment and public health. Examples of the effects include air pollution, deadly heatwaves, wildfires, powerful hurricanes, intensified droughts, and declining coral reefs as a result of ocean warming (Mann et al., 2017; Neumann et al., 2015; Yang et al., 2016; Hurteau et al., 2019). These circumstances necessitate a swift investigation of environmental friendly and commercially feasible, clean, and sustainable energy conversion technologies as soon as possible due to the worsening environmental and energy concerns.

Many attempts have been made over the past several decades. Electro-chemical energy conversion, like metal air batteries, conversion of CO_2, water splitting, N_2 reduction, and fuel cells had been regarded as one of the most efficient and capable methods to eventually replace conventional fossil fuels (Kwon et al., 2019; Mohanty et al., 2020; Peng et al., 2018; Zhang et al., 2019a; Zhang et al., 2021b,c). Rather than other fuels, hydrogen has a greater gravimetric energy density so it is a viable substitute to meet future and resent energy requirements in a variety of applications (Xu et al., 2020). For instance, ambient oxygen may be combined with hydrogen in fuel cells to generate electricity and water is obtained as the sole byproduct for use as transport fuels (like in automobiles). In a different context,

Photochemical Splitting of Water
DOI: https://doi.org/10.1016/B978-0-443-29064-0.00011-5

hydrogen is a crucial raw material used to produce a wide range of compounds with added value (such as hydrocarbons, pharmaceuticals, ammonia). Though production of hydrogen is not without its obstacles. In spite of being most abundant universe element, hydrogen is not easily obtained in nature due to its low volumetric energy density. Only 4% of the hydrogen generated globally comes from renewable sources such as thermochemical pyrolysis, photo/dark fermentation, water electrolysis, while on the other hand, about 96% hydrogen is obtained from fossil fuels like coal gasification, methane reforming, and naphtha reforming. Other sources include naphtha reforming, steam methane reforming, and coal gasification (Kumar and Himabindu, 2019; Sun et al., 2021; Renaudie et al., 2021; Geißler et al., 2015). Extensive research is conducted to develop sustainable pathways and make them as affordable as fossil fuel. Thus a kinetic overpotential of 0.4–0.6 V and 1.23 V thermodynamical potential is required for electrolysis of water.

Among the various techniques, photoelectrochemical (PEC) water splitting with semiconductor (SC) photoelectrodes has received the greatest interest. This method has three main advantages: (1) producing H_2 and O_2 at separate electrodes, which resolves the severance problem; (2) having the potential to operate under ambient stipulations; and (3) having the potential to construct a system that consists only of abundant and stable inorganic materials.

Designing of active catalysts for water splitting entails fast charge transfer, suitable band alignment and bandgap, effectual charge separation, and enduring stability in aqueous media. Catalysts used at the electrode can have a simultaneous impact on both efficiency and reaction rate, which is partially responsible for the energy conversion recital. So far, excellent activity and long-lasting stability have been observed in effective catalysts made of noble transition metals (IrO_2, RuO_2, and Pt/C) (Zhuang et al., 2020). Their limited availability and high purchasing prices hampered the broad applications of noble metal catalysts. It would therefore be possible to address this issue by reducing the amount of such metal used without compromising catalytic performance. Considerable research had been done on 2D materials in search of PEC catalysts with astonishing properties that might take the position of noble metals. These materials have promising characteristics in energy reaping field, storage, or conversion by virtue of their surface functionalization, optoelectronics characteristics, large surface area, and stack ability for heterostructure assembly (Chen et al., 2020b; Chen et al., 2020a). These characteristics enable them to play important roles in hydrogen evolution reaction (HER) and oxygen evolution reaction (OER) as light harvesters. In particular, 2D materials with ion exchanged sites, defects, synergistic composites, edge sites,and phase transition can significantly reduce free energy needed for surface reaction. Additionally, by forming junctions (multiheterojunction, Schottky junction, and pn junction) along the semiconducting photoelectrodes, they take on the role of charge transport mediators.

2D materials, like Mxenes (Chaves et al., 2020), layered double hydroxide (LDHs) (Mao et al., 2018a), graphene (Chia and Pumera, 2018), two-dimensional covalent organic framework (COF) (Long et al., 2019), transition metal dichalcogenides (TMDs) (Eom et al., 2021), and two dimensional metal organic framework (MOF) (Lee et al., 2021), have become center of scientific interest during the past few years. When we used 2D materials as supports with single atom site catalysts (SACs), the benefits of 2D materials enhanced more. 2D materials (2DM) with SACs provide adequate active sites and results a large area of surface. Structural and electrical characteristics of active centers can be changed by the strong metal support contact, which helps to increase catalytic stability and activity. Density functional theory (DFT) simulations combined with experimental methodologies provide a good platform to utilize the inherent mechanism of the catalytic process because of the homogenous coordination environment generated by basic structures. However, it is evident that there is still a lack of research on SACs/2DM at this time. More time and energy should be invested in investigating new, effective methods, uses, and precise mechanisms in order to give the development of SACs/2DM.

In order to get considerably stable and effective catalysts, we concentrate on 2DM and their attributes in this review. The special qualities and advantages of 2DM are presented. Next, 2DM for PEC water splitting are discussed. Benefits of SACs/2DM are spoken about. Lastly, a summary of the issues and viewpoints is provided depending on the state of development at the moment. Definitive objective of this review is to persuade the fabrication of more innovative, highly efficient 2D materials and SACs/2DM by offering a thorough and in-depth overview of the present state of this field of study.

14.2 2D materials

Graphene, an ultrathin material derived from graphite delamination, paved the way for developing the 2DM with remarkable anisotropic physical, chemical, and electronic characteristics. Therefore various kinds of 2DM are investigated including nitrides g-C_3N_4, LDHs, metal carbides, and TMDs (communally called MXenes) (Chia and Pumera, 2018). These materials typically comprised of one or more atomic layers connected by covalent bonds strongly, with inimitable interactions between the layers. The unusual structure endows 2DM with various features than parent bulk materials, like unique mechanical characteristics, remarkable conductivity, and an enhanced specific surface area (Tang et al., 2019). Due to such remarkable features, 2DM have emerged as a popular area of study for PEC water splitting applications. Some of the 2DM are shown in Fig. 14.1.

Without further processing, the majority of pure 2DM exhibit poor catalytic activity (Jin et al., 2018a). For instance, graphene's limited hydrogen adsorption capacity results in extremely low HER efficiency. As a result, several optimization strategies, including (1) defect engineering: it have been put forth to improve the catalytic efficiencies of 2DM. During catalytic process, true active sites are typically identified as the defect present in 2DM. As a result, it has been embraced as a highly effective approach for increasing the 2DM catalytic activities. (2) By controlling the lateral thickness and size may also enhance the activity of the 2DM. When the lateral thickness and size are reduced, more surface area is exposed, allowing more active sites to be accessed, and this also allows for the electronic structures to be adjusted for improved performance. (3) Strain engineering: it is another technique to overcome the low catalytic activity of such materials. The electronic states of 2DM may be efficiently tuned by lattice strain, and this technique has been shown to be a successful means of controlling the catalytic efficiency of 2DM. (4) Interface engineering: the interaction between two distinct materials at their interface changes their chemical and electrical characteristics, which enhances the catalytic activity. In order to create a unique

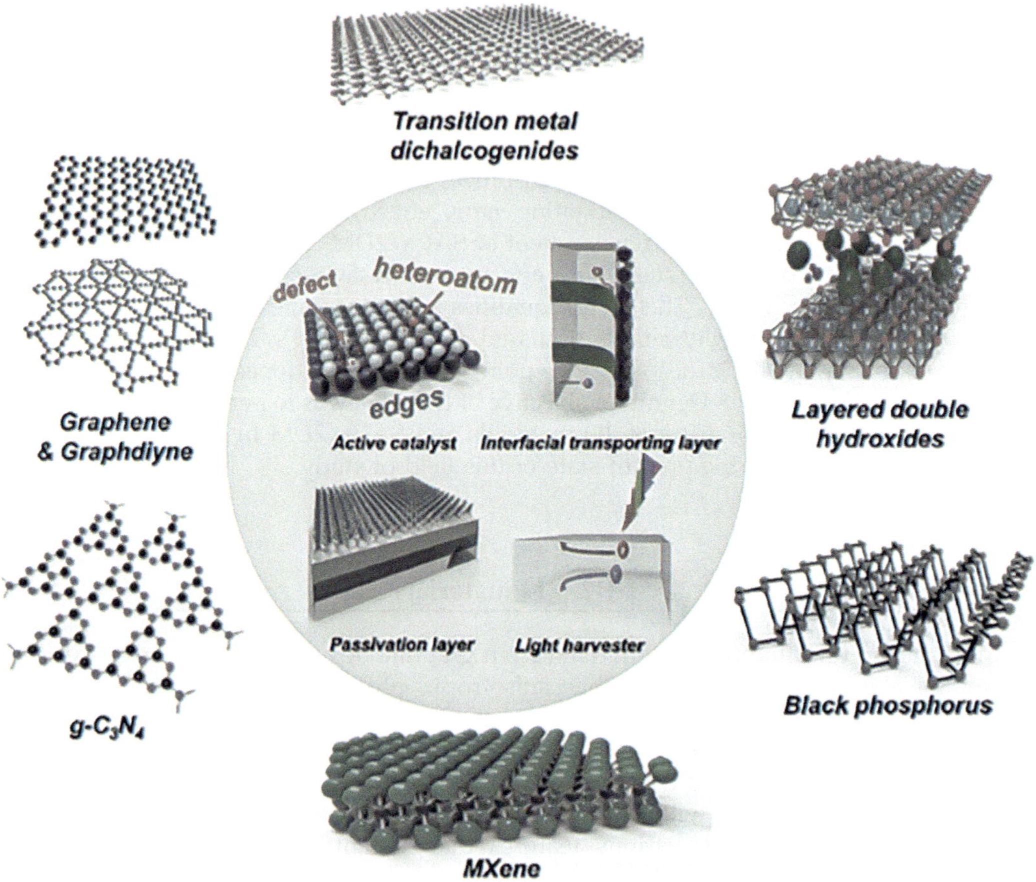

FIGURE 14.1 Various kinds of 2D materials (2DM) (Jun et al., 2023).

interface, this method often entails joining two distinct materials via physical contact and chemical bonding. (5) Doping of heteroatoms: heteroatom doping is a successful approach for raising catalytic efficiency of 2DM. (6) Doping of metal atoms: which is primarily employed in synthesis of SACs, have been extensively developed to improve efficiency for a variety reaction processes. When SACs were combined with 2DM, their vitality in various practical applications got enhanced because of the ease of the preparation procedure and the variety of the chosen metals (Jin et al., 2018a). The huge exposed surface area may offer a large number of mooring locales to attract the isolated atoms of metal and achieve an extraordinarily high metal stacking, which is challenging for 3D type substrates. For instance, nanosheets of TiO_2 are loaded by Pd and its value could reach to 1.5 wt.% (Liu et al., 2016), whereas Ag-SAC upheld on $g\text{-}C_3N_4$ may reach an astounding 10.21 wt.% of Ag loading (Xu et al., 2021). Furthermore, practically all of the active centers may be made accessible to the reagents by the ultrathin structure, achieving very high atom utilization a feat typically hindered by the stereo skeleton of 3D materials. However, 2DM with SACs support can interact with reactants more effectively to attain remarkable performance because they are often less coordinately saturated and because their electronic structures may be controlled via their interactions with 2D substrates. Additionally, because of the uniform and well defined active sites, SACs/2DM are more simpler and appropriate than 3D materials for studying the catalytic processes of various important heterogeneous reactions.

14.3 2D materials superiority

2DM layers interact with one another due to weak van der Waals forces, whereas within the planes strong covalent bonds exist (Chia and Pumera, 2018; Geim and Grigorieva, 2013). 2DM can be exfoliated into nanosheets or monolayers by utilizing weak interaction between the layers. The most notable feature of 2DM are their interactions, which provide them distinct advantages over materials with 0D, 1D, and 3D structures that include bandgap tuning (Chaves et al., 2020), large specific surface area (Long et al., 2016), surface functionalization (Martín et al., 2020), and heterostructures stackability (Li et al., 2018).

14.3.1 Heterostructure stackability

van der Waals interactions, in contrast to traditional way of bonding materials by covalent approach, which mechanically affect adjacent layers, enable 2DM to be layered onto any layer or substrate without causing atomic interdiffusion and crystal mismatch (Geim and Grigorieva, 2013). Specifically, to develop heterostructures with high quality at the atomic scale may be achieved by stacking several 2DM on top of one another (Li et al., 2018).

The most fascinating feature is that inherent surface states are not needed to achieve high-quality pn junctions, multiheterojunctions, or Schottky junctions within the energy bandgap (E_g) of 2DM. It prolongs the lifespan of charges passing over the heterointerface by lowering the probability of charges recombination. van der Waals forces might be utilized to redistribute the local electron density of 2DM, which can result in unpaired orbital electrons that are produced by altering the electron configuration (Li et al., 2018). This effect boosts the surface catalytic reaction which causes molecules to adsorb and desorb more easily. These factors allow them to be used with a variety of van der Waals heterostructure combinations, including 2D-2D, 1D-2D, 3D-2D, and 0D-2D (Fareza et al., 2022). Many applications, including sensors, optoelectronics, energy storage and conversion, and field-effect transistors, are made possible by these coupled dimensionality (Bhat et al., 2019; Lee et al., 2021; Long et al., 2019; Park et al., 2018; Wang et al., 2021).

14.3.2 Surface functionalization for electronic and mechanical characteristics

There are certain intrinsic active sites, such as defects and edges, in 2DM in their pristine state. However, distinct electrochemical, magnetic, electrical, and mechanical characteristics can be achieved by expanding these active locals or adding external impurities onto the surface. Functional units anchoring and chemical doping onto the 2DM result in unique features that enhance environmental stability, dispersibility, processability, and wettability (Martín et al., 2020). There are two forms of 2DM functionalization: noncovalent and covalent conjugation, depending on the nature of the interaction. The noncovalent approach regulates electrical and physical characteristics without deforming the structures of crystal, and physically adsorbed materials have an influence through electrostatic and van der Waals forces (Daukiya et al., 2019). Covalent modification, on the other hand, causes orbital rehybridization by assembling covalent chemical building blocks from 2DM and reactive molecules (Presolski and Pumera, 2016). Through dipole–dipole interactions or charge transfer, molecule building blocks are eventually used in both of them to amend electronic characteristic of 2DM (Brill et al., 2021).

14.3.3 Hefty specific surface area

2DM have a hefty specific surface area (SSA) because of their hefty lateral size along ultrathin thickness. Due to large SSA, the available atoms and active planes are maximized for catalytic purpose. Because of this, 2DM are widely useful for surface-active applications where reactants intermingle at the surface, such as batteries, gas sensors, energy conversion, and supercapacitors. Numerous active sites are available where the reactive components (OH, CO_2, and H^+) take part in the reaction for energy conversion usage such as OER, HER, and CO_2RR (Sun et al., 2017). More of the intended results are acquired with the large SSA as it facilitates the effective consumption of reactants. Zn^+, Na^+, and Li^+ are the transporting ions that are used in various kinds of batteries. Large surface area and reduced transfer paths cause the improved transportation and absorption of ions resulting in high rate capability and high energy density (Mao et al., 2018a). Large surface areas of 2DM offer multitude active sites for gas adsorption, which significantly increases sensitivity of sensor in gas-sensing applications (Eom et al., 2021).

14.3.4 Effect of energy band structure

The lowest energy required inciting an electron from ground state or valence band into conduction band is implied by the energy bandgap (Smith and Nie, 2010). A variety of 2DM are available with tunable energy bandgaps spanning from the ultraviolet to visible and infrared. Numerous investigations have shown that it is possible to tune the bandgaps of 2DM by alloying, chemical doping, intercalations, strain engineering, heterostructuring, and intercalations (Manzeli et al., 2015; Wan et al., 2016; Gong et al., 2014; Chen et al., 2013). For example, the crystal lattice's inversion symmetry is disrupted when a significant strain is applied to graphene, allowing gapless band structure to open up a bandgap (Gui et al., 2008; Ni et al., 2008). Moreover, bandgap bending based on W composition is seen in monolayer alloy of $Mo_{1-x}W_xS_2$. Devices that need optimal band alliance, such as light-emitting diodes, field-effect transistors, and photocatalysts/photoelectrodes, depend heavily on the mechanical characteristics of 2DM like thickness and composition-dependent bandgap. The corresponding light response of these materials spans the near infrared to ultraviolet (Gui et al., 2008; Ni et al., 2008). The adjustable band structure and the fluctuating electron distribution may both be used to control the interfacial desorption/adsorption-free energy between reactants and catalysts in addition to the light absorption characteristics (Wang et al., 2016).

14.4 Synthesis techniques of 2D materials

The two main processes for synthesizing 2D layered materials are thought to be bottom-up and top-down techniques. These strategies encompass a range of widely used techniques that are found on bottom up and top down techniques. In general, top down synthesis is viewed by external forces as a demixing process of layered bulk materials.

Wet ball milling (Peng et al., 2014), mechanical cleavage (Zheng et al., 2018), chemical etching (Seh et al., 2016), sonication (Rui et al., 2018), liquid/gas exfoliation (Ren et al., 2017), and shaking treatment (Kondo et al., 2013) are some of the methods for synthesizing 2DM through top-down techniques. One of the most widely used practices to manufacture multiple 2D layered materials is liquid exfoliation. This method's synthesis mechanism relies on interfacial debonding, which forms a steady sol through ultrasonic processing and deteriorating interactions between layers by adding doping (guest molecules) to increase interlayer space (Di et al., 2017).

Another common technique to synthesize 2D layered materials is mechanical cleavage. Jayasena and Subbiah (2011) exfoliated 2D graphene layers with an estimated area of $300 \times 900\ \mu m^2$ by using an ultrasharp diamond wedge aided via ultrasonic vibrations. The widely used Scotch tape process, which is thought to be among the most popular mechanical exfoliation techniques, caused a graphene discovery and may develop high-quality with desirable efficiency monolayered 2D materials (Chhowalla et al., 2013; Novoselov et al., 2004).

Using a certain ratio of zirconia spheres, solvents and materials are grinded into nanometer-sized slurry using the wet ball milling technique. Peng et al. synthesized ultrathin-layered MOFs with 1 nm thickness, high crystallinity, and large surface area by a procedure of wet ball milling and ultrasonic exfoliation, whereas methanol and propanol were utilized for ultrasonic exfoliation under the rotational speed of 60 rpm (Peng et al., 2014).

While top-down approaches are advantageous in that they produce 2DM with remarkable characteristics in an easy to use manner at a cheap cost (Kondo et al., 2013); they still have drawbacks that prevent the usage of 2DM in practical applications. These include low product yield, unstable nanosheet production, inconsistent layer counts, being restricted to layered materials, poor homogeneity, and the stripped nanosheets have propensity to break and restack. Therefore we think that fundamental research on top down 2DM synthesis is necessary, and that advancements in strategies and development will inevitably result in greater effort (Shifa et al., 2019), such as layer materials exfoliation containing molecules or ions among layers (Cai et al., 2018) and intercalation-assisted exfoliation and expansion (Cullen et al., 2017).

A common foundation for the bottom-up synthesis of layered 2DM is anisotropic assembly of tiny molecules with growth restrictions in vertical directions (Zhang et al., 2015b). Synthesis of 2DM may be achieved more controllably and more easily via bottom up wet chemical fabrication technique, which is also easier to implement for large-scale fabrication. These include chemical vapor deposition (CVD) (Geng and Yu, 2018), organic, inorganic lamellar (Adnan et al., 2020), template-assisted production (Wu et al., 2014), solvothermal/hydrothermal fabrication (Li et al., 2019a), inorganic surfactant assisted synthesis (Wang et al., 2020), and surfactant self-assembly (Sushko and Liu, 2011). The common procedures for synthesizing 2DM by bottom-up approaches will be covered in following discussion.

For large scale 2DM synthesis, CVD is a well liked process that offers superiority in controlling material thickness and size for practical applications (Han et al., 2019). Ji et al. synthesized 2D black phosphorus via CVD, with average areas typically greater than 3 μm^2. In Song's study, superior h-BN nanostructures with thicknesses generally ranging from two to five atomic layers were grown by CVD.

Surfactants and substrate materials are, nonetheless, frequently needed in conventional bottom-up synthesis methods. As a result, synthesized dispersed 2DM may be challenging, and removing the remaining surfactants from the products can be challenging, which may restrict their applications.

14.5 Characterization techniques for 2D materials

The characterization of 2DM and 2DM/SACs is critical for revealing their fundamental features and conducting mechanism investigations. The structural information about the supports of metal particles of 2DM/SACs may be obtained by recent developments in the atomic resolution characterization approach, known as aberration corrected scanning transmission electron microscopy (AC-STEM) (Qiao et al., 2011; Lu et al., 2012; Hackett et al., 2013). Specifically, when it comes to 2DM and 2DM/SACs, defects and structural disorder may expose crucial details via Raman spectroscopy to determine the SACs and 2DM phonon state. Several characterization methods for examining the operation of 2DM and 2DM/SACs as electro/photocatalyst for water splitting are pisted in this section and able to keep eye on the structural and activity of the catalyst under various reaction conditions. These techniques reveal details regarding the electrical characteristics, reaction intermediates, and surface chemistry of the catalyst. A few often employed techniques for characterizing the 2DM and 2DM/SACs are as follows:

1. Electron microscopic technique: Additionally, electron microscopic techniques may be used to determine the single atom's location on the support's surface as well as their degree of distribution (Sun et al., 2013; Lin et al., 2013; Yang et al., 2013). This methodology offers a reliable way to control the size and structure of SACs on the support materials. Moreover, electron energy loss spectroscopy and energy dispersive X-ray spectroscopy (EDS) might provide details on the compositional analysis of individual atoms. These additional methodologies did, however, speculate that it is challenging to clarify just individual atoms scattered over high-surface area supports, such as 2DM, because of the effect of electron beam induction and the detecting limitations of currently available equipment. Despite advances in microscopic technology, it remains difficult to properly characterize the SACs implanted on the surfaces of 2DM precisely (Zhao et al., 2011; Zhou et al., 2012).
2. Fourier transform infrared spectroscopy (FTIR): On the catalyst surface, the gases adsorption/desorption may be studied through FTIR technique. FTIR spectroscopy may detect the reaction intermediates and surface species that are produced throughout catalytic reaction process (Liu et al., 2019a). FTIR spectroscopy reveals the detection of the chemical interaction between the solid surfaces and absorbed molecules, and it has

played an important role in the SACs' characterization on supports (Ryczkowski, 2001; Chaiyakun et al., 2012). With adequate calibration, it is used to determine the chemical and physical characteristics of the active centers of catalysts. It is commonly used to identify defects and the effects of functionalization of graphene derivatives (Chaiyakun et al., 2012; Naebe et al., 2014). Both quantitative and qualitative information on 2DM/SACs in the catalysis processes may be obtained using the several forms of FTIR spectroscopy. It could be the most easiest technique to observe 2DM/SACs materials.

3. Transmission electron microscopy (TEM) and X-ray diffraction (XRD): The structural alterations of the synthesized material during reaction can be investigated by using TEM. TEM may reveal information regarding the size, morphology, and crystal structure during reaction conditions. The synthesis and development of reaction intermediates may also be studied using this method (Petersen and Weidenthaler, 2022). The typical TEM's detection limit is around 1 nm, whereas the XRD's is roughly 3 nm. Using these two methods, researchers anticipate to find no metal nanoparticles (NPs) when characterizing metal SACs. TEM, XRD, and energy dispersive X-ray (EDX) combination are often used to rule out the presence of metal NPs in the specified region. Using the high-resolution TEM, no significant NPs were found; however, this might be because the target metal element is absent from the region. In order to demonstrate the presence of required metals and the distribution of various elements, material's EDX mapping is essential. Because of the larger detection limit and increased requirements for metal loadings, the conclusion drawn from the XRD result is even less convincing. To verify that particular atoms are the only ones present, high angle annular darkfield scanning transmission electron microscopy (HAADF-STEM) is utilized. Agglomeration of metal atoms may be readily seen since the core atoms are often higher in size than the atoms in the support. Moreover, information on the composition of the SACs may be provided by the equipped EDS to demonstrate the existence of the required metal.
4. X-ray photoelectron spectroscopy (XPS): XPS is an efficient approach to examine electronic structure, surface elemental composition, and chemical characteristics of 2DM or 2DM/SACs. XPS may be performed to investigate how the synthesized materials surface chemistry changes as the reaction proceeds. Additionally, this method can reveal details about the adsorbed species and the catalyst's oxidation states throughout the reaction. 2DM or 2DM/SACs as electro/photocatalysts for water splitting may be used to better understand their working mechanism by examining the alterations in electronic characteristics, surface chemistry, and reaction intermediates throughout the reaction.
5. Raman spectroscopy: It is an effective technique for examining the structural, chemical, atomic vibrational details and evolution and formation of reaction intermediates of 2DM and 2DM/SACs during the reaction (Chen et al., 2021). Additionally, if reference data for the eminent material are available, it is a quick and simple technique to determine the characterization of fabricated 2DM. Specifically, it can offer valuable insights into the stacking order of graphitic materials, defects and in plane carbons sp^2 vibrations (Dresselhaus and Eklund, 2000; Jiang et al., 2005). Using resonant Raman scattering, the peak locations and intensity ratios might yield approximative information about the electronic structures and thickness. When the dopants are added to the 2DM, the Raman peak location may be changed. After the SACs are anchored, these data may be used to understand the electronic structure and doping state of the 2DM and can help to define

how the 2DM support and SACs interact. The Raman imaging approach makes it easy to find the anchoring sites by providing discontinuous variations in the peak placements or peak intensity ratio and reveals the doping states of 2DM. Therefore Raman spectroscopy is one of the useful techniques to understand the electronic structure of 2DM/SACs.

6. X-ray absorption fine structure (XAFS): XAFS provides information of both localized and overall structure on nonoccupied local electronic states as compared to the traditional electron microscopy which reveals information only about specific localization (Gaur et al., 2013). Therefore by employing X-ray technology, XAFS may offer a reasonably easy and useful method to clarify the chemical structure and local atomic structure for the chosen atomic species. Usually, XAFS is divided into two parts, extended region (EXAFS) (extends from ≈50 eV above absorption edge) and adjacent fringe region (XANES) (extends to ≈50 eV above absorption edge). Moreover, EXAFS is used to verify the distances, species of neighboring absorbed atoms, and coordination number, whereas XANES is very sensitive to coordination chemistry of the absorbing atom and formal oxidation state (Zhang et al., 2018a). Because of these benefits, XAFS is frequently used to confirm the absorbing SACs' chemical structure and also analyze the coordination states.
7. DFT computations: For the purpose of researching mechanisms behind catalytic reactions in water splitting of 2DM and SACs on supports, the DFT calculations are crucial. DFT simulations with simplified conditions of a single atom can provide fresh insights into the intrinsic properties and structures of these compounds in addition to predicting the best stable configuration for SACs on support materials. Recently, DFT simulations have been demonstrated for numerous cases, such as, (1) SACs/g-C_3N_4, (2) boron monolayer effects on CO_2 activation into anionic CO_2^- and H_2O molecules adsorption to lower reaction barrier via reaction intermediates, (3) atomic Ni species are trapped into graphene defects to develop active site (Ling et al., 2018a; Zhang et al., 2018c; Ling et al., 2018b; Mao et al., 2018b).

   DFT simulations may estimate the interaction among the catalyst performance and other synthetic factors, such as precursors, size, and methodologies utilized, without requiring any experimental work. It also provides information about the density of states of 2DM that may be influenced by SACs. It is easy to anticipate how the electronic structure of 2DM supports may change by doing variations in electron transfer and which cause the improvement catalytic performance (Liu et al., 2016). With the use of DFT studies, reaction systems, including catalyst design and reaction conditions, may be further optimized.
8. Other complementary technologies: To scrutinize the transfer dynamics and photogenerated charge separation in systems of PEC water splitting (WS), various analytical approaches have been used like Mott–Schottky plot (Xing et al., 2017), intensity modulated photocurrent spectroscopy (Zandi and Hamann, 2016), time-resolved transient photoluminescence (Zhang et al., 2021a), and open circuit potential decay profiles (Zhang et al., 2021a). These techniques offer insightful information on the efficiency and underlying mechanics of charge carrier processes. The potential of flat-band and acceptor or donor concentration of the photoelectrode may be estimated using the Mott Schottky plot. These analysis techniques may be used to get a thorough knowledge of the processes involved in photogenerated charge transfer and separation, which will help with the design and optimization of 2DM and 2DM/SAC based systems for PEC WS.

14.6 Mechanism of water splitting

The two methods of employing a catalyst to accelerate a chemical reaction are electrocatalysis and photocatalysis. Through the process of photocatalysis, a catalyst uses light energy to start a chemical reaction. In contrast, electrocatalysis uses a catalyst to assist an electrochemical process. The catalyst often accelerates the process by creating an additional route with lower energy barriers. Catalysts are essential for electrocatalysis and photocatalysis, boosting reactions, and enhancing efficiency. The deep discussion about the mechanism is given in the following:

14.6.1 Photoelectrochemical water splitting mechanism

In 1972 Honda and Fujishima showed the first effective PEC WS for H_2 production. Since then, a lot of work has been done to determine whether this field of study is commercially viable. The basic mechanism of PEC WS will be covered in this part, with an emphasis on the energetics needed to produce effective WS reactions. The principle behind PEC WS is the conversion of incoming photons with energy higher than the semiconductor's bandgap into electrochemical energy that may cause splitting of water straight into molecular O_2 and H_2.

The fundamental mechanism of this reaction is the production of minority charge-carriers at the time of illumination (holes for semiconductor of n-type and electrons for semiconductor p-type), which is followed by charge migration (Fig. 14.2). At the electrode surfaces, the migrating minority charge-carriers perform the required electrochemical processes, which result in the oxidation and reduction processes on n-type and p-type semiconductor, respectively. These processes might take place in the same semiconducting material or in various materials that are deposited on top of each other. Typically, PEC arrangement as shown in Fig. 14.2 comprises of a cathode and an anode, with either or both being photoactive. When bias potential is given to the system, it may increase efficiency of interfacial charge transfer and charge carrier separation. In an ideal scenario, photo WS would not require the use of external potential (like a PC device) because freely suspended semiconducting particles with loaded electrocatalysts would allow for the simultaneous HER and OER from photo-generated electrons and holes, respectively (Suen et al., 2017; Zeng and Li, 2015). Four photons are needed for each evolved oxygen molecule and every two evolved hydrogen molecules during WS.

When light is present, electrons are stimulated to conduction band due to which a hole is left in valence band which is positively charged. Builtin electric field, which is present in depletion region, causes the movement of photogenerated charge carriers toward the interface of electrolyte semiconductor to initiate the chemical reactions. In depletion region, this electric field is produced due to difference in potential between water's redox potential and semiconductor's Fermi level. A photoanode in a PEC system, photogenerated electrons go back to the contact via external circuit and water reduction occur to generate hydrogen. This process is referred to as HER. Conversely, OER is triggered when holes in semiconductor's valence band reach semiconductor's surface and oxidation of water take place to produce molecular oxygen (Harris-Lee et al., 2023).

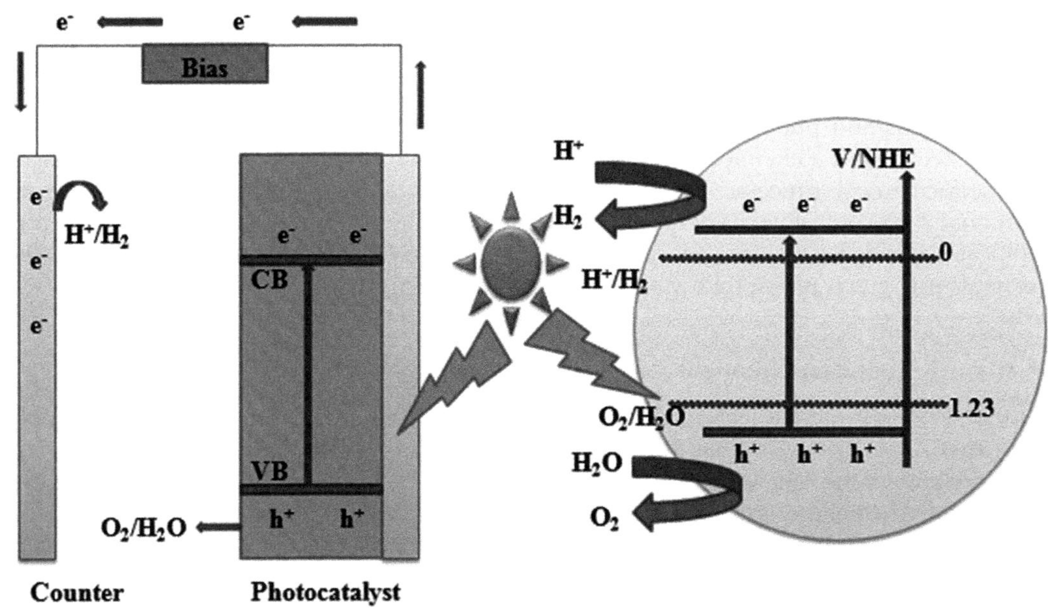

FIGURE 14.2 Schematic illustration of photoelectrochemical water splitting mechanism.

At every electrode, the following reactions are occurring as represented in Eqs. (14.1 and 14.2):

$$\text{(Photoelectrode) } 4h\upsilon \rightarrow 4e^- + 4h^+ \tag{14.1}$$

$$\text{(Anode) } 4h^+ + 2H_2O \rightarrow O_2\uparrow + 4H^+ \tag{14.2}$$

$$\text{(Cathode) } 4H^+ + 4e^- \rightarrow 2H_2\uparrow \tag{14.3}$$

$$\text{(Overall) } 2H_2O + 4h\upsilon \rightarrow O_2\uparrow + 2H_2\uparrow \tag{14.4}$$

Both hole and electron are remained in a "pair," attracted to one another and preferring to recombine to come back to lower energy state of electron in VB. Remaining charge-carriers recombine in bulk and thermally release their energy as phonons, which eventually reduce the photoelectrode's total activity. According to research, surface changes, heteroatomic doping, and nanostructuring have all been used to reduce other overpotential losses and recombination. However, the maximum solar-to-hydrogen (STH) efficiency for an independent PEC device has not been achieved by the materials currently in use (Ros et al., 2020; Walter et al., 2010; Alexander et al., 2008; Hisatomi and Domen, 2019).

A total of 237.2 kJ/mol is a Gibbs free energy analogous to 1.23 V versus reversible hydrogen electrode (RHE) thermodynamic potential having by the WS reaction. However, in reality, for WS process, photocorrosion and impedances cause the overpotential losses and in return the potential barrier is raised. Moreover, the water-oxidation process severely restricts the PEC device performance because of its slow kinetics for OER and greater energy needed for $4h^+$ and $4e^-$. As a result, the water oxidation reaction becomes a rate-determining step in the process of overall PEC WS.

Photoelectrode materials should be capable to produce ~1.2 V versus RHE and ~2.0 V versus RHE photovoltage for photocathode and photoanode, respectively. Thus using the right photoelectrode materials is crucial for hydrogen production via PEC WS.

14.6.2 Electrocatalytic water splitting mechanism

The first electrolytic cell driven through an electrostatic generator was constructed in 1789 by Deiman and Troostwijk. They discovered that the electric discharge would result in the evolution of gases, which would include two parts oxygen and one part hydrogen. Through several studies, the process of electrocatalytic WS progressively came to light. Efforts are still being made to lower the production cost and increase the efficiency of production of hydrogen by water electrocatalysis. In WS, the OER and HER are two significant surface processes. The electrolyte's pH level determines reaction's mechanism. The WS is mostly carried out in three types of medium like acidic, alkaline, or neutral. Schematic illustration of setup of electrochemical WS is shown in Fig. 14.3.

It involves utilizing catalyst and electrical energy to produce pure gases of hydrogen (H) and oxygen (O) at their respective potentials (Xiao et al., 2017). The whole WS reaction is represented by Eqs. (14.5–14.7).

$$2H_2O \rightarrow O_2 + 4H^+ + 4e^- E\ (\text{oxidation}) = 1.23\ \text{V} \tag{14.5}$$

$$4H^+ + 4e^- \rightarrow 2H_2 E\ (\text{reduction}) = 0.0\ \text{V} \tag{14.6}$$

$$2H_2O \rightarrow 2H_2 + O_2 E\ \text{cell} = 1.23\ \text{V (Overall reaction)} \tag{14.7}$$

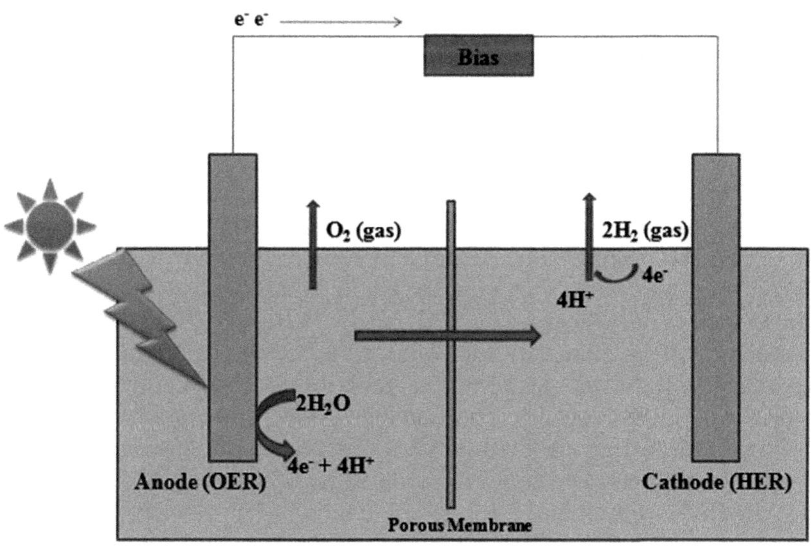

FIGURE 14.3 Schematic illustration of setup of electrochemical water splitting.

The OER is thought to be slower than the HER, which evolves two molecules of hydrogen with the same absorption of electrons and reason behind this phenomenon is that the OER needs the release of four electrons in order to form a single oxygen molecule (Kim et al., 2023).

14.6.2.1 OER and HER mechanisms in acidic media

In an acidic environment, two water molecules are probably oxidized to release four protons, one oxygen molecule, and four electrons during the OER process, whereas to generate two hydrogen molecules during HER, four protons are reduced by using four electrons. The electrolyte's pH level affects how these processes operate. The following equations describe how oxygen and hydrogen evolve in an acidic media.

OER in acidic media at anode is shown in Eqs. (14.8–14.11):

$$H_2O(l) + * \rightarrow H^+ + *OH + e^- \tag{14.8}$$

$$*OH \rightarrow H^+ + e^- + *O \tag{14.9}$$

$$H_2O(l) \rightarrow *O \rightarrow H^+ + *OOH + e^- \tag{14.10}$$

$$*OOH \rightarrow H^+ + * + e^- + O_2(g) \tag{14.11}$$

HER in acidic media at cathode is shown in Eqs. (14.12–14.14):

$$H^+ + e^- \rightarrow H_{ad} \tag{14.12}$$

$$H^+ + e^- + H_{ad} = H_2 \tag{14.13}$$

$$2H_{ad} \rightarrow H_2 \tag{14.14}$$

Where (g) and (l) stands for gas phase and liquid phase, OOH, O, OH are the free radicals adsorbed on the active site, and * corresponded to the catalyst's active sites (Liang et al., 2021). At the cathode, ions of hydrogen (H^+) are reduced to produce hydrogen gas (H_2) as part of the HER process. On the other hand, the OER process entails oxidation of water (H_2O) at anode to produce oxygen gas (O_2) (Kim et al., 2023).

Every stage can have a different metal ion oxidation state at the active site, either oxidizing or reducing the metal, or leaving the metal oxidation state unaltered. Additionally, the total oxidation state of $*O_2$ can be 0, -1, -2, based on the active site. The OER mechanism involves elementary steps that can either be nonelectroactive (nonelectron transfer) or electroactive (electron transfer), meaning that they do not involve the electron transfer (in this case, a electron–proton couple). The transfer of electron steps have more advantages than steps of nonelectron transfer because previous one requirements of activation energy (free energy) can be reduced by applying extraneous electric potential, since the other one (nonelectron transfer) undoubtedly hinge on energy attained from thermal resources, therefore they may not be preferred at room temperature even in the presence of external potentials.

14.6.2.2 HER and OER mechanism in alkaline media

At cathode, water is reduced to produce H_2 and OH as part of the HER process. At anode, hydroxide ions (OH^-) are oxidized to produce water (H_2O) and O_2 as part of the OER process (Zhang et al., 2017; Mayorga-Martinez et al., 2017; Jenjeti et al., 2016).

The following processes govern how oxygen and hydrogen evolve in an alkaline media (Eqs. 14.15–14.20):

$$H_2O + e^- \rightarrow H_{ad} + OH^- \tag{14.15}$$

$$H_2O + e^- + H_{ad} \rightarrow OH^- + H_2 \tag{14.16}$$

$$OH^- + * \rightarrow e^- + {*OH} \tag{14.17}$$

$$*OH + OH^- \rightarrow e^- + {*O} + H_2O(l) \tag{14.18}$$

$$OH^- + {*O} \rightarrow e^- + {*OOH} \tag{14.19}$$

$$OH^- + {*OOH} \rightarrow O_2(g) + e^- + H_2O(l) + * \tag{14.20}$$

14.7 2D materials for water splitting

14.7.1 Transition metal dichalcogenides

For PEC applications, materials with TMDs are most promising because of their special optical, electrical work, and catalytic qualities. Specifically, the flaws and edges function well as active sites and show superior PEC catalytic activity. MoS_2 (Zhou et al., 2018a), WS_2 (Kwon et al., 2017), WSe_2 (Seo et al., 2019), NbS_2 (Gnanasekar et al., 2019), $PtSe_2$ (Chung et al., 2021), and $ReSe_2$ (Huang et al., 2019b) are among the transition metal-based dichalcogenides that have recently been created for solar WS. These tactics include morphology control, heterostructuring, edge engineering, and thickness management.

14.7.1.1 Noble metal-based TMDs

$PtSe_2$ is deposited in two dimensions on Si photocathode (which is p-type) by employing CVD (Chung et al., 2021). $PtSe_2$ transforms from semimetal to semiconductive when its thickness reduces from bulk to monolayer. Furthermore, the transfer of photogenerated charges is facilitated by $PtSe_2$ enhanced electrical conductivity, which is achieved through thickness control. The energy barrier within the $PtSe_2$/p-Si interface and energy level of conduction band both drop with increasing thin film thickness.

14.7.1.2 Nonnoble metal-based TMDs

3D structured MoS_2 on p-Si substrate coated with TiO_2, which enhanced PEC catalytic activity of the silicon photocathode, was deposited by Andoshe et al. (2018). MoS_2 was deposited and developed into three-dimensional structures that were capable of vertical, head-on collision, and faulty bending. Growth kinetics in metal-organic chemical vapor deposition (MOCVD) governed their shape. Greater nucleation sites are produced by high precursor flow rates, which lead to the development of smaller grains and three-dimensional structures. Because of its exposed edge sites and three-dimensional structure, 3D MoS_2 exhibits increased PEC catalytic activity and decreased optical reflectance. With a 37 mA/cm^2 current density and 0.35 V onset potential versus RHE, 3D MoS_2/TiO_2/p-Si exhibits improved HER performance. Additionally, W-based TMDs have strong PEC catalytic capabilities.

The WS_2 was directly synthesized on p-Si through thermolysis by Hasani et al. (2019), and exhibited outstanding PEC activity with a 0.22 V onset potential versus RHE and at a 0 V the value of photocurrent density was 9.8 mA/cm^2 versus RHE. The synthesized sample demonstrated good stability, with the current density remaining constant even after 40 hours. With a nearly zero hydrogen adsorption Gibbs free energy, NbS_2 is a characteristic metallic TMD with a very active basal plane to hydrogen adsorption/desorption (Liu et al., 2017c). Gnanasekar et al. (2019) used a CVD technique to deposit NbS_2, demonstrating an effective and stable Si nanowire photocathode. The onset potential of NbS_2/Si NWs versus RHE was 0.34 V, and photocurrent density at 0 V versus RHE was 28 mA/cm^2.

14.7.2 2D metal oxide

Nonlayered nanosheets oxide (Fe_2O_3 and TiO_2) had been thoroughly scrutinized and contrasted with 2D layered nanosheets like TMD and graphene. The strong interaction between oxygen anions and metal cations makes the production of 2DM nonlayered more challenging. Nanosheets of 2D nonlayered materials, including WO_3, SnO_2, and TiO_2, have been produced and used in PEC WS (Liu et al., 2017b; Yao et al., 2017; Zhang et al., 2015a; Wang et al., 2014; Hou et al., 2014). TiO_2 based 2D nanosheets have been thoroughly explored because of their cost affectivity, nontoxicity, high permanence, and availability (Wang and Sasaki, 2014; Li et al., 2017; Edy et al., 2016).

A 3D $WO_3/C_3N_4/CoO_x$ was fabricated by Hou et al. (2014). It was composed of a branching WO_3 2D nanosheet which was heterojunction with 2D C_3N_4 that was decorated with nanoparticles of CoO_x. This demonstrated better PEC efficiency for oxidation of water. The hybrid system's interface charge transfer was investigated by electrochemical impedance spectroscopy (EIS) measurements. In comparison to WO_3-NA/C_3N_4-NS and WO_3-NA, this suggests reduced charge-transfer resistance and effective photogenerated electron–hole pair separation for 3D branched WO_3-NA/C_3N_4-NS/CoO_x under light irradiated and dark conditions. Porous structure consisting of Fe_2O_3NT/TiO_2NS was synthesized by Yao et al. (2017) and under visible light it exhibited a good PEC activity. As compared to bare TiO_2, its current density is six times higher 2.50 mA/cm^2. Wang et al. synthesized ultrathin coated films of TiO_2 on FTO substrate via layer by layer (LBL) assembly process. These TiO_2 films which were deposited via LBL process have improved PEC WS and thickness less than 1.5 nm was found to perform best.

14.7.3 Carbon allotropes

14.7.3.1 Graphene

A source of intrinsically layered 2DM, graphene is a individual atom-thick carbon sheet with sp^2 hybridization. It has garnered significant interest primarily in virtue of its exceptional physical and chemical characteristics, which include high thermal conductivity, high carrier mobility, effect of strong ambipolar electric field, mechanical strength, eccentric quantum Hall effect, and high transparency (Novoselov et al., 2004; Zhang et al., 2005; Balandin, 2011; Nair et al., 2008; Du et al., 2008; Lee et al., 2008; Li et al., 2016b). Graphene has been used extensively in many studies up to this point. It may also be used in PEC systems as a charge mediator, interfacial transporting layer, catalyst, molecular biology controller, and photon trapping layer (Jun et al., 2021; Xiao et al., 2014). By controlling the graphene layer, Sim et al. (2017) improved the photon to current efficiency of a silicon photocathode by adding high transmittance graphene. Double layer graphene was found to exhibit smallest work function from the ultraviolet photoelectron spectroscopy (UPS) test, suggesting that the conduction band edge is substantially bent toward H^+/H_2 redox pair. As a result, in depletion zone the strongest band bending produced highest anodic shift in linear sweep voltammetry (LSV) curve. After plasma treatment to introduce huge number of defects and edges, graphene/p-Si photoelectrode with double layer demonstrated improved PEC activity with 0.32% efficiency, which is 64 rank (times) larger than that of naked p-Si.

14.7.3.2 *Graphene oxide*

NiFe-LDH and reduced graphite oxide (rGO) on a TiO_2 nanorod array (NA) to fabricate photoanode with significantly improved PEC efficiency was reported by Ning et al. (2016). There was an increase in electrical conductivity because of the rGO. Because of rGO's exceptional electron mobility, migrating electrons go quickly to the current collector, significantly increasing the efficiency of separation of charge, whereas holes trapped in NiFe-LDH take part in OER.

14.7.3.3 *Graphdiyne*

Graphdiyne (GDY) is a recently discovered 2D allotrope of carbon that is made up of benzene ring moieties and butadiyne linkers, two types of hybridized carbons (Gao et al., 2019; Yu et al., 2019c). Because of its special sp-hybridized carbon atoms, GDY has exceptional qualities such as good chemical stability, large number of carbon bonds, and outstanding electrical conductivity. Zhang et al. (2018d) reported that adding a GDY nanowall to a Si heterojunction (SiHJ)-based photoanode improved its PEC characteristics. The synthesized GDY value of onset potential versus RHE was 1.082 V, with 39.1 mA/cm^2 saturated photocurrent density, and high conductivity of 2.516×10^{-4}/Sm and hole mobility of $1\times10^2\ cm^2/V/s$ which are comparable to silicon-based substrate. The tremendous catalytic efficiency, conductivity, and chemical bonding between GDY and NiO_x are responsible for the increased PEC-OER performance. Zhou et al. (2022) suggested a simple in situ encapsulation method employing hydrogen-substituted graphene (HsGDY) to activate water reduction and passivate the Cu_2O photocathode. At a 0 V, high photocurrent density of 12.88 mA/cm^2 versus RHE was demonstrated by the optimized HsGDY@Cu_2O NWs/CF, which is significantly greater than that of the pristine Cu_2ONWs/CF.

14.7.4 Black phosphorus

One of the phosphorus allotropes, black phosphorus (B) is formed when white or red phosphorus undergoes a phase change at a high pressure and temperature. It is a thermodynamically stable-layered semiconductor with a small bandgap (Liu et al., 2015). Only BP has been found to exhibit pressure-induced atomic structural changes, that is, from orthorhombic structure changing to cubic and rhombohedral at 10 and 5 GPa, respectively. BP has an energy bandgap that can be adjusted from 0.3 to 2.0 eV, which makes it a viable material for a variety of electrical and energy conversion applications. It also has a high carrier mobility.

Zheng et al. used BP quantum dot (BPQD) sensitization and defect engineering techniques (oxygen vacancy) to nanotube arrays of TiO_2 (NTAs) to produce first TiO_{2-x}/BP heterojunction photoanode (Zheng et al., 2020) images of TEM of TiO_{2-x}/BP with lattice fringes 0.297 nm have 040 planes of BP. PEC performance is improved by the combined action of oxygen vacancy-induced NTAs and synthesized BPQDs, which increase active sites and improve charge separation. At 1.23 V, a photocurrent density was 1.12 mA/cm^2 versus RHE and than that of pristine TiO_2, it is three times higher. By analyzing OEC/BP/$BiVO_4$, Zhang et al. (2019b) confirmed that charges trapping lifetime extends due to intrinsic p-type BP on the $BiVO_4$surface and helps to hole extract from the $BiVO_4$ photoanode. The density of states (DOS) of BP/$BiVO_4$ verified that BP nanosheets alter $BiVO_4$ valence band by overlapping the O 2p and P 2p orbitals, have no effect on the conduction band. Photocurrent density 1.5 times higher than NiOOH/$BiVO_4$ photoanode at 4.48 mA/cm^2 versus RHE. The increased charge separation efficiency amply demonstrated BP's effectiveness as a hole extraction material.

Heptazine-based crystalline nitride (KPHI) and BP composite catalysts fabricated via techniques of liquid phase exfoliation and molten salt which were ultrasound assisted had been synthesized by Xiao et al. (2023). The photocatalytic efficiency of composite KPHI/BP was examined by varying value of BP via loading mass ratio. KPHI/BP composite with 10% mass ratio demonstrated the maximum photocatalytic efficiency for HER 4.3 mmol/g/h and this value is three times greater than the pristine KPHI. This composite photocatalyst is highly effective, stable, simple, and cost effective and have better visible light absorption. Fig. 14.4A and B reveals the average HER rate for various values of KPHI/BP illumined by visible light and stability of KPHI/BP 10% for 5 successive cycles.

14.7.5 Layered double hydroxides

Anionic clays with a layered structure, known as LDHs, are characterized by their distinct atomic formula of $[M_{1-x}{}^{2+}M_x{}^{3+}(OH)_2](A^{n-})x/n.\ mH_2O$. A charge-balancing anion is A^{n-}, and the metal cations M^{3+} and M^{2+} are trivalent and divalent, respectively. Al, Cr, Ga, and Mn are utilized as M^{3+}(Han et al., 2013). Co, Mg, Ni, Zn, Fe, and Cu are employed as M^{2+}. Typically, LDHs are made up of layers which consists metallic cations and their ionic radii are comparable that are coupled by six atoms of oxygen in the $M^{2+}/M^{3+}(OH)_6$ octahedra structure (Khan and O'Hare, 2002) and via sharing edges, these octahedra form 2D sheets, which are then stacked together via hydrogen bonds among hydroxyl groups. Hydrothermal technique was used by Chong et al. (2017) to decorate CoAl-LDH, demonstrating a highly stable and effective a-Fe_2O_3 photoanode in the condition of neutral pH. They successfully synthesized the CoAl-LDH and then precisely integrated the synthesized LDH with α-Fe_2O_3. α-Fe_2O_3, Co–OH/α-Fe_2O_3, CoAl-LDH/α-Fe_2O_3, CoPi/a-Fe_2O_3, Al–OH/α-Fe_2O_3. At 1.23 V, photocurrent density was enhanced nine times versus RHE, and onset

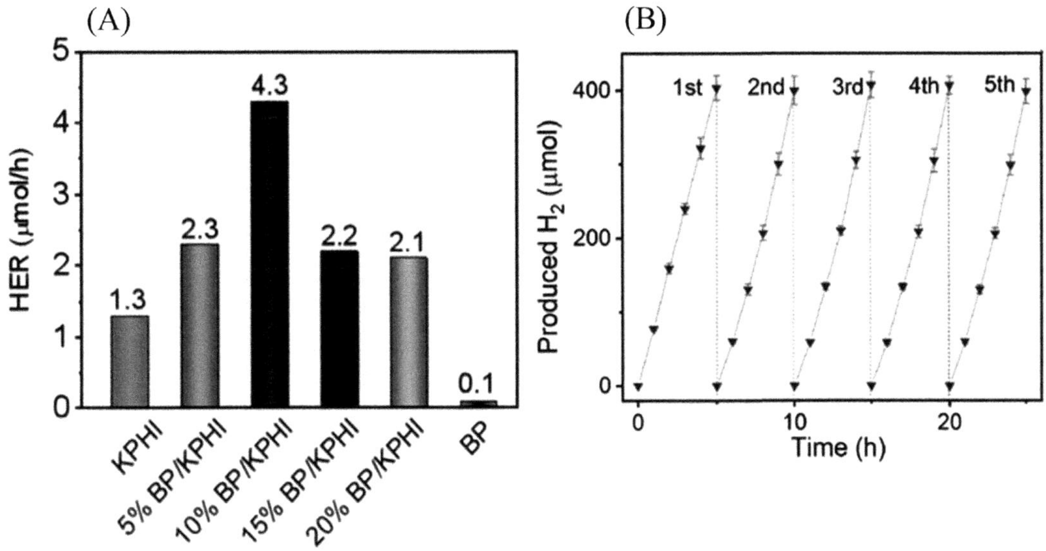

FIGURE 14.4 (A) Average HER rate for various values of KPHI/BP illumined by visible light and (B) stability of KPHI/BP 10% for 5 successive cycles (Xiao et al., 2023).

potential was moved 250 mV in the cathodic direction. CoAl-LDH/a-Fe_2O_3 performs better in high pH media as well as in neutral pH electrolyte. Choi et al. (2020) synthesized photoanode of Si with exceptional PEC activity by utilizing NiFe LDH and amorphous TiO_2 that was activated by cyclic voltammetry (CV).

To study the OER activity, Sun et al. (2024) synthesized the NiFe-LDH nanosheets and these nanosheets were doped by Ti_3C_2 and due to this, the charge transfer resistance decreased and its electrical conductivity increased to 49.5%. Therefore the NiFe-LDH/Ti_3C_2–60 mg composite electrocatalyst demonstrated a low value of Tafel slope 55.1 mV/dec and small value of overpotential of 334 mV and this value was 38.5% and 11.2% lower than the pristine NiFe/LDH. This approach offers a practical way to construct extremely effective electrocatalysts with nonnoble metals. Fig. 14.5A–D represents the morphological images and LSV curves and Tafel slope for the fabricated NiFe-LDH/Ti_3C_2 with various amounts of Ti_3C_2 (0, 15, 30, 45, 60, 75 mg).

Nasirpouri et al. (2024) discussed the electrochemical characteristics of prepared GO-NiAl/LDH. SEM studies revealed the successful GO-x% (x=5, 3, 1, 0.5, 0.3) hybridization with NiAl/LDH. At a 10 mA/cm^2, current density overpotential and onset potential were 443 mV and 0.61 V, respectively. Nanocomposite of GO-1%/NiAl/LDH exhibits improved water oxidation capability when exposed to a magnetic field. This is explained by an increase in electroactive sites and a decrease in oxygen bubble production on the surface of electrode brought on by Lorentzian motion. Under a magnetic field, it was shown that the synthesized nanocomposite with a GO of around 0.01 g (GO-1%/NiAl-LDH) acquired the lowest onset potential, roughly 0.61 V (vs SCE), to initiate water oxidation. Fig. 14.6A and B

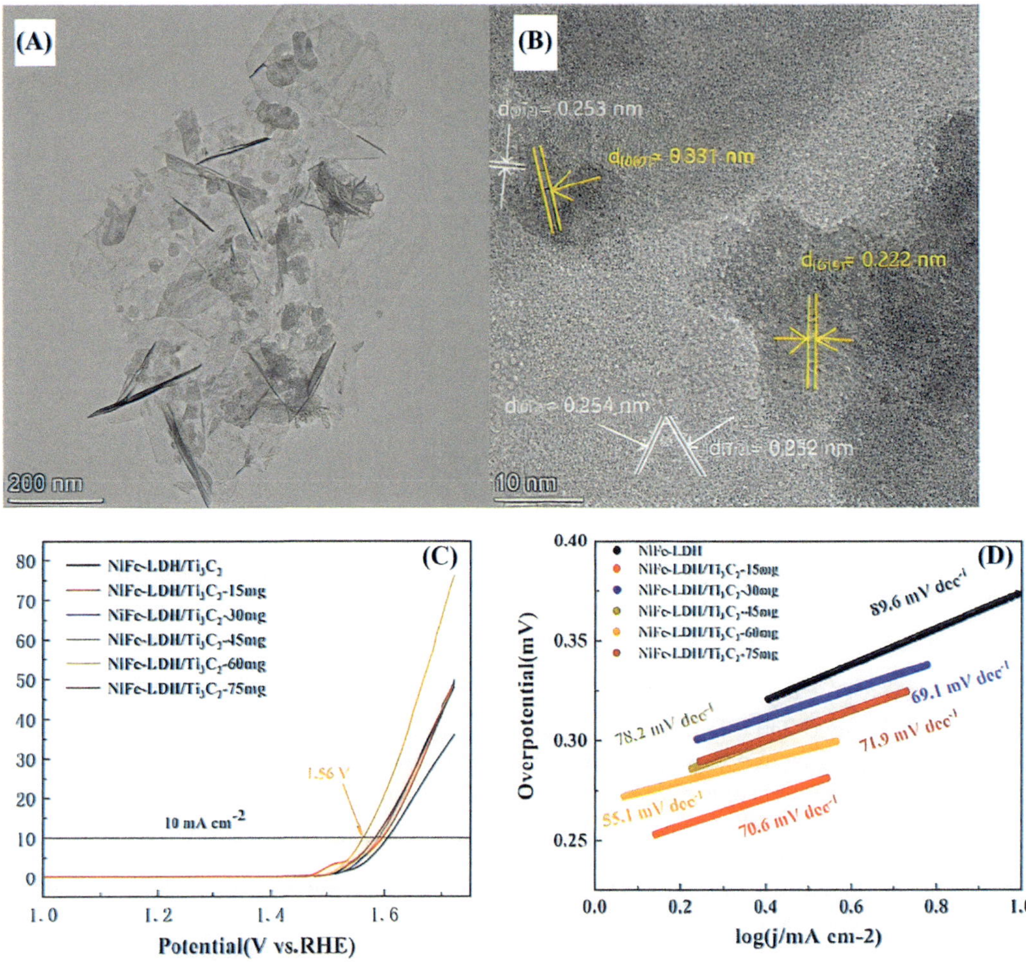

FIGURE 14.5 (A) Morphological image of 60 mg-Ti_3C_2/NiFe-LDH. (B) HRTEM image of 60 mg-Ti_3C_2/NiFe-LDH. (C, D) LSV curves and Tafel slope for the fabricated NiFe-LDH/Ti_3C_2 with various amounts of Ti_3C_2 (0, 15, 30, 45, 60, 75 mg), respectively (Sun et al., 2024).

illustrates the values of onset potential and overpotential (at current density 10 mA/cm) of the synthesized samples in the presence and absence of magnetic field (0 and 0.2 T).

Nanoparticles of CuO/Cu_2O were synthesized by Hameed et al. (2022) via HKUST-1 calcination during the CoMn/LDH nanosheets growth. The synthesized electrocatalyst exhibits high efficiency for OER activity, at 10 mA/cm current density only 297 mV value of overpotential was required for WS with Tafel slope 89 mV/dec. Furthermore, for OER stability is high for 15-hour period of time, only a minor current loss was observed. Fig. 14.6C and D depicts the LSV curves, Tafel slope of synthesized samples (CuO/Cu_2O, CoMn-LDH, CoMn-LDH@CuO/Cu_2O RuO_2).

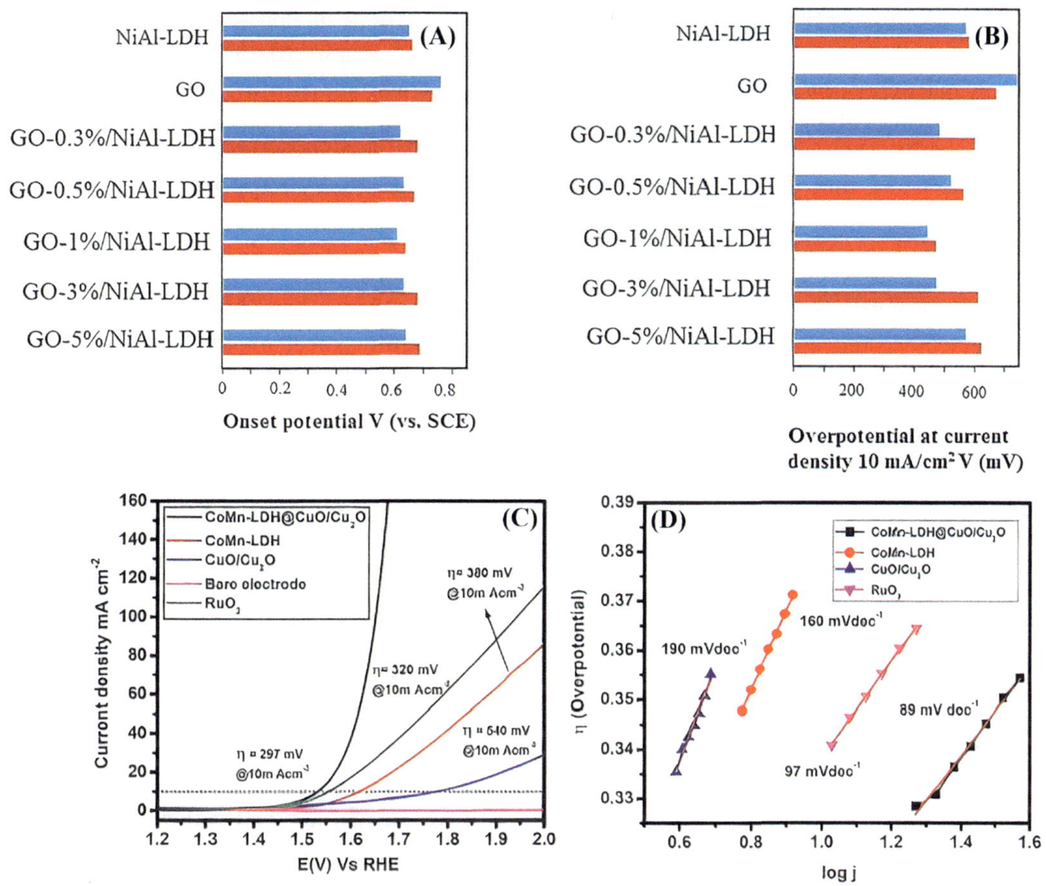

FIGURE 14.6 (A, B) Onset potential and overpotential (at current density 10 mA/cm) of the synthesized samples in the presence and absence of magnetic field (0 and 0.2 T) (Nasirpouri et al., 2024). (C) LSV curves of RuO_2, CuO/Cu_2O, CoMn-LDH, CoMn-LDH@CuO/Cu_2O and bare electrode. (D) Tafel slope of RuO_2, CuO/Cu_2O, CoMn-LDH, CoMn-LDH@CuO/Cu_2O (Hameed et al., 2022).

14.7.6 Graphitic carbon nitride

In photocatalysis and photoelectrochemistry, graphitic carbon nitride (g-C_3N_4) is regarded as a desirable material due to their outstanding bandgap suitability, physical and chemical stability, affordability, and pollution-free characteristic (Jiang et al., 2017). Still, certain limitations need to be addressed that include inadequate absorption of sunlight, a small surface area, and quick charge recombination. Until now, different approaches such as heterostructuring, elemental doping, increasing porosity, and morphological control have been used to produce high-performance g-C_3N_4 (Li et al., 2015; Bhat et al., 2020). Heterojunction of 3D ZnO/Ag/g-C_3N_4 (urchin-like) photocathode was synthesized by Wen et al. (2020) and this showed highly effective PEC WS. At 0.4 V versus RHE, ZnO/Au/g-C_3N_4 produced 0.91 mA/cm^2 photocurrent density. Furthermore, ZnO/Au/g-C_3N_4 demonstrated exceptional stability, yielding a value of 0.29 mA/cm^2 for a period of more than 36,000 seconds. An effective photoelectrode was synthesized with vertical alignment and hierarchically porous g-C_3N_4/CNT arrays by Jiang et al. (2021) through a vertical 3D printing technique. A clear redistribution of charges was noticed across the surface of the well-aligned g-C_3N_4/CNTs. In contrast, it was only seen at the ends of the sheets in the disordered g-C_3N_4/CNTs. It was established that the CNTs were intertwined and mutually supporting. The printed arrays yield a substantially improved photocurrent density than pure g-C_3N_4 and g-C_3N_4/CNT film.

Exfoliated g-CN was coated by a layer of porous carbon and as a result a heterostructure is formed in between them and then annealed under hydrogen atmosphere. The prepared sample was tested as photocatalyst for HER by Baranowska et al. (2022). The catalyst which was prepared at 400°C for 4 hours demonstrated the highest activity for HER. This material had 18.2 μmol/g rather than the ex-g-CN and its value was 0.8 μmol/g. Increased activity may result from (1) better separation and transfer of photogenerated carriers; (2) slowed down the process of recombination; (3) highly developed ECSA because of the layer of porous carbon. Largest photoresponse had been exhibited by the 400-4 g-CN photocatalyst and demonstrated excellent photogenerated carriers separation on the surface of photocatalyst. Fig. 14.7A–C illustrates photocatalytic HER production below simulated solar light by 400-4, and ex-g-CN, chronoamperometry spectra and comparison of evolved H_2 amount catalyzed by ex-g-CN modified by annealing under the H_2 atmosphere.

14.7.7 MXenes

The process of etching the "A" element from the ternary "MAX" structure yields 2D MXenes, which are made up of transition metal nitrides, carbides, and carbo-nitrides. "M" stands for a transition metal such as V, Mo, or Ti, "A" for Ga or Al, and "X" for N or C (Naguib et al., 2023). Due to its excellent conductivity, exceptional dispersibility, exposed active metal sites, and hefty surface area, which allows for a variety of applications, MXene has recently attracted a significant deal of attention (Huang et al., 2019a; Fan et al., 2019; Lim et al., 2021). Remarkably, special 2D MXene analogs such as $V_{0.2}Mo_{0.8}N_{1.2}$ solid solution and nitrogen-rich Mo_5N_6 nanosheets have been synthesized recently for electrochemical water splitting (Jin et al., 2022; Jin et al., 2018b; Jin et al., 2021). Because of their altered electronic state, they exhibited noticeably higher catalytic activity as compared to normal MXenes.

Ye et al. (2021) synthesized coated 2D Ti_3C_2-MXene through a simple hydrothermal and annealing procedure to successfully design highly active and stable α-Fe_2O_3 NRs. The photovoltaic cell's PEC performance increases with the atomic ratio of Ti, anode tends to expand, but if MXene is added in excess, the nanorods growth is impeded, resulting in the decline of PEC activity. Consequently, at 1.23 V versus RHE, α-Fe_2O_3/$MXene_{5/1}$ NRs generated 1.10 mA/cm^2 photocurrent density. $Ti_3C_2T_x$ MXene was deposited on Fe-doped nanorod arrays of TiO_2 by Yin et al. which is an effective photoanode. Electrodeposition was used to synthesize MXene, and hydrothermal was used to develop Fe–TiO_2 nanorod arrays (Yin et al., 2022). Nanosheets of $Ti_3C_2T_x$ adhered well to nanorods array of Fe–TiO_2 following electrodeposition. The 1.23 mA/cm^2 of photocurrent density is observed at 1.23 V

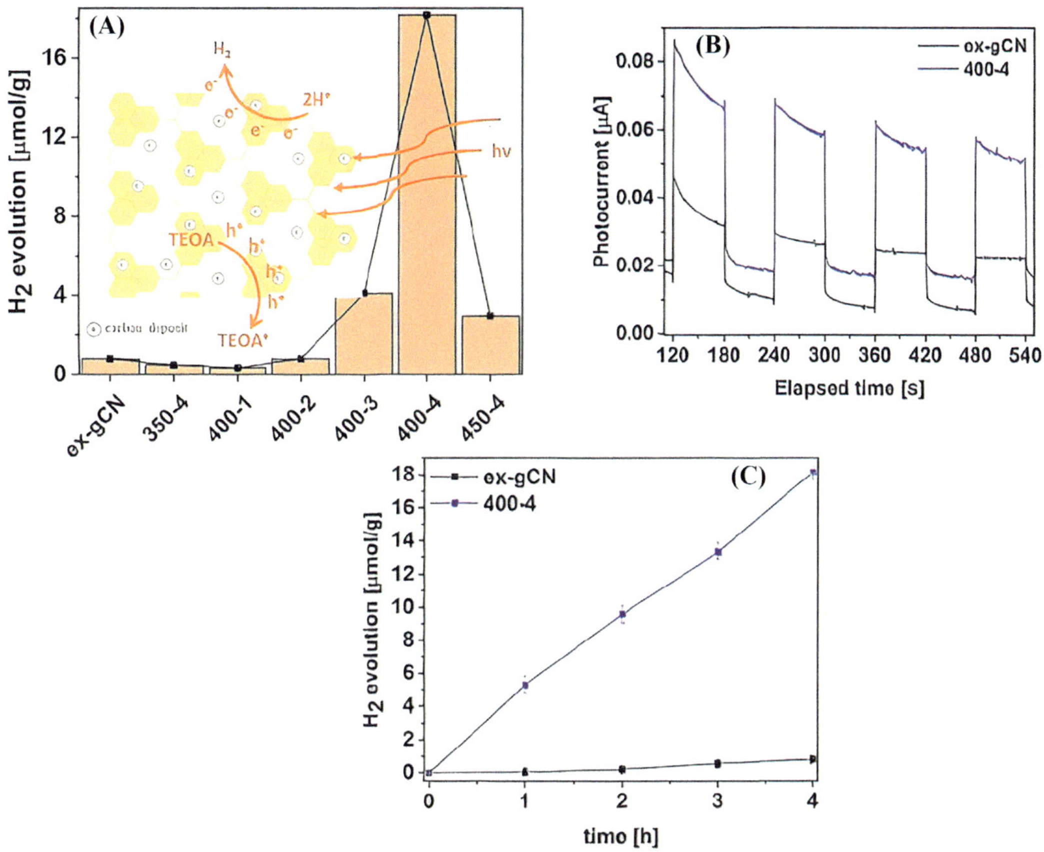

FIGURE 14.7 (A) Photocatalytic HER production below simulated solar light by 400-4 and ex-g-CN, (B) chronoamperometry spectra, (C) comparison of evolved H_2 amount catalyzed by ex-g-CN modified by annealing under the H_2 atmosphere (Baranowska et al., 2022).

versus RHE, greater than that of other TiO_2-based photoelectrodes. This outcome stems from $Ti_3C_2T_x$ nanosheets' superiority in enhancing Fe-TiO_2 water oxidation kinetics and charge transfer efficiency. Summary of 2D materials for WS and for HER production rate is briefly described in Tables 14.1 and 14.2.

14.8 SACs supported on diverse 2D materials

The discussions in this section will center on how 2DM/SACs are developed in relation to different substrate types. We will introduce SACs supported on LDHs, graphene, C_3N_4, 2D MOF, MXenes, TMDs, and 2D COF, along with their respective applications in the domain of WS.

TABLE 14.1 Summary of 2D materials for water splitting.

Catalyst	Electrolyte (M)	Type of reaction	Current density (mA/cm^2)	Overpotential (mV)	Tafel slope (mV/dec)	References
$Co_{0.4}Fe_{0.6}$ LDH/g-CNx	1M KOH	HER	10	280	29	Bhowmik et al. (2018)
NiFe (oxy)hydroxide	1M NaOH	HER	10	250	22.2	Sun et al. (2018)
$Ni_{2/3}Fe_{1/3}$-rGO	1M KOH	HER	10	230	42	Ma et al. (2015)
$NiSe_2$/CC	1M KOH	HER	10	1530	33.2	Zhou et al. (2018b)
FeCH@GDY/NF	1M KOH	HER	100	260	54.5	Hui et al. (2018)
ECT-$Co_{0.37}Ni_{0.26}Fe_{0.37}O$	1M KOH	HER	10	232	37.6	Chen et al. (2015)
NiS_2/V_2CT_x	1M KOH	HER	10	179	85	Kuang et al. (2019)
Ti_3 $CN(OH)_x@MoS_2$	0.5M H_2SO_4	HER	10	120	64	Jiang et al. (2023)
$CoNiFe_2O_4/Ti_3C_2T_x$	1M KOH	HER	10			Rasheed et al. (2023)
$MoS_2@Ti_3C_2T_x$ (Nano rolls)	0.5M H_2SO_4	HER	10	152	70	Liu et al. (2019b)
$Ni_3S_2@Ti_3C_2T_x$	1M KOH	HER	10	72	45	Tie et al. (2019)
$MoSe_2/NiSe_2MoSe_2$/CFP	0.5M H_2SO_4	HER	10	2494	46.959	Dong et al. (2015)
CoSe@NiFe LDH/NF	1M KOH	HER	10	–	89	Salarizadeh et al. (2019)
$Fe_{0.2}Ni_{0.8}Se$	1M KOH	HER	50	255	–	He et al. (2022)
NiCo-nitrides/$NiCo_2O_4$	1M KOH	HER		45	1.495	Zhang et al. (2022)
$MNi_{63}Co_{37}$/rGO		HER	10	115	45	Dong et al. (2021)
MoS_2/N–RGO		HER	10	168	32	Mei et al. (2022)
MoSSe/rGO		HER	10	258	98	Gowrisankar et al. (2021)
MoS2–BP		HER	10	85	209	Zhu et al. (2019)
MoS_2/Gr	0.5M H_2SO_4	HER	10	110	67.4	Li et al. (2021)

TABLE 14.2 Summary of 2D materials for HER production rate.

Photocatalyst	Source of light and its power	Preparation technique	Production rate of hydrogen (H_2) (μmol/g/h)	References
u/TiO_2@$Ti_3C_2T_x$	Xe lamp 300 W	HTo technique followed by photodeposition	860	Huang et al. (2020)
$LaFeO_3$/RGO	Xenon lamp 250 W	–	82,000	Samajdar et al. (2023)
Co_3O_4/$ZnIn_2S_4$	Xenon, 300 W	–	3844	Zhang et al. (2023)
Ti_3C_2/MoS_2/TiO_2	Xe arc lamp 300 W	Two-step hydrothermal technique	6425.29	Xiao et al. (2020)
$CdLa_2S_4$/Ti_3C_2 MXene	Xe lamp 300 W	Facile hydrothermal	11,182.4	Liu et al. (2020b)
$SnNb_2O_6$	Xe lamp 300 W (400 nm cuto filter)	–	264	Zhou et al. (2016)
MoO_2/Au/ $Mn_{0.5}Cd_{0.5}S$	Xe lamp 300 W	–	638.83	Liu et al. (2021)
PtO@Ti_3C_2/TiO_2	Xe lamp	Sonication	12,054	Peng et al. (2019)
CdS-MoS_2-MXenes	Xe lamp 300 W	Hydrothermal technique	9679	Zhang et al. (2020)
ZnCdS/TiO_2/Na-Ti_3C_2 MXene	Xe lamp 300 W	Hydrothermal	8436.58	Yu et al. (2019a)

14.8.1 Single atom site catalysts/graphene

Because of their broad surface area, stable chemical composition, and excellent conductivity, carbon-based materials were employed extensively as substrates to anchor single atoms (Wu et al., 2018; Liu et al., 2020a). In particular, graphene's excellent efficiency has garnered significant attention due to its monolayer planar shape. The graphene sheets' embedded functional groups and defects have the ability to modify surface and serve as anchoring points for each of the atoms, stabilizing them. Furthermore, the exceptional structure and high conductivity enable faster mass and electron transport kinetics, leading to enhanced electrocatalytic activity in the energy domain. Ru SAC supported on matrix of N-doped graphene was fabricated via annealing in an amount of NH_3 (Sun et al., 2019). Ru atoms' atomic dispersion form was determined using STEM methods. Concurrently, in NH_3 environment, doping of nitrogen and graphene oxide reduction were performed. Generated Ru-N/G-750 revealed excellent stability and strong activity for the oxygen reduction process (ORR), when given an appropriate voltage in acidic environments. When the stability test has run 10,000 times, 90% the activity was reserved for the preserved Ru single atomic species, indicating the material's excellent stability. Only very small deterioration was seen after 10,000 cycles.

It is vital to keep in mind that graphene's unique ordered structure makes it an ideal platform for assembling theoretical models to predict SAC performance. Fei et al. (2018) produced many 3D metal SACs (Ni, Fe, and Co) stacked on the frameworks of nitrogen-doped holey graphene (called M–NHGFs, where M=Ni, Fe, or Co) by using a generic and rational technique.

14.8.2 Single atom site catalysts/MXenes

Due to its hydrophilic characteristics, special layered structure, remarkable conductivity, and chemical stability have garnered significant interest in 2D transition metal nitrides or carbides, and MXenes have garnered significant attention for a lot of applications in energy sector (Jun et al., 2019). MXenes are appealing possibilities for permitting a wide range of electrochemical applications because of their distinctive qualities. Numerous energy conversion techniques have made substantial use of MXenes as substrates for single atom implantation, a use that has been thoroughly investigated. Zhang et al. studied synthesis of a Pt SAC built on the MXene nanosheets of double transition metal ($Mo_2TiC_2T_x$) with increased HER efficiency (Zhang et al., 2018b). During synthesis stage, a lot of surface Mo vacancies were produced by the electrochemical exfoliation method, which immobilized the individual Pt atoms. The final catalyst, $Mo_2TiC_2T_x$-PtSA, was synthesized as a result of single atom of Pt being attached by adjacent C atoms in MXene (Pt-C) and form covalent bonds. The resulting $Mo_2TiC_2T_x$-PtSA was equivalent to the Pt benchmark catalyst and showed remarkable catalytic efficacy for HER, requiring small potential of 77 and 30 mV to get 100 and 10 mA/cm^2, respectively. Almost no deterioration was seen during the 10,000 cycles (or 100 hours) of testing for stability, demonstrating the exceptional stability of $Mo_2TiC_2T_x$-PtSA. Specifically, Pt SAC's mass activity was almost 40 times more than commercial catalyst (Pt/C catalyst). In another study, Yang's group introduced SA-Zn MXene with a cathode of sulfur. High electronegativities of single Zn atoms on MXene enabled the catalysis of polysulfide conversion processes from Li_2S_4 to Li_2S_2/Li_2S and the achievement of a strong interaction with polysulfides as a result of the decreased energy barriers was intriguing.

14.8.3 Single atom site catalysts/C_3N_4

g-C_3N_4 is the most stable carbon nitride allotrope distinguished by its distinct structure, which consists of compositional motives with sp^{2}N and sp^{2}C atoms that form tri-s-triazine or s-triazine. This allotrope may be thought of as a particular instance of graphene. It has been the focus of much research recently because of its consistent N-coordinating cavities that trap isolated atoms and special optical characteristics, which have demonstrated that it is an excellent substrate for containing single atoms.

Li et al. introduced Pt atoms with high stability and dispersion into g-C_3N_4 by a straightforward impregnation process (Li et al., 2016a). FT-EXAFS curves show that rich Pt-N/C bonds have captured the solitary Pt atoms. The generated Pt SAC showed a remarkable high photocatalytic ability toward H_2 evolution reaction at 0.16 wt.% Pt loading.

It generated H_2 at a good rate of 318 μmol/h that is about 50 times more than bare g-C_3N_4 and 8.6 times than nanoparticles of Pt. Pt SAC has turnover frequency 775/h that is nine times greater compared to nanoparticles of Pt, demonstrating that SAC may significantly improve the process. The ultrafast transient absorption spectroscopy suggests that the g-C_3N_4 surface trap states may be modified by the inserted Pt atoms, which are accountable for observed improvement in efficiency. In a different study, Liu et al. (2018b) synthesized an atomically new form of Au_1N_x single-site enclosed in g-C_3N_4 that exhibits strong activity in photocatalysis as well as the OER and ORR processes. Because of the intrinsic structure of g-C_3N_4, doped heteroatoms may be used to stabilize individual atoms as well as alter the environment for coordination of isolated active metal catalysts. With a single Co_1-P_4 site inserted in nanosheets of g-C_3N_4, Liu et al. produced a unique Co SAC to split water overall using a straight forward photocatalysis approach (Liu et al., 2017a). The synthesis procedure involved two phases: (1) loading of single atoms of Co onto g-C_3N_4 and (2) phosphating under PH_3 environment for transferring active sites into centers of Co_1-P_4. These processes were identified via XPS spectra and XAS microscopy. With a maximum potential of 410.3 mmol/h/g under simulated solar irradiation, the as-prepared Co SAC outperformed the majority of known carbon nitride- and oxynitride-based materials in terms of WS performance and high progression rate of H_2. Lifespan of carriers may be increased 20 times rather than the pure g-C_3N_4 due to Co centers that prevent frequent charge recombination. This further improved water molecule activation and adsorption for reaction of oxygen evolution. This study suggests that treating a single active site with phosphidation is a viable approach to significantly enhance the total WS.

14.8.4 Single atom site catalysts/layered double hydroxides

Another common 2DM that shows potential for containing metal atoms is LDHs, which have a lot of OH^- groups and a stiff laminated structure. They typically exhibit minimal overpotential for OER due to their unique architectures, which has garnered them a lot of attention in recent years. In particular, research indicates that adding a third metal element will improve the OER process and result in even better performance. With the help of the specific interaction, the produced Ru/CoFe-LDHs with a Tafel slope of 39 mV/dec and a current density of 10 mA/cm^2 at a loading of 0.45 wt.% of Ru element only needed 198 mV. Stability test proved that current density might be sustained for up to 24 hours with very little decay (less than 1%), confirming the outstanding Ru/CoFe-LDHs performance, which significantly outperforms both commercial RuO_2 catalyst and pure CoFe-LDHs. Ru atoms functioned as active sites, which was verified by techniques of DFT+U and XAS and was crucial to the reaction. Additionally, as cocatalysts in the OER reaction, the CoFeLDHs sped up process by lowering kinetic barrier (Li et al., 2019b; Mori et al., 2017).

14.8.5 Single atom site catalysts/transition metal dichalcogenides

Researchers have focused a lot of interest on TMDs, particularly MoS_2, in recent decades due to their distinctive structures, electrical qualities, and enhanced actions toward HER; these compounds are thought to be a viable replacement for Pt (Feng et al., 2018). Many experiments have been conducted to enhance the activity, and it has been shown that restricting SACs is a successful strategy for reaching the goal. Single atoms can serve as auxiliaries to increase the HER activity, but in the realistic scenario, they are normally unable to function as active sites. It is shown that the unsaturated coordination number of the S atoms on their borders contain real active sites. In this regard, Deng and colleagues (Deng et al., 2015) originally described a unique way to induce the HER activity using a single-atom metal-doping technique. When Pt atoms were substituted for Mo atoms in the plane, the stability and activity of the resulting MoS_2 were noticeably higher than those of pure MoS_2. Doped adjacent Pt atoms were able to adjust the H atoms adsorption behavior on S atoms in plane, leading to an elevation in performance, as determined by the results of DFT simulations. This scheme may be applied to nonprecious metals in addition to atoms of noble metals. Wang et al. found that activity of HER may be raised in both acidic and alkaline environments by adding individual Ni atoms to pure MoS_2(Wang et al., 2018). Theoretical and experimental data demonstrated that the single Ni-doped atoms have a propensity to alter the way H atoms adsorb on nearby S atoms, which enhanced HER activity. To boost HER activity, single-atom restrictions are imposed by additional TMDs in addition to well-known MoS_2.

Han et al. fabricated monolayers of V SACs@1T-WS_2 via CVD technique which takes place directly in one step (Han et al., 2021). Raman spectroscopy and STEM results demonstrated that V species were atomically diffused on monolayers of 1TWS_2, on the other hand monolayers of 2H-WS_2 were formed without VCl_3 precursor under the same conditions. V SACs@1T-WS_2 demonstrated superior HER efficiency as compared to their 2H-WS_2 counterparts, at 100 mV the high value of TOF 3.01/sec and low value of Tafel slope 61 mV/dec and remarkable stability DFT simulations indicate that activated V-atom sites are responsible for improved HER performance.

14.8.6 Single atom site catalysts/2D metal organic framework

MOF is also a popular framework for making SACs because to its inimitable characteristics. Organic linkers combined together to form the skeleton which has a porous structure, large area of surface, and a superior conductivity, which contribute to the electrochemical effectiveness of derived SACs by improving atom trapping (Wan et al., 2021). These characteristics revealed that MOF is the ideal support for holding individual atoms in place. Recently, 2D MOF-supported SACs have gained popularity and are widely used in the energy conversion industry.

Zuo et al. synthesized PtSA-MNSs by utilizing Cu_2-$(COO)_4$ paddlewheel and PtII tetrakis porphyrin as linker to fabricate Pt SAC confined in ultrathin nanosheets of 2D MOF (Zuo et al., 2019). Loading of the resultant PtSA-MNSs using this approach may be as high as 12 wt.%. The ultrathin 2D structure of the resulting PtSA-MNSs allowed for faster transport of photon-generated carriers and plenty of accessible active sites, resulting in more effective photocatalytic efficiency in H_2 evolution reaction (11,320 mmol/g/h) via WS pursuant under irradiation of visible light as compared to other photocatalysts. Its catalytic activity still remained 93% after 4 cycles indicating its incredible durability. Using a wet chemistry approach devoid of pyrolysis, Lin et al. fabricated Co SAC confined on the 2D-conjugated organic framework of C_2N (Lin et al., 2020). It is possible to adjust the Co element loading up to 20.5 wt.%, which is nearly equal to the C_2N theoretical limit value. DFT computations revealed that the produced Co SAC exhibited remarkable bifunctional electrocatalytic properties toward ORR and OER.

14.8.7 Single atom site catalysts/2D covalent organic framework

A strong substrate to synthesize SACs which is the conjugated microporous polymers without metal is known as covalent organic framework (COF). Its distinct benefits include a hefty area of surface, the ability to control structure by altering the monomer choice, and a large number of substituted heteroatoms that can secure individual atoms through coordination. With the help of COF's synthetic diversity and distinct structure, it was possible to simply adjust the active sites' coordination environment and ultimately achieve exceptional catalytic performance.

Consequently, it has attracted a lot of attention and demonstrated great potential in last few years. In this regard, Zhong et al. created a Ni SAC supported on 2,2-bipyridine-based COF (Ni-TpBpy) that helped to reduce CO_2 to CO in a synergistic way when rendered to visible light (Zhong et al., 2019). Electrons from photosensitizer were transferred to distinct

TABLE 14.3 Electrocatalytic OER and HER activity of metal SACs confined in 2D supports.

Single atom	2D support	Reaction type	Electrolyte	Current density (mA/cm^2)	Tafel slope (mV/dec)	Overpotential (mV)	References
Pt	MXene	HER	0.5M H_2SO_4	10	30	30	Zhang et al. (2018b)
Mo	N-doped carbon	HER	0.5M H_2SO_4	10	86	154	Chen et al. (2017)
Ni	Graphdiyne	HER	0.5M H_2SO_4	10	45	88	Xue et al. (2018)
Pt	Fe-N-C	HER	0.1M	10	32	60	Zeng et al. (2018)
		OER	KOH		62	310	
Zn	MoS_2	HER	0.5M H_2SO_4	10	51		Shi et al. (2017)
Au	g-C_3N_4	OER	0.1M KOH	10	112	450	Liu et al. (2018b)
NiFe	g-C_3N_4	OER	1M KOH	10	67	326	Liu et al. (2018a)
Co	N-doped graphene	OER	1M KOH	10	80	402	Fei et al. (2018)
Ni	N-doped graphene	OER	1M KOH	10	63	301	Fei et al. (2018)

(*Continued*)

Table 14.3 (Continued)

Single atom	2D support	Reaction type	Electrolyte	Current density (mA/cm^2)	Tafel slope (mV/dec)	Overpotential (mV)	References
Fe	N-doped graphene	OER	1M KOH	10	164	488	Fei et al. (2018)
Mo	CN	HER	0.1M KOH	10	90	132	Chen et al. (2017)
Pd	MoS_2	HER	0.5M H_2SO_4	10	78	62–80	Luo et al. (2018)
Pd	Graphdiyne	HER	0.5M H_2SO_4	10	47	55	Yu et al. (2019b)

Ni sites throughout the reduction phase in order to produce CO. This produced 4057 μmol/g of CO with a 96% refinement over H_2 evolution in a 5-hour reaction time with good activity.

Nakanishi et al. developed Ru SAC attached on covalent triazine structure (Ru-CTF) in a separate study (Yamaguchi et al., 2017). Without any obvious OER process, this chemical showed clear selective alcohol oxidation activity. It was represented that in RU-CTF, the isolated atoms of Ru generated greatly reduced OER activity and enhanced the benzyl alcohol oxidation.

Some of the electrocatalytic OER and HER activity of metal SACs confined in 2D supports are enlisted in Table 14.3.

14.9 Conclusions

The sun's capacity to generate hydrogen energy without releasing carbon dioxide is one of the limitless energy sources that it may bestow onto the planet. In this regard, the development of affordable, good-quality photoelectrodes is essential for cost-effective hydrogen generation. Nevertheless, the photoelectrode alone is unable to prevent corrosion, speed up photogenerated charge transfer, and increase surface response. Therefore it is crucial to explore robust, catalytic, and highly conductive materials which are highly effective to synthesize the photoelectrodes with excellent PEC performance. In this study, we wrapped up the advantages of 2DM and discussed the use of several 2DM for PEC WS. Thickness of layers in 2DM have unique features due to van der Waals interactions, such as hefty area of surface, configurable bandgap, heterostructure stackability, and surface functionalization. They function as photoelectrodes in addition to supporting them. First of all, the surface reaction that produces oxygen and hydrogen is enhanced by their catalytic activity. A high photocurrent density and low overpotential are the result of specifically expanded active edge sites and defects, which help to reduce the amount of free energy needed for OER and HER. Furthermore, 2DM can outperform their inherent catalytic activity due to ion-exchanged sites, phase transition, and composites that work well together. 2DM which are chemically inert protect the silicon surface and photoelectrode which are oxide based and are vulnerable for alkaline and acidic electrolyte. Solar energy may be absorbed by 2DM with an energy bandgap, produces electron–hole pairs, and provide a photocurrent density in small amounts. Numerous 2DM have been discovered and used in photo/electrochemistry field up to this point. We divided them into carbon allotropes, LDHs, BP, MXene, TMDs, and g-C_3N_4. Even though 2DM have been successfully incorporated into a variety of semiconducting photoabsorbers, there are still problems that need to be solved in order to advance the technology. Firstly, for commercial use, 2DM must be directly fabricated on photoelectrodes in large quantities with homogeneity. We believe that scholars with an interest in 2DM, PEC catalysts, and renewable energy should find our comprehensive analysis on 2DM for PEC WS.

SACs supported on 2DM have drawn a lot of interest and have been widely used in a variety of energy conversion reactions in the recent years because of their distinctive structure and outstanding performance. Furthermore, interaction between 2D support and single atoms can result in unexpected properties. This means that in addition to the support's capacity to modify the electrical and geometrical configuration of individual active sites, thus boosting activity, furthermore, single atoms might operate as a supplement to increase the support's intrinsic activity, thereby improving performance. Furthermore, the SACs/2DM active sites homogeneity, which act as a link between heterogeneous and homogeneous catalytic systems, helps to create the perfect framework for a thorough examination of the fundamentals of heterogeneous catalysis, which will have a significant impact on the development of highly effective catalysts of the future. In this study, we also summarized the most current developments of SACs supported by different 2DM including 2D MOF, LDHs, 2D COF, MXenes, TMDs, graphene, and C_3N and their potential uses for several critical reactions in the sector of energy conversion. It is accepted that 2DM typically have large SSA, but increasing loading quantity of single atoms to fully use the surface area remains a gigantic difficulty. The development of more delicate and ubiquitous synthetic techniques for controlled fabrication of SACs/2DM is very desirable.

SACs/2DM durability is a key sign of whether or not SACs can be used for practical applications. During the catalytic process, homogeneous catalysis like reductive elimination and oxidative addition cycles may potentially diminish the strong contact that exists among 2D support and single atoms. In the ambient environment, certain nonnoble metals, such Ni, Co, and Fe are particularly quickly transformed into oxidative species. It could be challenging for SACs to find industrial use due to the ensuing activity deterioration. In addition, cost and intricacy of synthetic processes are other factors that should be considered. It will take time to implement SACs/2DM in its true sense because of these problems, hence more efforts has to be done to find early solutions.

References

Adnan, M., Baumberg, J.J., Vijaya Prakash, G., 2020. Linear and nonlinear optical probing of various excitons in 2D inorganic-organic hybrid structures. Scientific Reports 10, 2615.

Alexander, B.D., Kulesza, P.J., Rutkowska, I., Solarska, R., Augustynski, J., 2008. Metal oxide photoanodes for solar hydrogen production. Journal of Materials Chemistry 18, 2298–2303.

Andoshe, D.M., Jin, G., Lee, C.S., Kim, C., Kwon, K.C., Choi, S., et al., 2018. Directly assembled 3D molybdenum disulfide on silicon wafer for efficient photoelectrochemical water reduction. Advanced Sustainable Systems 2, 1700142.

Balandin, A.A., 2011. Thermal properties of graphene and nanostructured carbon materials. Nature Materials 10, 569–581.

Baranowska, D., Zielinkiewicz, K., Kedzierski, T., Mijowska, E., Zielinska, B., 2022. Heterostructure based on exfoliated graphitic carbon nitride coated by porous carbon for photocatalytic H_2 evolution. International Journal of Hydrogen Energy 47, 35666–35679.

Bhat, S.S., Pawar, S.A., Potphode, D., Moon, C.-K., Suh, J.M., Kim, C., et al., 2019. Substantially enhanced photoelectrochemical performance of TiO_2 nanorods/CdS nanocrystals heterojunction photoanode decorated with MoS_2 nanosheets. Applied Catalysis B 259, 118102.

Bhat, S.S., Jun, S.E., Lee, S.A., Lee, T.H., Jang, H.W., 2020. Influence of C_3N_4 precursors on photoelectrochemical behavior of TiO_2/C_3N_4 photoanode for solar water oxidation. Energies 13, 974.

Bhowmik, T., Kundu, M.K., Barman, S., 2018. CoFe layered double hydroxide supported on graphitic carbon nitrides: an efficient and durable bifunctional electrocatalyst for oxygen evolution and hydrogen evolution reactions. ACS Applied Energy Materials 1, 1200–1209.

Brill, A.R., Koren, E., De Ruiter, G., 2021. Molecular functionalization of 2D materials: from atomically planar 2D architectures to off-plane 3D functional materials. Journal of Materials Chemistry C 9, 11569–11587.

Cai, X., Luo, Y., Liu, B., Cheng, H.-M., 2018. Preparation of 2D material dispersions and their applications. Chemical Society Reviews 47, 6224–6266.

Chaiyakun, S., Witit-Anun, N., Nuntawong, N., Chindaudom, P., Oaew, S., Kedkeaw, C., et al., 2012. Preparation and characterization of graphene oxide nanosheets. Procedia Engineering 32, 759–764.

Chaves, A., Azadani, J.G., Alsalman, H., Da Costa, D., Frisenda, R., Chaves, A., et al., 2020. Bandgap engineering of two-dimensional semiconductor materials. NPJ 2D Materials and Applications 4, 29.

Chen, Y., Xi, J., Dumcenco, D.O., Liu, Z., Suenaga, K., Wang, D., et al., 2013. Tunable band gap photoluminescence from atomically thin transition-metal dichalcogenide alloys. ACS Nano 7, 4610–4616.

Chen, W., Pei, J., He, C.T., Wan, J., Ren, H., Zhu, Y., et al., 2017. Rational design of single molybdenum atoms anchored on N-doped carbon for effective hydrogen evolution reaction. Angewandte Chemie 129, 16302–16306.

Chen, W., Wang, H., Li, Y., Liu, Y., Sun, J., Lee, S., et al., 2015. In situ electrochemical oxidation tuning of transition metal disulfides to oxides for enhanced water oxidation. ACS Central Science 1, 244–251.

Chen, C.-J., Veeramani, V., Wu, Y.-H., Jena, A., Yin, L.-C., Chang, H., et al., 2020a. Phosphorous-doped molybdenum disulfide anchored on silicon as an efficient catalyst for photoelectrochemical hydrogen generation. Applied Catalysis B 263, 118259.

Chen, H., Wang, S., Wu, J., Zhang, X., Zhang, J., Lyu, M., et al., 2020b. Identifying dual functions of rGO in a $BiVO_4$/rGO/NiFe-layered double hydroxide photoanode for efficient photoelectrochemical water splitting. Journal of Materials Chemistry A 8, 13231–13240.

Chen, Z., Wang, X., Mills, J.P., Du, C., Kim, J., Wen, J., et al., 2021. Two-dimensional materials for electrochemical CO_2 reduction: materials, in situ/operando characterizations, and perspective. Nanoscale 13, 19712–19739.

Chhowalla, M., Shin, H.S., Eda, G., Li, L.-J., Loh, K.P., Zhang, H., 2013. The chemistry of two-dimensional layered transition metal dichalcogenide nanosheets. Nature Chemistry 5, 263–275.

Chia, X., Pumera, M., 2018. Characteristics and performance of two-dimensional materials for electrocatalysis. Nature Catalysis 1, 909–921.

Choi, S., Lee, S.A., Yang, H., Lee, T.H., Kim, C., Lee, C.W., et al., 2020. Stabilization of NiFe layered double hydroxides on n-Si by an activated TiO2 interlayer for efficient solar water oxidation. ACS Applied Energy Materials 3, 12298–12307.

Chong, R., Wang, B., Su, C., Li, D., Mao, L., Chang, Z., et al., 2017. Dual-functional CoAl layered double hydroxide decorated α-Fe_2O_3 as an efficient and stable photoanode for photoelectrochemical water oxidation in neutral electrolyte. Journal of Materials Chemistry A 5, 8583–8590.

Chung, C.-C., Yeh, H., Wu, P.-H., Lin, C.-C., Li, C.-S., Yeh, T.-T., et al., 2021. Atomic-layer controlled interfacial band engineering at two-dimensional layered $PtSe_2$/Si heterojunctions for efficient photoelectrochemical hydrogen production. ACS Nano 15, 4627–4635.

Cullen, P.L., Cox, K.M., Bin Subhan, M.K., Picco, L., Payton, O.D., Buckley, D.J., et al., 2017. Ionic solutions of two-dimensional materials. Nature Chemistry 9, 244–249.

Daukiya, L., Seibel, J., De Feyter, S., 2019. Chemical modification of 2D materials using molecules and assemblies of molecules. Advances in Physics 4, 1625723.

Deng, J., Li, H., Xiao, J., Tu, Y., Deng, D., Yang, H., Tian, H., Li, J., Ren, P., Bao, X., 2015. Triggering the electrocatalytic hydrogen evolution activity of the inert two-dimensional MoS_2 surface via single-atom metal doping. Energy and Environmental Science 8, 1594–1601.

Di, J., Xia, J., Li, H., Liu, Z., 2017. Freestanding atomically-thin two-dimensional materials beyond graphene meeting photocatalysis: opportunities and challenges. Nano Energy 35, 79–91.

Dong, H., Liu, C., Ye, H., Hu, L., Fugetsu, B., Dai, W., et al., 2015. Three-dimensional nitrogen-doped graphene supported molybdenum disulfide nanoparticles as an advanced catalyst for hydrogen evolution reaction. Scientific Reports 5, 17542.

Dong, J., Sun, T., Zhang, Y., Zhang, H., Lu, S., Hu, D., et al., 2021. Mesoporous NiCo alloy/reduced graphene oxide nanocomposites as efficient hydrogen evolution catalysts. Journal of Colloid and Interface Science 599, 603–610.

Dresselhaus, M., Eklund, P., 2000. Phonons in carbon nanotubes. Advances in Physics 49, 705–814.

Du, X., Skachko, I., Barker, A., Andrei, E.Y., 2008. Approaching ballistic transport in suspended graphene. Nature Nanotechnology 3, 491–495.

Du Plessis, A., Du Plessis, A., 2019. Climate change: current drivers, observations and impacts on the Globe's natural and human systems. Water as an Inescapable Risk: Current Global Water Availability, Quality and Risks with a Specific Focus on South Africa. pp. 27–53.

Edy, R., Zhao, Y., Huang, G.S., Shi, J.J., Zhang, J., Solovev, A.A., et al., 2016. TiO_2 nanosheets synthesized by atomic layer deposition for photocatalysis. Progress in Natural Science 26, 493–497.

Eom, T.H., Cho, S.H., Suh, J.M., Kim, T., Lee, T.H., Jun, S.E., et al., 2021. Substantially improved room temperature NO_2 sensing in 2-dimensional SnS_2 nanoflowers enabled by visible light illumination. Journal of Materials Chemistry A 9, 11168–11178.

Fan, X., Ding, Y., Liu, Y., Liang, J., Chen, Y., 2019. Plasmonic Ti_3C_2T x MXene enables highly efficient photothermal conversion for healable and transparent wearable device. ACS Nano 13, 8124–8134.

Fareza, A.R., Nugroho, F.A.A., Abdi, F.F., Fauzia, V., 2022. Nanoscale metal oxides–2D materials heterostructures for photoelectrochemical water splitting—a review. Journal of Materials Chemistry A 10, 8656–8686.

Fei, H., Dong, J., Feng, Y., Allen, C.S., Wan, C., Volosskiy, B., et al., 2018. General synthesis and definitive structural identification of MN4C4 single-atom catalysts with tunable electrocatalytic activities. Nature Catalysis 1, 63–72.

Feng, Y., Ding, L., Ji, D., Wang, L., Guo, W., 2018. Highly rectified ion transport through 2D WSe_2/MoS_2 bi-layered membranes. Chinese Chemical Letters 29, 892–894.

Gao, X., Liu, H., Wang, D., Zhang, J., 2019. Graphdiyne: synthesis, properties, and applications. Chemical Society Reviews 48, 908–936.

Gaur, A., Shrivastava, B.D., Nigam, H., 2013. X-ray absorption fine structure (XAFS) spectroscopy–a review. Proceedings of the Indian National Science Academy 921–966.

Geim, A.K., Grigorieva, I.V., 2013. Van der Waals heterostructures. Nature 499, 419–425.

Geißler, T., Plevan, M., Abánades, A., Heinzel, A., Mehravaran, K., Rathnam, R.K., et al., 2015. Experimental investigation and thermo-chemical modeling of methane pyrolysis in a liquid metal bubble column reactor with a packed bed. International Journal of Hydrogen Energy 40, 14134–14146.

Geng, D., Yu, G., 2018. Liquid catalysts: an innovative solution to 2D materials in CVD processes. Materials Horizons 5, 1021–1034.

Gnanasekar, P., Periyanagounder, D., Varadhan, P., He, J.-H., Kulandaivel, J., 2019. Highly efficient and stable photoelectrochemical hydrogen evolution with 2D-NbS_2/Si nanowire heterojunction. ACS Applied Materials and Interfaces 11, 44179–44185.

Gong, Y., Liu, Z., Lupini, A.R., Shi, G., Lin, J., Najmaei, S., et al., 2014. Band gap engineering and layer-by-layer mapping of selenium-doped molybdenum disulfide. Nano Letters 14, 442–449.

Gowrisankar, A., Sherryn, A.L., Selvaraju, T., 2021. In situ integrated 2D reduced graphene oxide nanosheets with MoSSe for hydrogen evolution reaction and supercapacitor application. Applied Surface Science Advances 3, 100054.

Gui, G., Li, J., Zhong, J., 2008. Band structure engineering of graphene by strain: first-principles calculations. Physical Review B 78, 075435.

Hackett, S.F., Brydson, R.M., Gass, M.H., Harvey, I., Newman, A.D., Wilson, K., et al., 2013. High-activity, single-site mesoporous Pd/Al_2O_3 catalysts for selective aerobic oxidation of allylic alcohols. Asian Journal of Control 15.

Hameed, A., Zulfiqar, F., Iqbal, W., Ali, H., Shah, S.S.A., Nadeem, M.A., 2022. Electrocatalytic water oxidation on CuO–Cu_2O modulated cobalt-manganese layered double hydroxide. RSC Advances 12, 28954–28960.

Han, J., Dou, Y., Zhao, J., Wei, M., Evans, D.G., Duan, X., 2013. Flexible CoAl LDH@ PEDOT core/shell nanoplatelet array for high-performance energy storage. Small 9, 98–106.

Han, W., Liu, K., Yang, S., Wang, F., Su, J., Jin, B., et al., 2019. Salt-assisted chemical vapor deposition of two-dimensional materials. Science China Chemistry 62, 1300–1311.

Han, A., Zhou, X., Wang, X., Liu, S., Xiong, Q., Zhang, Q., Gu, L., Zhuang, Z., Zhang, W., Li, F., Wang, D., Li, L.J., Li, Y., 2021. One-step synthesis of single-site vanadium substitution in 1T-WS_2 monolayers for enhanced hydrogen evolution catalysis. Nature Communications 12, 709.

Harris-Lee, T.R., Marken, F., Bentley, C.L., Zhang, J., Johnson, A.L., 2023. A chemist's guide to photoelectrode development for water splitting–the importance of molecular precursor design. EES Catalysis 1, 832–873.

Hasani, A., Van Le, Q., Tekalgne, M., Choi, M.-J., Choi, S., Lee, T.H., et al., 2019. Fabrication of a WS_2/p-Si heterostructure photocathode using direct hybrid thermolysis. ACS Applied Materials and Interfaces 11, 29910–29916.

He, J., Qian, T., Cai, C., Xiang, X., Li, S., Zu, X., 2022. Nickel-based selenides with a fractal structure as an excellent bifunctional electrocatalyst for water splitting. Nanomaterials 12, 281.

Hisatomi, T., Domen, K., 2019. Reaction systems for solar hydrogen production via water splitting with particulate semiconductor photocatalysts. Nature Catalysis 2, 387–399.

Hou, Y., Zuo, F., Dagg, A.P., Liu, J., Feng, P., 2014. Branched WO_3 nanosheet array with layered C_3N_4 heterojunctions and CoOx nanoparticles as a flexible photoanode for efficient photoelectrochemical water oxidation. Advanced Materials 26, 5043–5049.

Huang, H., Cui, J., Liu, G., Bi, R., Zhang, L., 2019a. Carbon-coated $MoSe_2$/MXene hybrid nanosheets for superior potassium storage. ACS Nano 13, 3448–3456.

Huang, J.-J., Liu, X.-Q., Meng, F.-F., He, L.-Q., Wang, J.-X., Wu, J.-C., et al., 2020. A facile method to produce $MoSe_2$/MXene hybrid nanoflowers with enhanced electrocatalytic activity for hydrogen evolution. Journal of Electroanalytical Chemistry 856, 113727.

Huang, W., Zhou, Q., Su, S., Li, J., Lu, X., Gao, X., et al., 2019b. Ion beam defect engineering on ReS_2/Si photocathode with significantly enhanced hydrogen evolution reaction. Advanced Materials Interfaces 6, 1801663.

Hui, L., Jia, D., Yu, H., Xue, Y., Li, Y., 2018. Ultrathin graphdiyne-wrapped iron carbonate hydroxide nanosheets toward efficient water splitting. ACS Applied Materials and Interfaces 11, 2618–2625.

Hurteau, M.D., Liang, S., Westerling, A.L., Wiedinmyer, C., 2019. Vegetation-fire feedback reduces projected area burned under climate change. Scientific Reports 9, 2838.

Jayasena, B., Subbiah, S., 2011. A novel mechanical cleavage method for synthesizing few-layer graphenes. Nanoscale Research Letters 6, 1–7.

Jenjeti, R.N., Austeria, M.P., Sampath, S., 2016. Alternate to molybdenum disulfide: a 2D, few-layer transition-metal thiophosphate and its hydrogen evolution reaction activity over a wide pH range. ChemElectroChem 3, 1392–1399.

Jiang, B., Huang, H., Gong, W., Gu, X., Liu, T., Zhang, J., et al., 2021. Wood-inspired binder enabled vertical 3D printing of g-C_3N_4/CNT arrays for highly efficient photoelectrochemical hydrogen evolution. Advanced Functional Materials 31, 2105045.

Jiang, J., Li, F., Bai, S., Wang, Y., Xiang, K., Wang, H., et al., 2023. Carbonitride MXene Ti_3CN (OH) x@ MoS_2 hybrids as efficient electrocatalyst for enhanced hydrogen evolution. Nano Research 16, 4656–4663.

Jiang, J., Saito, R., Samsonidze, G.G., Chou, S., Jorio, A., Dresselhaus, G., et al., 2005. Electron-phonon matrix elements in single-wall carbon nanotubes. Physical Review B 72, 235408.

Jiang, L., Yuan, X., Pan, Y., Liang, J., Zeng, G., Wu, Z., et al., 2017. Doping of graphitic carbon nitride for photocatalysis: a review. Applied Catalysis B 217, 388–406.

Jin, H., Guo, C., Liu, X., Liu, J., Vasileff, A., Jiao, Y., et al., 2018a. Emerging two-dimensional nanomaterials for electrocatalysis. Chemical Reviews 118, 6337–6408.

Jin, H., Liu, X., Vasileff, A., Jiao, Y., Zhao, Y., Zheng, Y., et al., 2018b. Single-crystal nitrogen-rich two-dimensional Mo_5N_6 nanosheets for efficient and stable seawater splitting. Acs Nano 12, 12761–12769.

Jin, H., Song, T., Paik, U., Qiao, S.-Z., 2021. Metastable two-dimensional materials for electrocatalytic energy conversions. Accounts of Materials Research 2, 559–573.

Jin, H., Yu, H., Li, H., Davey, K., Song, T., Paik, U., et al., 2022. MXene analogue: a 2D nitridene solid solution for high-rate hydrogen production. Angewandte Chemie 134, e202203850.

Jun, S.E., Choi, S., Choi, S., Lee, T.H., Kim, C., Yang, J.W., et al., 2021. Direct synthesis of molybdenum phosphide nanorods on silicon using graphene at the heterointerface for efficient photoelectrochemical water reduction. Nano-Micro Letters 13, 1–16.

Jun, B.-M., Kim, S., Heo, J., Park, C.M., Her, N., Jang, M., et al., 2019. Review of MXenes as new nanomaterials for energy storage/delivery and selected environmental applications. Nano Research 12, 471–487.

Jun, S.E., Lee, J.K., Jang, H.W., 2023. Two-dimensional materials for photoelectrochemical water splitting. Energy Advances 2, 34–53.

Khan, A.I., O'Hare, D., 2002. Materials feature article. Journal of Materials Chemistry 12, 3191–3198.

Kim, Y., Jun, S.E., Lee, G., Nam, S., Jang, H.W., Park, S.H., et al., 2023. Recent advances in water-splitting electrocatalysts based on electrodeposition. Materials 16, 3044.

Kondo, A., Tiew, C.C., Moriguchi, F., Maeda, K., 2013. Fabrication of metal–organic framework nanosheets and nanorolls with N-donor type bridging ligands. Dalton Transactions 42, 15267–15270.

Kuang, P., He, M., Zhu, B., Yu, J., Fan, K., Jaroniec, M., 2019. 0D/2D NiS_2/V-MXene composite for electrocatalytic H_2 evolution. Journal of Catalysis 375, 8–20.

Kumar, S.S., Himabindu, V., 2019. Hydrogen production by PEM water electrolysis–a review. Materials Science for Energy Technologies 2, 442–454.

Kwon, K.C., Choi, S., Hong, K., Andoshe, D.M., Suh, J.M., Kim, C., et al., 2017. Tungsten disulfide thin film/p-type Si heterojunction photocathode for efficient photochemical hydrogen production. MRS Communications 7, 272–279.

Kwon, K.C., Suh, J.M., Varma, R.S., Shokouhimehr, M., Jang, H.W., 2019. Electrocatalytic water splitting and CO_2 reduction: sustainable solutions via single-atom catalysts supported on 2D materials. Small Methods 3, 1800492.

Lee, C., Wei, X., Kysar, J.W., Hone, J., 2008. Measurement of the elastic properties and intrinsic strength of monolayer graphene. Science 321, 385–388.

Lee, S.A., Yang, J.W., Choi, S., Jang, H.W., 2021. Nanoscale electrodeposition: dimension control and 3D conformality. Exploration Wiley Online Library, 20210012.

Liang, Q., Brocks, G., Bieberle-Hütter, A., 2021. Oxygen evolution reaction (OER) mechanism under alkaline and acidic conditions. Journal of Physics 3, 026001.

Lim, H.-S., Choi, S.B., Kwon, H., Lim, J.-W., Han, C.J., Oh, J.-M., et al., 2021. Development of a highly flexible composite electrode comprised of Ti_3C_2-based MXene nanosheets and Ag nanoparticles. Electronic Materials Letters 17, 513–520.

Ling, C., Li, Q., Du, A., Wang, J., 2018a. Computation-aided design of single-atom catalysts for one-pot CO_2 capture, activation, and conversion. ACS Applied Materials and Interfaces 10, 36866–36872.

Ling, C., Niu, X., Li, Q., Du, A., Wang, J., 2018b. Metal-free single atom catalyst for N_2 fixation driven by visible light. Journal of the American Chemical Society 140, 14161–14168.

Lin, J., Wang, A., Qiao, B., Liu, X., Yang, X., Wang, X., et al., 2013. Remarkable performance of Ir1/FeO x single-atom catalyst in water gas shift reaction. Journal of the American Chemical Society 135, 15314–15317.

Lin, C., Zhang, H., Song, X., Kim, D.-H., Li, X., Jiang, Z., Lee, J.-H., 2020. 2D-organic framework confined metal single atoms with the loading reaching the theoretical limit. Materials Horizons 7, 2726–2733.

Liu, W., Cao, L., Cheng, W., Cao, Y., Liu, X., Zhang, W., et al., 2017a. Single-site active cobalt-based photocatalyst with a Long carrier lifetime for spontaneous overall water splitting. Angewandte Chemie 129, 9440–9445.

Liu, D., Ding, S., Wu, C., Gan, W., Wang, C., Cao, D., et al., 2018a. Synergistic effect of an atomically dual-metal doped catalyst for highly efficient oxygen evolution. Journal of Materials Chemistry A 6, 6840–6846.

Liu, H., Du, Y., Deng, Y., Peide, D.Y., 2015. Semiconducting black phosphorus: synthesis, transport properties and electronic applications. Chemical Society Reviews 44, 2732–2743.

Liu, Y., Kang, Z., Si, H., Li, P., Cao, S., Liu, S., et al., 2017b. Cactus-like hierarchical nanorod-nanosheet mixed dimensional photoanode for efficient and stable water splitting. Nano Energy 35, 189–198.

Liu, J., Liu, Y., Xu, D., Zhu, Y., Peng, W., Li, Y., et al., 2019b. Hierarchical "nanoroll" like MoS_2/Ti_3C_2Tx hybrid with high electrocatalytic hydrogen evolution activity. Applied Catalysis B 241, 89–94.

Liu, D., Shadike, Z., Lin, R., Qian, K., Li, H., Li, K., et al., 2019a. Review of recent development of in situ/operando characterization techniques for lithium battery research. Advanced Materials 31, 1806620.

Liu, L., Su, H., Tang, F., Zhao, X., Liu, Q., 2018b. Confined organometallic Au1Nx single-site as an efficient bifunctional oxygen electrocatalyst. Nano Energy 46, 110–116.

Liu, Y., Wang, L., Lv, H., Wu, X., Xing, X., Song, S., 2021. Constructing robust MoO_2/Au/Mn0. 5Cd0. 5S multiple heterojunctions for improved photocatalytic hydrogen evolution: an insight into the synergetic effect of MoO_2 and Au cocatalysts. Applied Surface Science 541, 148582.

Liu, H., Wang, C., Wang, G., 2020b. Photocatalytic advanced oxidation processes for water treatment: recent advances and perspective. Chemistry 15, 3239–3253.

Liu, Y., Wu, J., Hackenberg, K.P., Zhang, J., Wang, Y.M., Yang, Y., et al., 2017c. Self-optimizing, highly surface-active layered metal dichalcogenide catalysts for hydrogen evolution. Nature Energy 2, 1–7.

Liu, P., Zhao, Y., Qin, R., Mo, S., Chen, G., Gu, L., et al., 2016. Photochemical route for synthesizing atomically dispersed palladium catalysts. Science 352, 797–800.

Liu, G., Zhou, J., Zhao, W., Ao, Z., An, T., 2020a. Single atom catalytic oxidation mechanism of formaldehyde on Al doped graphene at room temperature. Chinese Chemical Letters 31, 1966–1969.

Li, X., Bi, W., Zhang, L., Tao, S., Chu, W., Zhang, Q., et al., 2016a. Single-atom Pt as co-catalyst for enhanced photocatalytic H_2 evolution. Advanced Materials 28, 2427–2431.

Li, C., Cao, Q., Wang, F., Xiao, Y., Li, Y., Delaunay, J.-J., et al., 2018. Engineering graphene and TMDs based van der Waals heterostructures for photovoltaic and photoelectrochemical solar energy conversion. Chemical Society Reviews 47, 4981–5037.

Li, S., Sun, J., Guan, J., 2021. Strategies to improve electrocatalytic and photocatalytic performance of two-dimensional materials for hydrogen evolution reaction. Chinese Journal of Catalysis 42, 511–556.

Li, P., Wang, M., Duan, X., Zheng, L., Cheng, X., Zhang, Y., et al., 2019b. Boosting oxygen evolution of single-atomic ruthenium through electronic coupling with cobalt-iron layered double hydroxides. Nature Communications 10, 1711.

Li, F.L., Wang, P., Huang, X., Young, D.J., Wang, H.F., Braunstein, P., et al., 2019a. Large-scale, bottom-up synthesis of binary metal–organic framework nanosheets for efficient water oxidation. Angewandte Chemie 131, 7125–7130.

Li, Y., Wang, R., Li, H., Wei, X., Feng, J., Liu, K., et al., 2015. Efficient and stable photoelectrochemical seawater splitting with TiO_2@ g-C_3N_4 nanorod arrays decorated by Co-Pi. The Journal of Physical Chemistry C 119, 20283–20292.

Li, C., Wang, H., Lu, D., Wu, W., Ding, J., Zhao, X., et al., 2017. Visible-light-driven water splitting from dyeing wastewater using Pt surface-dispersed TiO_2-based nanosheets. Journal of Alloys and Compounds 699, 183–192.

Li, X., Yu, J., Wageh, S., Al-Ghamdi, A.A., Xie, J., 2016b. Graphene in photocatalysis: a review. Small 12, 6640–6696.

Long, H., Harley-Trochimczyk, A., Pham, T., Tang, Z., Shi, T., Zettl, A., et al., 2016. High surface area MoS2/graphene hybrid aerogel for ultrasensitive NO_2 detection. Advanced Functional Materials 26, 5158–5165.

Long, M., Wang, P., Fang, H., Hu, W., 2019. Progress, challenges, and opportunities for 2D material based photodetectors. Advanced Functional Materials 29, 1803807.

Luo, Z., Ouyang, Y., Zhang, H., Xiao, M., Ge, J., Jiang, Z., et al., 2018. Chemically activating MoS_2 via spontaneous atomic palladium interfacial doping towards efficient hydrogen evolution. Nature Communications 9, 2120.

Lu, J., Aydin, C., Browning, N.D., Gates, B.C., 2012. Imaging isolated gold atom catalytic sites in zeolite NaY. Angewandte Chemie International Edition 24, 5842–5846.

Mann, M.E., Rahmstorf, S., Kornhuber, K., Steinman, B.A., Miller, S.K., Coumou, D., 2017. Influence of anthropogenic climate change on planetary wave resonance and extreme weather events. Scientific Reports 7, 1–12.

Manzeli, S., Allain, A., Ghadimi, A., Kis, A., 2015. Piezoresistivity and strain-induced band gap tuning in atomically thin MoS_2. Nano Letters 15, 5330–5335.

Mao, X., Ling, C., Tang, C., Yan, C., Zhu, Z., Du, A., 2018b. Predicting a new class of metal-organic frameworks as efficient catalyst for bi-functional oxygen evolution/reduction reactions. Journal of Catalysis 367, 206–211.

Mao, J., Zhou, T., Zheng, Y., Gao, H., Kun Liu, H., Guo, Z., 2018a. Two-dimensional nanostructures for sodium-ion battery anodes. Journal of Materials Chemistry A 6, 3284–3303.

Martín, N., Tagmatarchis, N., Wang, Q.H., Zhang, X., 2020. Chemical functionalization of 2D materials. Chemistry.

Mayorga-Martinez, C.C., Sofer, Z., Sedmidubsky, D., Huber, S., Eng, A.Y.S., Pumera, M., 2017. Layered metal thiophosphite materials: magnetic, electrochemical, and electronic properties. ACS Applied Materials and Interfaces 9, 12563–12573.

Ma, W., Ma, R., Wang, C., Liang, J., Liu, X., Zhou, K., et al., 2015. A superlattice of alternately stacked Ni–Fe hydroxide nanosheets and graphene for efficient splitting of water. ACS Nano 9, 1977–1984.

Mei, J., Liao, T., Sun, Z., 2022. 2D/2D heterostructures: rational design for advanced batteries and electrocatalysis. Energy and Environmental Materials 5, 115–132.

Mohanty, B., Jena, B.K., Basu, S., 2020. Single atom on the 2D matrix: an emerging electrocatalyst for energy applications. ACS Omega 5, 1287–1295.

Mori, K., Taga, T., Yamashita, H., 2017. Isolated single-atomic Ru catalyst bound on a layered double hydroxide for hydrogenation of CO_2 to formic acid. ACS Catalysis 7, 3147–3151.

Naebe, M., Wang, J., Amini, A., Khayyam, H., Hameed, N., Li, L.H., et al., 2014. Mechanical property and structure of covalent functionalised graphene/epoxy nanocomposites. Scientific Reports 4, 4375.

Naguib, M., Kurtoglu, M., Presser, V., Lu, J., Niu, J., Heon, M., et al., 2023. Two-dimensional nanocrystals produced by exfoliation of Ti_3AlC_2. MXenes. Jenny Stanford Publishing.

Nair, R.R., Blake, P., Grigorenko, A.N., Novoselov, K.S., Booth, T.J., Stauber, T., et al., 2008. Fine structure constant defines visual transparency of graphene. Science 320, 1308.

Nasirpouri, F., Jafari-Foruzin, L., Komari-Alaei, S., Farmani, A., Fazli-Shokouhi, S., Peighambardoust, N.S., 2024. Magneto-electrochemical water splitting performance of graphene oxide/nickel aluminum-layered double hydroxide nanocomposites. International Journal of Electrochemical Science 19, 100841.

Neumann, B., Vafeidis, A.T., Zimmermann, J., Nicholls, R.J., 2015. Future coastal population growth and exposure to sea-level rise and coastal flooding-a global assessment. PLoS ONE 10, e0118571.

Ning, F., Shao, M., Xu, S., Fu, Y., Zhang, R., Wei, M., et al., 2016. TiO_2/graphene/NiFe-layered double hydroxide nanorod array photoanodes for efficient photoelectrochemical water splitting. Energy and Environmental Science 9, 2633–2643.

Ni, Z.H., Yu, T., Lu, Y.H., Wang, Y.Y., Feng, Y.P., Shen, Z.X., 2008. Uniaxial strain on graphene: Raman spectroscopy study and band-gap opening. ACS Nano 2, 2301–2305.

Novoselov, K.S., Geim, A.K., Morozov, S.V., Jiang, D.-E., Zhang, Y., Dubonos, S.V., et al., 2004. Electric field effect in atomically thin carbon films. Science 306, 666–669.

Park, S.Y., Kim, Y.H., Lee, S.Y., Sohn, W., Lee, J.E., Shim, Y.-S., et al., 2018. Highly selective and sensitive chemoresistive humidity sensors based on rGO/MoS_2 van der Waals composites. Journal of Materials Chemistry A 6, 5016–5024.

Peng, Y., Li, Y., Ban, Y., Jin, H., Jiao, W., Liu, X., et al., 2014. Metal-organic framework nanosheets as building blocks for molecular sieving membranes. Science 346, 1356–1359.

Peng, Y., Lu, B., Chen, S., 2018. Carbon-supported single atom catalysts for electrochemical energy conversion and storage. Advanced Materials 30, 1801995.

Peng, C., Xu, W., Wei, P., Liu, M., Guo, L., Wu, P., et al., 2019. Manipulating photocatalytic pathway and activity of ternary Cu_2O/(001) TiO_2@ Ti_3C_2Tx catalysts for H_2 evolution: effect of surface coverage. International Journal of Hydrogen Energy 44, 29975–29985.

Petersen, H., Weidenthaler, C., 2022. A review of recent developments for the in situ/operando characterization of nanoporous materials. Inorganic Chemistry Frontiers 9, 4244–4271.

Presolski, S., Pumera, M., 2016. Covalent functionalization of MoS_2. Materials Today 19, 140–145.

Qiao, B., Wang, A., Yang, X., Allard, L.F., Jiang, Z., Cui, Y., et al., 2011. Single-atom catalysis of CO oxidation using Pt1/FeOx. Nature Chemistry 3, 634–641.

Qureshy, A.M., Ahmed, M., Dincer, I., 2019. Performance assessment study of photo-electro-chemical water-splitting reactor designs for hydrogen production. International Journal of Hydrogen Energy 44, 9237–9247.

Rasheed, T., Rasheed, A., Alzahrani, F.M.A., Ajmal, S., Warsi, M.F., Al-Buriahi, M., et al., 2023. Bifunctional electrocatalytic water splitting augmented by cobalt-nickel-ferrite NPs supported fluoride-free MXene as a novel electrocatalyst. Fuel 346, 128305.

Renaudie, M., Clion, V., Dumas, C., Vuilleumier, S., Ernst, B., 2021. Intensification and optimization of continuous hydrogen production by dark fermentation in a new design liquid/gas hollow fiber membrane bioreactor. Chemical Engineering Journal 416, 129068.

Ren, X., Zhou, J., Qi, X., Liu, Y., Huang, Z., Li, Z., et al., 2017. Few-layer black phosphorus nanosheets as electrocatalysts for highly efficient oxygen evolution reaction. Advanced Energy Materials 7, 1700396.

Ros, C., Andreu, T., Morante, J.R., 2020. Photoelectrochemical water splitting: a road from stable metal oxides to protected thin film solar cells. Journal of Materials Chemistry A 8, 10625–10669.

Rui, K., Zhao, G., Chen, Y., Lin, Y., Zhou, Q., Chen, J., et al., 2018. Hybrid 2D dual-metal–organic frameworks for enhanced water oxidation catalysis. Advanced Functional Materials 28, 1801554.

Ryczkowski, J., 2001. IR spectroscopy in catalysis. Catalysis Today 68, 263–381.

Salarizadeh, P., Askari, M.B., Seifi, M., Rozati, S.M., 2019. MoS_2 coating on different carbonaceous materials: comparison of electrochemical properties and hydrogen evolution reaction performance. Journal of Electroanalytical Chemistry 847, 113198.

Samajdar, S., Bera, S., Das, P.S., Finch, H., Dhanak, V.R., Chakraborty, S., et al., 2023. Exploration of 1D-2D $LaFeO_3$/RGO S-scheme heterojunction for photocatalytic water splitting. International Journal of Hydrogen Energy 48, 17838–17851.

Seh, Z.W., Fredrickson, K.D., Anasori, B., Kibsgaard, J., Strickler, A.L., Lukatskaya, M.R., et al., 2016. Two-dimensional molybdenum carbide (MXene) as an efficient electrocatalyst for hydrogen evolution. ACS Energy Letters 1, 589–594.

Seo, S., Kim, S., Choi, H., Lee, J., Yoon, H., Piao, G., et al., 2019. Direct in situ growth of centimeter-scale multi-heterojunction $MoS_2/WS_2/WSe_2$ thin-film catalyst for photo-electrochemical hydrogen evolution. Advanced Science 6, 1900301.

Shifa, T.A., Wang, F., Liu, Y., He, J., 2019. Heterostructures based on 2D materials: a versatile platform for efficient catalysis. Advanced Materials 31, 1804828.

Shi, Y., Zhou, Y., Yang, D.-R., Xu, W.-X., Wang, C., Wang, F.-B., et al., 2017. Energy level engineering of MoS_2 by transition-metal doping for accelerating hydrogen evolution reaction. Journal of the American Chemical Society 139, 15479–15485.

Sim, U., Moon, J., Lee, J., An, J., Ahn, H.-Y., Kim, D.J., et al., 2017. Double-layer graphene outperforming monolayer as catalyst on silicon photocathode for hydrogen production. ACS Applied Materials and Interfaces 9, 3570–3580.

Smith, A.M., Nie, S., 2010. Semiconductor nanocrystals: structure, properties, and band gap engineering. Accounts of Chemical Research 43, 190–200.

Suen, N.-T., Hung, S.-F., Quan, Q., Zhang, N., Xu, Y.-J., Chen, H.M., 2017. Electrocatalysis for the oxygen evolution reaction: recent development and future perspectives. Chemical Society Reviews 46, 337–365.

Sun, Z., Ma, T., Tao, H., Fan, Q., Han, B., 2017. Fundamentals and challenges of electrochemical CO_2 reduction using two-dimensional materials. Chem 3, 560–587.

Sun, X., Mi, Y., Jiao, F., Xu, X., 2018. Activating layered perovskite compound Sr_2TiO_4 via La/N codoping for visible light photocatalytic water splitting. ACS Catalysis 8, 3209–3221.

Sun, F., Qin, J., Wang, Z., Yu, M., Wu, X., Sun, X., et al., 2021. Energy-saving hydrogen production by chlorine-free hybrid seawater splitting coupling hydrazine degradation. Nature Communications 12, 4182.

Sun, Y., Wang, Z., Zhou, Q., Li, X., Zhao, D., Ding, B., et al., 2024. Ti_3C_2 mediates the NiFe-LDH layered electrocatalyst to enhance the OER performance for water splitting. Heliyon 10.

Sun, T., Xu, L., Wang, D., Li, Y., 2019. Metal organic frameworks derived single atom catalysts for electrocatalytic energy conversion. Nano Research 12, 2067–2080.

Sun, S., Zhang, G., Gauquelin, N., Chen, N., Zhou, J., Yang, S., et al., 2013. Single-atom catalysis using Pt/graphene achieved through atomic layer deposition. Scientific Reports 3, 1–9.

Sushko, M.L., Liu, J., 2011. Surfactant two-dimensional self-assembly under confinement. The Journal of Physical Chemistry B 115, 4322–4328.

Tang, L., Meng, X., Deng, D., Bao, X., 2019. Confinement catalysis with 2D materials for energy conversion. Advanced Materials 31, 1901996.

Tie, L., Li, N., Yu, C., Liu, Y., Yang, S., Chen, H., et al., 2019. Self-supported nonprecious MXene/$Ni3S_2$ electrocatalysts for efficient hydrogen generation in alkaline media. ACS Applied Energy Materials 2, 6931–6938.

Walter, M.G., Warren, E.L., Mckone, J.R., Boettcher, S.W., Mi, Q., Santori, E.A., et al., 2010. Solar water splitting cells. Chemical Reviews 110, 6446–6473.

Wang, Z., Liu, P., Ito, Y., Ning, S., Tan, Y., Fujita, T., et al., 2016. Chemical vapor deposition of monolayer $Mo1-xWxS_2$ crystals with tunable band gaps. Scientific Reports 6, 21536.

Wang, Y., Pang, J., Cheng, Q., Han, L., Li, Y., Meng, X., et al., 2021. Applications of 2D-layered palladium diselenide and its van der Waals heterostructures in electronics and optoelectronics. Nano-Micro Letters 13, 143.

Wang, L., Sasaki, T., 2014. Titanium oxide nanosheets: graphene analogues with versatile functionalities. Chemical Reviews 114, 9455–9486.

Wang, Z., Wang, G., Qi, H., Wang, M., Wang, M., Park, S., et al., 2020. Ultrathin two-dimensional conjugated metal–organic framework single-crystalline nanosheets enabled by surfactant-assisted synthesis. Chemical Science 11, 7665–7671.

Wang, D., Zhang, X.-T., Sun, P.-P., Lu, S., Wang, L.-L., Wei, Y.-A., et al., 2014. Enhanced photoelectrochemical water splitting on hematite thin film with layer-by-layer deposited ultrathin TiO_2 underlayer. International Journal of Hydrogen Energy 39, 16212–16219.

Wang, Q., Zhao, Z.L., Dong, S., He, D., Lawrence, M.J., Han, S., et al., 2018. Design of active nickel single-atom decorated MoS_2 as a pH-universal catalyst for hydrogen evolution reaction. Nano Energy 53, 458–467.

Wan, J., Lacey, S.D., Dai, J., Bao, W., Fuhrer, M.S., Hu, L., 2016. Tuning two-dimensional nanomaterials by intercalation: materials, properties and applications. Chemical Society Reviews 45, 6742–6765.

Wan, S., Wu, J., Wang, D., Liu, H., Zhang, Z., Ma, J., et al., 2021. Co/N-doped carbon nanotube arrays grown on 2D MOFs-derived matrix for boosting the oxygen reduction reaction in alkaline and acidic media. Chinese Chemical Letters 32, 816–821.

Wen, P., Sun, Y., Li, H., Liang, Z., Wu, H., Zhang, J., et al., 2020. A highly active three-dimensional Z-scheme ZnO/Au/g-C_3N_4 photocathode for efficient photoelectrochemical water splitting. Applied Catalysis B 263, 118180.

Wu, K., Chen, X., Liu, S., Pan, Y., Cheong, W.-C., Zhu, W., et al., 2018. Porphyrin-like Fe-N4 sites with sulfur adjustment on hierarchical porous carbon for different rate-determining steps in oxygen reduction reaction. Nano Research 11, 6260–6269.

Wu, X.J., Huang, X., Liu, J., Li, H., Yang, J., Li, B., et al., 2014. Two-dimensional CuSe nanosheets with microscale lateral size: synthesis and template-assisted phase transformation. Angewandte Chemie International Edition 53, 5083–5087.

Xiao, X., He, C.-T., Zhao, S., Li, J., Lin, W., Yuan, Z., et al., 2017. A general approach to cobalt-based homobimetallic phosphide ultrathin nanosheets for highly efficient oxygen evolution in alkaline media. Energy and Environmental Science 10, 893–899.

Xiao, F.-X., Miao, J., Liu, B., 2014. Layer-by-layer self-assembly of CdS quantum dots/graphene nanosheets hybrid films for photoelectrochemical and photocatalytic applications. Journal of the American Chemical Society 136, 1559–1569.

Xiao, Z., Wang, Y., Chen, H., Wang, H., Li, Y., Chen, Y., et al., 2023. Designing black phosphorus and heptazine-based crystalline carbon nitride composites for photocatalytic water splitting. Crystals 13, 312.

Xiao, R., Zhao, C., Zou, Z., Chen, Z., Tian, L., Xu, H., et al., 2020. In situ fabrication of 1D CdS nanorod/2D Ti_3C_2 MXene nanosheet Schottky heterojunction toward enhanced photocatalytic hydrogen evolution. Applied Catalysis B 268, 118382.

Xing, Z., Ren, F., Wu, H., Wu, L., Wang, X., Wang, J., et al., 2017. Enhanced PEC performance of nanoporous Si photoelectrodes by covering HfO_2 and TiO_2 passivation layers. Scientific Reports 7, 43901.

Xue, Y., Huang, B., Yi, Y., Guo, Y., Zuo, Z., Li, Y., et al., 2018. Anchoring zero valence single atoms of nickel and iron on graphdiyne for hydrogen evolution. Nature Communications 9, 1460.

Xu, H., Lee, U., Wang, M., 2020. Life-cycle energy use and greenhouse gas emissions of palm fatty acid distillate derived renewable diesel. Renewable and Sustainable Energy Reviews 134, 110144.

Xu, Q., Zhang, J., Wang, D., Li, Y., 2021. Single-atom site catalysts supported on two-dimensional materials for energy applications. Chinese Chemical Letters 32, 3771–3781.

Yamaguchi, S., Kamiya, K., Hashimoto, K., Nakanishi, S., 2017. Ru atom-modified covalent triazine framework as a robust electrocatalyst for selective alcohol oxidation in aqueous electrolytes. Chemical Communications 53, 10437–10440.

Yang, M., Allard, L.F., Flytzani-Stephanopoulos, M., 2013. Atomically dispersed Au–(OH) x species bound on titania catalyze the low-temperature water-gas shift reaction. Journal of the American Chemical Society 135, 3768–3771.

Yang, Y., Guan, H., Batelaan, O., Mcvicar, T.R., Long, D., Piao, S., et al., 2016. Contrasting responses of water use efficiency to drought across global terrestrial ecosystems. Scientific Reports 6, 23284.

Yao, H., Liu, L., Fu, W., Yang, H., Shi, Y., 2017. Fe_2O_3 nanothorns sensitized two-dimensional TiO_2 nanosheets for highly efficient solar energy conversion. FlatChem 3, 1–7.

Ye, R.-K., Sun, S.-S., He, L.-Q., Yang, S.-R., Liu, X.-Q., Fang, P.-P., et al., 2021. Surface engineering of hematite nanorods by 2D Ti_3C_2-MXene: suppressing the electron-hole recombination for enhanced photoelectrochemical performance. Applied Catalysis B 291, 120107.

Yin, H., Wang, Y., Ma, L., Zhang, S., Yang, B., Jiang, R., 2022. Effect of surface-deposited Ti_3C_2Tx MXene on the photoelectrochemical water-oxidation performance of iron-doped titania nanorod array. Chemical Engineering Journal 431, 134124.

Yu, H., Wang, Y., Jing, Y., Ma, J., Du, C.F., Yan, Q., 2019a. Surface modified MXene-based nanocomposites for electrochemical energy conversion and storage. Small 15, 1901503.

Yu, H., Xue, Y., Huang, B., Hui, L., Zhang, C., Fang, Y., et al., 2019b. Ultrathin nanosheet of graphdiyne-supported palladium atom catalyst for efficient hydrogen production. IScience 11, 31–41.

Yu, H., Xue, Y., Li, Y., 2019c. Graphdiyne and its assembly architectures: synthesis, functionalization, and applications. Advanced Materials 31, 1803101.

Zandi, O., Hamann, T.W., 2016. Determination of photoelectrochemical water oxidation intermediates on haematite electrode surfaces using operando infrared spectroscopy. Nature Chemistry 8, 778–783.

Zeng, M., Li, Y., 2015. Recent advances in heterogeneous electrocatalysts for the hydrogen evolution reaction. Journal of Materials Chemistry A 3, 14942–14962.

Zeng, X., Shui, J., Liu, X., Liu, Q., Li, Y., Shang, J., et al., 2018. Single-atom to single-atom grafting of Pt1 onto Fe N4 center: Pt1@ FeNC multifunctional electrocatalyst with significantly enhanced properties. Advanced Energy Materials 8, 1701345.

Zhang, Y., Chen, D., Li, N., Xu, Q., Li, H., Lu, J., 2023. Fabricating 1D/2D $Co_3O_4/ZnIn_2S_4$ core–shell heterostructures with boosted charge transfer for photocatalytic hydrogen production. Applied Surface Science 610, 155272.

Zhang, Y., Chou, J.B., Li, J., Li, H., Du, Q., Yadav, A., et al., 2019b. Broadband transparent optical phase change materials for high-performance nonvolatile photonics. Nature Communications 10, 4279.

Zhang, B., Fan, T., Xie, N., Nie, G., Zhang, H., 2019a. Versatile applications of metal single-atom@ 2D material nanoplatforms. Advanced Science 6, 1901787.

Zhang, L., Jia, Y., Gao, G., Yan, X., Chen, N., Chen, J., et al., 2018c. Graphene defects trap atomic Ni species for hydrogen and oxygen evolution reactions. Chem 4, 285–297.

Zhang, H., Liu, G., Shi, L., Ye, J., 2018a. Single-atom catalysts: emerging multifunctional materials in heterogeneous catalysis. Advanced Energy Materials 8, 1701343.

Zhang, Y., Tan, Y.-W., Stormer, H.L., Kim, P., 2005. Experimental observation of the quantum Hall effect and Berry's phase in graphene. Nature 438, 201–204.

Zhang, J., Wang, T., Liu, P., Liao, Z., Liu, S., Zhuang, X., et al., 2017. Efficient hydrogen production on $MoNi_4$ electrocatalysts with fast water dissociation kinetics. Nature Communications 8, 15437.

Zhang, J., Xing, C., Shi, F., 2020. MoS2/Ti3C2 heterostructure for efficient visible-light photocatalytic hydrogen generation. International Journal of Hydrogen Energy 45, 6291–6301.

Zhang, N., Yang, M.-Q., Liu, S., Sun, Y., Xu, Y.-J., 2015b. Waltzing with the versatile platform of graphene to synthesize composite photocatalysts. Chemical Reviews 115, 10307–10377.

Zhang, S., Yin, C., Kang, Z., Wu, P., Wu, J., Zhang, Z., et al., 2018d. Graphdiyne nanowall for enhanced photoelectrochemical performance of Si heterojunction photoanode. ACS Applied Materials and Interfaces 11, 2745–2749.

Zhang, L., Yuan, H., Li, X., Wang, Y., 2022. Hydrothermal synthesis of $NiCo_2O_4$@ $NiCo_2O_4$ core-shell nanostructures anchored on Ni foam for efficient oxygen evolution reactions catalysts. Coatings 12, 1240.

Zhang, B., Yu, S., Dai, Y., Huang, X., Chou, L., Lu, G., et al., 2021a. Nitrogen-incorporation activates NiFeOx catalysts for efficiently boosting oxygen evolution activity and stability of $BiVO_4$ photoanodes. Nature Communications 12, 6969.

Zhang, N., Zhang, X., Kang, Y., Ye, C., Jin, R., Yan, H., et al., 2021b. A supported Pd2 dual-atom site catalyst for efficient electrochemical CO_2 reduction. Angewandte Chemie 133, 13500–13505.

Zhang, N., Zhang, X., Tao, L., Jiang, P., Ye, C., Lin, R., et al., 2021c. Silver single-atom catalyst for efficient electrochemical CO_2 reduction synthesized from thermal transformation and surface reconstruction. Angewandte Chemie International Edition 60, 6170–6176.

Zhang, J., Zhang, P., Wang, T., Gong, J., 2015a. Monoclinic WO_3 nanomultilayers with preferentially exposed (002) facets for photoelectrochemical water splitting. Nano Energy 11, 189–195.

Zhang, J., Zhao, Y., Guo, X., Chen, C., Dong, C.-L., Liu, R.-S., et al., 2018b. Single platinum atoms immobilized on an MXene as an efficient catalyst for the hydrogen evolution reaction. Nature Catalysis 1, 985–992.

Zhao, L., He, R., Rim, K.T., Schiros, T., Kim, K.S., Zhou, H., et al., 2011. Visualizing individual nitrogen dopants in monolayer graphene. Science 333, 999–1003.

Zheng, L., Ye, X., Deng, X., Wang, Y., Zhao, Y., Shi, X., et al., 2020. Black phosphorus quantum dot-sensitized TiO_2 nanotube arrays with enriched oxygen vacancies for efficient photoelectrochemical water splitting. ACS Sustainable Chemistry and Engineering 8, 15906–15914.

Zheng, C., Yu, L., Zhu, L., Collins, J.L., Kim, D., Lou, Y., et al., 2018. Room temperature in-plane ferroelectricity in van der Waals In_2Se_3. Science Advances 4, eaar7720.

Zhong, W., Sa, R., Li, L., He, Y., Li, L., Bi, J., Zhuang, Z., Yu, Y., Zou, Z., 2019. A covalent organic framework bearing single Ni sites as a synergistic photocatalyst for selective photoreduction of CO_2 to CO. Journal of the American Chemical Society 141, 7615–7621.

Zhou, X., Fu, B., Li, L., Tian, Z., Xu, X., Wu, Z., et al., 2022. Hydrogen-substituted graphdiyne encapsulated cuprous oxide photocathode for efficient and stable photoelectrochemical water reduction. Nature Communications 13, 5770.

Zhou, W., Kapetanakis, M.D., Prange, M.P., Pantelides, S.T., Pennycook, S.J., Idrobo, J.-C., 2012. Direct determination of the chemical bonding of individual impurities in graphene. Physical Review Letters 109, 206803.

Zhou, B., Kong, X., Vanka, S., Chu, S., Ghamari, P., Wang, Y., et al., 2018a. Gallium nitride nanowire as a linker of molybdenum sulfides and silicon for photoelectrocatalytic water splitting. Nature Communications 9, 3856.

Zhou, J., Liu, Y., Zhang, Z., Huang, Z., Chen, X., Ren, X., et al., 2018b. Hierarchical $NiSe_2$ sheet-like nano-architectures as an efficient and stable bifunctional electrocatalyst for overall water splitting: phase and morphology engineering. Electrochimica Acta 279, 195–203.

Zhou, C., Zhao, Y., Shang, L., Shi, R., Wu, L.-Z., Tung, C.-H., et al., 2016. Facile synthesis of ultrathin $SnNb_2O_6$ nanosheets towards improved visible-light photocatalytic H_2-production activity. Chemical Communications 52, 8239–8242.

Zhuang, Z., Kang, Q., Wang, D., Li, Y., 2020. Single-atom catalysis enables long-life, high-energy lithium-sulfur batteries. Nano Research 13, 1856–1866.

Zhu, K., Li, C., Jing, Z., Liu, X., He, Y., Lv, X., et al., 2019. Two-dimensional transition-metal dichalcogenides for electrochemical hydrogen evolution reaction. FlatChem 18, 100140.

Zuo, Q., Liu, T., Chen, C., Ji, Y., Gong, X., Mai, Y., Zhou, Y., 2019. Ultrathin metal-organic framework nanosheets with ultrahigh loading of single Pt atoms for efficient visible-light-driven photocatalytic H_2 evolution. Angewandte Chemie (International ed. in English) 58 (30), 10198–10203. https://doi.org/10.1002/anie.201904058.

CHAPTER

15

Conjugated polymer materials for water-splitting

Benjamin Tawiah[1,2], *Emmanuel A. Ofori*[3], *Yang Ming*[2], *and Bin Fei*[2]

[1]School of Fashion and Textiles (SFT), The Hong Kong Polytechnic University, Hung Hom, Kowloon, Hong Kong SAR, P.R. China [2]Department of Industrial Art (Fashion and Textiles), Kwame Nkrumah University of Science and Technology, Kumasi, Ghana [3]Department of Materials Science and Engineering, Hubei University of Automotive Technology, Hubei, P.R. China

15.1 Introduction

The pressing need for renewable energy has prompted a great deal of research into cutting-edge clean energy generation and storage systems (Chu and Majumdar, 2012; Moriarty and Honnery, 2022). The electrochemical or photoelectrochemical breakdown of water into hydrogen and oxygen gases has become one of these technologies that shows the most promise for producing hydrogen; a clean and sustainable fuel for the future (Mansha et al., 2022; Zhao et al., 2020). However, the commercial implementation of water-splitting technologies has faced significant challenges, such as efficient and cost-effective catalysts, ineffective photoactive materials, and technical complexities in scalability (Dai et al., 2020; Rahman et al., 2020).

Congugated polymers (CPs) are macromolecules possessing alternating single and double bonds throughout their polymer chain. This feature enables electron delocalization and confers upon them distinct optical and electrical characteristics. CPs, with their unique electronic and optical properties, have garnered increasing attention as potential materials for water-splitting applications. These polymers possess extended π-conjugation along their backbone, enabling efficient light absorption and charge transport (Dai et al., 2020). Moreover, their molecular structures can be precisely tailored and functionalized, offering opportunities for optimizing their energy levels, bandgaps, and catalytic properties, thus making them ideal candidates for solar energy conversion and water-splitting (Dai et al., 2020; Wang et al., 2018). The mechanism of water-splitting and solar energy conversion by

Photochemical Splitting of Water
DOI: https://doi.org/10.1016/B978-0-443-29064-0.00007-3

CPs involves a series of interconnected processes comprising efficient light absorption, and generation of electron–hole pairs which facilitate charge separation and transport, enabling the catalytic reactions of water-splitting (Liu et al., 2021b). Typically, Wang et al. (2019) studied the mechanism of charge separation and transfer within these CPs and highlighted the influence of energy levels, bandgaps, and molecular structures on achieving efficient water-splitting for hydrogen production. In a similar study, the processes of charge separation and transport within the nanofibers were examined and the results emphasized the importance of morphology controlling and engineering interfaces to enhance the efficiency of water-splitting reactions (Zhao et al., 2020). Also, the relationship between morphology and performance in organic CPs for photoelectrochemical water-splitting has been examined detailing the importance of the polymer's nanoscale structure control and interfacial properties on the efficiency and stability of water-splitting reactions (Zhou et al., 2024). These and many other studies provide valuable insights into the mechanisms of water-splitting by CPs and offer strategies for improving their efficiency, stability, and scalability contributing to the ongoing research and development efforts in the field of clean energy technologies (Oka et al., 2021; Sun et al., 2019; Otep et al., 2020).

The utilization of CP materials in water-splitting systems offers several advantages over traditional materials. Firstly, their solution processability allows for low-cost and scalable fabrication techniques, such as spin coating, inkjet printing, and roll-to-roll processing, enabling the production of large-scale devices (Wang et al., 2019). Secondly, their versatile chemical structures and functionalization possibilities enable the incorporation of cocatalysts, dopants, and surface modifications to enhance catalytic activity, stability, and charge separation efficiency (Wang et al., 2019; Luo et al., 2020). Moreover, their compatibility with various electrode and photoelectrode materials opens opportunities for hybrid systems with improved performance and synergistic effects (Tentu and Basu, 2017).

This chapter provides a comprehensive overview of the utilization of CP materials for water-splitting applications and delves into the fundamental principles underlying the design and synthesis of CPs tailored specifically for water-splitting, including discussions on molecular engineering, bandgap control, and charge transport optimization. Furthermore, the various strategies employed to enhance the performance of conjugated polymer-based water-splitting systems, ranging from morphology control and interface engineering to the integration of cocatalysts and nanostructured materials.

Additionally, an in-depth analysis of recent advances in the fabrication of CP-based water-splitting devices, including photoelectrochemical cells and photocatalysts. Also, the key findings and challenges associated with these devices, with a focus on their efficiency, stability, and scalability are discussed and further highlight the emerging techniques and materials, such as perovskite-conjugated polymer hybrids and 2D nanomaterials, which hold great promise for further improving the performance of CP-based water-splitting systems.

Ultimately, this chapter offers a useful perspective on the potential of conjugated polymer materials as essential building blocks for the creation of effective and long-lasting water-splitting technologies. By highlighting the advantages, challenges, and recent advancements in this field, we hope to inspire further research and innovation towards the realization of a clean and renewable energy future.

15.2 The basics of water-splitting

Water-splitting is an essential process that produces hydrogen, a clean and sustainable energy source with many potential benefits. Using electricity from renewable sources (solar, wind, and hydroelectric) to split water into hydrogen gas (H_2 [g]), through the reaction $2H_2O \rightarrow 2H_2 + O_2$ is quite fascinating because 70% of the earth is covered by water (Buriak et al., 2018). After successful water-splitting, the H_2 can be used immediately as a fuel that burns cleanly or utilized to store energy chemically in the H–H bond for later use. Because this process is endothermic, energy must be supplied for it to continue. The energy needed can be thermal energy for thermochemical water-splitting or electrical energy for processes like water electrolysis (Abanades, 2019; Villafán-Vidales et al., 2017). Importantly, the fundamentals of water-splitting involve the photocatalyzed dissociation of water into its constituent elements, H_2 and oxygen (O_2), through a process that utilizes either artificial or natural light (Pai et al., 2012; McHugh et al., 2020). There are three methods to apply these fundamental ideas. First of all, water can be thermally broken down by solar concentrators and catalysts, but this process requires materials that are stable at high temperatures (Martin, 2015). The second method, which uses photovoltaic cells to photo-electrolyze water is more commonly utilized but at the moment, photovoltaic systems and electrolyzers are fairly priced (Woodhouse and Parkinson, 2009). As a third strategy, semiconductor materials are essential for absorbing solar energy and causing water-splitting reactions. Choosing the right semiconductor materials is essential for effective energy transfer and absorption. Due to its low cost, nontoxicity, great photocatalytic activity, self-cleaning qualities, and resilience against light corrosion, TiO_2 is a notable material. Its band gap aligns well with the redox potential of water, making it suitable for water-splitting applications (Chang et al., 2021). Essentially, the efficiency of water electrolysis is influenced by several factors, including the electrocatalysts used, the operating temperature, and the purity of the water feed (Chatenet et al., 2022; Sun et al., 2023).

15.2.1 Water-splitting's kinetics and thermodynamics

The thermodynamics and kinetics of the thermochemical cycle determine how quickly hydrogen is produced in this process. While kinetics controls the time needed to finish a cycle, thermodynamics determines how much hydrogen can be created in a single cycle (Asundi, 2015). The thermodynamic and kinetic mechanisms in water-splitting are studied through the underlying principle of density functional theory (DFT) calculations in determining the effect of the small gas molecules (Chen et al., 2021). The effectiveness and viability of water-splitting technologies are primarily determined by the thermodynamics and kinetics of this process when the photocatalyst's bandgap is occupied by both the proton reduction and water oxidation potentials (Bie et al., 2021). This can be described by the Gibbs free energy change (ΔG) and the enthalpy change of the reaction (refer to Fig. 15.1) (Bie et al., 2022). The ΔG is a measure of the favourability of a given reaction or process such that if ΔG is negative, it means the reaction or process is thermodynamically favored in the forward direction and is termed exergonic. In this case, the reaction or process will proceed spontaneously. If ΔG is positive, it means the reaction or process is thermodynamically favored in the reverse direction and is called endergonic. In this case,

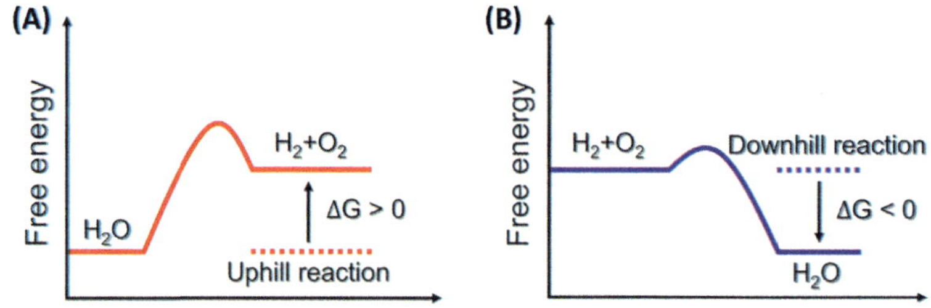

FIGURE 15.1 Change in Gibbs free energy (ΔG) for (A) the overall process of water-splitting, and (B) the interaction between H_2 and O_2 throughout that process (Bie et al., 2022). *Reproduced from an open-access article by Elsevier BV.*

the reaction or process will not occur spontaneously. The reaction or process is at equilibrium when ΔG is zero. When it comes to catalytic water-splitting, the photocatalyst's valence band (VB) top and conduction band (CB) bottom should be positioned below the water oxidation potential and above the proton reduction potential, respectively (Lin et al., 2020; Saunders et al., 2022), which shifts the reaction equilibrium and produces hydrogen and oxygen (Tsutsumi, 2009). Furthermore, the photoexcited state durations and charge-carrier mobilities of the catalyst must be sufficient for charge-carriers to move to the catalytic surface without recombination (Reza Gholipour et al., 2015; Sprick et al., 2016). Moreover, the catalyst's charge-carrier mobilities and photoexcited state durations must be long enough to permit charge-carrier migration to the catalytic surface without recombination (Bie et al., 2022). Crucially, the metal oxide system's thermodynamic characteristics, such as the reduction temperature, the partial pressure of oxygen, and the water vapor pressure, determine the overall effectiveness of the thermochemical water-splitting process (Gebremariam et al., 2023; Liang et al., 2024).

However, more voltage is needed to get beyond the kinetic and resistance barriers and to propel noticeable currents to flow for the oxygen evolution reaction (OER) and hydrogen evolution reaction (HER). This extra voltage is known as the overpotential, and it is the product of the kinetic overpotentials for each of the individual HER and OER half-reactions, the concentration, and the ohmic resistances in the electrolyzer (Xiang et al., 2016). Additionally, the use of catalysts, the optimization of material properties (e.g., porosity, surface area), and the design of efficient reactor systems. For example, the incorporation of catalysts, such as noble metals or transition metal oxides, enhances the reaction kinetics by providing alternative reaction pathways with lower activation barriers (Lyons et al., 2014; Li et al., 2023). In the same vein, the physical structure of the materials used in water-splitting, such as the particle size, morphology, and surface properties, impacts significantly, the reaction kinetics, mass, and heat transfer properties, leading to improved reaction rates under nano treatment (Nabgan et al., 2024; Chen et al., 2015).

15.2.2 Water-splitting mechanisms

The mechanisms underlying water-splitting are governed by the fundamental principles of electrochemistry and involve complex reactions at the electrode-electrolyte interface (Chen et al., 2023). An electrochemical cell, comprising an anode (positive electrode), a cathode (negative electrode), and an electrolyte solution or solid electrolyte, is where the water-splitting process takes place (Sen et al., 2022; Ifkovits et al., 2021), under an external voltage applied to the cell. At the corresponding electrodes, this propels the oxidation and reduction reactions. The photocatalysts-driven redox process has a positive ΔG. This suggests that external power, such as photons or electricity, which is subsequently converted into chemical energy, is needed for the water-spitting reaction and that it is not spontaneous (Chohan et al., 2022). The OER and the HER are the two half-reactions that can be discussed in relation to the water electrolysis process (see Fig. 15.2A) (Dau et al., 2010). The variations in these half-equations are contingent upon the pH at which the electrolysis is conducted. In essence, the process of water electrolysis requires a large amount of electrical energy. Standard conditions demand a minimum thermodynamic potential difference of 1.23 V to electrolyze water (McHugh et al., 2020). To meet thermodynamic criteria in this scenario, the band gap energy must be greater than 1.23 eV; nevertheless, most of the time, it should be greater than 2.0 eV in light of kinetic barriers. Exciton splitting permits charges to move to active sites, which in turn promotes proton reduction and water oxidation, which releases oxygen and hydrogen. When suitable semiconductors are dissolved in water and exposed to light, they catalyze water-splitting reactions that produce hydrogen and oxygen in a 2:1 ratio simultaneously. When light with sufficient energy shifts electrons from the photocatalyst's highest occupied molecular orbital (HOMO) to its lowest unoccupied molecular orbital (LUMO), it forms excitons, or electron–hole pairs (see Fig. 15.2B) (McQueen et al., 2022; Zhao et al., 2023). This leads to charge separation, and the crystal structure and crystallinity have a major influence on the migration of photogenerated carriers (Nosaka, 2011).

There are high-temperature water electrolysis systems that function at temperatures more than 100°C in addition to low-temperature ones. These systems, including solid oxide

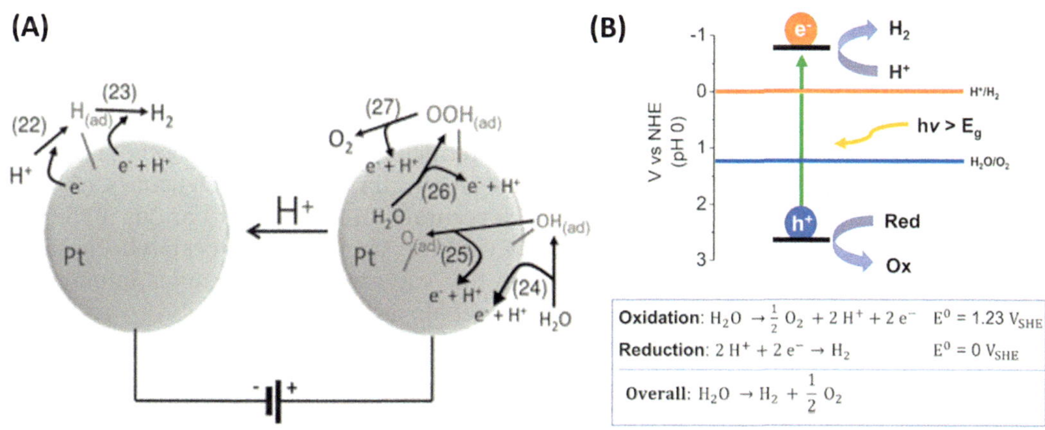

FIGURE 15.2 (A) Catalytic processes related to water electrolysis using electrochemistry. A battery-shaped external voltage drives the evolution of oxygen at the anode and hydrogen at the cathode (Dau et al., 2010). (B) The redox reaction and mechanism underlying photocatalytic water-splitting (McQueen et al., 2022). *(A) Reproduced with permission from John Wiley and Sons. Copyright © 2010 Weinheim, Germany's WILEY-VCH Verlag GmbH & Co. KGaA (B) Reproduced from an open-access article by MDPI under a CC license.*

electrolysis cells, have better kinetics at higher temperatures and lower internal resistance losses, which allow them to operate at higher efficiency levels (Min et al., 2022). Nevertheless, to accelerate the charge transfer of electrons and holes from the semiconductor to protons and/or water, metal catalysts are usually required (Aiga et al., 2013; Pan et al., 2016). CPs function as photocatalysts in situations where excited charged species are generated, sharing traits with semiconducting materials. The cocatalysts extract those new burn charges while simultaneously reducing or oxidizing the target substrates. The LUMO, also known as the negative CB, drives the electrolytic reaction of a photocatalyst toward the less negative side relative to the species' half-cell reduction potential; the HOMO, also known as the VB, drives the electrolytic reaction toward the more negative side relative to the other half-cell reduction potential of the species to be oxidized (Kosco et al., 2020).

15.2.3 Challenges in water-splitting

Water electrolysis and thermochemical water-splitting both provide many significant obstacles to overcome before the process can be implemented in a way that is both practical and economical. The intrinsic chemical and thermodynamic characteristics of water, as well as the shortcomings of the materials and technology currently in use in water electrolysis systems, are the sources of these difficulties (Horri and Hasan, 2024). Since photocatalytic overall water-splitting (OWS) is an energetically demanding process, the high stability of H_2O with strong O–H bonds presents a frequent thermodynamic difficulty (Bie et al., 2022) due to its highly exothermic nature. In addition to thermodynamic difficulties, kinetics presents difficulties in achieving OWS. This is because the production, separation, and transfer of photogenerated carriers typically take substantially longer (microseconds to seconds) during photocatalytic actions, and are frequently the step that limits the rate of reaction (Kranz and Wächtler, 2021). Likewise, there is a significantly higher overpotential and slower reaction kinetics (Serpone et al., 2016). Hence, optimizing the thermodynamic and kinetic properties of the materials, such as ceria, is crucial for improving the efficiency of the process (Samanta et al., 2022). Moreover, O_2 diffuses much more slowly than H_2 because of high oxygen affinity but low hydrogen affinity, which explains the low photocatalytic OWS efficiency (Bie et al., 2022). Similar to water electrolysis, scaling up thermochemical water-splitting systems to industrial levels and reducing the associated costs are major hurdles (Sun et al., 2023). Furthermore, because Ab Initio Molecular Dynamics (AIMD) simulations are prohibitively expensive and lack accurate force fields for various materials, accurate modeling of the water-splitting process, especially in photocatalytic systems requires sophisticated computational techniques that can capture the complex interactions at different lengths and accuracy scales (Samanta et al., 2022).

15.2.4 Importance of conjugated polymer materials in water-splitting

The alternating single and double bond structure of CPs, which is an organic macromolecule, permits electron delocalization along the polymer chain (refer to Fig. 15.3) (Mdluli et al., 2022). These polymers have emerged as a promising class of organic photocatalysts for photocatalytic hydrogen evolution in water-splitting processes due to their flexible chemical structures and electrical properties (Li et al., 2022; Ting et al., 2019; Sachs et al., 2018) in

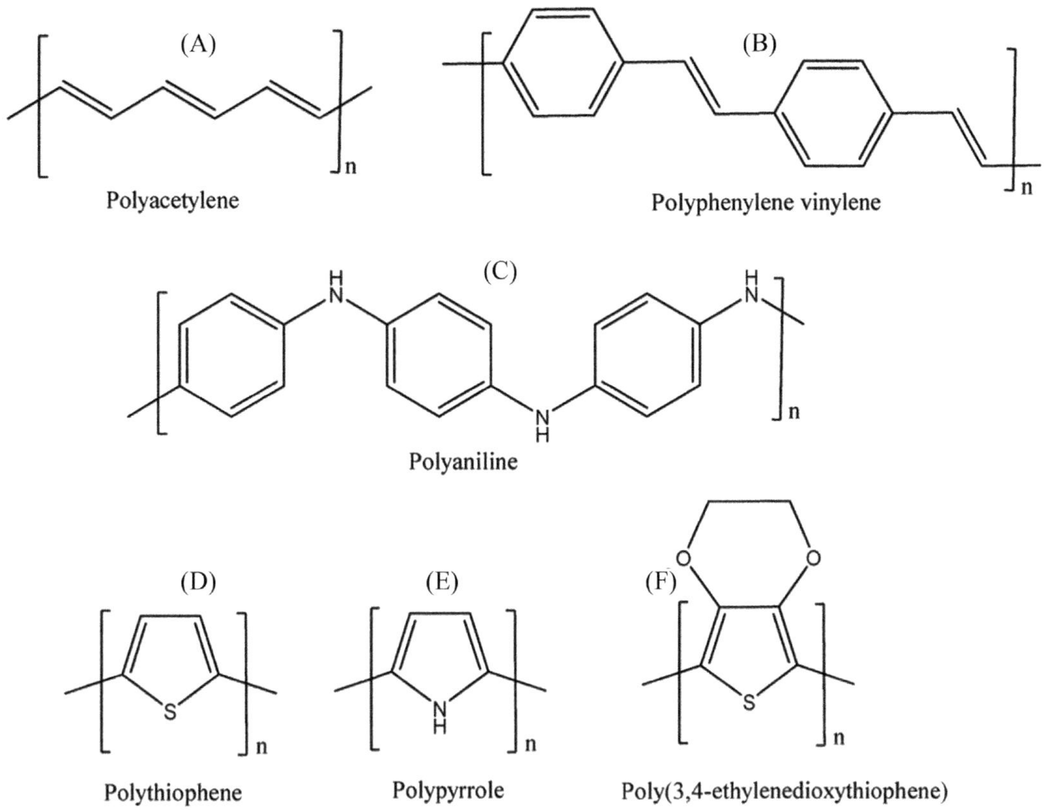

FIGURE 15.3 The chemical compositions of the most researched and utilized π-CPs. (A) Polyacetylene, (B) polyphenylene vinylene, (C) polyaniline, (D) polythiophene, (E) polypyrrole, and (F) poly(3,4-ethylenedioxythiophen) (Mdluli et al., 2022). *Reproduced from an open-access article by MDPI under a CC license.*

addition to their capacity for scalable and inexpensive manufacture (Yi and Abidian, 2016). Essentially, conjugated organic polymers are employed in water-splitting due to the ease in tunability (Sprick et al., 2015) for the desired electronic properties of photocatalysts (Wang et al., 2016; Sprick et al., 2016). This makes it possible of building frameworks and networks of photocatalytic polymers with large concentrations of soluble photocatalysts (McQueen et al., 2022; Woods et al., 2017; Wang et al., 2017). This makes possibilities for composite materials and large-area thin-film designs possible with applications in gadgets like organic light-emitting diodes (OLED), organic field effect transistors (OFET), and organic photovoltaics (OPV) (Pankow and Thompson, 2020). This is also true for CP materials, where the surface chemistry is changed by adding a hydrophilic adenine group to the side chain. The promotion of hydrogen bonding between the catalyst's surface and water molecules enhances the polymer photocatalyst's hydrophilicity and dispersity in the photocatalytic reaction solution (Li et al., 2022; Bai et al., 2022). With this wettability function, the degree of intermolecular contact between a liquid and a solid surface is dictated by surface shape and polarity (Darmanin and Guittard, 2014). CPs with network topologies provide small bandgaps that absorb visible light and promote charge transfer along the three-dimensional network due to their widely delocalized π-conjugated structures. Being highly porous, they could be perfect for large-scale water interactions (Oka et al., 2021). Additionally, CPs are nontoxic, affordable, and infinite resource materials that remain mechanically robust and chemically stable during extended periods of redox operation in water with a broad pH range (Oka et al., 2021).

In contrast to conventional metal-based catalysts, which frequently require costly and rare materials, CPs can be produced using simple renewable resources like biomass or feedstocks derived from petroleum (You et al., 2022). Furthermore, CP materials can be mixed with other materials to generate hybrid systems with improved performance, such as metal oxides or carbon-based nanomaterials (Gogoi et al., 2021). To optimize charge separation, charge transport, and catalytic activity—all of which are necessary to initiate photo-redox reactions for water-splitting, these hybrid systems can take advantage of the special qualities of each component (Dai and Liu, 2020).

15.3 CP material synthesis and characterization

To integrate CPs into a variety of applications, including energy conversion devices, their synthesis and characterization are essential (Al-Azzawi et al., 2023; Bini, 2018). CPs of all kinds, including "poly(pyrrole)s, poly(benzoazole)s, poly(thiophene)s, poly(cyclopenta-dithiophene)s, poly(fluorine)s, poly(carbazole)s, poly(arylene-ethynylene)s, poly(seleno-phene)s, poly(benzotriazole)s, poly(phenylene)s" and their derivatives, have been synthesized up to this point, and their possible uses in optical and electrical devices have been assessed (Amna et al., 2020). Typically, the synthesis process involves the controlled polymerization of monomers containing aromatic or heteroaromatic units, such as thiophene, phenylene, fluorene, and carbazole (Aravindu et al., 2018). A major obstacle in the conjugated polymer synthesis process is attaining accurate control over the molecular structure, which encompasses the topology, sequence, and regiochemistry of the polymer backbone (Ying et al., 2017). In essence, a multitude of different methods has been used to synthesize polymer photocatalysts (McQueen et al., 2022). The idea of sequence-defined CPs, in which the exact order of monomers along the polymer backbone is controlled, has surfaced more recently (Pandurangan and Porel, 2019). This is achieved through the use of orthogonal reactions, such as click chemistry and thiol-ene reactions, which enable the stepwise addition of monomers in a predetermined order (Long et al., 2023). It's significant that these CPs include novel links and π-conjugated systems, and their porosity enables the use of π-conjugated skeletons and pores for photocatalytic water-splitting (Chohan et al., 2022).

15.3.1 Synthetic methods for CPs

An extensive range of CPs can be readily synthesized thanks to the abundance of conjugated organic monomers and the variety of synthetic techniques available. These monomers serve as the basic building structural units that construct from which a range of CP polymers with tuneable properties can be synthesized (Cheng et al., 2009; Al-Azzawi et al., 2023). Furthermore, the monomers' adaptability allows for fine control over the resulting polymer for certain photocatalytic applications.

High-performance CPs can be produced by a variety of synthetic techniques, most notably the traditional cross-coupling polymerizations. Stille polycondensation reaction (see Fig. 15.4A) for donor–acceptor (D–A) type CPs which are typically carried out in batch reactors (Bini, 2018; Carsten et al., 2011; Bhattacherjee et al., 2021). Usually, an organotin monomer (the donor) and an organohalide monomer (the acceptor) engage in step-growth polymerization to accomplish this.

Nevertheless, this method causes a considerable variance in molecular weight and molecular weight dispersity (Đ) from batch to batch. Thus the creation of novel synthetic techniques to overcome this repeatability problem is a crucial step toward commercializing these materials and expanding the applications of these materials in highly efficient organic electrical and photonic devices (Kim et al., 2024). Among these techniques, the preparation of conjugated porous polymers and linear copolymers has been extensively employed using Pd-catalyzed Suzuki–Miyaura coupling polymerization (see Fig. 15.4B) between halogenated aromatics and diboronyl ester (Dai and Liu, 2020). Direct (hetero)arylation polymerization (DArP or DHAP) via C–H bond activation, catalyzed by transition metals, has become a potent and capable method for the synthesis of organic compounds recently (see Fig. 15.4C) (Mayhugh et al., 2022). These methods are preferred because they can produce materials with the necessary electrical and structural characteristics (Xu et al., 2023). Thus site-specific and highly reactive cross-coupling is created by monomer prefunctionalization, which opens the door to high molecular weight, low-defect materials.

15.3.1.1 *Chemical synthesis*

CPs are made by joining many monomers to generate a polymer chain that alternates between single and multiple bonds, resulting in a conjugated electron system. The main process in chemical synthesis is the creation of carbon–carbon (C–C) bonds between aromatic monomers, which can be achieved through cross-coupling reactions with catalysts made of transition metals (such as Pd, Ni, and Cu) to form π-conjugated compounds (Chua et al., 2021; Liu et al., 2021a). There are two primary forms of polymerization in the chemical synthesis method; thus organometallic coupling and treatment with an appropriate oxidant (Anguera and Sánchez-García, 2014). A few of the methods of polymerization reactions utilized for the chemical synthesis of CPs are discussed briefly.

15.3.1.1.1 Step-growth polymerization

Step-growth polymerization forms longer-chain polymers and bigger oligomers through the reaction of bi- or multifunctional monomers (Yokozawa and Ohta, 2016). This method is used to create a wide range of synthetic and natural polymers, including polyesters, polyamides, and polyurethanes. CPs, which have extraordinary optoelectronic qualities and a variety of applications, can also be made using step-growth polymerization (Aplan and Gomez, 2017). CPs with controlled topologies can be produced synthetically using three different techniques: step-growth polymerization, chain-growth polymerization, and iterative synthesis. Among these strategies, CPs with regulated chemical structures including

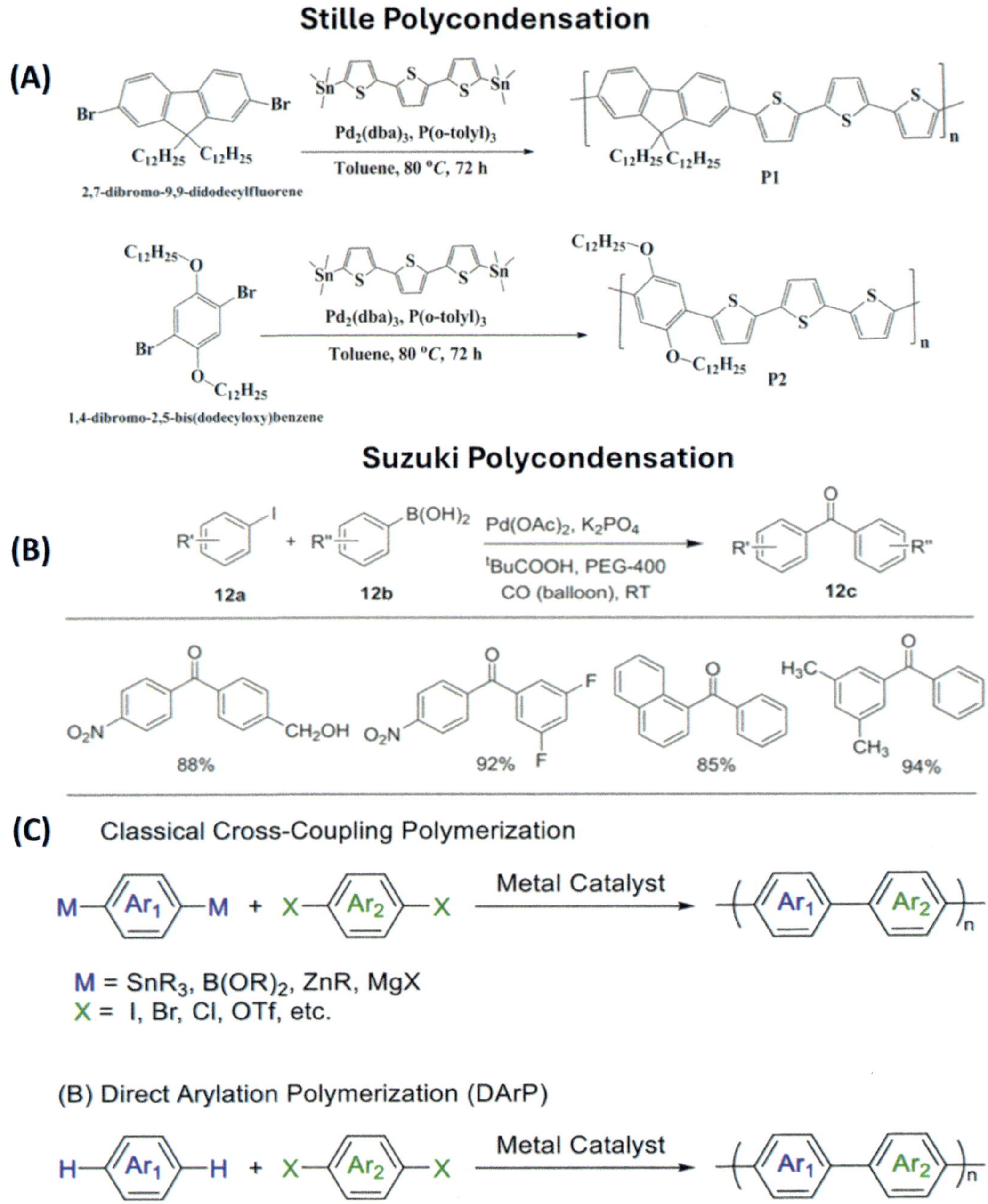

FIGURE 15.4 Synthetic approaches to CPs (A) Stille polycondensation. (B) Suzuki polycondensation of Pd-catalyzed carbonylative Suzuki coupling at ambient temperature and pressure (Han method). (C) Transition metal-catalyzed direct (hetero)arylation polymerization (DArP or DHAP) (Mayhugh et al., 2022). *(A) Reproduced with permission from John Wiley and Sons. Copyright © 2015 Weinheim, Germany's WILEY-VCH Verlag GmbH & Co. KGaA. (B) Reproduced from Bhattacherjee, D., Rahman, M., Ghosh, S., Bagdi, A.K., Zyryanov, G.V., Chupakhin, O.N., et al., 2021. Advances in transition-metal catalyzed carbonylative Suzuki–Miyaura coupling reaction: an update. Advanced Synthesis & Catalysis 363 (6), 1597–1624. https://doi.org/10.1002/adsc.202001509. Available from: https://onlinelibrary.wiley.com/doi/abs/10.1002/adsc.202001509 with permission from John Wiley and Sons. (C) Reproduced from an article licensed under CC-BY-NC-ND 4.0., Copyright © 2022, American Chemical Society.*

length, regioregularity, sequence, and topology; can be produced using step-growth polymerization.

One technique for achieving step-growth polymerization of CPs is the Suzuki–Miyaura catalyst-transfer polymerization (SCTP) (Yang et al., 2023). A variety of end-functionalized polymers can be accessed by the SCTP method's improved initiation system, increased functional group tolerance, and high polymerization efficiency under mild circumstances (Kim et al., 2022). In step-growth polymerization, a reaction occurs between the monomers' functional groups, resulting in the creation of a dimer, longer oligomers, and ultimately a long-chain polymer. To produce high molecular weight polymers, the reaction's extent is crucial. Note that step-growth polymerization can be difficult to produce large molecular weight since it needs a high degree of reaction. Nonetheless, CPs with high levels of purity in their end-functional groups can now be produced with the use of precision synthetic techniques like the SCTP. The overall mechanism of SCTP polymerization is depicted in Fig. 15.5A and B.

Also, typical instances of SCTP polymerization involving carbazole monomers of the triolborate type. Accordingly, "potassium 3-(6-bromo-9-(2-octyldodecyl)-9H-carbazole-2-yl) triolborate (M1) and potassium 2-(7-bromo-9-(2-octyldodecyl)-9H-carbazole-2-yl) triolborate (M2), as an effective and adaptable method for accurately synthesizing poly[9-(2-octyldodecyl)-3,6-carbazole] (3,6-PCz) and poly[9-(2-octyldodecyl)-2,7-carbazole] (2,7-PCz)", as shown in Fig. 15.6C and D respectively (Kobayashi et al., 2021). In general, CPs with regulated topologies can be synthesized using a flexible technique like step-growth polymerization.

The creation of precise synthetic techniques, such as the SCTP method, has made it possible to make end-functionalized CPs, which are crucial components needed to create soft, hybrid conjugated materials with intricate architectural patterns. Reactive intermediates are usually formed as the process moves forward, and these go through coupling reactions to produce the polymer chain (Nojima et al., 2016). Accurate control over conjugated polymer structure, regioregularity, sequence, and topology can be achieved using step-growth polymerization. To obtain a high degree of control over the structure of the resultant polymers, a variety of cross-coupling processes catalyzed by transition metals have been used (Bautista et al., 2021; Lee et al., 2021).

15.3.1.1.2 Chain-growth polymerization

Through the chain-growth polymerization process, unsaturated monomer molecules are incorporated one by one into the active site of a polymer chain that is expanding. CPs are commonly synthesized via chain-growth polymerization. By using this method, an active species such as a cation, anion, or radical triggers the polymerization of monomers (Verswyvel et al., 2011; Aplan and Gomez, 2017). Common techniques used in chain-growth polymerization include radical polymerization, anionic polymerization, and cationic polymerization (Yokozawa and Yokoyama, 2009). With regulated molecular weight and polydispersity, it makes the manufacture of high-molecular-weight CPs simple. This process involves three main types of reactions: initiation, propagation, and termination (Geng et al., 2010; Yokozawa and Yokoyama, 2009).

The initiation phase involves the formation of an active species, usually an ion or a radical, through the breakdown of an initiator molecule, as illustrated by the active

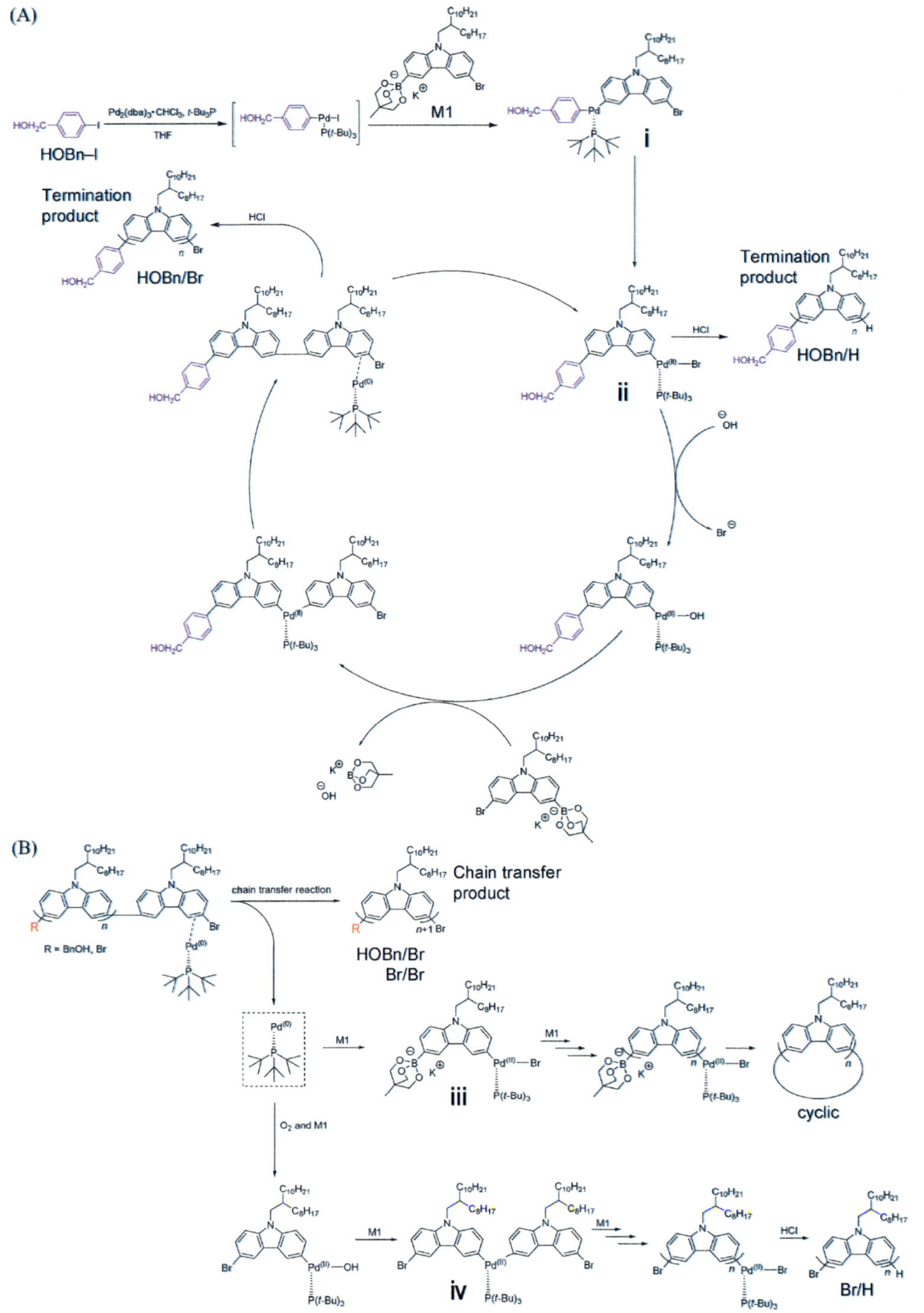

FIGURE 15.5 (A) The overall mechanism of 3,6-PCz Suzuki–Miyaura catalyst-transfer polymerization. (B) Process mechanisms most likely to give rise to cyclic 3,6-PCz and 3,6-poly[9-(2-octyldodecyl)-3,6-carbazoles] (PCzs) with unique chain end structures (Kobayashi et al., 2021). *3,6-PCz,* Poly[9-(2-octyldodecyl)-3,6-carbazole. *Reproduced from an open-access article by MDPI under a CC license.*

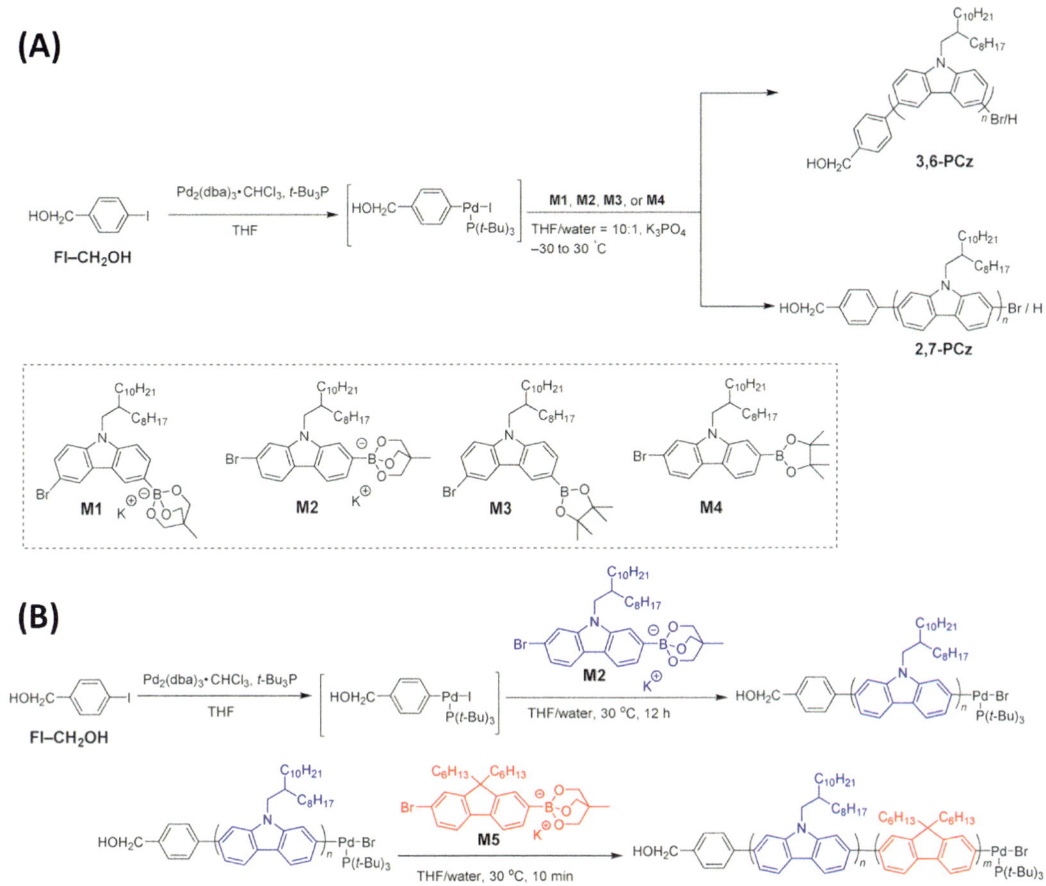

FIGURE 15.6 (A) Method design for the Suzuki–Miyaura catalyst-transfer polymerization of end-functionalized poly[9-(2-octyldodecyl)-3,6-carbazoles] (PCzs) based on the triolborate-type carbazole monomers. (B) Polymerization at random of M1/M2, M1/M5, and M2/M5 (Kobayashi et al., 2021). *M1*, Potassium 3-(6-bromo-9-(2-octyldodecyl)-9H-carbazole-2-yl)triolborate; *M2*, potassium 2-(7-bromo-9-(2-octyldodecyl)-9H-carbazole-2-yl) triolborate; *M5*, fifth monomer. *Reproduced from fFReproduced from an open-access article by MDPI under a CC license.*

dominant species initiation action in Fig. 15.7. This active species then reacts with a monomer molecule to start the conversion to the polymer through a process called propagation. The growing polymer chain retains the active site, allowing the addition of more monomers to further extend the chain (Yokozawa and Yokoyama, 2009). This process continues until the desired degree of polymerization is achieved. The chemical equation P + M → P-M, where P is the polymer chain, M is the monomer, and -M indicates the addition of a monomer unit to the polymer chain, can be used to explain the propagation stage in chain-growth polymerization (Geng et al., 2010). Finally, the termination step occurs when two polymer chains containing active centers react, deactivating the growth centers and resulting in the termination of chain propagation. Recombination and disproportionation are the two primary forms of termination. Recombination is the process by which two

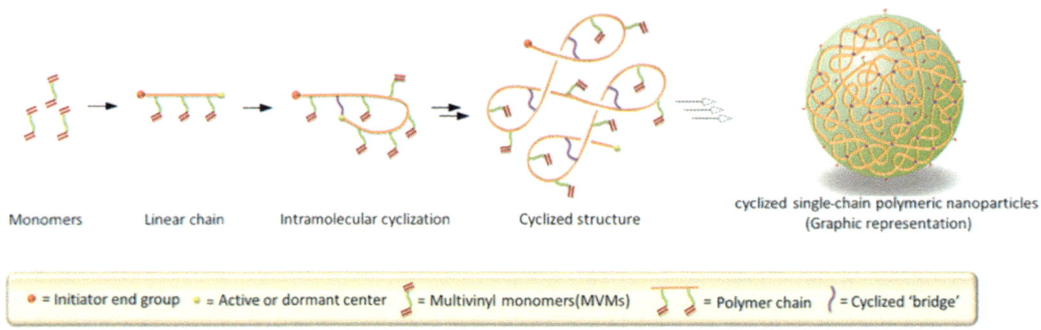

FIGURE 15.7 Diagrammatic free-radical polymerization of multivinyl monomers (Gao et al., 2017). *Published with authorization from John Wiley and Sons. Reserved all rights. Situated in Weinheim, Germany, Wiley-VCH Verlag GmbH & Co. KGaA.*

expanding polymer chains react to create a covalent link that unites them to form a single polymer molecule. Conversely, disproportionation results in the production of two distinct polymer molecules by moving a hydrogen atom from one chain to another (Xu et al., 2023; Yokozawa and Ohta, 2013).

Chain-growth polymerization has many benefits, including the capacity to produce polymers with certain molecular weights and shapes by adjusting the reaction conditions and the ratio of initiators to monomers (Ito, 2016; Jagadesan and Schanze, 2019). It also allows for the synthesis of polymers with tailored properties and structures (Yokozawa and Ohta, 2013, 2016). Polymers including polyethylene, polypropylene, polyvinyl chloride, and polymethyl methacrylate are produced by chain-growth polymerization (Aragrag et al., 2004). The molecular weight, structure, and other characteristics of the resultant polymers can be controlled by the chain-growth polymerization process.

15.3.1.1.3 Iterative synthesis

Iterative synthesis is the process of building up the polymer chain by adding monomer units repeatedly and step-by-step. By adding monomer units selectively in a preset order, this approach enables perfect control over the polymer structure and sequence (Xu et al., 2023). Iterative synthesis has demonstrated the effectiveness of transition metal-catalyzed cross-coupling processes, which allow for the production of complex CPs with regulated topology and regioregularity (Barnes et al., 2015). Through iterative synthesis, CPs can exhibit distinctive electrical, optical, and conducting features due to substantial π-electron delocalization along the polymer backbone. Repetitive linking of monomeric units to lengthen the polymer chain is the mechanism of CPs' iterative synthesis. This process typically utilizes a coupling reaction, such as a Suzuki coupling or a Stille coupling, to link the monomers together (Aiken et al., 2023).

The iterative process of the synthesis of diastereomers of a tetraboronic ester is described in Fig. 15.8. The iterative synthesis of CPs begins with the design and synthesis of monomeric units with suitable functional groups for polymerization. These monomers typically contain aromatic or heteroaromatic units that facilitate π-conjugation along the polymer backbone. The monomeric units are then subjected to polymerization reactions,

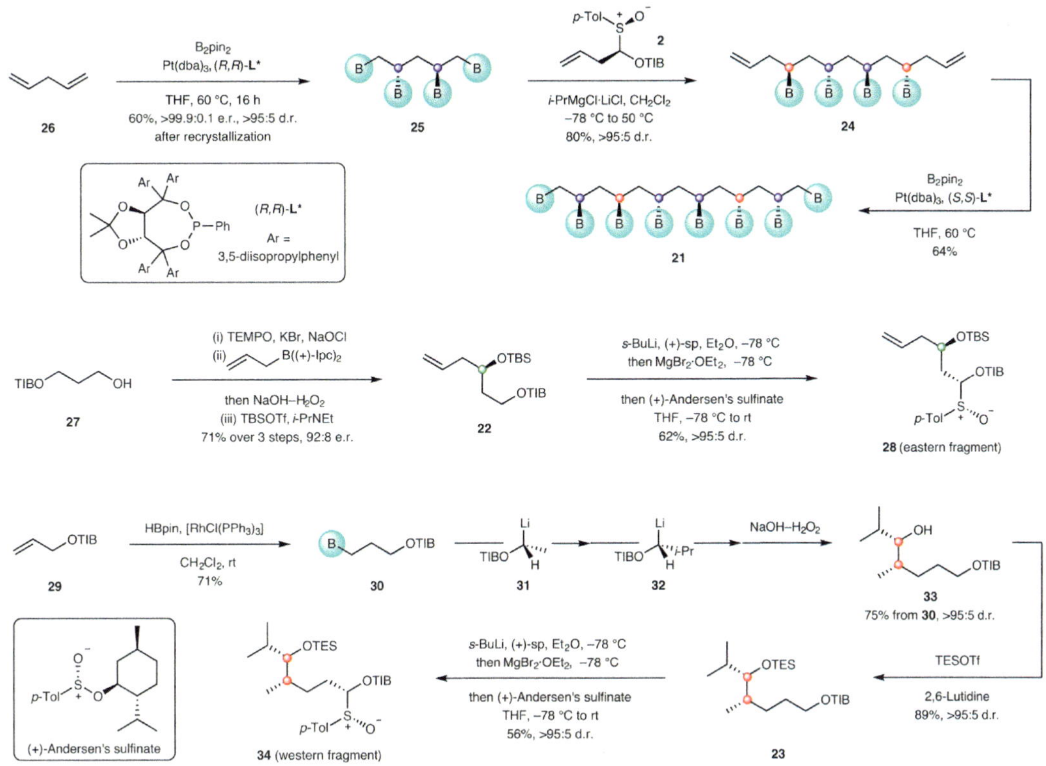

FIGURE 15.8 The synthesis process of eight diastereomers of a tetraboronic ester by iteration (Aiken et al., 2023). *Reproduced under license (Copyright © 2022) from Springer Nature.*

such as Suzuki–Miyaura coupling, Stille coupling, or Heck coupling, to form the first iteration of the polymer chain (Grisorio et al., 2010). These reactions often require catalysts and specific reaction conditions to achieve high yields and regioselectivity. After each iteration, the resulting polymer must be purified to remove any unreacted monomers or side products.

The molecular weight, structure, and purity of the polymer are examined using characterization methods such as mass spectrometry, gel permeation chromatography, and nuclear magnetic resonance (NMR) spectroscopy (Xu et al., 2023). In the repetitive iterations process, the purified polymer from the previous step serves as the starting material for subsequent iterations. Additional monomeric units are added using the same or different coupling reactions, allowing for the controlled growth of the polymer chain. This iterative process can be repeated multiple times to achieve the desired molecular weight and polymer length. Throughout the iterative synthesis, the properties of the CP can be fine-tuned by selecting specific monomeric units, controlling the polymerization conditions, and adjusting the polymer chain length (Miao et al., 2017; Higashihara et al., 2011). This allows for the customization of the polymer's electronic, optical, and conducting properties for various applications such as organic electronics, photovoltaics, and sensors. Because of their variable electrical characteristics and ease of processing, CPs have been used in OPV systems, OLED, OFET, and medicinal materials. The production of sophisticated functional materials for a range of technological applications is made possible by the powerful and adaptable method of conjugated polymer iterative synthesis, which can be used to customize the structure and properties of these materials.

The intended uses and characteristics of the conjugated polymer determine which specific synthesis method is used. Since each approach has benefits and drawbacks, the process is frequently improved to produce the desired polymer structure and qualities by modifying the synthesis conditions and monomer selection. It is crucial to note that CPs can be precisely synthesized to create polymers with desired shapes and characteristics. The creation of high-performance devices in a variety of industries, such as electronics, optoelectronics, energy storage, and sensors, has been made possible by the increased control over tandem polymer synthesis.

15.3.1.2 Electrochemical synthesis

To carry out electrochemical polymerization, a derivation of chemical synthesis, the right monomers can be oxidized on the right anode in a two-or three-electrode electrochemical cell (Liu et al., 2021a). The capacity to tailor the properties of the resultant polymer, control over film thickness and morphology, and simplicity are just a few benefits of the adaptable technology known as electrochemical polymerization. One unique feature of electrochemical polymerization is that it generates conjugated polymer films directly on the anode, making it simple to regulate the thickness and structure of the films. During the electrochemical polymerization process, counterions are usually added to the polymer film to balance the charge on the polymer backbone. This results in CPs that are often doped. The capacity to tailor the properties of the resultant polymer, control over film thickness and morphology, and simplicity are just a few benefits of the adaptable technology known as electrochemical polymerization. One unique feature of electrochemical polymerization is that it generates conjugated polymer films directly on the anode, making it simple to regulate the thickness and structure of the films. During the electrochemical polymerization process, counterions are usually added to the polymer film to balance the charge on the polymer backbone. This results in CPs that are often doped (Zhang et al., 2018; Pokhodenko and Krylov, 1994).

In the electrochemical polymerization process, a monomer is subjected to an electric potential in an electrolytic medium. The monomer is oxidized or reduced at the electrode surface, leading to the formation of polymer chains. The choice of monomer and electrolytic medium can affect the properties of the resulting polymer, such as its conductivity, band gap, and optical properties (Heinze et al., 2010). The terms anodic and cathodic electrochemical polymerization refer to two important types of polymerization processes. First, in the anodic electrochemical polymerization process, an oxidative monomer radical cation is produced. With an electrochemical stoichiometry of 2.07–2.6 F/mol, the monomer is utilized in the electro-oxidation process (Fomo et al., 2019). In polymer film manufacture, a monomer molecule typically needs two electrons or 2 F/mol. The cyclic voltammogram displays the oxidation of the oligomeric molecules due to the excess charge (Guay et al., 1992). Polymer synthesis and development are driven by the interaction of a radiation cation and a radical cation with a neutral monomer. A polymer can be formed by the radical cations reacting with neutral monomers or other nearby radical cations (Guyard et al., 1997).

Fig. 15.9A shows the radical cation-radical cation coupling step of the electrochemical polymerization process. The initial stage of the electrosynthesis process is the generation of a cation radical, which is brought about by an electron moving from the atom's sp3 orbital to the electrode surface during oxidation. The reduced state produces the most conducting polyaniline (PANI) (emeraldine), while aniline electropolymerizes to give a nonconducting PANI (leucoemeraldine state) (Fomo et al., 2019).

A comparatively low yield of conducting polymers is frequently produced by the cathodic electrochemical polymerization process, which employs reductive electrosynthetic techniques (Fomo et al., 2019). For this reason, conductive polymers are seldom produced via cathodic electrochemical polymerization. Fig. 15.8B depicts a typical cathodic electrochemical polymerization of anilines. Thus far, conducting polymers, or poly(p-phenylene-vinylenes), the building blocks of organic light-emitting devices, have only been produced by cathodic polymerization (Li and Li, 2008). Generally, CP electrochemical synthesis has been extensively used in many different domains, such as coatings, organic solar cells, OFETs, and sensors. The power conversion efficiency of organic solar cells can be greatly impacted by the CP selected and the polymerization process used (Pei, 2007). Similarly, the performance of OFETs depends critically on the charge carrier mobility of CPs, and electrochemical polymerization can be utilized to regulate the shape and enhance the polymer's charge transport characteristics (Yang et al., 2018).

During the polymerization process, distinct functional groups or side chains can be added to modify the characteristics of CPs (Liu et al., 2011). It has been demonstrated, for instance, that adding fluorine atoms or pyridine and selenophene moieties to the polymer backbone

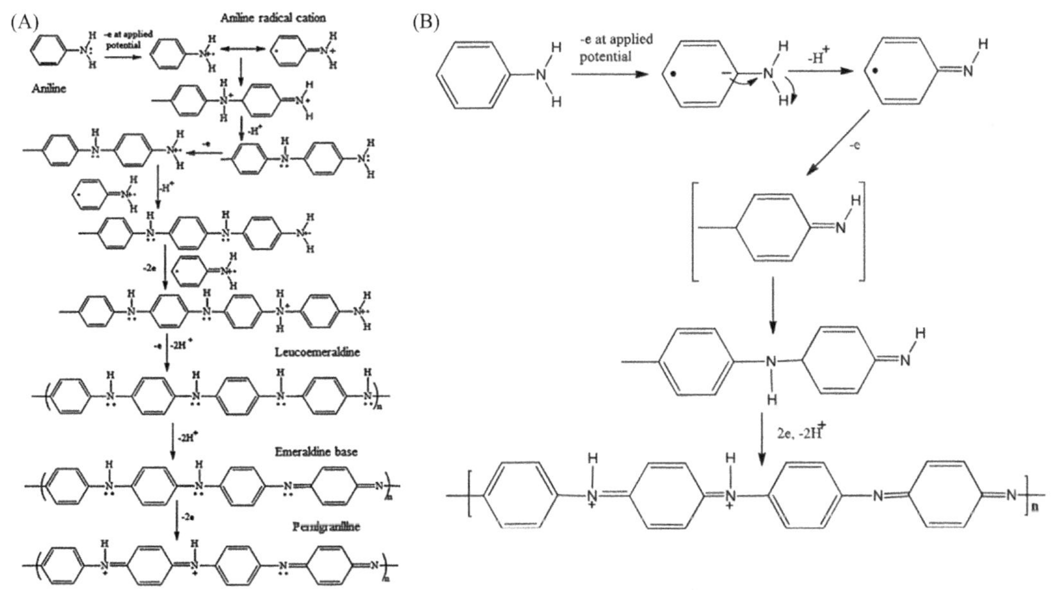

FIGURE 15.9 Aniline polymerization using the electrochemical process. (A) Mechanism for the anodic electrochemical polymerization of polyaniline (B) Mechanism of the cathodic electrochemical polymerization of aniline (Fomo et al., 2019). *Reproduced with permission from Springer Nature Switzerland AG, Copyright © 2019.*

increases the charge transfer mobility in OFETs (Zhang et al., 2016). Furthermore, there has been an increase in interest in the use of nonfullerene acceptors (NFAs) in organic solar cells. Electrochemical polymerization can be used to create polymers that, when combined with NFAs, have high power conversion efficiencies (Chatterjee et al., 2021). The electrochemical synthesis of CPs has found applications in various fields, including organic solar cells, OFETs, sensors, and coatings, contributing to the development of more efficient and versatile electronic and optoelectronic devices.

15.3.2 Characterization techniques for CPs

Characterization techniques play a critical role in understanding the properties of CPs and optimizing their performance in various applications. CPs can be characterized using a variety of general and specialized approaches. Since polymer science uses all of the approaches, only a brief synopsis of some of the most important characterization techniques will be provided.

15.3.2.1 Spectroscopic electrochemical methods

Electrochemical methods, like cyclic voltammetry (CV) and electrochemical impedance spectroscopy (EIS), are widely used to investigate CPs' redox behavior and charge transport properties. Spectro-electrochemical methods, such as infrared and UV–Vis absorption spectroscopy, can offer important insights into the structural and electrical alterations that result from electrochemical doping or oxidation/reduction processes in CPs (Aplan and Gomez, 2017). One method for figuring out CPs' molecular weight distribution and average molecular weight is size exclusion chromatography. It entails detecting the polymer concentration and sorting the polymer chains according to size using a porous column. For a more accurate analysis, the method can be combined with other detectors such as UV–Vis or refractive index detectors (Liu et al., 2011). Additionally, optical methods like fluorescence and UV–Vis absorption spectroscopy can offer important insights into the optical characteristics of CPs, such as their emission and absorption spectra. Additionally, the excited-state dynamics and fluorescence lifetimes of CPs can be investigated using methods such as time-resolved spectroscopy and photoluminescence spectroscopy.

15.3.2.2 Morphological and structural analysis

Structural analysis techniques such as X-ray diffraction and solid-state NMR are widely used to determine the crystalline structure, degree of order, and molecular packing topologies of CPs (Hildner et al., 2017). The topography and surface morphology of conjugated polymer films can be seen at the nanoscale using techniques like scanning electron microscopy and atomic force microscopy. Additionally, thermal analysis methods like thermogravimetric analysis and differential scanning calorimetry can reveal details regarding the thermal characteristics of CPs, such as their thermal stability and glass transition temperature (Wang et al., 2013).

15.3.2.3 Electrochemical analysis

Techniques for electrochemical investigation are essential for describing CPs' electrochemical behavior and characteristics. A brief description of a few popular techniques for electrochemical analysis is provided.

1. CV: CV is the most widely used electrochemical technique for analyzing the redox behavior of CPs. In CV, the current that is produced when the working electrode is subjected to a potential sweep is measured. CV gives details regarding the polymer's electrochemical stability, electron transfer rates, and oxidation and reduction potentials (Chatterjee et al., 2021).
2. Chronoamperometry: Using this technique, the working electrode is given a constant potential, and the resultant current is tracked over time. Chronoamperometry is a useful tool for researching CPs' electrochemical characteristics, including their capacity for charge storage and transmission (Gupta et al., 2023).
3. EIS: EIS is an effective method for researching the electrical characteristics of CPs and the interfaces between them. It entails putting the system under an AC voltage and monitoring the current response that results. EIS offers details on capacitance, charge transfer resistance, and other factors about the polymer's electrochemical behavior (Chang and Park, 2010).
4. Differential pulse voltammetry (DPV): Using a sequence of potential pulses applied to the working electrode, DPV is a voltammetric technique that measures the current that results. DPV helps investigate the relationship between CPs and analytes as well as for sensing applications since it may be used to identify and measure particular electroactive species in a solution (Tufan et al., 2014).
5. Scanning electrochemical microscopy (SECM): The spatially resolved information on CP electrochemical behavior is obtained by SECM, which integrates scanning probe microscopy with electrochemical techniques. It is useful for mapping the surface or polymer film's other characteristics, such as conductivity and local electroactivity (Lefrou and Cornut, 2010).

15.4 Photocatalytic water-splitting with CPs

Hydrogen and oxygen can be separated using light and a catalyst in a photocatalytic process. Water can be divided into hydrogen and oxygen to modify its usual Gibbs free energy of ΔG0, which is 237 kJ/mol as shown in Eq. (15.1). During this process, the lowest band gap can be determined using Eq. (15.2) if all of the energy released by splitting water is converted into electrical work. Simply put, the thermodynamic water-splitting condition requires semiconductors used as photocatalysts to have a band gap greater than 1.23 eV. Since a wider band gap would result in less efficient use of visible light (> 400 nm), the visible light response to the band gap of semiconductors must be smaller than 3.0 eV as represented in Eq. (15.3) (Xu et al., 2018).

$$H_2O \rightarrow H_2 + 1/2O_2, \Delta G^0 = 237 kJ/mol \quad (15.1)$$

$$\Delta G^0 = W_{f,\,max} = -nEF \quad (15.2)$$

$$E = h\nu = hc/\lambda \quad (15.3)$$

15.4.1 Mechanism of water-splitting by CPs

The basic process of photocatalysis is light absorption by a semiconductor photocatalyst, which produces electron–hole pairs. The necessary chemical changes are subsequently produced by these pairs reacting through redox reactions with target molecules (Dai et al., 2020). Many processes, such as light absorption, charge separation, and the catalytic production of hydrogen and oxygen, are usually involved in the mechanism of water-splitting by CPs as shown in Fig. 15.10A (Xu et al., 2018). The process begins with the CP absorbing photon energy, which transfers an electron from the VB to the CB, creating an electron–hole pair, as shown in Fig. 15.10A and B. Once the hole is consumed in another redox reaction, the excited electron can then be moved to a cocatalyst material, in this case, the CP (Mansha et al., 2022). The subsequent reactions involving water-splitting depend on this charge separation. Reactive oxygen species (ROS) produced by the heated polymer is one potential mechanism for water-splitting via CP. After that, these ROS species can oxidize water molecules, producing oxygen gas. As an alternative, the excited polymer can produce H_2 by reducing protons (H^+). Z-scheme water-splitting is a method wherein in certain instances, the oxidation and reduction reactions take place concurrently shown in Fig. 15.10C (Schumacher and Marschall, 2022). This method facilitates total water-splitting without the need for sacrificial chemicals by using an inorganic catalyst for water oxidation and a CP photocatalyst for proton reduction.

Cocatalysts, including platinum and other catalysts, can further increase the effectiveness of water-splitting by making it easier for protons or electrons to move over the polymer surface. To encourage the intended water-splitting events and inhibit the unwanted side reactions, these cocatalysts can function as electron sinks (Ruan, 2020). Several variables, such as the polymer's structure, content, morphology, and kind of cocatalyst, can affect how well CPs split water.

15.4.2 Role of CP materials in photocatalysis

In the field of photocatalysis, CP materials are essential, especially when it comes to splitting water to produce hydrogen. In photocatalysis, CP's primary function is *light harvesting*. The broad π-conjugated backbones of CPs allow for substantial light absorption in the visible and near-infrared regions of the spectrum. CPs are excellent choices for light-harvesting materials in photocatalytic applications due to their extensive absorption of solar radiation. Additionally, CP is a crucial *charge separation and transport medium* in photocatalysis. The delocalized π-electron system in CPs is responsible for this. This makes efficient charge transfer and separation possible, which is necessary for successful photocatalysis. Usually, CP can easily separate the excitons (electron–hole pairs) that are produced when light is absorbed, and the resultant charge carriers may then be efficiently transferred to the catalytic sites, improving the water-splitting process (Shit et al., 2017). Another important role of CP in the

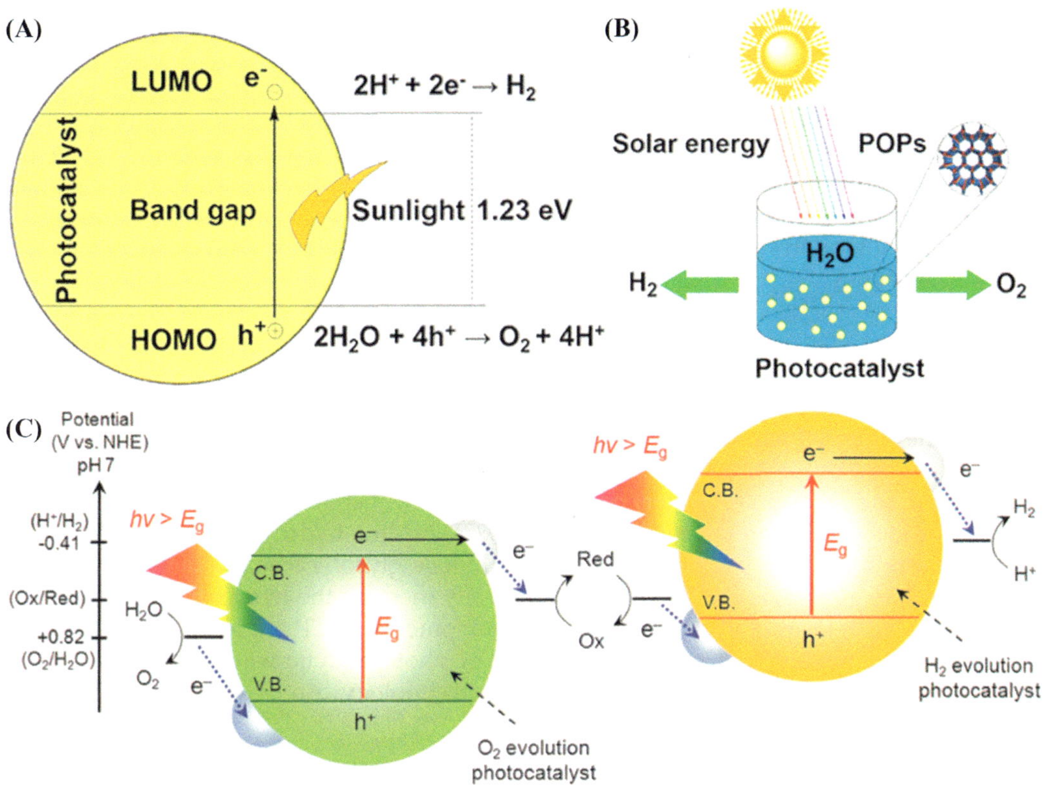

FIGURE 15.10 (A) The basic process of water-splitting using POPs. (B) The conventional method of using semiconductors for photocatalytic water-splitting (Xu et al., 2018). (C) An example of the CP Z-scheme for photocatalytic water splitting (Schumacher and Marschall, 2022). *POPs*, Porous organic polymers; *HOMO*, highest occupied molecular orbital; *LUMO*, lowest unoccupied molecular orbital. *Reproduced from Frontiers under CC BY 4.0 DEED. (C) From Copyright © 2022 by Springer Nature, under an open-access CC BY license.*

photocatalysis process is the *tuneable optoelectronic properties*. The chemical structure and composition of CPs can be tailored to fine-tune their optoelectronic properties, such as the bandgap, redox potentials, and charge transport characteristics. CPs can be tuned for particular photocatalytic reactions, such as water-splitting, CO_2 reduction, and the destruction of organic pollutants, thanks to this tunability. Furthermore, it is easy to combine the heterojunction generation in CPs with other materials to create heterojunction designs, including semiconductors, cocatalysts, and carbon-based materials. The photocatalytic performance of these heterojunctions is improved over individual components due to the effective charge separation and transfer that they enable (Chen and Cao, 2009). *Stability and durability* is another important role of CP in photocatalytic water-splitting. While CPs can suffer from stability issues under prolonged exposure to light and water, recent advancements in polymer design and engineering have led to improved stability and durability. Strategies like the incorporation of stabilizing groups, the use of cross-linking, and the development of protective coatings have helped to address the stability challenges (Freudenberg et al., 2018).

The *synergistic efficiency* of CP when combined with other materials with other materials, such as inorganic semiconductors, cocatalysts, or carbon-based materials has made CP the photocatalytic material of choice for hydrogen production. With the benefits of these complimentary materials and the special qualities of CPs, extremely effective and adaptable photocatalytic systems can be fabricated for possible commercial applications (Siemiaszko et al., 2021). The versatility, tuneable properties, and solution processability of conjugated polymer materials have made them a decidedly auspicious class of materials for advancing the field of photocatalysis, predominantly for solar-driven water-splitting and other clean energy applications.

15.4.3 Methods for improving CP photocatalytic performance

The main disadvantage of hydrophobic CPs in water-based processes is that they disperse unevenly, which means that an organic cosolvent must be added. Moreover, the two-electron process that creates hydrogen is inhibited by the high polymer chain aggregation and the poor water/polymer interfacial behavior, which limit electron transmission across the interface (Chang et al., 2022). Adding hydrophilic functional groups to the side-chain of hydrophobic CPs, such as ethylene glycol, carboxylic acid, amino, and ionic electrolyte groups, is another tactic to naturally increase the hydrophilicity of the CPs (Chen et al., 2020; Diao et al., 2019). Recent developments in CP engineering, design, and integration have made major strides in the field of photocatalytic water-splitting, opening the door to more effective and useful uses of this technology. An overview of some of the significant developments is provided.

1. Novel polymer structures: Researchers have been exploring the design of new CP structures with improved optoelectronic properties and photocatalytic performance. This includes the design of D–A-type polymers, which can enhance light absorption and charge separation. Examples include copolymers with electron-rich and electron-deficient units, as well as the incorporation of planar aromatic units to promote charge delocalization.
2. Heterojunction architectures: Improving photocatalytic water-splitting has needed the use of CPs and other materials, such as semiconductors or cocatalysts, to form heterojunction structures. These heterojunctions' efficient charge separation and transfer capabilities can hasten the synthesis of hydrogen (Zhao et al., 2020). Promising outcomes have been observed when CPs are coupled with metal-organic frameworks, carbon nitrides, or transition metal chalcogenides. The efficiency of water-splitting technologies can be significantly increased by employing CPs in heterojunction structures. Through meticulous material selection, structural optimization, and efficient synthesis techniques, scientists may create sophisticated photocatalytic devices that utilize solar radiation for the sustainable creation of hydrogen (Cho et al., 2021).
3. Cocatalyst integration: There has been much research done on the addition of cocatalysts to the surface of CPs, such as molecular catalysts or nanoparticles of noble metals. The total water-splitting efficiency can be increased by using these cocatalysts as active sites for the HER. Cocatalyst selection and optimization of their loading and distribution on

the polymer surface are crucial factors in this approach (Hayat et al., 2019a). Enhancing the effectiveness and stability of photocatalytic water-splitting systems requires the integration of cocatalysts with CPs. With their potential to enhance charge separation, reaction kinetics, and overall stability, catalysts significantly contribute to the performance enhancement of these systems, increasing their viability for the production of hydrogen sustainably (Hayat et al., 2019b).

4. Morphology control: The configuration and organization of the polymer can have a substantial impact on characteristics including light absorption, charge transfer, and catalytic activity. The photocatalytic efficiency of CPs can be greatly influenced by modifying their morphology and nanostructure (Zhang et al., 2023). Strategies like controlled polymerization, template-assisted synthesis, and surface engineering have been employed to create desirable morphologies, such as nanofibers, porous structures, and hierarchical architectures (Byun and Zhang, 2020). These anatomical characteristics have the potential to enhance light harvesting, simplify charge transmission, and offer easily accessible active areas for water-splitting. By carefully adjusting the structural properties of these materials, scientists can improve the catalytic efficiency overall, light absorption, and charge transport. This strategy opens the door to the development of more efficient and long-lasting photocatalytic hydrogen generation systems (Zhao et al., 2022).
5. Stability enhancement: Their poor stability, particularly when exposed to light and water for extended periods; is one of the main issues with conjugated polymer-based photocatalysts. For CPs to function better in water-splitting applications, their stability is crucial. Stable CPs can make a substantial contribution to the viability and efficiency of photocatalytic systems for the sustainable production of hydrogen by guaranteeing long operational lifetimes, reliable performance, and efficient charge dynamics. Exploring several ways to increase stability, like creating stronger polymer backbones, adding protective layers, and using encapsulating techniques has garnered significant research in recent years (Royakkers and Bronstein, 2021; Cui et al., 2024).
6. Spectral sensitization: In CP chemistry, spectral sensitization entails altering the optical characteristics of the polymers to increase their absorption of light into the visible and near-infrared portions of the spectrum (Li et al., 2024). This improvement is essential to raising the water-splitting's effectiveness. To improve solar spectrum usage, techniques like dye or quantum dot sensitization of CPs have been researched (Nayak et al., 2019; Zhao et al., 2024). These methods improve the polymers' capacity to extract a wider range of the solar spectrum for water-splitting by increasing the polymers' range of light absorption (Coghi and Coluccini, 2024).

15.5 Challenges and future perspectives

CPs show promise for water-splitting, which could lead to sustainable hydrogen production. To fully realize the promise of CPs in this industry, several obstacles must be addressed. Furthermore, future directions present stimulating chances to progress conjugated polymer applications in water-splitting.

15.5.1 Challenges of CPs for water-splitting

1. Efficiency and stability: CPs often suffer from limited photoconversion efficiency and stability under harsh water-splitting conditions. For potential future commercialization, it is imperative to improve the long-term stability of the polymer-based systems while simultaneously increasing the efficiency of light absorption, charge separation, and catalytic activity (Mansha et al., 2022).
2. Charge transport: Currently, charge transport within CP lacks a fast and efficient transport system to shuttle the photoexcited charges efficiently, hence, an effective charge transport within the conjugated polymer film is essential to prevent the recombination of photoexcited charges. The presence of defects, trap states, and limited mobility of charges within the polymer matrix hinders charge transport and reduces the overall efficacy of CP in large water-splitting ventures (Dai et al., 2020).
3. Catalyst integration: Research into the efficient integration of catalysts for both HER and OER is essential. Developing strategies to effectively couple catalysts with the CP matrix, optimizing their loading, and ensuring good charge transfer between the polymer and catalysts are key challenges (Zhao et al., 2020).
4. Scalability and manufacturing: CPs for water-splitting are thus far on the laboratory scale. Therefore scaling up the production of CP-based water-splitting systems while maintaining performance and controlling material properties presents challenges. Currently, all the synthesis processes have not been scaled up for commercial manufacturing, therefore developing scalable synthesis methods and manufacturing techniques for large-area devices with consistent performance is important for practical applications (Zhao et al., 2020; Zhou et al., 2024).

15.5.2 Future perspectives

Consistent with the recent calls for sustainable materials, energy, and clean production processes, the quest for further development of sustainable CP is essential and most of the development in this field will focus on the areas discussed below briefly.

1. Material design and synthesis: Tailoring the chemical structure and composition of CPs can lead to improved properties for water-splitting. Designing polymers with suitable energy levels, high absorption coefficients, and efficient charge transport properties can enhance overall performance.
2. Interface engineering: Future research efforts will focus on optimizing the interfaces between the conjugated polymer film, catalysts, and electrolytes can significantly influence the efficiency and stability of water-splitting systems. This will include developing efficient yet, sustainable strategies to control interfacial charge transfer processes and minimize losses that will result from interfacial resistance. This approach will be a promising avenue for the improvement of CP catalytic performance for water-splitting.
3. Tandem and multijunction systems: Future improvement efforts will focus on combining different CPs or integrating them with other materials in tandem or multijunction configurations can broaden the light absorption range, improve charge separation, and enhance overall efficiency. Although significant work has been done in this regard over

the past few years, however, the opportunities for further improvements in this regard are endless. Exploring novel architectures and materials combinations can lead to significant performance enhancements.

4. Advanced characterization techniques: As technology abounds, advancements in characterization techniques are set to be enhanced over time and this can provide detailed further insights into the structure, morphology, and charge dynamics of CPs. The creation of better materials and technologies can be guided by methods like ultrafast transient spectroscopy, imaging, and in situ spectroscopy, which can help clarify the underlying principles.
5. Hybrid systems and material combinations: Exploring hybrid systems that combine CPs with other materials, such as inorganic semiconductors, nanomaterials, or molecular catalysts, can offer synergistic effects and improved performance. Hybrid approaches can leverage the advantages of different materials to overcome individual limitations.

15.6 Case study: practical application of CP for water-splitting

Title: Photocatalytic water-splitting with poly(3-hexylthiophene)

Objective

To investigate the effectiveness of poly(3-hexylthiophene) (P3HT) in photocatalytic water-splitting under visible light irradiation.

Scope

Materials purchased for the experiment took a long time to arrive due to the complicated chemical procurement system. Moreover, due to time and the equipment available for the experiment, this case study could not be carried out in large quantities to fully understand the commercial potential of P3HT for hydrogen production.

Audience

This case study will benefit undergraduate students studying renewable energy, chemistry, and materials science.

Safety considerations

Ensure proper dispersion of the P3HT polymer before exposure to visible light. It is important to maintain the temperature within the vessel using a water jacket or temperature-controlled environment to avoid overheating during the reaction.

Actions taken

The first action was to procure materials and ensure the necessary, equipment and tools were made available for the experiment. The materials procured and the experimental processes are described briefly in this aspect.

Materials acquisition

P3HT was purchased (the structure is shown in Fig. 15.11A), DI water, visible light (LED with wavelength range 400–700 nm), and a quartz or glass container that allows light penetration were readily available in the laboratory. A setup to capture and measure evolved gases, such as a gas-tight syringe or gas chromatography (GC) system was set up.

Preparation

P3HT can be synthesized through oxidative polymerization in a laboratory. However, in this case study, commercially available P3HT was purchased and dispersed in distilled water using an

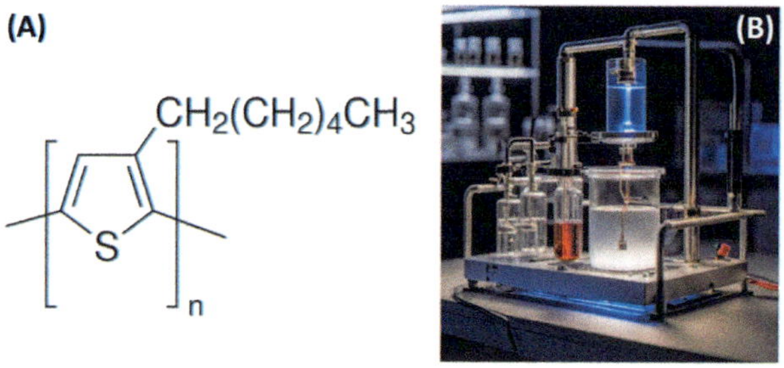

FIGURE 15.11 (A) Structure of commercially available poly(3-hexylthiophene), (B) laboratory set up for water-splitting (Note: Image for illustrative purposes only).

ultrasonicator to create a homogeneous suspension as shown in Fig. 15.11B setup. Typically, 150 g of P3HT was dispersed in 500 mL distilled water under mechanical stirring. The suspension was sonicated at 25°C for 40 minutes before and then transferred to the reaction vessel.

Experimental setup

Place the P3HT suspension in a reaction vessel and position it under the light source. Position the visible light source above the vessel to ensure uniform illumination. Maintain the temperature within the vessel using a water jacket or temperature-controlled environment to avoid overheating during the reaction. Connect the reaction vessel to a gas collection system, ensuring it is airtight to prevent gas escape. Install a GC system equipped with a thermal conductivity detector (TCD).

Measurements

Monitor hydrogen production using the GC system equipped with TCD to evaluate the composition of the gas. Take regular samples of the gas from the collection system at predetermined times (every 30 minutes, for example). As an alternative, track the amount of hydrogen created over time with a gas syringe or manometer.

Calculate the quantum efficiency (QE) of hydrogen production using Eq. (15.4).

$$QE = \frac{\text{Number of moles of hydrogen produced}}{\text{Number of moles of photons absorbed}} \times 100 \tag{15.4}$$

Determine the amount of light absorbed by measuring the absorbance spectrum of P3HT in the suspension using UV–Vis.

Challenges and solutions

The main challenge encountered is the stability of P3HT under reaction conditions. P3HT is a commercial polymer, hence, stability issues were not anticipated for further modifications before the experimental setup. Moreover, the experiment was not fitted with a metering pump for accurate quantification of the nitrogen gas generated. A possible solution in this case was to modify P3HT but that wasn't done for this case study.

Results

According to the study, P3HT efficiently facilitated the photolysis of water, yielding detectable amounts of H_2 over the length of the experiment. From the QE calculation, it was noted that P3HT is capable of producing sizable amounts of hydrogen under visible light. Electrons in P3HT are excited upon light absorption resulting in the creation of electron–hole pairs. Whereas the holes oxidize water to make oxygen, the electrons decrease it to form hydrogen.

Learning and knowledge outcomes

From this case study, CP P3HT with visible light absorption has proven to be a suitable photocatalyst for water-splitting. However, for large-scale applications, further study is needed.

15.7 Conclusion

Comparing CPs to conventional inorganic semiconductors, CPs show promise as a class of materials for water-splitting. Their tunable optoelectronic properties, high light absorption, and low-cost solution processability make them attractive for developing efficient and cost-effective water-splitting systems. Recent research has focused on designing novel conjugated polymer structures, incorporating cocatalysts, and optimizing device architectures to enhance the overall performance of these systems. Considerable advancements have been made in the attainment of visible-light-driven water-splitting, an essential component for realistic solar-to-fuel conversion. Nonetheless, there are still many obstacles to overcome, such as strengthening the polymers' stability, accelerating the production processes, and increasing charge separation and transmission. Realizing the full potential of CPs in the sustainable manufacture of clean hydrogen fuel will require ongoing research and development in this area. Generally, the unique properties of CPs have made them a promising candidate for advancing the field of solar water-splitting, with the potential to contribute to a more sustainable energy future.

References

Abanades, S., 2019. Metal oxides applied to thermochemical water-splitting for hydrogen production using concentrated solar energy. ChemEngineering 3 (3), 63. Available at https://doi.org/10.3390/chemengineering3030063.

Aiga, N., Jia, Q., Watanabe, K., Kudo, A., Sugimoto, T., Matsumoto, Y., 2013. Electron–phonon coupling dynamics at oxygen evolution sites of visible-light-driven photocatalyst: bismuth vanadate. The Journal of Physical Chemistry C 117 (19), 9881–9886. Available at https://doi.org/10.1021/jp4013027.

Aiken, S.G., Bateman, J.M., Liao, H.-H., Fawcett, A., Bootwicha, T., Vincetti, P., et al., 2023. Iterative synthesis of 1,3-polyboronic esters with high stereocontrol and application to the synthesis of bahamaolide A. Nature Chemistry 15 (2), 248–256. Available at https://doi.org/10.1038/s41557-022-01087-9.

Al-Azzawi, A.G.S., Aziz, S.B., Dannoun, E.M.A., Iraqi, A., Nofal, M.M., Murad, A.R., et al., 2023. A mini review on the development of conjugated polymers: steps towards the commercialization of organic solar cells. Polymers 15 (1). Available at https://doi.org/10.3390/polym15010164.

Amna, B., Siddiqi, H.M., Hassan, A., Ozturk, T., 2020. Recent developments in the synthesis of regioregular thiophene-based conjugated polymers for electronic and optoelectronic applications using nickel and palladium-based catalytic systems. RSC Advances 10 (8), 4322–4396. Available at https://doi.org/10.1039/c9ra09712k.

Anguera, G., Sánchez-García, D., 2014. Conjugated polymers: synthesis and applications in optoelectronics. Afinidad 71, 251–262.

Aplan, M.P., Gomez, E.D., 2017. Recent developments in chain-growth polymerizations of conjugated polymers. Industrial & Engineering Chemistry Research 56 (28), 7888–7901.

Aragrag, N., Castiglione, D.C., 2004. General procedures in chain-growth polymerization. The Practical Approach in Chemistry Series. pp. 43.

Aravindu, K., Cloutet, E., Brochon, C., Hadziioannou, G., Vignolle, J., Robert, F., et al., 2018. Poly (arylene vinylene) synthesis via a precursor step-growth polymerization route involving the Ramberg–Bäcklund reaction as a key post-chemical modification step. Macromolecules 51 (15), 5852–5862.

Asundi, A., 2015. Thermochemical Water Splitting by Ceria. Stanford University, USA. Available at http://large.stanford.edu/courses/2015/ph240/asundi1/.

Bai, Y., Li, C., Liu, L., Yamaguchi, Y., Bahri, M., Yang, H., et al., 2022. Photocatalytic overall water splitting under visible light enabled by a particulate conjugated polymer loaded with palladium and iridium. Angewandte Chemie (International Edition in English) 61 (26), e202201299. Available at https://doi.org/10.1002/anie.202201299.

Barnes, J.C., Ehrlich, D.J.C., Gao, A.X., Leibfarth, F.A., Jiang, Y., Zhou, E., et al., 2015. Iterative exponential growth of stereo-and sequence-controlled polymers. Nature Chemistry 7 (10), 810–815.

Bautista, M.V., Varni, A.J., Ayuso-Carrillo, J., Carson, M.C., Noonan, K.J.T., 2021. Pairing Suzuki–Miyaura cross-coupling and catalyst transfer polymerization. Polymer Chemistry 12 (10), 1404–1414.

Bhattacherjee, D., Rahman, M., Ghosh, S., Bagdi, A.K., Zyryanov, G.V., Chupakhin, O.N., et al., 2021. Advances in transition-metal catalyzed carbonylative Suzuki–Miyaura coupling reaction: an update. Advanced Synthesis & Catalysis 363 (6), 1597–1624. Available at https://doi.org/10.1002/adsc.202001509, https://onlinelibrary.wiley.com/doi/abs/10.1002/adsc.202001509.

Bie, C., Cheng, B., Fan, J., Ho, W., Yu, J., 2021. Enhanced solar-to-chemical energy conversion of graphitic carbon nitride by two-dimensional cocatalysts. EnergyChem 3 (2), 100051.

Bie, C., Wang, L., Yu, J., 2022. Challenges for photocatalytic overall water splitting. Chem 8 (6), 1567–1574. Available at https://doi.org/10.1016/j.chempr.2022.04.013, https://www.sciencedirect.com/science/article/pii/S245192942200211X.

Bini, K., 2018. Design, Synthesis and Characterization of Conjugated Polymers for Photovoltaics and Electrochromics. Chalmers Tekniska Hogskola, Sweden.

Buriak, J.M., Toro, C., Choi, K.-S., 2018. Chemistry of materials for water splitting reactions. Chemistry of Materials 30 (21), 7325–7327. Available at https://doi.org/10.1021/acs.chemmater.8b04419.

Byun, J., Zhang, K.A.I., 2020. Designing conjugated porous polymers for visible light-driven photocatalytic chemical transformations. Materials Horizons 7 (1), 15–31.

Carsten, B., He, F., Jung Son, H., Xu, T., Yu, L., 2011. Stille polycondensation for synthesis of functional materials. Chemical Reviews 111 (3), 1493–1528. Available at https://doi.org/10.1021/cr100320w.

Chang, B.-Y., Park, S.-M., 2010. Electrochemical impedance spectroscopy. Annual Review of Analytical Chemistry 3, 207–229.

Chang, J.-H., Kumar, M., Shen, S.-Y., 2021. Fundamentals of photoelectrochemical water splitting. Nanostructured Materials for Photoelectrochemical Water Splitting. IOP Publishing, London, United Kingdom 1-1-1-20. Available at https://doi.org/10.1088/978-0-7503-3699-4ch1 1-1-1-20.

Chang, C.-Li, Lin, W.-C., Ting, Li-Y., Shih, C.-H., Chen, S.-Y., Huang, T.-F., et al., 2022. Main-chain engineering of polymer photocatalysts with hydrophilic non-conjugated segments for visible-light-driven hydrogen evolution. Nature Communications 13 (1), 5460. Available at https://doi.org/10.1038/s41467-022-33211-1.

Chatenet, M., Pollet, B.G., Dekel, D.R., Dionigi, F., Deseure, J., Millet, P., et al., 2022. Water electrolysis: from textbook knowledge to the latest scientific strategies and industrial developments. Chemical Society Reviews 51 (11), 4583–4762. Available at https://doi.org/10.1039/d0cs01079k.

Chatterjee, S., Jinnai, S., Ie, Y., 2021. Nonfullerene acceptors for P3HT-based organic solar cells. Journal of Materials Chemistry A 9 (35), 18857–18886.

Cheng, Y.-J., Yang, S.-H., Hsu, C.-S., 2009. Synthesis of conjugated polymers for organic solar cell applications. Chemical Reviews 109 (11), 5868–5923. Available at https://doi.org/10.1021/cr900182s.

Chen, J., Cao, Y., 2009. Development of novel conjugated donor polymers for high-efficiency bulk-heterojunction photovoltaic devices. Accounts of Chemical Research 42 (11), 1709–1718.

Chen, S., Thind, S., Chen, A., 2015. Nanostructured materials for water splitting—state of the art and future needs: a mini-review. Electrochemistry Communications 63. Available at https://doi.org/10.1016/j.elecom.2015.12.003.

Chen, B., Yang, N., Wang, P., Xiang, Y., Chen, H., 2020. Post-side chain engineering of difluorinated benzothiadiazole-based conjugated microporous polymer for enhanced photocatalytic H_2 evolution. Applied Surface Science 499, 143865. Available at https://doi.org/10.1016/j.apsusc.2019.143865, https://www.sciencedirect.com/science/article/pii/S0169433219326819.

Chen, L., Li, Y., Liang, X., 2021. Ar/H_2/O_2-controlled growth thermodynamics and kinetics to create zero-, one-, and two-dimensional ruthenium nanocrystals towards acidic overall water splitting. Advanced Functional Materials 31 (11), 2007344. Available at https://doi.org/10.1002/adfm.202007344.

Chen, G., Wei, C., Zhu, Y., Huang, H., 2023. Emerging chemical driving force in electrocatalytic water splitting. EcoMat 5 (2), e12294. Available at https://doi.org/10.1002/eom2.12294.

Chohan, M., Ahmad, T., Ullah, N., Khan, S., Ashraf, M., Ali, S., et al., 2022. Photocatalytic water-splitting by organic conjugated polymers: opportunities and challenges. The Chemical Record 22. Available at https://doi.org/10.1002/tcr.202100336.

Cho, H.-H., Yao, L., Yum, J.-H., Liu, Y., Boudoire, F., Wells, R.A., et al., 2021. A semiconducting polymer bulk heterojunction photoanode for solar water oxidation. Nature Catalysis 4 (5), 431–438.

Chua, M.H., Mao Png, Z., Zhu, Q., Xu, J., 2021. Synthesis of conjugated polymers via transition metal catalysed C–H bond activation. Chemistry – An Asian Journal 16 (19), 2896–2919. Available at https://doi.org/10.1002/asia.202100749.

Chu, S., Majumdar, A., 2012. Opportunities and challenges for a sustainable energy future. Nature 488 (7411), 294–303.

Coghi, P., Coluccini, C., 2024. Literature review on conjugated polymers as light-sensitive materials for photovoltaic and light-emitting devices in photonic biomaterial applications. Polymers 16 (10), 1407.

Cui, X., Wu, M., Liu, X., He, B., Zhu, Y., Jiang, Y., et al., 2024. Engineering organic polymers as emerging sustainable materials for powerful electrocatalysts. Chemical Society Reviews.

Dai, C., Liu, B., 2020. Conjugated polymers for visible-light-driven photocatalysis. Energy & Environmental Science 13 (1), 24–52. Available at https://doi.org/10.1039/C9EE01935A.

Dai, C., Pan, Y., Liu, B., 2020. Conjugated polymer nanomaterials for solar water splitting. Advanced Energy Materials 10 (42), 2002474.

Darmanin, T., Guittard, F., 2014. Wettability of conducting polymers: From superhydrophilicity to superoleophobicity. Progress in Polymer Science 39 (4), 656–682. Available at https://doi.org/10.1016/j.progpolymsci.2013.10.003, https://www.sciencedirect.com/science/article/pii/S007967001300124X.

Dau, H., Limberg, C., Reier, T., Risch, M., Roggan, S., Strasser, P., 2010. The mechanism of water oxidation: from electrolysis via homogeneous to biological catalysis. ChemCatChem 2 (7), 724–761.

Diao, R., Ye, H., Yang, Z., Zhang, S., Kong, K., Hua, J., 2019. Significant improvement of photocatalytic hydrogen evolution of diketopyrrolopyrrole-based donor–acceptor conjugated polymers through side-chain engineering. Polymer Chemistry 10 (47), 6473–6480. Available at https://doi.org/10.1039/C9PY01404G.

Fomo, G., Waryo, T., Feleni, U., Baker, P., Iwuoha, E., 2019. Electrochemical polymerization. In: Mazumder, M.A.J., Sheardown, H., Al-Ahmed, A., (Eds.), Functional Polymers. Springer International Publishing, Cham, pp. 105–131.

Freudenberg, J., Jänsch, D., Hinkel, F., Bunz, Uwe H.F., 2018. Immobilization strategies for organic semiconducting conjugated polymers. Chemical Reviews 118 (11), 5598–5689.

Gao, Y., Newland, B., Zhou, D., Matyjaszewski, K., Wang, W., 2017. Controlled polymerization of multivinyl monomers: formation of cyclized/knotted single-chain polymer architectures. Angewandte Chemie International Edition 56 (2), 450–460. Available at https://doi.org/10.1002/anie.201608786, https://onlinelibrary.wiley.com/doi/abs/10.1002/anie.201608786.

Gebremariam, G.K., Jovanović, A.Z., Pašti, I.A., 2023. The effect of electrolytes on the kinetics of the hydrogen evolution reaction. Hydrogen 4 (4), 776–806. Available at https://doi.org/10.3390/hydrogen4040049.

Geng, Y.H., Huang, L., Wu, S.P., Wang, F.S., 2010. Kumada chain-growth polycondensation as a universal method for synthesis of well-defined conjugated polymers. Science China Chemistry 53, 1620–1633.

Gogoi, R., Dutt, S., Siril, P., 2021. Conjugated polymer-based nanocomposites as photocatalysts. In: Conjugated Polymer Nanostructures for Energy Conversion and Storage Applications. pp. 267–296.

Grisorio, R., Suranna, G.P., Mastrorilli, P., 2010. Chain-growth versus step-growth mechanisms for the Suzuki–Heck polymerisation of fluorenyldibromides with potassium vinyl trifluoroborate. Chemistry–A European Journal 16 (27), 8054–8061.

Guay, J., Kasai, P., Diaz, A., Wu, R., Tour, J.M., Dao, L.H., 1992. Chain-length dependence of electrochemical and electronic properties of neutral and oxidized soluble. alpha.,. alpha.-coupled thiophene oligomers. Chemistry of Materials 4 (5), 1097–1105.

Gupta, N., Halder, S., Behere, R.P., Singh, P., Kanungo, S., Dixit, M., et al., 2023. Side-chain modification in conjugated polymer frameworks for the electrocatalytic oxygen evolution reaction. ACS Applied Materials & Interfaces 15 (24), 29042–29051.

Guyard, L., Hapiot, P., Neta, P., 1997. Redox chemistry of bipyrroles: further insights into the oxidative polymerization mechanism of pyrrole and oligopyrroles. The Journal of Physical Chemistry. B 101 (29), 5698–5706.

Hayat, A., Raziq, F., Khan, M., Khan, J., Mane, S.K.B., Ahmad, A., et al., 2019a. Fusion of conjugated bicyclic co-polymer within polymeric carbon nitride for high photocatalytic performance. Journal of Colloid and Interface Science 554, 627–639.

Hayat, A., Shaishta, N., Mane, S.K.B., Khan, J., Hayat, A., 2019b. Rational Ionothermal copolymerization of TCNQ with PCN semiconductor for enhanced photocatalytic full water splitting. ACS Applied Materials & Interfaces 11 (50), 46756–46766.

Heinze, J., Frontana-Uribe, B.A., Ludwigs, S., 2010. Electrochemistry of conducting polymers persistent models and new concepts. Chemical Reviews 110 (8), 4724–4771.

Higashihara, T., Hayashi, M., Hirao, A., 2011. Synthesis of well-defined star-branched polymers by stepwise iterative methodology using living anionic polymerization. Progress in Polymer Science 36 (3), 323–375.

Hildner, R., Köhler, A., Müller-Buschbaum, P., Panzer, F., Thelakkat, M., 2017. π-Conjugated donor polymers: structure formation and morphology in solution, bulk and photovoltaic blends. Advanced Energy Materials 7 (16), 1700314.

Horri, B.A., Hasan, O., 2024. Green hydrogen production by water electrolysis: Current status and challenges. Current Opinion in Green and Sustainable Chemistry 47, 100932.

Ifkovits, Z.P., Evans, J.M., Meier, M.C., Papadantonakis, K.M., Lewis, N.S., 2021. Decoupled electrochemical water-splitting systems: a review and perspective. Energy & Environmental Science 14 (9), 4740–4759.

Ito, S., 2016. Chain-growth polymerization enabling formation/introduction of arylene groups into polymer main chains. Polymer Journal 48 (6), 667–677.

Jagadesan, P., Schanze, K.S., 2019. Poly (phenylene ethynylene) conjugated polyelectrolytes synthesized via chain-growth polymerization. Macromolecules 52 (10), 3845–3851.

Kim, H., Lee, J., Kim, T., Cho, M., Choi, T.-L., 2022. Precision synthesis of various low-bandgap donor–acceptor alternating conjugated polymers via living Suzuki–Miyaura catalyst-transfer polymerization. Angewandte Chemie 134 (31), e202205828.

Kim, Y., Lee, C., Seo, D., Kim, D., Kim, F.S., Peterson, G.I., et al., 2024. Advancing organic electronics: achieving reliable synthesis of conjugated polymers with various carrier polarities using a continuous flow reactor. Chemical Engineering Journal 480, 148016. Available at https://doi.org/10.1016/j.cej.2023.148016, https://www.sciencedirect.com/science/article/pii/S1385894723067487.

Kobayashi, S., Ashiya, M., Yamamoto, T., Tajima, K., Yamamoto, Y., Isono, T., et al., 2021. Suzuki–Miyaura catalyst-transfer polycondensation of triolborate-type carbazole monomers. Polymers 13 (23), 4168. Available at https://www.mdpi.com/2073-4360/13/23/4168.

Kosco, J., Moruzzi, F., Willner, B., McCulloch, I., 2020. Photocatalysts based on organic semiconductors with tunable energy levels for solar fuel applications. Advanced Energy Materials 10 (39), 2001935. Available at https://doi.org/10.1002/aenm.202001935.

Kranz, C., Wächtler, M., 2021. Characterizing photocatalysts for water splitting: from atoms to bulk and from slow to ultrafast processes. Chemical Society Reviews 50 (2), 1407–1437.

Lee, J., Kim, H., Park, H., Kim, T., Hwang, S.-H., Seo, D., et al., 2021. Universal Suzuki–Miyaura catalyst-transfer polymerization for precision synthesis of strong donor/acceptor-based conjugated polymers and their sequence engineering. Journal of the American Chemical Society 143 (29), 11180–11190.

Lefrou, C., Cornut, R., 2010. Analytical expressions for quantitative scanning electrochemical microscopy (SECM). Chemphyschem: A European Journal of Chemical Physics and Physical Chemistry 11 (3), 547–556.

Liang, H., Rehan, M.A., Li, J., Du, S., Zhai, Y., Li, G., 2024. Kinetic simulation of hydrogen production reaction parameters based on TiO_2 photocatalyst. Applied Thermal Engineering 239, 122134. Available at https://doi.org/10.1016/j.applthermaleng.2023.122134, https://www.sciencedirect.com/science/article/pii/S1359431123021634.

Lin, L., Hisatomi, T., Chen, S., Takata, T., Domen, K., 2020. Visible-light-driven photocatalytic water splitting: recent progress and challenges. Trends in Chemistry 2 (9), 813–824.

Liu, Y., Liu, Y., Zhan, X., 2011. High-mobility conjugated polymers based on fused-thiophene building blocks. Macromolecular Chemistry and Physics 212 (5), 428–443.

Liu, J., Baek, J.-B., Dai, L., 2021a. Conjugated polymer synthesis. In: Kobayashi, S., Müllen, K., (Eds.), Encyclopedia of Polymeric Nanomaterials. Springer, Berlin, Heidelberg, pp. 1–7.

Liu, Y., Li, B., Xiang, Z., 2021b. Pathways towards boosting solar-driven hydrogen evolution of conjugated polymers. Small (Weinheim an der Bergstrasse, Germany) 17 (34), 2007576.

Li, C., Li, Y., 2008. The process of functional conjugated organic polymers derived from triple-bond building blocks. Macromolecular Chemistry and Physics 209 (15), 1541–1552.

Li, R., Zhang, X., Zhang, C., Lu, J., Wang, J.-C., Cui, C.-X., et al., 2022. Adenine-functionalized conjugated polymer as an efficient photocatalyst for hydrogen evolution from water. International Journal of Hydrogen Energy 47 (69), 29771–29780. Available at https://doi.org/10.1016/j.ijhydene.2022.06.296, https://www.sciencedirect.com/science/article/pii/S0360319922029548.

Li, Z., Zhang, X., Ou, C., Zhang, Y., Wang, W., Dong, S., et al., 2023. Transition metal-based self-supported anode for electrocatalytic water splitting at a large current density. Coordination Chemistry Reviews 495, 215381. Available at https://doi.org/10.1016/j.ccr.2023.215381, https://www.sciencedirect.com/science/article/pii/S0010854523003703.

Li, S., Xie, Y., Yin, Y., Chen, J., Cao, Y., Ying, S., et al., 2024. Intramolecular sensitization assisting high-efficiency TADF conjugated polymers with accelerating exciton spin flip for solution-processed electroluminescent devices. Macromolecules.

Long, Y., Ma, M., Wang, Y., Zheng, C., Ye, S., Huang, J., et al., 2023. Sequence-controlled polymers constructed by alkyne-based polymerizations. Giant 15, 100168. Available at https://doi.org/10.1016/j.giant.2023.100168, https://www.sciencedirect.com/science/article/pii/S2666542523000309.

Luo, S., Zeng, Z., Zeng, G., Liu, Z., Xiao, R., Xu, P., et al., 2020. Recent advances in conjugated microporous polymers for photocatalysis: designs, applications, and prospects. Journal of Materials Chemistry A 8 (14), 6434–6470.

Lyons, M., Doyle, R., Fernandez, D., Godwin, I., Browne, M., Rovetta, A., 2014. The mechanism and kinetics of electrochemical water oxidation at oxidized metal and metal oxide electrodes. Part 1. General considerations: a mini review. Electrochemistry Communications 45, 60–62. Available at https://doi.org/10.1016/j.elecom.2014.04.018.

Mansha, M., Ahmad, T., Ullah, N., Akram Khan, S., Ashraf, M., Ali, S., et al., 2022. Photocatalytic water-splitting by organic conjugated polymers: opportunities and challenges. The Chemical Record 22 (7), e202100336.

Martin, D.J., 2015. Introduction: fundamentals of water splitting and literature survey. Investigation into High Efficiency Visible Light Photocatalysts for Water Reduction and Oxidation. Springer, Cham, pp. 1–53.

Mayhugh, A.L., Yadav, P., Luscombe, C.K., 2022. Circular discovery in small molecule and conjugated polymer synthetic methodology. Journal of the American Chemical Society 144 (14), 6123–6135. Available at https://doi.org/10.1021/jacs.1c12455.

McHugh, P.J., Stergiou, A.D., Symes, M.D., 2020. Decoupled electrochemical water splitting: from fundamentals to applications. Advanced Energy Materials 10 (44), 2002453. Available at https://doi.org/10.1002/aenm.202002453.

McQueen, E., Bai, Y., Sprick, R.S., 2022. Impact of interfaces, and nanostructure on the performance of conjugated polymer photocatalysts for hydrogen production from water. Nanomaterials (Basel) 12 (23). Available at https://doi.org/10.3390/nano12234299.

Mdluli, S.B., Ramoroka, M.E., Yussuf, S.T., Modibane, K.D., John-Denk, V.S., Iwuoha, E.I., 2022. π-Conjugated polymers and their application in organic and hybrid organic-silicon solar cells. Polymers 14 (4). Available at https://doi.org/10.3390/polym14040716.

Miao, L., Young, C.D., Sing, C.E., 2017. An iterative method for hydrodynamic interactions in Brownian dynamics simulations of polymer dynamics. The Journal of Chemical Physics 147 (2).

Min, G., Choi, S., Hong, J., 2022. A review of solid oxide steam-electrolysis cell systems: thermodynamics and thermal integration. Applied Energy 328, 120145. Available at https://doi.org/10.1016/j.apenergy.2022.120145, https://www.sciencedirect.com/science/article/pii/S0306261922014027.

Moriarty, P., Honnery, D., 2022. Renewable energy in an increasingly uncertain future. Applied Sciences 13 (1), 388.

Nabgan, W., Alqaraghuli, H., Owgi, A.H.K., Ikram, M., Vo, D.-V.N., Jalil, A.A., et al., 2024. A review on the design of nanostructure-based materials for photoelectrochemical hydrogen generation from wastewater: bibliometric analysis, mechanisms, prospective, and challenges. International Journal of Hydrogen Energy 52, 622–663. Available at https://doi.org/10.1016/j.ijhydene.2023.05.152, https://www.sciencedirect.com/science/article/pii/S0360319923024485.

Nayak, P.K., Suhas, M., Henry, J.S., 2019. Photovoltaic solar cell technologies: analysing the state of the art. Nature Reviews Materials 4 (4), 269–285.

Nojima, M., Kosaka, K., Kato, M., Ohta, Y., Yokozawa, T., 2016. Alternating intramolecular and intermolecular catalyst-transfer Suzuki–Miyaura condensation polymerization: synthesis of boronate-terminated π-conjugated polymers using excess dibromo monomers. Macromolecular Rapid Communications 37 (1), 79–85.

Nosaka, Y., 2011. Solar cells and photocatalysts. In: Andrews, D.L., Scholes, G.D., Wiederrecht, G.P., (Eds.), Comprehensive Nanoscience and Technology. Academic Press, Amsterdam, pp. 571–605 Chapter 17.

Oka, K., Winther-Jensen, B., Nishide, H., 2021. Organic π-conjugated polymers as photocathode materials for visible-light-enhanced hydrogen and hydrogen peroxide production from water. Advanced Energy Materials 11 (43), 2003724. Available at https://doi.org/10.1002/aenm.202003724.

Otep, S., Michinobu, T., Zhang, Q., 2020. Pure organic semiconductor-based photoelectrodes for water splitting. Solar RRL 4 (8), 1900395.

Pai, M.R., Banerjee, A.M., Tripathi, A.K., Bharadwaj, S.R., 2012. Fundamentals and applications of the photocatalytic water splitting reaction. In: Banerjee, S., Tyagi, A.K., (Eds.), Functional Materials. Elsevier, London, pp. 579–606 Chapter 14.

Pandurangan, N., Porel, M., 2019. Sequence-defined non-natural polymers: synthesis and applications. Polymer Chemistry 10. Available at https://doi.org/10.1039/C9PY00886A.

Pankow, R.M., Thompson, B.C., 2020. The development of conjugated polymers as the cornerstone of organic electronics. Polymer 207, 122874.

Pan, Z., Hisatomi, T., Wang, Q., Nakabayashi, M., Shibata, N., Pan, C., et al., 2016. Application of LaMg1/3Ta2/3O_2N as a hydrogen evolution photocatalyst of a photocatalyst sheet for Z-scheme water splitting. Applied Catalysis A: General 521, 26–33. Available at https://doi.org/10.1016/j.apcata.2015.10.034, https://www.sciencedirect.com/science/article/pii/S0926860X15302118.

Pei, Q., 2007. Light-emitting polymers. Material Matters 2 (3), 26–28.

Pokhodenko, V.D., Krylov, V.A., 1994. Electrochemistry of conjugated organic polymers. Theoretical and Experimental Chemistry 30 (3), 91–105.

Rahman, M., Tian, H., Edvinsson, T., 2020. Revisiting the limiting factors for overall water-splitting on organic photocatalysts. Angewandte Chemie 132 (38), 16418–16433.

Reza Gholipour, M., Dinh, C.T., Béland, F., Do, T.O., 2015. Nanocomposite heterojunctions as sunlight-driven photocatalysts for hydrogen production from water splitting. Nanoscale 7 (18), 8187–8208. Available at https://doi.org/10.1039/c4nr07224c.

Royakkers, J., Bronstein, H., 2021. Macrocyclic encapsulated conjugated polymers. Macromolecules 54 (3), 1083–1094.

Ruan, Q., 2020. The investigation into polymer semiconductor photoelectrodes for light-driven water splitting (Ph. D. thesis). UCL (University College London), London.

Sachs, M., Sprick, R.S., Pearce, D., Hillman, S.A.J., Monti, A., Guilbert, A.A.Y., et al., 2018. Understanding structure-activity relationships in linear polymer photocatalysts for hydrogen evolution. Nature Communications 9 (1), 4968. Available at https://doi.org/10.1038/s41467-018-07420-6.

Samanta, B., Morales-García, Á., Illas, F., Goga, N., Anta, J.A., Calero, S., et al., 2022. Challenges of modeling nanostructured materials for photocatalytic water splitting. Chemical Society Reviews 51 (9), 3794–3818. Available at https://doi.org/10.1039/D1CS00648G.

Saunders, B., Wilbraham, L., Prentice, A.W., Sprick, R.S., Zwijnenburg, M.A., 2022. The potential scarcity, or not, of polymeric overall water splitting photocatalysts. Sustainable Energy & Fuels 6 (9), 2233–2242.

Schumacher, L., Marschall, R., 2022. Recent advances in semiconductor heterojunctions and Z-schemes for photocatalytic hydrogen generation. Topics in Current Chemistry 380 (6), 53. Available at https://doi.org/10.1007/s41061-022-00406-5.

Sen, R., Das, S., Nath, A., Maharana, P., Kar, P., Verpoort, F., et al., 2022. Electrocatalytic water oxidation: an overview with an example of translation from lab to market. Frontiers in Chemistry 10. Available at https://doi.org/10.3389/fchem.2022.861604, https://www.frontiersin.org/articles/10.3389/fchem.2022.861604.

Serpone, N., Emeline, A.V., Ryabchuk, V.K., Kuznetsov, V.N., Artem'ev, Y.M., Horikoshi, S., 2016. Why do hydrogen and oxygen yields from semiconductor-based photocatalyzed water splitting remain disappointingly low? Intrinsic and extrinsic factors impacting surface redox reactions. ACS Energy Letters 1 (5), 931–948.

Shit, S.C., Khilari, S., Mondal, I., Pradhan, D., Mondal, J., 2017. The design of a new cobalt sulfide nanoparticle implanted porous organic polymer nanohybrid as a smart and durable water-splitting photoelectrocatalyst. Chemistry–A European Journal 23 (59), 14827–14838.

Siemiaszko, G., Niemirowicz-Laskowska, K., Markiewicz, K.H., Misztalewska-Turkowicz, I., Dudź, E., Milewska, S., et al., 2021. Synergistic effect of folate-conjugated polymers and 5-fluorouracil in the treatment of colon cancer. Cancer Nanotechnology 12, 1–24.

Sprick, R.S., Jiang, J.-X., Bonillo, B., Ren, S., Ratvijitvech, T., Guiglion, P., et al., 2015. Tunable organic photocatalysts for visible-light-driven hydrogen evolution. Journal of the American Chemical Society 137 (9), 3265–3270. Available at https://doi.org/10.1021/ja511552k.

Sprick, R.S., Bonillo, B., Clowes, R., Guiglion, P., Brownbill, N.J., Slater, B.J., et al., 2016. Visible-light-driven hydrogen evolution using planarized conjugated polymer photocatalysts. Angewandte Chemie International Edition 55 (5), 1792–1796. Available at https://doi.org/10.1002/anie.201510542.

Sun, H., Öner, I.H., Wang, T., Zhang, T., Selyshchev, O., Neumann, C., et al., 2019. Molecular engineering of conjugated acetylenic polymers for efficient cocatalyst-free photoelectrochemical water reduction. Angewandte Chemie International Edition 58 (30), 10368–10374.

Sun, H., Xu, X., Kim, H., Jung, W.C., Zhou, W., Shao, Z., 2023. Electrochemical water splitting: bridging the gaps between fundamental research and industrial applications. Energy & Environmental Materials 6 (5), e12441. Available at https://doi.org/10.1002/eem2.12441.

Tentu, R.D., Basu, S., 2017. Photocatalytic water splitting for hydrogen production. Current Opinion in Electrochemistry 5 (1), 56–62.

Ting, L.-Y., Jayakumar, J., Chang, C.-L., Lin, W.-C., Hammad Elsayed, M., Chou, H.-H., 2019. Effect of controlling the number of fused rings on polymer photocatalysts for visible-light-driven hydrogen evolution. Journal of Materials Chemistry A 7 (40), 22924–22929. Available at https://doi.org/10.1039/C9TA06425G.

Tsutsumi, A., 2009. Thermodynamics of water splitting. Energy Carriers and Conversion Systems (EOLSS). 1. pp. 136-145. Available online: https://www.eolss.net/Sample-Chapters/C08/E3-13-03-01.pdf (accessed on 26 July 2024).

Tufan, A.N., Baki, S., Guclu, K., Ozyurek, M., Apak, R., 2014. A novel differential pulse voltammetric (DPV) method for measuring the antioxidant capacity of polyphenols-reducing cupric neocuproine complex. Journal of Agricultural and Food Chemistry 62 (29), 7111–7117.

Verswyvel, M., Verstappen, P., De Cremer, L., Verbiest, T., Koeckelberghs, G., 2011. Development of a universal chain-growth polymerization protocol of conjugated polymers: toward a variety of all-conjugated block-copolymers. Journal of Polymer Science Part A: Polymer Chemistry 49 (24), 5339–5349.

Villafán-Vidales, H.I., Arancibia-Bulnes, C.A., Riveros-Rosas, D., Romero-Paredes, H., Estrada, C.A., 2017. An overview of the solar thermochemical processes for hydrogen and syngas production: reactors, and facilities. Renewable and Sustainable Energy Reviews 75, 894–908. Available at https://doi.org/10.1016/j.rser.2016.11.070, https://www.sciencedirect.com/science/article/pii/S1364032116308474.

Wang, H., Xu, Y., Yu, X., Xing, R., Liu, J., Han, Y., 2013. Structure and morphology control in thin films of conjugated polymers for an improved charge transport. Polymers 5 (4), 1272–1324.

Wang, L., Fernández-Terán, R., Zhang, L., Fernandes, D.L.A., Tian, L., Chen, H., et al., 2016. Organic polymer dots as photocatalysts for visible light-driven hydrogen generation. Angewandte Chemie International Edition 55 (40), 12306–12310. Available at https://doi.org/10.1002/anie.201607018.

Wang, L., Wan, Y., Ding, Y., Wu, S., Zhang, Y., Zhang, X., et al., 2017. Conjugated microporous polymer nanosheets for overall water splitting using visible light. Advanced Materials 29 (38), 1702428.

Wang, L., Zhang, Y., Chen, L., Xu, H., Xiong, Y., 2018. 2D polymers as emerging materials for photocatalytic overall water splitting. Advanced Materials 30 (48), 1801955.

Wang, Y., Vogel, A., Sachs, M., Sprick, R.S., Wilbraham, L., Moniz, S.J.A., et al., 2019. Current understanding and challenges of solar-driven hydrogen generation using polymeric photocatalysts. Nature Energy 4 (9), 746–760.

Woodhouse, M., Parkinson, B.A., 2009. Combinatorial approaches for the identification and optimization of oxide semiconductors for efficient solar photoelectrolysis. Chemical Society Reviews 38 (1), 197–210.

Woods, D.J., Sprick, R.S., Smith, C.L., Cowan, A.J., Cooper, A.I., 2017. A solution-processable polymer photocatalyst for hydrogen evolution from water. Advanced Energy Materials 7 (22), 1700479. Available at https://doi.org/10.1002/aenm.201700479.

Xiang, C., Papadantonakis, K.M., Lewis, N.S., 2016. Principles and implementations of electrolysis systems for water splitting. Materials Horizons 3 (3), 169–173.

Xu, C., Zhang, W., Tang, J., Pan, C., Yu, G., 2018. Porous organic polymers: an emerged platform for photocatalytic water splitting. Frontiers in Chemistry 6. Available at https://doi.org/10.3389/fchem.2018.00592, https://www.frontiersin.org/articles/10.3389/fchem.2018.00592.

Xu, C., Dong, J., He, C., Yun, J., Pan, X., 2023. Precise control of conjugated polymer synthesis from step-growth polymerization to iterative synthesis. Giant 14, 100154. Available at https://doi.org/10.1016/j.giant.2023.100154, https://www.sciencedirect.com/science/article/pii/S2666542523000164.

Yang, J., Zhao, Z., Wang, S., Guo, Y., Liu, Y., 2018. Insight into high-performance conjugated polymers for organic field-effect transistors. Chem 4 (12), 2748–2785.

Yang, H.-S., Choi, H.-N., Lee, I.-H., 2023. Recent progress on end-group chemistry of conjugated polymers based on Suzuki–Miyaura catalyst-transfer polymerization. Giant, 100152.

Ying, L., Huang, F., Bazan, G.C., 2017. Regioregular narrow-bandgap-conjugated polymers for plastic electronics. Nature Communications 8 (1), 14047. Available at https://doi.org/10.1038/ncomms14047.

Yi, N., Abidian, M.R., 2016. Conducting polymers and their biomedical applications. Biosynthetic Polymers for Medical Applications. Elsevier, Amsterdam, Netherlands, pp. 243–276.

Yokozawa, T., Ohta, Y., 2013. Scope of controlled synthesis via chain-growth condensation polymerization: from aromatic polyamides to π-conjugated polymers. Chemical Communications 49 (75), 8281–8310.

Yokozawa, T., Ohta, Y., 2016. Transformation of step-growth polymerization into living chain-growth polymerization. Chemical Reviews 116 (4), 1950–1968.

Yokozawa, T., Yokoyama, A., 2009. Chain-growth condensation polymerization for the synthesis of well-defined condensation polymers and π-conjugated polymers. Chemical Reviews 109 (11), 5595–5619.

You, F., Shi, W., Yan, X., Wang, X., Shi, X., 2022. Polymerization of bio-derived conjugated dienes with rare-earth-metal complexes. Chemistry—An Asian Journal 17. Available at https://doi.org/10.1002/asia.202200892.

Zhang, Q., Dong, H., Hu, W., 2018. Electrochemical polymerization for two-dimensional conjugated polymers. Journal of Materials Chemistry C 6 (40), 10672–10686.

Zhang, A., Xiao, C., Wu, Y., Li, C., Ji, Y., Li, L., et al., 2016. Effect of fluorination on molecular orientation of conjugated polymers in high performance field-effect transistors. Macromolecules 49 (17), 6431–6438.

Zhang, Y., Tian, J., Shaikh, H., MacKenzie, H.K., He, Y., Zhao, C., et al., 2023. Tailored energy funneling in photocatalytic π-conjugated polymer nanofibers for high-performance hydrogen production. Journal of the American Chemical Society 145 (41), 22539–22547.

Zhao, C., Chen, Z., Shi, R., Yang, X., Zhang, T., 2020. Recent advances in conjugated polymers for visible-light-driven water splitting. Advanced Materials 32 (28), 1907296.

Zhao, H., Dong, Y., Sun, P., Bai, Y., Ru, C., Wu, X., et al., 2022. Effect of D/A ratio on photocatalytic hydrogen evolution performance of conjugated polymer photocatalysts. ACS Applied Energy Materials 5 (4), 4631–4640.

Zhao, Y., Sheng, J., Zhao, X., Mo, J., Wang, J., Chen, Z., et al., 2023. Recent progress in conjugated polymers-based donor–acceptor semiconductor materials for photocatalytic hydrogen evolution from water splitting. Catalysts 13 (5). Available at https://doi.org/10.3390/catal13050850.

Zhao, T.-T., Zhang, Y.-Y., Wang, L.-Q., Shen, Y.-M., Qi, H., 2024. Visible-light-mediated cyclization of arylamines with aldehydes using CdSe/CdS quantum dots and titanium dioxide nanotube composite. Catalysis Letters 154 (4), 1855–1864.

Zhou, J., Cheng, H., Cheng, J., Wang, L., Xu, H., 2024. The emergence of high-performance conjugated polymer/inorganic semiconductor hybrid photoelectrodes for solar-driven photoelectrochemical water splitting. Small Methods 8 (2), 2300418.

CHAPTER

16

Role of cocatalyst in water splitting

Sumesh George[1], *and Jain Maria Thomas*[2]

[1]Department of Physics, St. George's College, Aruvithura, Kerala, India [2]Department of Chemistry, St. George's College, Aruvithura, Kottayam, Kerala, India

16.1 Introduction

The balance of universal power demand and the related environmental challenges are the major hurdle in the energy production sector. In the present scenario, solar energy is important since the earth absorbs more energy from the sun in just 1 hour than the entire world consumes in a year. Moreover, the solar energy is clean and renewable energy and can be converted into other forms by different means like solar cells and photoelectrochemical water splitting (Moniz et al., 2015; Robu and Flynn, 2017; Staffell et al., 2019). The major drawback of electricity produced using solar cell is its limitations on efficiency and also its difficulty in storage and transport. The production of chemical fuels such as hydrogen via photocatalytic processes is the most potential approach to capture sunlight (Colmenares & Luque, 2014; Kudo & Miseki, 2009; Liang et al., 2018; Walter et al., 2010). Hydrogen fuel is well known for its high energy density, emission-free, nontoxic nature, the reaction by-product is water, and it is carbon neutral. Hydrogen is regarded as the most suitable medium for effectively storing and transporting renewable energy. This affordable and environmentally sustainable method is considered an effective solution for addressing both the depletion of fossil fuels and the environmental problems they cause. The major breakthrough in solar energy harvesting is the Honda Fujishima effect (Fujishima & Honda, 1972) discovered by Honda and Fujishima in 1972. Since then hydrogen fuel production via photocatalytic water splitting has emerged as a promising method. In this method, the catalytic material absorbs photons (solar light) and generates a voltage to split water molecule. At the same time, under standard conditions, water can be reversibly electrolyzed at a potential of 1.23 eV.

$$H_2O + \text{solar light} = H_2 + \tfrac{1}{2}\, O_2$$

It occurs when the absorbed solar light energy is equal to or greater than the threshold energy (1.23 eV) for water splitting. The electrons in the valence band are excited to the conduction band via photon excitation, resulting in an excited state. This creates a hole in the valence band, and it can be filled with an electron produced with the oxygen generation.

Photochemical Splitting of Water
DOI: https://doi.org/10.1016/B978-0-443-29064-0.00019-X

The H_2 is produced via reduction reaction of electrons in the conduction band (CB), and holes in the valance band (VB) support the oxidation reaction (Colmenares & Xu, 2016; Landman et al., 2017; Subramanyam et al., 2022). Fig. 16.1 shows the basic process of photocatalysis using semiconductor (Maeda & Domen, 2010). The band gap energy of photocatalyst should be equal to or greater than 1.23 eV, the potential of conduction band bottom should be more negative than H^+/H_2, and the potential of valence band top should be more positive than O_2/H_2O (Kumar et al., 2022). These are the essential conditions for water splitting.

The basic processes of hydrogen production via photocatalytic water splitting are: (1) Absorption of solar energy (greater than the band gap of semiconductor) and thereby the generation of electron hole, (2) the photogenerated charge separation and migration without recombination, and (3) the hydrogen evolution and oxygen evolution reaction. The balance of thermodynamics and kinetics of the above three steps determines the efficiency of photocatalytic water splitting. In the last two decades, intensive research was conducted on the development of efficient photocatalytic semiconductor materials for water splitting.

It is reported that the band gap of semiconducting materials in the range of 2–3 eV can effectively absorb solar light and drive the oxygen evolution reaction (OER) and hydrogen evolution reaction (HER), but it is not the only condition (Tian et al., 2023). The recombination of charge carriers, slow surface reaction rate, and overpotentials limit the direct applications of semiconductors. The possibility of recombination is much higher due to the high activation energy and thereby limits the efficiency. Effective charge separation and transportation is the major hurdle in photocatalysis. The main objectives of this chapter are to summarize the research progress in engineering the photocatalytic materials, the role of cocatalyst, and the factors influencing the performance of cocatalyst in efficient photoelectrochemical water splitting (PECWS). This chapter also summarizes the different cocatalysts such as metal-based, nonmetal-based, carbon-based, and hybrid cocatalysts developed to date.

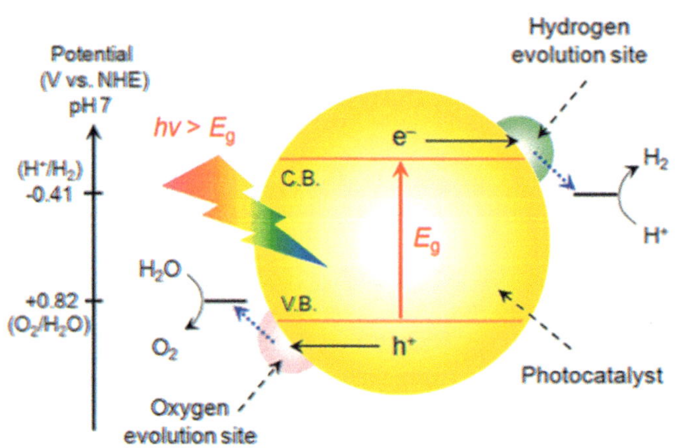

FIGURE 16.1 The basic process of hydrogen generation via photocatalytic water splitting. *Reproduced with permission from Maeda, K., Domen, K. 2020. Photocatalytic water splitting: recent progress and future challenges.* The Journal of Physical Chemistry Letters *118, 2655–2661. Copyright 2010 American Chemical Society.*

16.2 Approaches to enhance the photocatalytic activity

For efficient PECWS, following criteria must be met simultaneously (Ding et al., 2017): (1) effective sunlight utilization via broad light absorption; (2) efficient charge transfer; (3) quick utilization of photogenerated charges for the surface reaction; and (4) excellent robustness. To enhance the photocatalytic activity, various strategies like development of new photocatalytic materials, band gap engineering, and use of sacrificial agents and surface engineering were developed (Chen et al., 2012; Kubacka et al., 2012; Landman et al., 2017; Ran et al., 2014; Xiao et al., 2020; Yang et al., 2013; Zhou et al., 2019). It is reported that the defects in the materials serve as the recombination centers and hence reduce the efficiency of photoelectrochemical (PEC) (Li & Tsang, 2020). So, the selection of photocatalytic materials having high crystallinity is another strategy to improve the efficiency. Intensive works were done on the shape and surface customization, hybrid junction formation, and introduction of internal electric fields to improve the PECWS efficiency (Ran et al., 2014). The size of the semiconducting material particle also plays an important role, as reducing the size enhances the efficiency of PECWS. Reducing the particle size reduces the diffusion length for photogenerated electron-hole pairs (Maeda & Domen, 2016) and thereby increases the efficiency. It is reported that the electrolyte cations on the surface can influence the water splitting and backward reactions, and the anions may take part in the proton transfer processes (Li & Tsang, 2020; Ran et al., 2014). Hence, fine tuning of the electrolyte parameters is another strategy. The introduction of organic dyes and doping with cation and anions tailored the band structure in terms of band shift to enhance light energy harvesting (Etacheri et al., 2015). The visible light absorption can be enhanced by the creation of intermediate band levels by oxygen vacancies (Ran et al., 2014). In PEC water splitting, the major hurdle is the recombination of hydrogen–oxygen reaction which hinders the photocatalytic activity. In this context, the researchers were looking for the possibility of a second catalyst, which is called cocatalyst, to improve the PECWS efficiency of the photocatalytic process. The suitable quantity of cocatalysts decorated with semiconductor photocatalysts promotes forward H_2 or O_2 evolution and inhibits the backward reactions. The perfect stoichiometry of semiconductor catalyst with cocatalyst was verified to be the attractive means of improving the efficiency of photocatalytic activity. Moreover, it is reported that the modification of the interface between the cocatalysts and the semiconductor by the incorporation of interlayers enhances the charge separation and transfer (Ding et al., 2017). Various cocatalysts loading schemes have been developed to facilitate the surface redox reaction. Impregnation (Townsend et al., 2012), photo deposition (Wang et al., 2019b), electrostatic adsorption (Tian et al., 2023), growth of cocatalyst directly on the surface of photocatalysis using precursor, and direct loading of cocatalyst are the major cocatalyst loading methods. Cocatalyst loading site, loading quantity, and size of the cocatalyst also play a significant role in photocatalytic performance (Maeda & Domen, 2016). Recent reviews reveal that the researchers have developed different methods and cocatalysts in search of enhancing the PEC efficiency.

16.3 Significance of cocatalysts in water splitting

In PEC cells, the role of cocatalysts is to accelerate the kinetics of both the OER and HER (Zhao & Xu, 2021). The kinetics of these reactions increase electron-hole separation and prevent the recombination. Cocatalysts significantly enhance photocatalytic processes by improving charge separation, facilitating redox reactions, accelerating reaction kinetics, increasing stability, and creating additional active sites. Fig. 16.2 shows the role of a typical cocatalyst in water splitting. A cocatalyst helps in the charge separation process by providing an additional site or mechanism to capture and stabilize the charges (Colmenares & Luque, 2014). It can accept electrons from the primary photocatalyst or donate electrons to it, thereby supporting the charge separation. This reduces the likelihood of recombination and enhances the overall photocatalytic efficiency. The introduction of a suitable heterojunction between the cocatalyst and the semiconductor promotes the charge separation and transfer. Many of the visible-light sensitive catalysts are prone to oxidation by photogenerated holes, followed in their self-decomposition (Maeda & Domen, 2016). The decomposition can be hindered by the loading of cocatalysts on this catalyst. This may lead to extracting the photogenerated holes for O_2 evolution and thereby enhancing the robustness of semiconductors. The role of cocatalyst consisting of nanoparticles of transition metal and Cr on $SrTiO_3$ for photocatalytic water splitting (Kanazawa & Maeda, 2016) is depicted in Fig. 16.2. The cocatalyst loading results in the formation of interface between cocatalyst and semiconductor, and this depletion layer induces the electric field and provides a driving force for separation of the photogenerated electrons and holes to participate in the surface reduction and oxidation reactions (Xue et al., 2019; Zhang et al., 2015). Hence, the apt cocatalyst loading enhances the charge separation as well as provides active sites for proton reduction and water oxidation reactions.

The presence of cocatalyst on the surface of semiconductor modifies the energy bands of the semiconductor and thereby enhances the absorbed wavelength (Ran et al., 2014). It is reported

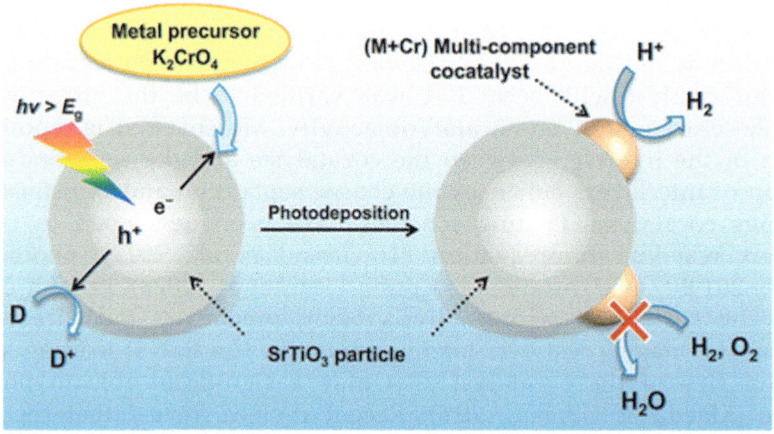

FIGURE 16.2 The role of cocatalyst in photocatalysis. *Reproduced with permission from Kanazawa, T., Maeda, K. 2016. Light-induced synthesis of heterojunctioned nanoparticles on a semiconductor as durable cocatalysts for hydrogen evolution.* ACS Applied Materials & Interfaces *8(11), 7165–7172. Copyright 2016 American Chemical Society.*

that the gold nanoparticles on titanium dioxide (TiO_2) enhanced the visible light absorption by a localized surface plasmon resonance (Maeda & Domen, 2016). This phenomenon involves the oscillation of conduction electrons on the metal nanoparticles upon exposure to light, which enhances the local electromagnetic field and increases light absorption. The heterojunction can be formed between cocatalyst and the main photocatalyst material by the incorporation of cocatalyst (Lv et al., 2010; Wang et al., 2012). These heterojunctions can create new electronic band alignments that enhance the light absorption properties and improve charge separation. For example, cocatalyst like cadmium sulfide (CdS) can form a heterojunction with titanium dioxide (TiO_2) (Ran et al., 2014). The composite can lead to a broader range of light absorption, thereby enhancing the photocatalytic activity. The introduction of cocatalysts may modify the surface properties such as surface area and surface charge. This can lead to improved interactions with light and enhance the photocatalyst's ability to absorb light.

16.4 Factors influencing the performance of cocatalyst

The cocatalysts capture photogenerated electrons from the photocatalyst and facilitate gas evolution by hosting active sites. Therefore the loaded cocatalyst played an important role in the overall efficiency of a photocatalytic system. The structural and intrinsic characteristics of a cocatalyst are important for H_2 evolution. There are several factors such as the cocatalyst loading quantity, particle size, and structure in the semiconductor-based photocatalytic water splitting that can influence the competence of H_2 and O_2 evolution (Ran et al., 2014). The incorporation of a cocatalyst onto a semiconductor could facilitate the charge collection and gas evolution reactions and thereby enhance the photocatalytic water splitting activity. The highest activity is achieved with the optimum value, and further increase in the loading amount decreases the activity (Maeda & Domen, 2016; Ran et al., 2014). The covering of active sites surface is the reason behind the reduction in the activity. The large amount of loading may lead to covering the incident light, and thus blocks the light absorption and generation of electron-hole pairs inside the semiconductor. Moreover, the charge recombination centers were formed with the high loading cocatalysts. The catalytic efficiency can be enhanced by the cocatalysts with smaller size. The smaller particle size leads to larger surface area and creates more active sites, which results in higher catalytic activity (Kumar et al., 2022; Tian et al., 2023). A cocatalyst significantly enhances the PECWS efficiency and intensive research is going on for developing novel earth-abundant, affordable, and effective cocatalysts for sustainable renewable energy.

16.5 Types of cocatalysts

16.5.1 Metal-based cocatalysts

Metal-based cocatalysts enhance the efficiency and reliability of photocatalytic water splitting by reducing activation energy. Metals such as Pt, Pd, Au, Fe, Co, Ni, Ru, and Rh are commonly used as H_2 evolution cocatalysts (Li & Tsang, 2020), since they have large work functions along with low Fermi levels, which create a Schottky barrier when in contact with semiconductors (Xi, 2021). This barrier induces band bending at the interface, aiding in charge carrier separation and improving electron conductivity (Xi, 2021). The holes remain in the semiconductor's VB, enabling efficient charge carrier separation. Photogenerated electrons transfer from the semiconductor's conduction band to the metal across the Schottky barrier until their Fermi levels align (Xiao et al., 2020).

16.5.1.1 Noble metals and alloy-based cocatalysts

Noble metals also offer a large number of surface trapping states and active sites, which speed up the hydrogen production (Leung et al., 2010). As cocatalysts, they significantly improve photocatalytic reactivity. Noble metals like Pt (Liu et al., 2018b; Ou et al., 2017; Wu et al., 2015a), Au (Chen et al., 2016; Liu et al., 2018a; Qian et al., 2018), Ag (Cheah et al., 2015; Chen et al., 2016; Qin et al., 2016), and Pd (Al-Azri et al., 2015; Mo et al., 2019) should be able to trap electrons more easily if they have a larger work function or a lower Fermi level. With the greatest work function among them, Pt is regarded as the ideal cocatalyst for the evolution of hydrogen (Xiao et al., 2020).

Incorporating 1%–3% noble metals as cocatalysts can drastically improve the catalytic performance of semiconductor photocatalysts compared to their bare form (Shoaib et al., 2016). Isolated single Pt atoms deposited on $g\text{-}C_3N_4$ significantly enhance photocatalytic H_2 evolution activity. Bare graphitic carbon nitride ($g\text{-}C_3N_4$) has a low rate of 6.5 μmol/h under visible light, but with 0.16 wt.% Pt loading, the rate jumps to 318 μmol/h. Loaded Pt atoms modify the surface state, increasing active sites and trapping more photogenerated electrons (Li et al., 2016c). Under visible light, $g\text{-}C_3N_4$ photocatalyst containing 18-nm Au nanospheres exhibits a high rate of H_2 evolution of about 540 μmol/g/h compared to the bare $g\text{-}C_3N_4$ (Guo et al., 2018). The plasmonic and electron sink properties of the Au nanocrystals are responsible for this increased activity. In another study, a high-temperature calcination is used to load Au, Ag, and Cu on TiO_2 [001] nanocrystals (Shoaib et al., 2016). An enhancement in photocatalytic activity of TiO_2 was observed due to the strengthening of the metal–semiconductor interface barrier, thereby reducing electron-hole recombination. Notably, Au/TiO_2 outperformed others due to localized surface plasmon resonance-induced accelerated electron-hole separation. Tong et al. (2018) explained the superior photocatalytic performance of single Pt, Pd, or Au atoms trapped in two-dimensional $g\text{-}C_3N_4$ bilayer ($BL\text{-}g\text{-}C_3N_4$) compared to the pristine $g\text{-}C_3N_4$. Pt and Pd atoms, trapped within $g\text{-}C_3N_4$ interlayers, bridge charge carriers and accelerate photoinduced electron migration. Au, on the other hand, prevents interlayer charge transfer since it loads only on the $g\text{-}C_3N_4$ surface. The d orbitals of noble atoms have a significant role in charge carrier distribution, band structure, and optical absorption in single-atom/$g\text{-}C_3N_4$ photocatalytic systems. Among all, Pt/BL- $g\text{-}C_3N_4$ has the highest optical absorption in visible and near ultraviolet light and the best charge carrier separation efficiency. Noble metals were hardly used in the large-scale industrial production of H_2 because of their high cost, unavailability, and corrosive nature.

Metal alloys have drawn increasing attention in recent years to be used as cocatalyst in photocatalytic water splitting. They usually demonstrate improved photocatalytic H_2 production compared to single-metal catalysts, largely due to the synergistic effects of the two metals (Liu et al., 2021). Plasmonic noble metals like Au and Ag, when used as cocatalysts, can improve the semiconductor photocatalysts' use of visible light through their strong surface plasmon resonance effect (Liu et al., 2021). Au and Ag alloys combine synergistic catalytic and plasmonic properties to generate double-plasmonic-metal cocatalysts. The impact of Ag/Pt alloy cocatalyst on the H_2 evolution efficiency of TiO_2 was studied and it was observed that the synthesized Ag/Pt@TiO_2 samples exhibit enhanced catalytic activity in hydrogen evolution (Jiang et al., 2014). This improvement is attributed to their short bulk diffusion length of charge carriers, low rate of electron-hole recombination, and high surface area. Moreover, Ag/Pt@TiO_2 nanoparticles demonstrate significantly higher activity compared to Ag@TiO_2 and Pt@TiO_2, owing to the optimal height of the Schottky barrier and synergistic effects. It has been proved that Pt-Pd is a functioning cocatalyst with great promise for improving H_2 evolution (Xiao et al., 2019). To investigate the various rates of H_2 generation by controlling the atomic arrangements, the Pt-Pd cocatalyst was deposited on the g-C_3N_4 surface. Thus the sample with the gas generation rate of 2885.0 μmol/g/h in the $Pt_{0.5}Pd_{0.5}$/g-C_3N_4 formula represents the ideal photocatalytic activity. Diagram illustrating the photocatalytic mechanism for overall water splitting with the PtPd/g-C_3N_4 photocatalyst is displayed in Fig. 16.3. Several other noble metal alloys like PtPd (Xiao et al., 2019) and AuPd (Han et al., 2017b; Han et al., 2015b) have also been manufactured and investigated for their function in photocatalytic H_2 evolution (Xiao et al., 2020). By acting as cocatalysts, these alloys improve the efficiency of charge separation and transfer by capturing electrons from the semiconductor. The role of noble metals and alloy-based cocatalysts on hydrogen production is summarized in Table 16.1.

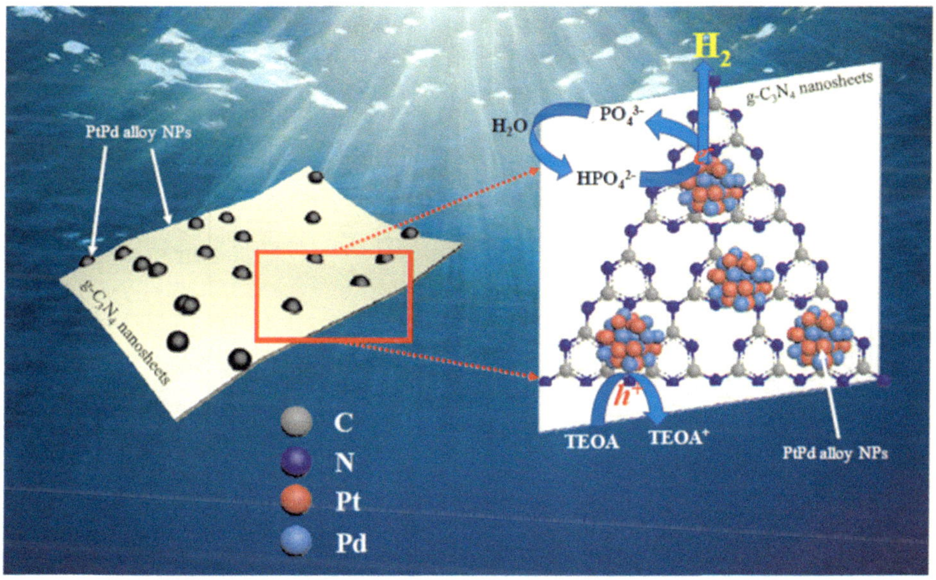

FIGURE 16.3 Photocatalytic mechanism for overall water splitting using the PtPd/g-C_3N_4 photocatalyst. *Reproduced with permission from Xiao, N., Li, S., Liu, S., Xu, B., Li, Y., Gao, Y., et al. 2019. Novel PtPd alloy nanoparticle-decorated g-C_3N_4 nanosheets with enhanced photocatalytic activity for H_2 evolution under visible light irradiation.* Chinese Journal of Catalysis *40(3), 352–361.*

TABLE 16.1 Summary of studies on the role of noble metals and alloy-based cocatalysts in hydrogen production.

Photocatalyst	Cocatalyst	Light source	Solution	H_2 evolution	References
$Pt/g\text{-}C_3N_4$	Pt	Xe lamp	200 mL 10 vol.% TEOA	318 μmol/h	Li et al. (2016c)
1 wt.% $Pt/g\text{-}C_3N_4$	Pt	300 W Xe lamp	100 mL 10 vol.% TEOA	34 μmol/h	Wu et al. (2015a)
0.2 wt.% $Pt/g\text{-}C_3N_4$	Pt	300 W Xe lamp	50 mL 10 vol.% TEOA	4875 μmol/h	Ou et al. (2017)
1 wt.% $Pt/g\text{-}C_3N_4$	Pt	350 W Xe lamp	80 mL 10 vol.% TEOA	4210.8 μmol/h/g	Liu et al. (2018b)
2 wt.% $Au/g\text{-}C_3N_4$	Au	300 W Xe lamp	40 mL 2.5 vol.% TEOA	540 μmol/h/g	Guo et al. (2018)
0.5% $Au/g\text{-}C_3N_4$	Au	300 W Xe lamp	200 mL 10 vol.% methanol	18.09 μmol/g/h	Chen et al. (2016)
Au/TiO_2	Au	300 W Xe lamp	20 mL 20 vol.% methanol	12.40 μmol/h/g	Liu et al. (2018a)
$Au/g\text{-}C_3N_4$	Au	300 W Xe lamp	100 mL 10 vol.% TEOA	84.1 μmol/h	Qian et al. (2018)
0.5% $Ag/g\text{-}C_3N_4$	Ag	300 W Xe lamp	200 mL 10 vol.% methanol	18.09 μmol/h/g	Chen et al. (2016)
20 wt.% Ag/MoS_2	Ag	300 W Xe lamp	4.5 mL 0.3 M Na_2S/0.3 M Na_2SO_3	179.5 μmol/g	Cheah et al. (2015)
3 wt.% $Ag/g\text{-}C_3N_4$	Ag	300 W Xe lamp	70 mL 10 vol.% TEOA	2014.2 μmol/h/g	Qin et al. (2016)
1 wt.% Pd/TiO_2	Pd	200 W SB-1000P/F lamp	20 mL 10 vol.% glycerol	47.5 mmol/h/g	Al-Azri et al. (2015)
0.5 wt.% $Pd/g\text{-}C_3N_4$	Pd	300 W Xe lamp	100 mL 10 vol.% TEOA	1208.6 μmol/h/g	Mo et al. (2019)
0.5 wt.% $AuPd/g\text{-}C_3N_4$	AuPd	300 W Xe lamp	100 mL 10 vol.% TEOA	326 μmol/h/g	Han et al. (2015b)
1.0 wt.% $AuPd/g\text{-}C_3N_4$	AuPd	300 W Xe lamp	100 mL 10 vol.% TEOA /K_2HPO_4	107 μmol/h	Han et al. (2017b)
0.2 wt % $Pt_{0.5}Pd_{0.5}/g\text{-}C_3N_4$	$Pt_{0.5}Pd_{0.5}$	300 W Xe lamp	100 mL 10 vol.% TEOA /K_2HPO_4	1600.8 μmol/g/h	Xiao et al. (2019)

16.5.1.2 Transition metals and alloy-based cocatalysts

Because of their low cost and abundance in the earth crust, transition metal-based materials especially those composed of Mo, Co, Ni, Fe, and Zn have been found as the most promising catalysts for the evolution of hydrogen (Guo et al., 2018; Liu et al., 2021; Tong et al., 2018). They also possess high work function and enhanced catalytic activity, which is comparable with noble metal cocatalysts (Liu et al., 2021). Typically, transition metal nanoparticles with high work functions, like Ni (Bi et al., 2015; Chen et al., 2014b; Chen et al., 2015; Kong et al., 2016; Vu et al., 2017), Co (Zhao et al., 2018), and Cu (Chen et al., 2018; Lin et al., 2011; Zhu et al., 2019) are chosen as cocatalysts for H_2 evolution. They are useful as sites for proton reduction as well as electron traps (Liu et al., 2021).

When ions such as Fe^{2+}, Cu^{2+}, Co^{2+}, Zn^{2+}, and Ni^{2+} are doped into g-C_3N_4, they intercalate into the nitrogen sites, altering the electronic properties and enhancing charge transfer due to the strong bond formation (Wen et al., 2017). This results in a reduced band gap and a shift of the absorption edge to higher wavelengths (Ding et al., 2011). It was observed that increasing Fe^{2+} concentration in Fe-doped g-CN decreases the band gap and shifts the absorption edge (Wang et al., 2009). It was reported that ultrathin Ni/CdS nanosheets significantly enhance H_2 generation and biomass intermediate coupling due to Fermi level of Ni and high work function, which accelerate charge separation (Han et al., 2017c). Similarly, NiCd/CdS nanorods exhibit synergistic photocatalytic activity, reaching a hydrogen evolution rate of 11.57 mmol/g/h, for H_2 much higher than pristine CdS, Cd/CdS, and Ni/CdS (Wang et al., 2019a). Replacing some Ni atoms with Cd ensures a high work function and strong adsorption of hydrogen atoms.

A non-semiconductor system using CuNi bimetal coupled with rGO was developed for hydrogen evolution. The $CuNi_{5:5}$/rGO (1 wt.%) combination showed the highest H_2 evolution rate of 1787 µmol/g/h. Ni and rGO serve as electron mediators to prevent electron-hole recombination, allowing more electrons to participate in the reaction (Zhang et al., 2017b). This photocatalyst also effectively utilizes near-infrared light, expanding its visible light absorption range. The PtCo bimetallic alloy cocatalyst was applied to g-C_3N_4 nanosheets using in situ chemical deposition. This approach helps reduce costs by replacing a portion of Pt with Co. Under visible light, the photocatalytic H_2 evolution rate of the 1.0 wt.% PtCo/g-C_3N_4 was 960 µmol/g/h, which is 2.9 times higher than that of the pristine bulk PtCo/g-C_3N_4 (Han et al., 2015a). Peng et al. (2018) observed that decorating g-C_3N_4 nanotubes with a PtNi bimetallic alloy significantly enhanced photocatalytic activity, achieving an H_2 generation rate of 104.7 µmol/g/, which is 47.6 times higher than pure g-C_3N_4 nanotubes and 1.1 times higher than Pt/g-C_3N_4. Thus PtCo and PtNi bimetallic alloys are promising, cost-effective cocatalysts for improved photocatalytic H_2 evolution. Fig. 16.4 depicts the mechanism of photocatalytic hydrogen evolution. The NiCd/CdS NRs sample enhanced photocatalytic H_2 production to 11.57 mmol/g/h, and it is about 64.8, 17.2, and 2.3 times higher than pristine CdS NRs, Cd/CdS, and Ni/CdS, respectively (Wang et al., 2019a). The role of transition metals and alloy-based cocatalysts on hydrogen production is summarized in Table 16.2.

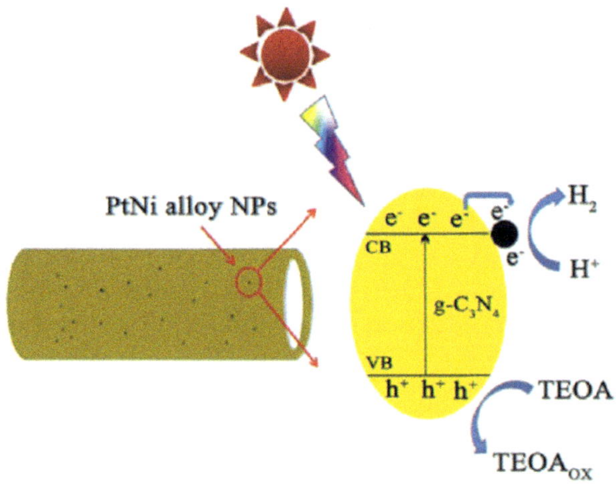

FIGURE 16.4 Schematic representation of the pathway for the evolution of photocatalytic hydrogen. *Reproduced with permission from Peng, W., Zhang, S.-S., Shao, Y.-B. & Huang, J.-H. 2018. Bimetallic PtNi/g-C_3N_4 nanotubes with enhanced photocatalytic activity for H_2 evolution under visible light irradiation.* International Journal of Hydrogen Energy *43 (49), 22215–22225* .

16.5.1.3 Metal oxides, hydroxides, sulfides, and phosphides-based cocatalysts

Metal oxides, hydroxides, sulfides, and phosphides are efficient cocatalysts that enhance H_2 photocatalytic activity by reducing the overpotential for H_2 generation and optimizing hole-trapping states to improve charge separation. Metal oxides like TiO_2 (Pan et al., 2019; Tan et al., 2018), Bi_2O_3 (Lakshmana Reddy et al., 2017), Cu_2O (Karthikeyan et al., 2018; Li et al., 2013; Li et al., 2015b), NiO (Li et al., 2015a), CuO (Kum et al., 2013; Yu et al., 2011b), hydroxides like $Ni(OH)_2$ (Mao et al., 2019a), $Co(OH)_2$ (Xu et al., 2017), sulfides like CuS and CdS (Hong et al., 2014; Yue et al., 2016), and phosphides have suitable standard potentials that make electron-hole separation easier and lower the energy threshold for photoinduced interband transition.

Metal oxides can also form heterojunctions with semiconductors, which enhances light absorption and electron transfer rates. However, their moderate conductivity can limit their overall efficiency, which can be improved by doping or creating composites with other materials. Copper oxides, CuO and Cu_2O, are effective cocatalysts for enhancing photocatalytic H_2 production. Cu_2O-loaded TiO_2 shows a 14.48-fold increase in H_2 evolution compared to pure P25 (Li et al., 2015b), while CuO-loaded TiO_2 reaches a H_2 evolution rate of 2061 μmol/h/g (Yu et al., 2011b). The quantum size effect of CuO influences its band edges and enhances electron transfer, improving photocatalytic activity by facilitating the transfer of electrons from CuO to H^+ for H_2 evolution. It was found that increasing the size of Cu_2O reduces the energy gap, improving electron transfer and optimizing H_2 generation (Karthikeyan et al., 2018). Wu et al. (2015b) synthesized an Ag_2O/g-C_3N_4 composite via a hydrothermal method, achieving a hydrogen evolution rate of 33.04 μmol/h under visible light. In this system, Ag_2O acts as a cocatalyst that facilitates charge separation; electrons from g-C_3N_4 are not transferred to Ag_2O, while holes from g-C_3N_4 move to Ag_2O and are consumed by TEOA, enabling efficient hydrogen production (Wu et al., 2015b).

Metal hydroxides can undergo oxidation to form oxyhydroxides under reaction conditions, further enhancing their catalytic activity. Their relatively lower electrical conductivity is often addressed by combining them with conductive substrates or integrating them into

TABLE 16.2 Summary of studies on the role of transition metals and alloy-based cocatalysts in hydrogen production.

Photocatalyst	Cocatalyst	Light source	Solution	H_2 evolution	References
0.5 wt.% Ni/TiO_2	Ni	100 W SB-100P/F lamp	20 mL 5 vol.% EtOH	24.3 mmol/h/g	Chen et al. (2015)
10% Ni/g-C_3N_4	Ni	500 W Xe arc lamp	100 mL 10 vol.% TEOA	8.41 μmol/h	Bi et al. (2015)
7.40% Ni/g-C_3N_4	Ni	300 W Xe lamp	10 mL 30 vol.% TEOA	4318 μmol/h/g	Kong et al. (2016)
3 wt.% Ni/$CdIn_2S_4$	Ni	100 W solar simulator	10 mL 10 vol.% TEOA	1060 μmol/h/g	Vu et al. (2017)
5 wt.% Ni/CdS	Ni	300 W Xe lamp	50 mL 50 vol.% lactic acid	3004.8 μmol/h	Chen et al. (2014b)
10% Co/g-C_3N_4	Co	300 W Xe lamp	15 mL 20 vol.% TEOA	11.48 μmol/h	Zhao et al. (2018)
2.5% Cu/TiO_2	Cu	300 W Xe lamp	100 mL 20 vol.% methanol	6046 μmol/h/g^1	Zhu et al. (2019)
0.3 wt.% Cu/TiO_2	Cu	300 W Xe lamp	65 mL 20 vol.% methanol	20.4 μmol/h/g	Chen et al. (2018)
1 wt.% Cu/Nb_2O_5	Cu	400 W medium pressure halide lamp	550 mL 16.7 vol.% methanol	1572 μmol/h/g	Lin et al. (2011)
1 wt.% CuNi/rGO	CuNi	300 W Xe lamp	70 mL 14.2 vol.% lactic acid	1787 μmol/h/g	Zhang et al. (2017b)

composite structures. Yu et al. demonstrated that $Ni(OH)_2$, synthesized via a simple precipitation method, serves as an effective cocatalyst on TiO_2 (Yu et al., 2011a), g-C_3N_4 (Yu et al., 2013), and CdS (Ran et al., 2011) for improved photocatalytic H_2 production under visible light (Yu et al., 2011a). Plate-like $Co(OH)_2$ was added to TiO_2 nanosheets as a cocatalyst to boost H_2 production, achieving a rate of 746.93 μmol/g/h with 0.5 mol% Co $(OH)_2$, which is 20 times greater than pure TiO_2, and it showed good stability over 14 hours of continuous irradiation (Xu et al., 2017). Bismuth vanadate ($BiVO_4$) shows promise in PEC water splitting but suffers from high charge recombination. Mo doping creates MoOx and oxygen vacancies that, with V^{4+} species, enhance hole transfer; without V^{4+}, these vacancies cause recombination. But CoOOH cocatalyst coating reduces recombination, boosting photocurrent by 37 times and shifting the onset potential by 500 mV (Yalavarthi et al., 2021).

Metal sulfide is a widely developed cocatalyst for photocatalytic water splitting, known for its excellent surface properties for binding water molecules and hydrogen protons, effectively promoting charge separation, being cost-effective, and having Gibbs free energy close to that of Pt (Huang et al., 2024). Its distinct structure and conductivity make it a highly efficient electrocatalyst (Huang et al., 2024). Various kinds of metal sulfides such as MoS_2, Cu_2S, FeS, and WS_2 are used as cocatalysts in H_2 generation. Transition-metal sulfides, with their large surface areas and abundant active sites, optimize H_2 evolution by improving charge separation and transfer, exposing sites for water dissociation, and trapping more electrons to prevent recombination (Han et al., 2017a). MoS_2, an effective cocatalyst, was loaded onto $ZnIn_2S_4$ by MoS_2, an effective cocatalyst, was loaded onto $ZnIn_2S_4$ using the hydrothermal method, resulting in a 28-fold increase in the H_2 evolution rate (Chen et al., 2014a). Novel bilayer junctions were created by chemically assembling metallic MoS_2 monolayers on oxygen-deficient $Bi_{12}O_{17}Cl_2$ monolayers using oxygen-vacancy chemistry, achieving a H_2 evolution rate of 33 mmol/h/g under visible light (Li et al., 2016a). A bifunctional core-shell CdS/Cu_2S heterojunction with self-optimization was synthesized using the solvothermal method, where Cu_2S serves as a cocatalyst. Under visible light, this system improved H_2 production by nearly 25 times compared to pure CdS, achieving a maximum rate of 14.4 mmol/h/g (Zhang et al., 2017a). The CoMoS cocatalyst enhances photocatalytic H_2 evolution in semiconductors like TiO_2 and g-C_3N_4 (Liu et al., 2018c). Its exceptional performance under visible light is attributed to the efficiency of CoMoS in facilitating electron transport for H_2 production. In recent years, metal sulfide has been employed as a cocatalyst to enhance the photocatalytic activity of g-C_3N_4. Its combination with g-C_3N_4 forms heterojunctions or ohmic contacts, including various types like Type-I, Type-II, Z-scheme, and S-scheme (Huang et al., 2024). A MoS_2/g-C_3N_4 hybrid photocatalyst with two-dimensional, flower-like hexagonal 2H-MoS_2 nanosheets, and a thin-layer structure was developed (Rao et al., 2021). The tight integration between MoS_2 and g-C_3N_4 significantly enhanced H_2 evolution. Compared to pure g-C_3N_4, MoS_2 nanoflowers increased light absorption across a wider spectrum. The unique flower-like nanostructure exposed more active edges and shortened electron transfer pathways, boosting photocatalytic activity. A new type of bimetallic sulfide (a-CoMoSx) based on molybdenum was developed and deposited on the CdS surface using a photoinduced electron-reduction method to enhance the interfacial H_2 production rate (Liu et al., 2018c). This cocatalyst significantly boosts the H_2 generation activity of CdS, showing improvements by factors of 26.3 and 3.1 compared to pure CdS and a-MoS_2/CdS, respectively. The unsaturated sulfur atoms in molybdenum sulfide act as efficient active sites for rapidly capturing H^+ ions and enhancing their direct reduction by photogenerated electrons. Fig. 16.5 shows a schematic depiction of the photocatalytic hydrogen evolution pathway of a-CoMoSx/CdS photocatalysts. The process involves three steps: (1) a-$CoMoS_x$ rapidly captures photogenerated electrons; (2) these electrons are effectively transferred from capture sites to active sites; and (3) at the a-CoMoSx active sites, efficient evolution of H_2 takes place.

Transition-metal phosphides (TMPs) are widely studied for their high stability, excellent activity, and abundance (Bi et al., 2016; Shen et al., 2018; Zeng et al., 2018). With H atoms, the P-ligand ensemble effect produces a modest binding strength, enhancing surface adsorption and desorption kinetics during water dissociation and increasing H_2 production rates (Pei et al., 2018). TMPs such as Ni_xP_y, Co_xP_y, Mo_xP_y, W_xP_y, and Fe_xP_y are widely used for H_2 production (Yin et al., 2024). The photocatalytic performance of TMPs is largely influenced by the atomic percentage of P, where negatively charged P atoms serve as active sites, while transition metals enhance conductivity. This makes phosphorus-based TMPs effective as integrated photocatalysts, balancing light absorption and cocatalytic activity (Ren et al., 2020a). Cobalt phosphide nanoparticles (CoP NPs) act as cocatalysts by accepting electrons and providing reactive sites for proton reduction. The Fermi level (EF) of CoP, relative to the normal hydrogen electrode is +0.173 eV (Liu et al., 2020a). This positioning of the Fermi level indicates that CoP NPs are effective in facilitating electron transfer processes essential for enhancing the hydrogen evolution reaction. By serving as electron acceptors, these nanoparticles improve the efficiency of charge separation and create additional active sites. This p-n heterojunction effectively promotes the spatial separation of photoinduced electrons and holes (Basavarajappa et al., 2020). By effectively separating these charge carriers, the electrons and holes have a longer lifetime and a reduced recombination rate (Basavarajappa et al., 2020). This enhanced separation significantly improves the efficiency of photocatalytic processes by allowing more electrons and holes to participate in the desired reactions, thereby boosting overall performance and stability. A key study on metal phosphides as cocatalysts investigated the scalability of photocatalysis of Ni_2P-decorated TiO_2, g-C_3N_4, and CdS (Chen & Qin, 2016). The results indicated that Ni_2P decoration

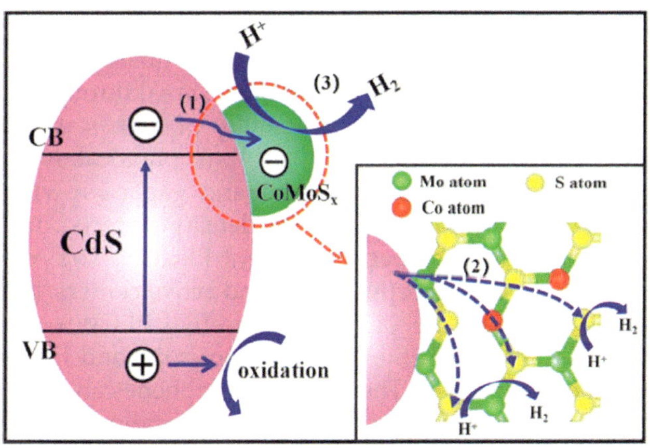

FIGURE 16.5 Schematic diagram depicting the photocatalytic hydrogen evolution mechanism of a-$CoMoS_x$/CdS photocatalysts. *Reproduced with permission from Liu, W., Wang, X., Yu, H., Yu, J. 2018c. Direct photoinduced synthesis of amorphous $CoMoS_x$ cocatalyst and its improved photocatalytic H_2-evolution activity of CdS.* ACS Sustainable Chemistry & Engineering *6 (9), 12436–12445. Copyright 2018 American Chemical Society.*

improved charge carrier transfer and optimized surface characteristics, enhancing the H_2 photo-reduction rate and quantum efficiency. Three crystalline phases of Ni_2P, $Ni_{12}P_5$, and Ni_3P were tested as cocatalysts hybridized with g-C_3N_4 for photocatalytic H_2 generation (Sun et al., 2017). It was observed that all three phases of Ni_xP_y significantly enhanced H_2 generation under visible light with g-C_3N_4. The highest H_2 generation was achieved with 5 wt.% Ni_2P/g-C_3N_4, reaching 0.162 mmol/h/g. This rate was about 34 times higher than pure g-C_3N_4 and nearly matched the 5 wt.% Pt/g-C_3N_4 rate of 0.168 mmol/h/g. Also, the photocatalyst demonstrated good stability after 9 hours of visible light exposure. Uniformly dispersed nanocrystalline Ni_2P particles were synthesized and used as cocatalysts to modify CdS, TiO_2, and C_3N_4 photocatalysts for hydrogen production (Chen & Qin, 2016). The Ni_2P cocatalyst boosted hydrogen production across all host photocatalysts, demonstrating its good catalytic properties and general applicability. The maximum H_2 generation rate of 9.38 mmol/h/g was attained by the Ni_2P/TiO_2 nanorods sample through optimization of the sacrificial reagents and Ni_2P loading concentration. This rate was around 85 times greater than the photocatalyst made of pure TiO_2 nanorods. Tiwari et al. (2022) demonstrated that using MoS_2/rGO as a cocatalyst with Cu_2O enhances the photoactivity by suppressing the recombination of photogenerated charge carriers and preventing photocorrosion. The ternary composite Cu_2O-MoS_2/rGO exhibited a significantly higher photoelectrochemical activity (8.46 mA/cm^2 at 0.95 V) compared to the bare Cu_2O (0.33 mA/cm^2) (Tiwari et al., 2022). The role of metal oxides, hydroxides, sulfides, and phosphides-based cocatalysts on hydrogen production is summarized in Table 16.3.

16.5.2 Nonmetal-based cocatalysts

Metal-free cocatalysts are typically more environmentally friendly and economically viable than their metal counterparts. They are increasingly attractive due to their abundance, stability, and lower toxicity. Various nonmetal-based cocatalysts for water splitting are discussed below.

16.5.2.1 Carbon-based cocatalysts

During the past decade, carbon has attracted much attention for its unique advantages in improving HER reaction. The discovery of carbon nanotubes (CNTs), carbon quantum dots (CQD) (Iijima, 1991), and graphene (Novoselov et al., 2004) are the major breakthrough in the material science research. This discovery opens the new possibility of high stability photocatalytic materials (Leary & Westwood, 2011; Shearer et al., 2014). In photocatalysis, the role of carbon materials ranges from supporting materials to band gap narrowing (Iijima, 1991). The specifications, such as size, number of walls, density defects, and interfacial area of the carbon-based materials, affect the performance of PECWS. The role of CNT and graphene is not limited to increasing the adsorption sites and active centers, but also to suppressing the recombination of electron-hole pairs. For effective PECWS, the essential condition is that the Fermi level is higher than the reduction potential and lower than the CB of coupled semiconductors (Cao & Yu, 2016). The CNT and graphene can act as a host platform to various photocatalytic semiconductors and prevent aggregation. It also extends the solar energy harvesting and promotes charge separation. Due to the high surface

TABLE 16.3 Summary of studies on the role of metal oxides, hydroxides, sulfides, and phosphides-based cocatalysts in hydrogen production.

Photocatalyst	Cocatalyst	Light source	Solution	H_2 evolution	References
TiO_2/g-C_3N_4	TiO_2	300 W Xe lamp	100 mL 10 vol.% TEOA	513 μmol/h/g	Tan et al. (2018)
B-TiO_2/g-C_3N_4	B-TiO_2	300 W Xe lamp	100 mL 10 vol.% TEOA	808.97 μmol/h/g	Pan et al. (2019)
0.4% Bi_2O_3/TiO_2	Bi_2O_3	Natural solar light	50 mL 5 vol.% glycerol	26.02 mmol/h/g	Lakshmana Reddy et al. (2017)
0.9 mol% Cu_2O/TiO_2	Cu_2O	300 W Xe lamp	40 mL 0.05 M glycerol	318 μmol/h	Li et al. (2013)
2.5 mol% Cu_2O/TiO_2	Cu_2O	500 W Xe lamp	450 mL 10 vol.% methanol	2048.25 μmol/h/g	Li et al. (2015b)
0.25 wt.% NiO/TiO_2	NiO	300 W Xe lamp	80 mL 25 vol.% methanol	337 μmol/h/g	Li et al. (2015a)
1.3 wt.% CuO/TiO_2	CuO	3 W UV-LEDs	80 mL 0.1 M glycerol	2061 μmol/h/g	Yu et al. (2011b)
$Ni(OH)_2$/CdS	$Ni(OH)_2$	300 W Xe lamp	50 mL 1 M $(NH_4)_2SO_3$	40.18 mmol/h/g	Mao et al. (2019a)
$Co(OH)_2$/TiO_2	$Co(OH)_2$	300 W Xe lamp	100 mL 25 vol.% methanol	746.93 μmol/h/g	Xu et al. (2017)
Ni/CdS/g-C_3N_4	Ni/CdS	300 W Xe lamp	100 mL 10 vol.% TEOA	1258.7 mmol/h/g	Yue et al. (2016)
ZnS/CuS/CdS	CuS/CdS	150 W Xe lamp	150 mL 60 mM Na_2S	837.6 μmol/h/g	Hong et al. (2014)
CdS/Co_2P	Co_2P	300 W Xe lamp	200 mL 10 vol.% lactic acid	10.8 mmol/h/g	Bi et al. (2016)
g-C_3N_4/Ni_2P-3.5 wt.%	Ni_2P	300 W high-pressure Xe lamp	100 mL 10 vol.% TEOA	474.7 mmol/h/g	Zeng et al. (2018)
Ni_2P/TiO_2	Ni_2P	300 W Xe lamp	50 vol.% methanol	9.38 mmol/h/g	Chen & Qin (2016)
10 wt.% Cu_3P/ g-C_3N_4	Cu_3P	300 W Xe lamp	80 mL 15 vol.% TEOA	159.41 μmol/h/g	Shen et al. (2018)

Reproduced with permission from Xiao, N., Li, S., Li, X., Ge, L., Gao, Y., Li, N. (2020). The roles and mechanism of cocatalysts in photocatalytic water splitting to produce hydrogen. *Chinese Journal of Catalysis* 41(4), 642–671.

area, the high electrical conductivity of carbon-based materials attracts the photogenerated electrons from the CB of a variety of semiconductors (Cao & Yu, 2016; Lian et al., 2015; Liu et al., 2014). Compared to CNT, the graphene is advantageous in terms of large specific surface area, better distribution, higher contact interface, and the augmented aromatic structure (Angaridis et al., 2014; Cao & Yu, 2016). Because of the unique property of electron acceptor and donor nature, the zerodimensional (0D) carbon-based nanocarbon materials (Angaridis et al., 2014; Cao & Yu, 2016) and CQD (Fernando et al., 2015) have also been widely investigated. Carbon fibers, activated carbon, and carbon black are the prominent members in the nanocarbon family and they were investigated in the design of efficient photocatalytic materials. The carbon-based nitrides are also investigated as a new class of photocatalytic material (Cao & Yu, 2016). The role of these materials is to support and enhance the structure stability. The carbon-based cocatalysts have several advantages, such as (1) the nanoparticles can be distributed and immobilized with large surface area, (2) they are chemically inert and thermally stable, (3) they can supply abundant nucleation sites, (4) they have a light specific weight for a supporting material, and (5) they are economic (Cao & Yu, 2016). Reports show that surface chemistry of carbon materials played an important role in the interaction between the photocatalytic nanoparticles and carbon materials (Cao & Yu, 2016). Carbon-based cocatalyst increases the adsorption and active sites. It is reported that the adsorption and active sites of a photocatalytic system can be increased with CNTs and graphene (Chen et al., 2013; Xiang & Yu, 2013). Theoretical studies suggested that the adsorption and dissociation of water molecule can be enhanced by the carbon materials, which is beneficial for hydrogen production (Politano et al., 2011; Rangel et al., 2009). For instance, the surface-treated CNT with $Zn_{0.83}Cd_{0.17}S$ is the best example (Wang et al., 2013). A larger specific surface area for CdS/graphene photocatalyst can be obtained with the uniform distribution of CdS clusters on the graphene surface. The ultrathin nanorods of Zn0.5Cd0.5S can stabilize and enhance the photocatalytic H_2 evolution by the utilization of reduced graphene oxide (RGO) (Wang et al., 2013). Reports show that the functional groups can act as anchoring sites toward the reagents and the carbon materials with surface functional groups such as COOH and OH can provide additional reaction centers to enhance the photocatalytic activity (Cao et al., 2014). Carbon-based cocatalyst also enhances the photocatalytic activity by electron acceptor and transport channel. As an example, for a carbon-based cocatalyst, the chemically oxidized multiwalled CNTs loaded onto The CdS/TiO_2/Pt composite loaded with multiwalled CNTs photocatalyst was investigated in the visible light (Khan et al., 2013). The electron-hole charge separation took place on the CdS/TiO_2 heterojunction. The separated electrons on the CB of TiO_2 further transferred to CNTs and Pt catalysts. The CNTs act as a cocatalyst together with Pt nanoparticles since the Fermi level of CNTs is lower than the conduction band of TiO_2. It is reported that the loading of Pt enhanced the performance by more than 50%. Literature reports that the graphene/MoS_2/TiO_2 composite shows a high H_2-production rate of 165.3 μmol/h. The capture of photoinduced electrons from the CB of TiO_2 was due to the high work function (Xie et al., 2013). The works on carbon-based cocatalyst, especially graphene, indicate that graphene can be a suitable substitute for the noble metals as a cocatalyst. It is worth noting that the photo-response of a wide band gap semiconductor photocatalyst can be extended by the carbon-based cocatalyst (Xie et al., 2013). Band gap narrowing effect was systematically investigated (Shaban & Khan, 2008), and it is reported that a specific chemical bonding was developed at the interface of carbon material and semiconductor (Li et al., 2014b). The main reason behind the band gap narrowing effect was the chemical bonding and, consequently, enhancing the photocatalytic efficiency. CNT, CQD, multiwalled carbon nanotubes (MWCNT), carbon nitride nanosheets (CNNS), and graphene are the major carbon-based cocatalysts. As an example, the interactions between CdS and graphene or CNT (Ye et al., 2012) result in the band gap narrowing in graphene or CNT-modified CdS (Ye et al., 2012). Xia et al. (2015) reported that the cocatalyst CNT-decorated CNNS gives an activity of 219.5 μmol/h/g, and CQDs-loaded TiO_2 results in an activity of 7.9 μmol/h. CNT-loaded $Zn_{0.83}Cd_{0.17}S$ and Ni_3S_2 result in an activity of 5.41 and 5.32 mmol/h^1/g, respectively. A total of 0.2 wt.% CNT/g-C_3N_4 shows an activity of 39.4 μmol/h (Chen et al. 2014c). Carbon-based materials such as CNT and graphene are the promising solution as cocatalyst for efficient PECWS. The role of carbon-based cocatalysts on the rate of hydrogen production is summarized in Table 16.4.

16.5.2.2 Phosphorus- and boron-based cocatalysts

Similar to graphene, black phosphorus (BP) is a two-dimensional material with excellent carrier mobility and a configurable band gap that offers possibilities for use in optoelectronics (Li et al., 2014a). BP modified with surface hydroxyl groups has been utilized as a cocatalyst, leading to a significant 341% increase in photocatalytic hydrogen evolution when combined with anatase TiO_2 under simulated sunlight (Wu et al., 2018). This system also achieved an impressive quantum efficiency of 18.23% at 365 nm. The enhanced photocatalytic performance is attributed to a Z-scheme mechanism, which facilitates more efficient charge separation and transfer, thereby boosting the overall hydrogen production. BP is an effective metal-free cocatalyst with thickness-dependent band structure, with a direct band gap ranging from 1.5 eV for monolayers to 0.3 eV for bulk, effectively bridging the gap between transition metal dichalcogenides and graphene (Ren et al., 2020a). When TiO_2 (P25) is treated with 1 wt.% of BP, the H_2 evolution rate increases four times, to 830 μmol/g/h under UV-Vis light. The use of CoP nanoparticles as a cocatalyst up to 2 wt.% increases H_2 generation to 7400 μmol/g/h, a 37-fold increase over that of pure TiO_2 (Provinciali et al., 2023). Since BP is a p-type semiconductor and TiO_2 is an n-type semiconductor, a type II p-n heterojunction is predicted at their interface. Energy band diagram and proposed photocatalytic reaction mechanism for TiO_2/BP/CoP under UV-Vis irradiation, with EF representing the Fermi level is displayed in Fig. 16.6. The valence band energy (EVB) was calculated using the equation EVB=ECB+Eg, where Eg represents the semiconductor's band gap. Measurements of photoluminescence and electrochemical impedance show that this heterostructure, when compared to pristine TiO_2, produces considerably more effective charge separation and lowers internal resistance. It is found that few-layer BP with a thickness of 1–5 nm can serve as an effective metal-free cocatalyst to enhance photocatalytic performance (Ran et al., 2017). A 1.0 wt.% few-layer BP mixed with CdS achieves a high photocatalytic H_2 production rate of 11,192 μmol/h/g and a 34.7% quantum yield at 420 nm, with theoretical calculations showing that CdS transfers photoexcited electrons to BP due to its higher CB and VB positions. Mao et al. (2019b) developed a BP-CdS-$La_2Ti_2O_7$ ternary composite that produced H_2 at rates of 0.96 mmol/h/g under solar light and 0.26 mmol/h/g in the near-infrared range using Na_2S/Na_2SO_3 aqueous solution. A 2D BP/g-C_3N_4 composite

TABLE 16.4 Summary of studies on the role of carbon-based cocatalysts in hydrogen production.

Photocatalyst	Cocatalyst	Light source	Solution	H_2 evolution	References
10 wt.% CNNS/CQDs	CQDs	1000 W Xe lamp($\lambda \geq 420$ nm)	10 mL 20 vol.% methanol	219.5 μmol/h/g	Xia et al. (2015)
CQDs/TiO_2	CQDs	300 W Xe lamp	60 mL 20 vol.% methanol	7.9 μmol/h	Sui et al. (2019)
CCTs/CQDs	CQDs	300 W Xe lamp ($\lambda \geq 400$ nm)	60 mL 20 vol.% methanol	3538.3 μmol/h/g	Wang et al. (2018)
1.0 wt.% graphene/g-C_3N_4	Graphene	350 W Xe lamp ($\lambda \geq 400$ nm)	80 mL 25% methanol	451 μmol/h/g	Xiang et al. (2011)
Cu_2O/RGO-2 wt.%	RGO	300 W Hg lamp ($\lambda \geq 400$ nm)	20 mL 20 vol.% methanol	64.3 μmol/h/g	Zhang et al. (2019)
0.1 wt.% MWCNT-TiO_2	MWCNT	Mercury vapor (Hg) lamp (200–800 nm)	50 mL 5 vol.% glycerol	8.8 mmol/h/g	Reddy et al. (2018)
$Zn_{0.83}Cd_{0.17}S$/CNTs-0.25 wt.%	CNTs	500 W Xe lamp	40 mL 0.1 M Na_2S/0.02 M Na_2SO_3	5.41 mmol/h/g	Yao et al. (2014)
CNT@Ni_3S_2	CNT	300 W Xe lamp ($\lambda \geq 420$ nm)	100 mL 15 vol.% TEOA and 200 mg ErY	5.32 mmol/h/g	Zhu et al. (2012)
0.25 wt.% CNT/$Cd_{0.1}Zn_{0.9}S$	CNT	300 W Xe lamp ($\lambda \geq 420$ nm)	80 mL 0.35 M Na_2S/0.25 M Na_2SO_3	78.2 μmol/h	Yu et al. (2012)
0.2 wt.% CNT/g-C_3N_4	CNT	300 W Xe lamp, ($\lambda \geq 420$ nm) 100	100 mL 10 vol.% TEOA	39.4 μmol/h	Chen et al. (2014c)
CNT/g-C_3N_4	CNT	10 LED lamps ($\lambda \geq 420$ nm)	50 mL 10 vol.% TEOA	654.8 μmol/h/g	Song et al. (2018)
$Zn_{0.83}Cd_{0.17}S$/CNTs	CNT	500 W Xe lamp	40 mL 0.1 M Na_2S/0.02 M Na2SO3	6.03 mmol/h/g	Wang et al. (2013)

Reproduced with permission from Xiao, N., Li, S., Li, X., Ge, L., Gao, Y., Li, N. (2020). The roles and mechanism of cocatalysts in photocatalytic water splitting to produce hydrogen. *Chinese Journal of Catalysis* 41(4), 642–671.

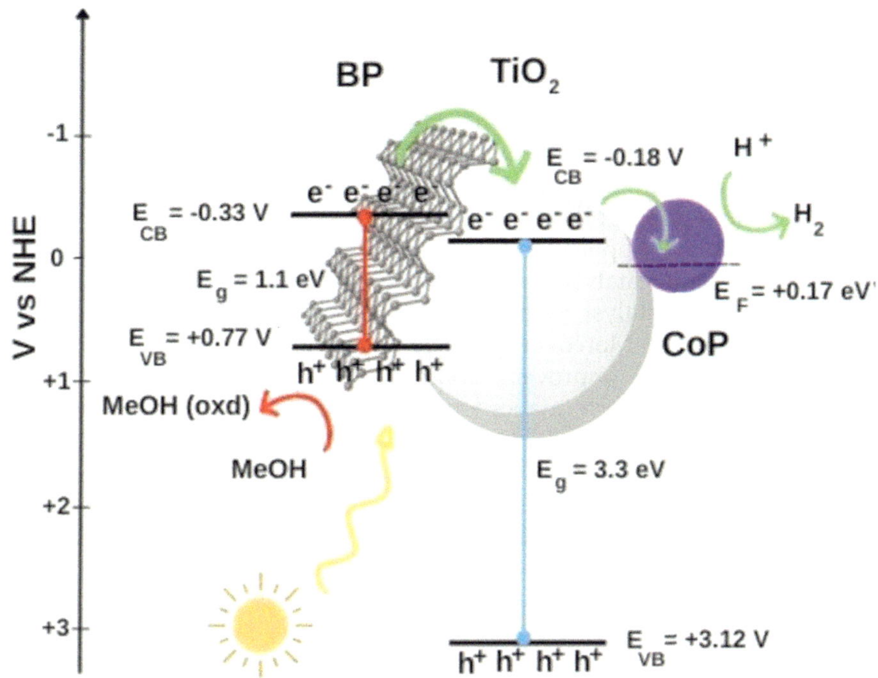

FIGURE 16.6 Schematic showing the energy band diagram and proposed photocatalytic hydrogen evolution mechanism for TiO_2/BP/CoP under UV-Vis irradiation (Provinciali et al., 2023).

with an optimal BP to g-C_3N_4 ratio of 1:4 demonstrated an H_2 evolution rate of 1.93 μmol under visible light irradiation for 3 hours in a methanol solution, maintaining stable performance even after 2 weeks. The composite forms a type-I heterojunction, where the P-N bond serves as an electron trap, effectively enhancing H_2 generation (Zhu et al., 2017). A novel BP-based photocatalyst with atomically dispersed cobalt on BP nanosheets (BP-Co) greatly enhances H_2 evolution from CdS under visible light, increasing production to 345.4 μmol/h compared to 18.5 μmol/h for pure CdS (Ren et al., 2020b).

Cobalt phosphate (Co-Pi), the cocatalyst, is introduced onto $ZnIn_2S_4$ by a simple in situ photodeposition technique (Wu et al., 2024). This aims to enhance the transport of photogenerated holes in $ZnIn_2S_4$, thereby aiding in the separation of electron-hole pairs. The results of the study demonstrate that the composite catalyst outperforms $ZnIn_2S_4$ alone in photocatalysis. For $ZnIn_2S_4$ loaded with 5% Co-Pi, the best photocatalytic activity is observed ($ZnIn_2S_4$/5%Co-Pi). This compound produces H_2 at a rate of 3593 μmol/g/h, which is almost twice as high as $ZnIn_2S_4$ alone. According to characterization data, the insertion of Co-Pi enhances $ZnIn_2S_4$ hole transfer, hence improving the separation efficiency of photogenerated carriers.

Transition metal borides (TMBs), made from abundant and inexpensive materials, are highly effective substitutes for Pt-based cocatalysts, yielding superior H_2 production in water splitting under visible light. TMBs, when used with NiCoB/CdS composites, achieved an impressive H_2 production rate of 144.8 mmol/h/g—36 times higher than CdS alone, with an apparent quantum efficiency of 97.42% at 500 nm, the highest reported for CdS photocatalysts (Li et al., 2016b). Porous boron nitride, composed of alternating boron and nitrogen atoms, is a focus of research due to its low density, large surface area, and exceptional thermal and chemical stability (Li et al., 2014a). Hexagonal boron nitride, known as "white graphene," is a 2D crystal with strong sp^2 covalent bonds, offering excellent chemical stability and thermal conductivity (Zhang et al., 2016). Its polar B-N bonds and high surface area allow for significant adsorption capacity, making it ideal to be used as cocatalyst in photocatalytic applications, including hydrogen production (Gao et al., 2020). Band gaps of hybrid photocatalysts can be effectively narrowed by combining BN with a visible-light-driven photocatalyst (Rahman et al., 2020). This method increases the effectiveness of light absorption. Moreover, photogenerated electron-hole pair creation, migration, and separation are all improved. Consequently, there is an improvement in the photocatalytic H_2 generation performance when exposed to visible light. The immigration of holes onto the surface of BN can lead to improved photocatalytic capabilities (Li et al., 2021). A disorder-engineered black H-TiO_2@BN photocatalyst was synthesized using BN nanosheets as the support. This material features Ti-B chemically bonded interfaces. The hybridization of TiO_2 with BN, combined with hydrogenated thermal treatment, enables photocatalytic H_2 production without noble metal cocatalysts. The H_2 yield was considerably enhanced when the BN content was more than 20%. The production of H_2 reached its maximum value when the BN concentration reached 40%. The hybrid photocatalyst enhances light absorption, reduces charge pair recombination, and facilitates interfacial electron transfer. It also lowers interfacial charge transfer resistance (He et al., 2018).

16.5.3 Metal-organic framework-based cocatalysts

Metal-organic frameworks (MOFs) are effective electrocatalysts and photocatalysts due to their flexible design, high surface area, and tunable pores that enhance active site accessibility and reduce photoexcited carrier transfer distance (Xiao et al., 2020). MOF-derived materials are promising for photocatalytic water splitting due to their ability to maintain MOF structure while enhancing conductivity. A stable MOF, UiO-66, shows intrinsic H_2 evolution activity (Xu et al., 2018). The CdS/UiO-66 composite combines CdS's visible light absorption with UiO-66 large surface area for better dispersion and increased photocatalytic sites. Electron transfer from excited CdS to the MOF creates a charge-separated state with electrons in UiO-66 and holes in the CdS valence band, reducing charge recombination and enhancing photocatalytic H_2 production efficiency. The hydrogen evolution mechanism of CdS/UiO-66 is schematically represented in Fig. 16.7. Zhang et al. (2018) developed NH_2-UiO-66/TpPa-1-COF (4:6) hybrid materials by combining TpPa-1-COF with stable NH2-UiO-66, resulting in high surface area, porosity, and crystallinity, and achieving a photocatalytic hydrogen production rate of 23.41 mmol/g/h, presenting a promising strategy for efficient hydrogen evolution photocatalysts.

16.5.4 Hybrid cocatalysts

Effectively removing charge carriers from semiconductors and transferring them to the surface is what makes a cocatalyst ideal. It must also offer numerous active sites and

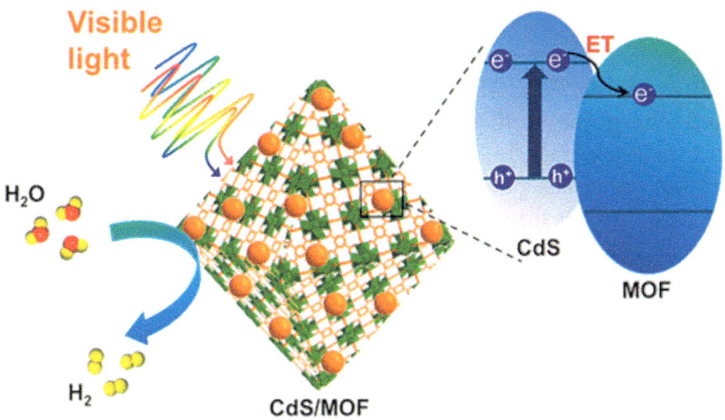

FIGURE 16.7 Schematical representation of hydrogen evolution mechanism of CdS/UiO-66. *Reproduced with permission from Xu, H.-Q., Yang, S., Ma, X., Huang, J., Jiang, H.-L. 2018. Unveiling charge-separation dynamics in CdS/metal-organic framework composites for enhanced photocatalysis.* ACS Catalysis *8 (12), 11615–11621. Copyright 2018 American Chemical Society.*

catalyze redox reactions with low activation energy and high selectivity (Bai et al., 2016). However, achieving all these features in a single cocatalyst is challenging. Lattice mismatch, hydrophilicity-hydrophobicity discrepancies, tiny contact areas, and interfacial defects are some of the reasons why some cocatalysts have trouble transferring charges across the surface (Lin et al., 2017). Their limited electrical conductivity prevents them from delivering captured electrons or holes in an efficient way. Charge transmission can be inefficient when there is an energy band mismatch between semiconductors and cocatalysts (Chang et al., 2016). Surface reaction kinetics also pose challenges. Slow reactions can occur when a cocatalyst has insufficient reaction sites for activation and adsorption. Others may promote unwanted side or back reactions alongside the target reactions (Hang Li et al., 2013). Additionally, many cocatalysts have poor product selectivity in multiproduct reactions (Bai et al., 2014). Low catalytic stability further limits some cocatalysts' performance (Jiang et al., 2016). Nano cocatalysts can aggregate or detach from semiconductors due to weak bonding and may also be susceptible to oxidation or corrosion when exposed to air, scavengers, or electrolyte. To address these issues, hybrid cocatalysts have been developed. These are composed of multiple components, at least one of which is nanoscale. The integration of different components helps overcome the limitations of single-component cocatalysts and creates synergistic effects, enhancing photocatalytic performance. When coupled with semiconductors, the architectural features of hybrid cocatalysts tremendously influence their functionalities (Bai et al., 2019).

Numerous outstanding works on element-doped cocatalysts have been reported. For instance, Chen et al. (2021) successfully replaced noble metal cocatalysts with an N-doped MoS_2 cocatalyst created using an effective calcination process. Together with an S-doped g-C_3N_4 photocatalyst, this N-doped MoS_2 allowed for photocatalytic water splitting, which produced hydrogen. After 4 hours of illumination, the system showed high photocatalytic activity, yielding 658.5 μmol/g/h. This efficiency outperformed the S-doped g-C_3N_4 by a factor of 23. From the results, it is evident that adding a cocatalyst enhanced electron conductivity and facilitated charge separation and transfer. The synergistic effect of dual cocatalysts is crucial for enhancing photocatalytic efficiency (Liu et al., 2020b). For instance, Moniruddin et al. (2019) developed a CdS/CoO_x/Co ternary composite photocatalyst by loading CdS nanorods (CdS NRs) with Co metal and CoO_x. Upon light stimulation, CoO_x captured most photogenerated holes, reducing carrier recombination, while Co metal absorbed electrons for the reduction reaction. This synergistic action between Co metal and CoO_x effectively promoted charge separation and transfer, leading to high photocatalytic hydrogen production efficiency. Fig. 16.8 illustrates the proposed mechanism by which adding CoOx and Co metal enhances photocatalytic H_2 evolution in CdS. Another widely used hybrid cocatalyst strategy, besides elemental doping, involves constructing heterojunctions to create an internal electric field within photocatalytic materials. In another study, a composite cocatalyst was developed with a $Ni_3B/Ni(OH)_2$ heterojunction structure (Lu et al., 2018). Analysis showed that its cocatalytic effect was nearly as effective as Pt. The $Ni_3B/Ni(OH)_2$ cocatalyst lowered the hydrogen evolution reaction overpotential of the photocatalyst, increased its active sites, and enhanced the visible light response. This enabled efficient visible-light catalysis of $g\text{-}C_3N_4$ and demonstrated excellent water splitting efficiency without needing a precious metal cocatalyst. This work emphasizes the potential of composite cocatalysts because of their synergistic effects and advances the development of high-activity, low-cost photocatalytic systems. A new catalyst consisting of $MoO_3/NiMoO_4$ heterostructured nanorods with Pt ultrafine nanoparticles has been developed for efficient hydrogen and oxygen evolution reactions in industrial water splitting (Luu Luyen Doan et al., 2024). Synthesized using a versatile method, it outperforms traditional noble-metal catalysts like Pt/C and RuO_2 at high current densities ($\geq 1000\ mA/cm^2$). When used as both cathode and anode, it delivers $10\ mA/cm^2$ at a low cell voltage of 1.55 V, maintaining excellent stability for over 50 hours. Density functional theory calculations reveal that the $MoO_3/NiMoO_4$ heterostructure is highly conductive, and its strong electronic interaction with Pt enhances charge transfer and optimizes the Gibbs free energies of intermediates

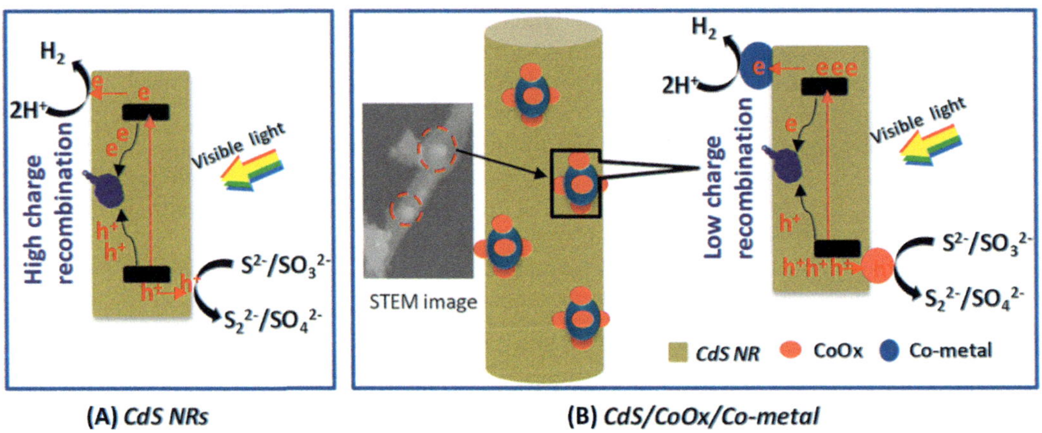

FIGURE 16.8 Mechanism to explain the proposed charge separation of CdS (A) and CdS/CoOx/Co-metal heterostructures, respectively (B). *Reproduced with permission from Moniruddin, M., Oppong, E., Stewart, D., McCleese, C., Roy, A., Warzywoda, J. et al. 2019. Designing CdS-based ternary heterostructures consisting of co-metal and CoOx cocatalysts for photocatalytic H_2 evolution under visible light.* Inorganic Chemistry *58 (18), 12325–12333. Copyright 2019 American Chemical Society.*

16.5.5 MXenes-based cocatalysts

MXenes, a recently discovered two-dimensional (2D) family of transition metal nitrides, carbides, and carbonitrides, have gained significant attention since their discovery in 2011 due to their exceptional chemical and physical properties (Isa et al., 2024). MXenes, pronounced as "Max-een," are a class of emerging 2D materials discovered by Anasori and Gogotsi (2019). The first MXene, titanium carbide ($Ti_3C_2T_x$), was synthesized by selectively etching the "A" layer (usually a Group III–VI element) from the ternary metal carbide Ti_3AlC_2, known as the MAX phase (Verger et al., 2019). MXenes have a general formula of $M_{n+1}X_nT_x$, where "M" is a transition metal, "X" is carbon, nitrogen, or both, and "T_x" represents surface groups like F, OH, O, or Cl formed during etching (Zahra et al., 2022). Their structures (e.g., M_2X, M_3X_2, M_4X_3) depend on the MAX phase composition and can consist of three, five, or seven atomic layers (Niu et al., 2022). Owing to their excellent electrical conductivity and unique surface chemistry, MXenes hold promise for applications in biosensing, energy storage, hydrogen production, supercapacitors, membrane separation, and more (Jatoi et al., 2023). The band gap of MXenes varies from 0.55 to 2.48 eV, depending on their composition. For instance, MXene oxides like Mo_2C/Ti_3C_2, Hf_2CO_2, Zr_2CO_2, and Ti_2CO_2 have band gaps of 2.2, 2.45, 2.13, and 1.15 eV, respectively (Naguib et al., 2014). Metal chalcogenide photocatalysts can produce high hydrogen yields but require further optimization for enhanced performance. For example, Tie et al. (2019) prepared ZnS/Ti_3C_2 using HF and solvothermal etching with lactic acid under a 300 W Xe lamp, achieving a hydrogen yield of 502.6 μmol/g/h. Similarly, Peng et al. (2021) synthesized 0D-CuS/2D-$Ti_3C_2T_x$ via an HF wet chemical process with Na_2SO_3 and Na_2S as sacrificial agents, resulting in a yield of 4.425 μmol/g/h. X. Liu et al. prepared ZnS/$Ti_3C_2T_x$ by etching Ti_3AlC_2 with NaF/HCl and ultrasonication with ethanol, achieving 212.0 μmol/g/h (Peng et al., 2021). Wang et al. (2023) prepared Au/Ti_3C_2/TiO_2 photocatalysts, achieving a hydrogen yield of 7989 μmol/g/h with Na_2S and Na_2SO_3 as sacrificial agents, which is about 13.7 times higher than that of bare Au/TiO_2. Ma et al. (2022) prepared a MoS_2/Ti_3C_2@CNFs hybrid material using LiF/HF etching and CVD growth. Under a 300 W xenon visible light lamp with DMF, the composite achieved a hydrogen evolution rate of 17.07 mmol/g/h (Ma et al., 2022). The MXene/C-TiO_2/g-C_3N_4 composite material, synthesized via a simple heat treatment by calcining melamine at 550°C, achieved a hydrogen yield of 1409 μmol/g/h under visible light irradiation at 420 nm (Han et al., 2020). A binary composite of P-doped tubular g-C_3N_4C3N4 and MXene (Ti_3C_2) was synthesized using HF etching, electrostatic hydrothermal, and self-assembly methods. The optimized g-C_3N_4/Ti_3C_2 composite achieved a hydrogen evolution rate of 565 mmol/h/g, which is 5.3 times higher than that of bare g-C_3N_4 under visible light from a 300 W Xe lamp (Huang et al., 2023).

16.6 Conclusion and future perspective

The global energy crisis and environmental concerns have made semiconductor-based photocatalytic water splitting an essential technique for converting solar energy into chemical energy. To enhance the rates of H_2 production, cocatalysts are often loaded onto semiconductor photocatalysts. They enhance light absorption, promote charge carrier separation and movement, increase active sites for gas generation, lower the energy barrier for water activation, and prevent charge recombination. This study reviews the function and the advancements in innovative cocatalysts in different categories. Noble metals are very effective, but their high cost and scarcity prevent widespread use. Therefore their use is not a favorable option for the sustainable industrial production of hydrogen. Consequently, research has focused on developing more abundant, cost-effective materials such as transition metal compounds and metal-free cocatalysts, which, though promising, often lack the efficiency and stability of noble metals. Metal-free cocatalysts show great promise due to their abundance, tunable electronic properties, cost-effectiveness, and environmental friendliness, making them strong candidates for future developments in water splitting technology. When comparing the overall performance, metal-free cocatalysts stand out for their cost-effectiveness, abundance, and lower environmental impact. The future of water splitting will focus on designing cocatalysts that offer high activity, durability, and cost-effectiveness. These innovations could lead to the development of efficient and sustainable water splitting systems. The future of water splitting will focus on designing cocatalysts that offer high activity, durability, and cost-effectiveness. Advances in various fields are expected to drive the discovery and optimization of new cocatalysts. Moreover, basic studies of the properties, electrochemical functions, and electron transfer techniques and their mechanisms involving loaded cocatalysts are limited. Such studies are crucial for advancing the field and enhancing our understanding of cocatalysts in photocatalytic applications. Ongoing research and innovation are essential to overcome existing challenges for a sustainable energy future.

References

Al-Azri, Z.H.N., Chen, W.-T., Chan, A., Jovic, V., Ina, T., Idriss, H., et al., 2015. The roles of metal co-catalysts and reaction media in photocatalytic hydrogen production: performance evaluation of M/TiO_2 photocatalysts (M=Pd, Pt, Au) in different alcohol–water mixtures. Journal of Catalysis 329, 355–367.

Anasori, B., Gogotsi, Û.G., 2019. 2D metal carbides and nitrides (MXenes). 2549 Springer.

Angaridis, P.A., Lazarides, T., Coutsolelos, A.C., 2014. Functionalized porphyrin derivatives for solar energy conversion. Polyhedron 82, 19–32.

Bai, S., Gao, C., Low, J., Xiong, Y., 2019. Crystal phase engineering on photocatalytic materials for energy and environmental applications. Nano Research 12 (9), 2031–2054.

Bai, S., Wang, X., Hu, C., Xie, M., Jiang, J., Xiong, Y., 2014. Two-dimensional g-C3N4: an ideal platform for examining facet selectivity of metal co-catalysts in photocatalysis. Chemical Communications 50 (46), 6094–6097.

Bai, S., Yin, W., Wang, L., Li, Z., Xiong, Y., 2016. Surface and interface design in cocatalysts for photocatalytic water splitting and CO_2 reduction. RSC Advances 6 (62), 57446–57463.

Basavarajappa, P.S., Patil, S.B., Ganganagappa, N., Reddy, K.R., Raghu, A.V., Reddy, C.V., 2020. Recent progress in metal-doped TiO_2, non-metal doped/codoped TiO_2 and TiO_2 nanostructured hybrids for enhanced photocatalysis. International Journal of Hydrogen Energy 45 (13), 7764–7778.

Bi, L., Xu, D., Zhang, L., Lin, Y., Wang, D., Xie, T., 2015. Metal Ni-loaded g-C_3N_4 for enhanced photocatalytic H_2 evolution activity: the change in surface band bending. Physical Chemistry Chemical Physics 17 (44), 29899–29905.

Bi, W., Zhang, L., Sun, Z., Li, X., Jin, T., Wu, X., et al., 2016. Insight into electrocatalysts as co-catalysts in efficient photocatalytic hydrogen evolution. ACS Catalysis 6 (7), 4253–4257.

Cao, S., Yu, J., 2016. Carbon-based H_2-production photocatalytic materials. Journal of Photochemistry and Photobiology C: Photochemistry Reviews 27, 72–99.

Cao, X., Tian, G., Chen, Y., Zhou, J., Zhou, W., Tian, C., et al., 2014. Hierarchical composites of TiO_2 nanowire arrays on reduced graphene oxide nanosheets with enhanced photocatalytic hydrogen evolution performance. Journal of Materials Chemistry A 2 (12), 4366–4374.

Chang, K., Hai, X., Pang, H., Zhang, H., Shi, L., Liu, G., et al., 2016. Targeted synthesis of 2H- and 1T-phase MoS_2 monolayers for catalytic hydrogen evolution. Advanced Materials 28 (45), 10033–10041.

Cheah, A.J., Chiu, W.S., Khiew, P.S., Nakajima, H., Saisopa, T., Songsiriritthigul, P., et al., 2015. Facile synthesis of a Ag/MoS_2 nanocomposite photocatalyst for enhanced visible-light driven hydrogen gas evolution. Catalysis Science & Technology 5 (8), 4133–4143.

Chen, D., Zhang, H., Liu, Y., Li, J., 2013. Graphene and its derivatives for the development of solar cells, photoelectrochemical, and photocatalytic applications. Energy & Environmental Science 6 (5), 1362–1387.

Chen, H., Nanayakkara, C.E., Grassian, V.H., 2012. Titanium dioxide photocatalysis in atmospheric chemistry. Chemical Reviews 112 (11), 5919–5948.

Chen, T., Quan, W., Yu, L., Hong, Y., Song, C., Fan, M., et al., 2016. One-step synthesis and visible-light-driven H_2 production from water splitting of Ag quantum dots/g-C_3N_4 photocatalysts. Journal of Alloys and Compounds 686, 628–634.

Chen, W.-T., Chan, A., Sun-Waterhouse, D., Moriga, T., Idriss, H., Waterhouse, G.I.N., 2015. Ni/TiO_2: a promising low-cost photocatalytic system for solar H_2 production from ethanol–water mixtures. Journal of Catalysis 326, 43–53.

Chen, W., Wang, Y., Liu, S., Gao, L., Mao, L., Fan, Z., et al., 2018. Non-noble metal Cu as a cocatalyst on TiO_2 nanorod for highly efficient photocatalytic hydrogen production. Applied Surface Science 445, 527–534.

Chen, G., Ding, N., Li, F., Fan, Y., Luo, Y., Li, D., et al., 2014a. Enhancement of photocatalytic H_2 evolution on $ZnIn_2S_4$ loaded with in-situ photo-deposited MoS_2 under visible light irradiation. Applied Catalysis B: Environmental 160-161, 614–620.

Chen, S., Chen, X., Jiang, Q., Yuan, J., Lin, C., Shangguan, W., 2014b. Promotion effect of nickel loaded on CdS for photocatalytic H_2 production in lactic acid solution. Applied Surface Science 316, 590–594.

Chen, Y., Li, J., Hong, Z., Shen, B., Lin, B., Gao, B., 2014c. Origin of the enhanced visible-light photocatalytic activity of CNT modified g-C_3N_4 for H_2 production. Physical Chemistry Chemical Physics 16 (17), 8106–8113.

Chen, Y., Qin, Z., 2016. General applicability of nanocrystalline Ni_2P as a noble-metal-free cocatalyst to boost photocatalytic hydrogen generation. Catalysis Science & Technology 6 (23), 8212–8221.

Chen, Y., Su, F., Xie, H., Wang, R., Ding, C., Huang, J., et al., 2021. One-step construction of S-scheme heterojunctions of N-doped MoS_2 and S-doped g-C_3N_4 for enhanced photocatalytic hydrogen evolution. Chemical Engineering Journal 404, 126498.

Colmenares, J.C., Luque, R., 2014. Heterogeneous photocatalytic nanomaterials: prospects and challenges in selective transformations of biomass-derived compounds. Chemical Society Reviews 43 (3), 765–778.

Colmenares, J.C., Xu, Y.-J., 2016. Heterogeneous Photocatalysis. Green Chemistry and Sustainable Technology.

Ding, C., Shi, J., Wang, Z., Li, C., 2017. Photoelectrocatalytic water splitting: significance of cocatalysts, electrolyte, and interfaces. ACS Catalysis 7 (1), 675–688.

Ding, Z., Chen, X., Antonietti, M., Wang, X., 2011. Synthesis of transition metal-modified carbon nitride polymers for selective hydrocarbon oxidation. ChemSusChem 4 (2), 274–281.

Etacheri, V., Di Valentin, C., Schneider, J., Bahnemann, D., Pillai, S.C., 2015. Visible-light activation of TiO_2 photocatalysts: advances in theory and experiments. Journal of Photochemistry and Photobiology C: Photochemistry Reviews 25, 1–29.

Fernando, K.A.S., Sahu, S., Liu, Y., Lewis, W.K., Guliants, E.A., Jafariyan, A., et al., 2015. Carbon quantum dots and applications in photocatalytic energy conversion. ACS Applied Materials & Interfaces 7 (16), 8363–8376.

Fujishima, A., Honda, K., 1972. Electrochemical photolysis of water at a semiconductor electrode. Nature 238 (5358), 37–38.

Gao, X., Yao, Y., Meng, X., 2020. Recent development on BN-based photocatalysis: a review. Materials Science in Semiconductor Processing 120, 105256.

Guo, Y., Jia, H., Yang, J., Yin, H., Yang, Z., Wang, J., et al., 2018. Understanding the roles of plasmonic Au nanocrystal size, shape, aspect ratio and loading amount in Au/g-C_3N_4 hybrid nanostructures for photocatalytic hydrogen generation. Physical Chemistry Chemical Physics 20 (34), 22296–22307.

Han, C., Lu, Y., Zhang, J., Ge, L., Li, Y., Chen, C., et al., 2015a. Novel PtCo alloy nanoparticle decorated 2D g-C_3N_4 nanosheets with enhanced photocatalytic activity for H_2 evolution under visible light irradiation. Journal of Materials Chemistry A 3 (46), 23274–23282.

Han, C., Wu, L., Ge, L., Li, Y., Zhao, Z., 2015b. AuPd bimetallic nanoparticles decorated graphitic carbon nitride for highly efficient reduction of water to H_2 under visible light irradiation. Carbon 92, 31–40.

Han, B., Liu, S., Zhang, N., Xu, Y.-J., Tang, Z.-R., 2017a. One-dimensional CdS@MoS_2 core-shell nanowires for boosted photocatalytic hydrogen evolution under visible light. Applied Catalysis B: Environmental 202, 298–304.

Han, C., Gao, Y., Liu, S., Ge, L., Xiao, N., Dai, D., et al., 2017b. Facile synthesis of AuPd/g-C_3N_4 nanocomposite: an effective strategy to enhance photocatalytic hydrogen evolution activity. International Journal of Hydrogen Energy 42 (36), 22765–22775.

Han, G., Jin, Y.-H., Burgess, R.A., Dickenson, N.E., Cao, X.-M., Sun, Y., 2017c. Visible-light-driven valorization of biomass intermediates integrated with H_2 production catalyzed by ultrathin Ni/CdS nanosheets. Journal of the American Chemical Society 139 (44), 15584–15587.

Han, X., An, L., Hu, Y., Li, Y., Hou, C., Wang, H., et al., 2020. Ti_3C_2 MXene-derived carbon-doped TiO_2 coupled with g-C_3N_4 as the visible-light photocatalysts for photocatalytic H_2 generation. Applied Catalysis B: Environmental 265, 118539.

Hang Li, Y., Xing, J., Jia Chen, Z., Li, Z., Tian, F., Rong Zheng, L., et al., 2013. Unidirectional suppression of hydrogen oxidation on oxidized platinum clusters. Nature Communications 4 (1), 2500.

He, Z., Kim, C., Jeon, T.H., Choi, W., 2018. Hydrogenated heterojunction of boron nitride and titania enables the photocatalytic generation of H_2 in the absence of noble metal catalysts. Applied Catalysis B: Environmental 237, 772–782.

Hong, E., Kim, D., Kim, J.H., 2014. Heterostructured metal sulfide (ZnS–CuS–CdS) photocatalyst for high electron utilization in hydrogen production from solar water splitting. Journal of Industrial and Engineering Chemistry 20 (5), 3869–3874.

Huang, K., Li, C., Zhang, X., Wang, L., Wang, W., Meng, X., 2023. Self-assembly synthesis of phosphorus-doped tubular g-C_3N4/Ti_3C_2 MXene Schottky junction for boosting photocatalytic hydrogen evolution. Green Energy & Environment 8 (1), 233–245.

Huang, X., Song, J., Wu, G., Miao, Z., Song, Y., Mo, Z., 2024. Recent progress on the photocatalytic hydrogen evolution reaction over a metal sulfide cocatalyst-mediated carbon nitride system. Inorganic Chemistry Frontiers 11 (9), 2527–2552.

Iijima, S., 1991. Helical microtubules of graphitic carbon. Nature 354 (6348), 56–58.

Isa, A.T., Hafeez, H.Y., Mohammed, J., Ndikilar, C.E., Suleiman, A.B., Kafadi, A.D.G., 2024. Photocatalytic performance of MXenes co-catalyst in hydrogen (H_2) production via photocatalytic water splitting: a review. Journal of Alloys and Compounds 1005, 175951.

Jatoi, A.S., Mubarak, N.M., Hashmi, Z., Solangi, N.H., Karri, R.R., Tan, Y.H., et al., 2023. New insights into MXene applications for sustainable environmental remediation. Chemosphere 313, 137497.

Jiang, W., Bai, S., Wang, L., Wang, X., Yang, L., Li, Y., et al., 2016. Integration of multiple plasmonic and co-catalyst nanostructures on TiO_2 nanosheets for visible-near-infrared photocatalytic hydrogen evolution. Small 12 (12), 1640–1648.

Jiang, Z., Zhu, J., Liu, D., Wei, W., Xie, J., Chen, M., 2014. In situ synthesis of bimetallic Ag/Pt loaded single-crystalline anatase TiO_2 hollow nano-hemispheres and their improved photocatalytic properties. CrystEngComm 16 (12), 2384–2394.

Kanazawa, T., Maeda, K., 2016. Light-induced synthesis of heterojunctioned nanoparticles on a semiconductor as durable cocatalysts for hydrogen evolution. ACS Applied Materials & Interfaces 8 (11), 7165–7172.

Karthikeyan, S., Kumar, S., Durndell, L.J., Isaacs, M.A., Parlett, C.M.A., Coulson, B., et al., 2018. Size-dependent visible light photocatalytic performance of Cu_2O nanocubes. ChemCatChem 10 (16), 3554–3563.

Khan, G., Choi, S.K., Kim, S., Lim, S.K., Jang, J.S., Park, H., 2013. Carbon nanotubes as an auxiliary catalyst in heterojunction photocatalysis for solar hydrogen. Applied Catalysis B: Environmental 142-143, 647–653.

Kong, L., Dong, Y., Jiang, P., Wang, G., Zhang, H., Zhao, N., 2016. Light-assisted rapid preparation of a Ni/g-C_3N_4 magnetic composite for robust photocatalytic H_2 evolution from water. Journal of Materials Chemistry A 4 (25), 9998–10007.

Kubacka, A., Fernandez-Garcia, M., Colon, G., 2012. Advanced nanoarchitectures for solar photocatalytic applications. Chemical Reviews 112 (3), 1555–1614.

Kudo, A., Miseki, Y., 2009. Heterogeneous photocatalyst materials for water splitting. Chemical Society Reviews 38 (1), 253–278.

Kum, J.M., Yoo, S.H., Ali, G., Cho, S.O., 2013. Photocatalytic hydrogen production over CuO and TiO_2 nanoparticles mixture. International Journal of Hydrogen Energy 38 (31), 13541–13546.

Kumar, M., Meena, B., Subramanyam, P., Suryakala, D., Subrahmanyam, C., 2022. Recent trends in photoelectrochemical water splitting: the role of cocatalysts. NPG Asia Materials 14 (1), 88.

Lakshmana Reddy, N., Emin, S., Valant, M., Shankar, M.V., 2017. Nanostructured Bi_2O_3@TiO_2 photocatalyst for enhanced hydrogen production. International Journal of Hydrogen Energy 42 (10), 6627–6636.

Landman, A., Dotan, H., Shter, G.E., Wullenkord, M., Houaijia, A., Maljusch, A., et al., 2017. Photoelectrochemical water splitting in separate oxygen and hydrogen cells. Nature Materials 16 (6), 646–651.

Leary, R., Westwood, A., 2011. Carbonaceous nanomaterials for the enhancement of TiO_2 photocatalysis. Carbon 49 (3), 741–772.

Leung, D.Y.C., Fu, X., Wang, C., Ni, M., Leung, M.K.H., Wang, X., et al., 2010. Hydrogen production over titania-based photocatalysts. ChemSusChem 3 (6), 681–694.

Li, L., Xu, L., Shi, W., Guan, J., 2013. Facile preparation and size-dependent photocatalytic activity of Cu_2O nanocrystals modified titania for hydrogen evolution. International Journal of Hydrogen Energy 38 (2), 816–822.

Li, L., Yu, Y., Ye, G.J., Ge, Q., Ou, X., Wu, H., et al., 2014a. Black phosphorus field-effect transistors. Nature Nanotechnology 9 (5), 372–377.

Li, X., Dai, Y., Ma, Y., Han, S., Huang, B., 2014b. Graphene/g-C_3N_4 bilayer: considerable band gap opening and effective band structure engineering. Physical Chemistry Chemical Physics 16 (9), 4230–4235.

Li, X., Zhang, J., Zhang, S., Xu, S., Wu, X., Chang, J., et al., 2021. Hexagonal boron nitride composite photocatalysts for hydrogen production. Journal of Alloys and Compounds 864, 158153.

Li, Y., Tsang, S.C.E., 2020. Recent progress and strategies for enhancing photocatalytic water splitting. Materials Today Sustainability 9, 100032.

Li, L., Cheng, B., Wang, Y., Yu, J., 2015a. Enhanced photocatalytic H_2-production activity of bicomponent NiO/TiO_2 composite nanofibers. Journal of Colloid and Interface Science 449, 115–121.

Li, Y., Wang, B., Liu, S., Duan, X., Hu, Z., 2015b. Synthesis and characterization of Cu_2O/TiO_2 photocatalysts for H_2 evolution from aqueous solution with different scavengers. Applied Surface Science 324, 736–744.

Li, J., Zhan, G., Yu, Y., Zhang, L., 2016a. Superior visible light hydrogen evolution of Janus bilayer junctions via atomic-level charge flow steering. Nature Communications 7 (1), 11480.

Li, L., Deng, Z., Yu, L., Lin, Z., Wang, W., Yang, G., 2016b. Amorphous transitional metal borides as substitutes for Pt cocatalysts for photocatalytic water splitting. Nano Energy 27, 103–113.

Li, X., Bi, W., Zhang, L., Tao, S., Chu, W., Zhang, Q., et al., 2016c. Single-atom Pt as co-catalyst for enhanced photocatalytic H_2 evolution. Advanced Materials 28 (12), 2427–2431.

Lian, Z., Xu, P., Wang, W., Zhang, D., Xiao, S., Li, X., et al., 2015. C60-decorated CdS/TiO_2 mesoporous architectures with enhanced photostability and photocatalytic activity for H_2 evolution. ACS Applied Materials & Interfaces 7 (8), 4533–4540.

Liang, S., Zhang, T., Zhang, D., Pu, X., Shao, X., Li, W., et al., 2018. One-pot combustion synthesis and efficient broad spectrum photoactivity of Bi/Bi OB r: Yb, Er/C photocatalyst. Journal of the American Ceramic Society 101 (8), 3424–3436.

Lin, H.-Y., Yang, H.-C., Wang, W.-L., 2011. Synthesis of mesoporous Nb_2O_5 photocatalysts with Pt, Au, Cu and NiO cocatalyst for water splitting. Catalysis Today 174 (1), 106–113.

Lin, H., Li, Y., Li, H., Wang, X., 2017. Multi-node CdS hetero-nanowires grown with defect-rich oxygen-doped MoS_2 ultrathin nanosheets for efficient visible-light photocatalytic H_2 evolution. Nano Research 10 (4), 1377–1392.

Liu, J., Zhang, H., Tang, D., Zhang, X., Yan, L., Han, Y., et al., 2014. Carbon quantum dot/silver nanoparticle/polyoxometalate composites as photocatalysts for overall water splitting in visible light. ChemCatChem 6 (9), 2634–2641.

Liu, B., Wang, Y., Shang, S., Ni, Y., Zhang, N., Cao, M., et al., 2018a. One-step synthesis of mulberry-shaped TiO_2-Au nanocomposite and its H2 evolution property under visible light. Colloids and Surfaces A: Physicochemical and Engineering Aspects 553, 203–209.

Liu, M., Xia, P., Zhang, L., Cheng, B., Yu, J., 2018b. Enhanced photocatalytic H_2-production activity of g-C_3N_4 nanosheets via optimal photodeposition of Pt as cocatalyst. ACS Sustainable Chemistry & Engineering 6 (8), 10472–10480.

Liu, W., Wang, X., Yu, H., Yu, J., 2018c. Direct photoinduced synthesis of amorphous $CoMoS_x$ cocatalyst and its improved photocatalytic H_2-evolution activity of CdS. ACS Sustainable Chemistry & Engineering 6 (9), 12436–12445.

Liu, Q., Huang, J., Tang, H., Yu, X., Shen, J., 2020a. Construction 0D TiO_2 nanoparticles/2D CoP nanosheets heterojunctions for enhanced photocatalytic H_2 evolution activity. Journal of Materials Science & Technology 56, 196–205.

Liu, Y., Li, Y.-H., Li, X., Zhang, Q., Yu, H., Peng, X., et al., 2020b. Regulating electron–hole separation to promote photocatalytic H_2 evolution activity of nanoconfined Ru/MXene/TiO_2 catalysts. ACS Nano 14 (10), 14181–14189.

Liu, Y., Sun, Z., Hu, Y.H., 2021. Bimetallic cocatalysts for photocatalytic hydrogen production from water. Chemical Engineering Journal 409, 128250.

Lu, X., Xie, J., Liu, S.-y, Adamski, A., Chen, X., Li, X., 2018. Low-cost Ni_3B/Ni(OH)2 as an ecofriendly hybrid cocatalyst for remarkably boosting photocatalytic H_2 production over g-C_3N_4 nanosheets. ACS Sustainable Chemistry & Engineering 6 (10), 13140–13150.

Luu Luyen Doan, T., Chuong Nguyen, D., Komalla, N., Hieu, N.V., Nguyen-Dinh, L., Dzade, N.Y., et al., 2024. Molybdenum oxide/nickel molybdenum oxide heterostructures hybridized active platinum co-catalyst toward superb-efficiency water splitting catalysis. Journal of Colloid and Interface Science 670, 12–27.

Lv, J., Kako, T., Li, Z., Zou, Z., Ye, J., 2010. Synthesis and photocatalytic activities of $NaNbO_3$ rods modified by In_2O_3 nanoparticles. The Journal of Physical Chemistry C 114 (13), 6157–6162.

Ma, S., Xu, Z., Jia, Z., Chen, L., Zhu, H., Chen, Y., et al., 2022. Facile fabrication of carbon fiber skeleton structure of MoS_2 supported on 2D MXene composite with highly efficient and stable hydrogen evolution reaction. Composites Science and Technology 222, 109380.

Maeda, K., Domen, K., 2010. Photocatalytic water splitting: recent progress and future challenges. The Journal of Physical Chemistry Letters 1 (18), 2655–2661.

Maeda, K., Domen, K., 2016. Development of novel photocatalyst and cocatalyst materials for water splitting under visible light. Bulletin of the Chemical Society of Japan 89 (6), 627–648.

Mao, L., Ba, Q., Jia, X., Liu, S., Liu, H., Zhang, J., et al., 2019a. Ultrathin $Ni(OH)_2$ nanosheets: a new strategy for cocatalyst design on CdS surfaces for photocatalytic hydrogen generation. RSC Advances 9 (3), 1260–1269.

Mao, L., Cai, X., Yang, S., Han, K., Zhang, J., 2019b. Black phosphorus-CdS-$La_2Ti_2O_7$ ternary composite: effective noble metal-free photocatalyst for full solar spectrum activated H_2 production. Applied Catalysis B: Environmental 242, 441–448.

Mo, Z., Xu, H., She, X., Song, Y., Yan, P., Yi, J., et al., 2019. Constructing Pd/2D-C_3N_4 composites for efficient photocatalytic H_2 evolution through nonplasmon-induced bound electrons. Applied Surface Science 467-468, 151–157.

Moniruddin, M., Oppong, E., Stewart, D., McCleese, C., Roy, A., Warzywoda, J., et al., 2019. Designing CdS-based ternary heterostructures consisting of co-metal and CoOx cocatalysts for photocatalytic H_2 evolution under visible light. Inorganic Chemistry 58 (18), 12325–12333.

Moniz, S.J.A., Shevlin, S.A., Martin, D.J., Guo, Z.-X., Tang, J., 2015. Visible-light driven heterojunction photocatalysts for water splitting – a critical review. Energy & Environmental Science 8 (3), 731–759.

Naguib, M., Mochalin, V.N., Barsoum, M.W., Gogotsi, Y., 2014. 25th Anniversary article: MXenes: A new family of two-dimensional materials. Advanced Materials 26 (7), 992–1005.

Niu, R., Han, R., Tang, S., Zhu, J., 2022. Microwave selective heating ultrafast construction of coral-like TiO_2-MXene/graphene hybrid architectures for high-performance lithium-ion battery. Journal of Power Sources 542, 231738.

Novoselov, K.S., Geim, A.K., Morozov, S.V., Jiang, D.-e, Zhang, Y., Dubonos, S.V., et al., 2004. Electric field effect in atomically thin carbon films. Science 306 (5696), 666–669.

Ou, M., Wan, S., Zhong, Q., Zhang, S., Wang, Y., 2017. Single Pt atoms deposition on g-C_3N_4 nanosheets for photocatalytic H_2 evolution or NO oxidation under visible light. International Journal of Hydrogen Energy 42 (44), 27043–27054.

Pan, J., Dong, Z., Wang, B., Jiang, Z., Zhao, C., Wang, J., et al., 2019. The enhancement of photocatalytic hydrogen production via Ti^{3+} self-doping black TiO_2/g-C_3N_4 hollow core-shell nano-heterojunction. Applied Catalysis B: Environmental 242, 92–99.

Pei, Y., Cheng, Y., Chen, J., Smith, W., Dong, P., Ajayan, P.M., et al., 2018. Recent developments of transition metal phosphides as catalysts in the energy conversion field. Journal of Materials Chemistry A 6 (46), 23220–23243.

Peng, C., Xie, X., Xu, W., Zhou, T., Wei, P., Jia, J., et al., 2021. Engineering highly active Ag/Nb_2O_5@Nb_2CT_x (MXene) photocatalysts via steering charge kinetics strategy. Chemical Engineering Journal 421, 128766.

Peng, W., Zhang, S.-S., Shao, Y.-B., Huang, J.-H., 2018. Bimetallic PtNi/g-C_3N_4 nanotubes with enhanced photocatalytic activity for H_2 evolution under visible light irradiation. International Journal of Hydrogen Energy 43 (49), 22215–22225.

Politano, A., Marino, A.R., Formoso, V., Chiarello, G., 2011. Water adsorption on graphene/Pt(111) at room temperature: a vibrational investigation. AIP Advances 1 (4), 042130.

Provinciali, G., Filippi, J., Lavacchi, A., Caporali, S., Banchelli, M., Serrano-Ruiz, M., et al., 2023. Black phosphorus as promoter for noble metal-free photocatalytic hydrogen production. ChemCatChem 15 (18), e202300647.

Qian, X.-B., Peng, W., Huang, J.-H., 2018. Fluorescein-sensitized Au/g-C_3N_4 nanocomposite for enhanced photocatalytic hydrogen evolution under visible light. Materials Research Bulletin 102, 362–368.

Qin, J., Huo, J., Zhang, P., Zeng, J., Wang, T., Zeng, H., 2016. Improving the photocatalytic hydrogen production of Ag/g-C_3N_4 nanocomposites by dye-sensitization under visible light irradiation. Nanoscale 8 (4), 2249–2259.

Rahman, M.Z., Kibria, M.G., Mullins, C.B., 2020. Metal-free photocatalysts for hydrogen evolution. Chemical Society Reviews 49 (6), 1887–1931.

Ran, J., Yu, J., Jaroniec, M., 2011. $Ni(OH)_2$ modified CdS nanorods for highly efficient visible-light-driven photocatalytic H2 generation. Green Chemistry 13 (10), 2708–2713.

Ran, J., Zhang, J., Yu, J., Jaroniec, M., Qiao, S.Z., 2014. Earth-abundant cocatalysts for semiconductor-based photocatalytic water splitting. Chemical Society Reviews 43 (22), 7787–7812.

Ran, J., Zhu, B., Qiao, S.-Z., 2017. Phosphorene co-catalyst advancing highly efficient visible-light photocatalytic hydrogen production. Angewandte Chemie International Edition 56 (35), 10373–10377.

Rangel, E., Ruiz-Chavarria, G., Magana, L.F., 2009. Water molecule adsorption on a titanium–graphene system with high metal coverage. Carbon 47 (2), 531–533.

Rao, S., Su, C., He, X., Zhang, L., Liu, Z., Qin, H., et al., 2021. Engineering MoS_2 cocatalysts as active sites over porous P-doped g-C_3N_4 nanosheets to enhance photocatalytic hydrogen production. Physica Status Solidi (RRL) – Rapid Research Letters 15 (12), 2000513.

Reddy, N.R., Kumari, M.M., Cheralathan, K., Shankar, M., 2018. Enhanced photocatalytic hydrogen production activity of noble metal free MWCNT-TiO_2 nanocomposites. International Journal of Hydrogen Energy 43 (8), 4036–4043.

Ren, X., Philo, D., Li, Y., Shi, L., Chang, K., Ye, J., 2020a. Recent advances of low-dimensional phosphorus-based nanomaterials for solar-driven photocatalytic reactions. Coordination Chemistry Reviews 424, 213516.

Ren, X., Shi, L., Li, Y., Song, S., Wang, Q., Luo, S., et al., 2020b. Single cobalt atom anchored black phosphorous nanosheets as an effective cocatalyst promotes photocatalysis. ChemCatChem 12 (15), 3870–3879.

Robu, V., Flynn, D., 2017. Outsmart supply dips in renewable energy. Nature 544 (7649), 161 -161.

Shaban, Y.A., Khan, S.U.M., 2008. Visible light active carbon modified n-TiO_2 for efficient hydrogen production by photoelectrochemical splitting of water. International Journal of Hydrogen Energy 33 (4), 1118–1126.

Shearer, C.J., Cherevan, A., Eder, D., 2014. Application and future challenges of functional nanocarbon hybrids. Advanced Materials 26 (15), 2295–2318.

Shen, R., Xie, J., Lu, X., Chen, X., Li, X., 2018. Bifunctional Cu_3P Decorated g-C_3N_4 nanosheets as a highly active and robust visible-light photocatalyst for H_2 production. ACS Sustainable Chemistry & Engineering 6 (3), 4026–4036.

Shoaib, A., Ji, M., Qian, H., Liu, J., Xu, M., Zhang, J., 2016. Noble metal nanoclusters and their in situ calcination to nanocrystals: precise control of their size and interface with TiO_2 nanosheets and their versatile catalysis applications. Nano Research 9 (6), 1763–1774.

Song, L., Kang, X., Zhang, S., 2018. CNT/g-C_3N_4 photocatalysts with enhanced hydrogen evolution ability for water splitting based on a noncovalent interaction. International Journal of Energy Research 42 (4), 1649–1656.

Staffell, I., Scamman, D., Velazquez Abad, A., Balcombe, P., Dodds, P.E., Ekins, P., et al., 2019. The role of hydrogen and fuel cells in the global energy system. Energy & Environmental Science 12 (2), 463–491.

Subramanyam, P., Meena, B., Biju, V., Misawa, H., Challapalli, S., 2022. Emerging materials for plasmon-assisted photoelectrochemical water splitting. Journal of Photochemistry and Photobiology C: Photochemistry Reviews 51, 100472.

Sui, Y., Wu, L., Zhong, S., Liu, Q., 2019. Carbon quantum dots/TiO_2 nanosheets with dominant (001) facets for enhanced photocatalytic hydrogen evolution. Applied Surface Science 480, 810–816.

Sun, Z., Zhu, M., Fujitsuka, M., Wang, A., Shi, C., Majima, T., 2017. Phase effect of NixPy hybridized with g-C3N4 for photocatalytic hydrogen generation. ACS Applied Materials & Interfaces 9 (36), 30583–30590.

Tan, Y., Shu, Z., Zhou, J., Li, T., Wang, W., Zhao, Z., 2018. One-step synthesis of nanostructured g-C_3N_4/TiO_2 composite for highly enhanced visible-light photocatalytic H_2 evolution. Applied Catalysis B: Environmental 230, 260–268.

Tian, L., Guan, X., Zong, S., Dai, A., Qu, J., 2023. Cocatalysts for photocatalytic overall water splitting: a mini review. Catalysts 13 (2), 355.

Tie, L., Yang, S., Yu, C., Chen, H., Liu, Y., Dong, S., et al., 2019. In situ decoration of ZnS nanoparticles with Ti_3C_2 MXene nanosheets for efficient photocatalytic hydrogen evolution. Journal of Colloid and Interface Science 545, 63–70.

Tiwari, S., Kumar, S., Ganguli, A.K., 2022. Role of MoS_2/rGO co-catalyst to enhance the activity and stability of Cu_2O as photocatalyst towards photoelectrochemical water splitting. Journal of Photochemistry and Photobiology A: Chemistry 424, 113622.

Tong, T., Zhu, B., Jiang, C., Cheng, B., Yu, J., 2018. Mechanistic insight into the enhanced photocatalytic activity of single-atom Pt, Pd or Au-embedded g-C_3N_4. Applied Surface Science 433, 1175–1183.

Townsend, T.K., Browning, N.D., Osterloh, F.E., 2012. Nanoscale strontium titanate photocatalysts for overall water splitting. Acs Nano 6 (8), 7420–7426.

Verger, L., Natu, V., Carey, M., Barsoum, M.W., 2019. MXenes: an introduction of their synthesis, select properties, and applications. Trends in Chemistry 1 (7), 656–669.

Vu, M.-H., Nguyen, C.-C., Sakar, M., Do, T.-O., 2017. Ni supported $CdIn_2S_4$ spongy-like spheres: a noble metal free high-performance sunlight driven photocatalyst for hydrogen production. Physical Chemistry Chemical Physics 19 (43), 29429–29437.

Walter, M.G., Warren, E.L., McKone, J.R., Boettcher, S.W., Mi, Q., Santori, E.A., et al., 2010. Solar water splitting cells. Chemical Reviews 110 (11), 6446–6473.

Wang, L., Yao, Z., Jia, F., Chen, B., Jiang, Z., 2013. A facile synthesis of $ZnxCd_1{-}xS$/CNTs nanocomposite photocatalyst for H_2 production. Dalton Transactions 42 (27), 9976–9981.

Wang, B., He, S., Zhang, L., Huang, X., Gao, F., Feng, W., et al., 2019a. CdS nanorods decorated with inexpensive NiCd bimetallic nanoparticles as efficient photocatalysts for visible-light-driven photocatalytic hydrogen evolution. Applied Catalysis B: Environmental 243, 229–235.

Wang, Q., Nakabayashi, M., Hisatomi, T., Sun, S., Akiyama, S., Wang, Z., et al., 2019b. Oxysulfide photocatalyst for visible-light-driven overall water splitting. Nature Materials 18 (8), 827–832.

Wang, X., Chen, X., Thomas, A., Fu, X., Antonietti, M., 2009. Metal-containing carbon nitride compounds: a new functional organic–metal hybrid material. Advanced Materials 21 (16), 1609–1612.

Wang, X., Xu, Q., Li, M., Shen, S., Wang, Y., Feng, Z., et al., 2012. Photocatalytic overall water splitting promoted by an α-β phase junction on Ga_2O_3. Angewandte Chemie 51 (52), 13089–13092.

Wang, Y., Fang, X., Zeng, J., Li, S., Wang, X., Zhang, B., 2023. 0D/2D/3D ternary Au/Ti_3C_2/TiO_2 photocatalyst based on accelerating charge transfer and enhanced stability for efficiently hydrogen production. Applied Surface Science 615, 156397.

Wang, Y., Liu, X., Liu, J., Han, B., Hu, X., Yang, F., et al., 2018. Carbon quantum dot implanted graphite carbon nitride nanotubes: excellent charge separation and enhanced photocatalytic hydrogen evolution. Angewandte Chemie 130 (20), 5867–5873.

Wen, J., Xie, J., Chen, X., Li, X., 2017. A review on g-C_3N_4-based photocatalysts. Applied Surface Science 391, 72–123.

Wu, J., Huang, S., Jin, Z., Chen, J., Hu, L., Long, Y., et al., 2018. Black phosphorus: an efficient co-catalyst for charge separation and enhanced photocatalytic hydrogen evolution. Journal of Materials Science 53 (24), 16557–16566.

Wu, M., Yan, J.-M., Zhang, X.-w, Zhao, M., 2015a. Synthesis of g-C_3N4 with heating acetic acid treated melamine and its photocatalytic activity for hydrogen evolution. Applied Surface Science 354, 196–200.

Wu, M., Yan, J.-M., Zhang, X.-W., Zhao, M., Jiang, Q., 2015b. Ag_2O modified g-C_3N_4 for highly efficient photocatalytic hydrogen generation under visible light irradiation. Journal of Materials Chemistry A 3 (30), 15710–15714.

Wu, Y., Wang, Z., Yan, Y., Wei, Y., Wang, J., Shen, Y., et al., 2024. Rational photodeposition of cobalt phosphate on flower-like $ZnIn_2S_4$ for efficient photocatalytic hydrogen evolution. Molecules 29 (2) Available online: [Accessed.

Xi, Z., 2021. Nanostructures design: the role of cocatalysts for hydrogen and oxygen generation in photocatalytic water splitting. Theoretical Physics Letters 6, 235–247.

Xia, X., Deng, N., Cui, G., Xie, J., Shi, X., Zhao, Y., et al., 2015. NIR light induced H_2 evolution by a metal-free photocatalyst. Chemical Communications 51 (54), 10899–10902.

Xiang, Q., Yu, J., 2013. Graphene-based photocatalysts for hydrogen generation. The Journal of Physical Chemistry Letters 4 (5), 753–759.

Xiang, Q., Yu, J., Jaroniec, M., 2011. Preparation and enhanced visible-light photocatalytic H_2-production activity of graphene/C_3N_4 composites. The Journal of Physical Chemistry C 115 (15), 7355–7363.

Xiao, N., Li, S., Li, X., Ge, L., Gao, Y., Li, N., 2020. The roles and mechanism of cocatalysts in photocatalytic water splitting to produce hydrogen. Chinese Journal of Catalysis 41 (4), 642–671.

Xiao, N., Li, S., Liu, S., Xu, B., Li, Y., Gao, Y., et al., 2019. Novel PtPd alloy nanoparticle-decorated g-C_3N_4 nanosheets with enhanced photocatalytic activity for H_2 evolution under visible light irradiation. Chinese Journal of Catalysis 40 (3), 352–361.

Xie, G., Zhang, K., Guo, B., Liu, Q., Fang, L., Gong, J.R., 2013. Graphene-based materials for hydrogen generation from light-driven water splitting. Advanced Materials 25 (28), 3820–3839.

Xu, H.-Q., Yang, S., Ma, X., Huang, J., Jiang, H.-L., 2018. Unveiling charge-separation dynamics in CdS/metal–organic framework composites for enhanced photocatalysis. ACS Catalysis 8 (12), 11615–11621.

Xu, H., Li, S., Ge, L., Han, C., Gao, Y., Dai, D., 2017. In-situ synthesis of novel plate-like $Co(OH)_2$ co-catalyst decorated TiO_2 nanosheets with efficient photocatalytic H_2 evolution activity. International Journal of Hydrogen Energy 42 (36), 22877–22886.

Xue, F., Si, Y., Wang, M., Liu, M., Guo, L., 2019. Toward efficient photocatalytic pure water splitting for simultaneous H_2 and H_2O_2 production. Nano Energy 62, 823–831.

Yalavarthi, R., Zbořil, R., Schmuki, P., Naldoni, A., Kment, Š., 2021. Elucidating the role of surface states of $BiVO_4$ with Mo doping and a CoOOH co-catalyst for photoelectrochemical water splitting. Journal of Power Sources 483, 229080.

Yang, J., Wang, D., Han, H., Li, C., 2013. Roles of cocatalysts in photocatalysis and photoelectrocatalysis. Accounts of Chemical Research 46 (8), 1900–1909.

Yao, Z., Wang, L., Zhang, Y., Yu, Z., Jiang, Z., 2014. Carbon nanotube modified Zn0. 83Cd0. 17S nanocomposite photocatalyst and its hydrogen production under visible-light. International Journal of Hydrogen Energy 39 (28), 15380–15386.

Ye, A., Fan, W., Zhang, Q., Deng, W., Wang, Y., 2012. CdS–graphene and CdS–CNT nanocomposites as visible-light photocatalysts for hydrogen evolution and organic dye degradation. Catalysis Science & Technology 2 (5), 969–978.

Yin, H., Rong, F., Xie, Y., 2024. A review of typical transition metal phosphides electrocatalysts for hydrogen evolution reaction. International Journal of Hydrogen Energy 52, 350–375.

Yu, J., Hai, Y., Cheng, B., 2011a. Enhanced photocatalytic H_2-production activity of TiO_2 by $Ni(OH)_2$ cluster modification. The Journal of Physical Chemistry C 115 (11), 4953–4958.

Yu, J., Hai, Y., Jaroniec, M., 2011b. Photocatalytic hydrogen production over CuO-modified titania. Journal of Colloid and Interface Science 357 (1), 223–228.

Yu, J., Wang, S., Cheng, B., Lin, Z., Huang, F., 2013. Noble metal-free $Ni(OH)_2$–g-C_3N_4 composite photocatalyst with enhanced visible-light photocatalytic H_2-production activity. Catalysis Science & Technology 3 (7), 1782–1789.

Yu, J., Yang, B., Cheng, B., 2012. Noble-metal-free carbon nanotube-Cd 0.1 Zn 0.9S composites for high visible-light photocatalytic H_2-production performance. Nanoscale 4 (8), 2670–2677.

Yue, X., Yi, S., Wang, R., Zhang, Z., Qiu, S., 2016. Cadmium sulfide and nickel synergetic co-catalysts supported on graphitic carbon nitride for visible-light-driven photocatalytic hydrogen evolution. Scientific Reports 6 (1), 22268.

Zahra, S.A., Waqas Hakim, M., Adil Mansoor, M., Rizwan, S., 2022. Two-dimensional double transition metal carbides as superior bifunctional electrocatalysts for overall water splitting. Electrochimica Acta 434, 141257.

Zeng, D., Xu, W., Ong, W.-J., Xu, J., Ren, H., Chen, Y., et al., 2018. Toward noble-metal-free visible-light-driven photocatalytic hydrogen evolution: monodisperse sub–15 nm Ni_2P nanoparticles anchored on porous g-C_3N_4 nanosheets to engineer 0D-2D heterojunction interfaces. Applied Catalysis B: Environmental 221, 47–55.

Zhang, F.-M., Sheng, J.-L., Yang, Z.-D., Sun, X.-J., Tang, H.-L., Lu, M., et al., 2018. Rational design of MOF/COF hybrid materials for photocatalytic H_2 evolution in the presence of sacrificial electron donors. Angewandte Chemie International Edition 57 (37), 12106–12110.

Zhang, L., Liu, Q., Aoki, T., Crozier, P.A., 2015. Structural evolution during photocorrosion of Ni/NiO core/shell cocatalyst on TiO_2. The Journal of Physical Chemistry C 119 (13), 7207–7214.

Zhang, J., Li, W., Li, Y., Zhong, L., Xu, C., 2017a. Self-optimizing bifunctional CdS/Cu_2S with coexistence of light-reduced Cu0 for highly efficient photocatalytic H_2 generation under visible-light irradiation. Applied Catalysis B: Environmental 217, 30–36.

Zhang, P., Song, T., Wang, T., Zeng, H., 2017b. Fabrication of a non-semiconductor photocatalytic system using dendrite-like plasmonic CuNi bimetal combined with a reduced graphene oxide nanosheet for near-infrared photocatalytic H_2 evolution. Journal of Materials Chemistry A 5 (43), 22772–22781.

Zhang, Y.-H., Cai, X.-L., Guo, D.-Y., Zhang, H.-J., Zhou, N., Fang, S.-M., et al., 2019. Oxygen vacancies in concave cubes Cu_2O-reduced graphene oxide heterojunction with enhanced photocatalytic H_2 production. Journal of Materials Science: Materials in Electronics 30, 7182–7193.

Zhang, Z., Liu, Y., Yang, Y., Yakobson, B.I., 2016. Growth mechanism and morphology of hexagonal boron nitride. Nano Letters 16 (2), 1398–1403.

Zhao, G., Xu, X., 2021. Cocatalysts from types, preparation to applications in the field of photocatalysis. Nanoscale 13 (24), 10649–10667.

Zhao, N., Kong, L., Dong, Y., Wang, G., Wu, X., Jiang, P., 2018. Insight into the crucial factors for photochemical deposition of cobalt cocatalysts on g-C_3N_4 photocatalysts. ACS Applied Materials & Interfaces 10 (11), 9522–9531.

Zhou, G., Shan, Y., Wang, L., Hu, Y., Guo, J., Hu, F., et al., 2019. Photoinduced semiconductor-metal transition in ultrathin troilite FeS nanosheets to trigger efficient hydrogen evolution. Nature Communications 10 (1), 399.

Zhu, J., Xiong, J., Cheng, G., Li, W., Dou, S., 2019. Promoting solar-to-hydrogen evolution on Schottky interface with mesoporous TiO_2-Cu hybrid nanostructures. Journal of Colloid and Interface Science 545, 116–127.

Zhu, M., Kim, S., Mao, L., Fujitsuka, M., Zhang, J., Wang, X., et al., 2017. Metal-free photocatalyst for H_2 evolution in visible to near-infrared region: black phosphorus/graphitic carbon nitride. Journal of the American Chemical Society 139 (37), 13234–13242.

Zhu, T., Wu, H.B., Wang, Y., Xu, R., Lou, X.W., 2012. Formation of 1D hierarchical structures composed of Ni_3S_2 nanosheets on CNTs backbone for supercapacitors and photocatalytic H_2 production. Advanced Energy Materials 2 (12), 1497–1502.

CHAPTER

17

Nanostructured materials for water-splitting

Suman Thakur, and Sharda Bharti

Department of Biotechnology, National Institute of Technology (NIT) Raipur, Raipur, Chhattisgarh, India

17.1 Introduction

The global use of energy is predicted to rise because of the fast growth of industrialization and population expansion. Presently, conventional, nonrenewable sources of fossil fuels account for 84% of the daily energy utilized globally, leading to an increase in greenhouse gas emissions (Ahmad et al., 2021). The average CO_2 concentration in the environment in 2019 was 410.5 parts per million, growing at a rate of about 2 ppm each year (Chen et al., 2023). Fossil fuel consumption harms the environment and people, and our world is quickly running out of resources. Hydrogen (H_2) has great potential as an energy carrier to give a long-term solution to the issue of ensuring a sustainable supply of cleaner, more ecologically acceptable fuel. Conventional H_2 generation systems emit greenhouse gases, posing a significant environmental threat (Carolin Christopher et al., 2021). Therefore developing an alternative and renewable method for H_2 production is essential. Because H_2 energy is a promising tool for achieving environmental sustainability and net-zero emissions, scientists closely monitor this field of study. A major move towards net-zero emission goals and environmental sustainability may be large-scale H_2 generation. H_2 may be generated utilizing both renewable and fossil fuels. Steam reforming, partial oxidation, autothermal oxidation, gasification, are all methods for producing H_2 from fossil fuels. Due to its carbon-free emissions, high gravimetric the density of energy, and global abundance, H_2 energy is foremost practical options in place of finite fossil fuels (Chen et al., 2021).

Significant promise exists for water splitting as an effective and sustainable source of energy. It is the procedure that divides water into its constituent elements, oxygen and H_2. The primary byproduct of water-splitting, H_2 is a flexible energy source that may be utilized in fuel cells for generating power. H_2 is also a greener fuel than traditional fuels because it doesn't release any greenhouse gases during combustion. However, the energy and cost associated with traditional water-splitting techniques, such electrolysis, restricts their

Photochemical Splitting of Water
DOI: https://doi.org/10.1016/B978-0-443-29064-0.00001-2

scalability. Researchers are using nanoengineering techniques to increase the effectiveness and affordability of water-splitting in an effort to get over these limitations. As an alternative, high-purity environmentally friendly H_2 may be produced via water electrolysis supported by solar or wind. It is recognized that using this technology will help materialize the H_2 economy in a sustainable and ecological manner (Chen et al., 2021). One of the most scalable and economical techniques is H_2 production through photoelectrochemical (PEC) water-splitting, which uses semiconductors as photoelectrodes (Li and Zhang, 2010). H_2 with high-energy density zero-carbon footprint and earth-abundance has shown to be an excellent energy source for sustainable energy systems, which are acknowledged as essential to reducing air pollution and mitigating climate change. In order to accomplish this, it is necessary to promote a dependable and reasonably priced H_2 economy in order to lower the price of producing renewable H_2 by electrochemical water-splitting (EWS; picture depicted in Fig. 17.1). Therefore it is crucial to create affordable catalysts for water electrolysis (Chen et al., 2021). Utilizing clean energy solutions can help achieve more sustainable development. Improvements in efficacy, resource utilization, cost efficiency, energy security, design analysis, and good environment are the main constituents of clean energy systems, according to Dincer and Zamfirescu (2012).

Recently, nanomaterials (NMs) or nanostructured materials have attracted a lot of interest. The primary reason comes from their distinct chemical and physical characteristics as compared with bulk materials, also their potential utility in diverse advancements, including energy conversion. Unique characteristics of nanostructured materials include greater surface area, catalytic activity, improved light adsorption, stability, durability, etc. Numerous systems have been investigated for photoelectrodes which is based on nanostructured semiconductor materials; these systems are often single-component nanomaterials like ZnO, WO_3, CdS, Bi_2S_3,

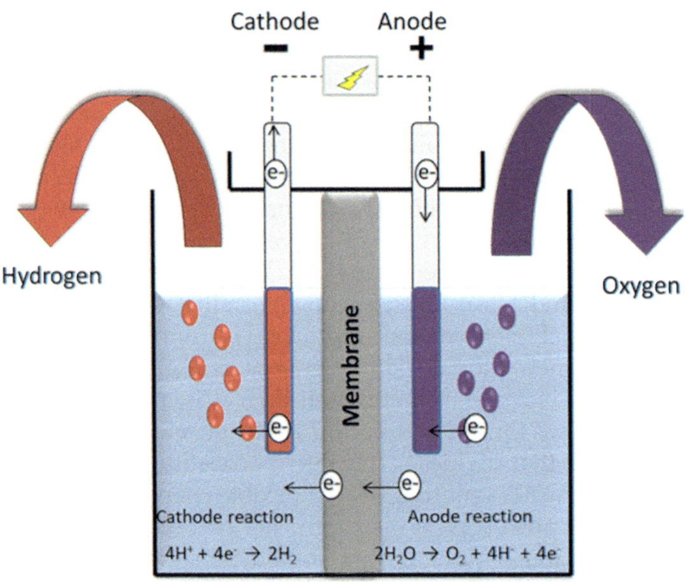

FIGURE 17.1 Electrolytic water-splitting.

Sb_2S_3, Fe_2O_3 and more frequently TiO_2. With various potential benefits, nanostructured semiconductor photoelectrodes possess the capacity to greatly alter PEC cell design and enhance the efficiency of conversion of solar to H_2 (Li and Zhang, 2010). Nanomaterials have a huge impact on enhancing H_2 generation and storage, but they are also starting to contribute significantly to the growth of new regions (Reddy et al., 2019).

The chapter discusses recent advancements in PEC water-splitting using nanostructured photoactive materials and explains the mechanism, chemical interactions, and factors affecting this process. This chapter explores various nanostructured materials, such as two-dimensional (2-D) materials, composite and hybrid materials, metal nanoparticles (MNPs), and metal-oxide nanoparticles (MONPs). It also covers the synthesis processes and properties of these materials and provides an overview of characterization techniques and electrochemical methodologies to understand the relationships between their structures and properties. It also addresses challenges and possible solutions in the field of nanostructured materials and focuses on the potential integration of nanomaterials into practical products and large-scale water-splitting systems. This chapter represents deep insights for scientists, scholars, and engineers engaged in diverse fields including energy-storage, conversion, and sustainable H_2 production, aimed at fostering a cleaner energy environment.

17.2 Conventional catalysts

One technique for splitting water is electrolysis, although it is restricted by the need for costly and uncommon catalysts like platinum (Pt). For application in electrocatalysis, researchers have been focusing on creating less expensive and more accessible catalyst substitutes, such as C-containing materials and various metal oxides. Research into the improvement and development of efficient electrocatalytic catalysts for water-splitting aims to discover greener and more cost-effective ways of creating H_2 for use in energy systems. Reaching this objective could reduce depending on nonrenewable resources and greenhouse gas emissions. Furthermore, the creation of effective catalysts may enable the generation of H_2 to be scaled up, providing a competitive substitute for fossil fuels. The main goal of study in this field is to build stable, long-lasting, and application-specific catalysts while addressing the drawbacks of traditional electrolysis, such as the need for costly and uncommon catalysts. Various categories of materials used for water-splitting are explained in the subsequent sections.

17.2.1 Metal-based catalysts

It has been demonstrated that catalysts like oxides of iridium and Ru, and Pt are extremely active and stable at diverse pH range (Jin et al., 2019). Their composites have demonstrated exceptional activity for hydrogen evolution reactions (HERs) and oxygen evolution reactions (OERs), in addition to the significant advancements in developing precious metal catalysts for EWS. Their high cost has limited their use in industry, and driving this process still requires cells with a high voltage and potential operating at large currents. Numerous attempts have been made since then to find alternatives. Increased catalytic efficacy (similar to that obtained using Pt, iridium- and Ru oxide), increased reliability in their operation (in acidic/basic/neutral media), ease of production, and cost-effectiveness based on market availability are required of the alternative choices.

It is well-known that in an alkaline atmosphere, transition metals of first row—Ni, Fe, Co, Cu, Mo, etc.—are active electrocatalytic materials for OER and HER. Given that Ni has a transition-metal electronic structure of $3d^8\ 4s^2$, it is clear that this metal is abundant in the earth and has remarkable catalytic activity. It has excellent thermal and electrical conductivity because of its unique electronic structure. For instance, Chu et al. (2019) created a combination of NiCu compounds for HER in a basic media. Based on the electrochemical reactions, Ni_2–xCux-P/NF are found to be effective electrocatalysts in HERs. Additionally, altering the transition-metal stoichiometry can enhance HER performance (Chu et al., 2019). In addition, since alkaline conditions contain a high concentration of OH^-, four important parameters that affect OER and HER catalyst performance in an alkaline atmosphere are water molecule adsorption on the surface, hydroxide molecule adsorption, H_2 or oxygen binding energy, and water molecule splitting.

17.2.2 Nobel metal-based catalysts

The majority of rare metals have extremely low binding energies for both OER and HER, and very low-energy inputs to initiate electrolysis. Pt-based compounds, exhibit good HER activity and function selectively to produce H_2. Pt-based catalysts, on the other hand, cannot function as well in an acidic environment as they do in an alkaline one because their exchange current density is very high in an acidic environment and abnormally low in an alkaline one, which leads to low EWS activity. Furthermore, Pt-based catalysts with nanoparticle (NP) sizes—especially sizes of a few nm to μm—are prone to agglomerating and coarsening during electrocatalytic processes. Pt-based catalysts also confront the issue of surface poisoning from chemically adsorbed intermediates generated during catalysis. To solve these limitations, Battiato et al. (2022) demonstrated that adding Pt to the Ni-P structure might improve its stability and the efficacy of HERs and OERs in basic conditions. They prepared Pt/Ni-P/Ni-foam samples in two stages: first, the Ni-P alloy was electrolessly deposited, and secondly, Pt was added to the framework by dip-coating. The integration of Pt atoms to the structure resulted in a strong synergy effect. When Pt atoms were added to the Ni-P structure, the HO-H bonds were equally distributed over the catalytic surface, preventing Pt active site poisoning and improving the overall structural conductivity (Battiato et al., 2022).

Another rare metal widely used as an electrocatalyst for OER and HER in acidic/basic medium is ruthenium (Ru). However, industrial use of the HER or OER process has been hindered by its scarcity, higher market value, and minimal reactivity in an alkaline state. Ru-based catalysts have binding energies that are comparable to Pt's, but Ru appears to be able to split water molecules more easily than Pt, and it is also less expensive on the market. Fan et al. (2022) created Ru nanoclusters that are anchored to N and B doped carbon nanotubes and examined how well they work in an alkaline electrolyte for both OER and HER. The HER efficiency of produced catalyst demonstrated that the Ru incorporation into the N, B codoped carbon structure led to a good performance in HER (Fan et al., 2022).

Palladium (Pd) serves as an active catalyst for EWS. Pd catalysts have garnered considerable attention owing to their increased catalytic performance and relatively reduced cost as compared to Pt, Ir, and Ru catalysts. Pd catalysts offer numerous benefits, including favorable thermodynamic energy for adsorption of OH^- or H^+ and lower binding energy for the adsorption/desorption of H-intermediates. Recently, Bhowmik et al. (2016) created a Pd/CNx based catalyst by reducing $PdCl_2$ which was then composited using CNx nanolayers. In an acidic environment, the as-prepared catalysts demonstrated strong performance against HER, exhibiting reduced values of onset potential and a lower Tafel slope, that is, 12 mV and 35 mV/dec, respectively. In an acidic environment, poor activity is considered for the values of initial potential > 60 mV and a Tafel slope > 110 mV/dec. It has been discovered that raising pH of the solution raises the binding energy of H_2, thereby lowering HER activity (Bhowmik et al., 2016).

17.2.3 Need for earth-abundant and sustainable catalyst alternatives

Apart from EWS materials connected to precious metals, it was previously believed that other earth-abundant catalytic materials (e.g., Cu, Fe, Co, etc.) were mostly utilized in the EWS process's basic solution. Several approaches have been developed for improving the EWS process and the catalytic efficiency of materials pertaining to TMs. Among the many known ways to improve TM performance, one of the most well-known is employing nonmetal dopants, which increase the catalytic activity of organic molecules. Nonmetal dopants such as S, N, and P can produce electronic structural modulation and vacancy creation in a crystalline structure because of imperfect doping. This, in turn, enhances the catalysis. For such cases, Li et al. revealed a simple, repeatable, and adjustable intercalation method for creating P-doped MoO_3 nanosheets, which have shown remarkable results for HER because of the synergistic effects of the P dopants and the vacancies that are created in the structure (Li et al., 2017).

17.2.4 Importance of nanostructured materials in addressing catalyst challenges

Nanotechnology, which offers engineered nanostructures with tremendous potential for producing things with noticeably higher performance, is a classic example of an emerging technology. Particularly when it comes to energy conversion and environmental applications, nanostructured materials are essential for tackling several catalyst-related issues. Several benefits that come with special qualities of nanostructures can greatly improve catalytic performance. Materials with structural components, molecules, crystallites, or clusters with sizes between 1 and 100 nm are referred to as nanostructured materials. Currently, nanomaterials are employed in electronics, cosmetics, athletic gear, sensors, energy-storage devices, nonscratch paints, and surface coatings on a commercial scale (Baig et al., 2021).

17.3 Synthesis and properties of nanostructured materials

Nanostructured materials are man-made materials with a deliberate arrangement of grains or phases on a scale smaller than 100 nm. They are manufactured utilizing a diverse array of mechanical, chemical, and physical methodologies. These materials can be created with several levels of complexity: one-dimensional (multilayer), 2-D (ultrafine-grained overlayers), 3-D (nanophase materials), or zero-dimensional (atom clusters, filaments, or tubules). All nanostructured materials shared three basic characteristics: they consisted of atomic domains (such as grains, layers, or phases) that were confined to a spatial scale of less than 100 nm.

17.3.1 Different approaches for synthesizing nanostructured catalysts

Different techniques are used in the synthesis of nanostructured catalysts to regulate the shape, size, and composition of nanomaterials, all of which affect their catalytic features. The following are a few typical methods for creating nanostructured catalysts. The synthetic methods of nanomaterials are represented in Fig. 17.2.

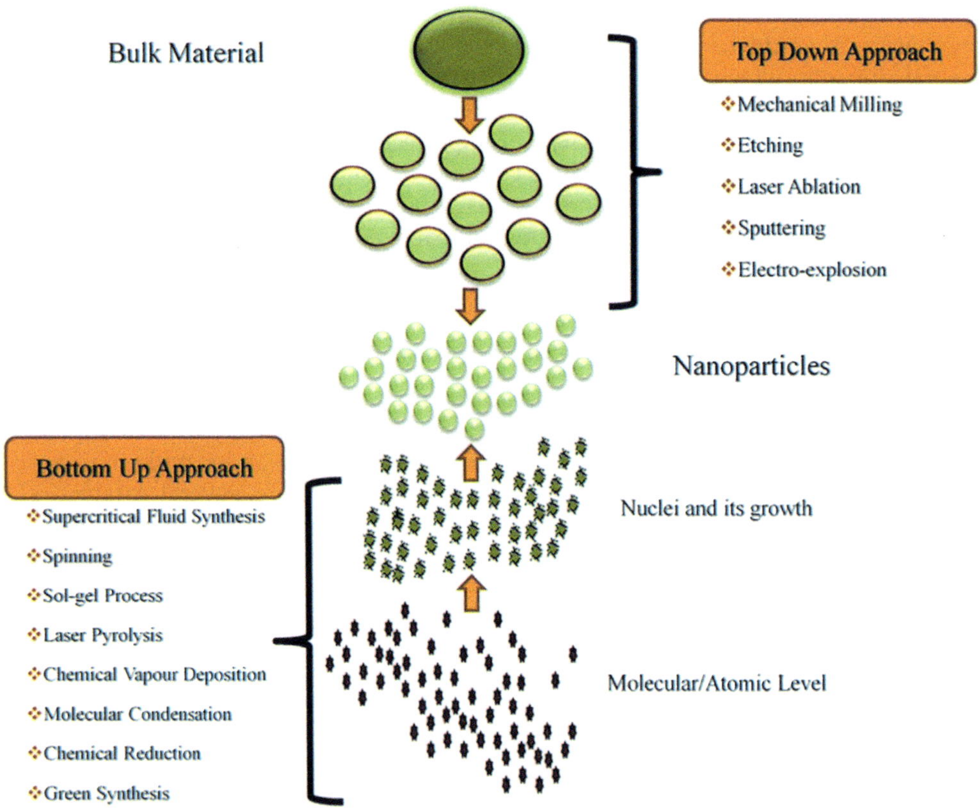

FIGURE 17.2 Synthesis approaches for nanomaterials (Khanna et al., 2019). Source: *Adapted with permission from Khanna, P., Kaur, A., Goyal, D., 2019. Algae-based metallic nanoparticles: synthesis, characterization and applications. Journal of Microbiological Methods 163, 105656. https://doi.org/10.1016/j.mimet.2019.105656, Copyright: ©2019, Elsevier B.V.*

17.3.1.1 Top-down synthesis approaches

Nanostructured materials are synthesized by employing top-down processes, which include the division of bulk materials. Mechanical milling, electro-explosion, sputtering, laser ablation, and etching, are all categorized as top-down procedures (Mukherji et al., 2019). Mechanical milling is an affordable technique of converting bulk materials into nanoscale materials. It works well for creating mixtures of various phases and is beneficial when creating nanocomposites (NCs). It is used to create Al/Ni/Mg/Cu-based nanoalloys, wear-resistant spray coatings, and a variety of other NCs materials (Yadav et al., 2012). Electrospinning is another simple method to create nanostructured materials. This is often utilized for producing nanofibers using diverse materials, predominantly polymers (Ostermann et al., 2011). To create nanoarchitectures a focused ray or electron beam can be utilized by lithography. Sputtering is considered a valuable approach for the production of thin films of NMs. It entails exposing solid surfaces to high-energy plasma/gas, to generate NMs (Baig et al., 2021). The arc discharge process is used to create carbon-based compounds such as few-layer graphene, carbon nanohorns, carbon nanotubes, fullerenes, and amorphous spherical carbon NPs (Zhang et al., 2019). NPs are produced via laser ablation synthesis, which uses a powerful laser beam to impact the desired material. The precursor or source material vaporizes during this process owing to irradiation with laser light of high intensity for producing NPs. Since it produces noble metal NPs without the need for extra chemicals or stabilizing agents, this process can be considered ecologically friendly (Amendola and Meneghetti, 2009).

17.3.1.2 Bottom-up synthesis approaches

The term "bottom-up synthesis" explains the procedure for creating complex structures or materials from simpler ones, starting at the atomic or molecular level. It entails putting together substances or molecules from their component atoms or smaller units. Nanomaterials with desired characteristics are frequently synthesized using this approach. They use molecules' intrinsic properties to form the structures you want. These approaches include Chemical vapor deposition (CVD), hydrothermal and solvothermal process (Thakur and Bharti, 2024), soft and hard template methods, sol-gel method (Thakur and Bharti, 2023), reverse micelle method. CVD methods are crucial to produce nanomedicines based on carbon fibers. A thin coating is created on the surface of the substrate by chemical reactions between vapor-phase precursors in CVD. It is deemed appropriate for CVD if it is highly volatile, nonhazardous, has a long shelf life, is stable during evaporation, is inexpensive, and has high chemical purity (Jones and Hitchman, 2008). The hydrothermal technique is widely used to create nanostructured materials. The hydrothermal technique creates nanostructured materials by the use of a heterogeneous reaction that takes place in a sealed vessel near the critical point of heat and pressure in an aqueous solution. The difference is that it occurs in a medium that is not water-based. Closed systems make use of both methods. The sol-gel method is a well-known wet chemical approach for developing NMs; it is used to synthesize several types of superior NMs based on metal oxides. The liquid precursor transforms a sol, and then that sol solidifies into a gel (Danks et al., 2016). Nanoporous material production heavily employs hard and soft template techniques. The soft template approach is an easy and classic way to make materials with nanostructures. Many people think the soft template approach is great since it allows for the creation of materials with different morphologies, is easy to use, and has relatively mild testing conditions. Using a well-designed solid material as a template, the hard template technique (nano-casting) creates nanostructures for the required applications. Filling the pores of the solid template with precursor molecules is the next step (Baig et al., 2021). The reverse micelle approach is another useful tool for making NMs of the right size and form. Micelles, in their standard form, are oil-in-water emulsions that include hydrophobic tails directing oil droplets towards a core (Malik et al., 2012).

17.3.2 Tailoring properties through size, shape, and composition

The functionality of nanomaterials is significantly influenced by their chemical composition and structure. Oxidation process, chemical potential, surface energy, and catalysis are some of the chemical properties that greatly impact NM composition. The physical properties of NMs are determined by the particles' color, morphology, shape, and size. The morphology, lattice parameter, size effect, and crystalline structure of NMs are outcomes of the interplay between particle size, shape, and arrangement. Light absorption, reflection, emission, and transmission are all optical properties of NMs that are dictated by their electronic structure, which is subsequently influenced by the surface atoms. When particle size is reduced to less than 10 nm, optical characteristics are employed for observing the size effect. High speed plasticity and high strength conditions are the results of mechanical behavior of NMs. Grain size, material structure, porosity hardness, adhesion, friction, and elastic modulus are all important aspects of NMs mechanical properties (Abid et al., 2022). In nanotechnology research, achieving exact control on the size, form, and functional features is crucial. Factors like synthesis techniques, raw materials, and processing parameters affect the controllability of NP and nanostructure composition, size, shape, and surface functions. Therefore getting NMs with desired properties and attaining targeted applications requires careful control over reaction circumstances and processing parameters as well as the selection of the best synthesis method (Harish et al., 2023).

17.3.3 Characterization methods for nanostructured materials

Nanomaterials possess several unique physicochemical qualities, that is, size, shape, composition, surface features, molecular weight, solubility, and sustainability. These characteristics are the main indicators of physiological performance. There are various methods for characterizing the NP used in energy devices. Common methods for evaluating NMs include zeta potential testing, thermal procedures, chromatography, electrophoresis, optical spectroscopy, surface scanning, electron microscopy, light scattering, X-ray scattering, mass spectrometry, magnetic resonance, and thermal techniques (Bharti et al., 2018; Mukherji et al., 2019). Nanomaterials are usually described by their size, surface charge, and form. The morphology of NMs is being studied using a variety of advanced microscopic techniques, that is, field emission gun-scanning electron microscopy, scanning tunneling microscopy, transmission electron microscopy (TEM), and scanning probe microscopy. Structure, shape, size, aggregation and dispersion are some of the NM properties examined by TEM and atomic force microscopy (AFM). Crystalline materials can have their phases identified using X-ray diffraction (XRD). Infrared spectroscopy is a crucial and significant characterization tool to comprehend molecular structure, chemical structure, and bonding patterns of a material (Salame et al., 2018). Raman spectroscopy, which makes use of Raman scattering, can detect vibrations at the molecular level, allowing for the identification of molecules. The material can be better identified with the use of data collected on its physical and molecular structures. Characteristics such as size distribution, size, and stability can be investigated utilizing zeta potential and particle size distribution (Bharti et al., 2020, 2021).

17.3.4 Factors affecting photocatalytic and photoelectrochemical water-splitting efficiency

The photoactivity of photocatalysts is primarily affected by the production and electrons and holes recombination along with other factors, such as band gap energy, light intensity, temperature effects, pH, and so on, as shown in Fig. 17.3 (Bharti et al., 2018). The orderliness, morphology and homogeneity of PEC water-splitting systems determine their overall performance and efficiency. Effective water-splitting process also depends on control over structure and shape, along with size reduction.

17.3.4.1 Band gap

The valence and conduction band (VB and CB, respectively) represent difference in energy level and are used to explain the electrical configuration of semiconductor-based photocatalysts. For thermodynamic considerations, the relative potential level of the acceptor must generally be lower than the semiconductor's CB. Furthermore, the conversion of H^+ ions to H_2 gas is favorably impacted by a larger negative value of the CB position (Fajrina and Tahir, 2019). A narrowing of the band gap is said to upgrade PEC features. This facilitates the wide spectrum of solar radiation absorption. A large band gap hinders effective water-splitting, but a 1.6–2.2 eV is favorable for efficient PEC splitting of water. The band gap may be lowered in a semiconductor by incorporating a donor-acceptor (Bharti et al., 2018).

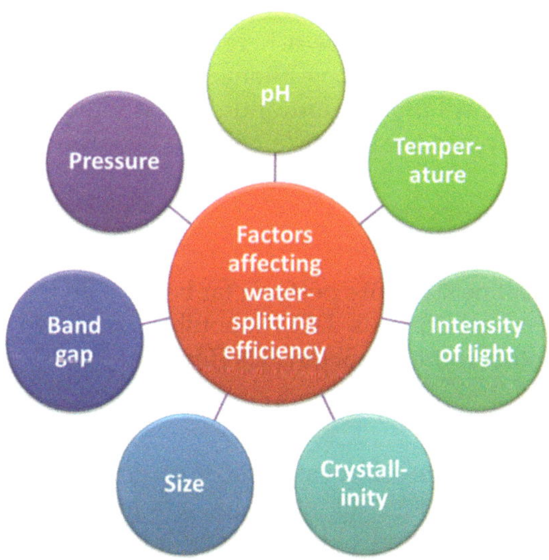

FIGURE 17.3 Representation of factors affecting water-splitting efficiency.

17.3.4.2 *Intensity of light*

When light with enough energy focuses on a surface, it can make the process of splitting water using light more efficient. There are two systems for splitting water using light, and they are determined by the amount of UV light used. In the first system, smaller amounts of light are used, and the chemical reactions make use of electron-hole pairs more quickly. In the second system, higher intensities of light are used, and the recombination reactions take over. The wavelength of light that the catalyst absorbs is important because it determines how fast the reaction happens during the water-splitting process, matching the band gap of the photocatalyst (Fajrina and Tahir, 2019). The source of light, its spectral distribution, and the intensity of the AM 1.5 G spectrum (whose integrated power is 1000 W/m^2) should all be provided. When semiconductors are doped with NPs, both visible and UV light can be absorbed (Alexander, 2016).

17.3.4.3 *Effect of temperature*

Thermodynamically, because temperature cannot produce electrons and holes in catalysts, it did not induce photocatalytic activity. Generally, temperature has the impact of promoting the H_2 release from photocatalyst surfaces, which raises the activity of photocatalytic water-splitting. Additionally, the temperature speeds up the reaction rate. An important factor in raising PEC efficiency is temperature. Experiments are typically carried out under hot conditions. A study has demonstrated that PEC efficiency can be increased using low temperature treatments. In lower temperature, the PEC efficiency increased by up to 95% (Nie et al., 2017).

17.3.4.4 *pH*

The rate of H_2 production from H_2O splitting is significantly affected by pH, that is, the concentration of H^+ ions in the solution. Optimal H_2 production occurs in a weakly basic solution as opposed to a stronger alkaline environment ($pH > 10$) or an acidic medium. Sinha et al. have fine-tuned the solution's pH to 8.6 to achieve a heightened rate of H_2 production in sacrificial agent-mediated photocatalytic reactions (Sinha, 2022). PEC cells operate at diverse pH values. The pH of electrolytic solution significantly affects the water-splitting reaction's equilibrium by determining the net total surface adsorbed charge (Joy et al., 2018).

17.3.4.5 *Crystallinity*

Highly organized materials possessing a crystalline structure perform far better than their amorphous materials. For instance, compared to amorphous TiO_2 nanotubes, annealed/crystalline TiO_2 nanotubes exhibit superior photocurrent characteristics (Joy et al., 2018).

17.3.4.6 *Size*

The PEC performance can be enhanced by the size of cocatalysts. Creating a well-designed catalyst is one way to increase PEC efficiency. Electrokinetics has a dominant role in smaller particles by increasing electron-hole recombination. Due to the larger particle size, band bending may occur, thereby facilitating improved extraction of charge at the electrode and electrolyte interface. Hence, larger cocatalysts is found suitable to improve the performance of PEC (Pokrant et al., 2017). Further considerations influencing the effectiveness of PEC water-splitting encompass adsorption and desorption kinetics (dependent on their pore structure), their grain shape and size (dependent on preparation strategy), lattice impurity, grain nucleation (based on doping), and the elucidation of specific details at the semiconductor/metal interfacial (Gupta, 2017).

17.4 Metal-based nanostructured materials for water-splitting

Water-splitting catalysts made of metal-based nanostructured materials are gaining popularity. These compounds speed up electrocatalysis and photocatalysis which divide water into H_2 and oxygen gases. Photocatalytic, PEC, and photovoltaic–electrolysis systems can divide water to produce H_2 from sunlight. We want photocatalysts with high solar light absorption, stability, and charge transfer. However, electrocatalysts are extensively used as cocatalysts to boost efficiency (Chen et al., 2016). Nanoscale catalyst design is significant because of its high catalytic activity, which improves reaction outcomes (Maduraiveeran et al., 2019). Nanoscale materials include rich surface textural patterns, many binding sites, extremely tiny sizes, and huge surface areas, which foster heterogeneous transportation thermodynamics and kinetics. Metal-based nanostructured materials can be prepared in many ways. Sol-gel, solvothermal, hydrothermal, electroless, physical, and electrochemical techniques can build metal-based nanostructures (Chen and Holt-Hindle, 2010).

17.4.1 Metal nanoparticles as efficient hydrogen evolution reaction catalysts

MNPs are promising catalysts for HER, which is an important phase in water-splitting to create H_2 gas. Protons are reduced to create molecular H_2 in the HER. Pt group metals like Pt, Ir, Rh, Pd and Rd catalyze HER better than other metals. According to Sabatier principle, reasonable M-H bond strength of Pt allows for optimum H_2 adsorption and desorption at its surface, making it an excellent HER catalyst. The catalysis at HER employing Pt nanomaterials as an electrocatalyst depends on their shape, size, and face. Overall, NP size determines catalytic activity with a decrease in the specific mass activity of Pt with a similar size range (Zhu et al., 2020). Table 17.1 and H_2 evolution for some of the nanomaterials is listed in Table 17.2. In 2022 Hua et al. proposed an electrode $Al_3/Ni_2/Ni$ to catalyze HER with excellent HER activity as compared to Pt including 83.4 mV lower overpotential and long-term stability over 36 hours in H_2SO_4 solution (0.5 M) (Hua et al., 2022).

17.4.2 Metal-oxide nanoparticles for improved oxygen evolution reaction performance

Several strategies for increasing the catalytic efficiency of metal oxides electrocatalysts have been identified. Doping with a specific foreign element is one way to enhance electronic properties and catalytic activity; nevertheless, this approach can bring about major flaws and surface conditions into catalysts, which can reduce their overall effectiveness. Several factors influence the efficiency with which MONPs catalyze the OER, including

TABLE 17.1 Electrocatalysts employed in different electrolytes with different performance.

Electrocatalyst	Synthesis method	Stability (mA/cm^2)	Application	Tafel slope (mV/dec)	References
m-WO_3 h-WO_3	Hydrothermal method, thermal treatment	–	HER	35	Yang et al. (2021)
$NiWO_4/WO_3$ fibers	Electrospinning	–	HER	50.27	Anis et al. (2017)
			OER	41.29	
Pt–Ni_3N	–	40 h (at 500 mA/cm^2)	HER	36.5	Wang et al. (2017)
Ru@CN	Simple copyrolysis of GAH (D-glucosamine hydrochloride)/melamine/$RuCl_3$ mixtures.	2000 cycles	HER	63	Wang et al. (2018)
PdP_2 @CB	–	10 h (at 27.6 mA/cm^2)	HER	29.5	Luo et al. (2018)
			OER	78.6	
Ni_3FeN nanoparticles	Thermal ammonolysis	9 h (at 10 mA/cm^2)	OER	46	Jia et al. (2016)
Au@Ni_2P	–	20 h (at 15 mA/cm^2)	OER	70	Cai et al. (2019)
Al_3Ni_2/Ni	One-step electrochemical reduction	36 h (at 10 mA/cm^2)	HER	40.7	Hua et al. (2022)
Fe/Ni doped graphene oxide @ $Ti_3C_2T_x$	Hummer's method, etching technique	50 cycles	OER	70	Raj et al. (2022)
			HER	110	
Co_2P/Ni_2P	–	40 h (at 10 mA/cm^2)	OER	69.9	Zhao et al. (2022)
			HER	32.3	

OER, Oxygen evolution reaction; *HER*, hydrogen evolution reactions.

surface form, crystal structure, and composition. The OER performance of CeO_2-RuO_2 and mixed CeO_2-RuO_2-IrO_2 was better than that of pristine RuO_2 and IrO_2. Due to its unique structural characteristics, cubic fluorite CeO_2 is a promising candidate for a variety of catalytic uses (Albu et al., 2021). The OER active sites corresponded to transition metals with oxidation levels of +2/+3/+4, making them the best OER electrocatalysts due to their diverse oxidation states. Extensive OER studies have been conducted on transition-metal oxides of the first row as well as mixed oxides of nickel, cobalt, manganese, and iron, and these materials have shown outstanding OER activity (Vazhayil et al., 2021). Some metal-oxide based NPs for improved OER are listed in Table 17.2.

17.4.3 Bimetallic and alloy nanostructures for enhanced activity

Numerous techniques have been utilized to increase overall catalytic activity; however, heteroatom doping and nanostructure engineering morphological control are two of the more feasible approaches (Li et al., 2019). In comparison to catalysts made of a single metal, bi-metal alloys engineered with nanostructures show much higher catalytic activity (Zhou et al., 2016). One such example is the electrocatalyst for OER and HER that Ghouri et al. (2021) developed and evaluated. It consisted of carbon nanowires and Co-Cu alloy NPs. The NMs were synthesized by electrospinning a sol-gel containing copper acetate, cobalt acetate, and polyvinyl alcohol. The mixture was subsequently calcined in an inert atmosphere. It was found that when the composition of metallic counterparts was varied, the electrochemical characteristics of the catalyst could be significantly enhanced.

Also, in an alkaline environment, the (CoCu@CNWs) show strong electrocatalytic activity for O_2 and H_2 evolution reaction due to unique nanowire structural features and morphology of the inserted Co-Cu alloy. The optimal composition involves a small overpotential η_{10} of ~285 mV (OER) and ~160 mV (HER), with a corresponding Tafel slope of 92 mV/dec and 172 mV/dec in comparison to the reversible H_2 electrode. Furthermore, in a two-electrode system, there was minimal activity loss after a thousand cycles, and 1.58 V cell voltage at 10 mA/cm^2 (1.72 V) at 50 mA/cm^2 (Ghouri et al., 2021). Janani et al. (2020) investigated the

ABLE 17.2 Metal-oxide based nanoparticles for improved OER performance.

Metal-oxide electrocatalysts for OER	Overpotential value (mV)	References
$Fe_xNi_{1-x}O$	297	Fominykh et al. (2015)
MnO_x film	–	Huynh et al. (2015)
α-Fe_2O_3 and γ-Fe_2O_3	275	Xu et al. (2020)
Fe_2O_3@CuO NTs/CF	398	Gao et al. (2020)
ZnO-CoO_x	276	Xiong et al. (2016)
Co–CoO/Rh	277	Tran et al. (2021)
Fe-CrOOH/Fe_3O_4/NF	198	Li et al. (2022)

OER, Oxygen evolution reaction.

water electrolysis capabilities of bimetallic oxide, selenide, and sulfide nanostructures such as spinel-type, phase-pure $NiCo_2O_4$, $NiCo_2S_4$, and $NiCo_2Se_4$ in terms of their bifunctional activity (Janani et al., 2020).

17.5 2D materials and heterostructures

The global energy challenges and ecological concerns have created need for alternate energy sources. Electrochemical complete water-splitting is a highly promising approach for transforming power from renewable energy sources into H_2 and O_2, which can be easily transported and stored. For commercial water-splitting to be successful, highly effective and affordable electrocatalysts are needed to produce H_2 and O_2. 2D electrocatalysts are widely used in water-splitting applications due to their increased activity, high surface area, stability, and their potential for structural and functional customization. The electrocatalytic activity is improved by the combination of efficient mass transport and rapid e^- transfer occurring at the surface flaws of these structures. The water-splitting method for H_2 production focuses on utilizing 2D materials such as polymers, oxides, carbons, MXenes (nitrides, carbides, and carbonitrides), metal-organic frameworks, and chalcogenides (Ertekin and Ozer, 2022).

The 2D materials, known as transition-metal dichalcogenides (TMDCs) have unique characteristics that make them excellent photocatalytic tools. TMDCs were found relatively recently. They are commonly used as cocatalysts with other semiconductor materials because the charge transfer from photogenerated electrons and holes is made possible by the junction that forms between them. Chalcogen elements (Se, S, or Te) and transition metals (Mo, Zr, W, Ti, Nb, Hf, etc.) make up the standard formula for TMDCs, which is MX_2. The MX_2 general formula describes TMDCs that are abundant on Earth. These include pyrite phase organized TMDCs (e.g., $CoSe_2$, CoS_2, NiS_2, FeS_2, etc.) and 2D-layered materials (e.g., WS_2, MoS_2, $MoSe_2$, TaS_2, WSe_2, TiS_2, etc.). The preparation of materials for water-splitting is progressing at a rapid pace. The atomically thin, 2D-layered MX_2 nanosheets exhibit distinct physical, electrical, and chemical properties that set them apart from their bulk equivalents. As a rule, these structures have X, M, and X atoms linked together in three layers. Furthermore, nearby MX_2 layers are linked by weak van der Waals interactions, which results in the formation of enormous crystals (Wang et al., 2015).

Since graphene's discovery, molybdenum-based dichalcogenides have been the most talked about TMD-based nanomaterials due to their potential in HER catalysis (Wang et al., 2015). The Lewis et al. research is also commendable in terms of uncovering active spots. By employing an such methodology, a porous $MoSe_2$ layer was directly synthesized on a glassy carbon electrode. HER performance was also superior. To further improve HER performance, it is usual practice to increase MoS_2 conductivity either through atomic level enhancement or by combining it with other conductive materials. The performance of HERs can be improved with hybrid constructions based on molybdenum dichalcogenides by increasing their conductivity. Among the single-component catalysts, MoO_3/MoS_2 core-shell nanowires, MoS_2/3D-CFs, MoS_2/CNT-graphene, MoS_2/N-doped carbon nanotube, MoS_2/Au, MoS_2/rGO, $MoSe_2$/graphene, $MoSe_2$/ SnO_2, and $MoSe_2$/rGO demonstrated the highest level of electrocatalytic activity in the HER.

Poor OER stability and unsatisfactory HER activity are barriers to its practical implementation. Here, we developed a plan to build a Co and Mo_2C heterostructure supported on carbon (Co/Mo_2C@C) for increasing the electrocatalytic performance of Mo_2C. The synergetic effects of Co and Mo_2C led to significant improvements, as demonstrated by experimental and theoretical calculations. Electrons from Co can move to Mo_2C and build up on Mo. This enhanced the HER by promoting the adsorption of H_2O and the desorption of H. By way of a "self-sacrifice" effect, on the other hand, the Co may prevent the oxidation and dissolution of Mo_2C, greatly enhancing the OER stability of Mo_2C. The catalyst that was created had exceptional electrocatalytic activity in an alkaline electrolyte, thanks to the synergetic effects of Co and Mo_2C. It also provided low overpotentials of 254 mV and 98 mV at 10 mA/cm^2 for catalyzing the OER and HER, respectively (Yuan et al., 2022).

TMD-based nanomaterials are utilized to augment the OER in water-splitting process. OER, an essential process in water-splitting, entails intricate electron transfer reactions characterized by sluggish kinetics caused by the presence of energy barriers. Consequently, a significant overpotential (η) is required to drive the evolution of oxygen. Important characteristics of OER activity encompass the enthalpy change during the conversion from a lower to a higher oxide, the number of 3d e^- (electrons) in transition-metal ions in bulk, the energy required for oxygen to bind to the surface, and the utilization of Yang Shao-Horn's principle, which highlights the nearly complete occupation of surface transition-metal cations. Recent progress in cobalt-based electrocatalysts includes a wide range of materials, which includes basic metal oxides, metals, hydro(oxy)oxides, phosphates, chalcogenides, and perovskites.

17.5.1 Challenges and opportunities in utilizing 2D materials

Planar 2D nanomaterials may have few catalytic active sites. Effective exposure to active areas is difficult. 2D nanomaterials degrade over time, especially in harsh reaction conditions. For practical application, these materials must be sturdier and more robust. Functional integration of 2D nanomaterials with conventional water-splitting system components like electrodes and catalyst supports is problematic. Synthesizing high-quality 2D nanomaterials and increasing output while preserving regularity is tough. It is necessary to develop techniques for production on a wide scale. Synergistic effects from 2D NPs and other nanomaterials or supporting elements boost catalytic performance. PEC water-splitting systems can use graphene and TMDs, 2D materials with great light absorption efficiency. 2D nanomaterials have many catalytic active sites due to their enormous surface areas. This attribute could boost water-splitting efficiency. Some 2-D NMs can act as bifunctional catalysts, addressing the OER and HER. Surface functionalization of 2D nanomaterials introduces selective catalytic sites, improving water-splitting selectivity and activity.

However, the potential toxicity of these substances presents risks to both human and environmental well-being, thus requiring cautious management and proper disposal protocols. Although complex approaches like heteroepitaxy and self-sacrificial templates have been employed, a universally applicable and uncomplicated method for producing 2D nanosheets has not yet been established. 2D nanosheets provide a solution to the problem of using only one semiconductor for a high-quality photoelectrode in PEC water-splitting. They offer the advantage of being able to use different materials, and the presence of nanoscale heterojunctions improves the performance of PEC devices. The rational design of 2D nanosheets at the nanoscale offers the potential for enhancing artificial photosynthesis by creating junctions, modifying surface catalysis, adding charge collectors, burying junctions, and including light absorbers. Incorporating hybrid materials into 2D nanosheets offers the potential for substantial performance improvements in PEC systems (Zhou et al., 2013).

17.6 Nanostructured materials for photoelectrochemical water-splitting

Water-splitting process is a highly intriguing method for extracting H_2. Several methodologies, including PEC, photocatalytic, photobiological, radiolysis, and thermal breakdown, have been documented for water-splitting. Radiolysis generates nuclear waste as a secondary outcome. Conventional photosynthesis is transformed into photobiological water-splitting using an algae bioreactor, which efficiently produces high levels of O_2 and low levels of H_2. The primary drawbacks of thermal decomposition are the elevated temperature threshold and the restricted output of H_2. PEC and photocatalytic water-splitting are the most efficient, economical, and eco-friendly methods for producing H_2. Fujishima and Honda developed the PEC water-splitting technique, utilizing a semiconducting material that demonstrates both high efficiency and low cost. PEC water-splitting is based on the principle of turning solar energy into H_2 gas employing photovoltaic materials containing a redox pair in an electrolyte. One of the redox pair components is a semiconductor material that can absorb light on applying an external bias. The power is utilized to perform water electrolysis. One notable characteristic of semiconductors is their capacity to function as photocatalysts. Therefore it is important to initiate the chemical process of oxidation and reduction on light irradiation. These photoelectrodes can gather light, hence providing energy to the operations. An externally supplied electrical or chemical bias is used to supply the additional voltage needed to conduct a reaction. Joy et al. (2018) found several essential parameters affecting the performance of PEC water-splitting significantly, which include efficiency of absorbing light (η A), efficiency of separating charges (ηCS), efficiency of transporting charges (ηCT), and efficiency of collecting/reacting charges (ηCR) (Joy et al., 2018). For enhanced photocatalytic activity, the NPs can be suspended or disseminated in the reaction media (water) or coated on the electrode. The MNPs were concentrated on the interface between semiconductor and electrolyte, where water-splitting efficiency was found to increase. The minority charge transmission from the electrode to the electrolyte is the cause of this.

The most popular material for investigations on H_2 production and water spitting is TiO_2. Although bulk TiO_2 is stable and inexpensive, its band gap of 3.03–3.18 eV results in decreased efficiency in absorption of sunlight. This could be useful for water-splitting. Among TiO_2NMs, TiO_2 nanotube is the better choice. Low quantum efficiency is caused by faster electron-hole recombination that takes place in bulk TiO_2. Materials with narrow band gaps were also coated with TiO_2 to lessen this effect. Black TiO_2 NPs, a substitute for other materials used to decorate TiO_2, were created and exhibit more effective photocatalytic activities (Isimjan et al., 2012). Optimizing the size and form of nanoscale TiO_2 can substantially increase the transformation efficacy (Joy et al., 2018).

Photoactive materials, particularly thin films made of semiconductors, have a crucial role to play in PEC based water-splitting. The photoexcitation of semiconductor films achieves the fulfillment of 1.23 V redox potential gap through their interactions with light and matter. In such case, the charge carriers need to have a sufficient overpotential to enable the reactions at both the photoanode and photocathode. In controlled experimental conditions, the semiconductor can generate a photovoltage of approximately 400 mV lower than its band gap. Therefore to effectively harness solar energy for the facilitation of the OER and HER, a semiconductor material must possess a band gap that falls within the ideal range of 1.6–2.6 (eV) (Kumar et al., 2022). The band-edge position of most n-type semiconductors, including TiO_2, CdS, $SrTiO_3$, g-C_3N_4, and ZrO_2, is highly favorable for the OER. P-type semiconductors with narrow band gaps, such as Cu_2O, Cu_2S, Sb_2Se_3, p-Si, and $CuInS_2$, possess CB edges that are well-suited for photocathodes. Additionally, their band gaps are ideal for efficient absorption of light. In contrast, n-type semiconductors such as, $BiVO_4$, WO_3, and Fe_2O_3 possess favorable visible light absorption properties and viable VB edges for the OER.

17.6.1 Mechanism of photoelectrochemical water-splitting

Patil et al. (2021) utilized a self-template sequential solvothermal and cation-exchange technique with inorganic-organic hybrid, that is, $ZnS(en)_{0.5}$ to produce a unique ZnO NR-$Zn_{1-x}Cd_xS$-PNS dual nanostructure on zinc foil, marking the first instance of its creation (Patil et al., 2021). However, the permeable structure is crucial for absorbing solar radiation and reducing charge recombination resistances. The data shows a rate of H_2 production as 250.65 μmol/cm^2 and a photocurrent density of 6.60 mA/cm^2 for 3 hours. This explains the mechanism of charge production and their transer at the ZnO NR-$Zn_{1-x}Cd_xS$-PNS 160 C photoanode process during H_2 production using the PEC process. When the photoanodes are exposed to light, the electrons that are generated can follow two different paths: (1) the e^- in CB of ZnO-NR and $Zn_{1-x}Cd_xS$-PNS move directly towards the zinc substrate, and (2) the e^- in the CB of NR-$Zn_{1-x}Cd_xS$-PNS transfer to CB of ZnO-NRs because of the electric field present in the space-charge region in ZnO NR-$Zn_{1-x}Cd_xS$-PNS heterojunctions. The electron produced by the photochemical reaction is subsequently sent to the Pt electrode through the external circuit, where it undergoes reduction to become H_2 gas ($H_2\uparrow$), indicating a highly efficient pathway for electronic transportation through the porous $Zn_{1-x}Cd_xS$-PNS and ZnO-NR structure's ideal surface area. As another possibility, the photogenerated hole at the porous $Zn_{1-x}Cd_xS$-PNS nanosheet and ZnO-NR valance band further moves to the photoanode surface, where it interacts with the electrolyte. The reactions at the ZnO NR-$Zn_{1-x}Cd_xS$-PNS160C photoanode are represented in Fig. 17.4.

In another research, pure and Mo-doped CuO thin films were formed using the DC/RF sputtering method on Indium Tin Oxide to produce H_2 via PEC water-splitting process which showed photocurrent density of around 2.25 mA/cm^2 at room temperature. Mo–CuO photoelectrodes work exceptionally well for producing green H_2. A variety of activities, including electron-hole pair formation, photon irradiation, charge carrier movement and interfacial interactions were studied using the electrolyte (0.3 M). In short, when light contacts the semiconductor photocathode's surface, e^--h^+ pairs are created at the VB and instantly split. To finish the H_2 reduction process, the photogenerated e^- diffuse at the interface of photocathode and electrolyte to reach CB. Moreover, h^+ present in the VB lead

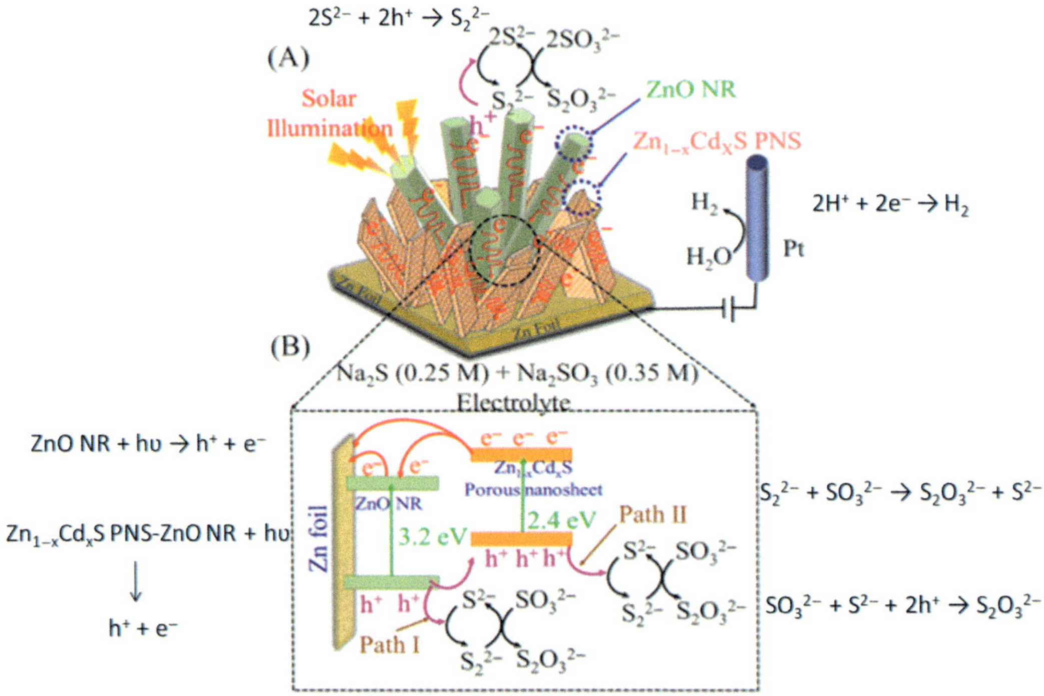

FIGURE 17.4 (A) Schematics of PEC H_2 production system and (B) Mechanism of charge transfer during PEC H_2 production over ZnO NR-$Zn_{1-x}Cd_xS$-PNS160C photoanode. *PEC*, Photoelectrochemical. Source: *Adapted with permission from Patil, R.P., et al., 2021. Self-templated fabrication of 2-D dual nanoarchitecture $Zn_{1-x}Cd_xS$ porous nanosheet and ZnO nanorod for photoelectrochemical hydrogen production, Applied Surface Science 539, 148267. https://doi.org/10.1016/j.apsusc.2020.148267.*

water oxidation process directly at the electrolyte/ counter electrode' interface. H_2 and O_2 bubbles within the PEC cell were assessed, which is taken into consideration as proof that the PEC water-splitting process was completed effectively. Schematics shown in Fig. 17.4 (Qutb et al., 2024).

17.6.2 Composite and hybrid nanostructured materials

PEC catalysis shows significant potential for sustained expansion. A synergistic effect of ZrO_2 catalyst and cobalt phosphate was demonstrated on anodized TiO_2 nanotubes (TNTs), leading to a substantial improvement in the PEC efficiency for water-splitting using visible light. The ZrO_2/CoPi composite was introduced into the TNTs (titanium nanotubes) by a sequential process involving electrodeposition and photoassisted electrodeposition techniques. Substituting Zr^{4+} with Ti^4 led to formation of O_2 vacancies, which enhanced e^- capture, thereby reducing recombination of e^- and h^+, and thus enhancing charge transfer efficacy. For the TNTs/ZrO_2/CoPi photoanode, the CoPi water oxidation catalyst (WOC) working as a hole-transfer relay improved the efficiency of water-splitting. The integration of ZrO_2/CoPi accelerates the surface reaction kinetics of TNTs, resulting in considerable improvements in charge transfer efficiency (ηCT = 90%), photocurrent density (0.86 mA/cm^2 at 1.23 voltage versus reversible hydrogen electrode (VRHE)), and durability. Furthermore, the mechanistic investigation demonstrated that the synthesized materials exhibited enhanced charge transfer and surface conductivity, as evidenced by the impedance measurements. The proposed method has significant promise for constructing electrodes from unconventional materials, hence facilitating the development of solar fuels (Shaddad et al., 2022). The researchers created an enhanced PEC water-splitting system. The structure comprises a two-dimensional array of TiO_2 nanosheets that are adorned with CoOx NPs and overlaid with SnS_2 absorbers. When subjected to simulated sunlight, the $TiO_2/SnS_2/CoO_x$ hybrid photoanodes exhibited a photoconversion efficiency that was 3.6 times greater than that of the naked TiO_2 nanosheet arrays photoanode. Additionally, the TiO_2/SnS_2 hybrid photoanodes had a photoconversion efficiency that was 2.0 times higher.

$CuO/NiO/ZrO_2$ (CNZr) mixed metal-oxide composites (MMOCs) were created using a simple hydrothermal method, and its EWS and photocatalytic H_2 generation were studied by Kannan et al. (2023). An average 31 nm crystallite size was achieved for CNZr (monoclinic, cubic, and tetragonal) structures. The H_2 production rate in the CNZr600 composite was 14.27 mmol/g/h. The CNZr600/NF in KOH (Potassium hydroxide) electrolyte exhibited low overpotentials of 218 and 372 mV for HER and 400 and 540 mV for OER at current densities of 10 and 50 mA/cm^2, respectively. This is due to its numerous catalytically active sites and efficient charge transfer capacities. The CNZr600/NF alkaline electrolyser requires a working voltage of 1.69 V to efficiently generate H_2. The CNZr/NF catalysts exhibited a low overpotential (η) of 218 mV/10 mA/cm^2 and a Tafel slope of 114.9 mV/dec. The MMOCs exhibited exceptional catalytic activity due to their structural characteristics, extensive electroactive surface areas, rapid electron transport kinetics, and robust chemo-electrical interaction among CuO, NiO, and ZrO_2.

In 2023 Praragusan et al used a simple hydrothermal technique and prepared nano-needle-like morphology of $BaTiO_3$-$CoFe_2O_4$ photocatalyst and characterized their optical, structural, morphological, and electrochemical characteristics for potential use in PEC water-splitting. They observed PEC behavior of 12 mA/cm^2 at an applied potential of 1.7 V (Parangusan et al., 2023).

In 2023 Ramesh et al. demonstrated the water-splitting capabilities of GaN nanocolumns (NCs) coated with reduced graphene oxide (rGO) and pristine on flexible niobium (Nb) metal foils using PEC approaches. In a 1 M oxalic acid electrolyte, the PEC measurement of pure GaN-NCs under standard (1 Sun) circumstances revealed an efficient photocatalysis showing photocurrent density of ~60 μA/cm^2 at 0.8 V (vs Ag/AgCl). For rGO/GaN-NCs, this value rises to ~110 μA/cm^2. The effective charge separation and transporting of photogenerated carriers is responsible for the efficient PEC properties of rGO/GaN-NCs. Moreover, hybrid and pristine GaN-NCs on flexible metal foil in an aqueous solution exhibited a quick and steady photoresponse, according to transient photocurrent measurement. The creation of this PEC device based on hybrid nitride nanostructure on flexible metal foils opens the door for making scalable PEC based systems for H_2 generation required in various applications (Ramesh et al., 2023).

Effective PEC over gold over anatase-sodium borohydride-titanium (Au@A-B-TiO_2) has been shown by Shi et al. (2018). In comparison to the state-of-the-art Au-TiO_2 photoelectrodes for PEC water-splitting, the photoelectrode shows a much higher 2.82 mA/cm^2 photocurrent density at 1.23 V versus reversible H_2 electrode (RHE) under AM 1.5G illumination. The synergetic interaction between Au NPs and A-B-TiO_2 under solar light for PEC water-splitting can be explained as numerous electronic transitions (e.g. VB to CB, VB to O_2 vacancy levels, or O_2 vacancy levels to CB) are caused by light irradiation of A-B-TiO_2. Concurrently, the Au NPs capture resonant photons and use the localized surface plasma resonance excitation to produce energetic hot electrons that move to the adjacent A-B-TiO_2 CB and O_2 vacancies states. A H_2O reduction process was carried out by the e^- that collected on the A-B-TiO_2 and were transported via an external circuit to the counter Pt photocathode (Shi et al., 2018).

17.7 Commercial prospects and technological challenges

The preparation of stable and efficient catalysts with modified nanostructures for specific water-splitting reactions (OER and HER) is a significant technological challenge. Finding and producing a large supply of ecologically acceptable materials with the right catalytic qualities for nanostructured catalysts is a recurring problem. Some of the problems that need to be fixed when incorporating nanostructured catalysts into water electrolysis systems include stability, compatibility, and overall system efficiency. One of the obstacles to the commercialization of nanostructured catalysts is the scaling up from the lab to industrial application. On a broader scale, it might be challenging to keep catalysts functioning and retaining their characteristics. Technological advancements in manufacturing and synthesis can help bring down costs, making these systems more competitive when compared to other renewable energy sources. Ensuring that the catalysts do not degrade while they are being used is very important. Maintaining a stable adhesion of the catalysts on the conducting substrates has been identified as another general difficulty in enabling the stable working of the electrodes, in addition to enhancing the catalytic activity. Historically, binders and additives have been applied to help electrocatalyst NPs adhere to substrates. Nonetheless, the electrodes continue to be unstable due to a complex production process, material aggregation, and peeling off, which significantly restricted their long-term operations. Another significant challenge is the mass production of reliable, high-performance electrodes at a reasonable cost on a huge scale.

17.8 Case study on nanostructured materials for water-splitting

17.8.1 Case study title

Application of monoclinic phase tungsten oxide (m-WO_3) synthesized using plasma routes for solar water-splitting (de Respinis et al., 2013).

17.8.2 Objective

This study synthesized oxide of tungsten, WO_3 (in the monoclinic phase) nanostructure using an effective plasma-processing method for Solar Water-Splitting. The nanostructure possessed strong photocatalytic activity and enhanced electrical conductivity and was utilized as photoactive anodes after oxidation for the solar water-splitting (de Respinis et al., 2013).

17.8.3 Scope

This study offers a novel processing technique that utilizes low-energy helium ions to treat tungsten substrates. This method offers a productive and efficient approach to produce a porous metallic nanostructure, which was employed as a photoactive anode for solar water-splitting. Metal-oxide-based semiconductors are especially promising for various realistic purposes owing to cost-effectiveness, low toxicity, availability, and resistance to corrosion. This work employed monoclinic-WO_3 as a photoactive material for solar water-splitting, demonstrating its cost-effectiveness and low toxicity. Additionally, this can be used for several other purposes, like photocatalysis and environmental clean-up.

17.8.4 Audience

This case study can be valuable for scientists, scholars, and enterprises involved in a variety of sectors such as energy-storage, conversion, and sustainable H_2 generation, as well as carbon capture and usage targeted at promoting a cleaner energy environment. It is feasible to scale up the methods described in the article for use in industries. Several international research institutions, including the University of California, San Diego, Delft University of Technology in the Netherlands, the Dutch Institute for Fundamental Energy Research in the Netherlands, and the Institute for Solar Fuels in Germany, have published findings related to this study.

17.8.5 Rationale

One of the main obstacles in the current scenario is finding ways to fulfill the enormous energy requirements at the terawatt level by utilizing renewable energy sources. The utilization of solar energy into a usable form of energy that can be stored for further applications. This can be achieved via the PEC system using inexpensive and widely available semiconductors and catalysts. Metal-oxide-based semiconductors have the potential to be used in different applications, including addressing global energy demand, due to their affordability, lack of toxicity, easy availability, and enhanced stability. This makes them a sustainable material from an environmental standpoint.

Despite many experiments conducted over the past four decades, the effectiveness of employing metal oxides for solar water-splitting remains low, and scalability is an additional challenge. The diminished effectiveness is mainly attributed to e^--h^+ recombination in semiconductors based on metal oxides. Furthermore, the limited light penetration and sluggish reaction kinetics surfaces of metal-oxide nanostructures, especially for oxidation of water, are further constraints. To address these difficulties, one possible solution is to alter the structure of the material by introducing porous nanostructures and nanowires. This modification can lead to an increased surface area to volume ratio, thereby providing a larger area for photocatalytic reactions to take place. Additionally, it can also reduce the distance that charged particles need to travel for efficient transport.

17.8.6 Expected results and deliverables

This case study introduced a very effective plasma-processing technique that allows for precise control over the structure's morphology. This control is crucial for solar water-splitting. Additionally, the researchers utilized tungsten, a metal known for its exceptional photocatalytic properties, making it the ideal choice for their study. Nanostructured materials provide promising prospects for use in energy-storage and conversion.

The synthesis and transformation of WO_3 are the main topics of this study, although a wide variety of characterization techniques might improve the understanding of characteristics of the prepared material. Furthermore, it is necessary to do long-term stability testing. Examining the long-term performance of the material would provide crucial insights about its durability. By conducting tests on WO_3 under different environmental variables, including temperature, visible light intensities, and various pH values, useful insights can be gained regarding its adaptability. This will enhance the comprehension of the material's behavior in many contexts, which is crucial for practical applications.

17.8.7 Safety considerations

Controlled temperature is utilized to facilitate the crystallization of nanomaterials to achieve specific desirable characteristics. Higher temperatures promote fast crystallization, but they also create stronger reducing conditions.

The tungsten samples were subjected to annealing in either one or two phases while being exposed to an O_2 flow, following the plasma method. The initial stage is accomplished within a time frame of two hours at a temperature of 550°C to generate the desired m-WO_3. To avoid the possibility of forming a substoichiometric WO_{3-x} phase, the sample was then subjected to complete oxidation for 4 hours at a temperature of 450°C (considered low) before being cooled down to room temperature.

17.8.8 Actions taken/Workflow/Tools used/Simulations and analyses

The m-WO_3 nanostructure was produced utilizing a plasma-processing approach, as illustrated in the schematic diagram in Fig. 17.5.

After subjecting the WO_3 to low-energy helium ions, its surface structure exhibits a nanostructured morphology consisting of nanometric filaments with an interconnected and open structure. The filament has a diameter ranging from 100 to 200 nm at a temperature of 1500°C. Tungsten, a nanostructured material with 90% porosity, reflects between 94–97% of visible light, which is significantly higher than polished targets that only reflect more than 50% of the light. The presence of O_2 vacancies is shown by the blue color of the samples that were annealed at 550°C using a single-step process. After undergoing a 4-hours annealing process at a temperature of 450°C, the sample exhibits a yellow hue. This observation is in line with the color of stoichiometric WO_3, suggesting that it is completely oxidized.

Based on the diffuse reflectance curves, the polished material m-WO_3 had a band gap value of 2.6 eV, while the porous material m-WO_3 had a band gap value of 2.7 eV. The WO_3 nanostructure created in this work displayed enhanced properties in comparison to the polished sample obtained after annealing at 700°C. The oxidation efficiency was six times

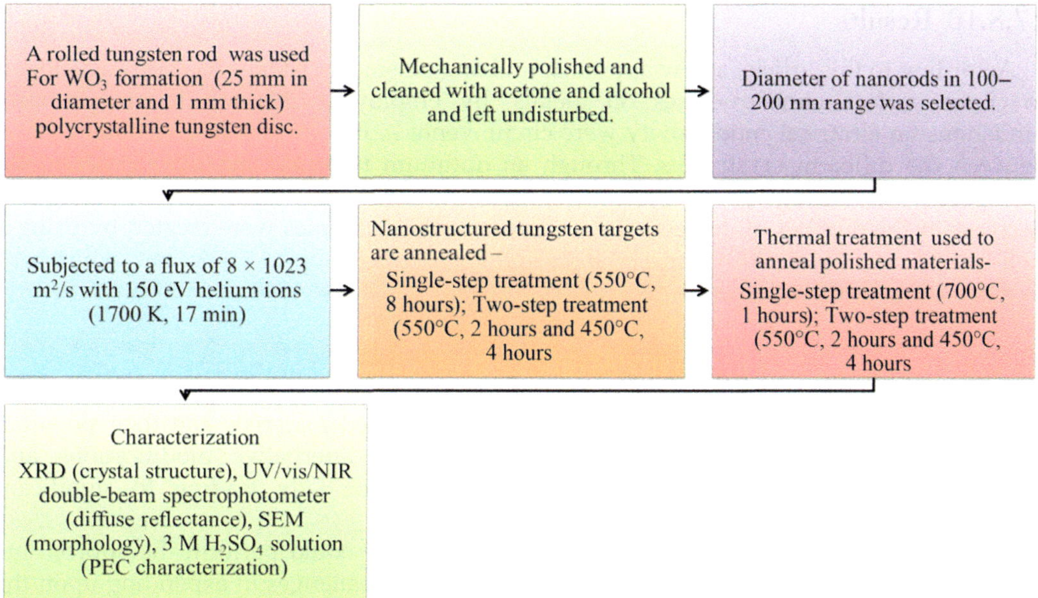

FIGURE 17.5 Schematic representation demonstrating synthesis of WO_3 via plasma-processing method.

greater at 1.0 V and three times higher at 2.5 V compared to the RHE. The augmented generation of photocurrent by the WO_3 nanostructures can be ascribed to three primary variables, namely, surface area, heightened absorption, and their dimensions. The catalytic activity of the NPs for water oxidation was enhanced by their higher specific surface area. This improvement was attributed to the efficient separation of charges at higher potentials, which led to faster surface reaction kinetics. Additionally, the increased supply of photogenerated holes also contributed to this enhancement. The nanostructured material exhibits enhanced absorption, as indicated by the lower reflectance (i.e., higher absorbance) of the nanoporous WO_3 compared to the polished WO_3. Ultimately, the nanostructure's modest size, similar to the minority carrier diffusion length (LD ~150 nm), enables the efficient transportation of photogenerated holes to the semiconductor/electrolyte interface, preventing their recombination with electrons.

Quantum efficiency studies have confirmed that the porous WO_3 exhibits a higher photoresponse over the entire spectrum of light above the band gap. The effectiveness achieves a value of 22% and reaches its highest point at a wavelength of 370 nm. The Tauc-plot revealed a band gap of 2.65 eV, which is consistent with the value obtained from the diffuse reflectance spectra and the published values for m-WO_3 in the literature.

17.8.9 Challenges and solutions

The main challenge faced was identifying the optimal form and improved electrical conductivity during the creation of nanostructured photoelectrodes with exceptional efficiency.

17.8.10 Results

According to this article, a new approach utilizing plasma assistance has been developed to create surface nanostructures on metals. By employing this top-down approach, limitations on electrical conductivity were circumvented, thereby ensuring optimal contact between the different crystallites. Through an optimum two-step annealing process, the mesoporous metal tungsten transforms into nanostructured monoclinic-WO_3, resulting in a decrease in the number of defects. The nanoporous WO_3 samples were created by using a polished W disc. These WO_3 films exhibited photocurrents of up to 1 mA/cm^2, which was five times higher than the dense films.

17.8.11 Learning and knowledge outcomes

The findings facilitate further research on nanostructured materials and their potential applications in solar energy. Researchers can explore alternative modifications and enhancements to the plasma-processing technique, potentially leading to significant developments in solar energy technology.

This study lacks detailed parameter optimization. They could perform the appropriate optimization for each step. Future researchers may have challenges in expanding upon the findings due to a scarcity of information.

17.9 Conclusion and future perspectives

Concerns regarding the expansion of nuclear power plants, the exhaustion of fossil fuel reserves, and contamination of the environment have all focused heavily on the use of H_2 gas as a clean and efficient fuel for energy production. Hence, using nanostructured materials in PEC water-splitting is an emerging method for H_2 production in a more convenient, affordable, and sustainable way. For PEC water-splitting, nanomaterial-based photoelectrodes provide several benefits, including an optical band gap that can be adjusted, a greater surface area, a reduction in light reflection loss, a shorter carrier diffusion length, and an electrical structure. Nanostructured materials can adjust the size and shape of their molecules to tailor their bandwidth. For PEC, a variety of nanomaterials, including nanosheets, nanotubes, nanowires, and nanorods, were focused. This study concentrates on the latest advancements in nanomaterial-based PEC water-splitting methods, along with several approaches to enhance H_2 production. To fully understand the complexities of multielectron/hole water-splitting reactions, how charge carriers function, and how photocatalysts and cocatalysts interact with solutions, it is essential to use both theoretical and experimental approaches. In addition, efforts should be concentrated on determining the band structure and offering suggestions for the development of photocatalysts that have a high level of performance. Furthermore, the improvement of water-splitting systems necessitates additional investigation into nanostructured semiconductors as photocatalysts, which are distinguished by their expansive surface areas and distinctive topologies. The objective of this strategy is to reduce the diffusion paths of minority charger carriers and enhance the kinetic reaction. The presence of nanostructures at the interface between water and nanomaterials is crucial for the transfer of charges and the facilitation of the processes of oxidation and reduction. Despite the great potential of nanostructured materials, issues related to scalability, durability, and procuring abundant components continue to exist. Conquering these obstacles promotes continuous investigation and advancement, propelling the development of sustainable methods for producing H_2. The progress made in this field shows potential for reducing carbon emissions in different sectors, such as transportation and industry, by employing nanostructured catalysts and renewable energy sources. Future proposals should give priority to resolving issues related to the implementation of nanostructured materials for solar-mediated water-splitting. These challenges should specifically focus on achieving scalability, durability, and sustainable sourcing of components. Continued research collaboration among academia, industry, and policymakers is essential for maximizing industrial processes and advancing material science and renewable energy technologies. Collaborative endeavors across several fields of study and worldwide support are crucial for progressing effective, economically feasible systems and encouraging the wider acceptance of environmentally friendly H_2 technology.

References

Abid, N., et al., 2022. Synthesis of nanomaterials using various top-down and bottom-up approaches, influencing factors, advantages, and disadvantages: a review. Advances in Colloid and Interface Science 300, 102597. Available at https://doi.org/10.1016/j.cis.2021.102597.

Ahmad, K., Raza, W., Khan, M.Q., 2021. Water splitting: design, synthesis and fabrication of nanostructured materials based efficient electrodes for water splitting applications. Handbook of Greener Synthesis of Nanomaterials and Compounds. Synthesis At the Macroscale and Nanoscale. 2. Elsevier, pp. 549–564. Available at https://doi.org/10.1016/B978-0-12-822446-5.00024-1.

Albu, Z., et al., 2021. Improving water oxidation performance by implementing heterointerfaces between ceria and metal-oxide nanoparticles. Journal of Colloid and Interface Science 587, 39–46. Available at https://doi.org/10.1016/j.jcis.2020.12.006.

Alexander, J.C., 2016. Principles of Photo-Electrochemical Cells. Surface Modifications and Growth of Titanium Dioxide for Photo-Electrochemical Water Splitting. Springer, pp. 47–92. Available at https://doi.org/10.1007/978-3-319-34229-0_3.

Amendola, V., Meneghetti, M., 2009. Laser ablation synthesis in solution and size manipulation of noble metal nanoparticles. Physical Chemistry Chemical Physics 11 (20), 3805. Available at https://doi.org/10.1039/b900654k.

Anis, S.F., et al., 2017. Electrospun nickel–tungsten oxide composite fibers as active electrocatalysts for hydrogen evolution reaction. Journal of Materials Science 52 (12), 7269–7281. Available at https://doi.org/10.1007/s10853-017-0964-2.

Baig, N., Kammakakam, I., Falath, W., 2021. Nanomaterials: a review of synthesis methods, properties, recent progress, and challenges. Materials Advances 2 (6), 1821–1871. Available at https://doi.org/10.1039/D0MA00807A.

Battiato, S., et al., 2022. Superior performances of electroless-deposited Ni–P films decorated with an ultralow content of Pt for water-splitting reactions. ACS Applied Energy Materials 5 (2), 2391–2399. Available at https://doi.org/10.1021/acsaem.1c03880.

Bharti, S., Mukherji, S.S., Mukherji, S.S., 2018. Photocatalytic action of silver-titanium dioxide nanocomposite against pathogenic bacteria'. Journal of Indian Chemical Society 95 (March), 1–8.

Bharti, S., Mukherji, S., Mukherji, S., 2020. Extracellular synthesis of silver nanoparticles by thiosphaera pantotropha and evaluation of their antibacterial and cytotoxic effects. 3 Biotech 10 (6), 237. Available at https://doi.org/10.1007/s13205-020-02218-0.

Bharti, S., Mukherji, S., Mukherji, S., 2021. Antiviral application of colloidal and immobilized silver nanoparticles. Nanotechnology 32 (20), 205102. Available at https://doi.org/10.1088/1361-6528/abe489.

Bhowmik, T., Kundu, M.K., Barman, S., 2016. Palladium nanoparticle–graphitic carbon nitride porous synergistic catalyst for hydrogen evolution/oxidation reactions over a broad range of pH and correlation of its catalytic activity with measured hydrogen binding energy. ACS Catalysis 6 (3), 1929–1941. Available at https://doi.org/10.1021/acscatal.5b02485.

Cai, C., et al., 2019. Direct observation of yolk-shell transforming to gold single atoms and clusters with superior oxygen evolution reaction efficiency. ACS Nano 13 (8), 8865–8871. Available at https://doi.org/10.1021/acsnano.9b02135.

Carolin Christopher, F., et al., 2021. A review on critical assessment of advanced bioreactor options for sustainable hydrogen production. International Journal of Hydrogen Energy 46 (10), 7113–7136. Available at https://doi.org/10.1016/j.ijhydene.2020.11.244.

Chen, A., Holt-Hindle, P., 2010. Platinum-based nanostructured materials: synthesis, properties, and applications. Chemical Reviews 110 (6), 3767–3804. Available at https://doi.org/10.1021/cr9003902.

Chen, S., Thind, S.S., Chen, A., 2016. Nanostructured materials for water splitting - state of the art and future needs: a mini-review. Electrochemistry Communications 63, 10–17. Available at https://doi.org/10.1016/j.elecom.2015.12.003.

Chen, W.-H., et al., 2023. A critical and systematic review of sustainable hydrogen production from ethanol/bioethanol: steam reforming, partial oxidation, and autothermal reforming. Fuel 333, 126526. Available at https://doi.org/10.1016/j.fuel.2022.126526.

Chen, Z., Wei, W., Ni, B.J., 2021. Cost-effective catalysts for renewable hydrogen production via electrochemical water splitting: recent advances. Current Opinion in Green and Sustainable Chemistry 27, 100398. Available at https://doi.org/10.1016/j.cogsc.2020.100398.

Chu, S., et al., 2019. Holey Ni-Cu phosphide nanosheets as a highly efficient and stable electrocatalyst for hydrogen evolution. Applied Catalysis B: Environmental 243, 537–545. Available at https://doi.org/10.1016/j.apcatb.2018.10.063.

Danks, A.E., Hall, S.R., Schnepp, Z., 2016. The evolution of "sol–gel" chemistry as a technique for materials synthesis. Materials Horizons 3 (2), 91–112. Available at https://doi.org/10.1039/C5MH00260E.

Dincer, I., Zamfirescu, C., 2012. Potential options to greenize energy systems. Energy 46 (1), 5–15. Available at https://doi.org/10.1016/j.energy.2011.11.061.

Ertekin, Z., Ozer, D., 2022. 2D Materials for overall water splitting. Handbook of Energy Materials. Springer Nature Singapore, Singapore, pp. 1–26. Available at https://doi.org/10.1007/978-981-16-4480-1_72-1.

Fajrina, N., Tahir, M., 2019. A critical review in strategies to improve photocatalytic water splitting towards hydrogen production. International Journal of Hydrogen Energy 44 (2), 540–577. Available at https://doi.org/10.1016/j.ijhydene.2018.10.200.

Fan, M., et al., 2022. Highly dispersed Ru nanoclusters anchored on B,N co-doped carbon nanotubes for water splitting. Inorganic Chemistry Frontiers 9 (5), 968–976. Available at https://doi.org/10.1039/D1QI01672E.

Fominykh, K., et al., 2015. Iron-doped nickel oxide nanocrystals as highly efficient electrocatalysts for alkaline water splitting. ACS Nano 9 (5), 5180–5188. Available at https://doi.org/10.1021/acsnano.5b00520.

Gao, Y., et al., 2020. Construction of Fe_2O_3 @CuO heterojunction nanotubes for enhanced oxygen evolution reaction. ACS Applied Energy Materials 3 (1), 666–674. Available at https://doi.org/10.1021/acsaem.9b01866.

Ghouri, Z.K., et al., 2021. Theoretical and experimental investigations of Co-Cu bimetallic alloys-incorporated carbon nanowires as an efficient bi-functional electrocatalyst for water splitting. Journal of Industrial and Engineering Chemistry 96, 243–253. Available at https://doi.org/10.1016/j.jiec.2021.01.027.

Gupta, N.M., 2017. Factors affecting the efficiency of a water splitting photocatalyst: a perspective. Renewable and Sustainable Energy Reviews 71, 585–601. Available at https://doi.org/10.1016/j.rser.2016.12.086.

Harish, V., et al., 2023. Cutting-edge advances in tailoring size, shape, and functionality of nanoparticles and nanostructures: a review. Journal of the Taiwan Institute of Chemical Engineers 149, 105010. Available at https://doi.org/10.1016/j.jtice.2023.105010.

Hua, Z., et al., 2022. One-step controllable fabrication of 3D structured self-standing Al_3Ni_2/Ni electrode through molten salt electrolysis for efficient water splitting. Chemical Engineering Journal 427 (August 2021), 131743. Available at https://doi.org/10.1016/j.cej.2021.131743.

Huynh, M., et al., 2015. Nature of activated manganese oxide for oxygen evolution. Journal of the American Chemical Society 137 (47), 14887–14904. Available at https://doi.org/10.1021/jacs.5b06382.

Isimjan, T.T., Rohani, S., Ray, A.K., 2012. Photoelectrochemical water splitting for hydrogen generation on highly ordered TiO_2 nanotubes fabricated by using Ti as cathode. International Journal of Hydrogen Energy 37 (1), 103–108. Available at https://doi.org/10.1016/j.ijhydene.2011.04.167.

Janani, G., et al., 2020. Enhanced bifunctional electrocatalytic activity of Ni-Co bimetallic chalcogenides for efficient water-splitting application. Journal of Alloys and Compounds 846, 156389. Available at https://doi.org/10.1016/j.jallcom.2020.156389.

Jia, X., et al., 2016. 'Ni3FeN nanoparticles derived from ultrathin NiFe-layered double hydroxide nanosheets: an efficient overall water splitting electrocatalyst. Advanced Energy Materials 6 (10), 1–6. Available at https://doi.org/10.1002/aenm.201502585.

Jin, H., et al., 2019. Recent progress in bifunctional electrocatalysts for overall water splitting under acidic conditions. ChemElectroChem 6 (13), 3244–3253. Available at https://doi.org/10.1002/celc.201900507.

Jones, A.C., Hitchman, M.L., (Eds.), 2008. Chemical Vapour Deposition. Royal Society of Chemistry, Cambridge. Available at https://doi.org/10.1039/9781847558794.

Joy, J., Mathew, J., George, S.C., 2018. Nanomaterials for photoelectrochemical water splitting – review. International Journal of Hydrogen Energy 43 (10), 4804–4817. Available at https://doi.org/10.1016/j.ijhydene.2018.01.099.

Kannan, K., et al., 2023. Hydrothermally synthesized mixed metal oxide nanocomposites for electrochemical water splitting and photocatalytic hydrogen production. International Journal of Hydrogen Energy 48 (93), 36412–36426. Available at https://doi.org/10.1016/j.ijhydene.2023.06.042.

Khanna, P., Kaur, A., Goyal, D., 2019. Algae-based metallic nanoparticles: synthesis, characterization and applications. Journal of Microbiological Methods 163, 105656. Available at https://doi.org/10.1016/j.mimet.2019.105656.

Kumar, M., et al., 2022. Recent trends in photoelectrochemical water splitting: the role of cocatalysts. NPG Asia Materials 14 (1), 88. Available at https://doi.org/10.1038/s41427-022-00436-x.

Li, H., et al., 2019. Transition metal tuned semiconductor photocatalyst CuCo/β-SiC catalyze hydrolysis of ammonia borane to hydrogen evolution. International Journal of Hydrogen Energy 44 (16), 8307–8314. Available at https://doi.org/10.1016/j.ijhydene.2019.02.034.

Li, L., et al., 2017. P doped MoO_{3-x} nanosheets as efficient and stable electrocatalysts for hydrogen evolution. Small 13 (25). Available at https://doi.org/10.1002/smll.201700441.

Li, L., et al., 2022. Constructing the Fe/Cr double (oxy)hydroxides on Fe_3O_4 for boosting the electrochemical oxygen evolution in alkaline seawater and domestic sewage. Applied Catalysis B: Environmental 302, 120847. Available at https://doi.org/10.1016/j.apcatb.2021.120847.

Li, Y., Zhang, J.Z., 2010. Hydrogen generation from photoelectrochemical water splitting based on nanomaterials. Laser and Photonics Reviews 4 (4), 517–528. Available at https://doi.org/10.1002/lpor.200910025.

Luo, F., et al., 2018. Palladium phosphide as a stable and efficient electrocatalyst for overall water splitting. Angewandte Chemie - International Edition 57 (45), 14862–14867. Available at https://doi.org/10.1002/anie.201810102.

Maduraiveeran, G., Sasidharan, M., Jin, W., 2019. Earth-abundant transition metal and metal oxide nanomaterials: synthesis and electrochemical applications. Progress in Materials Science 106, 100574. Available at https://doi.org/10.1016/j.pmatsci.2019.100574.

Malik, M.A., Wani, M.Y., Hashim, M.A., 2012. Microemulsion method: a novel route to synthesize organic and inorganic nanomaterials. Arabian Journal of Chemistry 5 (4), 397–417. Available at https://doi.org/10.1016/j.arabjc.2010.09.027.

Mukherji, S., et al., 2019. Synthesis and characterization of size- and shape-controlled silver nanoparticles. Physical Sciences Reviews 4 (1), 20170082. Available at https://doi.org/10.1515/psr-2017-0082.

Nie, Q., et al., 2017. Interface optimization of ZnO nanorod/CdS quantum dots heterostructure by a facile two-step low-temperature thermal treatment for improved photoelectrochemical water splitting. Chemical Engineering Journal 325, 151–159. Available at https://doi.org/10.1016/j.cej.2017.05.021.

Ostermann, R., et al., 2011. Metal–organic framework nanofibers viaelectrospinning. Chemical Communications 47 (1), 442–444. Available at https://doi.org/10.1039/C0CC02271C.

Parangusan, H., et al., 2023. Engineering the structural, optical and photoelectrochemical properties of $BaTiO_3$-$CoFe_2O_4$ nanocomposite for photoelectrochemical water splitting. Electrochimica Acta 464, 142849. Available at https://doi.org/10.1016/j.electacta.2023.142849.

Patil, R.P., et al., 2021. Self-templated fabrication of 2-D dual nanoarchitecture $Zn_{1-x}Cd_xS$ porous nanosheet and ZnO nanorod for photoelectrochemical hydrogen production. Applied Surface Science 539, 148267. Available at https://doi.org/10.1016/j.apsusc.2020.148267.

Pokrant, S., et al., 2017. Size effects of cocatalysts in photoelectrochemical and photocatalytic water splitting. Materials Today Energy 5, 158–163. Available at https://doi.org/10.1016/j.mtener.2017.06.005.

Qutb, S.R., et al., 2024. Superior photoelectrodes of nanostructured Mo-doped CuO thin film for green hydrogen generation from photoelectrochemical water-splitting. International Journal of Hydrogen Energy 76, 190–201. Available at https://doi.org/10.1016/j.ijhydene.2024.03.200.

Raj, S.K., Bhadu, G.R., Upadhyay, P., 2022. 'ScienceDirect Three-dimensional Ni/Fe doped graphene oxide @ MXene architecture as an efficient water splitting electrocatalyst. International Journal of Hydrogen Energy 47 (99), 41772–41782. Available at https://doi.org/10.1016/j.ijhydene.2022.05.204.

Ramesh, C., et al., 2023. Hybrid reduced graphene oxide/gan nanocolumns on flexible niobium foils for efficient photoelectrochemical water splitting. ACS Applied Nano Materials 6 (3), 1898–1909. Available at https://doi.org/10.1021/acsanm.2c04864.

Reddy, N.L., et al., 2019. A review on frontiers in plasmonic nano-photocatalysts for hydrogen production. International Journal of Hydrogen Energy 44 (21), 10453–10472. Available at https://doi.org/10.1016/j.ijhydene.2019.02.120.

de Respinis, M., et al., 2013. Efficient plasma route to nanostructure materials: case study on the use of m-WO_3 for solar water splitting. ACS Applied Materials & Interfaces 5 (15), 7621–7625. Available at https://doi.org/10.1021/am401936q.

Salame, P.H., Pawade, V.B., Bhanvase, B.A., 2018. Characterization tools and techniques for nanomaterials. Nanomaterials for Green Energy. Elsevier, pp. 83–111. Available at https://doi.org/10.1016/B978-0-12-813731-4.00003-5.

Shaddad, M.N., et al., 2022. Exploiting the synergistic catalytic effects of CoPi nanostructures on Zr-doped highly ordered TiO_2 nanotubes for efficient solar water oxidation. International Journal of Energy Research 46 (9), 12608–12622. Available at https://doi.org/10.1002/er.8030.

Shi, L., et al., 2018. A synergistic interaction between isolated Au nanoparticles and oxygen vacancies in an amorphous black TiO_2 nanoporous film: toward enhanced photoelectrochemical water splitting. Journal of Materials Chemistry A 6 (27), 12978–12984. Available at https://doi.org/10.1039/C8TA04621B.

Sinha, S.K., et al., 2022. Production of hydrogen via water splitting using photocatalytic and photoelectrocatalytic route. Clean Energy Technologies - Hydrogen and Gasification Processes. IntechOpen. Available at https://doi.org/10.5772/intechopen.106915.

Thakur, S., Bharti, S., 2023. The synergy of nanoparticles and metal-organic frameworks in antimicrobial applications: a critical review. Journal of Environmental Chemical Engineering 11 (6), 111458. Available at https://doi.org/10.1016/j.jece.2023.111458.

Thakur, S., Bharti, S., 2024. Unlocking the potential of metal–organic frameworks: a review on synthesis, characterization, and multifaceted applications. Journal of Inorganic and Organometallic Polymers and Materials. Available at https://doi.org/10.1007/s10904-024-03174-5 [Preprint].

Tran, P.K.L., et al., 2021. Highly effective freshwater and seawater electrolysis enabled by atomic Rh-modulated Co-CoO lateral heterostructures. Small 17 (50). Available at https://doi.org/10.1002/smll.202103826.

Vazhayil, A., et al., 2021. A comprehensive review on the recent developments in transition metal-based electrocatalysts for oxygen evolution reaction. Applied Surface Science Advances 6, 100184. Available at https://doi.org/10.1016/j.apsadv.2021.100184.

Wang, F., et al., 2015. Recent advances in transition-metal dichalcogenide based nanomaterials for water splitting. Nanoscale 19764–19788. Available at https://doi.org/10.1039/c5nr06718a.

Wang, J., et al., 2018. Highly uniform Ru nanoparticles over N-doped carbon: PH and temperature-universal hydrogen release from water reduction. Energy and Environmental Science 11 (4), 800–806. Available at https://doi.org/10.1039/c7ee03345a.

Wang, Y., et al., 2017. Superb Alkaline Hydrogen Evolution and Simultaneous Electricity Generation by Pt-Decorated Ni3N Nanosheets. Advanced Energy Materials 7 (2), 1–7. Available at https://doi.org/10.1002/aenm.201601390.

Xiong, S., et al., 2016. Enhanced catalytic performance of ZnO-CoOx electrode generated from electrochemical corrosion of Co-Zn alloy for oxygen evolution reaction. Electrochimica Acta 222, 999–1006. Available at https://doi.org/10.1016/j.electacta.2016.11.068.

Xu, Q.-Q., et al., 2020. Crystal phase determined Fe active sites on Fe_2O_3 (γ- and α-Fe_2O_3) yolk-shell microspheres and their phase dependent electrocatalytic oxygen evolution reaction. Applied Surface Science 533, 147368. Available at https://doi.org/10.1016/j.apsusc.2020.147368.

Yadav, T.P., Yadav, R.M., Singh, D.P., 2012. Mechanical milling: a top down approach for the synthesis of nanomaterials and nanocomposites. Nanoscience and Nanotechnology 2 (3), 22–48. Available at https://doi.org/10.5923/j.nn.20120203.01.

Yang, J., et al., 2021. From hexagonal to monoclinic: engineering crystalline phase to boost the intrinsic catalytic activity of tungsten oxides for the hydrogen evolution reaction. ACS Sustainable Chemistry and Engineering 9 (16), 5642–5650. Available at https://doi.org/10.1021/acssuschemeng.1c00485.

Yuan, S., et al., 2022. Dual synergistic effects between Co and Mo2C in Co/Mo2C heterostructure for electrocatalytic overall water splitting. Chemical Engineering Journal 430, 132697. Available at https://doi.org/10.1016/j.cej.2021.132697.

Zhang, D., et al., 2019. Controllable synthesis of carbon nanomaterials by direct current arc discharge from the inner wall of the chamber. Carbon 142, 278–284. Available at https://doi.org/10.1016/j.carbon.2018.10.062.

Zhao, H., Liang, J., Zhao, Y., 2022. Construction of hierarchical Co_2P/Ni_2P heterostructures on Ni foam as efficient bifunctional electrocatalyst for overall water splitting. Journal of Alloys and Compounds 907, 164479. Available at https://doi.org/10.1016/j.jallcom.2022.164479.

Zhou, M., Lou, X.W., Xie, Y., 2013. Two-dimensional nanosheets for photoelectrochemical water splitting: possibilities and opportunities. Nano Today 8 (6), 598–618. Available at https://doi.org/10.1016/j.nantod.2013.12.002.

Zhou, W., et al., 2016. $CoSe_2$ nanoparticles embedded defective carbon nanotubes derived from MOFs as efficient electrocatalyst for hydrogen evolution reaction. Nano Energy 28, 143–150. Available at https://doi.org/10.1016/j.nanoen.2016.08.040.

Zhu, J., et al., 2020. Recent advances in electrocatalytic hydrogen evolution using nanoparticles. Chemical Reviews 120 (2), 851–918. Available at https://doi.org/10.1021/acs.chemrev.9b00248.

CHAPTER

18

Photovoltaic electrochemical integration for water-splitting

Princemon Philip[1], *Ajesh K. Zachariah*[2], *and Anju K. Nair*[2]

[1]St. Alosyius College, Edathua, Alappuzha, Kerala, India [2]Department of Physics, Mar Thoma College, Tiruvalla, Kerala, India

18.1 Introduction

The world's energy supply is largely dependent on finite resources such as petroleum, coal, and natural gas. In recent decades, rising public concern over the challenges linked to fossil fuel usage and the large-scale CO_2 emissions as a greenhouse gas has spurred the demand for innovative energy conversion methods that utilize green and renewable sources (Vesborg and Jaramillo, 2012; Perera, 2017; Lewis and Nocera, 2006; Armaroli and Balzani, 2007). Solar energy is the sole renewable resource capable of replacing fossil fuels and addressing the rising environmental needs (Asif and Muneer, 2007; Lewis and Nocera, 2006). Even though solar energy can be harnessed to generate electricity with the most frequently utilized solar panels [or photovoltaic (PV) cells, abbreviated as PV], the intermittent nature of solar energy due to natural cycles poses a significant challenge to a sustainable energy supply. To address this issue, many researchers are working on converting solar energy into chemical compounds that can be generated when the sun shines and stored for use on demand (Yin et al., 2020).

The most promising method to utilize solar energy is using it to produce hydrogen through photocatalytic water-splitting. Currently, over 95% of H_2 gas is produced from fossil fuels, leading to significant environmental concerns (Bak et al., 2002a). As an alternative, water-splitting powered by solar energy can generate high-purity, eco-friendly hydrogen (Bak et al., 2002b). This method is considered a green and sustainable technology that can support the hydrogen economy (Momirlan & Veziroglu, 2005). Hydrogen is the purest energy source, useful for powering eco-friendly vehicles, fuel cells, home heating, and airplanes (Dorner et al., 2010). Additionally, using hydrogen for energy storage and transfer offers a long-term strategy to reduce global CO_2 emissions by converting CO_2 into valuable hydrocarbons (Jessop et al., 1995; Ashley et al., 2009).

Photochemical Splitting of Water
DOI: https://doi.org/10.1016/B978-0-443-29064-0.00016-4

In theory, PV-electrolysis systems could achieve 90%–95% of the efficiency of PV cells, resulting in approximately 57% efficiency for a 3-junction cell and around 62% for a 4- or 5-junction cell. These figures suggest there is considerable potential for enhancing the performance of PV-electrolysis system prototypes. In photoelectrochemical (solar) water-splitting devices, PEC cells typically include at least one light-absorbing electrode, which is usually a single semiconductor or a combination of two semiconductors in a heterojunction. PEC tandem cells, however, feature dual light absorbers, either connected by wires or operating wirelessly, and utilize the photovoltage generated by both photoelectrodes (Walter, 2010; Chen, 2019; Nam, 2018; Zeng, 2017; Bai, 2016). Fig. 18.1 illustrates a sustainable strategy for advancing the hydrogen economy by integrating renewable energy sources with electrochemical water-splitting (Zhang et al., 2020).

In this book chapter, we offer a comprehensive overview of the principles underlying the integration of PV systems with electrochemical cells for water-splitting. We also discuss the significance and potential of this integration in the context of renewable energy production and sustainable hydrogen generation. Additionally, we explore the mechanisms involved in water-splitting using integrated PV-electrochemical systems and examine the materials used for PV cells and electrochemical catalysts, with a focus on their efficiency, stability, and cost-effectiveness.

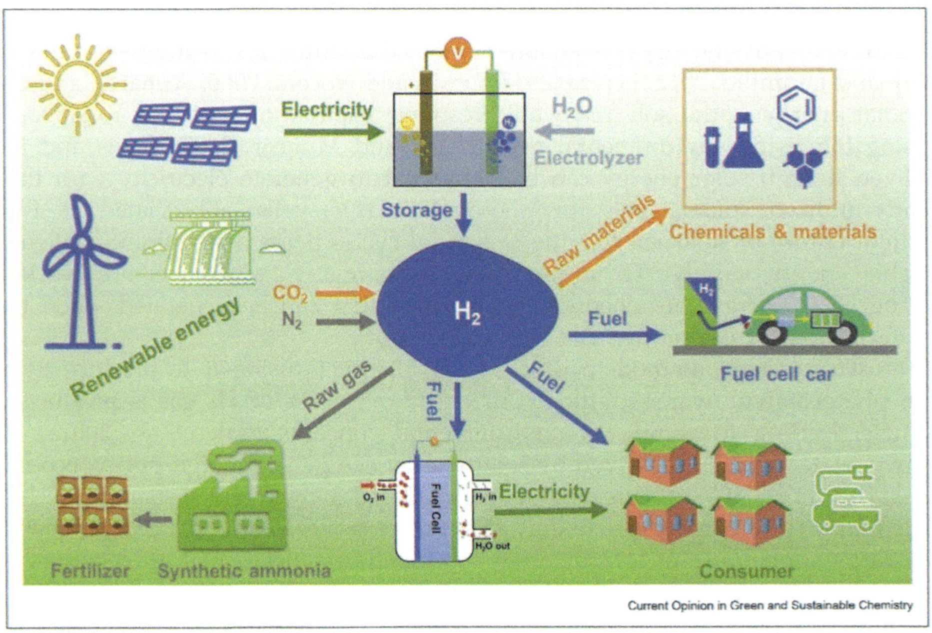

FIGURE 18.1 A sustainable approach to supporting the hydrogen economy involves integrating renewable energy sources with electrochemical water-splitting. (Zhang et al., 2020). *Source: Copyright reserved to the Wiley (Creative Commons Attribution 4.0 Unported License).*

18.2 Water-splitting

From a thermodynamic perspective, splitting water is an uphill process that does not occur spontaneously and necessitates external energy to proceed, as the reverse reaction happens readily. Fig. 18.2 summarizes various water-splitting methods. These include thermolysis (e.g., thermochemical cycles), electrolysis (e.g., electrocatalysis), photolysis (e.g., photocatalysis), and biolysis (e.g., dark fermentation). Hydrogen production can also be achieved using a combination of energy forms. Key hybrid energy systems include thermal + electrical (e.g., thermoelectrolysis, high-temperature electrolysis), electrical + photonic (e.g., photoelectrolysis, PV electrolysis), and photonic + biochemical (e.g., biophotolysis, photofermentation). Hybrid systems are typically more thermodynamically advantageous than nonhybrid systemsbecause they incorporate cheaper or renewable resources for part of the required energy, which lowers overall operational costs, reduces activation barriers, improves the speed of chemical reactions and increases the rate of hydrogen production.

18.2.1 Basics of water-splitting

The overall process of water electrolysis consists of two half-cell reactions: the water reduction reaction (hydrogen evolution reaction [HER]) and the water oxidation reaction (oxygen evolution reaction [OER]). The cathodic (HER: Eqs. 18.1 and 18.3) and anodic (OER: Eqs. 18.2 and 18.4) reactions vary depending on the electrolyte conditions (pH of the solution) and can be represented as follows, with standard hydrogen electrode (SHE) indicating the standard hydrogen electrode (Spöri et al., 2017, Mahmood et al., 2018)

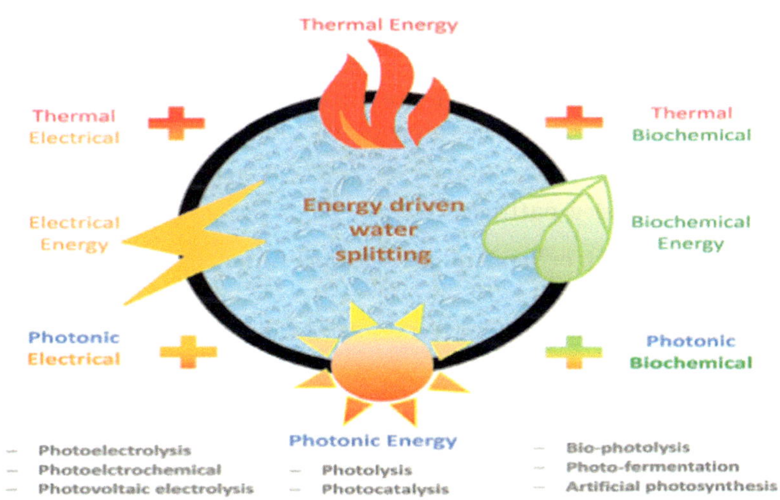

FIGURE 18.2 Various energy driven water-splitting routes by using thermal, electrical, biochemical and photonic energy or their combinations (Si et al., 2017). *Source: Copyright reserved to the Wiley (Creative Commons Attribution 4.0 Unported License).*

In acidic condition (pH 0),

$$2H^+(aq) + 2e^- \rightarrow H_2(g) \quad (HER, E^\circ = 0.00V \text{ versus SHE}) \tag{18.1}$$

$$H_2O(l) \rightarrow \tfrac{1}{2}O_2(g) + 2H^+(aq) + 2e^-(OER, E^\circ = 1.23V \text{ versus SHE}) \tag{18.2}$$

In alkaline conditions (pH 14),

$$2H_2O(l) + 2e^- \rightarrow H_2(g) + 2OH^-(aq)(HER, E^\circ = -0.83V \text{ versus SHE}) \tag{18.3}$$

$$2OH^-(aq) \rightarrow H_2O(l) + \tfrac{1}{2}O_2(g) + 2e^-(OER, E^\circ = 0.40V \text{ versus SHE}) \tag{18.4}$$

Overall reaction can be written as,

$$H_2O(l) \rightarrow H_2(g) + \tfrac{1}{2}O_2(g)(\Delta E = 1.23V) \tag{18.5}$$

The water-splitting reaction is a thermodynamical uphill process, requiring a net Gibbs free energy input of 237 kJ/mol.

The Volmer-Tafel and Volmer-Heyrovsky pathways describe different mechanisms of the HER (Shinagawa et al., 2015). These pathways detail the steps involved in reducing protons to produce hydrogen gas (H_2) on an electrode surface. Let's break down each pathway:

18.2.2 Volmer-Tafel pathway

18.2.2.1 Volmer step (electrochemical adsorption)

o $H^+ + e^- \rightarrow H_{ads}$

o A proton from the electrolyte gains an electron from the electrode and adsorbs onto the electrode surface as a hydrogen atom (H_{ads}).

18.2.2.2 Tafel step (chemical desorption)

$H_{ads} + H_{ads} \rightarrow H_2$

Two adsorbed hydrogen atoms combine to form a hydrogen molecule, which then desorbs from the electrode surface into the electrolyte or gas phase.

18.2.3 Volmer-Heyrovsky pathway

18.2.3.1 Volmer step (electrochemical adsorption)

$H^+ + e^- \rightarrow H_{ads}$

A proton from the electrolyte gains an electron from the electrode and adsorbs onto the electrode surface as a hydrogen atom (H_{ads}).

18.2.3.2 Heyrovsky step (electrochemical desorption)

$H_{ads} + H^+ + e^- \rightarrow H_2$

An adsorbed hydrogen atom reacts with another proton (H^+) from the electrolyte and gains an electronto form a hydrogen molecule (H_2), which then desorbs from the electrode surface.

In an alkaline electrolyte, the HER begins with the reduction of a water molecule at the active site of the catalysts. This process forms a proton at the active site. Consequently, more energy (or overpotential) is typically needed compared to acidic media to generate protons by breaking down water molecules. Unlike the HER, the OER requires a four-electron transfer process to produce 1 mole of oxygen gas (O_2), making it a bottleneck in the entire water electrolysis system. Although four different reaction mechanisms for the OER have been proposed (Matsumoto and Sato 1986).

18.2.4 Electrochemical oxide pathway

This pathway involves the formation and involvement of higher oxidation state intermediates through direct electrochemical steps.

18.2.4.1 Formation of metal-OH (M-OH)

$M + H_2O \rightarrow M\text{-}OH + H^+ + e^-$

A water molecule adsorbs onto the catalyst (M), forming a hydroxyl group (M-OH) with the release of a proton and an electron.

18.2.4.2 Formation of metal-O (M-O)

$M\text{-}OH \rightarrow M\text{-}O + H^+ + e^-$

The M-OH undergoes further oxidation to form an oxo group (M-O), releasing another proton and electron.

18.2.4.3 Formation of metal-OOH (M-OOH)

$M\text{-}O + H_2O \rightarrow M\text{-}OOH + H^+ + e^-$

The M-O reacts with another water molecule to form a hydroperoxo group (M-OOH), with the release of a proton and an electron.

18.2.4.4 Formation of oxygen gas

$M\text{-}OOH \rightarrow M + O_2 + H^+ + e^-$

The M-OOH is further oxidized, releasing O_2 and regenerating the catalyst site, with the release of a proton and an electron.

18.2.5 Oxide pathway

The oxide pathway involves the formation of bulk or surface oxides, which then participate in the OER process.

18.2.5.1 Formation of metal hydroxide

$M + H_2O \rightarrow M\text{-}OH + H^+ + e^-$

A water molecule adsorbs onto the metal surface, forming a MOH, releasing a proton and an electron.

18.2.5.2 Formation of metal oxide and water release

o $MOH \rightarrow MO + H_2O$
o The metal hydroxide (MOH) loses a water molecule, forming a metal oxide (MO).

18.2.5.3 Formation of higher oxidation state oxide

o $MO + H_2O \rightarrow MO_2 + 2\ H^+ + 2e^-$
o The MO is further oxidized by another water molecule, forming a higher oxidation state oxide (MO_2), releasing protons and electrons.

18.2.6 Oxygen evolution

$MO_2 \rightarrow M + O_2$

The higher MO_2 decomposes, releasing O_2 and regenerating the catalyst surface.

In the OER mechanism in alkaline media, the initial step involves the adsorption of a hydroxide ion on the catalyst's active site. Subsequently, the evolution of O_2 can proceed via either a radical oxo coupling pathway or a superoxo intermediate pathway (Song et al., 2018). Fig. 18.3 depicts the mechanism of electrochemical water-splitting reactions in an acidic aqueous solution (Minoh et al., 2022).

18.3 Photovoltaic systems for water-splitting

18.3.1 Basic principle of the photovoltaic system

When light hits a photovoltaic PV cell, it converts the energy into electricity. The word "photovoltaic" is derived from "photo," which refers to light, and "voltaic," which pertains to electricity. When the cell is illuminated, it captures photons from the sunlight, which excites electrons and causes them to move away from their atoms, creating holes. When the PV cell is linked to a load, the voltage difference causes charges to shift from the n-side to the p-side of the cell, resulting in electricity generation.

Generally, a single PV cell generates approximately 0.5 V, while the amount of current produced depends on factors such as weather conditions, sunlight intensity, and the cell's surface area (Ahmad et al., 2023, Jiang et al., 2017, Chan et al., 2016). To improve the performance of PV cells, they are usually designed into larger modules or panels with cells connected in series or parallel arrangements. Different configurations provide different enhancements; series connections increase voltage, while parallel connections boost current. For certain applications, PV cells can be arranged in a hybrid configuration to create a PV array customized to fulfill specific requirements.

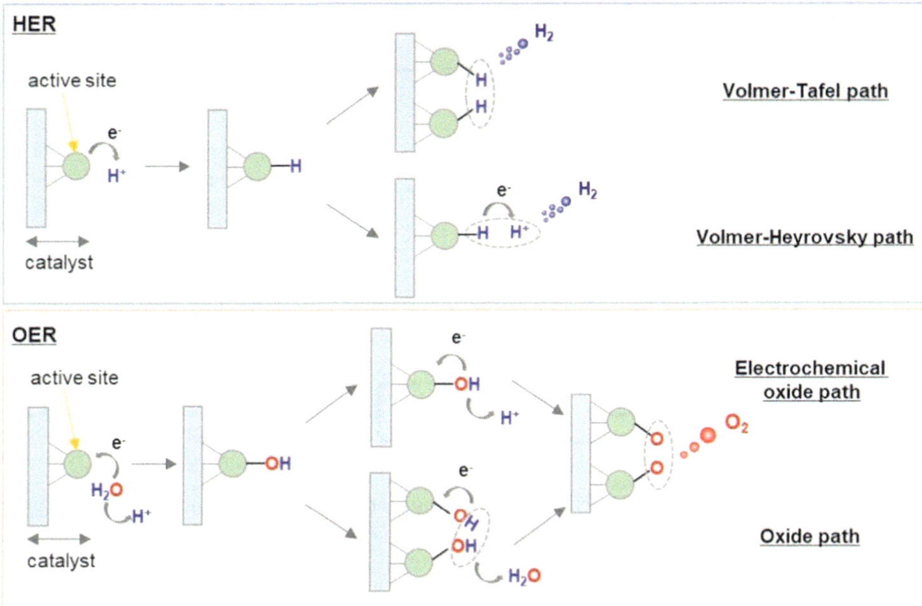

HER: $2\ H^+(aq) + 2e^- \rightarrow H_2(g)$ (in acidic medium)

$2\ H_2O(l) + 2e^- \rightarrow H_2(g) + 2\ OH^-$ (in alkaline medium)

OER: $H_2O(l) \rightarrow \frac{1}{2}\ O_2(g) + 2\ H^+(aq) + 2e^-$ (in acidic medium)

$2\ OH^-(aq) \rightarrow \frac{1}{2}\ O_2(g) + 2\ H_2O(l) + 4e^-$ (in alkaline medium)

Total reaction: $2\ H_2O(l) \rightarrow 2\ H_2(g) + O_2(g)$ $\quad \Delta E = 1.23\ V$

FIGURE 18.3 Mechanism of electrochemical water-splitting reactions in an acidic aqueous solution. (Minoh et al., 2022). *Source: Copyright reserved to the Chemistry Europe (Creative Commons Attribution 4.0 Unported License).*

18.3.2 Types of solar photovoltaic technology

Solar PV technology has captured considerable attention within communities and society, establishing itself as one of the leading renewable energy sources. Currently, three generations of solar PV technologies have been developed, as illustrated in Fig. 18.4 (Samsudin, 2023).

18.3.3 First generation: traditional silicon-based solar cells

The initial generation of solar PV technology is mainly composed of crystalline silicon solar cells, available in both monocrystalline and polycrystalline forms. These cells have led the market for many years, thanks to their high-efficiency and dependable performance. Monocrystalline silicon cells are known for their high-efficiency, typically ranging between 15% and 20% in commercial products, with laboratory efficiencies reaching around 26%–27% (Allouhi et al., 2022; Samsudin et al., 2020). Polycrystalline silicon cells, while slightly less efficient, offer a more cost-

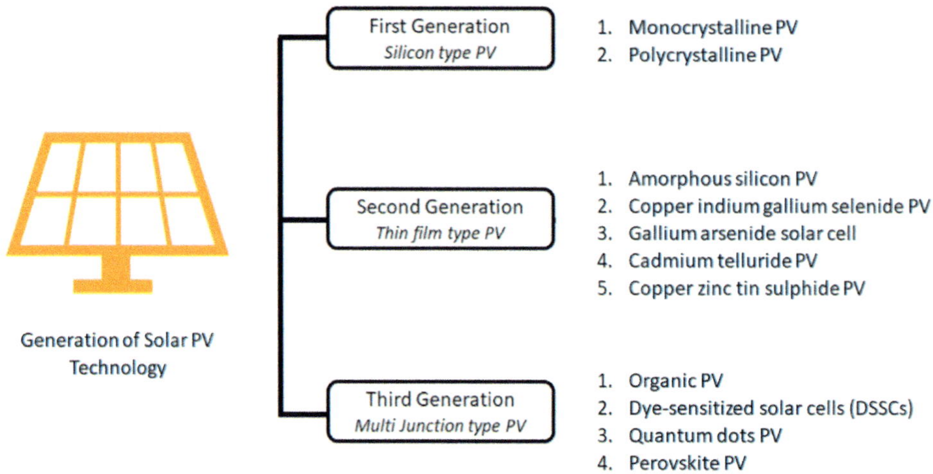

FIGURE 18.4 Generations of solar PV technologies that has been developed (Samsudin, 2023). *Source: Copyright reserved to the energies (Creative Commons Attribution 4.0 Unported License).*

effective alternative. Monocrystalline silicon PV technology utilizes PV cells composed of a single silicon crystal derived from sand. The manufacturing of these cells involves a complicated, multistage process that includes extracting, purifying, and heating silicon to elevated temperatures, followed by the use of the Czochralski method (Saga, 2010). Due to the high manufacturing costs associated with monocrystalline PV systems, polycrystalline silicon PV technology has been developed as a more cost-effective alternative (Allouhi et al., 2022; Saga, 2010). While both technologies operate on the same basic principles, polycrystalline PV systems differ slightly in their crystalline silicon structure. In polycrystalline systems, the silicon consists of multiple crystal grains, which introduce additional recombination sources that affect the total efficiency of the PV cells. Polycrystalline PV technology generally attains efficiencies between 15% and 20%, varying based on the particular technology implemented and the environmental circumstances in which the system functions (Kenu, 2020; Pastuszak and Węgierek, 2022).

18.3.4 Second generation: thin-film solar cells

The next generation of solar PV technology brought forth thin-film solar cells, created to minimize material consumption and lower production costs. These cells are made up of single or multiple layers of PV elements. This generation includes various PV systems such as amorphous silicon, copper indium gallium selenide, gallium arsenide (GaAs), cadmium telluride, and copper zinc tin sulfide (Huang and Wang, 2018; Radue and van Dyk, 2010; Mateo et al., 2022). Among these, amorphous silicon PV is the most advanced thin-film option, featuring a p-i-n or n-i-p junction configuration that creates an internal electric field and offers a superior absorption capacity of 1.1–1.7 eV, surpassing the first generation's 1.1 eV (Mikolasek et al., 2019). However, its low structural homogeneity affects electron and hole movement, impacting absorption. To address this, researchers have developed heterojunctions of amorphous and crystalline silicon (heterojunction technology [HJT] cells), which improve efficiency, open-circuit voltage, and thermal stability (Zeng et al., 2023).

GaAs and CdTe PV systems are also notable in this generation. GaAs is preferred for space applications due to its radiation resistance. CdTe, with an energy gap of 1.4 eV, can capture a broad spectrum of light intensities and has seen efficiency improvements from over 10% in the early 1980s to more than 15% in recent years (Bonnet, 2012).

18.3.5 Third generation: emerging and advanced photovoltaic technologies

Third-generation solar PV technology represents a significant leap in solar energy conversion, aiming to address the shortcomings of previous generations and attain high-efficiency while keeping production costs low. This generation includes innovative approaches such as dye-sensitized solar cells (DSSCs), organic photovoltaics (OPVs), perovskite solar cells, quantum dot solar cells, and tandem solar cells, among others. DSSCs use organic dye molecules to capture sunlight and produce electrons. They are recognized for their affordable production costs and their effectiveness in low-light environments. The architecture typically includes a photosensitizer (dye), a semiconductor (usually titanium dioxide), and a liquid electrolyte. OPVs use organic molecules or polymers totransform solar energy into electrical power. They provide benefits including mechanical flexibility, light weight, and the possibility of cost-effective production via roll-to-roll manufacturing. Recent progress in materials and device engineering has notably enhanced the power conversion efficiency of OPV cells. Perovskite solar cells have attracted considerable interest because of their excellent efficiency and relatively straightforward manufacturing process. These cells employ a perovskite-structured material as the layer responsible for capturing light. These cells use a perovskite-structured compound as the light-harvesting active layer. They have rapidly increased in efficiency, reaching over 25% in laboratory settings, making them one of the most promising emerging PV technologies. Quantum dot solar cells incorporate nanoscale semiconductor materials (quantum dots) that have unique electronic properties due to quantum mechanics. These cells can be designed to capture various segments of the solar spectrumby adjusting the size of the quantum dots, potentially leading to higher efficiencies (Solak and Irmak, 2023; Alami et al., 2022).

18.4 Electrochemical systems for water-splitting

Electrochemical water-splitting, which produces hydrogen of high-purity with no CO_2 emissions, holds great promise for achieving a sustainable energy future (Seh et al., 2017; Grigoriev et al., 2020). In contrast to coal gasification and steam-methane reforming, electrochemical water-splitting has benefits owing to the abundant availability of basic inputs like electricity and water (Kou et al., 2021).

18.4.1 Overview of electrolyzers

Various types of water electrolyzers, such as Proton Exchange Membrane (PEM), alkaline, Anion Exchange Membrane (AEM), and Solid Oxide electrolyzers, are defined based on their operating temperatures and the electrolytes employed (Li and Baek, 2021; Sapountzi et al., 2017; Zakaria and Kamarudin, 2021). Solid oxide electrolyzers operate at elevated temperatures (700°C–800°C), whereas the other three operate at much lower temperatures (typically below 100°C). Despite having a theoretical efficiency of up to 100%, solid oxide electrolyzers face challenges due to high operating temperatures causing safety and durability issues (Sapountzi et al., 2017). In PEM, alkaline, and AEM water electrolyzers, two electrodes are submerged in a water-based electrolyte and linked to an external circuit. This setup facilitates the OER at the anode and the HER at the cathode (Zakaria and Kamarudin, 2021; Lagadec and Grimaud, 2020).

Alkaline water electrolyzers employ an alkaline electrolyte, that is, constantly circulated through the electrolysis stack, where it generates hydrogen and oxygen gases. They have several advantages, such as fast start-up, high-purity hydrogen production, compact design, and low resistance losses. The primary element is the diaphragm, which facilitates ion movement between the anode and cathode. Additionally, heat exchangers are employed to ensure the electrolyte remains at the ideal temperature.

PEM water electrolyzers involve a membrane electrode assembly containing catalysts and the PEM. The ionomer within the system manages liquid or gas flow, ensures ionic conductivity, and provides mechanical stability for the catalysts (Chen et al., 2022).

AEM water electrolyzers are recognized as a technology of the third generation, combine features of both alkaline and PEM electrolyzers. They use transition-metal-based materials and a solid polymer electrolyte structure. While alkaline electrolyzers use concentrated KOH/NaOH solutions, AEM electrolyzers can use distilled water or low-concentration alkaline solutions (Hassan et al., 2022). AEM electrolyzers with hydrocarbon AEM and low-cost electrodes are promising for hydrogen production but need further research on applicable catalysts and designs.

The efficiency and yield of water electrolyzers depend on the selection and design of membranes, support materials, binders/ionomers, gas diffusion, operational conditions, local pH values, and impurities (Table 18.1).

18.4.2 Catalysts for hydrogen and oxygen evolution reactions

18.4.2.1 Hydrogen evolution reaction electrocatalysts

HER electrocatalysts generally fall into two principal types: those that use noble-metals and those that employ nonnoble metals.

18.4.2.2 Noble-metal-based electrocatalysts

Noble metals, like those in the Pt group, exhibit excellent catalytic performance for HER. Nonetheless, their widespread use is restricted due to their rarity and high expense. To overcome this issue, it is essential to develop catalysts with minimal metal content and efficient metal use. Combining Pt with transition metals can greatly improve the efficiency of Pt usage and modify electronic environments, boosting HER electroactivity. In alkaline conditions, HER activity is typically reduced compared to acidic conditions because water dissociation on the Pt surface is less effective (Zhu et al., 2019). To improve this, Pt is often coupled with water dissociation promoters (Xie et al., 2019). Among various Pt-based electrocatalysts, regulating the metal composition on the surface is essential for improving

TABLE 18.1 Comparison of the main characteristics of alkaline, PEM and AEM water electrolyzers (Sun et al., 2023).

	Alkaline water electrolyzer	PEM water electrolyzer	AEM water electrolyzer
Separator	Diaphragm (Zirfon Perl 500 μm)	Nafion 117	AEM (20–100 μm)
Catalysts	Anode:Ni, Ni-Co alloys Cathode: Ni, Ni-Moalloys	Anode: Ru/IrO_2 Cathode: Pt, Pt-Pd	Anode: Ni, Fe, Cooxides Cathode: Ni and Ni alloys
Electrolyte	KOH (20–40 wt.%)	High-purity deionized water (18 MΩcm)	Alkaline solution or deionized water
Ion transport	OH^-	H^+	OH^-
Temperature (°C)	50–80	50–80	50–70
Pressure(bar)	1–30	30–76	1–30
Current density (Acm^{-2})	0.2–0.6	0.6–2.0	0.2–0.4
Cellvoltage (V)	1.8–2.4	1.8–2.2	1.8–2.2
Efficiency (%)	60–70	70–80	–
Gas purity (vol%)	> 99.5	> 99.9999	> 99.99
Developing status	Mature technology	Mature for small scale	Under development
Advantages	Nonnoble metal catalysts, great durability, low capital cost	Compact design, fast response or start-up, high-purity production	Combined advantages of Alkaline and PEM water electrolyzers.
Disadvantages	Slow dynamics, corrosive electrolyte, further purification stage	Noble-metal catalysts, poor durability, high capital cost	Low OH^-conductivity in polymeric membranes

PEM, Proton exchange membrane; *AEM*, anion exchange membrane.

electrocatalytic performance. Additionally, doping Pt-based materials with different metals improves catalytic performance while reducing Pt usage.

18.4.2.3 Nonnoble-metal-based electrocatalysts

Transition-metal carbides have garnered significant attention for developing nonnoble metal electrocatalysts. For instance, Mo_2C and WC demonstrate high catalytic activity for HER. In addition to their outstanding electrical conductivity, their hydrogen adsorption characteristics and d-band electronic density resemble those of Pt, which is believed to be a major factor in their high HER activity (Levy and Boudart, 1973; Kitchin et al., 2005; Vrubel and Hu, 2012).

Research into transition-metal phosphides (TMPs) is progressing quickly in the creation of electrocatalysts that offer both high catalytic activity and stability in acidic and basic environments. The P atoms in TMPs are believed to play a crucial role because of their outstanding conductivity and distinctive electronic configuration. Since 2005, Ni_2P has been recognized as one of the most effective practical catalysts for HER (Liu and Rodriguez, 2005). Ni_2P exceeds the performance of bulk Pt and Ni due to the strong binding of hydrogen, produced during HER, to the hollow sites of the metal. This strong bonding increases the energy required for hydrogen to desorb from the surface of the metal. The inclusion of P atoms decreases the number of highly active Ni sites, resulting in a more moderate interaction with intermediates and products, a phenomenon known as the "ensemble effect" (Popczun et al., 2013). Nevertheless, the stability of the Ni_2P/Ti electrode, particularly in alkaline electrolytes, was inadequate. The $NiCo_2Px$ catalyst, on the other hand, shows exceptional longevity and consistent stability across different electrolytes, establishing it as a highly effective pH-universal catalyst for HER (Zhang et al., 2017).

Transition-metal chalcogenides (sulfides and selenides): In 2005, density functional theory (DFT) calculations conducted by Nørskov et al. demonstrated that the free energy of atomic hydrogen binding to the MoS_2 edge is similar to that of Pt. This finding underscores MoS_2's potential as an effective electrocatalyst for HER (Hinnemann et al., 2005). To pinpoint the precise active site of MoS_2, Chorkendorf et al. synthesized triangular MoS_2 single crystals of various sizes on an Au(111) substrate. Their research revealed that HER activity is directly proportional to the number of edge sites on the MoS_2 catalyst, which are highly active for catalysis (Jaramillo et al., 2007). Building on this insight, several strategies, including nanostructure design and morphology optimization, have been proposed to enhance HER activity by exposing these active sites (Xie et al., 2013). In a different study, Jin et al. successfully prepared CoS_2 with tunable morphologies, including film, microwire, and nanowire (NW) forms. They systematically analyzed the structures, activities, and stabilities of these different morphologies, revealing that both activity and stability are enhanced depending on the morphology. Among the three types, CoS_2 NWs showed the greatest HER catalytic efficiency and stability, which is credited to their extensive effective electrode surface area and the effective removal of gas bubbles from the surface of the electrode (Faber et al., 2014). Additionally, doping with heteroatoms has been shown to be a successful approach for enhancing HER activity.

18.4.3 Oxygen evolution reaction electrocatalysts

OER electrocatalysts can be divided into two main categories: those derived from noble-metals and those derived from nonnoble metals.

18.4.3.1 Noble-metal-based electrocatalysts

Noble-metal and MO electrocatalysts, like RuO_2 and IrO_2, have traditionally been considered top materials for the OERdue to their superior performance. However, their high cost and significant dissolution issues necessitate modifications to optimize their composition, structure, and morphology. Several approaches, such as heteroatom doping and mixing with other transition metals, have been explored to enhance the activity, stability, and reduce costs (Su et al., 2018; Wang et al., 2019). Surface structure modifications, especially in creating hollow nanoparticles like nanocages and nanoframes, have been successful in boosting catalytic activity (Park et al., 2017). Other noble-metals like Rh, Au, Pt, and Pd are also being recognized as potential OER electrocatalysts. Designing bi- or tri-functional catalysts with these metals can improve performance for OER, oxygen reduction reaction, and HER. These catalysts are typically evaluated in alkaline solutions due to their lower dissolution resistance in acidic electrolytes. Controlling the morphology and composition is crucial for attaining the best electrocatalytic performance (Lu et al., 2016).

18.4.3.2 Nonnoble metal-based electrocatalysts

Nonnoble metal-based electrocatalysts have garnered significant research attention due to their low-cost and widespread availability. Attention has increasingly shifted towards identifying efficient noble-metal-free OER electrocatalysts. Zhang et al. (2016) developed a gelled FeCoW oxyhydroxide (W, Fe-doped CoOOH, G-FeCoW) with uniform metal distribution using the sol–gel method. This FeCoW oxyhydroxide exhibited an overpotential of 191 mV at a current density of 10 mA/cm^2 and demonstrated 500 hours of cycle stability, outperforming the benchmark Ni–Fe based catalyst (Zhang et al., 2016). Defect engineering is a useful approach for modifying the structural and electronic characteristics of electrocatalysts. Improving OER activity can be accomplished by adjusting the intermediate adsorption energy, which can occasionally lead to the discovery of unforeseen active sites (Xu et al., 2016). Beyond intrinsic changes for optimal intermediate species adsorption energy, the capacity to modify electron transport is crucial for enhancing OER activity (Lu et al., 2014; Zhou et al., 2018).

Most nonnoble metal catalysts are MOs and (oxy)hydroxides. Lately, a range of promising electrocatalysts, including TMPs, sulfides, and selenides, have exhibited remarkable catalytic performance for OER. Nonetheless, these compounds often show limited durability in highly oxidative environments within alkaline solutions. Therefore understanding the chemical nature of the true active sites has become a major focus in developing OER-related catalysts. Hu and colleagues synthesized nanostructured nickel iron diselenide ($Ni_xFe_{1-x}Se_2$) as a template precursor for in situ generation of a highly active nickel iron oxide catalyst. This catalyst demonstrated outstanding OER activity with an overpotential of only 195 mV at 10 mA cm^2 (Xu et al., 2016).

18.5 Efficiency of water-splitting

The efficiency of water-splitting, particularly through electrolysis, can vary depending on the specific method, the materials used for the electrodes, and the operating conditions. Table 18.2 estimates for the efficiency of different water-splitting technologies.

TABLE 18.2 Efficiency of different water-splitting technologies.

Sl. no.	Water-splitting process	Efficiency	Reference
1.	Alkaline water electrolysis	60%–80%	Ursua et al. (2012)
2.	Proton exchange membrane electrolysis	70%–85%.	Ayers et al. (2010)
3.	Solid oxide electrolysis cells	exceed 85% (above 800°C)	Graves et al. (2011)
4.	Photoelectrochemical (PEC) water-splitting	10%–20%,	Lewis and Nocera (2006)
5.	Thermochemical water-splitting	30%–40%.	Agrafiotis et al. (2015)

18.6 Photovoltaic electrocatalysis

Photo-electrochemical water-splitting (PEC) can be considered as an ideal method, because it is environmentally friendly and renewable in nature. This method efficiently transforms sunlight into chemical energy, enabling the oxidation or reduction of substances to produce valuable materials, including hydrogen (Gopinath et al., 2022).

Hydrogen fuel, which is environmentally friendly, storable, and renewable, can be effectively produced by splitting water, a resource, that is, abundantly available. However, the high voltage required for water electrolysis (> 1.23 V) poses a challenge to cost-effectiveness. To address this, PEC cells are used to lower the voltage. When powered by solar-driven PV devices, PEC cells create an all-clean, efficient technology that relies solely on renewable sources, making this an area of significant research interest (Chatterjee et al, 2022a,b). Photovoltaic electrocatalysis (PV-EC) mitigates the limitations of both particulate photocatalysis and electrocatalysis. This method involves connecting the output of a PV module to an electrolyzer. Consequently, water-splitting occurs in a two-step process: (1) photon-to-electric conversion via the PV module and (2) electric-to-chemical energy conversion using electrolyzers with electrodes made from electrocatalytic materials. By employing PEMs between electrodes, H_2 and O_2 are generated separately.

In a study, Gimm et al. introduced an equivalent circuit model to evaluate the steady-state performance of photoelectrochemical cells. This model compares devices situated in various locations with differing average illumination levels and tilt angles. The findings indicate that using actual illumination data significantly affects the efficiency of PV-EC devices. The average annual solar-to-hydrogen (STH) efficiency is notably lower than the ideal value and is influenced by the tilt angle, with the optimal angle for latitudes similar to Europe's being approximately 40° (Grimm et al., 2022).

Landman and colleagues introduced a novel concept of cell separation within the PEC system, which decouples the PEC system into distinct oxygen and hydrogen cells. In this decoupled PEC setup, the researchers utilized the first generation of silicon PV modules to facilitate photoelectrocatalytic hydrogen production in a separate cell arrangement. The silicon PV module exhibited a short-circuit current of 164.5 mA and an open-circuit voltage of 5.4 V. Notably, charge carrier separation in both individual cells was achieved by incorporating battery-grade nickel hydroxide into each cell compartment (Chatterjee et al., 2022a,b; Lublow and Schedel Niedrig, 2020). Another work explores the use of perovskite/silicon tandem solar cells integrated with a thermal exchanger, which significantly improves performance for scalable direct water-splitting. The study highlights the development of a modular approach to scale up the concept and discusses the stability and degradation mechanisms of these devices (Maragno et al., 2024). In other work, the efficiency of PV-assisted electrochemical water-splitting systems has been achieved using low-cost light-absorbing materials to attain high STH conversion rates. It demonstrates a PV-electrolysis system with an impressive STH efficiency, highlighting the potential for scalable hydrogen production (Jia et al., 2016). Finger et al. (2020) examine the use of silicon-based multijunction solar cells in integrated PV-EC systems, focusing on achieving high open-circuit voltages and efficient hydrogen production.

18.7 Case study

18.7.1 Title

Photovoltaic electrochemical integration for water-splitting.

18.7.2 Objective

The case study on "Photovoltaic electrochemical integration for water-splitting" typically aims to achieve the following objectives:

1. **Efficiency improvement**: To enhance the efficiency of converting solar energy into chemical energy through water-splitting, leveraging the integration of PV and electrochemical systems.
2. **Sustainable hydrogen production**: To develop a sustainable and clean method for producing hydrogen from water using solar energy, which can serve as a renewable energy source.
3. **System integration**: To explore and optimize the integration of PV cells with electrochemical water-splitting processes, ensuring that the system operates efficiently and effectively.
4. **Cost reduction**: To reduce the overall cost of hydrogen production by improving the performance and economic viability of the PV-electrochemical systems.
5. **Technical feasibility**: To evaluate the technical feasibility of combining PV and electrochemical technologies, including the assessment of material compatibility, system design, and operational stability.

18.7.3 Scope

Project limitations

1. **Limited time period**: Research projects often face constraints due to the academic calendar or funding timelines. This may affect the depth of experimentation and data collection.
2. **Access to materials/tools**: Availability of specialized materials (like high-efficiency PV cells) and tools (such as advanced electrochemical testing equipment) can be a limiting factor.

3. **Technical expertise**: The complexity of integrating PV systems with electrochemical water-splitting may require a high level of technical knowledge, which can be a constraint if the team lacks experience in certain areas.

Intended workflow (in brief)

1. **Literature review**: Start with a comprehensive review of existing technologies and integration methods for PV water-splitting systems.
2. **Design and simulation**: Develop theoretical models and simulations to predict the performance of different integration schemes.
3. **Material selection**: Choose appropriate PV cells and electrochemical catalysts based on performance criteria.
4. **System integration**: Construct the prototype by integrating the PV system with the electrochemical water-splitting setup.
5. **Testing and optimization**: Conduct experiments to test the system's efficiency, stability, and scalability. Optimize parameters based on initial results.
6. **Analysis and reporting**: Analyze the data, compare results with theoretical predictions, and prepare a comprehensive report.

Budget and costs

1. **Materials**: Costs will vary based on the quality and quantity of PV cells, electrochemical catalysts, and other components.
2. **Equipment**: Expenses for testing equipment, such as potentiostats or solar simulators, can be significant.
3. **Labor**: Consider costs related to the team's time and expertise, including any hiring of additional staff or consultants.
4. **Miscellaneous**: Budget for unexpected expenses, such as repairs, additional testing, or unforeseen challenges.

Planning carefully and addressing these factors will help in managing the project effectively and achieving meaningful results.

Audience

1. **Researchers and scientists:**
   - **Who will find it useful**: Those working in the fields of renewable energy, materials science, and electrochemistry. Researchers focused on improving STH conversion efficiencies and developing sustainable energy solutions.
   - **Ability to follow**: They should have a strong background in PVs, semiconductor materials, and electrochemical processes. Familiarity with current challenges in water-splitting technology is essential.
2. **Graduate and postgraduate students**:
   - **Who will find it useful**: Students pursuing degrees in energy systems, chemical engineering, or environmental science. Those interested in solar energy and hydrogen production would benefit from understanding the integration techniques discussed.
   - **Ability to follow**: They should have foundational knowledge in electrochemistry, renewable energy technologies, and an understanding of the principles behind water-splitting and PV systems.

3. **Energy industry professionals**:
   - **Who will find it useful**: Engineers, technical consultants, and professionals involved in the development of renewable energy projects, particularly those focused on hydrogen production and solar energy integration.
   - **Ability to follow**: Industry professionals with experience in PVs, electrochemical systems, and energy storage would be able to follow the case study, though they may require some background in the latest research developments.
4. **Policy makers and environmental advocates**:
   - **Who will find it useful**: Those involved in developing policies for renewable energy adoption and sustainability goals. Advocates promoting clean energy solutions could use the case study to support policy changes or advocacy campaigns.
   - **Ability to follow**: While not necessarily experts, they should have a general understanding of renewable energy technologies and the significance of hydrogen as a clean energy carrier.
5. **Funding agencies and investors**:
   - **Who will find it useful**: Organizations and individuals interested in investing in cutting-edge renewable energy technologies or funding research in this area.
   - **Ability to follow**: They would need a general understanding of the technology's potential, economic implications, and how it fits into the broader energy landscape.

Participants in the case study

- **Lead researchers**: Specialists in PVs and electrochemical processes who conducted the experimental work and analysis.
- **Collaborating institutions**: Universities, research centers, or companies specializing in renewable energy technologies.
- **Funding bodies**: Organizations that provided financial support for the research, including government agencies, private foundations, or industry partners.
- **Technical staff**: Engineers and laboratory technicians who supported the experimental setup and data collection.
- **Peer reviewers**: Academics and experts who reviewed the case study to ensure the validity and significance of the findings before publication.

18.7.4 Rationale

The case study on "PV Electrochemical Integration for Water-Splitting" addresses the critical need for sustainable energy solutions. The rationale for this activity is rooted in several key motivations:

1. **Energy crisis and environmental concerns**: The growing global demand for energy, coupled with the detrimental effects of fossil fuel consumption, has necessitated the exploration of renewable energy sources. Water-splitting, which produces hydrogen—a clean fuel—is seen as a promising solution. However, traditional methods of hydrogen production are energy-intensive and rely heavily on nonrenewable resources.
2. **Photovoltaic technology**: PV technology, which converts sunlight into electricity, offers a sustainable and abundant source of energy. By integrating PV with electrochemical

water-splitting, the aim is to harness solar energy to produce hydrogen efficiently. This integration could potentially lead to a self-sustaining, low-cost, and environmentally friendly method of hydrogen production.

3. **Technological innovation**: The main issue being addressed is the inefficiency and cost associated with current hydrogen production methods. The experiment seeks to explore how effectively PV cells can be coupled with electrochemical systems to enhance the efficiency of water-splitting. The focus is on optimizing the materials, system design, and operational conditions to maximize hydrogen yield while minimizing energy loss.
4. **Sustainable energy transition**: The experiment was held to contribute to the broader goal of transitioning to a hydrogen-based economy. By demonstrating the feasibility and scalability of PV-electrochemical integration, the study aims to provide insights that could accelerate the adoption of hydrogen as a clean energy carrier.

In summary, the activity was driven by the need to find efficient, sustainable, and cost-effective methods for hydrogen production, leveraging the potential of PV technology in the process.

Expected results and deliverables

1. **Efficient solar-to-hydrogen conversion**: The case study is expected to demonstrate advancements in PV electrochemical integration for water-splitting, focusing on achieving higher STH conversion efficiency. This would contribute to the development of sustainable and cost-effective methods for hydrogen production.
2. **Optimization of materials and processes**: By analyzing various materials and processes used in PV and electrochemical cells, the study aims to identify the most efficient combinations. This optimization can lead to improved performance and stability, benefiting future research and industrial applications.
3. **Scalability and practical implementation**: The study will explore the scalability of the integrated system, assessing its potential for large-scale implementation. Insights gained from this can guide the design of commercial hydrogen production systems.
4. **Environmental impact assessment**: The case study will provide an analysis of the environmental benefits of using renewable energy sources for hydrogen production. This includes reducing carbon emissions and minimizing the ecological footprint of the energy production process.
5. **Cost-benefit analysis**: A detailed cost-benefit analysis will be conducted to evaluate the economic viability of the integrated system. This will help stakeholders understand the financial implications and potential returns on investment in PV electrochemical water-splitting technologies.

Reader benefits

1. **Comprehensive understanding**: Readers will gain a comprehensive understanding of the latest advancements in PV electrochemical integration for water-splitting, including the underlying principles, materials, and processes involved.
2. **Practical insights**: The case study will provide practical insights into the optimization and implementation of such systems, offering guidance for researchers, engineers, and industry professionals.

3. **Informed decision-making**: By presenting a detailed cost-benefit analysis and environmental impact assessment, the case study will enable readers to make informed decisions about investing in or adopting these technologies for sustainable energy production.
4. **Future research directions**: The study will highlight areas where further research is needed, helping to shape the direction of future studies and innovations in the field of renewable energy and hydrogen production.

18.7.5 Safety considerations

When integrating PV and electrochemical systems for water-splitting, safety considerations are crucial to the project's success. Here are key safety aspects that the reader must be aware of:

1. **Electrical safety**
   - **High voltage risks:** The combination of PV panels and electrochemical cells can generate high voltages, especially under optimal sunlight conditions. Proper insulation, grounding, and circuit protection (like fuses and circuit breakers) are essential to prevent electric shock or short circuits.
   - **Overcurrent protection:** Given the variability of solar energy, the system must include overcurrent protection to safeguard against potential surges that could damage equipment or cause fires.
2. **Chemical safety**
   - **Handling of electrolytes:** Electrolytes used in water-splitting systems can be corrosive or toxic. Proper storage, handling, and disposal of these chemicals are essential. Personal protective equipment (PPE) such as gloves, goggles, and lab coats should be used when dealing with electrolytes.
   - **Gas management:** The process generates hydrogen and oxygen gases, which can be explosive if mixed in the wrong proportions. Ensuring proper ventilation and gas separation is critical. The design should include pressure relief valves and fail-safes to avoid gas buildup.
3. **Thermal safety**
   - **Heat generation:** Both PV panels and electrochemical cells can produce significant amounts of heat during operation. This necessitates adequate cooling systems to prevent overheating, which could lead to thermal runaway or equipment failure.
   - **Fire prevention:** The combination of electrical and chemical components poses a fire risk. Fire suppression systems should be in place, and all materials used should be flame-retardant.
4. **Structural integrity**
   - **Material compatibility:** Ensure that materials used in the construction of the PV and electrochemical systems are compatible and resistant to corrosion, especially in humid or outdoor environments.
   - **Load bearing:** The structure supporting the PV panels and electrochemical cells should be capable of handling the weight and any dynamic forces, such as wind or seismic activity.

5. **Environmental and health safety**
   - **Waste management:** Proper protocols must be in place for the disposal of chemical waste and decommissioned components, considering the environmental impact.
   - **Exposure to sunlight:** Prolonged exposure to sunlight during installation or maintenance can pose risks such as heatstroke or UV damage to workers. Adequate shade, hydration, and sunscreen should be provided.
6. **Regulatory compliance**
   - **Adherence to standards:** The system must comply with local, national, and international safety standards, including electrical codes, chemical handling regulations, and environmental laws.
   - **Permitting and inspections:** Before starting the project, ensure that all necessary permits are obtained, and regular inspections are scheduled to maintain safety standards throughout the project's lifecycle.

These safety considerations are critical to minimizing risks and ensuring the safe and successful implementation of a PV electrochemical water-splitting system.

18.7.6 Actions taken/Workflow/Tools used/Simulations and analyses

Photovoltaic electrochemical integration for water-splitting is an advanced method combining PV cells with electrochemical systems to convert sunlight into chemical energy, primarily in the form of hydrogen through water-splitting. This case study explores the actions taken, workflow, tools used, simulations, and analyses performed in a research setting aimed at optimizing this technology.

1. Understanding the problem statement

Objective:
The primary goal is to integrate PV cells with an electrochemical system for efficient water-splitting to generate hydrogen. The study focuses on optimizing the efficiency of the system and analyzing the performance under different conditions.

2. Literature review and initial planning

Actions taken:
- Reviewed existing literature on PV-EC systems, focusing on the types of PV cells (e.g., silicon-based, perovskite) and electrochemical catalysts used for water-splitting.
- Identified key parameters influencing system efficiency, including light absorption, charge carrier dynamics, and catalyst efficiency.

Tools used:
- **Zotero** for managing research papers and references.
- **Mendeley** for collaborative reference management.

3. System design and selection of materials
Step 1: Selection of photovoltaic cells
- Evaluated different PV materials for their efficiency, cost, and compatibility with electrochemical systems.

- Selected a high-efficiency perovskite solar cell based on its superior light absorption and ease of integration with electrochemical cells.

Step 2: Selection of electrochemical catalysts

- Explored various catalysts like platinum (Pt), iridium oxide (IrO_2), and nickel-iron (NiFe) for the OER and HER.
- Chose NiFe as the catalyst due to its cost-effectiveness and relatively high efficiency.

Tools used:

- **Material Studio** for computational material selection and initial simulations.
- **COMSOL Multiphysics** for modeling the integration and simulation of electrochemical processes.

4. Simulation and modeling

Step 3: Photovoltaic cell simulation

- Simulated the PV cell's performance under standard sunlight conditions to determine the voltage and current characteristics.

Step 4: Electrochemical cell simulation

- Modeled the electrochemical reactions within the cell to optimize the electrode design and catalyst placement.

Step 5: Integrated system simulation

- Combined the PV and EC models to simulate the integrated system's performance.
- Analyzed the effect of varying light intensity, temperature, and electrolyte concentration on system efficiency.

Tools used:

- **MATLAB®/Simulink** for system-level simulation and optimization.
- **Ansys** for finite element analysis to understand the thermal and stress behavior within the system.

Workflow:

1. **Initial setup:** Define the problem and gather necessary materials.
2. **Material selection:** Choose the appropriate PV material and electrochemical catalysts.
3. **Simulation:** Perform individual simulations of PV and EC systems.
4. **Integration:** Combine PV and EC systems into a single model.
5. **Optimization:** Refine system parameters to maximize efficiency.
6. **Validation:** Compare simulated results with experimental data.

5. Experimental validation

Step 6: Prototype development

- Constructed a small-scale prototype based on the optimized simulation model.
- Tested the prototype under controlled lab conditions to measure hydrogen production efficiency.

Step 7: Data collection and analysis

- Collected data on voltage, current, hydrogen production rate, and system stability.
- Compared the experimental results with the simulated predictions to validate the model.

Tools used:

- **LabVIEW** for data acquisition and real-time monitoring.
- **OriginPro** for data analysis and graphing.

6. **Results and discussion**
   - **Efficiency analysis:** Discussed the overall efficiency of the integrated PV-EC system, highlighting the key factors that contributed to the observed performance.
   - **Challenges and solutions:** Identified challenges such as system stability, degradation of materials, and energy losses, along with proposed solutions.
   - **Future Work:** Suggested areas for further research, including exploring new materials and optimizing system design.
7. **Conclusion**
   The study successfully demonstrated the integration of PV cells with electrochemical systems for water-splitting, achieving a significant improvement in hydrogen production efficiency. The use of advanced simulation tools and careful material selection were key to the project's success.

   Flowchart

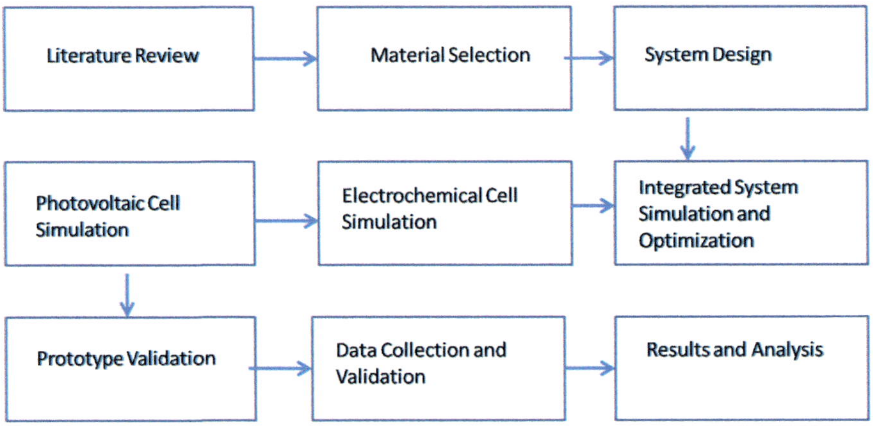

This detailed case study should provide a comprehensive understanding of the steps taken in the PV-EC integration process for water-splitting.

18.7.7 Challenges and solutions

Photovoltaic electrochemical water-splitting is a promising technology for sustainable hydrogen production. The process involves using sunlight to drive the electrolysis of water, producing hydrogen and oxygen. However, the integration of PV and electrochemical systems for efficient and cost-effective water-splitting faces several challenges. Below are some of the key challenges and potential solutions that have been explored in case studies and research:

18.7.8 Challenges

1. **Efficiency mismatch**:

- **Challenge**: The efficiency of PV cells and the electrochemical processes often do not match. PV cells might produce more electricity than the electrochemical cell can use efficiently, leading to losses.
- **Solution**: Researchers are exploring tandem solar cells, which combine different materials to absorb a broader range of the solar spectrum, increasing overall efficiency. Power management strategies that involve buffering and matching the power output of the PV cells to the needs of the electrochemical process are also being developed.

2. **Stability and durability**:
   - **Challenge**: The materials used in both the PV and electrochemical systems can degrade over time, especially under harsh operating conditions such as high temperatures or corrosive environments. This degradation reduces the system's lifespan and efficiency.
   - **Solution**: The development of more robust materials, such as corrosion-resistant catalysts and encapsulation techniques to protect PV cells, is crucial. Advances in material science are also focusing on improving the stability of photoelectrodes used in the process.
3. **Cost**:
   - **Challenge**: The high cost of materials, such as catalysts for the electrochemical reactions and high-efficiency PV cells, makes the overall system expensive.
   - **Solution**: Cost reduction can be achieved through the use of earth-abundant materials, improving manufacturing processes, and scaling up production. Researchers are also looking into cheaper, more efficient catalysts, such as those based on transition metal oxides instead of Pt.
4. **Integration complexity**:
   - **Challenge**: Integrating PV and electrochemical systems requires complex engineering to ensure that the two components work together efficiently. This includes challenges in power management, heat management, and system optimization.
   - **Solution**: Integrated system designs that optimize the interaction between PV and EC components are being developed. Computational models and simulations are also used to predict and improve system performance before physical prototypes are built.
5. **Intermittency of solar energy**:
   - **Challenge**: The intermittent nature of solar energy (e.g., night-time or cloudy days) makes it difficult to achieve consistent hydrogen production.
   - **Solution**: Hybrid systems that integrate energy storage solutions, such as batteries or other forms of energy storage, can smooth out the supply and ensure continuous operation. Additionally, the development of more efficient energy storage methods for hydrogen itself is an ongoing area of research.

18.7.9 Solutions uncovered

While these challenges are significant, ongoing research has led to several promising solutions:

- **Use of tandem cells**: Tandem cells that combine perovskite with silicon have shown higher efficiency in converting sunlight to electricity, which can then be used for water-splitting.
- **Advanced catalysts**: The development of nonprecious metal catalysts, such as NiFe alloys, has reduced costs while maintaining high catalytic activity for water-splitting.
- **System integration**: Improved design of integrated PV-EC systems, with better thermal management and more efficient power electronics, has enhanced overall system performance.
- **Energy storage**: The integration of energy storage solutions like batteries or hydrogen storage tanks helps in dealing with the intermittency of solar power.

These solutions are still in development, and large-scale implementation will require further advancements in technology, cost reductions, and more extensive field testing. However, the progress so far has been promising, with several pilot projects demonstrating the feasibility of PV-EC water-splitting as a sustainable hydrogen production method.

18.7.10 Results

Photovoltaic electrochemical integration for water-splitting is a cutting-edge approach that combines PV cells with electrochemical systems to directly convert sunlight into chemical energy, particularly for splitting water into hydrogen and oxygen. The results of case studies on this technology typically focus on efficiency, stability, and scalability.

18.7.11 Outcomes of PV-EC integration for water-splitting

1. **Efficiency**:
   - **Expected**: The primary goal of PV-EC systems is to achieve high STH efficiency. Theoretically, STH efficiencies of 15–20% or higher are considered highly desirable.
   - **Actual**: Many case studies report STH efficiencies ranging from 10% to 19%, depending on the specific materials and design used. This is generally within expected ranges, though ongoing research aims to push these efficiencies higher.
2. **Stability and durability**:
   - **Expected**: Long-term stability under operational conditions is crucial for practical applications. A stable system would operate effectively over months or years without significant degradation.
   - **Actual**: Case studies often report stability challenges, particularly in the harsh conditions required for water-splitting. Some systems maintain stability for thousands of hours, while others experience rapid degradation, which can be a limiting factor.
3. **Scalability**:
   - **Expected**: The ability to scale up the technology for industrial use is a significant consideration. Laboratory-scale efficiencies may not always translate directly to larger systems.
   - **Actual**: Case studies show that while small-scale systems demonstrate promising results, scaling up can introduce new challenges, such as increased costs and reduced

efficiency. However, some studies report successful scale-up with minimal losses in performance.

4. **Cost**:
   - **Expected**: Reducing the cost of materials and manufacturing processes is critical to making PV-EC systems commercially viable.
   - **Actual**: While some case studies indicate that costs are still high, ongoing research and development are focused on using cheaper materials and more efficient manufacturing techniques. Some success has been reported in reducing costs, though significant challenges remain.

18.7.12 Were the outcomes expected?

- **Partially**: Many outcomes align with theoretical predictions, particularly in terms of efficiency and initial performance. However, challenges related to long-term stability, scalability, and cost have been more difficult to overcome than expected. Researchers are continuing to refine materials and system designs to address these issues.

In summary, while the outcomes of PV-EC integration for water-splitting are promising, they also highlight the need for continued research to address the challenges that remain before this technology can be widely adopted.

18.7.13 Learning and knowledge outcomes

1. **Learning and knowledge outcomes:**
   - **Integration benefits**: The case study likely highlighted the synergy between PV cells and electrochemical water-splitting. Combining PV systems with electrochemical cells can enhance the efficiency of converting solar energy into hydrogen, a clean fuel.
   - **System efficiency**: Insights into how the efficiency of PV electrochemical systems can be optimized, including the impact of different materials, system designs, and operational conditions on hydrogen production rates.
   - **Material compatibility**: Understanding the importance of selecting appropriate materials for both the PV cells and the electrochemical cell components to ensure long-term stability and performance.
   - **Economic and environmental impact**: The potential for reducing costs and environmental impacts compared to traditional energy sources, as well as the scalability of these systems for practical applications.
2. **What could have been done differently:**
   - **Experimental design**: More rigorous experimental setups and controls might have been used to isolate variables and accurately measure system performance under varying conditions.
   - **Long-term testing**: Extended testing over longer periods could provide a better understanding of the durability and stability of the integrated system.
   - **Broader material testing**: Exploring a wider range of materials for both the PV cells and the electrochemical components might reveal more efficient or cost-effective alternatives.

3. **What mustn't be changed:**
   - **Focus on integration**: The emphasis on integrating PV systems with electrochemical cells should remain, as it is central to the innovation and potential of this approach.
   - **Sustainability goals**: The focus on sustainability and clean energy solutions is crucial and should continue to drive future research and development efforts.
4. **Recommendations for future investigation:**
   - **Advanced materials research**: Investigate new and improved materials that could enhance efficiency, reduce costs, or improve the longevity of the integrated systems.
   - **System optimization**: Focus on optimizing system designs for different scales, from small-scale applications to larger, industrial-scale systems.
   - **Economic analysis**: Conduct comprehensive economic analyses to better understand the cost-effectiveness and market viability of PV electrochemical integration.
   - **Integration with other technologies**: Explore how this integration can be combined with other renewable energy technologies or energy storage systems to create more comprehensive and resilient energy solutions.

These aspects will help in refining the current understanding of PV electrochemical integration for water-splitting and drive advancements in the field.

18.8 Conclusion

The integration of photovoltaic technology with electrochemical water splitting offers a sustainable and efficient approach to hydrogen production. By harnessing solar energy, this method decreases dependence on fossil fuels and reduces greenhouse gas emissions, making a substantial contribution to renewable energy efforts. Advances in PV materials and electrochemical catalysts have significantly improved the overall efficiency of the water-splitting process. High-efficiency PV cells combined with effective electrocatalysts ensure better solar-to-hydrogen conversion rates, making the technology more viable for large-scale applications. The use of renewable solar energy for hydrogen production significantly reduces the environmental footprint compared to conventional methods. By minimizing carbon emissions and reducing dependence on non-renewable resources, PV-assisted water splitting bolsters worldwide initiatives to address climate change and advance sustainable development. Ongoing research is focused on further improving the efficiency and durability of both PV cells and electrochemical catalysts. Innovations in materials science, such as the development of new semiconductor materials and novel catalyst designs, are expected to drive further advancements. Additionally, integrating PV-assisted water splitting with other renewable energy systems, like wind and biomass, could enhance overall energy efficiency and storage capabilities. Although considerable advancements have been achieved, challenges persist regarding energy conversion efficiency, system stability, and economic competitiveness. Addressing these challenges requires continued interdisciplinary research and collaboration among scientists, engineers, and policymakers. Incentives for renewable energy adoption and supportive regulatory frameworks will also play crucial roles in overcoming these barriers.

References

Agrafiotis, C., Roeb, M., Sattler, C., 2015. A review on solar thermal syngas production via redox pair-based water/carbon dioxide splitting thermochemical cycles. Renewable and Sustainable Energy Reviews 42, 254–285.

Ahmad, I., Zou, Y., Yan, J., Liu, Y., Shukrullah, S., Naz, M.Y., et al., 2023. Semiconductor photocatalysts: a critical review highlighting the various strategies to boost the photocatalytic performances for diverse applications. Advances in Colloid and Interface Science 311, 102830.

Alami, A.H., Aljaghoub, H., Ayoub, M., Mdallal, A., Hasan, R., 2022. Third-Generation Photovoltaics: Organic Photovoltaics (OPV).

Allouhi, A., Rehman, S., Buker, M.S., Said, Z., 2022. Up-to-date literature review on solar PV systems: technology progress, market status and R&D. Journal of Cleaner Production 362, 132339.

Armaroli, N., Balzani, V., 2007. The future of energy supply: challenges and opportunities. Angewandte Chemie International Edition 46, 52–66.

Ashley, A.E., Thompson, A.L., O'Hare, D., 2009. Angewandte Chemie International Edition 48, 9839–9843.

Asif, M., Muneer, T., 2007. Renewable and Sustainable Energy Reviews 11, 1388–1413.

Ayers, K.E., Dalton, L.T., Yang, C., 2010. An evaluation of scale-up methods for the production of low-cost, high-performance PEM electrolyzers. Journal of the Electrochemical Society 157 (3), B352–B356.

Bai, X., 2016. A Cu_2O/Cu_2S-ZnO/CdS tandem photoelectrochemical cell for self-driven solar water splitting. Journal of Alloys and Compounds 698, 133–138. Available at https://doi.org/10.1016/j.jallcom.2016.12.261.

Bak, T., Nowotny, J., Rekas, M., Sorrell, C., 2002a. International Journal of Hydrogen Energy 27, 991–1022.

Bak, T., Nowotny, J., Rekas, M., Sorrell, C., 2002b. International Journal of Hydrogen Energy 27, 19–26.

Bonnet, D., 2012. CdTe Thin-Film PV Modules, 2nd ed. Elsevier Ltd., Amsterdam.

Chan, C.K., Tüysüz, H., Braun, A., Ranjan, C., La Mantia, F., Miller, B.K., et al., 2016. Advanced and in situ analytical methods for solar fuel materials. Springer: Berlin/Heidelberg, Germany, Volume 371, ISBN 9783319230986.

Chatterjee, P., Ambati, M., Chakraborty, A., Chakrabortty, S., Biring, S., Ramakrishna, S., et al., 2022a. Energy conversion and management. Energy Conversion and Management 261, 115648.

Chatterjee, P., Krishna, S., Chakraborty, A.K., Chakraboratty, S., Kumar, A., Biring, S., et al., 2022b. Energy Conversion and Management 261 (1), 115648.

Chen, Z., 2019. Metal Oxide-Based Tandem Cells for Self-Biased Photoelectrochemical Water Splitting. ACS Energy Letters 5, 844–850. Available at https://doi.org/10.1021/acsenergylett.9b02620.

Chen, Z., Guo, L., Pan, L., Yan, T., He, Z., Li, Y., et al., 2022. Advanced Energy Materials 12, 2103670.

Dorner, R.W., Hardy, D.R., Williams, F.W., Willauer, H.D., 2010. Energy & Environmental Science 3, 884–890.

Faber, M.S., Dziedzic, R., Lukowski, M.A., Kaiser, N.S., Ding, Q., Jin, S., 2014. High-performance electrocatalysis using metallic cobalt pyrite (CoS_2) micro- and nanostructures. Journal of the American Chemical Society 136 (28), 10053–10061.

Finger, F., Welter, K., Urbain, F., Smirnov, V., Kaiser, B., Jaegermann, W., 2020. Photoelectrochemical water splitting using adapted silicon based multi-junction solar cell structures: development of solar cells and catalysts, upscaling of combined photovoltaic-electrochemical devices and performance stability. Zeitschrift für Physikalische Chemie 234 (6), 1055–1095.

M.GopinathA review on solar energy-based indirect water-splitting methods for hydrogen generationInternational Journal of Hydrogen Energy2022

Graves, C., Ebbesen, S.D., Mogensen, M., Lackner, K.S., 2011. Sustainable hydrocarbon fuels by recycling CO_2 and H_2O with renewable or nuclear energy. Renewable and Sustainable Energy Reviews 15 (1), 1–23.

Grigoriev, S.A., Fateev, V.N., Bessarabov, D.G., Millet, P., 2020. International Journal of Hydrogen Energy 45, 26036.

Grimm, A., Marie, A.S., Kramer, G.J., Gazzani, M., 2022. International Journal of Hydrogen Energy 47 (23), 11764–11777.

Hassan, N., Zheng, Y., Kohl, P., Mustain, W.E., 2022. Journal of the Electrochemical Society

Hinnemann, B., Moses, P.G., Bonde, J., Jørgensen, K.P., Nielsen, J.H., Horch, S., et al., 2005. Atomic-scale insight into the hydrogen evolution reaction on MoS_2 and why transition metal dichalcogenides are efficient catalysts. Journal of the American Chemical Society 127 (15), 5308–5309.

Huang, C., Wang, L., 2018. Simulation study on the degradation process of photovoltaic modules. Energy Conversion and Management 165, 236–243.

Jaramillo, T.F., Jørgensen, K.P., Bonde, J., Nielsen, J.H., Horch, S., Chorkendorf, I., 2007. Identification of active edge sites for electrochemical H_2 evolution from MoS_2 nanocatalysts. Science 317 (5834), 100–102.

Jessop, P.G., Ikariya, T., Noyori, R., 1995. Chemical Reviews 95, 259–272.

Jiang, C., Moniz, S.J.A., Wang, A., Zhang, T., Tang, J., 2017. Photoelectrochemical devices for solar water splitting-materials and challenges. Chemical Society Reviews 46, 4645–4660.

Jia, J., Seitz, L., Benck, J., et al., 2016. Solar water splitting by photovoltaic-electrolysis with a solar-to-hydrogen efficiency over 30%. Nature Communications 7, 13237.

Kenu, E.S., 2020. A review of solar photovoltaic technologies. International Journal of Engineering Research 9, 741–749.

Kitchin, J.R., Nørskov, J.K., Barteau, M.A., Chen, J.G., 2005. Trends in the chemical properties of early transition metal carbide surfaces: a density functional study. Catalysis Today 105 (1), 66–73.

Kou, T., Wang, S., Li, Y., 2021. ACS Materials Letters 3, 224.

Lagadec, M.F., Grimaud, A., 2020. Nature Materials 19, 1140–1150.

Levy, R., Boudart, M., 1973. Platinum-like behavior of tungsten carbide in surface catalysis. Science 181 (4099), 547–549.

Lewis, N.S., Nocera, D.G., 2006. Powering the planet: chemical challenges in solar energy utilization. Proceedings of the National Academy of Sciences 103 (43), 15729–15735.

Liu, P., Rodriguez, J.A., 2005. Journal of the American Chemical Society 127 (42), 14871–14878.

Li, C., Baek, J.-B., 2021. Nano Energy 87, 106162.

Lublow, M., Schedel-Niedrig, T., 2020. Water oxidation catalysts from waste metal resources: a facile metal-organic electrochemical approach. Zeitschrift für Physikalische Chemie 234 (6), 1097–1113.

Lu, A., Peng, D.-L., Chang, F., Skeete, Z., Shan, S., Sharma, A., et al., 2016. ACS Applied Materials & Interfaces 8 (31), 20082–20091.

Lu, Z., Xu, W., Zhu, W., Yang, Q., Lei, X., Liu, J., et al., 2014. Chemical Communications 50 (49), 6479–6482.

Mahmood, N., Yao, Y., Zhang, J.W., Pan, L., Zhang, X., Zou, J.J., 2018. Advanced Science 5, 1700464.

Maragno, A.R.A., Morozan, A., Fize, J., Pellat, M., Artero, V., Charton, S., et al., 2024. Thermally integrated photoelectrochemical devices with perovskite/silicon tandem solar cells: a modular approach for scalable direct water splitting. Sustainable Energy & Fuels 8, 3726.

Mateo, C., Hernández-Fenollosa, M.A., Montero, Seguí-Chilet, S., 2022. Ageing and seasonal effects on amorphous silicon photovoltaic modules in a mediterranean climate. Renewable Energy 186, 74–88.

Matsumoto, Y., Sato, E., 1986. Materials Chemistry and Physics 14, 397–426.

Mikolasek, M., Kemeny, M., Chymo, F., Ondrejka, P., Huran, J., 2019. Amorphous silicon PEC-PV hybrid structure for photoelectrochemical water splitting. Journal of Electrical Engineering 70, 107–111.

Minoh, L., Haas, S., Smirnov, V., Merdzhanova, T., Rau, U., 2022. Scalable photovoltaic-electrochemical cells for hydrogen production from water-recent advances. ChemElectroChem 9 e202200838.

Momirlan, M., Veziroglu, T.N., 2005. International Journal of Hydrogen Energy 30, 795–802.

Nam, W., 2018. Solar water splitting with a hydrogenase integrated in photoelectrochemical tandem cells. Angewandte Chemie International Edition 57, 10595–10600. Available at https://doi.org/10.1002/anie.201805027.

Park, J., Sa, Y.J., Baik, H., Kwon, T., Joo, S.H., Lee, K., 2017. ACS Nano 11 (6), 5500–5509.

Pastuszak, J., Węgierek, P., 2022. Photovoltaic cell generations and current research directions for their development. Materials 15, 5542.

Perera, F., 2017. Pollution from fossil-fuel combustion is the leading environmental threat to global pediatric health and equity: solutions exist. International Journal of Environmental Research and Public Health 15, 16.

Popczun, E.J., McKone, J.R., Read, C.G., Biacchi, A.J., Wiltrout, A.M., Lewis, N.S., et al., 2013. Journal of the American Chemical Society 135 (25), 9267–9270.

Radue, C., van Dyk, E.E., 2010. A comparison of degradation in three amorphous silicon PV module technologies. Solar Energy Materials and Solar Cells 94, 617–622.

Saga, T., 2010. Advances in crystalline silicon solar cell technology for industrial mass production. NPG Asia Materials 2, 96–102.

Samsudin, M.F.R., Frebilot, C., Kaddoury, Y., Sufian, S., Ong, W.J., 2020. Bifunctional Z-scheme $Ag/AgVO_3/g\text{-}C_3N_4$ photocatalysts for expired ciprofloxacin degradation and hydrogen production from natural rainwater without using scavengers. Journal of Environmental Management 270, 110803.

Samsudin, M.F.R., 2023. Photovoltaic-assisted photo(electro)catalytic hydrogen production: a reviewEnergies, 161520235743.

Sapountzi, F.M., Gracia, J.M., Weststrate, C.J., Fredriksson, H.O.A., Niemantsverdriet, J.W., 2017. Progress in Energy and Combustion Science 58, 1–35.

Seh, Z.W., Kibsgaard, J., Dickens, C.F., Chorkendorff, I., Nørskov, J.K., Jaramillo, T.F., 2017. Science 355 eaad4998.

Shinagawa, T., Garcia-Esparza, A.T., Takanabe, K., 2015. Scientific Reports 5, 13801.

Si, Y.T., Yin Win, K., Siang Teo, W., Koh, L.-D., Liu, S., Teng, C.P., et al., 2017. Recent progress in energy-driven water splitting. Advanced Functional Materials 4, 1600337.

Solak, E.K., Irmak, E., 2023. Advances in organic photovoltaic cells: a comprehensive review of materials, technologies, and performance. RSC Advances 13, 12244–12269.

Spöri, C., Kwan, J.T.H., Bonakdarpour, A., Wilkinson, D.P., Strasser, P., 2017. Angewandte Chemie International Edition, 56, pp. 5994-6021; Angewandte Chemie, 129, pp. 6088-6117

Sun, H., Xu, X., Kim, H., Jung, W.C., Zhou, W., Shao, Z., 2023. Electrochemical water splitting: bridging the gaps between fundamental research and industrial applications. Energy & Environmental Materials 6, e12441.

Song, F., Bai, L., Moysiadou, A., Lee, S., Hu, C., Liardet, L., Hu, X., 2018. Transition Metal Oxides as Electrocatalysts for the Oxygen Evolution Reaction in Alkaline Solutions: An Application-Inspired Renaissance. Journal of the American Chemical Society 140 (25), 7748–7759. Available at https://doi.org/10.1021/jacs.8b04546.

Su, J., Ge, R., Jiang, K., Dong, Y., Hao, F., Tian, Z., et al., 2018. Advanced Materials 30 (29), 1801351.

Ursua, A., Gandia, L.M., Sanchis, P., 2012. Hydrogen production from water electrolysis: current status and future trends. Proceedings of the IEEE 100 (2), 410–426.

Vesborg, P.C.K., Jaramillo, T.F., 2012. Addressing the terawatt challenge: scalability in the supply of chemical elements for renewable energy. RSC Advances 2, 7933–7947.

Vrubel, H., Hu, X., 2012. Molybdenum boride and carbide catalyze hydrogen evolution in both acidic and basic solutions. Angewandte Chemie International Edition 51, 12703–12706.

Walter, M., 2010. Solar water splitting cells. Chemical Reviews 110, 6446–6473. Available at https://doi.org/10.1021/cr1002326.

Wang, Y., Zhang, L., Yin, K., Zhang, J., Gao, H., Liu, N., et al., 2019. ACS Applied Materials & Interfaces 11 (43), 39728–39736.

Xie, Y., Cai, J., Wu, Y., Zang, Y., Zheng, X., Ye, J., et al., 2019. Advanced Materials 31 (16), 1807780.

Xie, J., Zhang, H., Li, S., Wang, R., Sun, X., Zhou, M., et al., 2013. Defect-rich MoS_2 ultrathin nanosheets with additional active edge sites for enhanced electrocatalytic hydrogen evolution. Advanced Materials 25 (40), 5807–5813.

Xu, L., Jiang, Q., Xiao, Z., Li, X., Huo, J., Wang, S., et al., 2016. Angewandte Chemie International Edition 128 (17), 5363–5367.

Yin, J., Molini, A., Porporato, A., 2020. Nature Communications 11, 4781.

Zakaria, Z., Kamarudin, S.K., 2021. International Journal of Energy Research 45, 18337–18358.

Zeng, Q., Guo, G., Meng, Z., Gao, L., Meng, H., Zhou, L., 2023. Improvement of amorphous silicon/crystalline silicon heterojunction solar cells by light-thermal processing. Materials Science in Semiconductor Processing 154, 107192.

Zeng, Y., 2017. A low-cost photoelectrochemical tandem cell for highly-stable and efficient solar water splitting. Nano Energy 41, 225–230. Available at https://doi.org/10.1016/j.nanoen.2017.09.032.

Zhang, H., Maijenburg, A.W., Li, X., Schweizer, S.L., Wehrspohn, R.B., 2020. Bifunctional heterostructured transition metal phosphides for efficient electrochemical water splitting. Advanced Functional Materials 30, 20032.

Zhang, R., Wang, X., Yu, S., Wen, T., Zhu, X., Yang, F., et al., 2017. Advanced Materials 29 (9), 1605502.

Zhang, B., Zheng, X., Voznyy, O., Comin, R., Bajdich, M., García-Melchor, M., et al., 2016. Science 352 (6283), 333–337.

Zhou, Q., Chen, Y., Zhao, G., Lin, Y., Yu, Z., Xu, X., et al., 2018. ACS Catalysis 8 (6), 5382–5390

Zhu, J., Hu, L., Zhao, P., Lee, L.Y.S., Wong, K.-Y., 2019. Chemical Reviews 120 (2), 851–918.

CHAPTER

19

Challenges for photoelectrochemical water-splitting technology for the near-future

Gaweł Sołowski, and Fethi Ahmet Özdemir

Department of Molecular Biology and Genetics, Faculty of Science and Art, Bingöl University, Bingöl, Turkey

19.1 Introduction

Fossil fuel depletion and the greenhouse effect force governments and researchers to develop new consumption of surrounding sources ideas. The leading idea is a zero-waste policy converging life cycle assessment that claims to close the loop of material from appearing to be recycled with no waste different than natural. To reduce the greenhouse effect the lifestyle needs to reduce or stop generating energy with carbon dioxide emission. Therefore for 50 years, we have developed energy and daily life product generation methods for finding approaches without harming surrounding nature and recovering their level to the preindustrial age. There are many programs across the world with the major Paris Agreement of 2015–International Climate Change Treaty.

European Union aims to receive a zero-waste policy up to 2040 and neutrality from fossil fuels. The pledge was considered the Paris Agreement, where China promised to reach this level in 2060 (Yan and Jiang, 2023). Iran introduced the Sixth Development Plan in which from 1% in 2018 and 5% in 2023 is going to reach 100% renewable energy with a majority of solar and wind power, however, nowadays, hydroelectric energy leads among Iranian renewable energies that developing improperly (using freshwater resources) provides to environment troubles in neighbor countries like Iraq (Othman et al., 2023; Saedi and Ahmadi, 2023). One of the parts of zero-waste is restoring freshwater resources. Moreover, We should improve water quality by reducing the influence of pesticides and photoelectrochemical water-splitting combined with bioprocesses, and wind power technology could be important tools (Sinisterra-Solís et al., 2023).

Hydrogen is considered one of the cleanest energy carriers and fossil-fuel-independent industrial methods of obtaining hydrogen are sought (C. Liu et al., 2023). According to Anaya et al. (2023) hydrogen production methods are evaluated as follows:

Photochemical Splitting of Water
DOI: https://doi.org/10.1016/B978-0-443-29064-0.00017-6

1. Process design/simulation in which feed and equipment are considered.
2. Net energy ratio- if a method would give positive energy income.
3. Hydrogen cost estimation. How long investment would take to return costs and earning?
4. Sensitivity and uncertainty analysis.
5. If greenhouse gases are emitted or not.

Many decomposition methods use energy (solar, wind power, thermal or microwave) to obtain from hydrogen-containing methods (like pyrolysis, photolysis, photoelectrochemical, gasification, and electrolysis) that are faster but require more materials than much slower biological methods (like biophotolysis, dark fermentation, microbial electrolysis cells, or photofermentation).

Photoelectrochemical water-splitting develops achievements of photocatalysis, and electrolysis for enhancing the efficiency of hydrogen generation. The topic was without greater research interest 3 publications per year until 2003 and a successful hydrogen generation approach (Miller et al., 2003). Then, from 2003 to 2014 there was a gradual increase in interest after overcoming of corrosive effect of metal oxides by supporting silica electrodes with the deposition of titanium oxide (Seger et al., 2013). Then the growth by 2023 has reached 8458 according to Scopus. In the last decade, China outpassed the USA by 3273 publications (like Ma et al., 2023; Meng et al., 2022) 1273 (including T. Liu et al., 2023) majorly with a researcher in cooperation with China (Pan et al., 2023) or performed by Chinese researchers (Xiao et al., 2023). The huge interest is in marine countries of Asia like Korea (Goddati et al., 2023) or Taipei (Chiou et al., 2023) and less in Japan (Chen et al., 2023) with a dominant specialist (Pihosh et al., 2023) and relevant Indian influence (Kumar Biswas et al., 2022). The development of hydrogen production is weak despite USA resolution from 2020 and governmental support therefore (Ansari et al., 2021; Thakur et al., 2020). Water and dihydrogen oxide are applied in the method voltage of 1.23 V by photoelectrochemical electrodes for decomposing the compound. The approach utilizes the most widespread hydrogen compound but requires more energy than some less common ones (Gou et al., 2021; Price et al., 2023). According to Zdonek et al. (2023) renewable sources of energy are unsupported by communities due to high costs, still lower yields than traditional energy sources, and low authorities' interest in popularizing these technologies including hydrogen production by photoelectrochemical water-splitting.

The work highlights major problems of the method as follows: Potential material for electrodes and combinations, potential recovery of used photoelectrochemical circuits and their elements, water selection, irradiance, and government and citizen support. In the research very rapidly developing recent achievements in the presented challenges were discussed. All mentioned thresholds in this water-splitting we compared and extended in further parts.

19.2 Methodology

The chapter selected 70 positions from 8458 that had been published according to Scopus after 1971 and until August 29, 2023. The Scopus did not detect publications before 1971.

Next, we compared in July 2024 to discern development in the research area to 9351 records. In July 2024, 643–702 were published, covering August 2023.

We cited articles with recent outstanding outcomes and mostly influencing the research subject. For better comparison with other fields, we used 19 other publications to demonstrate thresholds and potentials of photoelectrochemical water-splitting and possible progress.

19.3 Discussion

19.3.1 Challenges and solutions

In the section, thresholds of photoelectrochemistry of splitting water. The problems were divided into four sections and a discussion. The overall summary and recommendation we put in Learning and Knowledge Outcomes. In the one year time, the four leading scientists have not changed their position in the plays. Only Professor Liu Zhifeng from Tianjin Key Laboratory of Building Green Functional Materials, Tianjin, China increased the number of records above 100 publications decreasing the distance with Professor Domen, Kazunari from Faculty of Natural Science and Technology, Okayama University, Kita-ku, Okayama, Japan. In the first 10 positions from 10 to 5th most publishing authors changed. China outpassed other countries increasing the gap between activities. Similarly, Korea, the United States, and India scientific activity is higher than the rest countries making these four countries scientific leaders. We observed that after the peak in research in 2022 the research activity is lowered. The lower activity is caused by the high potential required for water-splitting and seeking other waste. The other reason can be less interest of grant donators due to failures in finding more common metal suitable metal or ligands and high prices of conjugated polymers and silicon. The war between Russia and Ukraine in 2022 and the change of power in African countries like Niger closed or highly obstructed access for potential water-splitting catalysis.

19.3.2 Mechanisms of water-splitting understanding

Mechanism of water electrolysis can be divided due to temperature like alkaline or ion (proton or anion) exchange membrane (PEM or AEM) electrolyzers applying ambient or near ambient condition or solid oxide electrolysis applying temperature above 400°C or/with special pressure solid oxide electrolysis where water molecule bonding is weaker. (Bilgiç and Ozturk 2023; Prosser et al., 2023). The general mechanism of water-splitting can be classified due to stages such as conventional water electrolysis, coupled electrolysis, and decoupled. In coupled electrolysis the water is one of the components of the mixture allowing wastewater with components often like ammonia or urea more easily decomposed than water thus using energy from splitting them to enhance water electrolysis. In decoupled water decomposition is stimulated using a specially introduced redox mediator.

Important thresholds are still low splitting to hydrogen factor, that is, the efficiency of electrodes. This is the conversion of photon potential into electron-hole pairs (eÿ $_{CB}$ + $h_{+\ VB}$) if this transfer is higher or equal to the band gap, it can occur water decomposition due to solar irradiance (Vilanova et al., 2024). If the band gap is narrow the photocatalysts are designed in a Z-scheme (Lin et al., 2024). Photocatalysts can lead to photosplitting or photoreforming (Clarizia et al., 2023). Photovoltaic electrochemical systems can reach even 32% of solar-to-hydrogen energy efficiency (STH) while the photoelectrochemical system according to Qureshi and Tahir (2024) can reach even 19% without an explanation difference in structure. Photocatalytic water-splitting works on two or three stages process on the cathodes where occurs hydrogen evolution reaction (HER) can occur using a mediator in case of decoupled water electrolysis. The HER is cathodic three-staged: Volmer, Heyrowski, and Tafel. (Li et al., 2024). There parallelly on anode occurs oxygen evolution reaction with usually four electron coupled proton transfer working on alkaline or acidic pathway. Generally, vast photocatalytical materials belong to metal elements of IV and V groups with different combinations. The common research trend is limiting to metals that due with seeking an approach for increasing more common elements to overcome limitations more than employing commonly occurring silica, graphene, and semiconductive polymers (Budama et al., 2023; Shiva Kumar and Lim, 2022). The limitations caused that the practical energy needed for splitting is 2 eV.

19.3.3 Potential material for electrodes and combinations

The choice of material for the electrode depends on the selection of a water solution for more and less corrosive resistance. Other parameters are surface charge, HER rate, STH, and proceeding oxidation reaction rate. The material should have high power density and high resistance to coagulation and poisoning. The potential photosensitive materials are transition and rare earth metals, metal oxides or ligands; semiconductive polymers, and silicon. A special type but not photosensitive electrode is combined with a microbial electrolysis cell; their combination improves electrochemical characteristics. The material should absorb a wide spectrum that features a low band gap. The photoelectrode is classified as poor if STH is below 1%, marginal 2%–4%, acceptable 4%–10%, and highly efficient above 10% (Li et al., 2024; Thakur et al., 2020; Yang et al., 2024). Reaching proper band gap obtaining sufficient reaction rate, current and resistance density, and noncorrosiveness are features of the optimal electrode. Obtaining the desired potential electrode is hard to obtain using a single particle or polymer. Therefore there are many attempts to implement nanoparticles of various types from nanorods, and nanovoids to nanoporous with macromolecular nets of other compounds like carbon polymer, silica, or metal. According to Shrivastava et al. (2023) doping, grafting, drop-casting, inkjet printing, encapsulation dipping, pumping, spraying, layer-by-layer pump approach flowing, spin-coating and linking are the most common approaches to filling intermolecular spaces to obtain more π-bond. The deposition of CoP to TiO_2 nanotubes lowered band gaps from 3.1 eV to 2.9% and 5% of current density which was a significant curtail of thresholds (Xu et al., 2023). The tailoring of the network is well-explained in conjugated polymers, which possess the most suitable potential band gap and wider spectrum than metals or silica. Unfortunately, the synthesis of these polymers uses mostly fossil fuels and is much more expensive than other types. The problems and characteristics of semiconductivity are explained below focusing on polymers. The porosity of photocatalysts improves surface area and thus the efficiency of water-splitting (Zhao et al., 2020). Potential materials for electrodes and combinations are shown in Table 19.1.

TABLE 19.1 Potential material for electrodes and combinations.

Photosensitive material or nanomaterial for electrodes	Achievements	Disadvantages	Combinations with another type	Combination with microbial electrolysis cells	STH	References
Metals/Metal oxides	– Wider spectrum and absorptivity than silica. – Rare earth metals feature especially high efficiencies and low corrosiveness. – Transition metals nanorods combination are efficient but expensive.	– Rare earth metals and compounds are expensive. – Generation unless recycled, increases contamination risk, especially increasing illegal mining, for example, gold. – Transition metals change fast charge recombination causing insufficient light utilization. – Usually n-type semiconductors with donor-type defects. – Beside oxysulfides hardly achieve a band gap of 2.0 eV.	Combinations of rare earth elements like vanadium, help reduce the corrosiveness sensitiveness of cheaper and fragile zinc nanoporous materials. Silver addition promotes separation and transfer charge on the surface.	The connection, for example, with bismuth oxide nanosheets allows to split water without fossil sources charging. The method allowed to extension of MEC splitting to water.	– Rare like bismuth or i/InGaP/GaAs reaches 8%. – Rh/TiO 2/AlInP/GaInP/GaInAs/GaAs reached 19,2%	– Chiou et al. (2023); Pahari and Chen (2023); Sajjadizadeh et al. (2024a)
Semiconductive polymers	– Wide spectrum of solar light absorption. – High efficiency. – Research is used for improving biomarkers in medicine.	– The green method of synthesis is low yield if microbial while pyrolysis or gasification requires much heat. – Need exact and restrictive plans of reuse or recycling to prevent an increase in microplastic/plastic pollution of the environment.	– They expand easily photoelectrochemical properties after shifting the spectrum analyzed for biomarkers. – Metals doping of semiconductive polymers enhances photoelectrochemical water-splitting by increasing oxygen evolution reaction proceeding hydrogen evolution reactions.	MEC decreases resistance, increasing current and power density.	– $CH_3NH_3PbI_3$ based perovskite solar cells reached 15%. – Tiophene polymers reached 9%. The special linking thiophane- or pyrrole-based backbone improved by 14 %.	Kausar et al. (2023); Shokrollahi et al. (2023); Takayama et al. (2024); Xu et al. (2024),

(Continued)

Table 19.1 (Continued)

Photosensitive material or nanomaterial for electrodes	Achievements	Disadvantages	Combinations with another type	Combination with microbial electrolysis cells	STH	References
Silica	– Safe. – Cheap. – Widespread. – Quite a high efficiency for such a low price. – Silica wastes besides silicosis are without known negative impact on the environment.	– Less absorptivity than other materials. – Low research of nano-silica compounds and silica hazard of powder forming risk, extensive simulation needed to prevent similar events to plastic disaster. – Nanoparticle of silica hooked in acidic $HClO_4$ achieved photoelectrochemical properties of gold.	– Deposition of copper sulfate on silica improved their current density and gradient filling of the vias – suitable for basic pH conditions. – Nanoporous silica with polypyrroles stimulated irradiance and photoelectrochemical properties.	It allows splitting water range in a cheap manner.	Up 20% thus twice higher than most metallic compounds	Brinker et al. (2022); Fehr et al. (2023); Nabgan et al. (2023); Zhao et al. (2018)

Conjugated polymers are groups of polymers that in chain have alternating single and double bonds. The double bond contains one σ bond and one π-bond. σ-bond lays always along the internucleus axis. The single bonds have only σ bonds. A shared electron in a σ-bond should be found between two carbon atoms The π bonds contain the electrons above and below each carbon atom – appearing at p-orbitals. The σ bond and one π-bond at double bond don't intersect their orbitals. The π-bond electrons are usually placed just above and below the central axis and are formed by overlapping p-orbitals of two carbon nuclei. Conjugated polymers have a σ-bond backbone of overlapping sp^2- orbital (Tvingstedt, 2007) In the case of nonconjugated polymers the backbone is composed of mers which are σ-banded. Some nonconducting polymers may have polystyrene π-bonding on the side groups. According to Tvingstedt (2007). The high electron energy band gaps in σ-bonded polymers, E'(σ), causes these materials electrically insulating, and almost nonabsorbing to visible light. For example, in polyethylene with a monomeric repeat unit, the optical band gap is 8 eV (Kumar et al., 2008). The band gap suitable for optical purposes according to Hiorns is between 1 and 3 eV (Moliton and Hiorns, 2004). One of the relevant aims of the design of photoelectrochemical electrodes with a proper band gap. Band gap is the difference in energy between the highest occupied molecular orbital (HOMO) and lowest unoccupied molecular orbital levels. Much lower band gap semiconducting polymers comprise oxidative properties and higher energy HOMO, they can easily be doped. Every such particle can dissociate in some conditions to the polarons and bipolarons which are believed to be the most desired quasiparticle to be formed for the case of effective broad-range absorption (Moliton and Hiorns, 2004). The charge-bearing species are śelf-localized' (Salaneck et al., 1999) or self -trapped'. Hence' Self-trapping provides local changes in atomic spectral structure, and localized changes in structure appear. Polaron is a radical anion or cation, that is, associated with a spectral relaxation self-localized state. Bipolaron - a stable entity formed from a combination of two mobile polarons is a radical dication quasiparticle spinless, charge carrier, and main charge storage species. Around bipolarons, strong relaxation appears. Every quasiparticle is integrated from a coupled charge-latter deformation entity. Such species are generated and influenced by the absorption of the optical spectrum in a neutral or doped state. This generation leads to the fission of conjugated polymers to degenerated and nongenerated species. Degenerated state polymers are represented mainly by trans-polyacetylene and polysulfur nitride. They can form only quasiparticle- soliton. Solitons feature the lowest energy eigenstates of the coupled electron (hole) lattice systems. They are responsible for the unusual electrical, magnetic, and linear as well as nonlinear optical properties. So the limited ability to form quasiparticles is caused by alternating double and single bonds 'duplicating' the identical ground-state geometry (Beaujuge et al., 2010). The other theory is the frontier molecular orbital theory principles is a theory that tries to explain the deviation of the donor-acceptor theory (Zhao et al., 2023). Here donor-acceptor macromolecules are considered to destabilize the electrons occurring in the framework. Shifted orbitals give energy a unique energy band diagram. Theory is a subject that leads to different opinions. The work of Salzner assumed an attitude that the designation of polymer energy band gap only by donor-acceptor is insufficient and is necessary to correct by use of perturbation energy. On the other hand, these results (Salzner, 2021) confirmed the special case of the donor-acceptor theory (Reynolds et al., 1991). Experiments by Salzner et al. (1998) have shown that small dispersion provides to minimize curtail in band gap upon an increase in chain lengths. Wide valence bandwidths seem not to be a direct result of a low band gap. Small valence bandwidths can elucidate the lower conductivities of donor-acceptor systems upon doping regarding homopolymers and for low n-type conductivity dedoping-return from oxidating states to neutral being result of reduction by n-reductant using alkali metals. Those properties allow both metals and polymers to be efficient electrodes. Additionally unlike metals polymers easier obtain apparent quantum efficiency of 30% sufficient for solar conversion above 11.4% (Fang et al., 2022). Semiconductive polymers absorb a wider spectrum of solar radiation than silica ones or metals. Thin-coated polymeric carbon nitride on $BiVO_4$ anode boosted seven times photocurrent density (Feng et al., 2017). After finding their applicability in cancer and cardiac biomarkers, the mechanisms of shifting spectrum and irradiance are better explained and more feasible for designing photoelectrochemical systems (Feldner et al., 2023; Shokrollahi et al., 2023). The complex Z-scheme of $Sm_2Ti_2O_5S_2$ deposited on $BiVO_4$ improved STH by only 0.22 while it is required above 10% to be acceptable (Lin et al., 2024; Wang et al., 2023).

Chiou et al. (2023), Goddati et al. (2023), and Kang et al. (2023) modified surface (forestation) and linked nanorods of iron titanium, and zinc with gold (Chiou et al., 2023) extending slightly spectrum with increasing costs. Metal oxide transition metals besides rare metals need support from other noble metals such as Pt to prevent the recombination of electrons and holes making them more expensive (Zwara et al., 2020). Silica nanoparticles in wafer form in $HClO_4$ had a change of length of surface during photoelectrolysis equal to nanoporous gold (Brinker and Huber, 2022). Silica and polymers have the advantage of being widespread while the most efficient metal photocatalysts like bismuth 8% of STH and vanadium 4% STH are random elements on Earth and full consumption of bismuth after 17 years while vanadium after 244 years (Hillenbrand et al., 2024). The increase of STH using conjugated polymers and silicon has recently more expensive but technological development especially by using biomass waste can provide proper linking faster acceptable water-splitting rate (Guo et al., 2024). Thus usage of them required full closing of life cycle analysis and reusing more than in using silica and conjugated polymers. In polymerization of nanoporous silica with artificial muscle material (polypyrroles) gained significant growth in electrochemical properties (Brinker et al., 2020). Deposition of different elements to conductive is preferred since doping that shifts a more efficient spectrum can risk point defects or dislocation of added elements especially in void shape (Lei et al., 2022). In conjugated polymers, metal coordinate combinations like Ni or Cu improve charged surface areas facilitating electron transfer via microporous areas (Das et al., 2023). An enormous problem is the achievement of polymers by green chemistry methods like fermentation with enough efficiency for use there (Anjum et al., 2023; Zatta et al., 2023). The potential is also energy-consuming pyrolysis/gasification (Alazaiza et al., 2023) of organic wastes and black liquor (Parot et al., 2022; Peixoto et al., 2023). Black liquor microbial conversion to polymers is very slow and genetic engineering seems a potential tool for improving the rate (Li et al., 2023; X. Wang et al., 2020; Yoo et al., 2020).

Combinations of metal oxides with polymers, silica, or microbial eletrolysis cell (MEC) reduce their defect occurrence thus due to their strength and high power density can be considered in the design of photoelectrochemical electrodes. Defects desired are those that are induced by unsaturated metal atoms and are efficient in water-splitting and binding to carbon dioxide and defect engineering aims in building such structures. Zhang et al. (2023) considered organic linkers of semiconductive polymers or graphene as an optimal tool for obtaining morphology and defect structure optimal for water-splitting (Zhang et al., 2023).

19.3.4 Lifetime of electrodes

A relevant issue of modern engineering is the exact determining lifetime of a designed product usually called life cycle assessment (Johanning et al., 2023). Similarly, photocatalysts need exact restrictions to protect the environment from the pollution of their components. The removed electrode should be recycled with the proposed, well-documented 'closing of loop' thus no waste is generated. The potential pathway of photocatalyst is pictured in Fig. 19.1.

Photoelectrochemical electrodes can enhance wastewater treatment beside water and increase after use in water-splitting can be applied dark fermentation. This allows optimal use potential and allows obtaining zero-waste level (Phan et al., 2023). Another method of recycling electrodes can be approached by molding bacteria like *Aspergillus niger*, brown rot,

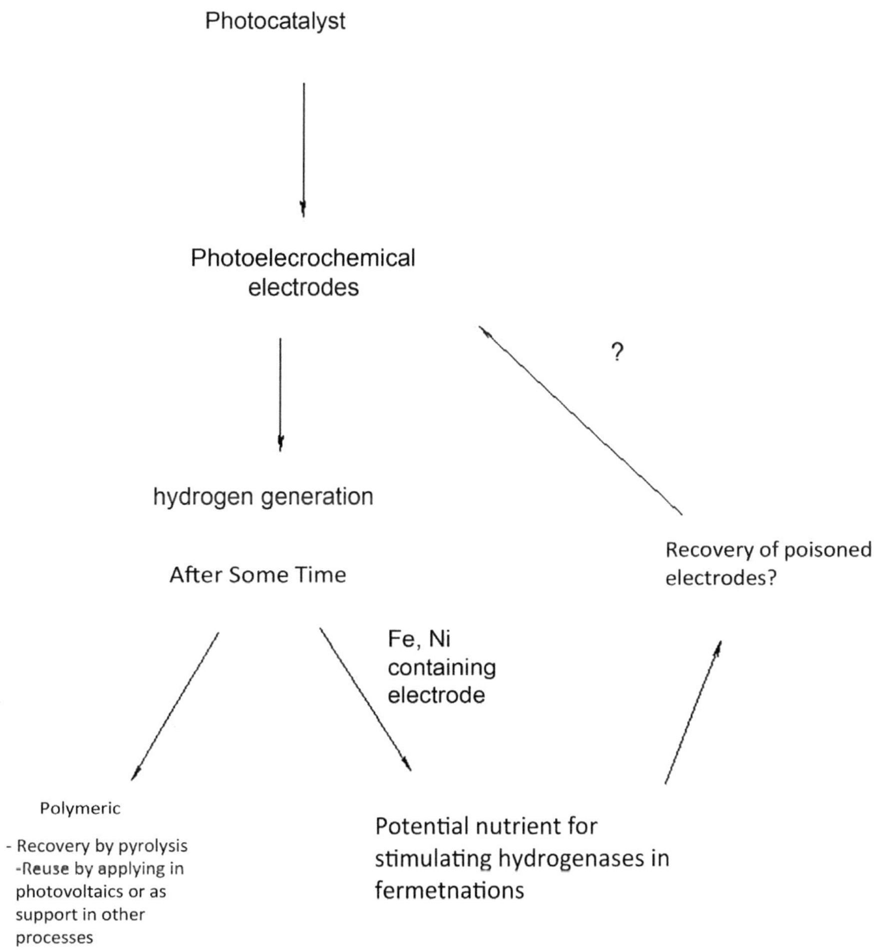

FIGURE 19.1 Problem after the use of photoelectrochemical electrodes.

or nematodes similar to printed circuit boards (Kaur et al., 2022). Kaur et al. (2022) method can even lead to the reuse of the recovered metals for photoelectrochemical purposes.

Silica electrodes and wafers seem safe unless their powder form can risk silicosis and much modeling and research for preventing hazards are required (Martínez-López et al., 2023). Implementing new technologies we need to test and assess scaling up very carefully remembering plastic history (Chamas et al., 2020; Worch and Dove, 2020) and proceeding with the side effects due to overusing it (Sołowski et al., 2022). Both silica and polymers can support microbial reactors by enhancing the potential (De La Cruz-Noriega et al., 2023; Segundo et al., 2023).

19.3.5 Water sources

The relevant problem is the proper water origin source from the surroundings. The water divides into freshwater or saline. Both types of water can be grouped into classes due to pollution. Fig. 19.2 shows possible types of water for being fed water-splitting. The water resources especially clean freshwater need proper conservation and more reasonable consumption (Velthuis et al., 2023). Energy generation is preferred for waste origin. Wastewater contains many compounds like butyric acid needed for utilization with lower required voltage thus potential hydrogen can be higher (Chen et al., 2021). On the other hand polluted wastewater forces more frequent replacement of electrodes due to their poisoning (Pecoraro et al., 2023).

Photoelectrochemical splitting beside water can split other compounds simultaneously and can support the utilization of wastes including winery, livestock with energy generation, and chemicals, especially in cooperation with microbial methods (Gomes et al., 2023).

Another challenge blocking scaling up all biohydrogen methods is the storage of fuel. Their storage tanks are expensive (Sorrenti et al., 2023). Hydrides are expensive while ammonia storage is relatively cheap (Lamb et al., 2019; G. Wang et al., 2020). The ammonia splitting by microbial electrolysis cells or/within photoelectrochemically and then restoring it using the Haber method is only particularly (G. Wang et al., 2020) applied without spreading throughout the globe. The ammonia method is an option for the large energy consumption required for long-tube trailers and compressors (Yan and Jiang, 2023). The ammonia and water could be especially beneficial for countries with high livestock density and a high need for nitrogen and phosphorus recovery like the Netherlands which can be combined with already existing wind power plants (de Vries et al., 2023). Metal oxides on cathodes and microbial electrolysis cells operate efficiently on wastewater with low resistance and long-term (De La Cruz-Noriega et al., 2023; Segundo et al., 2023), if well designed it is even possible to produce of butanol -relevant biodiesel compounds (Romans-Casas et al., 2024).

Payá et al. (2023) developed S-doped carbon sponge capacitors, especially for electrodes. These capacitors feature high-energy storing and discharge rates, and applying them to magazine generation of photoelectrodes would prevent the system from fluctuating periods in unstable weather periods (Payá et al., 2023). For example, the application of the photoelectrochemical electrodes with MEC can allow enhanced cleaner agrovoltaics sufficient in the subarctic region for obtaining fresh vegetables (Jamil and Pearce, 2023).

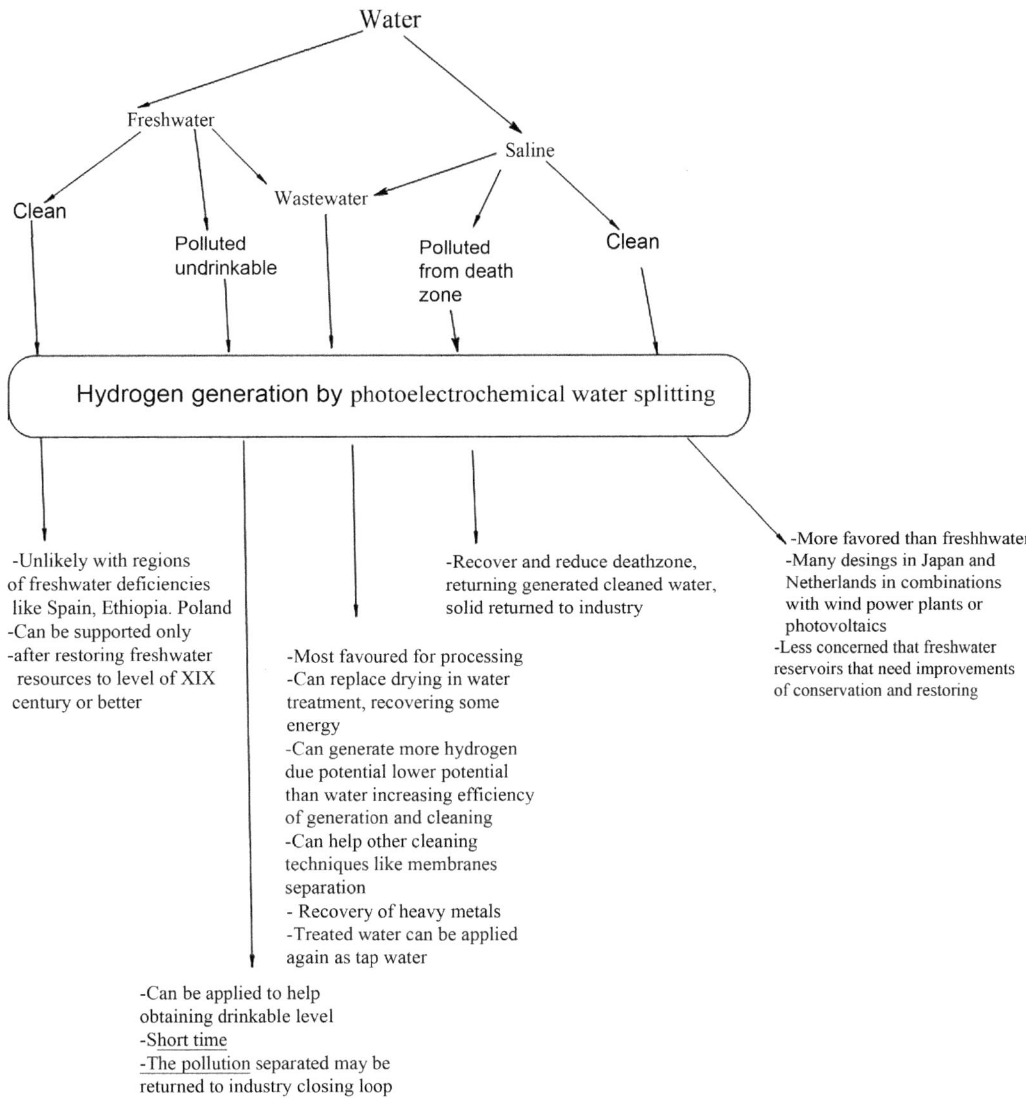

FIGURE 19.2 Water feed options for photoelectrochemical electrodes.

Clean or polluted water, even wastewater can be in different conditions and be preferred for hydrogen production. Grid inhomogeneous solvation theory points out that in some solutions water molecules split more easily, for example, protein solutions (Mukherjee and Schäfer, 2022). Water-splitting energy yields are less than acceptable for dark fermentation (Carrillo-Reyes et al., 2019) but they capture solar light that offers much higher energy than biomass and, therefore can fulfill more energy demands the advancement of methods is comparable without waste utilization unlike biological hydrogen production methods

(Algayyim et al., 2024; Hackula et al., 2024; Sajjadizadeh et al., 2024a). Therefore similar simulations of other water solutions can indicate potential optimal water solution content optimal for hydrogen production.

19.3.6 Irradiance measurement

Both wind power and photovoltaics including photoelectrochemical electrolysis challenge strong fluctuation in power due to changing weather and demand accurate forecasts for proper control of on- or –off-grid systems (Kojima et al., 2023). The proper position of photoelectrodes due to current sun irradiance frequency is still a relevant issue. Between 2014 and 2023 more devices were developed like the sunphotometer (Raptis et al., 2023) supported with improved models using an unsupervised artificial neural neutwork (ANN) model (Pevtsov et al., 2023) in comparison to (Engerer 2015; Pandey and Katiyar, 2013). Short-circuit loss is less due to the implementation of perovskite but is still expensive disabling common usage (Caprioglio et al., 2023). The enormous threshold is sunspot number and group number calculation for predicting irradiance, especially in unobserved times, There is still only one physical model of variability of solar irradiance on timescale and we are unable to establish of magnitude of solar-cycle irradiance variability in the near-infrared and the near-ultraviolet (Pevtsov et al., 2023). The proper control is sufficient for the efficient prevention of system failure due to thermal shock. Thus during wider employing, we set up photoelectrochemical and photovoltaic devices with high error, making unreliable more efficient usage of sun energy, and measurement with modeling lasting several years can give more promising approaches for solving that.

19.3.7 Government and citizen support

The development of renewable energy methods is possible if it is supported by governments and communities that would support their implementation independently of ruler change or not. The enormous trouble is low education level, especially in the scientific branches; the war in Ukraine showed a great knowledge gap in education levels of math and physics between Ukrainian and French pupils (Shyyan and Shyyan, 2023; St. John, 2023). The higher common scientific knowledge is essential for convincing people of the harmful negatives of Earth's climate change in recent years of human origin. The programs for improvement of zero-waste policy would not be reached without the long-term and continuous support of the majority of voters. The voters and leaders can support only if the scientific outcome is clear thus proper education level is crucial. The globe in conference Paris claimed promising programs that are slowly provided by only some national governments (Fartash and Ghorbani, 2023; Wu et al., 2023; Yang et al., 2023). The photoelectrochemical splitting develops mostly in China, Korea, and Taiwan more than in the USA, and besides scientific grants governmental policy is also due to better science education (Guo et al., 2023; Xia et al., 2023). East Asia level of real sciences in high and primary schools enables more local universities to educate more real science specialists like mathematicians, physicians, or chemists and common civil better understanding surrounding environment and the necessity of changes. Technical problems can be solved efficiently only if people use and they will apply them commonly. This is only possible if political leaders and societies permit the change.

19.4 Learning, knowledge outcomes, and conclusions

Photoelectrochemical water-splitting is an efficient method of hydrogen production but requires much development due to the design of a structure suitable for projected water conditions. The expected split water characteristic is sufficient for implementing electrodes combined from metals, semiconductive polymers, silicon, and microbial electrolysis cells. The needed technologies should fulfill green chemistry requirements for the preparation of metal and polymer electrodes. Semiconductive polymer productions from nonfossil materials are in the early stages. Physical models of a physical model of the variability of solar irradiance on a timescale are needed for improving planning and controlling a plant for efficient 'harvesting' of solar radiation that produces hydrogen. Hydrogen can be stored at relatively low expenses as an energy carrier in the form of ammonia or other hydrides and seems the most suitable method of storing this item.

The technological achievements in photoelectrochemical water-splitting help people prevent 'selfboiling' of themselves only if properly supported by them and their states.

The combination of microbial electrolysis cells with photoelectrochemical cells can improve and extend potential water and can be more used for wastewater, being a more favored substrate due to recovery, utilization, and energy generation not only hydrogen production.

Metal suitable for electrodes should originate from waste to prevent further pollution of the environment by mining. Primarily photoelectrochemical installation should prefer saline water to freshwater to prevent increasing freshwater deficiency unless used as part of a water cleaning system.

The improvement of electrolysis should obtain the highest possible STH ratio using electrode common and cheap material. Furthermore, photoelectrode should absorb all spectrums of solar irradiance emitted to the earth with optimum efficiency and without burning itself. The photoelectrodes are believed as potential solar charmers using all irradiated energy according to Earth's living demands. The process should enable us to utilize waste besides generating energy. The catalyst source should be waste.

References

Alazaiza, M.Y.D., Albahnasawi, A., Eyvaz, M., Al Maskari, T., Nassani, D.E., Abu Amr, S.S., Abujazar, M.S.S., Bashir, M.J.K., 2023. An overview of green bioprocessing of algae-derived biochar and biopolymers: synthesis, preparation, and potential applications. Energies 16 (2), 791. https://doi.org/10.3390/en16020791.

Algayyim, S.J.M., Saleh, K., Wandel, A.P., Fattah, I.M.R., Yusaf, T., Alrazen, H.A., 2024. Influence of natural gas and hydrogen properties on internal combustion engine performance, combustion, and emissions: a review. Fuel 362, 130844.

Anaya, K., Olufemi Oni, A., Kumar, A., 2023. Investigating the techno-economic and environmental performance of chemical looping technology for hydrogen production. Sustainable Energy Technologies and Assessments 56.

Anjum, S., Aslam, S., Hussain, N., Bilal, M., Boczkaj, G., Smułek, W., et al., 2023. Bioreactors and biophoton-driven biohydrogen production strategies. International Journal of Hydrogen Energy.

Ansari, A.A., Thakur, V.K., Chen, G., 2021. Functionalized upconversion nanoparticles: new strategy towards FRET-based luminescence bio-sensing. Coordination Chemistry Reviews 436, 213821.

Beaujuge, P.M., Amb, C.M., Reynolds, J.R., 2010. Spectral engineering in π-conjugated polymers with intramolecular donor–acceptor interactions. Accounts of Chemical Research 43, 1396–1407.

Bilgiç, G., Ozturk, B., 2023. Modeling of artificial neural networks for hydrogen production via water electrolysis. El-Cezerî Journal of Science and Engineering 2023 (10), 137–146.

Brinker, M., Huber, P., 2022. Wafer-scale electroactive nanoporous silicon: large and fully reversible electrochemo-mechanical actuation in aqueous electrolytes. Advanced Materials 34.

Brinker, M., Huber, P., Dittrich, G., Richert, C., Huber, N., Lakner, P.H., et al., 2020. Electrochemical actuation in porous silicon. ECS Meeting Abstracts MA2020-02 1216.

Brinker, M., Thelen, M., May, M., Rings, D., Krekeler, T., Lakner, P., et al., 2022. How nanoporous silicon-polypyrrole hybrids flex their muscles in aqueous electrolytes: in operando high-resolution x-ray diffraction and electron tomography-based micromechanical computer simulations. *Physical Review Materials* 6.

Budama, V.K., Rincon Duarte, J.P., Roeb, M., Sattler, C., 2023. Potential of solar thermochemical water-splitting cycles: a review. Solar Energy 249, 353–366.

Caprioglio, P., Smith, J.A., Oliver, R.D.J., Dasgupta, A., Choudhary, S., Farrar, M.D., et al., 2023. Open-circuit and short-circuit loss management in wide-gap perovskite p-i-n solar cells. Nature Communications 14.

Carrillo-Reyes, J., Tapia-Rodríguez, A., Buitrón, G., Moreno-Andrade, I., Palomo-Briones, R., Razo-Flores, E., et al., 2019. A standardized biohydrogen potential protocol: an international round robin test approach. International Journal of Hydrogen Energy 44, 26237–26247.

Chamas, A., Moon, H., Zheng, J., Qiu, Y., Tabassum, T., Jang, J.H., et al., 2020. Degradation rates of plastics in the environment. ACS Sustainable Chemistry & Engineering 8, 3494–3511.

Chen, S., Nandy, S., Vequizo, J.J.M., Hisatomi, T., Nakabayashi, M., Pan, Z., et al., 2023. Promoted utilization of charge carriers in $La_5Ti_2Cu_{0.9}Ag_{0.1}O_7S_5$-based photocatalyst sheets for efficient Z-scheme overall water splitting. ACS Catalysis 13, 3285–3294.

Chen, Y., Yin, Y., Wang, J., 2021. Influence of butyrate on fermentative hydrogen production and microbial community analysis. International Journal of Hydrogen Energy 46 (53). https://doi.org/10.1016/j.ijhydene.2021.05.185.

Chiou, S.-H., Ho, H.-C., Liao, H.-T., Tsai, F.-Y., Tsao, C.-W., Hsu, Y.-J., et al., 2023. Plasmonic gold nanoplates-decorated ZnO branched nanorods@TiO_2 nanorods heterostructure photoanode for efficient photoelectrochemical water splitting. Journal of Photochemistry and Photobiology A: Chemistry 443, 114816.

Clarizia, L., Nadagouda, M.N., Dionysiou, D.D., 2023. Recent advances and challenges of photoelectrochemical cells for hydrogen production. Current Opinion in Green and Sustainable Chemistry 41 (6), 100825. https://doi.org/10.1016/j.cogsc.2023.100825.

Das, S.K., Chowdhury, A., Bhunia, K., Ghosh, A., Chakraborty, D., Das, M., ... Bhaumik, A., 2023. Ni (II) and Cu (II) grafted porphyrin-pyrene based conjugated microporous polymers as bifunctional electrocatalysts for overall water splitting. Electrochimica Acta 459, 142553.

De La Cruz-Noriega, M., Benites, S.M., Rojas-Flores, S., Otiniano, N.M., Sabogal Vargas, A.M., Alfaro, R., et al., 2023. Use of wastewater and electrogenic bacteria to generate eco-friendly electricity through microbial fuel cells. Sustainability 15.

Engerer, N.A., 2015. Minute resolution estimates of the diffuse fraction of global irradiance for southeastern Australia. Solar Energy 116, 215–237.

Fang, Y., Hou, Y., Fu, X., Wang, X., 2022. Semiconducting polymers for oxygen evolution reaction under light illumination. Chemical Reviews 122 (3), 4204–4256. https://doi.org/10.1021/acs.chemrev.1c00686.

Fartash, K., Ghorbani, A., 2023. From a promising technological niche to an established market niche: solar photovoltaic niche formation in Iran. Energy for Sustainable Development 74, 50–65.

Fehr, A.M.K., Agrawal, A., Mandani, F., Conrad, C.L., Jiang, Q., Park, S.Y., et al., 2023. Integrated halide perovskite photoelectrochemical cells with solar-driven water-splitting efficiency of 20.8%. Nature Communications 14.

Feldner, A., Völkle, J., Lieberzeit, P., Fruhmann, P., 2023. Conductive molecularly imprinted polymers (cMIPs): rising and versatile key elements in chemical sensing. Chemosensors 11 (5), 299. https://doi.org/10.3390/chemosensors11050299.

Feng, C., Wang, Z., Ma, Y., Zhang, Y., Wang, L., Bi, Y., 2017. Ultrathin graphitic C_3N_4 nanosheets as highly efficient metal-free cocatalyst for water oxidation. Applied Catalysis B 205, 19–23.

Goddati, M., Nguyen, H., Kang, S., Gicha, B.B., Tufa, L.T., Nwaji, N., et al., 2023. Rugged forest morphology of magnetoplasmonic nanorods that collect maximum light for photoelectrochemical water splitting. Small 19, 2302980. https://doi.org/10.1002/smll.202302980.

Gomes, A., Borges, A., Peres, J.A., Lucas, M.S., 2023. Bioenergy production from agro-industrial wastewater using advanced oxidation processes as pre-treatment. Catalysts 13, 1186.

Gou, J., Xian, B., Wang, X., Cheng, P., Thye Shen, A., Xu, Y., et al., 2021. Realizing quinary charge states of solitary defects in two-dimensional intermetallic semiconductor. National Science Review 9 (2), nwab070. https://doi.org/10.1093/nsr/nwab070.

Guo, J., Jiang, Y., Lin, S., Pang, W., 2023. Individualism and collectivism as predictors of creative potentials and real-life creativity in China and US. Creativity Research Journal 1–12.

Guo, B., Zhao, J., Xu, Y., Wen, X., Ren, X., Huang, X., et al., 2024. Noble metal phosphides supported on CoNi metaphosphate for efficient overall water splitting. ACS Applied Materials & Interfaces 16, 8939–8948.

Hackula, A., O'Shea, R., Murphy, J.D., Wall, D.M., 2024. Developing distillery biorefineries through dark fermentation of whiskey production by-products: the effect of organic loading rate on decarbonisation pathways. Energy Conversion and Management 301, 118064.

Hillenbrand, M., Helbig, C., Marschall, R., 2024. Supply risk considerations for photoelectrochemical water splitting materials. Energy & Environmental Science 17, 2369–2380.

Jamil, U., Pearce, J.M., 2023. Maximizing biomass with agrivoltaics: potential and policy in Saskatchewan Canada. Biomass 3, 188–216.

Johanning, M., Widenmeyer, M., Escobar Cano, G., Zeller, V., Klemenz, S., Chen, G., et al., 2023. Recycling process development with integrated life cycle assessment–a case study on oxygen transport membrane material. Green Chemistry 25, 4735–4749. https://doi.org/10.1039/D3GC00391D.

St John, E.P., 2023. Higher education in post-neoliberal times: building human capabilities in the emergent period of uncertainty. Education Sciences 13.

Kang, S., Goddati, M., Hwang, D., Nguyen, H.-Q., Gicha, B.B., Tufa, L.T., et al., 2023. Tuning assembly of Ni-doped Au@FexOy magnetoplasmonic nanowires for optimizing photochemical water splitting. Journal of Alloys and Compounds 967, 171417.

Kaur, P., Sharma, S., Albarakaty, F.M., Kalia, A., Hassan, M.M., Abd-Elsalam, K.A., 2022. Biosorption and bioleaching of heavy metals from electronic waste varied with microbial genera. Sustainability 14.

Kausar, A., Ahmad, I., Zhao, T., Maaza, M., Bocchetta, P., 2023. Green nanocomposite electrodes/electrolytes for microbial fuel cells—cutting-edge technology. Journal of Composites Science 7 (4), 166. https://doi.org/10.3390/jcs7040166.

Kojima, H., Nagasawa, K., Todoroki, N., Ito, Y., Matsui, T., Nakajima, R., 2023. Influence of renewable energy power fluctuations on water electrolysis for green hydrogen production. International Journal of Hydrogen Energy 48 (12), 4572–4592. https://doi.org/10.1016/j.ijhydene.2022.11.018.

Kumar Biswas, N., Srivastav, A., Saxena, S., Verma, A., Dutta, R., Srivastava, M., et al., 2022. Role of varying ionic strength on the photoelectrochemical water splitting efficiency. Solar Energy 247, 543–552.

Kumar, A., Bokria, J.G., Buyukmumcu, Z., Dey, T., Sotzing, G.A., 2008. Poly (thieno [3, 4-b] furan), a new low band gap polymer: experiment and theory. Macromolecules 41, 7098–7108.

Lamb, K.E., Dolan, M.D., Kennedy, D.F., 2019. Ammonia for hydrogen storage; a review of catalytic ammonia decomposition and hydrogen separation and purification. International Journal of Hydrogen Energy 44, 3580–3593.

Lei, W., Yu, Y., Zhang, H., Jia, Q., Zhang, S., 2022. Defect engineering of nanostructures: Insights into photoelectrochemical water splitting. Materials Today 52 (January/February), 133–160. https://doi.org/10.1016/j.mattod.2021.10.028.

Lin, L., Ma, Y., Vequizo, J.J.M., Nakabayashi, M., Gu, C., Tao, X., et al., 2024. Efficient and stable visible-light-driven Z-scheme overall water splitting using an oxysulfide H_2 evolution photocatalyst. Nature Communications 397, 15. https://doi.org/10.1038/s41467-024-44706-4.

Liu, T., Li, W., Wang, D.Z., Luo, T., Fei, M., Shin, D., et al., 2023. Low catalyst loading enhances charge accumulation for photoelectrochemical water splitting. Angewandte Chemie International Edition 62, e202307909. https://doi.org/10.1002/ange.202307909.

Liu, C., Zhang, N., Li, Y., Fan, R., Wang, W., Feng, J., et al., 2023. Long-term durability of metastable β-Fe_2O_3 photoanodes in highly corrosive seawater. Nature Communications 14, 4266.

Li, Y., Meng, X., Meng, R., Cai, T., Pu, Y., Zhao, Z.-M., et al., 2023. Valorization of homogeneous linear catechyl lignin: opportunities and challenges. RSC Advances 13, 12750–12759.

Li, Z., Sun, L., Zhang, Y., Han, Y., Zhuang, W., Tian, L., et al., 2024. Coupled and decoupled electrochemical water splitting for boosting hydrogen evolution: a review and perspective. Coordination Chemistry Reviews 510 (1 July), 215837. https://doi.org/10.1016/j.ccr.2024.215837.

Martínez-López, A., Candel, S., Tyrkalska, S.D., 2023. Animal models of silicosis: fishing for new therapeutic targets and treatments. European Respiratory Review 32 (169), 230078;. https://doi.org/10.1183/16000617.0078-2023.

Ma, M., Ruan, M., Nie, W., Lei, E., Liu, Z., 2023. $BiVO_4/Bi_2S_3$ heterojunction decorated by Pt and FeOOH double cocatalysts to enhance photoelectrochemical degradation of Rhodamine B. Journal of Materials Science: Materials in Electronics 34.

Meng, Y., Li, Z., Liu, Z., 2022. Enhanced electrochromic properties of inorganic–organic tungsten oxide and Prussian blue core–shell film. Journal of Materials Science: Materials in Electronics 33, 24995–25005.

Miller, E.L., Rocheleau, R.E., Deng, X.M., 2003. Design considerations for a hybrid amorphous silicon/photoelectrochemical multijunction cell for hydrogen production. International Journal of Hydrogen Energy 28, 615–623.

Moliton, A., Hiorns, R.C., 2004. Review of electronic and optical properties of semiconducting π-conjugated polymers: applications in optoelectronics. Polymer International 53, 1397–1412.

Mukherjee, S., Schäfer, L.V., 2022. Spatially resolved hydration thermodynamics in biomolecular systems. Journal of Physical Chemistry B 126 (20), 3619–3631. https://doi.org/10.1021/acs.jpcb.2c01088.

Nabgan, W., Alqaraghuli, H., Owgi, A.H.K., Ikram, M., Vo, D.V.N., Jalil, A.A., et al., 2023. A review on the design of nanostructure-based materials for photoelectrochemical hydrogen generation from wastewater: bibliometric analysis, mechanisms, prospective, and challenges. International Journal of Hydrogen Energy 52 (C), 622–633. https://doi.org/10.1016/j.ijhydene.2023.05.

Othman, A.A., Ali, S.S., Salar, S.G., Obaid, A.K., Al-Kakey, O., Liesenberg, V., 2023. Insights for estimating and predicting reservoir sedimentation using the RUSLE-SDR approach: a case of Darbandikhan Lake Basin, Iraq–Iran. Remote Sensing 15.

Pahari, S.K., Chen, Y.-T., 2023. In situ observation of photoelectrochemical water oxidation intermediates for selective biomass upgrading with simultaneous hydrogen production. Chemical Engineering Journal 473, 145232.

Pandey, C.K., Katiyar, A.K., 2013. Solar radiation: models and measurement techniques. Journal of Energy 2013, 1–8.

Pan, Y., Zhang, H., Zhang, B., Gong, F., Feng, J., Huang, H., et al., 2023. Renewable formate from sunlight, biomass and carbon dioxide in a photoelectrochemical cell. Nature Communications 14.

Parot, M., Rodrigue, D., Stevanovic, T., 2022. High purity softwood lignin obtained by an eco-friendly organosolv process. Bioresource Technology Reports 17, 100880.

Payá, S., Díez, N., Sevilla, M., 2023. Biomass-derived carbon sponges for use as sodium-ion capacitor electrodes. Sustainable Energy & Fuels 7 (10), 2378–2389. https://doi.org/10.1039/D3SE00273J.

Pecoraro, C.M., Di Franco, F., Bellardita, M., Loddo, V., Santamaria, M., 2023. Enhancing H_2 production rate in PGM-free photoelectrochemical cells by glycerol photo-oxidation. International Journal of Hydrogen Energy 49 (C), 322–336. https://doi.org/10.1016/j.ijhydene.2023.08.011.

Peixoto, G., Mockaitis, G., Moreira, W.K., Lima, D.M.F., de Lima, M.A., Ferreira, F.V., et al., 2023. Acidogenesis of pentose liquor to produce biohydrogen and organic acids integrated with 1G–2G ethanol production in sugarcane biorefineries. Waste 1, 672–688.

Pevtsov, A.A., Nandy, D., Usoskin, I., Pevtsov, A.A., Corti, C., Lefèvre, L., et al., 2023. Long-term solar variability: ISWAT S1 cluster review for COSPAR space weather roadmap. Advances in Space Research. https://doi.org/10.1016/j.asr.2023.08.034.

Phan, T.P., Nguyen, T.L., Nguyen, P.K.T., 2023. Evaluation and enhancement of hydrogen production from Vietnamese macroalgae in the single-stage of dark fermentation and microbial electrolysis cell. Biomass Bioenergy 175, 106885.

Pihosh, Y., Nandal, V., Higashi, T., Shoji, R., Bekarevich, R., Nishiyama, H., et al., 2023. Tantalum nitride-enabled solar water splitting with efficiency above 10%. Advanced Energy Materials 13, 2301327. https://doi.org/10.1002/aenm.202301327.

Price, R., MacDonald, L., Gillies, N., Day, A., Brightman, E., Li, J., 2023. Utilisation and valorisation of distillery whisky waste streams via biomass electrolysis: electrosynthesis of hydrogen. Faraday Discussions 247, 268–288. https://doi.org/10.1039/D3FD00086A.

Prosser, J.H., James, B.D., Murphy, B.M., Wendt, D.S., Casteel, M.J., Westover, T.L., et al., 2023. Cost analysis of hydrogen production by high-temperature solid oxide electrolysis. International Journal of Hydrogen Energy 49 (C), 207–227. https://doi.org/10.1016/j.ijhydene.2023.07.084.

Qureshi, F., Tahir, M., 2024. Photoelectrochemical water splitting with engineering aspects for hydrogen production: recent advances, strategies and challenges. International Journal of Hydrogen Energy 69 (5), 760–776. https://doi.org/10.1016/j.ijhydene.2024.05.039.

Raptis, I.P., Kazadzis, S., Fountoulakis, I., Papachristopoulou, K., Kouklaki, D., Psiloglou, B.E., et al., 2023. Evaluation of the solar energy nowcasting system (SENSE) during a 12-months intensive measurement campaign in Athens, Greece. Energies 16 (14), 5361. https://doi.org/10.3390/en16145361.

Reynolds, J.R., Ruiz, J.P., Child, A.D., Nayak, K., Marynick, D.S., 1991. Electrically conducting polymers containing alternating substituted phenylenes and bithiophene repeat units. Macromolecules 24, 678–687.

Romans-Casas, M., Feliu-Paradeda, L., Tedesco, M., Hamelers, H.V.M., Bañeras, L., Balaguer, M.D., et al., 2024. Selective butyric acid production from CO_2 and its upgrade to butanol in microbial electrosynthesis cells. Environmental Science and Ecotechnology 17.

Saedi, A.H., Ahmadi, A., 2023. Life cycle assessment of Iran energy portfolio: renewable energy replacement approach. Energy & Environmental Science 2, 1–20.

Sajjadizadeh, H.-S., Goharshadi, E.K., Karimi-Nazarabad, M., 2024a. Highly efficient photoanode in visible light water splitting through development of Z-scheme structure between compositing TiO_2 with GQDs and Ba doped VO_2 (m) with smart selection of Ag nanoparticles sites. Fuel 355, 129544.

Salaneck, W.R., Friend, R.H., Brédas, J.L., 1999. Electronic structure of conjugated polymers: consequences of electron–lattice coupling. Physics Reports 319, 231–251.

Salzner, U., 2021. Optoelectronic properties of diketopyrrolopyrrole homopolymers compared to donor–acceptor copolymers. The Journal of Chemical Physics 154 (5), 054309. https://doi.org/10.1063/5.0038284.

Salzner, U., Lagowski, J.B., Pickup, P.G., Poirier, R.A., 1998. Comparison of geometries and electronic structures of polyacetylene, polyborole, polycyclopentadiene, polypyrrole, polyfuran, polysilole, polyphosphole, polythiophene, polyselenophene and polytellurophene. Synthetic Metals 96, 177–189.

Seger, B., Tilley, D.S., Pedersen, T., Vesborg, P.C.K., Hansen, O., Grätzel, M., et al., 2013. Silicon protected with atomic layer deposited TiO_2: durability studies of photocathodic H_2 evolution. RSC Advances 3, 25902–25907.

Segundo, R.F., Benites, S.M., De La Cruz-Noriega, M., Vives-Garnique, J., Otiniano, N.M., Rojas-Villacorta, W., et al., 2023. Impact of dragon fruit waste in microbial fuel cells to generate friendly electric energy. Sustainability 15.

Shiva Kumar, S., Lim, H., 2022. An overview of water electrolysis technologies for green hydrogen production. Energy Reports 8, 13793–13813.

Shokrollahi, P., Omidi, Y., Cubeddu, L.X., Omidian, H., 2023. Conductive polymers for cardiac tissue engineering and regeneration. Journal of Biomedical Materials Research Part B: Applied Biomaterials 111 (11), 1979–1995. https://doi.org/10.1002/jbm.b.35293.

Shrivastava, R., Fekar, D., Mahilang, M., 2023. Up-conversion phosphors: A short review. Ferroelectrics 614 (1), 112–118.

Shyyan, O., Shyyan, R., 2023. Teacher education in Ukraine: surfing the third wave of change. In: Kowalczuk-Walędziak, M., Valeeva, R.A., Sablić, M., Menter, I. (Eds.), The Palgrave Handbook of Teacher Education in Central and Eastern Europe. Springer International Publishing, Cham, pp. 527–551.

Sinisterra-Solís, N.K., Sanjuán, N., Ribal, J., Estruch, V., Clemente, G., 2023. From farm accountancy data to environmental indicators: assessing the environmental performance of Spanish agriculture at a regional level. Science of The Total Environment 894, 164937.

Sorrenti, I., Rasmussen, T.B.H., Xydis, G., Enevoldsen, P., You, S., 2023. Correlations between component size green hydrogen demand and breakeven price for energy islands. Renewable and Sustainable Energy Reviews 183, 113439.

Sołowski, G., Shalaby, M., Özdemir, F.A., 2022. Plastic and waste tire pyrolysis focused on hydrogen production-a review. Hydrogen 3 (4), 531–549. https://doi.org/10.3390/hydrogen3040034.

Takayama, T., Iwase, A., Kudo, A., 2024. Enhancing photocathodic performances of particulate-$CuGaS_2$-based photoelectrodes via conjugation with conductive organic polymers for efficient solar-driven hydrogen production and CO_2 reduction. ACS Applied Materials & Interfaces 16 (28), 36423–36432. https://doi.org/10.1021/acsami.4c06083.

Thakur, A., Ghosh, D., Devi, P., Kim, K.H., Kumar, P., 2020. Current progress and challenges in photoelectrode materials for the production of hydrogen. Chemical Engineering Journal 397 (1 October), 125415. https://doi.org/10.1016/j.cej.2020.125415.

Tvingstedt, K., 2007. Light trapping and alternative electrodes for organic photovoltaic devices.

Velthuis, M., Teurlincx, S., van Dijk, G., Smolders, A.J.P., de Senerpont Domis, L.N., 2023. Salinisation effects on freshwater macrophyte growth and establishment in coastal eutrophic agricultural ditches. Freshwater Biology 68, 547–560.

Vilanova, A., Dias, P., Lopes, T., Mendes, A., 2024. The route for commercial photoelectrochemical water splitting: a review of large-area devices and key upscaling challenges. Chemical Society reviews 53, 2388–2434. https://doi.org/10.1039/D1CS01069G.

de Vries, W., Kros, J., Voogd, J.C., Ros, G.H., 2023. Integrated assessment of agricultural practices on large scale losses of ammonia, greenhouse gases, nutrients and heavy metals to air and water. Science of the Total Environment 857.

Wang, X., Chen, H., Feng, X., Zhang, Q., Labbé, N., Kim, K., et al., 2020. Isolation and characterization of lignocellulosic nanofibers from four kinds of organosolv-fractionated lignocellulosic materials. Wood Science and Technology 54, 503–517. https://doi.org/10.1007/s00226-020-01167-4.

Wang, G., Mitsos, A., Marquardt, W., 2020. Renewable production of ammonia and nitric acid. AIChE Journal 66, 1–9.

Wang, G., Zheng, J., Duan, W., Yang, J., Mahmud, M.A., Lian, Q., et al., 2023. Molecular engineering of hole-selective layer for high band gap perovskites for highly efficient and stable perovskite-silicon tandem solar cells. Joule 7, 2583–2594.

Worch, J.C., Dove, A.P., 2020. 100th anniversary of macromolecular science viewpoint: toward catalytic chemical recycling of waste (and future) plastics. ACS Macro Letters 9, 1494–1506.

Wu, W., Hu, Y., Wu, Q., 2023. Subsidies and tax incentives-does it make a difference on TFP? Evidences from China's photovoltaic and wind listed companies. Renewable Energy 208, 645–656.

Xiao, Y., Kong, X., Vanka, S., Dong, W.J., Zeng, G., Ye, Z., et al., 2023. Oxynitrides enabled photoelectrochemical water splitting with over 3,000 hrs stable operation in practical two-electrode configuration. Nature Communications 14.

Xia, C., Li, X., Cao, S., 2023. Challenges for the government-controlled higher education system in China. International Journal of Educational Development 97, 102721.

Xu, Y., Yan, T., Chen, N., Chen, H., Xue, A., Zhang, X., 2024. Conjugated porous polymers regulated by thiophene and polycyclic aromatic hydrocarbons for photocatalytic water splitting toward hydrogen production. International Journal of Hydrogen Energy 51, 1417–1428.

Xu, Y., Zhang, Z., Lin, Q., Ahmed, R., Zhang, S., Sun, Y., et al., 2023. Photoelectrochemical water oxidation through a hybrid architecture nanoparticles/nanotubes. Applied Surface Science 635.

Yang, W., Hu, Y., Ding, Q., Gao, H., Li, L., 2023. Comprehensive evaluation and comparative analysis of the green development level of provinces in eastern and western China. Sustainability 15.

Yang, Y., Sun, B., Sun, Z., Xue, J., He, J., Wang, Z., et al., 2024. Recent advances and strategies of metal phosphides for accelerating polysulfide redox and regulating Li plating. Coordination Chemistry Reviews 510 (1 July), 215836. https://doi.org/10.1016/j.ccr.2024.215836.

Yan, R., Jiang, Z., 2023. Energy-saving and emission-reduction potential of fuel cell heavy-duty trucks in China during the fuel life cycle. Environmental Science and Pollution Research 30, 80559–80572. https://doi.org/10.1007/s11356-023-28085-9.

Yoo, C.G., Meng, X., Pu, Y., Ragauskas, A.J., 2020. The critical role of lignin in lignocellulosic biomass conversion and recent pretreatment strategies: a comprehensive review. Bioresource Technology 301, 122784.

Zatta, P.H.S., Jacob-Furlan, B., Mirabile, R.C., Gonçalves, R.S.R., Da Silva, P.A.S., Taher, D.M., et al., 2023. Microalgae biopolymers: a review. Revista de Engenharia Térmica 21, 44.

Zdonek, I., Mularczyk, A., Turek, M., Tokarski, S., 2023. Perception of prosumer photovoltaic technology in Poland: usability, ease of use, attitudes, and purchase intentions. Energies 16, 4674.

Zhang, Y., Zheng, X., Guo, X., Zhang, J., Yuan, A., Du, Y., et al., 2023. Design of modified MOFs electrocatalysts for water splitting: high current density operation and long-term stability. Applied Catalysis B 336, 122891.. https://doi.org/10.1016/j.apcatb.2023.122891.

Zhao, Y., Niu, Z., Zhao, J., Xue, L., Fu, X., Long, J., 2023. Recent advancements in photoelectrochemical water splitting for hydrogen production. Electrochemical Energy Reviews 6, 14.

Zhao, M., Shadman, F., Keswani, M., 2018. Gradient filling of copper in porous silicon using a non-contact electrochemical method. Applied Surface Science 445, 505–511.

Zhao, T., Wang, Y., Karuturi, S., Catchpole, K., Zhang, Q., Zhao, C., 2020. Design and operando/in situ characterization of precious-metal-free electrocatalysts for alkaline water splitting. Carbon Energy 2 (4), 582–613. https://doi.org/10.1002/cey2.79.

Zwara, J., Pancielejko, A., Paszkiewicz-Gawron, M., Łuczak, J., Miodyńska, M., Lisowski, W., et al., 2020. Fabrication of ILs-assisted $AgTaO_3$ nanoparticles for the water splitting reaction: the effect of ILs on morphology and photoactivity. Materials 13 (18), 1.

Index

Note: Page numbers followed by "*f*," "*t*,' and "*b*" refer to figures, tables, and boxes, respectively.

A

Aberration-corrected scanning transmission electron microscopy (AC-STEM), 326–328
Absorbed photon-to-current conversion efficiency (APCE), 126
Absorption of photons, 34
Acidic environment, 332, 430–431
Acidic media, OER and HER mechanisms in, 332
Activated carbon, 408–411
Activation energy, 258
Additive manufacturing (AM), 14–16
  role in water electrolysis, 14–16, 15*f*
Adsorption, 30–31, 296, 309
Advanced colloidal nanofabrication techniques, 254–255
Advanced electrolysis system designs, 70
Air pollution, 1
Algae, 62
Alkaline, 465–466
  electrolyte, 461
  electrolyzers, 55, 466
  environment, 439
  HER and OER mechanism in alkaline media, 333
  water electrolyzers, 466
Alloy-based cocatalysts
  noble metals and, 400–401, 402*t*
  transition metals and, 403, 405*t*
Aluminum, 12–13, 260
Ammonia (NH_3), 7*b*, 26
  splitting, 496
Amorphous carbon, 271
Amorphous silicon (a-Si), 135–136
Anion exchange membrane (AEM), 465–466, 489
Anionic clays, 336–337
Announced pledges scenario (APS), 1
Anode, 54
Anodic electrochemical polymerization, 376–377
Anodic reaction, 28, 30–31
  factors affecting, 31
Apparent quantum efficiency (AQE), 196–197
Applied bias photon-to-current efficiency (ABPE), 126
Arc discharge, 277–279
Aromatic monomers, 369
Artificial intelligence (AI), 14
  role of artificial intelligence and additive manufacturing in water electrolysis, 14–16, 15*f*
Artificial photosynthesis, 158
Aspergillus niger, 495–496
Automated systems, 235–236
Autothermal reforming (ATR), 4

B

Balanced chemical equation, 54
Bandgap energy, 56
Band structure engineering, 133–134
Batch-type cells, 106–107
Bimetallic metal oxides, 199
Bimetallic nanostructures for enhanced activity, 439–440
Biological methods, water-splitting, 309
Biological water splitting, 62–63
  algae and cyanobacteria, 62
  biological systems with synthetic components, 63
  bioreactors and cultivation systems, 63
  enzymatic catalysis, 63
  genetic engineering, 63
Biomass wastes, 272–273
Bioreactors, 63
Bipolaron, 493–494
Bismuth vanadate ($BiVO_4$), 56, 97–98, 129–130, 155, 404–406
Black phosphorus (BP), 335–336, 411–413
Black phosphorus quantum dot (BPQD), 336
Boron-based cocatalysts, 411–414
Bottom-up synthesis approaches, 433–434
Bottom-up wet chemical fabrication technique, 325
Butler–Volmer equation, 195

C

Cadmium materials, 296
Cadmium selenide (CdSe), 219–220
Cadmium sulfide (CdS), 154, 219–220, 398–399
Cadmium sulfide nanorods (CdS NRs), 415–416
Cadmium telluride (CdTe), 168
Calcium ferrite ($CaFe_2O_4$), 128–129

Carbon
allotropes, 334–335
atoms, 271, 283
black, 408–411
capture technology, 4
carbon-based catalyst, 286, 310
carbon-based cocatalysts, 408–411, 412*t*
carbon-based nitrides, 408–411
fibers, 408–411
nanomaterials, 270
Carbon-based materials, 270, 285–286
carbon nanomaterials, 270–286
case study, 286–289
actions taken/workflow/tools used/simulations and analyses, 288
audience, 287
challenges and solutions, 288
expected results and deliverables, 288
learning and knowledge outcomes, 289
objective, 287
rationale, 287
results, 289
safety considerations, 288
scope, 287
title, 286
Carbon-carbon bonds (C-C bonds), 369
Carbon dioxide (CO_2), 3, 193–194
Carbon nanotubes (CNTs), 228–229, 270, 277–282, 280*t*, 408–411
Carbon nitride nanosheets (CNNS), 408–411
Carbon quantum dots (CQD), 270, 408–411
Catalysis in water splitting, 166–168
electrolysis, 166–167
photocatalysis, 167
photoelectrochemical processes, 167–168
Catalysts, 28–29, 34–36, 166, 194, 212, 309–310, 320, 329, 380, 439, 468–469
activity of, 29
dispersion, 185
for hydrogen and oxygen evolution reactions, 466–468
importance of nanostructured materials in addressing catalyst challenges, 431
introducing magnetism in, 212
Catalytic activity, 37–38, 149, 448–449
Cathode, 54
Cathodic electrochemical polymerization, 376–377
Cathodic reaction, 28, 31
factors affecting, 32
Cell
cell setup in photochemical water-splitting reaction, 86–91
design, 78–83, 155–156
monilithic photoelectrochemical cell, 79
PEC/PEC tandem cells, 81
PEC/PV tandem cells, 81–83, 82*f*, 82*t*
working principle for photoelectrochemical water splitting, 83
Ceramic oxide, 55
Cerium metal-organic framework, metal doping in, 298
Cerium oxide (CeO_2), 58
cycle, 58
Cesium dinickel selenide ($CsNi_2Se_2$), 199–203
Cetyltrimethylammonium chloride (CTAC), 173
Chalcogenides, 197, 199–203, 204*t*, 205*t*
Chemical bonding via Raman spectroscopy, analyzing, 310
Chemical bonds, 310
Chemical energy, 143–144
Chemical industry, 26
Chemical precipitation, 307
Chemical reactions, 10
Chemical synthesis, conjugated polymer materials, 369–376
chain-growth polymerization, 371–374
iterative synthesis, 374–376
step-growth polymerization, 369–371
Chemical vapor deposition (CVD), 277–279, 325, 433–434
Chronoamperometry, 379
Clean energy, 1
sources, 193
Closed system, 106–107
Coal, 75–76
gasification, 5
Cobalt (Co), 11, 32, 173, 281–282
Cobalt-based metal-organic framework, 298
Cobalt-based MOFs, 300, 305–307
Cobalt iron (CoFe), 69–70
Cobalt phosphate (Co-Pi), 413
Cobalt phosphide nanoparticles (CoP NPs), 407–408
Cobalt Sulfide (CoS), 154
Cocatalysts, 256, 258, 380, 382–383, 397
extract, 365–366
factors influencing performance of cocatalyst, 399
significance of cocatalysts in water splitting, 398–399
types of cocatalysts, 399–417
Commercial water-splitting, 440
Complementary analysis, 305
Composite nanostructured materials, 444–446
Conduction band (CB), 34–36, 77, 121–122, 199, 219, 253–254, 363–364
Conduction band minimum (CBM), 101
Conductive polymers, 377
Conference of Parties (COP26), 3
Conjugated organic polymers, 366–368

Conjugated polymers, 493–494
basics of water-splitting, 363–368
case study, 385–387
challenges and future perspectives, 383–385
challenges of CPs for water-splitting, 384
future perspectives, 384–385
characterization techniques for, 378–379
electrochemical analysis, 379
morphological and structural analysis, 378
spectroscopic electrochemical methods, 378
CP material synthesis and characterization, 368–379
importance of conjugated polymer materials in water-splitting, 366–368
photocatalytic water-splitting with, 379–383
mechanism of water-splitting by, 380
methods for improving CP photocatalytic performance, 382–383
role of CP materials in photocatalysis, 380–382
practical application of CP for water-splitting, 385–387
synthetic methods for, 368–378
chemical synthesis, 369–376
electrochemical synthesis, 376–378
Conventional catalysts, 429–431
Conventional fossil fuels, 1, 143–144
Conventional fossil source, 3
Conventional metal-based catalysts, 368
Conventional photosynthesis, 442
Conventional water electrolysis, 53
Copper (Cu), 11
copper-based chalcogenides, 257
Copper bismuth oxide ($CuBi_2O_4$), 128–129
Copper indium gallium selenide (CIGS), 168
Copper oxide (CuO), 11, 404
Copper Sulfides (CuS), 154
Copper vanadate ($Cu_3V_2O_8$), 130
Core-shell structures, 187
Cost-benefit analysis, 474
Counter electrode, role of, 89–90
Coupled electrolysis, 489
Covalent modification, 323–324
Covalent organic framework (COF), 320–321, 347
COF-based electrocatalysts, 297–298
Crystal defects, 99–100
Crystallinity, 436
Cultivation systems, 63
Custom-designed reaction cells, 106–107
Cutting-edge energy source, 1–2
Cyanobacteria, 62
Cyclic voltammetry (CV), 173, 336–337, 378–379
Cyclic voltammogram, 376–377

D

Density functional theory (DFT), 320–321
Density of states (DOS), 336
Differential pulse voltammetry (DPV), 379
Dipole–dipole interactions, 248–249
Direct arylation polymerization (DArP), 369
Direct current (DC), 25–26, 54
Direct electron transfer (DET), 249
Direct hetero polymerization (DHAP), 369
Dodecanethiol, 256–257
Domen's group, 109
Donor–acceptor (D–A) CPs, 369
Donor–acceptor (D–A) macromolecules, 493–494
Durability, 380–382
Dye-sensitized solar cells (DSSCs), 259, 465

E

Earth-abundant alternatives, need for, 431
Electrical conductivity, 414–415
Electrical contact, 288
Electrical energy, 32
Electric field, creating, 147
Electricity, 363, 395
Electrocatalysis, 329, 437
Electrocatalysts, 163, 437
Electrocatalytic water splitting mechanism, 331–333
HER and OER mechanism in alkaline media, 333
OER and HER mechanisms in acidic media, 332
Electrochemical assessments, 307
Electrochemical cell, 364–365
simulation, 476–478
Electro-chemical energy conversion, 319–320
Electrochemical exfoliation method, 344
Electrochemical impedance spectroscopy (EIS), 307, 378–379
Electrochemical methanol reformation (ECMR), 175
Electrochemical methods, 309
Electrochemical oxide pathway, 461
formation
of metal-O, 461
of metal-OH, 461
of metal-OOH, 461
of oxygen gas, 461
Electrochemical polymerization, 376–378
Electrochemical systems, for water-splitting, 465–469
Electrochemical water splitting, 194–195
exchange current density, 195
overpotential, 194–195
Tafel slope, 195
turn over frequency, 195
Electrode(s), 79, 185, 330
considerations, 83–86
design, 185

lifetime of electrodes, 495–496
material, 27, 172
potential material for electrodes and combinations, 490–494, 491*t*
reactions, 25–26
Electrokinetics, 436–437
Electrolysis, 7–8, 63, 166–167, 499
comparison of mechanisms of splitting water by electrolysis method and photochemical water splitting, 38, 39*t*
factors affecting kinetics of water splitting by electrolysis process, 28–29
kinetic aspects of water splitting by, 29–30
mechanism of water splitting by electrolysis process, 30–32
overall reaction of water splitting by electrolysis process, 32
process, 25–27, 54, 55*f*
techniques for optimizing water splitting by electrolysis process, 32–33
Electrolyte, 32, 53, 122
electrolyte–semiconductor, 329
material and electrolyte selection, 156
role of, 130–133
selection, 185
transporter of, 54
Electrolytic splitting water technique, kinetics of splitting water by, 28–30
Electrolyzer(s), 53, 55, 166–167, 465–466, 467*t*
types of, 55
Electromagnetic field, 244
Electronic band, 211–212
Electronic characteristics, surface functionalization for, 323–324
Electronic configuration of noble metals, 170
Electronic structure, 178
Electron microscopic technique, 326–328
Electrons (e-), 36–37, 89–90, 129, 347–348, 395–396, 441
energy loss spectroscopy, 326–328
engineering materials for, 251–252
transfer, 414
Electrospinning, 433
Emission gun-scanning electron microscopy, 434–435
Endergonic, 363–364
Energy, 119
band gap, 271–272
conversion, 123–125
crisis, 473–474
demands, 97
effect of energy band structure, 324
efficiency, 63
industry professionals, 68
input problem, 65
Energy-dispersive X-ray spectroscopy (EDX), 235, 326–328
Engineering materials for electron acceptors, 251–252
Enthalpy (ΔH), 165
Environmental concerns, 473–474
Environmental impact assessment, 474
Enzymatic catalysis, 63
Enzyme-based devices, 154
European Union, 487
Extraction
extraction of heavy metals using hydrogen evolution reaction, 309–310
types of water-splitting methods for extraction of heavy metals, 307–309
Extrinsic catalytic activity, 206

F
Faradaic efficiency, 125–126, 151–153, 249
Fermi level, 407–408
Few-layered graphene (FLG), 272
Flat band potential (E_{fb}), 91
determination of, 92
Flow system, 108
Fluorine (FTO), 230–231
Fossil fuels, 3–4, 110, 164, 193–194, 427
depletion, 487
Fourier transform-infrared spectroscopy (FT-IR spectroscopy), 111, 326–328
Four photon (D4), 81
Fowler's law, 248
Freshwater, 68–69
Fuels, 319–320
Fullerenes-based materials, 282–284, 285*t*

G
Gallium arsenide (GaAs), 219–220, 464
Gallium phosphide (GaP), 219–220
Gas chromatography (GC), 385
Gel permeation chromatography, 375–376
Genetic engineering, 63
Gibbs free energy (ΔG), 98, 121–122, 151, 165, 363–364
Global energy
challenge, 1–3, 2*f*
demands, 157
Global warming potential (GWP), 6*b*
Gold (Au), 164, 169–170, 228
gold-based photocatalysts, 179–180
Gold nanoparticles (AuNPs), 179–180
Graphdiyne (GDY), 335
Grapheme, 408–411
Graphene, 270–273, 320–321, 334–335
nanosheets, 271–272
single atom site catalysts, 343–344

Graphene oxide (GO), 270, 273–275, 335, 445
Graphitic carbon nitride (g-C_3N_4), 203–206, 339–340, 400
Graphitic carbon nitride QDs (g-C_3N_4 QDs), 255
Green energy, 197
  systems, 144–145
Greenhouse effect, 487
Greenhouse gas, 4
  emissions, 1
Green hydrogen, seawater to, 67–71
Grid inhomogeneous solvation theory, 497–498

H

Haber method, 496
Half-cell reactions, 77
Heavy metal extraction
  advances in metal-organic framework photocatalysts for efficient water-splitting, 298–307, 301*t*
  analyzing chemical bonding via Raman spectroscopy, 310
  case study, 310–313
    actions taken/workflow/tools used/simulations and analyses, 312
    audience, 311
    challenges and solutions, 312
    expected results and deliverables, 311
    learning and knowledge outcomes, 312
    objective, 311
    rationale, 311
    recommendations for future investigation, 313
    safety considerations, 311
    scope, 311
  different metal-organic frameworks used in water splitting, 296–298
  extraction of heavy metals using hydrogen evolution reaction, 309–310
  types of water-splitting methods for extraction of heavy metals, 307–309
Heavy metals, 295, 309
  heavy metal-free quantum dots, building of, 255–256
  pollutants, 307
  via metal-organic framework-assisted water-splitting, 303–307, 306*t*
Heck coupling, 374–375
Hefty specific surface area, 324
Helmholtz layer, 122
Hematite (Fe_2O_3), 167
Heptazine-based crystalline nitride, 336
Heteroatom doping, 211–212, 321–323
Heterogeneous photocatalysts, 105–106
Heterojunction photoelectrodes, 135
Heterostructures, 440–442
  stackability, 323
Hexagonal boron nitride, 413–414
Heyrovsky reactions, 29–30, 198
Heyrovsky step, 8–9
High-angle annular darkfield scanning TEM (HAADF-STEM), 326–328
High energy consumption, 70
  potential solution, 70
Highest occupied molecular orbital (HOMO), 364–365, 493–494
High-resolution transmission electron microscopy (HRTEM), 176
High surface area, 170
High-temperature steam electrolysis (HTSE), 60–62
  advantages of, 61
  challenges and limitations, 62
  electrochemical reactions, 61, 61*f*
  external electrical power, 61
  high operating temperature, 60
  hydrogen and oxygen separation, 61
  recent advancements in, 62
  solid oxide electrolysis cells, 60
Hong Kong University of Science and Technology (HKUST-1), 302
Hot carrier injection, 248
Hybrid catalysts, 164
Hybrid cocatalysts, 414–416
Hybrid nanostructured materials, 444–446
Hybrid systems, 459
  and storage solutions, 67
Hydergen evolution reaction (HER), 77
Hydrides, 496
Hydrogen (H), 1–4, 51–52, 56, 64, 76, 79, 104, 107, 119–120, 143–144, 163, 171, 193–194, 217, 235, 269–270, 319–320, 331, 379–380, 427, 457, 487–488, 499
  colors of, 4–6, 5*f*
  economy, 3–6
    methods adapted for hydrogen generation, 4–6, 5*f*
    properties of, 3–4
  evolution, 228–229
  fuel, 395, 470
  gas, 51
  generation, 251
  hydrogen-oxygen reaction, 397
  hydrogen produced at industry, 10
  industrialization of hydrogen generation from photocatalytic water splitting, 109
  potential source of, 6–14
  production, 3, 75–76, 119–120, 219
  separation, 61
  splitting of water into, 7–10
Hydrogenases, 154–155
Hydrogen evolution photocatalyst (HEP), 102–103

Hydrogen evolution process, 173, 176
Hydrogen evolution reaction (HER), 6–9, 52, 99–100, 119–120, 143–144, 163, 193–194, 274, 301, 320, 364, 396, 429–430, 459, 489–490
catalysts for, 466–468
electrocatalysts, 466
extraction of heavy metals using, 309–310
factors affecting, 172
electrode material, 172
overpotential, 172
pH of solution, 172
hydrogen evolution reaction electrocatalysts, 466
noble-metal-based electrocatalysts, 466
nonnoble-metal-based electrocatalysts, 467–468
iridium-based catalysts for, 176–178
mechanism, 171
in alkaline media, 333
in acidic media, 332
metal nanoparticles as efficient hydrogen evolution reaction catalysts, 437, 438*t*, 439*t*
noble metals in, 171–179
platinum metal-based catalysts for, 172–175
ruthenium-based catalysts for, 178–179
Hydrogen farm project (HFP), 109–110
Hydrogen gas (H_2), 23–24, 332
Hydrogen iodide (HI), 58
Hydrogen ions (H), 25–26, 31, 57
Hydrogen-substituted graphene (HsGDY), 335
Hydrothermal process, 433–434
Hydrothermal technique, 336–337
Hydroxides, 404–408, 409*t*

I

Imperishable energy sources, 24
Incident photon-to-current efficiency (IPCE), 126
Indian Institute of Technology Madras (IIT Madras), 67
prototype by IIT Madras scientists, 67–71
Indium tin oxide (ITO), 253, 443–444
Industrialization of hydrogen generation from photocatalytic water splitting, 109
Industrial-scale hydrogen production, 109–110
Infrared spectroscopy, 378, 434–435
Inorganic semiconductors, 384–385
In situ fabrication process, 253
In situ synthesis method, 111
Intrinsic catalytic activity, 206
Ion, 53, 84–85, 403
exchange, 308
Iridium (Ir), 164, 169–170
iridium-based catalysts for hydrogen evolution reaction, 176–178
nanoparticles, 178
Iridium oxide (IrO_2), 31, 476–478
Iron (Fe), 11, 154–155, 173, 212
Iron (III) oxide (Fe_2O_3), 219–220
Iron oxide nanoparticles, 279–281
Iron Sulfide (FeS), 154

J

j–V analysis, 91–92

K

Kinetics
of photochemical processes, 151
of photochemical technique of splitting water, 34–36
of water splitting, 166
by electrolysis process, factors affecting, 28–29

L

Lanthanum iron oxide ($LaFeO_3$), 128–129
Laser ablation, 277–279
Layer by layer (LBL), 334
Layered double hydroxide (LDHs), 320–321, 336–338
single atom site catalysts, 345
Lead (Pb^2), 313
Life cycle assessment, 495
Light absorption
and management, 156
and scattering, 247–248
Light sources, 57
Liquid-phase plasma (LPP), 224
Liquid precursor transforms, 433–434
Living organisms, 62
Local enhancement of electric field (LEF), 246
Localized surface plasmon resonance (LSPR), 179, 244
Lowest unoccupied molecular orbital (LUMO), 364–365
Low-temperature electrolysis (LTE), 56

M

Magnetic nanocatalysts, 212
Mass spectrometry, 375–376
Membrane filtration, 309
Mercaptoacetic acid (MAA), 253
Mercury (Hg), 91, 313
Mercury oxide (HgO), 91
Mesoporous electron acceptor materials, 252
Metal, 295, 493–494
alloys, 400–401
atoms, 321–323
chalcogenide photocatalysts, 417
metal-based catalysts, 429–430
metal-based nanostructured materials for water-splitting, 437–440
metal-free cocatalysts, 408
metal-semiconductor Schottky barrier, 250

Metal-based cocatalysts, 399–408
metal oxides, hydroxides, sulfides, and phosphides-based cocatalysts, 404–408, 409*t*
noble metals and alloy-based cocatalysts, 400–401, 402*t*
transition metals and alloy-based cocatalysts, 403, 405*t*
Metal hydroxide (MOH), 404–406, 462
formation of, 461
Metal ions, 302
case studies, 232–236
actions taken/workflow/tools used/simulations and analyses, 235
audience, 234
challenges and solutions, 235–236
expected results and deliverables, 234
objectives, 232
rationale, 234
results, 236
safety considerations, 234
scope, 232–234
learning and knowledge outcomes, 236
future investigation, 236
learned, 236
mechanism of metal oxide-assisted photochemical water splitting, 219–220, 220*f*
TiO_2-based photocatalysts, 221–226, 222*t*
WO_3-based photocatalysts, 229–232, 233*t*
ZnO-based photocatalysts, 226–229, 227*t*
Metallic elements, 295
Metallic nanoparticles, 247–248
Metallic oxides, 252
Metal nanoparticles (MNPs), 243–244, 250, 326–328, 429
as efficient hydrogen evolution reaction catalysts, 437, 438*t*
Metal-organic chemical vapor deposition (MOCVD), 333–334
Metal-organic frameworks (MOFs), 252, 296, 320–321, 414
advances in MOFs photocatalysts for efficient water-splitting, 298–307, 301*t*
exploring catalytic potential and synthesis strategies of metal-organic framework, 301–303, 301*t*
extracting heavy metals via metal-organic framework-assisted water-splitting, 303–307, 306*t*
different MOFs used in water splitting, 296–298
cobalt-based, 298
metal doping in cerium, 298
metal-organic framework as photocatalysts, 297
mixed metal-metal-organic framework, 298, 299*t*
nanoporous metal/covalent organic framework-based electrocatalysts, 297–298
nickel phosphonate, 298
two-dimensional MOF, 297
MOF-based cocatalysts, 414
Metal oxide (MO), 84–85, 153, 199, 200*t*, 201*t*, 206–211, 217–219, 218*t*, 404–408, 409*t*, 462, 494, 496
component, 187
formation of, 462
mechanism of metal oxide-assisted photochemical water splitting, 219–220, 220*f*
MO-based semiconductors, 447
Metal-oxide nanoparticles (MONPs), 429
for improved oxygen evolution reaction performance, 437–439
Metal sulfide, 154–155, 406–407
Methane, 5
pyrolysis, 5
Microbial electrolysis cells, 499
Microscopy, 307
Microwave heating method, 111
Microwire, 468
Mixed metal-metal-organic framework, 298, 299*t*
Mixed metal-oxide composites (MMOCs), 445
Molecular catalysts, 384–385
Molybdenum-based dichalcogenides, 440
Molybdenum sulfide (MoS_2), 154
Monilithic photoelectrochemical cell, 79, 79*f*
Monocrystalline silicon cells, 463–464
Monolithic cell, 79
Monomer, 368–369, 376–377
Mott-Schottky plot, 326–328
Multijunction systems, 384–385
Multiwalled carbon nanotubes (MWCNTs), 252, 272, 408–411
Mxenes, 320–321, 340–341
Mxenes-based cocatalysts, 417
single atom site catalysts, 344

N

Nafion, 55
Nano cocatalysts, 414–415
Nanocolumns (NCs), 445
Nanocomposites (NCs), 433
Nanoengineering techniques, 427–428
Nanofiltration, 309
Nanomaterials (NMs), 196–197, 271, 288, 384–385, 428–429, 434–435
Nanoparticles (NPs), 221–224, 246, 271, 326–328, 338, 430
Nanoporous frameworks, 297–298
Nanoporous metal organic framework-based electrocatalysts, 297–298
Nanoporous silica, 494
Nanorod array (NA), 335
Nanorods (NRs), 246
Nanoscale materials, 437
Nanosheets, 340–341

Nanostructured electrodes, 28–29
Nanostructured materials, 37–38, 172, 428–429
  2D materials and heterostructures, 440–442
  case study on nanostructured materials for water-splitting, 446–450
  characterization methods for nanostructured materials, 434–435
  commercial prospects and technological challenges, 446
  conventional catalysts, 429–431
  different approaches for synthesizing nanostructured catalysts, 432–434
  electrolytic water-splitting, 428*f*
  factors affecting photocatalytic and photoelectrochemical water-splitting efficiency, 435–437
    band gap, 435
    crystallinity, 436
    effect of temperature, 436
    intensity of light, 436
    pH, 436
    size, 436–437
  future perspectives, 450–451
  metal-based nanostructured materials for water-splitting, 437–440
  nanostructured materials for photoelectrochemical water-splitting, 442–446
  synthesis and properties of nanostructured materials, 431–437
  tailoring properties through size, shape, and composition, 434
Nanostructured photoelectrodes, 134
Nanostructures, 186
Nanotechnology, 157, 431
Nanotube arrays of TiO_2 (NTAs), 336
Nanowires (NWs), 226, 468
Natural photosynthesis, 259
N-doped graphene quantum dots (NGQD), 137
Near-infrared range (IR range), 77–78
Near-infrared region (IR region), 101
Nernst equation, 121–122, 194
Net zero emissions (NZE), 1
Nickel (Ni), 11, 32, 154–155, 173
Nickel Cobalt Organic Framework, 301
Nickel-encapsulated carbon tubules (Ni/NCT), 69–70
Nickel-iron (NiFe), 476–478
Nickel phosphonate MOFs, 298, 300
Niobium (Nb), 445
Nitrides, 197, 203–206, 207*t*, 208*t*
Nitrogen (N), 107, 203
Nitrogen-doped carbon nanotubes (N-CNTs), 69–70, 281–282
Noble metals, 99–100, 164, 168–171, 181, 400–401, 402*t*
  alloys, 400–401
  catalysts, 168
  comparative analysis of noble metals *vs.* noble metal-free catalysts, 182–184, 183*t*
    availability, 183–184
    cost, 183
    selectivity, 183
    stability, 183
  development of cost-effective alternatives, 182
    transition metal-based catalysts, 182
  future prospects for noble metals in water-splitting technologies, 187–188
  in hydrogen evolution reaction, 171–179
    mechanism of hydrogen evolution reaction, 171
  mechanistic pathways, 171–172
  noble metal-based catalysts, 430–431
  noble metal-based photocatalysts, 179–181
  noble metal-based TMDs, 333
  noble metal-free
    alternatives, 181–182
    catalysts, 182–183, 194
  properties of, 170–171
    electronic configuration of, 170
    high surface area and thermal stability, 170
    resistance to corrosion, 170–171
  types of, 168–169
Nonconjugated polymers, 493–494
Noncovalent approach, 323–324
Nonfullerene acceptors (NFAs), 377–378
Nonlayered nanosheets oxide, 334
Nonmetal-based cocatalysts, 408–414
Nonmetal dopants, 431
Nonnoble metal-based TMDs, 333–334
Non-semiconductor system, 403
Normal hydrogen electrode (NHE), 121–122
Novel designs, 187
n-type photoelectrode, 120
n-type semiconductor, 81–84, 156, 251–252
Nuclear magnetic resonance spectroscopy (NMR spectroscopy), 375–376

O

Oleic acid, 256–257
Oleylamine, 256–257
Onion-like carbons (OLCs), 285
Open-circuit voltage (V_{OC}), 123
Organic field effect transistors (OFET), 366–368
Organic light-emitting diodes (OLED), 366–368
Organic photovoltaics (OPV), 366–368, 465
Osmium (Os), 169–170
Osmium tetroxide (OsO_4), 169
Overall water-splitting (OWS), 366

Overpotential, 28
Oxidation
  efficiency, 448–449
  process, 28, 434
Oxide pathway, 461–462
  formation
    of higher oxidation state oxide, 462
    of metal hydroxide, 461
    of metal oxide and water release, 462
Oxygen (O_2), 51–52, 56, 79, 104, 107, 163, 193–194, 331, 363, 379–380
  evolution of, 34–36, 462
  formation of oxygen gas, 461
  separation, 61
  splitting of water into, 7–10
Oxygen evolution photocatalyst (OEP), 102–103
Oxygen evolution reaction (OER), 6–7, 10, 52–53, 77, 99–100, 119–120, 147, 164, 193–194, 274, 298, 320, 364, 396, 429–430, 459
  anode, 54
  catalysts for, 466–468
  cathode, 54
  electrocatalysts, 468–469
  electrolysis process, 54, 55*f*
  electrolyte, 53
  kinetics, 70
    potential solution, 70
  mechanism
    in acidic media, 332
    in alkaline media, 333
  metal-oxide nanoparticles for improved oxygen evolution reaction performance, 437–439
  noble-metal-based electrocatalysts, 468–469
  nonnoble metal-based electrocatalysts, 469
  process, 198
  recent advancements in water electrolysis, 56
  transporter of electrolyte, 54
  types of electrolyzers, 55
Oxygen gas (O_2), 23–24, 332
Oxygen reduction process (ORR), 343–344
Oxygen-vacancy control, 206–211
Oxynitrides, 206, 209*t*, 210*t*
  catalysts, 197

P

Palladium (Pd), 164, 168, 170, 430–431
  Pd-based alloys, 175–176
  Pd metal-based catalysts, 175–176
  Pd metal-based catalysts for hydrogen evolution reaction, 172–175
Palladium nanoparticles (Pd-NP), 175
Paris Agreement, 487
Particulate photocatalytic system (PPS), 86
Perovskite-based solar cells (PSC), 12
Perovskite-conjugated polymer hybrids, 362
Perovskite oxide catalysts, 184
Perovskite oxynitrides, 206
Perovskite solar cells, 465
Personal protective equipment (PPE), 41, 111, 475–476
Petroleum-based motor fuels, 1–2
Petroleum products, 75–76
Phosphides-based cocatalysts, 404–408, 409*t*
Phosphorus allotropes, 335–336
Phosphorus-cocatalysts, 411–414
Photoactive materials, 153–156, 443
  metal oxides, 153
  metal sulfides, 154–155
  semiconductors, 155
Photoanode, 129–130, 132*t*, 136, 329
  based on quantum dots, 251–256
  design, 156
Photocatalysis, 144–145, 167, 179, 269–270, 339–340, 380, 408–411, 437
  materials and mechanisms, 167
  role of CP materials in, 380–382
Photocatalyst, 56, 101, 104, 279–281, 395–396, 403, 489–490, 495
  metal-organic framework as, 297
  photocatalysts-driven redox process, 364–365
  types of photocatalysts for water splitting, 101–102, 102*f*
Photocatalyst system (PC), 119–120
Photocatalytic activity, 279–281
Photocatalytic efficiency, 336, 382–383
Photocatalytic HER, 196–197
Photocatalytic materials, 397
Photocatalytic metal-organic frameworks, 297
Photocatalytic overall water-splitting, 366
Photocatalytic properties, 283
Photocatalytic system (PC), 119–120
Photocatalytic technology, 97–98
Photocatalytic water splitting, 27, 76, 97, 108, 196–197, 196*f*, 231–232, 279
  basics of semiconductor-mediated photocatalytic water splitting, 98–100, 99*f*
  case study, 110–111
    actions taken/workflow/tools used/simulations and analyses, 111
    audience, 110
    challenges and solutions, 111
    expected results and deliverables, 110
    learning and knowledge outcomes, 111
    objective, 110
    rationale, 110
    results, 111
    safety considerations, 111
    scope, 110

title, 110
constraints in, 112
with CPs, 379–383
efficiency, factors affecting, 435–437
material design for, 100–106
importance of visible light-driven photocatalytic water-splitting systems, 102–103
photocatalytic hydrogen evolution from water splitting, 103–105, 103*f*
photocatalytic oxygen evolution from water splitting, 105–106
types of photocatalysts for water splitting, 101–102, 102*f*
principle of, 76–77, 78*f*
process, 57, 76–77
reaction setup of water-splitting devices, 106–108
recent progress in, 108–109, 108*f*
industrialization of hydrogen generation from photocatalytic water splitting, 109
selection, design, and development of efficient photocatalytic materials, 109
Photocathode, 84, 128–129, 131*t*, 257, 287
based on quantum dot, 257–258
design, 156
Photochemical cells
design considerations for, 155–157
integration of components, 156–157
light absorption and management, 156
material and electrolyte selection, 156
photoanode and photocathode design, 156
devices, 156–157
system, 156
Photochemical reactions, 37–38
fundamentals of, 145–151, 148*f*
kinetics of photochemical processes, 151
photoelectrochemical process, 147–149, 150*f*
thermodynamics of water-splitting, 150–151
Photochemical splitting of water, 33–38
criteria for material selection, 151–153
factors affecting, 37–38
kinetics of photochemical technique of, 34–36
mechanism of photochemical technique of, 36–37
photoactive materials, 153–155
Photochemical water splitting, 27, 33, 56–58, 158, 217–218, 235
bandgap energy, 56
comparison of mechanisms of splitting water by, 38, 39*t*
light sources, 57
photocatalyst, 56
photocatalytic water-splitting process, 57
process, 36–37
reaction, cell setup in, 86–91
recent advancements in, 57–58
recent advancements in photochemical water splitting, 57–58
Photo-corrosion, 286
Photocurrent, 258
density, 92
Photodegradation mechanism, 287
Photoelectrocatalytic cells (PEC cells), 286, 470
Photoelectrochemical process (PEC process), 78–79, 81–83, 82*t*, 86, 119–121, 130–133, 147–149, 150*f*, 166–168, 193–194, 499
devices, 83–84
electrodes, 495–496
materials and systems, 168
splitting, 496
splitting develops, 498
tandem cells, 81
Photoelectrochemical water photolysis system (PEC water photolysis system), 229–230
Photoelectrochemical water splitting (PECWS), 120–126, 194–195, 320, 396, 427–428, 470, 499
advancements in, 133–136
basis of, 120–122
case study, 136–139
actions taken/workflow/tools used/simulations and analyses, 138
audience, 137
challenges and solutions, 138
expected results and deliverables, 137
learning and knowledge outcomes, 138–139
objective, 137
rationale, 137
results, 138
safety considerations, 138
scope, 137
title, 137
challenges and solutions, 489
develops, 488
efficiency, factors affecting, 435–437
exchange current density, 195
government and citizen support, 498
irradiance measurement, 498
learning, knowledge outcomes, 499
lifetime of electrodes, 495–496
materials for, 126–130, 127*f*
photoanodes, 129–130, 132*t*
photocathodes, 128–129, 131*t*
mechanism, 329–331
of water-splitting understanding, 489–490
methodology, 488–489
nanostructured materials for, 442–446
composite and hybrid nanostructured materials, 444–446

mechanism of photoelectrochemical water-splitting, 443–444
overpotential, 194–195
photoelectrochemical water splitting efficiency analysis, 123–126, 124*f*
potential material for electrodes and combinations, 490–494, 491*t*
problem after use of photoelectrochemical electrodes, 495*f*
role of electrolyte, 130–133
strategies to enhance photoelectrochemical efficiency, 133–135
alternative approaches, 135
band structure engineering, 133–134
morphology control of photoelectrode, 134
silicon-based electrodes for PEC water splitting, 135–136
surface treatment of semiconductor, 134
Tafel slope, 195
turn over frequency, 195
water feed options for photoelectrochemical electrodes, 497*f*
water sources, 496–498
working mechanism of photoelectrochemical cell, 122–123, 123*f*
working principle for, 83
Photoelectrochemistry, 339–340, 489
Photoelectrode (PE), 81, 122, 428–429, 442, 498–499
materials, 330–331
morphology control of, 134
Photoexcited electron microscopy (PEC), 256–257
plasmonic materials for, 244
system based on quantum dots, 250
Photogenerated electrons, 259
Photoluminescence spectroscopy (PL spectroscopy), 111, 281–282
Photon, 84, 120–121
absorption, 147–148
absorption of, 34
photon-to-current, 259
photon-to-current efficiency, 334–335
Photosynthetic organisms, 62
Photovoltage, 123
Photovoltaic electrocatalysis (PV-EC), 470–471
outcomes of PV-EC integration for water-splitting, 480–481
Photovoltaic electrochemical system (PV-EC), 12, 119–120, 476–478, 489–490
Photovoltaic energy converter (PEC), 148
Photovoltaics (PV), 81–83, 498
cells, 81–83, 363, 457
cell simulation, 476–478
deployment, 1
PV-electrolysis systems, 457–458
PV-photoelectrochemical hydrogen production, 119–120
tandem cells, 81–83, 82*t*
technology, 473–474
for water-splitting, 462–465
basic principle of photovoltaic system, 462
emerging and advanced photovoltaic technologies, 465
thin-film solar cells, 464–465
traditional silicon-based solar cells, 463–464
types of solar photovoltaic technology, 463
Plasma-processing technique, 448
Plasmonic effects, 181
Plasmonic materials, 243–244, 261*t*, 262
for PEC, 244
Plasmonic nanoparticles, 243–244
Plasmonic noble metals, 400–401
Plasmon-induced charge-transfer transition (PICTT), 249
Plasmon-induced hot electron injection (PHET), 249
Plasmon-induced resonant energy transfer (PIRET), 248–249
Platinum (Pt), 11, 32, 87, 143–144, 164, 168, 170, 195, 288, 380, 429, 476–478
platinum-based counter electrode, 89–90
platinum-based photocatalysts, 181
Platinum-cobalt-ruthenium alloy nanoassemblies (PtCoRu NAs), 173
Platinum group metals (PGMs), 168
Platinum nanoparticles (Pt NPs), 173
Polaron, 493–494
Poly(3-hexylthiophene) (P3HT), 385
Polycrystalline silicon cells, 463–464
Polydopamine (PDA), 312
Polyethylene, 493–494
Polygonal deposition method, 221–224
Polymer electrolyte membrane electrolyzer (PEM), 55
Polymeric carbon nitride (PCN), 275–276
Polymerization reactions, 374–375
Polymers, 361–362, 366–368, 374, 493–494
Porous boron nitride, 413–414
Porous structure, 301
Potassium hydroxide (KOH), 53
Power consumption, 75–76
Propagation, 371–374
Proton exchange membrane (PEM), 167, 176–178, 465–466, 489
Protons, 437
p-Silicon Nanowires (p-SiNWs), 138
Pure water, 54

Q

Quantum competence, 34–36
Quantum dots (QDs), 243–244, 261*t*
  PEC system based on, 250
  photoanode based on, 251–256
    adjusting light harvesting, 252–253
    building of heavy metal-free quantum dots, 255–256
    customizing quantum dot stability and charge dynamics, 253–254
    engineering materials for electron acceptors, 251–252
    modifying quantum dot shape, 254–255
  photocathode based on, 257–258
    electron transfer engineering, 258
    modifying hole transfer process, 257
  quantum dots assembly, 258
  solar cells, 465
  surface customization of, 256–257
    choice of surface ligands, 256–257
    cocatalysts, 256
    passivating layer, 256
Quantum efficiency (QE), 386
  studies, 449
Quantum yield (AQY), 225–226

R

Radiolysis, 442
Ralstonia eutropha, 13
Raman spectroscopy, 307, 326–328, 346, 434–435
  analyzing chemical bonding via, 310
Rare earth elements, 184
Rate-determining step (RDS), 198
Rate laws, 151
Reactive oxygen species (ROS), 380
Recombination, 371–374
Redox reactions, 34
Reduced graphene oxide (RGO), 270, 275–277, 277*t*, 278*t*, 408–411
Reduced graphite oxide (rGO), 335
Reduction process, 28
Reduction reaction, 54
Reference electrode, role of, 90–91
Reference half-cell electrode (RHE), 305
Renewable energy, 75–76, 361
  integration of renewable energy sources, 64
  methods, 498
  sources, 1–2
  technologies, 158, 217
Renewable hydrogen generation, 102
Renewable integration, 65
Renewable sources, 488
Resistance to corrosion, 170–171
Reverse osmosis, 309
Reversible hydrogen electrode (RHE), 84, 128–129, 445–446
Rhodium (Rh), 169–170
Rhodium-graphene oxide (Rh-GO), 274
Robust storage materials, 235–236
Ruthenium (Ru), 164, 169–170, 430
  ruthenium-based catalysts for hydrogen evolution reaction, 178–179
Ruthenium oxide (RuO_2), 31

S

Sacrificial reagents, 219
Saturated calomel electrode (SCE), 91
Scalable synthesis methods, 235–236
Scanning electrochemical microscopy (SECM), 379
Scanning electron microscopy (SEM), 235
Scanning probe microscopy, 434–435
Scanning tunneling microscopy, 434–435
Seawater, 70
  challenges and solutions, 70–71
  corrosion and scaling, 70
    high energy consumption, 70
    potential solution, 70
  electrolysis, 70
  environmental impact, 70
    future prospects, 71
    learning and knowledge outcomes, 71
    oxygen evolution reaction kinetics, 70
    potential solution, 70
  to green hydrogen, 67–71
    audience, 68
    expected results and deliverables, 68–69
    objective, 67
    rationale, 68
    scope, 68
    workflow, 69–70
Selenides, 468
Semiconductive polymers, 493–494
Semiconductivity, 490
Semiconductors (SC), 27, 34–36, 81, 121–122, 155, 199, 243–244, 320, 382–383, 404, 442
  materials, 130–133, 148–149, 180, 363
  SC-based photo-electrode, 155–156
  SC-mediated photocatalytic water splitting, 98–100
  surface treatment of, 134
Si heterojunction (SiHJ), 335
Silica electrodes, 496
Silica nanoparticles, 494
Silicon (Si), 135
  silicon-based electrodes for PEC water splitting, 135–136
Silicon carbide (SiC), 219–220

Silver (Ag), 11–12, 91, 170, 180
metal, 228
silver-based photocatalysts, 180–181
Silver and zinc oxide (Ag-ZnO), 180–181
Silver chloride (AgCl), 91
Single-atom catalysts (SACs), 186–187, 225–226
Single-atom silver-incorporated g-C_3N_4 (SAAg-g-C_3N_4), 143–144
Single atom site catalysts (SACs), 320–321
supported on diverse 2D materials, 341–348
2D covalent organic framework, 347–348, 347*t*
2D metal organic framework, 346
C_3N_4, 344–345
graphene, 343–344
layered double hydroxides, 345
MXenes, 344
transition metal dichalcogenides, 345–346
Single-crystal phosphides, 128
Single-element carbon dots, 255
Sodium hydroxide (NaOH), 53
Solar
cells, 12
energy, 119, 143–144, 217–218, 310, 395, 447, 457
irradiance, 498
types of solar photovoltaic technology, 463
Solar to hydrogen (STH), 7, 85, 100–101, 125, 151–153, 330, 470
conversion, 474
Sol-gel process, 274
Solid oxide electrolysis cells (SOECs), 60
Solid oxide electrolyzers, 465–466
Solid oxide fuel cell (SOFC), 60
Solid-state NMR, 378
Solid-state synthetic methods, 302–303
Solvothermal techniques, 303, 433–434
Specific surface area (SSA), 324
Spectro-electrochemical methods, 378
Spectroscopic analyses, 272
Splitting water, 23–24
comparison of mechanisms of splitting water by electrolysis method and photochemical water splitting, 38, 39*t*
key challenges of splitting water technique, 27
kinetics of photochemical technique of, 34–36
kinetics of splitting water by electrolysis or electrolytic splitting water technique, 28–30
activity of catalysts, 29
factors affecting kinetics of water splitting by electrolysis process, 28–29
kinetic aspects of water splitting by electrolysis, 29–30
mechanism of water splitting by electrolysis process, 30–32
overall reaction of water splitting by electrolysis process, 32
photochemical splitting of water, 33–38
techniques for optimizing water splitting by electrolysis process, 32–33
kinetics of splitting water techniques, 27–38
mechanism of photochemical technique of, 36–37
Standard hydrogen electrode (SHE), 90–91
Steam methane reforming (SMR), 4, 10–11
Stille coupling, 374–375
Strontium titanate ($SrTiO_3$), 56
Sulfide ions, 258
Sulfides, 404–408, 409*t*, 468
Sulfur dioxide (SO_2), 58
Sulfuric acid (H_2SO_4), 58
Sulfur-iodine cycle, 58
Surface back reactions (SBR), 104
Surface plasmon polaritons (SPP), 244
Surface plasmon resonance (SPR), 244–249
hot carrier injection, 248
light absorption and scattering, 247–248
mechanisms, 246
plasmon-induced resonant energy transfer, 248–249
shape, 246
size, 246
Surface treatment of semiconductor, 134
Sustainable energy, 196–197
Sustainable hydrogen production, 471
Suzuki–Miyaura catalyst-transfer polymerization (SCTP), 371
Suzuki–Miyaura coupling, 374–375
Synergistic effects, 188
Synthesis methods, 188

T

Tafel reaction, 29–30, 198
Tafel slope, 195
Tandem cell, 93, 479–480
Tandem quantum dot-based PEC architecture, 259
Tandem solar cells, 465
Tandem systems, 384–385
Tantalum pentoxide (Ta_2O_5), 137
Thermal analysis, 307
methods, 378
Thermal conductivity detector (TCD), 386
Thermal gravimetric analysis, 310
Thermal stability, 170
Thermochemical water splitting, 58–59, 65, 163–164, 366
Thermodynamics, 165, 363–364
minimum, 25–26
of water splitting, 150–151, 165–166
Thin-film solar cells, 464–465
Thiol groups, 258

Three-dimensional printing, 14–16
Three-electrode configuration, 87
Tin dioxide (SnO_2), 219–220
Titanium dioxide (TiO_2), 56, 97–98, 129–130, 143–144, 167, 219–224, 260, 281, 398–399
  case study for TiO_2-based photoanode, 260
  TiO_2-based photocatalysts, 221–226, 221*f*, 222*t*
Titanium nanotubes (TNTs), 444–445
Top-down synthesis approaches, 325, 433
Traditional electrolysis, 61
Traditional energy production methods, 97
Traditional silicon-based solar cells, 463–464
Traditional water purification methods, 311
Transition metals, 403, 405*t*
  carbides, 467
  chalcogenides, 468
  complexes, 149
  sulfides, 406–407
  transition metal-based catalysts, 182
  transition metal-based perovskites, 12–13
Transition metal borides (TMBs), 413–414
Transition metal chalcogenides (TMCs), 199–203
Transition metal dichalcogenides (TMDs), 320–321, 333–334, 440
  noble metal-based TMDs, 333
  nonnoble metal-based TMDs, 333–334
  single atom site catalysts, 345–346
Transition metal oxides (TMOs), 194, 197, 199
Transition-metal phosphides (TMPs), 407–408, 468
Transition metals, 182, 194, 225
Transmission electron microscopy (TEM), 326–328, 434–435
Trioctylphosphine oxide, 256–257
Tungsten disulfide (WS_2), 154
Tungsten metal ion, 225
Tungsten trioxide (WO_3), 56, 97–98, 130, 219–220
  WO_3-based photocatalysts, 229–232, 233*t*
Turnover frequency (TOF), 194
Two-dimension (2D)
  carbonaceous materials, 271–272
  covalent organic framework, 320–321
    single atom site catalysts, 347–348, 347*t*
  family, 417
  metal organic framework, 320–321
    single atom site catalysts, 346
  metal oxide, 334
  MOF, 297
  nanomaterials, 362
Two-dimensional materials (2-D materials), 320–323, 429, 440–442
  challenges and opportunities in utilizing 2D materials, 441–442
  characterization techniques for 2D materials, 326–328
  mechanism of water splitting, 329–333
  SACs supported on diverse 2D materials, 341–348
  superiority, 323–324
    effect of energy band structure, 324
    hefty specific surface area, 324
    heterostructure stackability, 323
    surface functionalization for electronic and mechanical characteristics, 323–324
  synthesis techniques of 2D materials, 325–326
  for water splitting, 333–341, 342*t*
Two-electrode, 87
  configuration, 87–89
  j–V measurement, 92–93

U

Ultrafast transient absorption spectroscopy, 344–345
Ultrafiltration, 309
Ultrathin graphitic carbon nitride nanoplatelets (UGCNPs), 285
Ultraviolet (UV), 57, 76, 101
  based systems, 102–103
Ultraviolet-Vis absorption spectroscopy, 378
United Nation (UN), 3
Uphill reaction, 98
US Geological Survey, 295–296

V

Valence band (VB), 34–36, 77, 105, 121–122, 199, 219, 253–254, 363–364
Valence band energy (EVB), 411–413
Valence band maximum (VBM), 101
Van der Waals interactions, 323
Visible light-driven photocatalytic water-splitting systems, importance of, 102–103
Volmer–Heyrovsky mechanism, 171–172
Volmer–Heyrovsky pathway, 460–461
  Heyrovsky step, 460–461
  Volmer step, 460
Volmer reaction, 8–9
Volmer–Tafel mechanism, 172
Volmer–Tafel pathway, 460
  Tafel step, 460
  Volmer step, 460

W

Waste water, 496
Water (H_2O), 6–14, 23–24, 51–52, 104, 147, 163, 332
  case study, 38–44, 471–482
    actions taken/workflow/tools used/simulations and analyses, 41, 476–478
    audience, 39–38
    challenges, 478–479
    challenges and solutions, 41–43

expected results and deliverables, 39–40
learning and knowledge outcomes, 43–46, 481–482
objective, 38, 471
rationale, 39, 473–475
results, 480
safety considerations, 40–41, 475–476
scope, 38, 471–473
solutions, 478
solutions uncovered, 479–480
title, 471
cell design, 78–83
Water splitting (WS), 24, 150–151, 163, 165, 179, 197, 234, 269–270, 326–328, 459–462
2D materials for, 333–341, 342*t*
2D metal oxide, 334
black phosphorus, 335–336
carbon allotropes, 334–335
graphitic carbon nitride, 339–340
layered double hydroxides, 336–338
MXenes, 340–341
transition metal dichalcogenides, 333–334
advances in metal-organic framework photocatalysts for efficient water-splitting, 298–307, 301*t*
background of, 145–151
fundamentals of photochemical reactions, 145–151, 148*f*
basics of, 363–368
challenges in water-splitting, 366
importance of conjugated polymer materials in, 366–368
water-splitting mechanisms, 364–366
water-splitting's kinetics and thermodynamics, 363–364
basics of, 459–460
case study for TiO_2-based photoanode, 260
case study on nanostructured materials for, 446–450
actions taken/workflow/tools used/simulations and analyses, 448–449
audience, 447
case study title, 446
challenges and solutions, 449
expected results and deliverables, 448
learning and knowledge outcomes, 450
objective, 446
rationale, 447
results, 450
safety considerations, 448
scope, 447
catalysis in, 166–168
cell setup in photochemical water-splitting reaction, 86–91
challenges and limitations, 184–186
cost and resource scarcity, 184
electrode design and catalyst dispersion, 185
electrolyte selection, 185
long-term performance and durability, 185
scalability and practical implementation, 184
stability and long-term performance issues, 185–186
challenges of, 63–64
comparative study of, 65, 66*t*
corrosion and durability, 64
costs, 64
energy efficiency, 63
environmental impact, 64
integration of renewable energy sources, 64
scaling up, 64
storage and transportation, 64
challenges of CPs for, 384
comparative analysis of noble metals *vs.* noble metal-free catalysts, 182–184, 183*t*
concept of, 25–27
applications of, 26
comparison of mechanisms of splitting water by electrolysis method and photochemical water splitting, 38, 39*t*
important processes and reactions, 25–26
key challenges of splitting water technique, 27
kinetics of splitting water techniques, 27–38
types of water splitting, 27
current scenario, 10–14
hydrogen produced at industry, 10
laboratory research, 11–14
design, 250
design considerations for photochemical cells, 155–157
determination of flat band potential, 92
different metal-organic frameworks used in, 296–298
efficiency, 300–301
efficiency of water-splitting, 469, 470*t*
electrochemical and photo-electrochemical water splitting, 194–195
electrochemical oxide pathway, 461
electrochemical systems for water-splitting, 465–469
electrode considerations, 83–86
electrolysis, 5, 52, 366, 489
efficiency, 56
energy yields, 497–498
factors affecting kinetics of water splitting by electrolysis process, 28–29
formation of water release, 462
fundamentals of, 165–166
kinetics of, 166
thermodynamics of, 165–166
future aspects of, 65–67
environmental sustainability, 67

hybrid systems and storage solutions, 67
improved efficiency and cost reduction, 65
market growth and commercial viability, 67
policy support and collaboration, 67
renewable energy integration, 65
scaling up for commercialization, 65
seawater to green hydrogen, 67–71
technological innovations, 67
future perspective and recommendations, 157–158
advanced material innovations, 157
artificial photosynthesis, 158
economic and environmental considerations, 158
integration of nanotechnology, 157
system integration and scalability, 158
future prospects for noble metals in, 187–188
j–V analysis, 91–92
kinetic aspects of water splitting by electrolysis, 29–30
material selection for photochemical splitting, 151–155
mechanism by CPs, 380
mechanism by electrolysis process, 30–32
mechanism of, 197–206, 329–333
mechanisms of water-splitting understanding, 489–490
metal-based nanostructured materials for water-splitting, 437–440
bimetallic and alloy nanostructures for enhanced activity, 439–440
metal nanoparticles as efficient hydrogen evolution reaction catalysts, 437, 438*t*, 439*t*
metal-oxide nanoparticles for improved oxygen evolution reaction performance, 437–439
molecules, 63
noble metal-based photocatalysts, 179–181
noble metal-free alternatives, 181–182
noble metals, 168–171
in hydrogen evolution reaction, 171–179
outcomes of PV-EC integration for, 480–481
overall reaction of water splitting by electrolysis process, 32
oxide pathway, 461–462
oxygen evolution, 462
PEC system based on quantum dots, 250
photoanode based on quantum dots, 251–256
photocatalytic hydrogen evolution from, 103–105
photocatalytic oxygen evolution from, 105–106
photocatalytic water splitting, 196–197, 196*f*
photocathode based on quantum dot, 257–258
photochemical splitting of, 33–38
photovoltaic electrocatalysis, 470–471
photovoltaic systems for water-splitting, 462–465
plasmonic materials for PEC, 244
practical application of CP for, 385–387
principle of photocatalytic water splitting, 76–77, 78*f*
quantum dots assembly, 258
reactions, 151–153
reaction setup of water-splitting devices, 106–108, 107*f*
recent advancements in water electrolysis, 56
recent trends and future directions, 186–187
renewable energy and hydrogen production, 75–76
role of artificial intelligence and additive manufacturing in water electrolysis, 14–16, 15*f*
significance of cocatalysts in, 398–399
sources, 496–498
splitting, 7, 119–120
objectives, 24
of water into H_2 and O_2, 7–10
strategies to improve water splitting, 206–212
surface customization of quantum dots, 256–257
surface plasmon resonance, 244–249
tandem quantum dot-based PEC architecture, 259
techniques for optimizing water splitting by electrolysis process, 32–33
technologies, 51–63, 52*f*
thermodynamics of, 150–151
transition metal oxides, chalcogenides, nitrides, oxynitrides catalysts, 197
synthesis methodologies, 197
two-electrode j–V measurement, 92–93
types of photocatalysts for water splitting, 101–102
types of water-splitting methods for extraction of heavy metals, 307–309
water-oxidation process, 330–331
Wet ball milling, 325
White graphene, 413–414
Wind power, 498
Working electrode, 79

X

X-ray absorption fine structure (XAFS), 326–328
X-ray diffraction (XRD), 111, 235, 307, 326–328, 378
X-ray photoelectron spectroscopy (XPS), 111, 326–328

Z

Zerodimensional carbon-based nanocarbon materials (0D carbon-based nanocarbon materials), 408–411
Zinc oxide (ZnO), 130, 219–220, 226
ZnO-based photocatalysts, 226–229, 227*t*
Zinc Sulfide (ZnS), 154

9780443290640